ADVANCES IN ENERGY, ENVIRONMENT AND CHEMICAL ENGINEERING
VOLUME 1

Advances in Energy, Environment and Chemical Engineering collects papers resulting from the conference on Energy, Environment and Chemical Engineering (AEECE 2022), Dali, China, 24–26 June, 2022. The primary goal is to promote research and developmental activities in energy technology, environment engineering and chemical engineering. Moreover, it aims to promote scientific information interchange between scholars from the top universities, business associations, research centers and high-tech enterprises working all around the world.

The conference conducts in-depth exchanges and discussions on relevant topics such as energy engineering, environment technology and advanced chemical technology, aiming to provide an academic and technical communication platform for scholars and engineers engaged in scientific research and engineering practice in the field of saving technologies, environmental chemistry, clean production and so on. By sharing the research status of scientific research achievements and cutting-edge technologies, it helps scholars and engineers all over the world comprehend the academic development trend and broaden research ideas. So as to strengthen international academic research, academic topics exchange and discussion, and promote the industrialization cooperation of academic achievements.

PROCEEDINGS OF THE 8TH INTERNATIONAL CONFERENCE ON ADVANCES IN ENERGY, ENVIRONMENT AND CHEMICAL ENGINEERING (AEECE 2022), DALI, CHINA, 24–26 JUNE 2022

Advances in Energy, Environment and Chemical Engineering

Volume 1

Edited by

Ahmad Zuhairi Abdullah
School of Chemical Engineering, Universiti Sains Malaysia, Malaysia

Azlin Fazlina Osman
Universiti Malaysia Perlis, Malaysia

CRC Press
Taylor & Francis Group
Boca Raton   London   New York   Leiden

CRC Press is an imprint of the
Taylor & Francis Group, an **informa** business

A BALKEMA BOOK

First published 2023
by CRC Press/Balkema
4 Park Square, Milton Park / Abingdon, Oxon OX14 4RN / UK
e-mail: enquiries@taylorandfrancis.com
www.routledge.com – www.taylorandfrancis.com

CRC Press/Balkema is an imprint of the Taylor & Francis Group, an informa business

Library of Congress Cataloging-in-Publication Data
A catalog record has been requested for this book

ISBN: 978-1-032-42618-1 (SET Hbk)
ISBN: 978-1-032-42619-8 (SET Pbk)

ISBN Volume 1: 978-1-032-36083-6 (Hbk)
ISBN Volume 1: 978-1-032-36085-0 (Pbk)
ISBN Volume 1: 978-1-003-33016-5 (eBook)
DOI: 10.1201/9781003330165

ISBN Volume 2: 978-1-032-42623-5 (Hbk)
ISBN Volume 2: 978-1-032-42627-3 (Pbk)
ISBN Volume 2: 978-1-003-36354-5 (eBook)
DOI: 10.1201/9781003363545

Typeset in Times New Roman
by MPS Limited, Chennai, India

Advances in Energy, Environment and
Chemical Engineering – Abdullah & Osman (Eds)
© 2023 The Editor(s), ISBN: 978-1-032-36083-6

Table of contents

VOLUME 1

Energy application efficiency and energy saving and emission reduction technology

Research and development of renewable energy structure and performance

Environmental engineering and geological and hydrological structure research

*Advances in Energy, Environment and
Chemical Engineering – Abdullah & Osman (Eds)
© 2023 The Editor(s), ISBN: 978-1-032-36083-6*

Preface

Due to the prevention and control of COVID-19 and travel restrictions worldwide, 2022 8th International Conference on Advances in Energy, Environment and Chemical Engineering (AEECE 2022) was adjusted to online (video) Conference, which held on June 24–26, 2022. AEECE 2022 is to bring together innovative academics and industrial experts in the field of energy engineering, electrical engineering and materials application to a common forum. The primary goal of the conference is to promote research and developmental activities in energy engineering, electrical engineering and power system and another goal is to promote scientific information interchange between researchers, developers, engineers, students, and practitioners working all around the world.

The conference model was divided into three sessions, including oral presentations, keynote speeches, and online Q&A discussion. In the first part, some scholars, whose submissions were selected as the excellent papers, were given about 5-10 minutes to perform their oral presentations one by one. Then in the second part, keynote speakers were each allocated 30-45 minutes to hold their speeches. There were 300 individuals who attended this on-line conference, represented many countries including Malaysia, France, Italy, Australia and China.

In the second part, we invited four professors as our keynote speakers. Prof. Azlin Fazlina Osman, Universiti Malaysia Perlis, Malaysia. Her research interests are in the field of biomedical polymer, nanotechnology, nanocomposites, biocomposites, chemistry and properties of materials. And then we had Prof. Ahmad Zuhairi Abdullah, Universiti Sains Malaysia, Malaysia. His research works mostly involve the use of ordered porous materials in oleochemical reactions, renewable energy, waste treatment, and waste valorization. Associate Professor Md Wasikur Rahman, Jashore University of Science and Technology/Department of Chemical Engineering, Bangladesh. His research area: Nanotechnology, Energy storage, Polymer synthesis, Food packaging, Water treatment, etc. Lastly, we were glad to invite ASSOC. PROF. ChM. DR. ONG MENG CHUAN, Faculty of science and marine science, Universiti Malaysia Terengganu. His field of expertise includes marine pollution specifically on heavy metals pollution. Their insightful speeches had triggered heated discussion in the third session of the conference. Every participant praised this conference for disseminating useful and insightful knowledge.

The proceedings are a compilation of the accepted papers and represent an interesting outcome of the conference. Topics include but are not limited to the following areas: Energy, Power, Electrical, Environmental, Chemical and other related topics. All the papers have been through rigorous review and process to meet the requirements of international publication standard.

We sincerely hope that the AEECE 2022 turned out to be a forum for excellent discussions that enabled new ideas to come about, promoting collaborative research. We are sure that the proceedings will serve as an important research source of references and knowledge, which will lead to not only scientific and engineering findings but also new products and technologies.

The Committee of AEECE 2022

Committee members

Conference Chairs
Prof. Chong Kok Keong, *Universiti Tunku Abdul Rahman, Malaysia*
Prof. Hongwei Li, *Southwest Petroleum University, China*

Conference General Chair
Prof. Chong Kok Keong, *Universiti Tunku Abdul Rahman, Malaysia*

Publication Chairs
Prof. Hongwei Li, *Southwest Petroleum University, China*

Program Committee
Prof. Tianhuan Huang, *South China University of Technology, China*
Prof. Yaoku Guo, *Southeast University, China*
A. Prof. Jianbo Yang, *Henan academy of sciences institute of geography, China*
Dr. Weili Zhang, *Henan, University Of Economics And Law, China*
Dr. Xijun Yao, *Inner Mongolia Land Surveying and Planning Institute, China*
Dr. Wei Luo, *Hohai University, China*
Dr. Xiaoguang Huang, *Guangzhou Baiyunshan Pharmaceutical Co. Ltd, China*
Dr. Abdul Latif Bin Mohd Tobi, *Universiti Tun Hussein Onn Malaysia, Malaysia*

International Technical Committee
Prof. Ahmad Baharuddin Abdullah, *Universiti Sains Malaysia, Malaysia*
Prof. Norhaliza Abdul Wahab, *Universiti Teknologi Malaysia, Malaysia*
Prof. Keishi Matsuda, *Matsuyama University, Japan*
Prof. Ikuo IHARA, *Nagaoka University of Technology, Japan*
Prof. Wen-Tsai Sung, *National Chin-Yi University of Technology, Taiwan, China*
Prof. Laurent Ruhlmann, *University of Strasbourg, France*
Prof. Luigi Fortuna, *University of Catania Department of Electrical, Italy*
Prof. Marcello De Falco, *University "Campus Bio-Medico" of Rome, Italy*
Prof. Jyhjong Lin, *Department of Information Management, Ming Chuan University, Taiwan, China*
A. Prof. Chien Chiang Lin, *Shih Hsin University, Taiwan, China*
Dr. Yanhui Huang, *Rensselaer Polytechnic Institute, USA*
Dr. Narottam Das, *University of Southern Queensland, Australia*

Energy application efficiency and energy saving and emission reduction technology

Advances in Energy, Environment and Chemical Engineering – Abdullah & Osman (Eds)
© 2023 The Author(s), ISBN: 978-1-032-36083-6

An energy consumption evaluation method for coal-fired power plant

Yuekai Cai, Yanjie Zhou, Wangen Jia*, Zhongxu Zhu & Hongxing Wang
Shangan Power Plant of Huaneng Power International, Inc., Shijiazhuang, China

ABSTRACT: Each system of a coal-fired power plant is coupled with each other, and the change of operating parameters impacts boiler efficiency, cold source loss and auxiliary power. In order to comprehensively evaluate the influence of operating parameters on the energy consumption level of coal-fired power plants, this paper establishes the quantitative relationship between a single variable and unit energy consumption level according to the unit performance test and calculates the energy consumption level of each system through the real-time data of SIS system. The results show that when the air-cooling fan is at 45 Hz, the comprehensive energy consumption of fan power consumption and unit vacuum is the lowest, and the air-cooling fan has the optimum frequency. The comprehensive evaluation method can effectively guide the operation of coal-fired power plants and has a certain reference significance for the energy conservation and consumption reduction of similar units.

1   INTRODUCTION

In China's energy structure, coal is the most important primary energy, which plays an important role in China's power production. According to the 2020 China Statistical Yearbook, the coal consumption of coal-fired power plants accounts for about 54% of the total industrial coal consumption. Reducing the energy consumption level of coal-fired power plants is significant in achieving the goal of "carbon peaking and carbon neutralization."

During the normal operation of coal-fired power plants, there are many operating parameters of units. In order to ensure the economic operation of units and reduce coal consumption, most power plants adopt the assessment method of main economic indicators, including main steam temperature, reheat steam temperature, exhaust gas temperature, fly ash combustibles, vacuum, auxiliary power consumption rate and other parameters. This method is dominated by a single small index with certain limitations, one-sidedness and lag. As the unit of a coal-fired power plant is a complex dynamic operation system, adjusting one parameter to change in the positive direction will cause the other parameter to develop in the negative direction. For example, reducing the boiler air supply will reduce the auxiliary power consumption rate of the unit, but it will cause the carbon content of boiler fly ash to rise [1–3]; Increasing the fan frequency of direct air cooling unit will reduce the exhaust pressure of the unit, but it will cause the increase of auxiliary power consumption rate of the unit [4, 5].

Some scholars evaluate the energy consumption level of some aspects of the unit based on the SIS. Yang Yunx [6] established the relationship between the 20 variables that have the most impact on coal consumption and power generation efficiency based on the method of random forest theory with an analytic hierarchy process and obtained the influence degree of adjustable parameters on energy consumption. Li Zhe [7] used the SIS system of the power plant to monitor the power consumption of each main auxiliary machine in real-time, and analyzed the collected data. Through the horizontal comparison with the power consumption of different units and power plants, he mastered the change of the auxiliary power consumption rate and analyzed the reasons for the change.

---

*Corresponding Author: sahuaneng@163.com

However, the single index assessment method cannot maximize the unit economy. How to use a method to unify and visualize the parameters with strong mutual coupling has become an urgent demand. Based on the performance test of a 600MW direct air cooling unit, this paper obtains the influence level of single variable change on energy consumption, establishes the functional relationship through data fitting, simplifies the problem of complex coupling system, and has stronger operability.

2 OVERVIEW

The 600MW supercritical unit of Shangan Power Plant adopts single-furnace, primary reheat, double flue, balanced ventilation, dry-bottom, all-steel frame, full suspension structure, semi-open layout and supercritical variable pressure once through the coal-fired boiler. The turbine is supercritical, with primary reheat, three cylinders four exhausts, a single shaft, direct air-cooled condensing type.

Based on the supervisory information system (SIS) of the power plant, this paper introduces the concept of comprehensive energy consumption index of boiler side and turbine side, and finally converts the changes of parameters of boiler side and steam turbine side into the impact on coal consumption of power generation through formula calculation. Through SIS online real-time display and real-time adjustment, the economy of the unit is maximized.

3 ANALYTICAL METHOD

The thermal efficiency of the boiler refers to the percentage of effective heat in the heat sent by fuel. Boiler thermal efficiency is one of the important indexes reflecting boiler energy efficiency. There are generally two methods for the measurement and accounting of boiler thermal efficiency.

3.1 *Positive balance method*

The method of calculating thermal efficiency by the ratio of the heat used by the boiler to the total heat released by the fuel is called the positive balance method, also known as the direct measurement method. The calculation formula of positive balance thermal efficiency can be expressed by the following formula:

$$\eta = (Q'/Q) * 100\% \tag{1}$$

$$Q' = Q_1' * (Q_2' - Q_3') \tag{2}$$

$$Q = Q_1 * Q_2 \tag{3}$$

Where,

η represents boiler thermal efficiency.

Q' represents the heat used.

Q represents the heat released by all fuel entering the boiler.

Q_1' represents boiler steam flow, kg/h.

Q_2' is the enthalpy of steam, kJ/kg.

Q_3' represents the enthalpy of water entering the boiler, kJ/kg.

Q_1 represents the amount of coal entering the boiler, kg/h.

Q_2 represents the Low Calorific Value of Coal, kJ/kg.

The thermal efficiency formula does not consider the influence of steam humidity, blowdown and steam consumption, and is suitable for the rough calculation of the thermal efficiency of small steam boilers. It is not used in most current boilers.

3.2 *Counter balance method*

By measuring and calculating the heat loss of the boiler, the method to obtain thermal efficiency is called the counterbalance method, also known as the indirect measurement method. This method is conducive to comprehensively analyzing the boiler, finding various factors affecting the thermal efficiency, and putting forward ways to improve the thermal efficiency. The thermal efficiency can be calculated by the following formula.

$$\eta = 100\% - q_1 - q_2 - q_3 - q_4 - q_5 - q_6 \tag{4}$$

Where,

q_1 represents the ratio of exhaust gas heat for the boiler to the total heat entering the boiler.

q_2 represents the ratio of incomplete combustion heat loss of gas to the total heat entering the boiler.

q_3 represents the ratio of incomplete combustion heat loss of solid to the total heat entering the boiler.

q_4 represents the ratio of heat loss of boiler heat dissipation to the total heat entering the boiler.

q_5 represents the ratio of the heat content of ash to the total heat entering the boiler.

In addition, q_1, q_2 and q_3 are adopted as boiler energy consumption efficiency indicators. q_4 and q_5 in the counterbalance method have little impact on boiler thermal efficiency and cannot be monitored online in real-time, so we ignore them.

4 COMPREHENSIVELY ENERGY CONSUMPTION INDEX

4.1 *Energy consumption formula related to boiler side*

The comprehensive energy consumption evaluation method for coal-fired power plants is following,

$$\lambda = T_{yx} * \alpha + C_{th} * \beta + C_c/100 * i + \gamma * \theta + \varepsilon * \theta \tag{5}$$

Where,

T_{yx} represents the exhaust gas temperature for the boiler, °C.

C_{th} represents the carbon content of fly ash.

α represents the change of coal consumption for power generation affected by the change of boiler exhaust temperature of the power plant by 1°C, g/kWh.

β represents the change of coal consumption for power generation affected by the change of carbon content in boiler fly ash by 1%, g/kWh.

C_c represents carbon monoxide concentration.

i represents the change of coal consumption for power generation affected by every 100 mg/m^3 increase of carbon monoxide and co value in boiler flue gas, g / kWh.

θ represents the change of coal consumption of power generation affected by the change of power consumption rate of unit equipment by 1%, g / kWh.

γ represents the power consumption rate of a fan, including supply fans, induced draft fans and primary air fans.

ε represents the power consumption rate of coal mill.

The coefficient is obtained from Shangan Power Plant Unit 5 test,

$$\alpha = 0.16,$$

$$\beta = 1.248,$$

$$i = 0.096,$$

$$\theta = 3.2$$

Other variables are taken from the SIS system. The power consumption rate of the air-cooling island is calculated in real-time. The data are taken from the SIS system, and the calculation results can be obtained by voltage and current.

4.2 *Energy consumption formula related to the turbine side*

As the unit of the plant is a direct air cooling unit, there is a coupling relationship between the power consumption and exhaust pressure, and the parameters of the other turbine side are relatively simple. In order to simplify the calculation method of comprehensive energy consumption at the turbine side, in this paper, the comprehensive energy consumption at the turbine side is converted by extracting only two indicators of air cooling island exhaust pressure and power consumption, and other units can add real-time indicators according to the actual situation.

This paper introduces the comprehensive energy consumption index of the turbine side.

$$\mu = p * X + q * d \qquad (6)$$

Where,

μ represents comprehensively energy consumption index of turbine side.

p represents the exhaust pressure of the turbine, kPa.

X represents the coefficient of coal consumption change affected by 1kPa exhaust pressure change under different electric loads.

q represents the power consumption rate of the air-cooling island.

d represents the change of coal consumption of power generation affected by the change of power consumption rate of unit equipment by 1%, g/kWh.

Let $X = f(x)$

Where,

x represents generating the load of units, MW.

$f(x)$ represents the functional relationship between coal consumption change affected by 1kPa exhaust pressure change under different electric loads and generating the load of units.

$f(x)$ can be obtained by data fitting according to the test data of the steam turbine air-cooling island.

Through experiments, we get the functional relationship,

$$f(x) = 1.12 * 10^{-5} * P^2 - 136.5105 * 10^{-4} * P + 5.16381586 \qquad (7)$$

P represents generating the load of units, MW.

The power consumption rate of the air-cooling island is calculated in real-time. The data are taken from the SIS system, and the calculation results can be obtained by voltage and current.

5   APPLICATION ANALYSIS

5.1 *Energy consumption index of unit 5 at 313MW*

When the load of unit 5 is 313mw, the real-time value of comprehensive energy consumption at the boiler side and steam turbine side is shown in Table 1.

It can be seen from Table 1, the comprehensive energy consumption index at the boiler side is 35.10g/kWh, and the comprehensive energy consumption index at the turbine side is 17.58g/kWh. The standard value can be set according to what is required.

From Table 1, the coal consumption converted from the exhaust gas temperature, carbon content in fly ash and CO emission value at the boiler side is less than the coal consumption converted from the standard value, which is in a better economic state; the power consumption rate of the fan at the boiler side and the coal mill is greater than the coal consumption converted from the standard value, which is in a poor economic state; the comprehensive energy consumption index

Table 1. Display value of comprehensive energy consumption of SIS system.

Unit 5 Parameters	313MW		
	Real time value	Standard value	Coal consumption difference
Comprehensive energy consumption for boiler	35.1	34.4–35.4	−0.3
Comprehensive energy consumption for turbine	17.58	20.43	−2.85

at the boiler side is within the range of the reference value as a whole, that is, the boiler side is in a better operating state as a whole; The comprehensive energy consumption at the turbine side is better than the converted coal consumption of the standard value, and is in a better operation state. The unit operator can make directional adjustments according to the real-time consumption difference.

5.2 *550MW working condition optimization*

When the stable load of the unit is 550MW, the economic optimization test of the unit is carried out to determine the best operation mode for the unit. For convenience, the comprehensive energy consumption index of the unit is taken as an example. Table 2 is the screen before parameter adjustment.

From Table 2, the air cooling fan is 50 Hz, the real-time value of the exhaust pressure of the unit is 11.47kpa, the real-time value of the power consumption rate of the air cooling fan is 0.65%, and the difference between the comprehensive index of the turbine side and the standard value is 2.09. The economy of the unit operation is poor.

Table 2. Display value of comprehensive energy consumption of SIS system.

Unit 5 Parameters	550MW		
	Real time value	Standard value	Coal consumption difference
Comprehensive energy consumption for boiler	36.81	34.4–35.4	1.41
Exhaust pressure	11.47	9	2.47
Air cooling fan	0.65%	0.8%	−0.15
Comprehensive energy consumption for turbine	14.05	11.96	2.09

The first adjustment test is shown in Table 3. The frequency of the air cooling fan is reduced from 50Hz to 45Hz. After stabilizing for 10 minutes, the exhaust pressure of the unit is 11.92 kPa, the real-time value of the power consumption rate of the air cooling fan is 0.48%, and the difference between the comprehensive index on the machine side and the standard value is 2.02, which is better than the 50Hz operation state of the air cooling fan.

The second adjustment test is shown in Table 4. The frequency of the air cooling fan is reduced from 45Hz to 40Hz. After stabilizing for 10 minutes, the exhaust pressure of the unit is 12.63 kPa, the real-time value of the power consumption rate of the air cooling fan is 0.35%, and the difference between the comprehensive index on the machine side and the standard value is 2.34, which is worse than the 50Hz and 45Hz operation state of the air cooling fan.

It can be concluded that when the current temperature and unit load are 550MW, a lower unit exhaust pressure is not necessarily better. There is an optimal speed for the frequency of the air cooling fan. The above is only an application diagram, and multiple tests can be used to carry out relevant optimization work.

Table 3. Display value of comprehensive energy consumption of SIS system.

Unit 5 Parameters	550MW		
	Real time value	Standard value	Coal consumption difference
Comprehensive energy consumption for boiler	36.91	34.4–35.4	1.51
Exhaust pressure	11.92	9	2.47
Air cooling fan	0.48%	0.8%	−3.2
Comprehensive energy consumption for turbine	13.97	11.95	2.02

Table 4. Display value of comprehensive energy consumption of SIS system.

Unit 5 Parameters	550MW		
	Real time value	Standard value	Coal consumption difference
Comprehensive energy consumption for boiler	36.98	34.4–35.4	1.58
Exhaust pressure	12.63	9	3.63
Air cooling fan	0.35%	0.8%	−4.5
Comprehensive energy consumption for turbine	14.30	11.96	2.34

6 CONCLUSION

In this paper, through specific scientific methods, the parameters with strong coupling at the boiler and steam turbine are unified into the impact on coal consumption for power generation, which can objectively and accurately evaluate the comprehensive energy consumption of thermal power plants as a whole and can better guide the regulation parameters of the thermal power plant to really reduce coal consumption for power generation.

Using the online real-time display function of the power plant level monitoring information system of the coal-fired power plant, the comprehensive energy consumption results are displayed on the SIS screen in real-time. The operators can carry out real-time directional adjustment and guidance according to the system, carry out the horizontal comparison between the same units through the system, or carry out the economic test under relevant loads through the system to obtain the optimal economical operation mode of the generator unit.

REFERENCES

[1] Duan, D W. (2017) *Assessment of Energy-saving Potential and Energy Diagnosis on 600mw Subcritical Air Cooling Coal-fired Power Unit*. North China Electric Power University.
[2] Li, Q. Wei, H. He, J F. Wang, F. (2021) Discussion on Energy Saving of Operation in Thermal Power Plant. *Chemical Management*, 24:7–8.
[3] Li, Z. (2020) Research and Application of the Assessment System of Thermal Power Plant Auxiliary Power Consumption Based on SIS Data. *Jilin Electric Power*, 48(05):37–39
[4] Yang, Q F. (2020) Comparison of Boiler Efficiency Calculated by DL/T 904—2015 and GB/T 10184—2015. Shanxi Electric Power, 03:55–58.
[5] Yang, Y X. (2021) *Research on Data-driven Based Energy Consumption Diagnosis Method and Analysis for Coal-fired Power Plants*. North China Electric Power University
[6] Yang, Y Z. Li, L M. (2014) Power Consumption Analysis of Air Cooling Fan of 330mw Direct Air Cooling Generator Unit. *Resources Economization & Environmental Protection*, 12:17.
[7] Zhou, S Q. (2018) *Analysis of Energy Consumption Characteristics and Optimal Load Dispatch in Thermal Power Unit*. Southeast University.

Advances in Energy, Environment and
Chemical Engineering – Abdullah & Osman (Eds)
© 2023 The Author(s), ISBN: 978-1-032-36083-6

Current exploration status and utilization direction of geothermal resources in Hainan Province

Xiaolan Lv & Cheng Ma*
Development Research Center of China Geological Survey, Beijing, China

Lin Zhang
Haikou Marine Geological Survey Center of China Geological Survey, Haikou, Hainan Province, China
Oil and Gas Resources Survey Center of China Geological Survey, Beijing, China

Beibei Yang
Development Research Center of China Geological Survey, Beijing, China

ABSTRACT: In 2021, eight ministries of China jointly issued several opinions on promoting the development and utilization of geothermal energy. In this paper, with the comparative method, the development status of the domestic and foreign geothermal energy industry is analyzed, the geological and hydrogeological background of Hainan Province in recent 30 years is combed, and the key geological problems restricting exploration in Hainan Province are analyzed. Five conclusions are put forward to support the development of free trade ports in services and the national strategy of "dual carbon targets."

1 INTRODUCTION

In September 2021, 8 ministries jointly issued several opinions on promoting the development and utilization of geothermal energy, which proposed that "by 2025, the heating (cooling) area of geothermal energy will increase by 50% compared with 2020, and the installed capacity of geothermal power generation in China will double compared with 2020". In March 2022, the National Development and Reform Commission and the National Energy Administration issued the "14th Five-year Plan for Modern Energy System," proposing that "by 2025, the proportion of non-fossil energy consumption should increase to around 20% and the proportion of non-fossil energy power generation to around 39%". To take full advantage of the abundant geothermal resources in Hainan Province, a good exploration and development plan can effectively support the realization of relevant objectives. As one of the five non-carbon-based energy sources, compared with other renewable energy sources, geothermal energy has the characteristics of sustainable and stable energy supply, efficient recycling, and renewable. It has unique advantages in the energy revolution and is expected to become a new direction of energy structure transformation.

2 DEVELOPMENT STATUS OF DOMESTIC AND FOREIGN THERMAL ENERGY INDUSTRY

2.1 *Current situation of geothermal energy utilization abroad*

Geothermal energy utilization can be divided into power generation utilization and direct utilization, and geothermal power generation has become an important way of geothermal energy utilization. Affected by high-temperature geothermal resources and the economy, global geothermal power

---

*Corresponding Author: 32210591@qq.com

generation is growing slowly (Lund & Toth 2020; Ma et al. 2021; Wang et al. 2020; Zhao et al. 2020). The global installed capacity of geothermal power generation was 9.9GW in 2010 and increased to 15.95GW in 2020. The annual installed capacity and growth rate are shown in Figure 1. Geothermal capacity had exceptional growth in 2021, with 1.6 GW added (The International Renewable Energy Agency (IRENA). (2022).

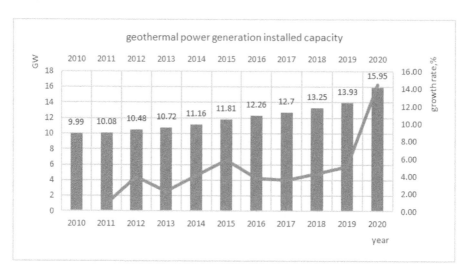

Figure 1.   Installed capacity of geothermal power generation in the world and China from 2010 to 2020.

In 2020, the top 10 countries for geothermal power generation are the United States, Indonesia, and other countries. The Americas and the Asia Pacific region dominate (Lund & Toth 2020; Ma et al. 2021; Wang et al. 2020; Zhao et al. 2020), as shown in Figure 1. Geothermal capacity had exceptional growth in 2021, with 1.6 GW added. The United States increased capacity by 1.3 GW in 2021, and other expansions occurred in Indonesia (+146 MW), Turkey (+63 MW), Italy (+30 MW), and Mexico (+25 MW) (The International Renewable Energy Agency (IRENA). (2022).

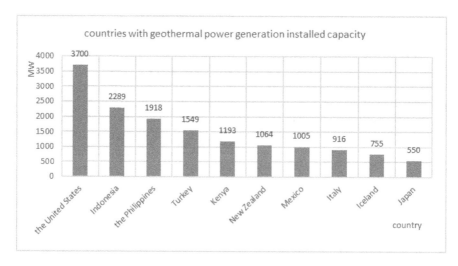

Figure 2.   Top 10 countries with geothermal power generation installed capacity in 2020.

2.2 Current situation of geothermal energy development and utilization in China

2.2.1 Classification of geothermal resources

According to geothermal energy (Nb/T 10097-2018), geothermal resources are "geothermal energy, geothermal fluid, and their useful components." Geothermal resources are mainly divided into three categories. The characteristics (China Geological Survey 2016; Li et al. 2022; Zhou 2022) are shown in Table 1.

Table 1. Characteristics of main types of geothermal resources in China.

Type	Depth	Temperature	Resources (equivalent to standard coal)	Annual recoverable amount of geothermal resources (equivalent to standard coal)	Main purpose
Shallow geothermal energy	Less than 200m	$T \leq 25°C$	9.5 billion tons	700 million tons	Building heating and cooling
Hydrothermal geothermal resources, including	Less than 4000m	$T > 25°C$	1250 billion tons	1.9 billion tons	
High-temperature geothermal resources		$T \geq 150°C$	14.1 billion tons	18 million tons	Heating and power generation
Medium temperature geothermal resources		$90 \leq T < 150°C$	1230 billion tons	1.85 billion tons	Heating, planting, breeding, tourism and recuperation, industrial production, mineral water
Low-temperature geothermal resources		$T < 90°C$			
Dry hot rock resources	3000–10000 km	$T > 150°C$	856 trillion tons		Electricity generation

2.2.2 Geothermal energy industry has initially taken shape in China

There are abundant geothermal resources in China. Currently, medium and low-temperature geothermal resources are mostly used, mainly distributed in sedimentary basins, such as Bohai Bay Basin, Songliao Basin, Ordos Basin, etc. (China Geological Survey. (2016). China is one of the earliest countries in the world to develop and utilize geothermal energy resources. The use of Lishan soup and other hot springs can be traced back to the pre-Qin period. Since the 1950s, China began to use hot springs on a large scale, and hundreds of hot spring sanatoriums have been established. In the early 1970s, China's development and utilization of geothermal energy resources began to enter the stage of various utilization modes.

The utilization of geothermal energy resources in China is mainly direct. By the end of 2020, the heating and cooling area of geothermal energy in China has reached 1.39 billion square meters, ranking first in the world. It can replace 41 million tons of standard coal and reduce 108 million tons of carbon dioxide yearly.

Geothermal power generation in China began in the 1970s, but the progress is relatively slow. The total installed capacity of geothermal power generation has changed from 24 MW in 2010 to 27.8MW in 2019, with little change in the recent ten years.

2.2.3 Geothermal development and utilization policy in China

Over the past two decades, as clean and renewable energy, geothermal energy has attracted the attention of the state. China has issued a series of policies to point out the direction for development, and the national legal framework and management system have been preliminarily established. After 2000, the water and renewable energy laws further regulated the development and utilization of geothermal resources. A series of administrative documents have also been issued at the national level, such as planning, outline, and guidance (Li & Zhao 2019). Tianjin, Heilongjiang, and more than ten provinces have issued policies and put forward the development goals during the 14th Five Year plan period.

2.3 Current situation of geothermal resources exploration in Hainan Province

There are abundant geothermal resources in Hainan Province, but the proven rate and utilization degree of resources are low (Report on the investigation results. (2016). The current survey data are mainly hydrothermal type, and the dry, hot rock type geothermal is basically blank. The geothermal temperature is mainly medium and low, and the temperature of the Qixianling geothermal field in Baoting County is the highest, reaching 94°C. Other areas are mainly distributed at $36 \sim 77$°C.

3 GEOLOGICAL BACKGROUND OF HAINAN PROVINCE

Hainan Province is located in the southernmost tropical and subtropical regions of China, with a total land area of 33,900km^2 and a sea area of about 2 million km^2. It includes 19 cities and counties. The main geological background (Hainan Geological Brigade of Guangdong Geological Bureau. (1981); Geological Survey of Hainan Province. (2012); Hainan Geological Survey Institute. (2006); Hainan Provincial Department of land, environment, and resources. (2007) is as follows.

3.1 Neotectonics movement

The geological conditions of Hainan Province are relatively complex. Under the Himalayan movement, the neotectonics movement in Hainan Island since Neogene is active and frequent, with various manifestations. Under the background of the intermittent rise of the earth's crust, strong fault activity occurred, and volcanic platform and volcanic cone landform, marine landform along the coastal zone, and fluvial landform on both banks of the river were formed in northern Hainan.

3.2 Crustal movement

The crustal movement generally has the characteristics of intermittent upward movement, resulting in multiple parallel unconformity contacts between Neogene and Quaternary and within Quaternary. The differential crustal movement leads to the widespread parallel unconformity contact between Pliocene and Pleistocene Beihai formation and Basuo formation. The Middle Pleistocene Beihai formation is distributed in the coastal and piedmont plain areas, becoming a marine tertiary terrace with an elevation of more than 20m. Basuo formation of Upper Pleistocene is distributed on the second terrace of Haicheng formation, with an elevation of 5–15m. The Holocene Yandun formation is distributed in the coastal zone, a first-class marine terrace, and can be divided into first-class sand embankment deposition (elevation 3–13m) and first-class terrace floodplain deposition (elevation 0–2m).

3.3 *Fault activity*

The fault activity of Hainan Island since Quaternary is mainly developed in the Qiongbei area, and the fault activity in other areas is mainly the inherited activity of old faults. Fault activity, especially the inherited activity of old faults, directly controls the occurrence of seismicity, volcanic eruption, and geothermal activity. According to the existing regional geological survey data, the active faults in the Qiongbei area mainly include the East-West Wangwu Wenjiao fault zone, northwest Yanchunling Daoya fault, Rongshan Lingnan fault, Qionghua Liantangcun fault, northeast Gongtang fault, Baifen Huangyan fault and North-South South Dujiang fault. The activity of the Wangwu culture education fault zone is the most obvious and has a wide range of influence.

3.4 *Seismicity*

Hainan Province, adjacent to the famous Circum-Pacific Seismic Belt, is an area with relatively active seismic activity. According to the seismic intensity zoning map of Hainan Province drawn by the Seismological Bureau of Guangdong Province, the seismic intensity of Hainan Island is high in the north and low in the south. The seismic intensity in the south is magnitude VI or below, and the seismic activity is not active; the earthquake intensity in the north is magnitude VII and VIII, which is still in the active period of seismic activity.

3.5 *Volcanic activity*

Volcanic activity is developed in the north of the Wangwu Wenjiao fault zone. Volcanic activity is frequent and has the characteristics of multi-stage eruptions. In terms of its distribution scale, from morning to night, volcanic activity shows the characteristics of weak-strong-weak and tends to exhaustion. In the Holocene, volcanic activity was localized, occurred only in the area of Eman and Shishan, and erupted in a central group. The arrangement direction of its crater was also NW, which was restricted by NW trending faults.

3.6 *Geothermal activity*

The direct manifestation of geothermal activity is the development of hot springs. There are many hot springs on Hainan Island, which are widely distributed. There are hot springs exposed in Danzhou Lanyang farm and other places. Danzhou Lanyang hot spring, Qionghai Guantang hot spring, Wanning Xinglong hot spring, and Sanya Nantian hot spring have become tourist attractions.

4 MAIN PROBLEMS IN THE EXPLORATION AND DEVELOPMENT OF GEOTHERMAL RESOURCES

The formation of geothermal resources is closely related to fault activity, magmatic activity, formation lithology, and hydrogeological conditions. On the whole, the geothermal resources in Hainan Province are widely distributed, but the understanding of geological conditions and resource potential is insufficient. In particular, geological exploration mainly focuses on implementing hydrothermal geothermal, and the exploratory research of dry, hot rock and the exploration of medium-high-temperature geothermal resources are lacking. In terms of geophysical exploration and drilling, most of them are shallow geophysical exploration, and the implementation of structural faults is not clear. The Jiusuo Lingshui fault in the South has a low degree of integration of geothermal basic geological data, which can't form a systematic cross block and Basin wide basic geological data set and map of geothermal resources, which restricts the overall evaluation of hydrothermal geothermal resources. The development and utilization need the basis of the investigation and evaluation of geothermal resources. Geothermal exploration mainly focuses on exposed hot springs and shallow geothermal, and the geophysical exploration method is relatively single. The current identification technology mainly involves shallow low-temperature hydrothermal geothermal resources, and the identification of medium deep, medium-high-temperature geothermal resources and dry, hot rocks is not ideal. Lack of geophysical identification technology combined with multiple means. At the same time, geothermal resource utilization is low, mainly used for tourism purposes such as hot springs, and rarely used in high-end aspects such as medical treatment and power generation. In terms of hot spring development and utilization, most of them have problems of improper or destructive exploitation, which can easily cause environmental damage and resource depletion. Therefore, the development of thermal mineral water resources and resource and environmental protection are a unified whole. The development must pay attention to resource and environmental protection to prevent man-made resource loss, waste, and damage to resources and the environment (Hainan Provincial Department of land, environment, and resources. (2007).

5 CONCLUSION

This paper adopts a comparative method to conduct research on the exploration and utilization of geothermal resources in Hainan Province. The main conclusions can be summarized as follows: (1) To develop geothermal resources in Hainan Province, it is urgent to strengthen basic geological surveys and find the resources. Through geophysical exploration combined with multiple means, the scope of geothermal anomaly, the boundary of geothermal field, and the spatial distribution of thermal reservoir are delineated. The basement fluctuation of the geothermal field and the spatial distribution of hidden faults are determined, which promotes the formation of "one map" of geothermal resources. (2) To realize the sustainable development of geothermal resources, we should adhere to the principle of "scientific and rational development, cascade comprehensive utilization and equal emphasis on development and protection," pay attention to the relationship between development and conservation and realize the unity of economic benefits, resource benefits, and environmental benefits. (3) The development and utilization of geothermal resources involve many fields. Each utilization field and mode must realize the unity of resources, environment, and economic benefits. Different utilization fields were selected, and intensive geothermal demonstration projects were established. In the process of management, the demonstration engineering technology is popularized and applied, which can play a role in point to an area and overall improvement. (4) Following development and utilization, the mining shall be determined by quantity, and the development shall be carried out within the allowable mining amount. At the same time, the recharging technology of the geothermal field in the porous layered thermal reservoir in Hainan Province is explored, and the combination of recharge and mining is implemented. Through Reinjection Technology, more heat in the thermal reservoir is taken to maintain the balanced and stable hot water head pressure of reserves to realize the sustainable utilization of resources. (5) By implementing the pilot basic geological survey of geothermal resources, multiple departments jointly explore and develop geothermal resources in Hainan. By further accelerating the exploration and development of geothermal resources and connecting with international standards, we can effectively support energy transformation and achieve the dual carbon goal. In terms of future work, the above approach should be carried out to enhance the role of geothermal resources in the economic and social development of Hainan, even the national strategy of "dual carbon targets."

ACKNOWLEDGMENTS

This work was financially supported by the funding of the following projects: Special Project of Geological Mineral Resources and Environment Investigation "geological survey standardization and standard formulation and revision (2019–2021)" (No.: DD20190470) and "geological survey standardization and standard formulation and revision (2022–2025)" (No.: DD202 21826).

REFERENCES

China Geological Survey. (2016) *Investigation report on geothermal resources in China.*

Geological survey of Hainan Province. (2012) *Regional geological records of Hainan Province.*

Hainan Geological Brigade of Guangdong Geological Bureau. (1981) Report on 1:200000 regional hydrogeological survey of Hainan Island.

Hainan Geological Survey Institute. (2006) Comprehensive research report on eco-environmental geology of Hainan Island.

Hainan Provincial Department of land, environment, and resources. (2007) Geothermal Resources Exploration and Utilization Plan of Hainan Province.

Li Tianshu, Wang Huimin, Huang Jiachao. (2022) Analysis of the current situation and development opportunities of geothermal energy utilization in China. In: *Journal of petrochemical Management Cadre College*, 22 (3): 62–66.

Li Yang, Zhao Wanyu. (2019) Analysis report on Industrial Technology in the geothermal energy field. In: *High Tech and Industrialization*, (9): 46–51.

Lund J W, Toth A N. (2020) *Direct utilization of geothermal energy 2020 worldwide review.* In: Proceedings World GeothermalCongress 2020, Reykjavik, Iceland, April 26–May 2.

Ma Bing, Jia Lingxiao, Yu Yang. (2021) *Current situation and Prospect of geothermal energy development and utilization in the world.* In: Geology of China, 48 (6): 1734–1747.

Report on the results of investigation. (2016) *Evaluation and zoning of geothermal resources in Hainan Province.*

The International Renewable Energy Agency (IRENA). (2022) Renewable capacity highlights.

Wang Shejiao, Chen Qinglai, Yan Jiahong. (2020) Development trend of geothermal energy industry and technology and suggestions for oil companies. In: *Petroleum Science and Technology Forum*, 39 (3): 9–16.

Zhao Xu, Yang Yan, Liu Yuhong, et al. (2020) Current situation and technical development trend of global geothermal industry. In: *World Petroleum Industry*, 27 (2): 53–57.

Zhou Borui. (2022) Current situation and future trend of geothermal energy development and utilization in China. In: *Energy*, (2): 77–80.

Advances in Energy, Environment and Chemical Engineering – Abdullah & Osman (Eds)
© 2023 The Author(s), ISBN: 978-1-032-36083-6

A study on time-frequency analysis and automatic classification of microseismic and blasting signals underground mine

Rui Dai*
BGRIMM Technology Group, Beijing, China
Institute of Geology and Geophysics, Chinese Academy of Sciences, Beijing, China

Yibo Wang
Institute of Geology and Geophysics, Chinese Academy of Sciences, Beijing, China

Da Zhang & Hu Ji
BGRIMM Technology Group, Beijing, China

ABSTRACT: Microseismic monitoring technology is an important means to realize the stability evaluation of ground stress, while the rapid warning of a microseismic monitoring system for rock mass stability depends greatly on the efficiency of effective microseismic signal identification. So far, the classification between blasting events and microseismic events is manually completed by data processors. This means the experience and subjectivity of the data processor would have a great influence on the effective microseismic data and the process is inefficient. Aimed at this problem and based on the monitoring data from a mine, this paper uses many blasting and microseismic signals for the study. The results show that the energy of the blasting signal is mainly concentrated in the high-frequency band (300–1000 Hz), while the energy of the microseismic signal is mainly concentrated in the low-frequency band (0-300 Hz). By analyzing the energy proportion of blasting and microseismic signal in different frequency bands, this paper provides a way to realize the automatic identification and classification of two kinds of signals, which promotes the actual realization of automatic classification of collected signals underground mine.

1   INTRODUCTION

In the process of mining, roadway and goaf are formed, which destroys the stress balance of original rock underground mine, and the internal stress of stope, surrounding rock and ore body is redistributed, forming secondary stress field, leading to displacement and deformation of pillar, roof, and surrounding rock of working face, and even destruction. In the process of mining, the mining disturbance will change the stress distribution of rock mass in a large range. The rock mass fracture in the stress concentration area and the sudden release of rock mass elastic energy will form a mine earthquake, which will induce roof caving, pillar fracturing, surrounding rock cracking, floor heave, slab wall and other ground pressure disasters. Microseismic monitoring technology can realize the spatial and temporal location of regional mine earthquakes and the stability evaluation of rock mass, and become an important means of ground pressure monitoring (Feng Xiating 2013).

   Through the use of signal filtering noise reduction, the signal-to-noise ratio of the seismic signal can be improved through the acquisition of signals containing valid microseismic signals, blasting, power frequency interference, and background noise (Jia Ruisheng et al. 2015; Lu Caiping et al. 2005), but blasting seismic signals often accompanied by a rock burst, Similar to natural earthquakes, large earthquakes are often accompanied by smaller aftershocks. Manual processing and

*Corresponding Author: dairui@bgrimm.com

DOI 10.1201/9781003330165-3

identification of blasting events and microseismic events are greatly affected by the experience and subjectivity of data processors, and it is time-consuming and laborious, so it is of great significance to study the automatic classification method of blasting signals and microseismic signals (Dong Longjun et al. 2016; Li Wei 2017; Shang Xueyi et al. 2016; Zhu Quanjie et al. 2012).

This paper aims at the technical problem that the blasting signal and the microseismic signal are difficult to distinguish automatically in the time domain in the microseismic monitoring data acquisition system, through time and frequency analysis of signals and extracting blasting seismic signals in frequency domain characteristics of effective through the typical seismic signal and the signal in the frequency domain of blasting energy ratio analysis, implement automatic distinguish between two types of signals. There are some suggestions for automatic classification of the signals collected in underground mines.

2 METHOD

Time-frequency analysis of the collected signals can distinguish the main characteristic information as far as possible, and ensure that the same kind of information is not too dispersed. Time-frequency analysis plays a very important role in time-varying signal analysis, which can reveal the variation of signal energy distribution characteristics with frequency. Different kinds of signals generally have different frequency band distribution ranges of energy concentration, and their energy-frequency band characteristics can be used to distinguish signal types. In this paper, the time-frequency analysis of microseismic and blasting signals is carried out by a fast Fourier transform.

The principle of the fast Fourier transform is as follows:

$$X_F(\omega) = \frac{1}{2\pi} \int_{+\infty}^{-\infty} X(t) e^{-iwt} dt \tag{1}$$

Where $X(t)$ is the signal function in the time domain; e^{-iwt} is the basis function for the Fourier transform. It can be seen from the above equation that any mutation of the signal in the time domain will affect the frequency domain of the whole function. Therefore, the part of the signal energy concentration in the time domain will have a corresponding change in the frequency domain.

The energy information of microseismic data is mapped to different frequency bands by vertical leaf transform. According to Equation (2), the signal energy can be defined as:

$$E_j(w_j) = \int_T |f_j(w_j)|^2 dt = \sum_{k=1}^{m_j} |x_{j,k}|^2 \tag{2}$$

Where $E_j(w_j)$ is the energy of the j-th frequency band; m_j is the total number of discrete sampling points of the j-th frequency band; $x_{j,k}$ is the amplitude of discrete sampling points of the k-th microseismic data of the j-th frequency band.

When the original signal $X(t)$ is converted from a time domain to a frequency domain, its total energy is equal to the sum of the signal energy of each frequency band, which can be expressed as:

$$E = \sum_{j=0}^{f_m} E_j(w_j) = \sum_{j=0}^{f_m} \sum_{k=1}^{m_j} |x_{j,k}|^2 \tag{3}$$

Where E is the total energy, $E_j(w_j)$ is the energy of the j-th frequency band, m_j represents the number of discrete sampling points in the j-th frequency band, f_m is the Nyquist frequency, $x_{j,k}$ is the amplitude of the k-th microseismic data in the j-th frequency band.

The percentage of signal energy in the j-th frequency band of microseismic data to the total energy is as follows:

$$P_j = \frac{E_j(w_j)}{E} \tag{4}$$

17

According to the distribution characteristics of the percentage of the energy in each frequency band of the microseismic data to the total energy, the frequency band range of the energy concentration of the two types of signals can be solved, and the energy-frequency band characteristics can be used to distinguish the two types of signals.

3 ANALYSIS OF ACTUAL MINE SEISMIC DATA

To explore the effect of automatic classification of microseismic and blasting signals by the above method, the frequency band energy characteristics of blasting wave and rock fracture microseismic wave shape of a mine were quantitatively analyzed. The pillarless sublevel caving mining method is adopted as the main mining method in the mining area, and the shallow hole retention mining method is adopted for the thin orebody at the corner of the orebody. With the development of mining, the surrounding rock of partial structure is broken, the surrounding rock is muddled seriously, and the joint surface is developed. The microseismic online monitoring system is established to monitor the caving production process in real-time through monitoring data, monitor and warn potential hazard sources, and guide the mine safety products based on the analysis results of microseismic monitoring data. The above-automated classification method is used to analyze 1,060 blasting data and 1,081 microseismic event data. Figure 1 depicts on-site monitoring of the original blasting

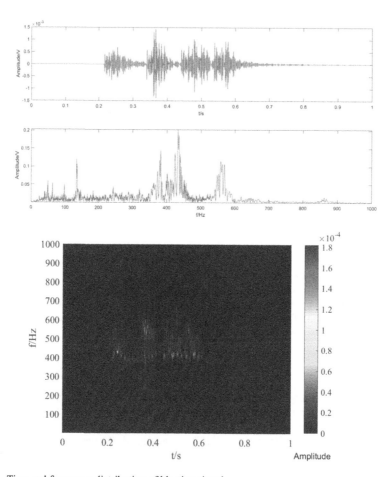

Figure 1.   Time and frequency distribution of blasting signal.

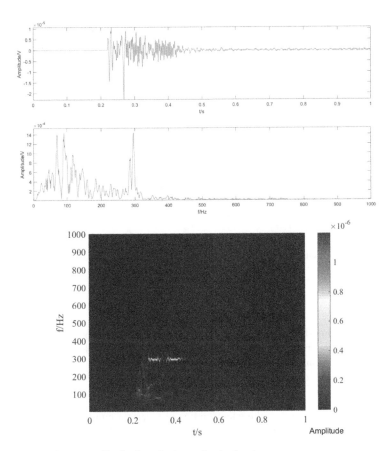

Figure 2. Time and frequency distribution of micro-seismic signal.

signal in the time domain and frequency domain distribution. Figure 2 depicts primitive microtremor signals in the time domain and frequency domain distribution. Through the comparative analysis of the two kinds of signal frequency band energy distribution, we found that the energy of the blasting signal is mainly concentrated in the high-frequency band, while the energy of the microseismic signal is mostly concentrated in the low-frequency band.

To analyze the energy proportion of blasting signal and microseismic signal in different frequency bands in detail, the main frequency band is divided into 10 broadband bands: 0–100 Hz, 100–200 Hz, 200–300 Hz, 300–400 Hz, 400–500 Hz, 500–600 Hz, 600–700 Hz, 700–800 Hz, 800–900 Hz, 900–1000 Hz. Blasts with microseismic signals frequency domain energy distribution as shown in Figure 3. The red energy corresponds to the blasting event, the blue energy corresponds to the seismic event, and the blue energy corresponds to the seismic event. The blasting signal energy is concentrated in the high-frequency band, about 88% of the total energy ratio, whereas the seismic signal energy is primarily concentrated in the low frequency band (0 to 300 Hz). Based on the statistical analysis of the energy proportion of blasting signals and microseismic signals in different frequency bands, the percentage of energy distribution in the 0-300 Hz low-frequency band is taken as the characteristic index. Based on the statistics of 1,081 microseismic events and 1,060 blasting events, the threshold index is set as 80%. When the energy proportion exceeds the threshold in the 0–300 Hz frequency band, it is automatically classified as microseismic events, which can preliminarily realize the identification and classification of blasting signals and microseismic signals.

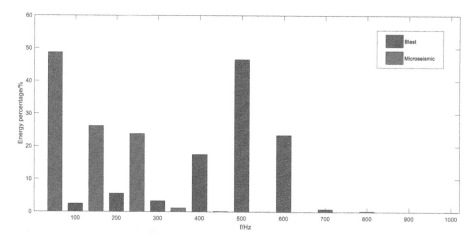

Figure 3. Comparison of energy distribution between blasting and micro-seismic signals in the frequency domain.

4 CONCLUSIONS

Through the study of time-frequency analysis and automatic classification method of microseismic and blasting signals, this paper makes a comparative analysis of actual mine data and draws the following main conclusions.

The energy-frequency characteristics of blasting signals and microseismic signal are quite different, which can be used to distinguish the two signal types.

(1) Through statistical analysis of the energy ratios of a certain number of blasting signals and microseismic signals in different frequency bands, the energy-frequency characteristics of the two kinds of signals are quite different. The energy of blasting signals is mainly concentrated in the high-frequency band (300–1000 Hz), while the energy of microseismic signals is mainly concentrated in the low-frequency band (0–300 Hz).

(2) The percentage of the energy distribution in the low-frequency band (0–300 Hz) in the total energy is taken as the characteristic index. Through the setting of the threshold index, the energy proportion in the low-frequency band (0–300 Hz) exceeding the threshold can be automatically classified as microseismic events.

(3) In this paper, the automatic classification of microseismic signals and blasting signals is carried out, and the automatic classification of other signals generated in the actual mining process needs to be further studied.

ACKNOWLEDGEMENT

This study was supported by the National Key Research and Development Program of China (Grant Nos. 2018YFE0121000 and 2020YFE0202800).

REFERENCES

Dong Longjun, Sun Daoyuan, Li Xibing, et al. *Statistical identification method of microseismic and blasting events and its engineering application.*, 2016, 35 (7): 1423–1433.

Feng Xiating. *Mechanism, Early Warning and Dynamic Control of Rockburst Incubation Process* [M]. Science Press, 2013.

Jia Ruisheng, Zhao Tongbin, Sun Hongmei, et al. Microseismic signal denoising based on EMD and INDEPENDENT component analysis. *Chinese Journal of Geophysics*, 2015, 58 (2): 1013–1023.

Li Wei. Mining microseismic signal Feature Extraction and classification Method based on LMD and Pattern Recognition [J]. *Journal of China Coal Society*, 2017, 42 (5): 1156–1164.

Lu Caiping, Dou Linming, Wu Xinggrong, et al. *Chinese Journal of Geotechnical Engineering*, 2005, 27 (7): 772–775. (in Chinese)

Shang Xueyi, Li Xibing, Peng Kang, et al. Mining Microseismic and blasting Signal Feature Extraction and Classification Method based on EMD_SVD [J]. *Chinese Journal of Geotechnical Engineering*, 2016, 38 (10): 1849–1858.

Zhu Quanjie, Jiang Fuxing, Yu Zhengxing, et al. *Journal of Rock Mechanics and Engineering*, 2012, 31 (4): 723–730.

Advances in Energy, Environment and
Chemical Engineering – Abdullah & Osman (Eds)
© 2023 The Author(s), ISBN: 978-1-032-36083-6

Research on power system scheduling mode considering dynamic carbon metering

Jiaxing Chen* & Chunming Liu*
North China Electric Power University, Beijing, China

ABSTRACT: At present, facing the pressure of emission reduction brought by the greenhouse effect, the inadequacy of carbon reduction methods and the accuracy of carbon emission measurement of power generation enterprises have become critical issues. Therefore, this paper proposes a low-carbon economic dispatch model of power system considering dynamic carbon measurement. Firstly, a first-stage optimization model for low-carbon power generation operation of a combination of day-ahead units is established to determine the optimal output plan of power generation. Secondly, a second-stage optimization model combining the carbon market and power market is established to calculate the power generation under the spot market conditions and reduce the total carbon emission of the system by using the carbon market transaction cost. Finally, the power system with and without carbon market is analyzed, and the power generation, total carbon emission, and total cost of the system are calculated by a simulation study. The results show that the proposed model can calculate the carbon emissions of power generators in real-time, reduce the total carbon emissions of the system, and validate the rationality and effectiveness of the proposed method.

1 INTRODUCTION

With the increasingly serious problem of greenhouse gas emissions, protecting the environment has become the focus of social attention. The power industry is the main coal-consuming industry, and its carbon dioxide emissions account for a large proportion of the total national carbon emissions. In September 2020, China put forward the "3060" double carbon target for the first time, and the power generation industry is facing a serious situation of carbon emission reduction. Therefore, it is very important for power generation companies to realize carbon emission reduction and to accurately measure the carbon emission of power plants.

The combination of the electricity market and carbon emission trading rights market (hereinafter referred to as the carbon market) is an effective measure to promote carbon emission reduction in the power generation industry. Electricity trading will cause different carbon emissions from power plants, and the change in carbon price will affect the output and dispatch pattern of power plants, which can reduce the carbon emissions of the whole power system while increasing the activity of carbon market trading. Research on the integration of the carbon market and power system can be divided into two types. 1) low carbon economy based on the ladder carbon price to plan the power grid, mostly using the grid planning cost model or the operation cost model combined with the ladder carbon price model. 2) considering the change of carbon trading price, mostly using the carbon "penalty" cost to stimulate new energy consumption.

1) Reference (Tian et al. 2012) established a price prediction model for carbon emissions, added the predicted carbon cost to the cost of grid planning, and studied the impact of carbon price

*Corresponding Authors: maxandtiger@163.com and liuchunming@ncepu.edu.cn

DOI 10.1201/9781003330165-4

fluctuations on grid operation. However, the paper uses a fixed carbon emission factor and does not address the intraday market study of electricity. References (Chen et al. 2021; 2021; Qiu et al. 2021; Zhang et al. 2020) introduced a stepped carbon cost model in the operating costs of integrated energy systems to control total carbon emissions.

2) Reference (Zhang et al. 2014) proposed a low carbon economic dispatch method for wind power systems based on an uncertain scenario probabilistic simulation method but did not consider the impact of combining the spot power market and carbon market. Reference (Zhang et al. 2020) proposed a virtual power plant trading method under a carbon trading mechanism to address the abandoned wind and light behavior in virtual power plants. The study shows that a certain level of carbon trading price can increase the overall revenue of virtual power plants and reduce carbon emissions. Reference (Li & Liu 2021) uses a power dispatch model that considers carbon trading and wind power load forecast errors and combines a stepped carbon price cost model for analysis. Still, carbon emissions are calculated based on the day-ahead dispatch output without considering the combination of the real-time carbon market and spot power market and the fact that power plant output can cause changes in carbon emission factors.

The above-mentioned articles do not provide a thorough analysis of the combination of the electricity market and the carbon market, and the calculations regarding emissions are not accurate enough. Reference (Yu et al. 2018) provides a preliminary analysis of the emission characteristics of power plants by calculating CO_2 emissions under different unit types, unit capacities, and fuel conditions, and the results show that the emission factors are different under various conditions. Currently, the existing dynamic emission measurement models in China require a large amount of data and are not suitable for short-term rapid calculations and analysis in combination with the electricity spot market.

Therefore, this paper proposes a step carbon emission measurement model to simulate the trend of carbon emission factors, divides the power plant's power output into multiple intervals, calculates the total carbon emission of the power plant based on the emissions in different intervals, and solves the unit start/stop schedule and power generation of the power plant using GUROBI in combination with the two-stage economic dispatch model of power. Finally, the overall carbon emissions of the system under the traditional economic dispatch model and the dynamic carbon emission measurement model are compared and analyzed to verify the validity and reasonableness of the proposed model.

2 PROBLEM STATEMENT

This section describes the problems addressed by the article and the corresponding measures.

1) To improve the current way of calculating carbon emissions from power plants using constant carbon emission factors, a dynamic real-time carbon emission measurement model is constructed in this paper to improve the real-time accuracy of carbon emission accounting for power plants. See Section 3.1 for details.
2) To meet the emission reduction demand of power generation enterprises, this paper incorporates carbon market trading into the power trading market (two-stage dispatch model of power). The carbon market trading model is introduced into the day-ahead power trading and intraday power trading models to improve the carbon emission reduction effect of power plants. See Chapter 4 for details.

After case verification, the dynamic carbon measurement model of the unit and the economic dispatch model of the two-stage power system proposed in this paper can realize real-time and accurately calculate the carbon emission of the unit, which can effectively reduce the carbon emission of the power system and has certain guiding significance for the development of China's carbon market, as detailed in Chapters 6 to 7.

3 CARBON METERING MODEL

Most of the existing literature and policies are based on a constant carbon dioxide emission factor to calculate the carbon emissions of power plants. The CO_2 emission factor of the fuel is calculated as follows (Tan 2018).

The emission factor for the *i-th* fossil fuel is calculated as follows:

$$EF_i = CC_i \times OF_i \times 44/12 \tag{1}$$

Where CC_i is carbon content per unit calorific value of the *i-th* fossil fuel, tC/TJ; OF_i is the carbon oxidation rate of the *i-th* fossil fuel, %; $44/12$ is the ratio of the molecular weight of carbon dioxide to carbon.

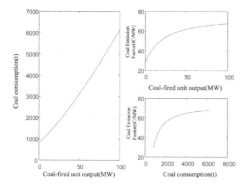

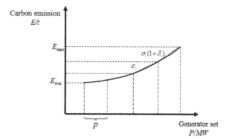

Figure 1.   Carbon emission factor change chart.

Figure 2.   Schematic diagram of dynamic carbon emission model.

Using traditional emission factors to calculate the carbon emissions of power generation enterprises requires a long accounting period and has certain requirements for the accuracy and scope of data collection, which is not suitable for real-time accounting of power generation enterprises. As shown in Figure 1, when coal consumption by power plants increases, the carbon dioxide emission factor increases, and vice versa. If the constant carbon emission factor is used to calculate the carbon emissions of enterprises, the accuracy is insufficient. Therefore, to simplify the analysis and facilitate the real-time accounting of carbon emissions of power plants, this paper proposes a dynamic carbon emissions measurement model.

3.1   *Dynamic carbon emission model*

The dynamic carbon emission model divides carbon emissions into different intervals based on the generation capacity of the unit. The higher the generation capacity of the unit, the more carbon emission factor intervals there are and the higher the carbon emissions of the plant. The dynamic carbon emission measurement model is as follows:

$$E_{(i,t)} = \begin{cases} \sigma_1(P_{(i,t)} - P_0), \min\{P\min_i, P_0\} \le P_{(i,t)} \le P_0 + p \\ \sigma_1(P_{(i,t)} - P_0) + \sigma_1(P_{(i,t)} - P_0 - p)\delta_1, P_0 + p \le P_{(i,t)} \le P_0 + 2p \\ \sigma_1(P_{(i,t)} - P_0) + \sigma_1(P_{(i,t)} - P_0 - p)\delta_1 + \sigma_1(P_{(i,t)} - P_0 - 2p)\delta_1, P_0 + 2p \le P_{(i,t)} \le P_0 + 3p \\ \sigma_1(P_{(i,t)} - P_0) + \ldots + \sigma_1(P_{(i,t)} - P_0 - (n-1)p)\delta_1, P_0 + (n-1)p \le P_{(i,t)} \le \min\{P_0 + np, P\max_i\} \end{cases} \tag{2}$$

Where, $E_{(i,t)}$ is the carbon emission of the *i-th* thermal power unit at moment t, t; p is the interval length of the output of the generator set, $p.u$; σ_1 is the base value of the carbon emission factor, taking $40tC/p.u$; δ_1 is the growth rate of the carbon emission factor, taken as 5%; P min and P max are respectively the upper and lower output limits of the generator set, $p.u$

3.2 The stepped carbon price model

The trading mechanism of the carbon market is mainly divided into two forms (Cui et al. 2021), the quota-based carbon trading model and the stepped carbon trading model. In this paper, the stepped carbon trading model is used, and the stepped carbon trading is as follows:

$$
F_{p(i,t)} =
\begin{cases}
\sigma_2(E_{(i,t)} - E_c), E_{(i,t)} \leq E_c + p \\
\sigma_2(E_{(i,t)} - E_c) + \sigma_2(E_{(i,t)} - E_c - p)\delta_2, E_c + p \leq E_{(i,t)} \leq E_c + 2p \\
\sigma_2(E_{(i,t)} - E_c) + \sigma_2(E_{(i,t)} - E_c - p)\delta_2 + \sigma_2(E_{(i,t)} - E_c - 2p)\delta_2, E_c + 2p \leq E_{(i,t)} \leq E_c + 3p \\
\cdots
\end{cases}
$$

$$(3)$$

Where, p is the regional span of carbon emissions in the carbon market; δ_2 is the growth rate for carbon trading price, taken as 5%; E_c is the given carbon allowance.

4 TWO-STAGE DISPATCHING MODEL OF ELECTRIC POWER

4.1 Day-ahead unit combination optimization model

The main content of the day-ahead dispatching model is to develop a unit combination scheme (Chang et al. 2011). The traditional day-ahead unit combination optimization model mainly considers the minimum system energy consumption, but when combined with the carbon market, the energy consumption cost and carbon market transaction cost need to be considered comprehensively. The objective function is as follows.

$$
\min \sum_{t=0}^{T} \sum_{i=1}^{G} (C_i(P_{(i,t)}) + S_i + F_{p(i,t)})
$$

$$(4)$$

where: G is the number of units; T is the scheduling time; $C_i(P_{(i,t)})$ is the operating cost of the unit; S_i is the start and stop costs of the unit.

Restrictions:

1) System power balance constraints

$$
\sum_{i=1}^{G} P_{(i,t)} = P_{load}^{t} \quad \forall t \in T
$$

$$(5)$$

Where, $P_{(i,t)}$ is the output of the unit at moment t.

2) Spinning reserve constraints

$$
\sum_{i=1}^{G} v_{(i,t)} P \max_i \geq P_{load}^{t} + P_r^{t} \quad \forall t \in T
$$

$$(6)$$

Where, $v_{(i,t)}$ is the unit state variable; 0 is shutdown; 1 is startup; P_r^{t} is the spinning reserve capacity.

3) Unit capacity limit

$$
v_{(i,t)} P \min_i \leq P_{(i,t)} \leq v_{(i,t)} P \max_i \quad \forall t \in T, \forall i \in G
$$

$$(7)$$

4) Minimum start and stop time constraints of the unit

$$
\begin{cases}
(v_{(i,t-1)} - v_{(i,t)})(M_{(i,t-1)} - UT_i) \geq 0 \\
(v_{(i,t)} - v_{(i,t-1)})(- M_{(i,t-1)} - DT_i) \geq 0
\end{cases}
\quad \forall t \in T, \forall i \in G
$$

$$(8)$$

Where, $M_{(i,t-1)}$ is the time the unit has been running, UT_i and DT_i are the minimum start time and stop time of the unit.

5) Unit ramping constraints

$$\begin{cases} P_{i,t-1} - P_{i,t} \leq Pdown_i \\ P_{i,t} - P_{i,t-1} \leq Pup_i \end{cases} \forall t \in T, \forall i \in G \tag{9}$$

Where, Pup_i and $Pdown_i$ are the ramping upper limit and lower climbing limit of the unit.

6) Unit startup and shutdown cost constraints

$$\begin{cases} S_i = C_i^{on} + C_i^{off} \\ C_i^{on} \geq \lambda_i^{on}(v_{(i,t)} - v_{(i,t-1)}) \\ C_i^{off} \geq \lambda_i^{off}(v_{(i,t-1)} - v_{(i,t)}) \end{cases} \tag{10}$$

Where, C_i^{on} and C_i^{off} are the unit startup cost and unit stop cost; λ_i^{on} and λ_i^{off} are the single start and stop costs of the unit.

4.2 Real-time generation scheduling model

The first stage model seeks the optimal combination of generation cost and carbon cost based on the system constraints and finally obtains the start-stop schedule of the unit. The second stage of the real-time generation dispatch model is based on the existing day-ahead unit start/stop schedule. Under the condition of ensuring the safety of system power flow, the energy consumption of power generation and the cost of real-time carbon market transactions are minimized. This objective function is:

$$\min \sum_{t=0}^{T} \sum_{i=1}^{G} (a_i(P_{(i,t)})^2 + b_i(P_{(i,t)}) + c_i v_{(i,t)} + F_{p(i,t)}) \tag{11}$$

Where, a_i, b_i and c_i are the consumption parameters of the generator set; $P_{(i,t)}$ is the output of generator sets in the spot market; $v_{(i,t)}$ is the state variable of the unit, obtained according to the one-stage model; $F_{p(i,t)}$ is the transaction cost for the real-time carbon trading market.

$$F_{p(i,t)} = E_{(i,t)} \times \sigma_t \tag{12}$$

Where, $E_{(i,t)}$ is real-time carbon emissions for units; σ_t is real-time carbon price in carbon trading.

In addition to the constraints (5), (7), and (9), the line power flow safety constraints need to be considered in the intraday real-time scheduling model:

$$Pline_{l,\min} \leq Pline_{(l,t)} \leq Pline_{l,\max} \quad \forall l \in L \tag{13}$$

Where, $Pline_{(l,t)}$ is the current flowing on line l at moment t; $Pline_{l,\min}$ and $Pline_{l,\max}$ are the upper and lower limits of the line flow.

5 OPTIMIZATION PROCESS ANALYSIS

The two-stage model is solved using the GUROBI solver, and the flowchart for the solution of the model is shown in Figure 3.

6 EXAMPLE ANALYSIS

In this section, a modified IEEE14-node system is used for analysis, and the model is solved by invoking the GUROBI solver under the MATLAB software platform. As shown in Figure 4, there are five generators and 21 transmission lines in the system. The basic parameters of the units are shown in Tables 1–2, the 24-hour day-ahead load is shown in Figure 5, and the coal price is 50 ¥/t.

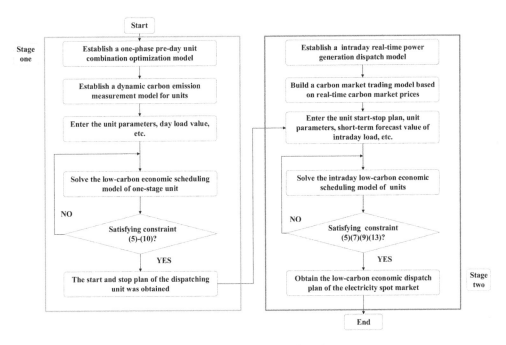

Figure 3.   Low-carbon dispatching of two-stage power system flow chart.

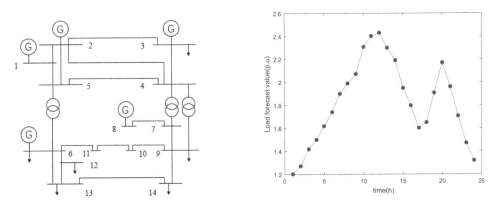

Figure 4.   Improved IEEE14 node system diagram.   Figure 5.   Day-ahead load.

6.1   One-stage model results analysis

According to the day-ahead unit combination optimal dispatch model and the day-ahead unit low-carbon economy combination optimal dispatch model, the generation schedule and carbon emissions of each unit are shown in Figures 6–9.

From Figure 6, it can be concluded that without considering the carbon market transaction cost, most of the load is carried by Unit 1 because it has the largest adjustment range and high shutdown cost. Compared with other units, Unit 2 and Unit 3 consume the least amount of coal. The load that exceeds the upper limit of Unit 1 will be shared by Units 2 and 3 when the load is met. During the day, only three units will be started considering meeting the load and unit constraints. The generation of the units will cause a change in the carbon emissions of the system. Figure 7 shows

Table 1. Basic parameters of system generator (1).

Generator number	Generator serial number	Active upper limit (p.u)	Active lower limit (p.u)	Generator cost characteristic factor		
				a (t/(p.u)^2)	b(t/(p.u))	c(t)
1	1	2.0	0.5	0.00375	2	0
2	2	0.8	0.2	0.00175	1.75	0
3	3	0.5	0.15	0.0625	1	0
4	6	0.35	0.1	0.00834	3.25	0
5	8	0.3	0.1	0.0025	3	0

Table 2. Basic parameters of system generator (2).

Generator number	Generator ramp rate (p.u/h)	Generator minimum start and stop time (h)	Generator single start cost ($)	Generator single stop cost ($)
1	0.375	2	176	50
2	0.3	2	187	60
3	0.15	2	75	30
6	0.2	2	100	85
8	0.15	2	50	52

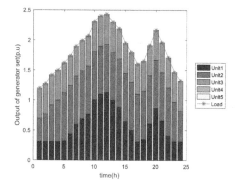

Figure 6. Unit mix plan and output excluding carbon market transaction costs.

Figure 7. Unit carbon emissions excluding carbon market transaction costs.

that the higher the generation capacity of the units, the more carbon emissions and the greater the carbon cost.

Figure 8 shows how the day-ahead scheduling plan will change when the carbon market's transaction cost is factored in. Even though Units 1, 2, and 3 are "primary generators," Unit 1 will cut some of the output from the initial condition due to the cost of carbon market transactions. Unit 2 and a portion of the output of Unit 3 are also substituted by other units in the system to minimize the total cost of the system. In this scenario, the system's five generators will collaborate. Carbon emissions will fluctuate according to changes in the generator set's output. As a result, as shown in Figure 9, the output of Unit 1 falls in this model, and carbon emissions fall as well.

Units 4 and 5 are kept at their highest outputs during peak power consumption hours, as illustrated in Figure 10, to guarantee that consumers' electricity demands are met. As a result, Unit 4's overall output level is higher than Unit 5's. Unit 5's carbon emissions will likewise be greater. Because Unit 1 and Unit 2 produce about the same quantity of power, carbon emissions are nearly identical.

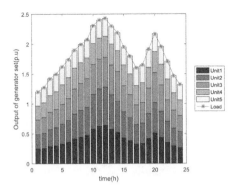

Figure 8. Unit mix plan and output considering carbon market transaction.

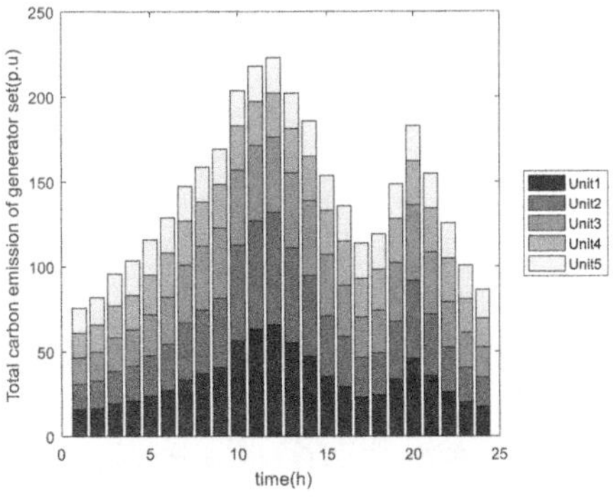

Figure 9. Carbon emissions of units.

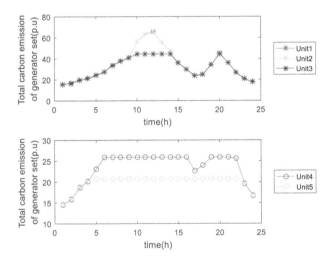

Figure 10. Carbon emissions of 5 generator sets taking into account carbon trading costs.

Because the highest power output limit for Units 4 and 5 is lower than the upper limit for other units, the overall carbon emissions are lower.

6.2 Analysis of the results of the two-stage model

Because intraday real-time scheduling must account for unit ramping limits, the scheduling time period for intraday real-time scheduling is 45 minutes, while the carbon market is cleared and settled every 15 minutes. The intraday carbon market price in the model is set at 70/t to simplify the study. The power distribution plan of the unit is developed using short-term forecast data, combined with the start-stop plan of the unit combination optimization model, and under the condition of meeting the safety restrictions. The results of the scheduling model for 1 h–12 h are shown in the diagram below.

The overall difference between the intraday unit output and the day-ahead unit output plan is minor because the intraday load prediction results are not considerably different from the load forecast results in the day-ahead dispatch model. The carbon emissions of the generator sets will be slightly different than the carbon emissions originally projected due to the combined effect of the energy market and the carbon market. For example, due to the impact of high carbon pricing, the overall electricity output of Unit 1 grew at a slower rate in the first 12 hours than planned.

29

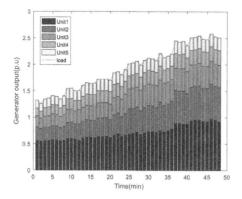

Figure 11. Real-time generator output change chart within a day (1 h–12 h).

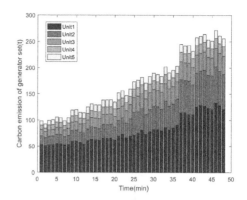

Figure 12. Intraday real-time generator carbon emissions change graph (1 h-12h).

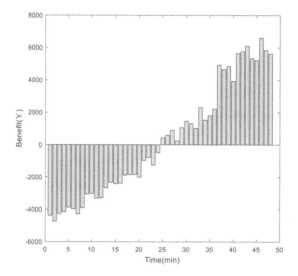

Figure 13. Intraday total system cost change graph (1 h-12 h).

Figure 13 shows the total cost of the spot market and the carbon market. For the period 0 h-6 h of low electricity consumption, the generator set's output will decrease, and the carbon emission will also decrease. The transaction cost of the carbon market is related to the carbon quota, so the difference in carbon emissions that is lower than the carbon allowance can be sold for additional income. Starting from 7:00, the demand for system load continues to rise, the carbon emissions of power plants will increase accordingly, and the carbon cost of the system will also increase accordingly, and corresponding allowances need to be purchased from the carbon market.

7 CONCLUSION

To guide the precise low-carbonization of the power system and let the power plants actively take responsibility for "carbon emission reduction," this paper proposes a two-stage power economic dispatch model based on the dynamic carbon emission measurement model and verifies the rationality of the model through simulation examples. The main conclusions are as follows:

(1) Compared with the carbon accounting of the constant carbon emission factor, the dynamic carbon emission measurement model can reflect the changing trend of the emission factor caused by the output of the unit and conduct more accurate carbon emission accounting for the unit, which is conducive to the development of the carbon market and the electricity market.

(2) Based on the clearing price of the carbon market, the intra-day real-time scheduling in the second stage can reasonably distribute the load of the units according to the optimization results of the first stage to meet the overall low-carbon demand of the system.

The proposed model reduces the total carbon emissions and provides a more feasible reference scheme for the power dispatching method under a low-carbon background. On this basis, follow-up research plans to further study the "carbon responsibility" sharing mechanism, so that power plants and users share the responsibility for carbon emissions.

ACKNOWLEDGMENTS

This work was financially supported by the Science and Technology Project of State Grid of Shanxi Electric Power Company (520531220003), based on the key technology of "electricity-carbon" metering in the process of achieving the "double carbon" target.

REFERENCES

Chang Yong, Wang Hongjiang, Chen Li, Gui Junping. Optimal dispatching of power system considering the combined action of multiple low-carbon factors [J] *Journal of Wuhan University*. 2018, 51(12)

Chen Xi, Yuan Mengling, Wang Song, Mu Jingxia, Peng Honghua, Wang You. Optimal operation of integrated energy system considering the impact of carbon trading on wind power consumption [J/OL]. *Journal of Chongqing University of Technology* (NATURAL SCIENCE):1–9 [2021-12-28]

Chen Jinpeng, Hu Zhijian, Chen Jiabin, Chen Yingguang, Gao Mingxin, Lin Mingrong. Optimal scheduling of integrated energy system considering stepped carbon trading and flexible dual response of supply and demand [J] *High Voltage Technology*. 2021, 47(09):30943106.

Cui Yang, Zeng Peng, Zhong Wuzhi, Cui Wenli, Zhao Yuting. Low carbon economic scheduling of electricity gas heat integrated energy system considering stepped carbon trading [J] *Power Automation Equipment*, 2021, 41(03).

Li Jiayao, Liu Weina. Low carbon economic dispatch of power system considering carbon trading and wind load prediction error [J] *Zhejiang Electric Power*, 2021, 40.

Qiu Bin, Song Shaoxin, Wang Kai, Yang Zhen. Optimal operation of regionally integrated energy system considering demand response and stepped carbon trading mechanism [J/OL] *Journal of Power System and Automation*:1–16 [2021-12-28].

Tian Kuo, Qiu Liuqing, Zeng Ming. Power grid planning model based on dynamic carbon emission price [J] *Chinese Journal of Electrical Engineering*, 2012, 32(04):57–64+18.

Tan Chao. *Study on Monitoring Methods of Carbon Emission in Coal-Fired Power Plants* [D] South China University of Technology 2018.

Wang Nan. *Study on Optimization Model and Method of Generation Dispatching* [D] North China Electric Power University, 2011.

Yu Honghai, Yin Zhuochao, Qu Litao, Li Chao. Analysis of carbon dioxide emission characteristics of coal-fired power plants based on empirical fitting method [J] *Energy and Environmental Protection*, 2018, 40.

Zhang Xiaohui, Liu Xiaoyan, Zhong Jiaqing. Integrated energy system planning considering reward and punishment stepped carbon trading and power heat transfer load uncertainty [J] Chinese *Journal of Electrical Engineering*. 2020, 40(19):6132–6142.

Zhang Xiaohui, Yan Keke, Lu Zhigang, Zhong Jiaqing. Multi-objective low carbon economic dispatch of wind power system based on scenario probability.[J]*Power Grid Technology*, 2014, 38(07).

Zhang Lihui, Dai Guyu, Nie Qingyun, Tong Zihao. Economic dispatch model of virtual power plant considering electricity consumption under a carbon trading mechanism.[J] *Power System Protection and Control*, 2020, 48.

Advances in Energy, Environment and
Chemical Engineering – Abdullah & Osman (Eds)
© 2023 The Author(s), ISBN: 978-1-032-36083-6

Study on reservoir types and characteristics of Shuguang buried hill in Liaoxi depression

Tianye Li*, Yachun Wang & Dandan Wang
School of Earth Science, North East Petroleum University, Daqing, China

Lei Zhang, Xiangxiong Jin & Yinglin Liu
School of Petroleum Engineering, Chongqing University of Science and Technology, Chongqing, China

ABSTRACT: Buried hill belt has been a favorable accumulation area of oil and gas since ancient times. In order to clarify the reservoir type and oil-water distribution law of Shuguang buried hill belt in Western Liaohe depression, through previous research results and fine anatomy of discovered reservoirs, divide and identify reservoir types, and then analyze the law of oil and water distribution. The results show that five types of reservoirs are mainly developed in this area: unconformity shielding reservoir, fault reservoir, buried hill inner reservoir, stratigraphic overlap reservoir, and fault unconformity shielding reservoir. Weathering crust and interior are the main accumulation parts. Oil and gas are distributed along the fault strike on the plane and can be accumulated vertically in the high, medium, and low parts of the structure. Clarifying this reservoir type plays a supporting role in the development of the buried hill reservoir in Liaoxi sag.

1   INTRODUCTION

From the discovery of the Xinglongtai buried hill reservoir in 1972 to the successful drilling of the Shugu 1 well in 1979, it set off a climax of buried hill exploration in the western depression (Leng et al. 2008; Liu 2007; Shan 2007). With the continuous improvement of the degree of exploration and development, the research shows that the reservoir types inside the Shuguang buried hill in the western depression of Liaohe are complex and diverse (Gao et al. 2009; Li et al. 2009; Li 2018; Yang et al. 2016; Zang et al. 2009). Because different types of reservoirs should have different production methods, understanding the reservoir types is the prerequisite for the efficient development of oil fields, and grasping the distribution law of oil and water is conducive to improving the exploration efficiency and effect. The fine dissection of the discovered reservoirs, the classification and identification of reservoir types, and the analysis of oil-water distribution law provide the geological basis for further exploration and development and target prediction of favorable areas in this area, which is important petroleum geological significance.

2   REGIONAL GEOLOGICAL BACKGROUND

Liaohe fault depression is located in the northeast corner of the Bohai Bay Basin. It is a Meso Cenozoic rift basin. The western depression is the main secondary negative structural unit of Liaohe fault depression. The study area covers an area of about 314 km^2. The buried hill was formed due to the uplift, denudation, and sedimentary discontinuity of the Bohai Bay Basin caused by the Caledonian movement in the Late Ordovician. According to the drilling data, the strata in

---

*Corresponding Author: lty_0925@163.com

DOI 10.1201/9781003330165-5

the study area are Archean, Proterozoic, Paleozoic, Mesozoic, and Cenozoic from bottom to top. The main target strata of this study are Paleozoic and Proterozoic buried hill strata. The main oil sources of the buried hills in this area are the dark mudstone and oil shale in the fourth and third members of the Shahejie formation. Metamorphic rocks and Middle Upper Proterozoic carbonate rocks are the main reservoirs, mudstone and slate are the caprocks, and multiple sets of spacers are developed.

The study area is mainly controlled by NE trending faults. Under the cutting of NE trending and near EW trending faults, the whole buried hill zone is divided into multiple buried hill fault blocks. The long axis direction of each buried hill is mainly NE trending, distributed in NE trending strips, with three buried hill zones of high, medium, and low. The top shape of the buried hill is characterized by high in the north, low in the south, high in the West, and low in the East. Since sedimentation, it has experienced multi-stage tectonic cycles. It has been transformed by the formation of reverse torsional compression, along with torsional tension and multi-stage faulting. In the study area, the fault activity is strong and frequent in the Mesozoic early and middle Paleogene, showing obvious inheritance and regeneration. The fault activity gradually weakens in the middle and late Paleogene, and there is basically no new fault development in the Neogene, showing the stage of structural evolution.

3 RESERVOIR TYPE

Reservoir type has always been one of the important contents of petroleum geology research, which is of great significance to the selection of oil and gas exploration direction and the way of oilfield development. Buried hill reservoir is not a single reservoir type, but the general name of reservoirs existing in the crystalline basement. Buried hill reservoirs can be divided into many types according to the structural (evolution) characteristics, trap characteristics, reservoir characteristics, and reservoir forming combinations of buried hill reservoirs. Liaohe buried hill reservoir has experienced the initial discovery stage, weathering crust control stage, inner oil and gas reservoir, and bedrock oil and gas reservoir. According to the shape of the buried hill and the main controlling factors of its formation, and through the fine anatomy of the discovered buried hill reservoirs, the reservoirs in this area are mainly divided into five categories. Unconformity barrier reservoir, fault reservoir, buried hill inner reservoir, stratigraphic overlap reservoir, and fault unconformity barrier reservoir (Figure 1).

3.1 *Stratigraphic unconformity barrier reservoir*

The stratigraphic unconformity shielding trap is formed by taking the impermeable stratum above the stratigraphic unconformity surface as the shielding condition and is located below the stratigraphic unconformity surface. Oil and gas accumulation in a stratigraphic unconformity trap is a stratigraphic unconformity reservoir. According to whether there is a structural form under unconformity in Shuguang buried hill of Western Sag, the stratigraphic unconformity shielding reservoir can be divided into latent denudation protrusion reservoir and latent denudation monoclinic reservoir. Among them, the latent denuded monoclinic reservoir is divided into the downhill latent denuded monoclinic reservoir, and the reverse slope latent denuded monoclinic reservoir.

3.1.1 *Latent denudation protrusion reservoir*

This kind of reservoir refers to the trap formed by the paleotopographic protrusion (no obvious structural form) covered by the overlying impermeable stratum, in which oil and gas are accumulated. As shown in Figure 2a, the S101 well in the study area is such an oil reservoir. Generally, traps formed by unconformity or unconformity fault joint sealing above and in the upward direction of the reservoir form massive oil and gas reservoirs. The high point of the ground level structure is a favorable reservoir forming position.

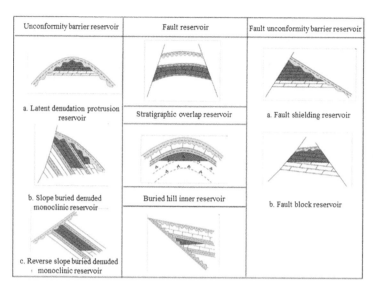

Unconformity barrier reservoir	Fault reservoir	Fault unconformity barrier reservoir
a. Latent denudation protrusion reservoir	Stratigraphic overlap reservoir	a. Fault shielding reservoir
b. Slope buried denuded monoclinic reservoir	Buried hill inner reservoir	b. Fault block reservoir
c. Reverse slope buried denuded monoclinic reservoir		

Figure 1.   Reservoir types of Shuguang buried hill in western sag.

3.1.2   *Slope buried denuded monoclinic reservoir*

This type of oil and gas reservoir is mainly distributed inside the buried hill slope area. Figure 3B shows this kind of reservoir at well s125 (Figure 2b) in the study area. The reservoir is an inside monocline, the top of which is covered by the unconformity surface of the buried hilltop, and the two sides are clamped by the internal interlayer of the buried hill. There are also reverse slope buried denuded monoclinic reservoirs, distinguished by the formation tendency opposite to the unconformity surface.

3.2   *Fault reservoir*

As shown in Figure 2c, well s54 in the study area is a reservoir. The single fault or double fault structural fault block trap type formed on the monoclinic gentle slope landform background can form a reservoir in the high part of the barrier and fault due to the shielding of the inner barrier.

3.3   *Buried hill inner reservoir*

In the exploration stage, the reservoir of well SG1 has been proved to be an internal reservoir of the buried hill, which is an oil and gas reservoir type with the fracture system inside the buried hill as the reservoir and the internal tight layer as the shielding. As shown in Figure 2d, s605 in the study area is an oil reservoir. The inner oil reservoir of the buried hill is not in direct contact with the unconformity surface of the buried hill, and the inner barrier forms a barrier. In the formation of oil and gas reservoirs, the tight rock layer acts as an interlayer, which divides the oil and gas reservoirs into layered distribution, and reservoirs with different oil-water interface heights can be formed on the same paleogeomorphic buried hill.

3.4   *Stratigraphic overlap reservoir*

Stratigraphic overlying reservoir refers to the reservoir formed by the reservoir overlying the impermeable unconformity surface and being covered by the impermeable stratigraphic overlying. As shown in Figure 2e, well s56 in the study area mainly takes PZ limestone as the reservoir, develops a large set of mudstone as the caprock, and develops multiple sets of the thicker interlayer. The trap condition is that the Paleozoic stratum laterally overlaps the Proterozoic top unconformity surface, and the top denudation pinches out, forming top unconformity shielding and lateral stratum overlying unconformity shielding.

3.5 *Fault unconformity barrier reservoir*

Due to the long-term activity of faults, the closure amplitude of buried hills is increasing, and then fault unconformity shielding reservoirs are formed. Such reservoirs are generally divided into two types: (1) As shown in Figure 2f, well S107 belongs to the single-side Mountain Fault shielding reservoir in the study area. (2) As shown in Figure 2, well sg102 in the study area is a fault block reservoir. It is mainly formed by fault activity. The bedrock block is lifted under the influence of fault activity to form a fault block buried hill. The main feature is that the fault edge is the highest part, and the occurrence of the overlying stratum is parallel to the erosion surface. This buried hill was eroded before it became a mountain, and its erosion surface is mostly slope.

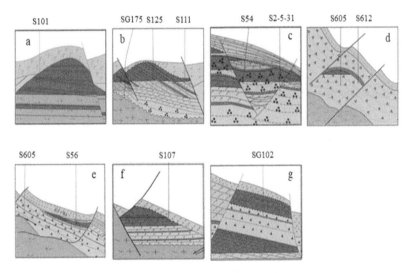

Figure 2.   Main hydrocarbon reservoir types in Shuguang buried-hill of western sag.

4   RESERVOIR AND OIL-WATER DISTRIBUTION CHARACTERISTICS

4.1 *Reservoir distribution characteristics*

Reservoirs are mainly distributed in Paleozoic strata. Oil and gas are accumulated in high, medium, and low buried hills on the plane, and the formed reservoirs are distributed along the fault strike. There are differences in reservoir size. As shown in Figure 3, reservoirs can be formed at the high and low parts of the structure in the vertical buried hill, with weathered crust and the same zone inside.

4.2 *Vertical distribution characteristics of oil and water*

Vertically, in each set of inner strata, different degrees of the oil layer, oil-bearing water layer, water-bearing oil layer, and oil-water same layer are developed. The main factors are high buried hills → low buried hills. The distance between the lowest oil layer and the top of the buried hill gradually increases. That is, the closer to the oil source area, the higher the oil column height. The lowest oil-bearing layer in the sg131 block of the medium-high buried hill is 13m away from the top of the buried hill. The lowest oil-bearing layer in block sg98 is 22.5 ~ 140.96m away from the top of

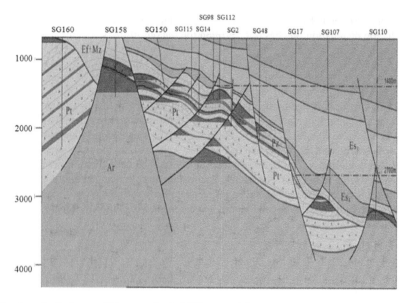

Figure 3.   Longitudinal map of Shuguang buried-hill reservoir in western sag.

the buried hill. The lowest oil-bearing layer in block sg112 is 22.8 ~ 194.1m away from the top of the buried hill. The lowest oil-bearing layer in low buried hill block dg79 is 49.84 ~ 279.93m away from the top of the buried hill. The distance between the lowest oil-bearing layer in block S103 and the top of the buried hill is 28 ~ 387.19m. The distance between the lowest oil-bearing layer in block S107 and the top of the buried hill is 64.43 ~ 378.5m. The distance from the lowest oil-bearing layer in the s56 block to the top of the buried hill is 139.96m. The minimum oil-bearing layer in block s125 is 47 ~ 475.96m away from the top of the buried hill. The minimum distance from the buried hilltop of SG 31192.92 is 38 ~ 36m.

The oil-water interface increases with the buried depth of the buried hill. If the low buried hill is closer to the oil source, the oil-gas filling is more thorough (Figure 4). Due to the relationship between reservoir and interlayer in the formation of the buried hill stratum, when the oil-gas reservoir is formed, the tight rock stratum acts as an interlayer, dividing the oil-gas reservoir into layered distribution with multiple independent oil-water systems. Reservoirs with different heights of the oil-water interface are formed on the same paleogeomorphic buried hill, thus causing the difference in oil and gas distribution.

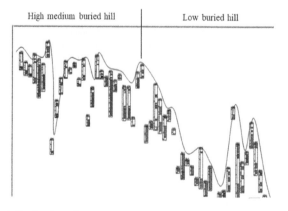

Figure 4.   Longitudinal distribution relationship of oil and water.

5 CONCLUSION

There are five types of buried hill reservoirs developed in the study area, which are ① unconformity shielding reservoirs, including underlying denudation protrusion reservoir, along slope latent denudation monoclinic reservoir and reverse slope latent denudation monoclinic reservoir; ② fault reservoir; ③ buried hill reservoir; ④ stratigraphic overlap reservoir; ⑤ fault unconformity shielding oil reservoir, including single side mountain fault shielding oil reservoir and fault block oil reservoir. However, it is mainly buried hill inner reservoir and stratigraphic overlap and unconformity shielding reservoir. Clarifying this reservoir type plays a supporting role in developing the buried hill reservoir in Liaoxi sag.

Oil and gas are distributed along the fault direction on the plane, and high, medium, and low buried hills can form reservoirs. Vertically, the weathering crust and the inner part form reservoirs, which can be formed in the high and low parts of the structure. The oil-water interface increases with the buried depth of the buried hill. If the low buried hill is closer to the oil source, the oil-gas filling is more thorough, and the oil column height is higher. However, because the tight rock layer acts as an interlayer, the reservoir has multiple independent oil-water systems, resulting in a difference in oil-gas distribution.

REFERENCES

Gao Ping, Jiang Suhua, Wang Zhiying. Formation mechanism and accumulation model of buried hill reservoir in Jiyang Depression [J]. *Marine Geology Dynamics*, 2009, 25(03): 1–6.

Hu Yang, Liu Huimin, Hao Xuefeng. Characteristics and controlling factors of glutenite reservoir in steep slope zone of faulted lacustrine basin: A case study of Paleogene in Dongying Sag [J]. *Geological Review*, 2019, 65(S1): 151–152.

Huo Jin, Zhi Dongming, Zheng Menglin, et al. Characteristics and main controlling factors of shale reservoirs in Lucaogou Formation, Jimusar Sag, Junggar Basin [J]. *Petroleum Geology & Experiment*, 2020, 42(04): 506–512.

Leng Jigao, Pang Xiongqi, Li Xiaoguang, et al. Main controlling factors of hydrocarbon accumulation in western sag of Liaohe Fault Depression [J]. *Journal of Palaeogeography*, 2008, (05): 473–480.

Li Donghao. *Analysis of accumulation characteristics and main controlling factors of buried hill reservoir in Shulu sag, Jizhong Depression* [D]. 2018, Chengdu: Southwest Petroleum University.

Li Xiaoguang, Liu Baohong, Cai Guogang. Genetic analysis of metamorphic buried hill inside oil reservoir in Liaohe Depression [J]. *Special Oil and Gas Reservoirs*, 2009, 16(04):1–5+12+104.

Liu Xingzhou. Identification and reservoir characteristics of buried hill reservoirs in western Liaohe Oilfield and Damuntun sag [J]. *Petroleum Geology and Engineering*, 2007, (04): 5–8.

Shan Junfeng, Zhang Juxiang, Zhang Zhuo, et al. Study on hydrocarbon accumulation conditions of buried hill in west slope of Western sag of Liaohe Fault Depression Basin[J]. *Special Oil and Gas Reservoirs*, 2004, (04): 39–42+124–125.

Shan Junfeng. *Study on reservoir forming conditions inside metamorphic buried hill in Liaohe Depression* [D]. 2007, Beijing: China University of Geosciences.

Shen Weibing, Pang Xiongqi, Chen Jianfa, et al. Types and distribution characteristics of deep oil and gas reservoirs in Tarim Basin[J]. *Geological Journal of China Universities*, 2017, 23(02): 324–336.

Sun Yu. *Study on high-resolution sequence stratigraphy and lithologic reservoir formation in fluvial delta system* [D]. 2010, Daqing: Northeast Petroleum University.

Tian Shifeng, Zha Ming, Wu Kongyou, et al. Distribution and enrichment of buried hill oil and gas in Raoyang depression[J]. *Marine Geology & Quaternary Geology*, 2009, 29(04): 143–150.

Xing Yawen, Zhang Yiming, Jiang Shuanqi, et al. Characteristics and distribution of oil and gas reservoirs in Wulanhua Sag, Erlian Basin [J]. *China Petroleum Exploration*, 2020, 25(06): 68–78.

Xu Kun, Fu Wan. Tectonic movement of Liaohe Basin[J]. *Journal of Liaoning Normal University* (Natural Science Edition), 1996, (01): 58–63.

Yan Zijie, Jiang Nengdong, Ji Shuangwen.Classification of buried-hill types and hydrocarbon accumulation in northeast Jiyang Depression[J]. *China Petroleum Exploration*, 2008, 13(06): 15–18+80.

Yang Keji, Qi Jiafu, Yu Yixin, et al. Differential evolution and accumulation conditions of buried hills in Liaodong Bay area [J]. Natural gas geoscience, 2016, 27(06): 1014–1024.

Yao Jifeng, Liao Xingming, Yu Tianxin. Structural analysis of Liaohe Basin[J]. *Fault-block Oil and Gas Field*, 1995, (05): 21–26.

Zang Mingfeng, Wu Kongyou, Cui Yongqian, et al. Types of buried hills and hydrocarbon accumulation in Jizhong Depression[J]. *Journal of Oil and Gas Technology*, 2009, 31(02): 166–169+6.

Zeng Rong Cai, Hu Cheng, Dong Xia. Analysis of the inner structure and accumulation conditions of buried hill in Liaoxi Depression[J]. *Lithologic reservoirs*, 2009, 21(04): 10–18.

Zhang Shanwen. Hydrocarbon accumulation characteristics and exploration prospect of unconformity in the marginal strata of Junggar Basin[J]. *Petroleum Geology & Experiment*, 2013, 35(03): 231–237+248.

Zhang Ting, Wang Ke, Luo Anxiang, et al. Formation and model of Chang 7 tight oil reservoir in Triassic Yanchang Formation, Ordos Basin [J]. *Mineral Exploration*, 2021, 12(02): 295–302.

Zhao Huimin, Liu Xuesong, Meng Weigong, et al. Subtle reservoirs and their accumulation dynamics in Shuguang Lejia area[J]. *Journal of Jilin University* (Earth Science Edition), 2011, 41(01): 21–28.

Advances in Energy, Environment and
Chemical Engineering – Abdullah & Osman (Eds)
© 2023 The Author(s), ISBN: 978-1-032-36083-6

Study on evolution characteristics of 2410 fully mechanized caving face in Yuhua coal mine by numerical simulation of FLAC^{3D}

Kun Zhang, Sen Zhang* & Man Wang
School of Architecture and Civil Engineering, Xi'an University of Science and Technology, Xi'an, Shaanxi, China

Dong Yue, Xiaoquan Huo & Zengyun Yuan
Shaanxi Coal Tongchuan Mining Co., Ltd., Tongchuan, China

ABSTRACT: The roof water disaster of the coal seam in the western mining area has become a great threat to mine safety production. In order to study the failure evolution characteristics of mining disturbed rock, FLAC^{3D} numerical simulation software was used to study the mining-induced rock "three zones" height of 2410 fully mechanized caving face in the Yuhua Coal Mine. The dynamic changes of stress evolution, fracture distribution, and displacement of strata in the mining field after mining are obtained by simulation analysis. The height of the bending subsidence zone is 446.4 m, the height of the fracture zone is 153.6, and the height of the caving zone is 22 m. Through the comparative analysis of the empirical formula and the numerical simulation results and the field measured "three zones" height values after mining in the mining area, the numerical simulation results are more consistent with the measured values, which indicates that the numerical simulation method of mining rock "three zones" can be used as an effective means for the study of mining rock "three zones" in the working face, and provides the necessary technical support for the safe and efficient mining of fully mechanized caving face in coal mines.

1   INTRODUCTION

After the coal seam mining, with the advance of the coal mining face, the coal seam strata under the influence of mining have regular deformation, fracture, movement, subsidence, and large-area caving, forming fracture zone, bending subsidence zone, and caving zone (hereinafter referred to as "three zones") (Qian & Xu 2019; Zhang & Wang 2014). Studying the evolution law and distribution characteristics of the "three zones," especially the determination of the height of the caving zone and fault zone, has important guiding significance for determining the mining upper limit, the location of high-level borehole end holes, the layout horizon of high-level drainage roadway, the location of high-level borehole end holes of gas, preventing roof water inrush and optimizing coal mining technology (Gao 2013; Yang et al. 2020).

At present, many scholars have carried out a lot of research on the evolution characteristics of the "three zones" of mining overburden and achieved fruitful results. The research methods adopted can be roughly divided into three categories: theoretical calculation (Gong & Jin 2004; Han et al. 2019; Xiang et al. 2017), simulation analysis (Huang 2018; Li et al. 2012; Lei et al. 2015; Wang et al. 2009) and field measurement (Sun et al. 2013; Yang 2016; Yang et al. 2019). Field measurement has the advantages of authenticity and reliability, but it is difficult to comprehensively analyze the mechanism because of its large workload, long monitoring cycle, high cost, and complex influencing factors. The theoretical calculation has high applicability, but complex geological factors are often quantified or ignored, which is prone to engineering problems with large errors and cannot simulate the dynamic change process of overburden (Liu et al. 2012; Zhang & Li

---

*Corresponding Author: zhanghbf97@163.com

DOI 10.1201/9781003330165-6

2011). Simulation analysis includes physical similarity simulation tests and numerical simulation research. Through simulation analysis, the dynamic change process of overburden deformation and failure can be observed and studied indirectly. It is one of the important methods to study the evolution characteristics of "three zones" (Yang 2014).

Taking 2410 fully mechanized top coal caving face of Yuhua Coal Mine in Jiaoping mining area as the research object, according to the measured physical and mechanical parameters of overburden of 4-2 coal seam, and using the research methods of numerical simulation and empirical formula, this paper analyzes the evolution characteristics of cracks in the deformation and failure process of overburden under the influence of mining and simulates the height value of "three zones," which provides a basis for determining the distribution law of "three zones" of overburden in the fully mechanized top coal caving face of this coal mine.

2 PROJECT OVERVIEW

Yuhua Coal Mine is located in the Jiaoping mining area, Tongchuan City, Shaanxi Province. The 4-2 coal seam mined in 2410 fully mechanized top coal caving face in panel two of the coal seam has a stable occurrence, the buried depth is $524.2 \sim 748.4$ m, the thickness of the coal seam is about 8 m, the hardness is grade $2 \sim 3$, and the fractures are relatively developed. The coal seam adopts the comprehensive mechanized mining method of the inclined layered strike, long wall, full span, and large mining height, and the roof is managed by the full caving method. The pit of 2410 working face is located in the West Wing of panel two, which is the fifth working face in panel two, with an altitude of $1674.14 \sim 1447.09$ m. The overall shape of the coal seam shows a NE-SW trend, and the wide and gentle wavy monoclinic structure shows a SE trend. The lithology and engineering geological characteristics of the coal seam roof and floor strata in the Yuhua Coal Mine are shown in Table 1:

Table 1. Styles stratum characteristics of coal seam roof and floor in yuhua coal mine.

Stratigraphic system				Formation thickness	
Group	System	Unification	Formation	(m)	Lithologic characteristics
Mesozoic	Cretaceous K	Lower unification K1	Luohe formation K1l	417.00	The southern part of the coalfield is occupied by a large number of conglomerates and glutenite mixed with coarse-grained sandstone. The middle part is a coarse-grained rock with $5 \sim 7$ layers of conglomerates. The rock strata in the north and West are mostly coarse-grained sandstone.
			Yijun formation K1y	65.00	It is mostly composed of massive conglomerate with coarse gravel diameter in the South and fine gravel diameter in the north, filled with sand and argillaceous, cemented with calcium and iron.
	Jurassic J	Middle unification J2	Anding Formation J2a	123.00	It is composed of mudstone and sandy mudstone, intercalated with medium and coarse-grained sandstone and glutenite.
			Zhiluo Formation J2z	20.00	It is composed of medium coarse-grained arkose, trace mudstone, and sandy mudstone.
			Yan'an Formation J2y	115.00	It is mostly composed of mudstone, sandy mudstone, siltstone, and fine-grained sandstone, mixed with medium and coarse-grained sandstone.

Combined with the geological drilling data of the Yuhua Coal Mine, the coal seam roof overburden structure is classified into the soft type and hard type, as shown in Figure 1. The direct top of the fully mechanized top coal caving face is fine siltstone, with a general thickness of 5 m and a hardness of grade 4 ~ 5. It belongs to a medium-solid-to-very-weak roof. The main roof is mainly fine sandstone and medium coarse sandstone, with a general thickness of 12.9 m and a hardness of grade 5 ~ 7. It belongs to a moderately stable roof. The coal seam floor is mostly composed of carbonaceous mudstone and root-soil rock, with a general thickness of 1.1 m, belonging to an unstable to the extremely unstable floor. In order to make the simulation results closer to the engineering practice, rock blocks and cores are taken from each rock layer overlying the coal seam, and the rock physical and mechanical property parameters of each rock layer are obtained through indoor experiments, as shown in Table 2.

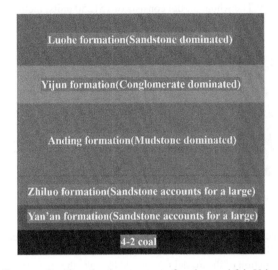

Figure 1.   Schematic diagram of roof overburden structure of coal seam 4-2 in Yuhua Coal Mine.

Table 2.   4-2 Physical and mechanical property parameters of overlying strata of coal seam.

Stratum name	Thickness/m	Density/ (kg·m^{-3})	Cohesion/ MPa	Internal friction angle/(°)	Tensile strength/ MPa	Bulk modulus (GPa)	Shear modulus (GPa)
Sandstone	30	2050	42.01	42	2.81	9.82	5.46
Interbedding of conglomerate and medium sandstone	280	2330	34.23	41	1.24	6.14	4.87
Sand shale interbed	90	2350	32.04	35	2.03	4.97	2.18
Siltstone	130	2350	40.21	40	3.12	9.95	6.52
Medium coarse sandstone	60	2580	61.13	38	2.51	8.24	5.87
Fine siltstone	10	2390	34.06	41	1.24	8.14	4.87
4-2 coal	8	1370	4.81	17	0.43	0.57	0.37
Carbonaceous mudstone	5	2080	32.07	28	0.81	9.43	6.56
Dark red mudstone	40	2050	25.22	23	1.13	8.43	6.16
Gray packsand	160	2350	51.01	39	2.23	8.24	6.87
Medium fine sandstone	200	2330	61.03	38	2.51	8.22	5.87

3 FLAC³ᴰ NUMERICAL SIMULATION OF "THREE ZONES" EVOLUTION CHARACTERISTICS OF MINING OVERBURDEN IN 2410 FULLY MECHANIZED TOP COAL CAVING FACE OF YUHUA COAL MINE

3.1 Establishment of longitudinal mining calculation model

The numerical calculation model of 2410 fully mechanized top coal caving face is established by using FLAC³ᴰ three-dimensional finite difference software, and the size is long (x) × Width (y) × Height (z) = 3500 m × 1 m × 1008 m. Mohr-Coulomb constitutive model and large strain deformation mode are adopted. The working face and its roof and floor rock stratum are composed of a hexahedral block grid. The bottom boundary of the model is set as the fully constrained boundary, and the horizontal displacement constraints are imposed on the x-direction boundary and Y-direction boundary. The whole model consists of 151250 units, including 305244 nodes. The three-dimensional numerical calculation model is shown in Figure 2.

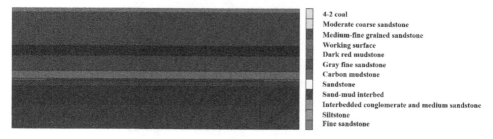

4-2 coal
Moderate coarse sandstone
Medium-fine grained sandstone
Working surface
Dark red mudstone
Gray fine sandstone
Carbon mudstone
Sandstone
Sand-mud interbed
Interbedded conglomerate and medium sandstone
Siltstone
Fine sandstone

Figure 2. Diagram of numerical calculation model of longitudinal mining.

When simulating the mining stage of the working face, step-by-step excavation is adopted, the step length is 2 m, and the maximum advancing length is 1900 m. Before the excavation calculation, the initial balance calculation of the model is required to make the rock stratum in the original rock stress state.

3.2 Analysis of simulation calculation results of longitudinal mining

3.2.1 Characteristic analysis of maximum principal stress

In the process of coal seam mining, the stress state of the original rock of the rock stratum is constantly disturbed, the stress balance is broken, the stress is redistributed, and the fracture of the surrounding rock of the stope is induced, resulting in the collapse of the rock stratum (Tang 2003). The goaf makes the stress of coal seam and its surrounding rock change dynamically, forming additional stress and causing mining influence. The distribution characteristics of the maximum principal stress of the overlying strata at different advancing distances of the working face are shown in Figure 3:

It can be seen from Figure 3 that when the working face is advanced for 500 m, the maximum principal stress is axisymmetrically distributed with the central line of the goaf, and the direct roof has appeared small-area caving, and with the advancement of the working face, the basic roof separation layer develops rapidly upward and produces large vertical cracks. When the working face advances 1100 m, the range of tensile stress area above the goaf expands gradually, and the maximum height of the fracture zone increases gradually, but the change of stress is not obvious. When the working face advances 1900 m, the goaf is filled with the collapsed rock stratum of the overlying rock stratum. Under the action of compressive stress, a high compressive stress area is formed in the middle of the rock stratum.

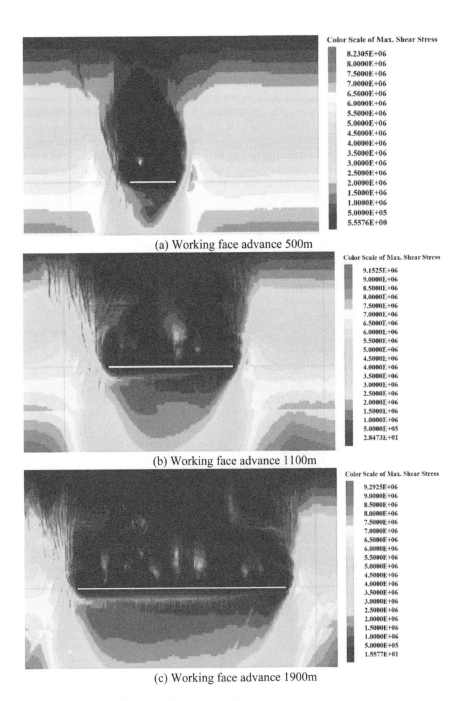

(a) Working face advance 500m

(b) Working face advance 1100m

(c) Working face advance 1900m

Figure 3.   Diagram of distribution maximum principal stress.

3.2.2   *Characteristic analysis of plastic zone*

When the working face advances to 500 m, the failure area is distributed in the shape of a ladder. The tension failure area expands rapidly within 85 m above the direct roof of the coal seam, and the thickness of the rock stratum in the bending subsidence zone reaches about 70 m, which develops to the surface of the bedrock with the advance of mining. When the working face advances 1100 m, the plastic failure zone develops to both sides of the roof rock above the working face, and the bending subsidence zone increases to about 210 m, accompanied by fracture and separation in the growth process. When the working face advances to 1500 m, the overburden failure height above the working face will no longer develop upward, and the scope of the overburden failure area will still extend to the mining direction. When the working face advances 1900 m, the compaction degree of the collapsed overburden is further strengthened, and the change in the height of the failure area tends to be stable. Since the maximum height of the plastic zone is equivalent to the maximum value of the fracture zone, the development height of the fracture zone can be determined by the range of tensile failure zone and tensile fracture zone in the plastic deformation zone of overburden during excavation. The whole rock stratum bends and sinks, and the range of unidirectional tension on both sides increases. At this time, the height of the fracture zone is 153.6 m, and the ratio of the height of the fractured zone to the mining height of the working face is 19.2.

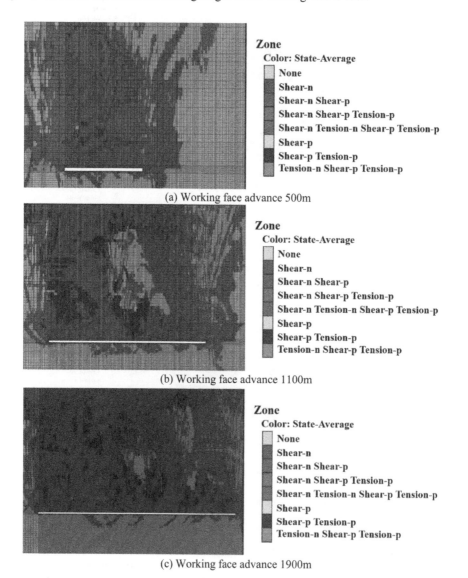

(a) Working face advance 500m

(b) Working face advance 1100m

(c) Working face advance 1900m

Figure 4.   Diagram of distribution plastic zone.

3.2.3 *Characteristic analysis of vertical displacement*

It can be seen from Figure 5 that the displacement of the upper overburden in the gob in the vertical direction gradually increases with the advance of mining. When the working face is advanced for 500 m, the vertical displacement of the direct top above the working face is about 63 cm. When advancing to 1100 m, the maximum subsidence of the roof is 1.5 m, and the deformation within 12 m away from the overburden of the coal seam roof increases; that is, all the strata in this range collapse. When the working face continues to advance to 1500 m, the deformation increases greatly, and the vertical displacement of the upper part of the goaf continues to extend to the advancing direction of the working face. At this time, the maximum subsidence of the roof is 3 m, about 19 m away from the roof of the coal seam. When the final mining is 1900 m, the subsidence reaches about 4 m. At this time, it is 22 m away from the coal seam roof, and the overburden is sunk and compacted within 295 m. The overburden deformation within 25 m from the coal seam roof tends to be stable. Therefore, 22 m is defined as the maximum height of the overburden caving zone in the fully mechanized top coal caving face.

4 DETERMINATION OF OVERBURDEN "THREE ZONES" HEIGHT OF 2410 FULLY MECHANIZED TOP COAL CAVING FACE OF YUHUA COAL MINE

4.1 *Empirical formula prediction of overburden "Three Zones" height*

According to the geological survey report of the Yuhua Coal Mine, the cumulative mining thickness of coal seam 4-2 in 2410 fully mechanized top coal caving face is 8 m. According to the calculation formula of fracture zone height given in the code for coal pillar reservation and coal pressure mining of buildings, water bodies, railways, and main shafts and roadways and the code for hydrogeological engineering geological exploration of the mining area, the fracture zone height of coal seam 4-2 after mining is expected to be:

$$H_{li} = 30\sqrt{\sum M} + 10 \tag{1}$$

$$H_{li} = \frac{100M}{2.4n + 2.1} + 11.2 \tag{2}$$

H_{li} – Maximum height of fracture zone, m;
M – Cumulative mining thickness of coal seam, m;
n – Number of coal seam layers.

The calculation formula for the height of mining overburden caving zone in coal seam mining (Zhang et al. 2017):

$$H_c = \frac{M}{(k - 1)\cos\alpha} + 3.73 \tag{3}$$

H_c – Maximum height of caving zone, m;
M – Cumulative mining thickness of coal seam, m;
α – Dip angle of the coal seam, (°), taken as $\alpha = 0°$;
k – Expansion coefficient of roof rock in goaf, taken as $k = 1.3$.

According to formula (1), the height of the fracture zone is 94.85 m, and the ratio of the height of the fractured zone to the mining height is 11.85. According to formula (2), the height of the fracture zone is 188.97 m, and the ratio of the height of the fractured zone to the mining height is 23.62. Therefore, the average height of the overburden fracture zone of coal seam 4-2 in 2410 fully mechanized top coal caving face of Yuhua Coal Mine is 141.91 m, and the average ratio of the height of the fractured zone to the mining height is 17.73. The height of the caving zone calculated from formula (3) is 30.4 m. The height of the bending subsidence zone is from the top boundary of the fracture zone to the surface to form a continuous and overall moving rock stratum. After full mining, the bending subsidence zone will develop to the surface (Huang 2013). The calculation results of the "three zones" empirical formula are shown in Figure 6.

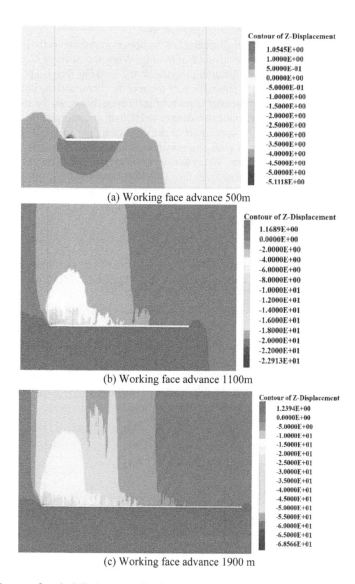

(a) Working face advance 500m

(b) Working face advance 1100m

(c) Working face advance 1900 m

Figure 5.   Diagram of vertical displacement distribution.

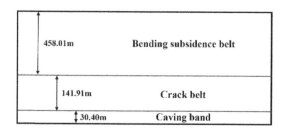

Figure 6.   Based on the empirical formula, the height distribution diagram of "three zones" of mining overburden in 2410 fully mechanized top coal caving face is predicted.

4.2 Comparison and analysis of "Three Zones" height results of overburden

Based on the above FLAC numerical simulation analysis, when the simulated mining reaches 1900 m of the working face, the maximum height of the fracture zone is about 153.6 m, the height of the overburden bending subsidence zone is about 446.4 m, the maximum height of the caving zone is about 22 m, and the ratio of the height of the fractured zone to the mining height is 19.2. See Figure 7 for the results of the FLAC^{3D} numerical simulation of "three zones" height.

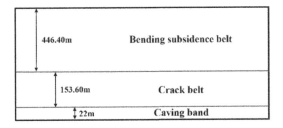

Figure 7. Based on FLAC^{3D} numerical simulation, the height distribution diagram of mining overburden "three zones" in 2410 fully mechanized top coal caving face is predicted.

Table 3. Statistics of mining overburden fracture zone height of 2410 fully mechanized top coal caving face in 4-2 coal seam of yuhua coal mine.

Computing method of the height of "three belts"	Empirical formula	FLAC^{3D} numerical simulation	Measured data of mining area	Error%
Fracture zone height/m	141.91	153.6	156	9.03/1.54
Height of caving zone/m	30.4	22	25.6	15.8/14.1
Height of bending subsidence zone/m	458.01	446.4	449.8	1.8/0.8
the ratio of the height of the fractured zone to the mining height	17.73	19.2	19.5	9.1/1.5

Table 3 shows the height results of "three zones" measured by different methods in fully mechanized top coal caving face of seams 4-2 in the Yuhua Coal Mine. Compared with the field measured value of the "three zones" height of the overburden under the influence of mining in 4-2 coal seam, it is found that the error of the calculation result of the empirical formula is slightly larger. The predicted value of the "three zones" height obtained by the 1:1 model of the numerical simulation method is closer to the field measured value, which is suitable for the calculation of the "three zones" of the mining overburden in the fully mechanized top coal caving face.

5 CONCLUSION

(1) With the advance of the working face, the range of tensile stress area above the goaf is gradually expanded, part of the overlying strata collapse, and a high compressive stress area is formed in the middle of the strata. The plastic failure zone develops along both sides of the roof rock above the working face, increasing the failure height of the overburden above the working face. With the continuous development of the mining of the working face, the failure height of the

overburden no longer develops upward, and the scope of the failure zone continues to extend to the mining direction.

(2) When the compaction degree of the collapsed overburden reaches certain mining progress, it is strengthened, and the changes in the height and displacement of the failure area tend to be stable.

(3) The height of the mining overburden fracture zone obtained by numerical simulation is 153.6 m, the caving zone is 22 m, the bending subsidence zone is 446.4 m, and the ratio of the height of the fractured zone to the mining height is 19.2. Compared with the calculated value of the empirical formula, the numerical simulation result is more in line with the field measured value.

(4) In the numerical simulation study of mining-induced rock "three zones," under the condition of sufficient research time, a three-dimensional mining model can be established to more intuitively explore the evolution law of "three zones" and improve the accuracy of "three zones" height determination.

ACKNOWLEDGMENT

The research was supported by the National Natural Science Foundation of China (No. 11872299,12072259), Coal Joint Fund of Shaanxi Natural Science Foundation of China (2019JLP-01).

REFERENCES

Gao, H., Yang, H. W., An, X. M. (2013) Study on the development law of "three zones" of hard roof in Gushuyuan mine. *J. Coal Eng.*, 45: 80–82.

Gong, P. L., Jin, Z. M. (2004) Study on the structure characteristics and movement laws of overlying strata with large mining height. *J. China Coal Soc.*, 29: 7–11.

Han, G., Li, X. D., Qu, X. C., Zhu, L., Liu, L. X. (2019) Study on correlation between spatial fracturing of overlying strata and distribution of mining stress field in stope. *J. Coal Sci Technol.*, 47: 53–58.

Huang, X. M. (2013) Application and determination method of vertical three regions of working face. *J. Coal Sci Technol*, (S2): 48–50.

Huang, Z. Z. (2018) Experimental study of top coal size effect on fully-mechanized caving mining under abutment pressure. *J. Coal Sci Technol.*,46: 63–67.

Lei, W. J., Feng, Y. J., Wang, Z.F. (2015) Simulation of cover rock caving zone and fractured distribution in fully mechanized caving mining by strength increase finite element method. *J. Min Safe Eng.*, 32: 623–627.

Li, W. S., Li, W., Yi, S. X. (2012) Study on development height of water flow crack zone in roof above fully mechanized one passing full seam mining face. *J. Coal Sci Technol*, 40: 104–107.

Liu, H. T., Ma, N. J., Li, J., Zhang, W. W., Zhang, S. P. (2012) Evolution and distribution characteristics of roof shallow fissure channel. *J. China Coal Soc.*, 37: 1451–1455.

Qian, M. G., Xu J. L. (2019) Behaviors of strata movement in coal mining. *J. China Coal Soc.*, 44: 973–984.

Sun, Q. S., Mou, Y., Yang, X. L. (2013) Study on "two zones" height of overlying of fully-mechanized technology with high mining height at Hongliu coal mine. *J. China Coal Soc.*, 38: 283–286.

Tang, C. A. (2003) *Numerical test of mining rock mass fracture and strata movement.* Jilin University Press, The Jilin.

Wang, Z. G., Zhou, H. W., Xie, H. P. (2009) Research on fractal characterization of mined crack network evolution in overburden rock stratum under deep mining. *J. Rock Soil Mech*, 30: 2403–2408.

Xiang, P., Sun, L. H., Ji, H. G. (2017) Dynamic distribution characteristics and determination method of caving zone in overburden strata with large mining height. *J. Min Safe Eng.*, 34: 861–867.

Yang, J. Z. (2016) Study on development law of water conducted zone in fully—mechanized mining face with 7 m mining height. *J. Coal Sci Technol*, 46: 61–66.

Yang, J. Z., Hu, B. W., Wang, Z. R. (2020) Study on distribution characteristics of collapse zone, fissure zone and curved subsidence zone and layered settlement of overburden on 8.8 m super-large mining height coal mining face. *J. Coal Sci Technol*, 48: 42–48.

Yang, M. D., Guo, W. B., Zhao, G. B., Tan, Y., Yang, W. Q. (2019) Height of water-conducting zone in longwall top coal caving mining under thick alluvium and soft overburden. *J. China Coal Soc.*, 44: 3308–3316.

Yang, P. (2014) Similar simulation of mining cracking evolution law for overburden strata above coal mining face. *J. Coal Sci Technol.*, 42: 121–124.

Zhang, J., Wang, J. P. (2014) Similar simulation and practical research on the mining overburden roof strata "three zones" height. *J. Min Safe Eng.*, 31: 249–254.

Zhang, Y. H., Bo, Y. L., Gao, Y. K. (2017) The "Three Zones" Division of overlying strata and the position determination of gas drainage in Pingshan coal mine. *J. Min Res Devel.*, 37: 44–48.

Zhang, Y. J., Li, F. M. (2011) Monitoring analysis of fissure development evolution and height of overburden failure of high tension fully-mechanized caving mining. *Chin. J. Rock Mech. Eng.*, (S1): 2994–3001.

*Advances in Energy, Environment and
Chemical Engineering – Abdullah & Osman (Eds)
© 2023 The Author(s), ISBN: 978-1-032-36083-6*

Research on reservoir characteristics of Chang 2 in W oilfield in Northern Shaanxi

Yingmin Cui*
Changqing Industrial Group Co., Ltd., Geological Technology Research Institute, Xi'an, China

Shuai Yang*, Jun Jiang*, Tengteng Zhuang*, Xiangqian Yu* & Binhong Jia*
Changqing Oilfield Branch No. 9 Oil Production Plant, Yinchuan, China

ABSTRACT: To determine the reservoir performance of Chang 2 oil reservoir in W oilfield in northern Shaanxi, analysis of casting thin section was done based on scanning electron microscope, porosity, permeability, etc., petrology, pore structure, diagenesis, physical properties, and heterogeneity. The results show that the Chang 2 reservoir in the W oilfield belongs to the delta front-plain sedimentary system, and its lithology is mainly gray, fine-grained feldspar sandstone. The interstitial materials are mainly chlorite and calcite. The muddle layer is the main one. The main storage spaces are intergranular pores and cuttings dissolution pores. The main types of diageneses are compaction, cementation, and dissolution. In the middle diagenetic stage A, compaction and cementation lead to tight reservoirs, and dissolution improves the physical properties of the reservoirs to a certain extent. The porosity is concentrated at 13%–15%, and the permeability is mainly concentrated at 3–5 mD. Intra-layer, inter-layer, and plane heterogeneity are strong. The above results can provide a reference for determining favorable areas in the study area and the next exploration and development.

1 INTRODUCTION

The W oilfield is located in the northern Shaanxi region of the Ordos Basin of China, which comprises beam-shaped hills and gully area of the Loess Plateau. The surface is covered by Quaternary loess with a thickness of 100–200 m, and the terrain is relatively complex. The ground elevation is 1233–1809 m, and the relative height difference is about 576 m. The Yanchang Formation in the Ordos Basin can be divided into ten oil reservoir groups (Liu 2017). Among them, the Chang 2 oil layer group developed a large area of braided rivers, rivers, and delta plains, which laid a favorable foundation for the enrichment of oil and gas (He & Guo 2019; Xing 2019). In particular, the W oilfield developed delta fronts and plains, and the sedimentary grain size was relatively coarse and gradually became an important oil and gas accumulation area. However, researchers have not had a clear and unified view of the distribution law of Chang 2 oil reservoirs over the years, resulting in poor exploration results.

The reservoir is one of the main controlling factors affecting the distribution of oil and gas and is the basis for finding oil and gas enrichment areas and developing oil fields (Li 2011; Pang 2010; Shi 2020). Therefore, it is very important to conduct reservoir research and comprehensive evaluation. The porosity of the reservoir determines the storage capacity of the reservoir, and the permeability determines the seepage capacity and productivity of the reservoir (Huang et al. 2004; Shi et al. 2013). This paper takes Chang 2 reservoir in W oilfield in northern Shaanxi as the research object and carries out comprehensive research and evaluation of the reservoir mainly from the aspects of

---

*Corresponding Authors: cym2_cq@petrochina.com.cn; yangshuai_cq@petrochina.com.cn;
jiangj01_cq@petrochina.com.cn; zttteng01_cq@petrochina.com.cn; yxq_cq@petrochina.com.cn and
jiabinhong_cq@petrochina.com.cn

DOI 10.1201/9781003330165-7

petrological characteristics, diagenesis, physical properties, and reservoir heterogeneity. It provides a strong theoretical basis for the exploration and development of oil fields.

2 CHANG 2 RESERVOIR CHARACTERISTICS

2.1 *Petrological features*

The Chang 2 reservoir in the W oilfield can be divided into three sub-layers, Chang 21, Chang 22, and Chang 23, according to cycles. The rock type is mainly gray, fine-grained feldspar sandstone. The main mineral components are quartz and feldspar. Quartz accounted for 47.3%, and feldspar accounted for 45.5%. The average total content of reservoir interstitials is 9.9%, mainly chlorite and calcite. The clay minerals are dominated by chlorite and montane mixed layers (Figure 1A). The cementation types are mainly pore-filling and pore-contacting (Figure 1B). Chlorite exists in the form of ring edges in this area, which not only improves the mechanical strength and compaction resistance of the rock but also reduces the nucleation amount of authigenic quartz on the detrital particles and inhibits the secondary increase of quartz, giving rise to sandstone. Primary and secondary pores are preserved.

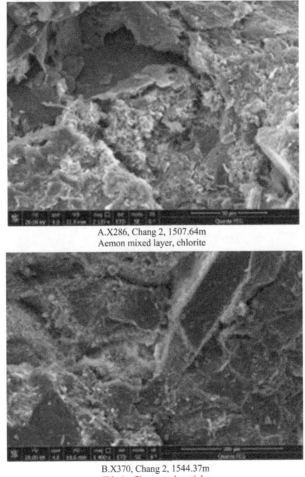

A.X286, Chang 2, 1507.64m
Aemon mixed layer, chlorite

B.X370, Chang 2, 1544.37m
Chlorite film-coated particles

Figure 1.   Schematic diagram of the scanning electron microscope of Chang 2 reservoir in W oilfield.

51

2.2 Pore structure features

The pore types of Chang 2 reservoir include intergranular pores and dissolved pores (intergranular, feldspar, cement, and detrital dissolved pores) (Figure 2), of which intergranular pores and detrital dissolved pores are the main storage spaces. The Chang 2 reservoir layer has a porosity of 3.3%. The mercury intrusion data show that the Chang 2 reservoir belongs to the low displacement pressure-small pore and fine throat type. The capillary pressure curve has an obvious plateau, the sorting coefficient and homogeneity coefficient are small, the mercury removal efficiency is high, and the storage performance and permeability are good.

2.3 Diagenesis characteristics

The various physical, chemical, and biological processes experienced in the long geological history after sediment deposition and burial until undergoing metamorphism are collectively referred to as diagenesis. There are various types of diageneses in Chang 2 reservoirs in the W oilfield.

A.X270, Chang 2, 1457.81 m
Intergranular pores and feldspar dissolved pores

B.X270, Chang 2, 1463.92 m
Feldspar dissolved pores

Figure 2.　Schematic diagram of pore types of Chang 2 reservoir in W oilfield.

The formation and development of secondary pores directly affect the physical properties and heterogeneous distribution of the reservoir. Therefore, it is very important to study diagenesis.

Through the observation of cast thin sections and scanning electron microscope experiments, the diagenesis of Chang 2 reservoir in W oilfield mainly includes the three types: compaction, cementation, and dissolution (Figure 4). In the middle diagenetic stage A, compaction and cementation lead to tight reservoirs, and dissolution improves the physical properties of the reservoirs to a certain extent.

2.4 Physical characteristics

The porosity of the Chang 2 reservoir is concentrated in 13%–15% (Figure 3), and the permeability is mainly distributed in 3–5 mD (Figure 4, Table 1).

Table 1. Classification of physical properties of Chang 2 sandstone reservoirs in W oilfield.

Horizon	Parameter variety	Porosity (%)	Permeability (mD)	Types of reservoir
Chang 21	Maximum	21.2	18.69	
	Minimum	5.8	0.13	
	Average	13.3	3.91	
Chang 22	Maximum	20.66	13.23	
	Minimum	7.8	0.29	Low Hole Special Low Seep Type
	Average	13.8	4.01	
Chang 23	Maximum	21.4	15.57	
	Minimum	7.5	0.11	
	Average	14.4	4.70	

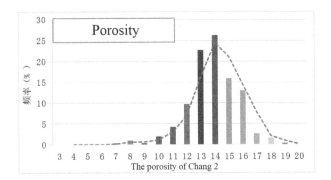

Figure 3. Normal distribution of porosity of Chang 2 reservoir in W oilfield.

2.5 Heterogeneous features

Under the combined influence of sedimentation, diagenesis, and later tectonic processes, oil and gas reservoirs have different spatial distributions in plane and vertical directions, and their internal components and structures show different characteristics and uneven changes in various properties. The basic properties of the reservoir include the inhomogeneity of lithology, physical properties, oil content, and microscopic pore structure in three-dimensional space.

1) Intralayer heterogeneity

Intra-layer heterogeneity refers to the internal vertical variation of reservoir properties within a single sand layer scale. Intra-layer heterogeneity is the internal cause of intra-layer contradiction

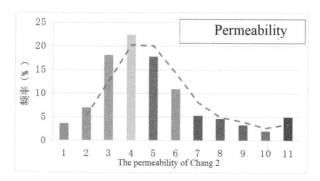

Figure 4.   Normal distribution of permeability of Chang 2 reservoir in W oilfield.

in production. Preliminary analysis shows that different sedimentary microfacies in the study area have different deposition modes, so the sand bodies between different facies have different intralayer heterogeneity characteristics. Intra-layer heterogeneity is mainly affected by turbulent depositional environment, unstable provenance, and diagenesis and is mainly expressed by the heterogeneity parameter of intra-layer permeability, which is embodied in particle size, cementation, dissolution, and other aspects. According to the classification standard of heterogeneity, the heterogeneity in each sub-layer is determined to be strong (Table 2).

Table 2.   Statistical table of heterogeneity division of Chang 2 reservoir in W oilfield.

Horizon	Chang 21	Chang 22	Chang 23
Permeability	0.13–18.69	0.28–13.23	0.11–15.57
average	3.91	4.01	4.70
Coefficient of variation	0.75	0.56	0.6
Dash factor	4.8	3.3	3.1
Difference	148.3	45.8	140.3
Degree of heterogeneity	Strong	Strong	Strong

2) Interlayer heterogeneity

The interlayer heterogeneity is mainly controlled by the depositional environment and depositional cycles and is mainly reflected in the sandstone density, stratification coefficient, and interlayer distribution. The Yan'an Formation in the study area mainly develops a delta front plain, and the distribution of channel sand bodies is uneven. The lower the sandstone density, the higher is the stratification coefficient and the stronger is the interlayer heterogeneity. Statistics show that the ratio of stratification coefficient and sandstone density shows that the heterogeneity of the Chang 21 sub-layer is stronger than that of other layers.

3) Planar heterogeneity

Planar heterogeneity is caused by the geometric shape, scale, top-bottom undulation, porosity, and permeability of the reservoir sand bodies. These factors directly affect the oil displacement efficiency and swept volume in the plane water.

According to the analysis of the plane distribution characteristics of the reservoir parameters, it is found that the distribution of each parameter in the plane has strong heterogeneity, and the distribution characteristics are different. However, there is certain regularity between the parameters of each sub-layer: that is, the area with large sandstone thickness generally has good porosity, permeability, and high oil saturation, and the opposite is true in thin sandstone areas. The plane distribution heterogeneity of reservoir parameters is controlled by the development degree of sand bodies and sedimentary microfacies.

a) Porosity plane distribution

The porosity at the intersection of underwater distributary channels in Chang 2 reservoir of W oilfield is relatively high, and the porosity in the central and southern parts varies greatly, mainly distributed over 12%. The porosity is flake-like and lens-like, and the overall plane has strong heterogeneity. The high-porosity areas are distributed in the small layers of the study area and form small flakes and lenses locally. The porosity of each sub-layer has a large vertical variation and strong heterogeneity.

b) Planar distribution of permeability

The permeability of the Chang 2 reservoir in the W oilfield is relatively high, and the permeability is mainly distributed above 0.3 mD. The high-permeability area in the northeast is concentrated in the distribution, and the south is distributed in lakes or lenses. The plane has strong heterogeneity. The high-permeability areas are dispersed in the small layers of the study area and form small flakes and lenses locally, mainly concentrated at the intersection of river channels. The permeability of each sub-layer has a large vertical variation and strong heterogeneity.

3 CONCLUSIONS

(1) The Chang 2 reservoir in the W oilfield belongs to the delta front-plain sedimentary system, and the lithology is mainly gray, fine-grained feldspar sandstone. The interstitial materials are mainly chlorite and calcite. The clay minerals are all chlorite. The muddle layer is the main one.

(2) The main storage spaces of Chang 2 reservoir in W oilfield are intergranular pores and cuttings dissolution pores. The main types of diageneses are compaction, cementation, and dissolution. In the middle diagenetic stage A, compaction and cementation lead to tight reservoirs, and dissolution improves the physical properties of the reservoirs to a certain extent.

(3) The porosity of the Chang 2 reservoir is concentrated in 13%–15%, and the permeability is mainly distributed in 3–5 mD. Intra-layer, inter-layer, and plane heterogeneity are strong.

REFERENCES

He Yawei, Guo Yonghong. Characteristics and comprehensive evaluation of Chang 2 reservoir in Qingpingchuan Oilfield[J]. *Journal of Yan'an University* (Natural Science Edition),2019,38(01):99–105.

Huang Sijing, Zhang Meng, Zhu Shiquan, Wu Wenhui, Huang Chenggang. Controlling effect of sandstone pore origin on porosity/permeability relationship: A case study of the Triassic Yanchang Formation in the Longdong area of the Ordos Basin[J]. *Journal of Chengdu University of Technology* (Natural Science) Edition), 2004(06):648–653.

Li Keyong. *Study on the sedimentary system and reservoir characteristics of the Yanchang Formation in the Yellow Exploration Area of the Ordos Basin*[D]. Northwestern University, 2011.

Liu Xianyang. *Accumulation characteristics and exploration potential analysis of tight oil in Yanchang Formation, Ordos Basin*[D].Chengdu University of Technology, 2017.

Pang Jungang. *Reservoir characteristics of Yanchang Formation in Zichangshijiapan area, Ordos Basin*[D]. Northwest University, 2010.

Shi Hui, Liu Zhen, Pan Gaofeng, Hu Xiaodan. Analysis of porosity evolution model of clastic rock formations in sedimentary basins: Taking Yanchang Formation in Ordos Basin as an example [J]. *Geological Science*, 2013, 48(03): 732–746.

Shi Jian. *Study on the characteristics and controlling factors of low permeability sandstone reservoirs of Yanchang Formation in Hujianshan area, Ordos Basin*[D]. Northwestern University, 2020.

Xing Fen. *Reservoir characteristics and main controlling factors of Chang 2 reservoir group in Qingyangcha area of Jingbian Oilfield*[D]. Chang'an University, 2019.

*Advances in Energy, Environment and
Chemical Engineering – Abdullah & Osman (Eds)
© 2023 The Author(s), ISBN: 978-1-032-36083-6*

Quantitative evaluation of fault lateral sealing of Shuangtaizi gas storage group

Zhiming Hu & Mingming Jiang*
School of Earth Science, North East Petroleum University, Daqing, China

Zhongshun Min, Guoguang Zhao, Bin Li & Jie Liu
Exploration and Development Research Institute of Liaohe Oilfield, Panjin, China

ABSTRACT: Underground gas storage plays an important role in storing oil and gas, which requires that the gas storage must have a good sealing capacity, but the research on fault sealing of underground gas storage is still lacking. Taking the control circle fault of the Shuangtaizi gas storage group as the quantitative evaluation research object, the fault shale content (SGR) of reservoir units on the fault plane is counted. At the same time, the fault pressure difference (AFPD) on both sides is calculated according to the oil, gas, water density and buried depth on both sides of the fault. The results are calibrated to fit the quantitative relationship between fault shale content (SGR) and fault pressure difference (AFPD). Finally, it is determined that the minimum SGR value of fault rock sealing of the Shuangtaizi gas reservoir is 19.7%, and the minimum sealing pressure difference and minimum breakthrough pressure of each block are calculated according to this relationship. This method can provide the necessary theoretical basis and practical reference for ensuring the safety of gas storage.

1   INTRODUCTION

Underground gas storage is the "warehouse" for underground natural gas storage. It injects natural gas storage in the off-season of consumption and exploits natural gas at the peak of heating, which plays the role of seasonal peak shaving and strategic reserve. Gas storage has become the most important way to store natural gas globally. It is of great significance to the strategic adjustment of natural gas storage and emission reduction in the northeast of Liaoning and Taizi, which is an important part of China's natural gas storage and regulation of climate change, and plays an important role in promoting the safety of natural gas storage and emission reduction in the northeast of China. As Shuangtaizi oilfield belongs to the western slope of the Liaohe depression, there are many types of faults involved, and the sealing capacity of different faults is different, especially because of the perennial injection and production characteristics of gas storage; all have high requirements for the sealing strength of gas storage faults. Therefore, it is necessary to quantitatively evaluate the fault sealing strength of Sha 1 and Sha 2 in the gas storage group in the study area, so as to clarify the pressure boundary of gas storage group operation, and provide a necessary theoretical basis for better operation of gas storage (Wang 2021).

2   REGIONAL GEOLOGICAL OVERVIEW

Shuangtaizi gas storage is located in the west of the Liaohe depression. The geological structure of this area is complex. The fault system of the Shuangtaizi structural belt has inheritance and

---

*Corresponding Author: jiangmingming1997@163.com

DOI 10.1201/9781003330165-8

regeneration. The plane distribution direction is mainly near the east-west, mostly South falling normal faults. A total of 28 faults are developed, which fall gradually to the center of the depression under the action of gravity. The development period is Shayi, Erhe and Dongying period, and the maximum fault distance of circle controlling faults is about 400m. Most fault distances are concentrated in the range of 20 ~ 220m, which mainly controls the structural pattern and oil, gas and water distribution in this area, complicating the long axis anticline structure (Lu 2016). According to the statistics of regional traps, it is found that eight traps of different sizes are formed under the control of faults and structures as a whole (Table 1).

Table 1. Trap elements of Shuangtaizi structural belt.

Trap name	Trap type	Trap area (km^2)	Trap height	Trap control factors	Trap overflow point (m)
S6	Broken nose	3.21	260	Fault controlled overflow point	−2540
S67	Broken nose	2.78	200	Fault controlled overflow point	−2500
S7	lithology	1.31	60	Lithology control	−2580
S51	anticline	5.95	160	Dual control of structural form and fault	−2530
S602	Broken nose	1.92	80	Construct dual control	−2700
S9	anticline	4.72	100	Fault controlled overflow point	−2872
S601	anticline	2.05	100	Fault controlled overflow point	−2400
S31	Broken nose	3.75	120	Dual control of structural form and fault	−2470

3 QUANTITATIVE EVALUATION OF FAULT SEALING

3.1 *Lateral sealing mechanism and sealing type of fault*

The essence of fault lateral sealing research is to study the permeability between the fault zone and its surrounding rock. From the perspective of causing permeability differences, faults can be roughly divided into juxtaposition seal, fault rock seal and cemented seal.

3.2 *Sealing type and evaluation principle of ring-controlled faults in gas storage*

The overlying mudstone of the Xinglongtai oil layer group is thick and stably distributed in the whole area, with a thickness of 200 m-400 m. The fault distance of the loop control fault is mainly distributed between 20 m-220 m. Since most gas reservoirs are located in the rising wall of the loop control fault, the upper reservoir is easy to form butt sealing with the thick mudstone of the opposite wall, while most oil-water units that are far away from the caprock vertically are difficult to form butt sealing with the caprock, The thickness of most interlayer and the oil-water unit is less than the maximum fault distance of the control circle section, which is completely staggered by the fault, thus forming fault rock sealing (Ding & Li 2002).

According to the relevant basic geological data, it can be concluded that the section of the loop control fault of this trap develops different sealing modes. By comparing the relationship between fault distance, sandstone thickness and mudstone thickness: 1. When the fault distance is greater than the thickness of the sandstone layer but less than the thickness of the mudstone layer, the sand layer and mud layer are completely connected. The butt sealing is formed at this time, and the sealing is good. 2. When the fault distance is less than the thickness of sandstone, the sand body is not completely staggered at this time, and there is a sand butt joint on the fault plane. At this time, the fault cannot seal the fluid in the hanging wall, and the sealing capacity of the footwall also depends on the degree of sand mud butt joint. 3. When the fault distance is greater than the thickness of sandstone and mudstone, the sand layer is butted with the sandstone of different layers, and the leakage risk is large. At this time, the sealing capacity is directly and positively correlated with the fault argillaceous content SGR. At this time, the sealing capacity can be evaluated by extracting the fault plane SGR value.

3.3 *Quantitative evaluation of fault lateral sealing in gas storage*

Under still water conditions, the displacement pressure difference between fault and reservoir determines the sealing capacity of the fault, and this cross fault pressure difference is affected by the shale content of the fault plane. Generally, the greater the shale content, the greater the displacement pressure difference. Therefore, it is extremely important to accurately predict the shale content in the fault zone and clarify the relationship between shale content and cross fault pressure difference for sealing evaluation. In principle, the value of SGR can be corrected by the water pressure in the fault zone, but these data are difficult to obtain directly in reality, so it can be replaced by measuring the hydrocarbon water pressure passing through the fault in the reservoir. Then, the fault gouge ratio (SGR) and fault pressure difference (AFPD) are counted. There will be different shale content and fault pressure difference at different depths. An envelope line can be fitted along the outer trend of all points, which is the effective closure envelope line of GFPD-SGR in the Shuangtaizi area, as shown in Figure 1.

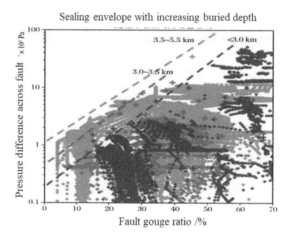

Figure 1.   Orientation map of the fault gouge ratio SGR and crossing fault pressure difference AFPD in sand reservoir collocation area.

In order to build a quantitative characterization model of fault lateral sealing capacity, after determining the differential pressure across the fault actually supported by the fault at different depths of the drilled reservoir, it is also necessary to calculate the SGR parameters representing the content of fault gouge. The calculation of SGR mainly uses three-dimensional seismic data volume and single well data (shale content data, lithology, stratification and well deviation) to build a numerical simulation model to calculate the fault offset distribution and formation shale content distribution. Then, according to the SGR calculation formula, the SGR value of each position on the fault plane can be obtained (Figure 2), and the SGR value corresponding to the oil-bearing section can be extracted (Figure 3) after determining the actual support fault differential pressure and section SGR value corresponding to the oil-bearing section of the control fault, the fault differential pressure is combined with a series of SGR values of corresponding depths on the fault plane to ensure that GFPD corresponds to a series of SGRs of corresponding depths, so as to obtain the SGR-GFPD intersection diagram of the trap (Wang 2015).

The same method is applied here, and the same SGR-GFPD relationship calibration method is adopted for other oil-bearing traps in Shuangtaizi gas storage. Then we put these data points in the same coordinate system, fit the outer envelope of these points based on all points (Figure 4), and

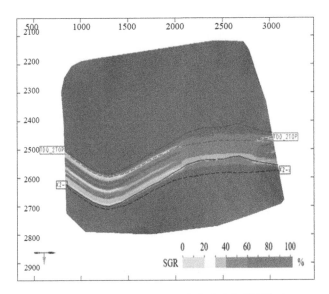

Figure 2.   SGR distribution of F17 fault plane of control circle fault in S31 block.

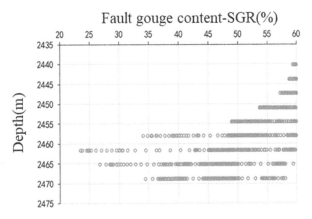

Figure 3.   SGR distribution scatter diagram of the footwall of F17 fault in control circle fault in S31 block.

obtain the formula (1) of this envelope. It can be seen that the left critical value of the envelope is 19.7%, which is the sealing critical value of the fault rock of Shuangtaizi gas storage. Below this value, the leakage risk of fault is great. With the increase in SGR value, the pressure difference across the fault that can be supported by the fault shows an increasing trend (Fu et al. 2021; 2013). When the pressure difference across the fault supported by the fault is equal to the floating pressure of oil and gas, the fault reaches the limit of closed hydrocarbon column height. Based on this, the quantitative relationship formula (3) between the height of the oil column closed by the fault and section SGR can be obtained by combining formula (1) and formula (2). Using this relationship formula, the sealing capacity of other faults in the study area can be evaluated.

$$\text{AFPD} = 0.278 \ln (\text{SGR}) - 0.856 \tag{1}$$

$$P_F = (\rho_w - \rho_o)gH \tag{2}$$

$$\text{H} = \frac{0.278 \ln (\text{SGR}) - 0.856}{(\rho_w - \rho_o)g} \tag{3}$$

59

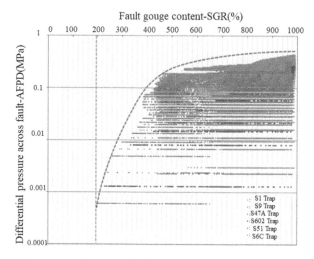

Figure 4. SGR-AFPD intersection diagram of fault trap oil-water unit section of Shuangtaizi gas storage.

Where the differential pressure across the fault is expressed as AFPD, Mpa; the floating pressure of oil and gas is expressed as , the height of the oil column is expressed as H, m; the acceleration of gravity is expressed as g, m/s^2; and represent the density of water and hydrocarbon under formation conditions, kg/m^3.

3.4 *Fault development characteristics and quantitative evaluation of fault breakthrough pressure*

Shuang9 block is located in the rising wall of circle control fault F11. F11 fault is NW trending, with an inclination of 55° ~ 65°, and the extension length of the fault is 4km. It can be seen from the figure that the joint between the fault and the upper part of fault f141-43 is developed, and the joint between the fault and the upper part of fault f141-43 is developed. Through the fault, it can be seen that the joint between the fault and the upper part of fault f141-43 is developed, and the sand and mudstone are closed (Knipe 1989; Pei et al. 2015). By calculating the SGR distribution map of the fault plane, it can be concluded that the minimum SGR of the section of F11 fault in the Shuang9 block is 25.04% (Figure 5). According to the SGR-GFPD relationship, it can be calculated that the minimum sealing pressure difference of the section of F11 fault is 0.11Mpa (Figure 6). Its minimum breakthrough pressure needs to be added with its hydrostatic pressure to obtain that the minimum breakthrough pressure of F11 fault is 29.15 Mpa.

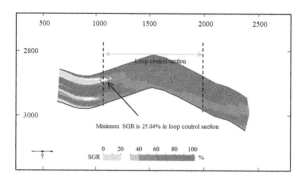

Figure 5. SGR distribution of F11 section of control circle fault in Shuang9 Block (side view).

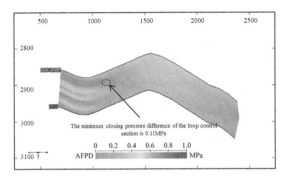

Figure 6.   Closed differential pressure distribution of F11 section of control circle fault in Shuang9 Block (side view).

The same method is used to obtain the minimum closing pressure difference and minimum breakthrough pressure data of other hydrocarbon traps, as shown in Table 2.

Table 2.   Data summary of minimum sealing differential pressure and minimum breakout pressure.

Name of gas storage	Minimum closed differential pressure/MPa	Minimum breakthrough pressure/Mpa
S9	0.11	29.15
S602	0.217	28.11
S7	7.11	32.13
S67	0.245	23.7
S6	0.005	23.01
S51	0.16	24.9
S601	0.22	24.45
S1	0.085	24.62

4   CONCLUSION

The provenance direction of the hei258 area is southwest, and the river pushes into the lake from the southwest. The Conclusion

(1) The lateral sealing capacity of the fault has a direct impact on the storage capacity of the gas storage. The stronger the sealing capacity of the loop controlling fault, the stronger the capacity of the trap involved to store oil and gas.

(2) By quantitatively evaluating the lateral sealing ability of the control circle fault of the gas storage, the relationship of SGR-AFPD is fitted as AFPD = 0.278 ln (SGR) – 0.856. Based on this, it is determined that the critical SGR value of the lateral sealing of the fault of Shuangtaizi gas storage is 19.7%, and the minimum sealing pressure difference and minimum breakthrough pressure of each block are calculated according to the relationship.

REFERENCES

Ding Guosheng, Li Wenyang. Current situation and development trend of underground gas storage at home and abroad [J] *International Petroleum Economy*, 2002 (06): 23–26 + 63.

Fu Xiaofei, Li Kun, Dong Rou, Li Jinku. Review of fault sealing research [J] *Energy Technology and Management*, 2021,46 (03): 24–26 + 171.

Fu Xiaofei, Shang Xiaoyu, Meng Lingdong. Internal structure of low porosity rock fracture zone and its relationship with oil and gas accumulation [J] *Journal of Central South University* (NATURAL SCIENCE EDITION), 2013,44 (06): 2428–2438.

Knipe R J. Deformation mechanisms — Recognition from natural tectonites[J]. *Journal of Structural Geology*, 1989, 11(1–2):127–146.

Lu Zhengguang. Main problems and Countermeasures of underground gas storage in China [J] *Sino Foreign Energy*, 2016, 21 (06): 15–19.

Wang Haibo. Development status of underground gas storage in China and application of geological guidance [J] *Western Exploration Engineering*, 2021,33 (07): 84–86.

Wang Xia Exploration and development analysis of Shuangtaizi block in Liaohe Oilfield [J] *China Chemical Trade*, 2015, 000 (008): 186–186.

Yangwen, Pei, Douglas, et al. A review of fault sealing behavior and its evaluation in siliciclastic rocks[J]. *Earth Science Reviews*, 2015.

Advances in Energy, Environment and Chemical Engineering – Abdullah & Osman (Eds)
© 2023 The Author(s), ISBN: 978-1-032-36083-6

Energy-saving analysis of typical traditional residence in coastal and inland area

Junxue Wang & Peng Chen*
Zhejiang University of Water Resources and Electric Power, Hangzhou, China

Siwen Wang
China United Engineering Corporation Limited, Hangzhou, China

Jiacheng Xu & Qiangqiang Wu
Zhejiang University of Water Resources and Electric Power, Hangzhou, China

ABSTRACT: Traditional residential houses are formed by people in different places to adapt to the long-term development of the local natural environment, with deep regional characteristics. In this paper, Zhoushan and Turpan are selected as typical coastal and inland areas, and the similarities and differences of traditional residence are investigated and discussed through the aspects of the natural environment, settlement layout and residential characteristics. In addition, the effective application of traditional houses in low-carbon technology and their reference to energy conservation and emission reduction in modern architectural design are also analyzed in the paper.

1 INTRODUCTION

In the context of the Chinese government's commitment to carbon peak and carbon neutrality, energy conservation and emission reduction of buildings have attracted more and more attention from different regions in China. In the low level of building energy conservation technology, in order to adapt to the unique local climate, terrain and resource characteristics and obtain good living comfort, ancient Chinese residents adopted a series of clever and low-carbon technical means in architectural design and construction technology, reflecting the ancient Chinese philosophy of harmony between nature and man (Gao 2020). Up to now, these traditional technical and philosophical thoughts are still of great significance to the development of green buildings.

Existing research on building energy conservation and emission reduction focuses on the use of clean energy, optimization of thermal insulation materials, energy recycling and other emerging technologies. In contrast, the reference and innovation significance of low-carbon technology means that traditional residential design is less involved. Therefore, this paper selected Zhoushan and Turpan as the typical coastal cities and inland cities, respectively, carried out field research on the traditional local residence, analyzed the effective application of traditional dwellings in low-carbon technology, and proposed its reference and thinking for modern architectural design of energy conservation and emission reduction.

---

*Corresponding Author: chenp@zjweu.edu.cn

2 DESCRIPTION OF CHARACTERISTICS OF TRADITIONAL RESIDENCE IN ZHOUSHAN

2.1 Natural conditions of Zhoushan

Zhoushan belongs to a monsoon Marine climate area on the southern edge of the northern subtropical region, and the annual precipitation changes greatly (Zhang et al. 2013). Affected by the special geographical environment, Zhoushan has a small rainfall area and few flowing rivers. Therefore, Zhoushan lacks freshwater resources.

According to the characteristics of monthly precipitation distribution, the most precipitation in Zhoushan is June and August, of which the precipitation in June is caused by the plum rain season, and August is the month with the most frequent typhoon impact and the most precipitation.

2.2 Main characteristics of traditional local residence

2.2.1 Site selection and layout according to local conditions

In order to make it easy for island fishermen to go to the sea, coastal villages generally grow and expand along the outline of the coastline around the seaport. They often make decisions between the U-shaped bays facing the sea and backed by the mountains. Houses are distributed step by step along the terrain. On the one hand, they could avoid typhoons and easily resist the invasion of typhoons; On the other hand, the concave port is convenient for fishing boats to berth and supply.

2.2.2 Energy saving of traditional local residence

Due to the monsoon climate of Zhoushan island, there are many typhoons in summer and autumn. Due to the windproof performance and durability of the houses, coupled with the island being rich in stone resources, but the wood is relatively scarce, the residents on the island have built many stone houses, as shown in Figure 1, according to local conditions.

For the sake of lighting and ventilation, the ancestors carved many stone Windows with an artistic atmosphere similar to Figure 2. The terrain height difference between the building and the building makes up for the lighting problem of the hollow-out designed stone window and can minimize the impact of typhoons, wind and heavy rain in summer. The relatively tight window structure can also reduce heat loss in winter and maintain the indoor temperature.

Figure 1.   Stone house in Zhoushan.

Zhoushan island is hilly with a small catchment area, which can not form a large runoff, and a rich groundwater supply cannot be obtained. Therefore, freshwater resources are extremely scarce.

Figure 2.   Stone windows with art flavor.

However, the island residents who depend on the sky draft have complex feelings about the typhoon, which not only brings a lot of precipitation but also ravages the island every year.

Figure 3.   Reservoir roof of a traditional local residence in Zhoushan.

In order to avoid the waste of "sky and rain" and effectively collect rainwater, the eaves of traditional dwellings do not drop water as freely as the eaves of inland dwellings, but mostly adopt organized drainage methods. It is common practice to place four tanks in each of the four corners of the patio, and then the rainwater collected from the adjacent eaves of Figure 3 is discharged into the tank via the downpipe for daily use (Zhu 2009). Wells can also be drilled to preserve valuable water resources. For the irrigation of crops, people generally build canals to lead the mountains into the field reservoir to irrigate the crops.

2.2.3   Utilization of environmental resources

As an island, Zhoushan is surrounded by the sea. The ancient ancestors introduced the seawater into the dug evaporation beach during the rising tide, concentrated the seawater with the rich wind energy brought by sunlight and southeast monsoon, and then introduced it into the crystal pool.

Continuing exposure to the sun, the seawater would become a saturated solution and then salt would be precipitated.

Figure 4.   Salt pan in Zhoushan.

3   DESCRIPTION OF CHARACTERISTICS OF TRADITIONAL RESIDENCE IN TURPAN

3.1   *Natural conditions of Turpan*

Turpan is located in an east-west intermountain basin in the eastern Tianshan Mountains, shaped around a lake, forming a three-ring topographic distribution, high and low in the south, wide and narrow to the east, with geographical coordinates between north latitude 41°12′ to 43°40′ and east longitude 87°16′ to 91°55′. At the same time, as a transportation hub in Xinjiang, Turpan is an important hub in China's inland region that connects Xinjiang, Central Asia and southern and northern Xinjiang (Tian 2021).

The climate of Turpan is characterized by abundant sunshine, abundant heat, extreme dryness, frequent strong winds and low rainfall, especially during periods of a large temperature difference between day and night. As the saying is, "Wear a cotton-padded jacket in the morning and wear gauze at noon; eat watermelon around the stove." The mean temperature in July and August was 25.7°C, while the mean temperature in January was −15.2°C. Extreme temperatures range from 47.8°C to 41.5°C. The four seasons here are unevenly distributed; in which the winter is cold and long.

3.2   *Main characteristics of traditional local residence*

3.2.1   *Building insulation*

Turpan is hot in summer and cold in winter. And large temperature difference existed between day and night here. In order to adapt to the environment, local houses need good insulation and insulation. Because the local soil winning soil has a good slow heat release effect, local houses are built from local materials, and residential buildings are often constructed by soil structure.

Local dwellings often build half underground bedroom structures. In the summer, residents can utilize the underground environment to adjust the indoor temperature, which makes these dwellings even do not use modern electrical appliances to create a comfortable indoor environment and achieve the purpose of energy saving. The half underground bedroom is designed in the front house, with the room door open or closed, which can inhibit heat energy transfer and enhance indoor temperature suitability.

On the outside of the residence, high scaffolding is set up in such ways as roof, cantilever and column support. The high scaffolding was erected in the second-story building in Figure 5, which can not only provide shade, but also provide a good leisure place for people. Combining lighting, ventilation and vision, it broadens the space for people's production activities.

Figure 5.　Local traditional residence in Turpan.

3.2.2　*Water resources energy saving*

Due to the low rainfall and strong transpiration in Xinjiang, the available surface water resources are scarce. Therefore, the construction of Karez can effectively solve the problem of the lack of local freshwater resources. Karez can not only protect water resources from transpiration loss, but also use the cool underground environment to keep the water cool, and then adjust the temperature of residential buildings to achieve energy conservation.

3.2.3　*Utilization of environmental resources*

Due to its unique geographical location, Turpan is rich in wind energy resources. The characteristic local building named "Grape House," shown in Figure 6, makes full use of these wind energy resources to dry the grapes in the house. While avoiding direct sunlight, convection is formed by using the vent of the grape room to fully exchange the external dry and hot air with the wet air in the room to keep the environment in the grape room dry, so as to achieve the effect of quickly drying the grapes without easy deterioration of the grapes due to direct sunlight.

Large solar power panels can often be seen in Turpan because Turpan has a dry climate, less cloudy and rainy weather, high atmospheric transparency, more annual average sunny days, and abundant solar energy resources. The vast area and many desertification areas make it possible to build a large area into a solar power plant. Therefore, Turpan has become a high-quality solar power generation place.

4　DISCUSSION ON THE REFERENCE SIGNIFICANCE OF TRADITIONAL RESIDENCE TO MODERN ARCHITECTURE

4.1　*Protection of traditional residence*

For areas prone to natural disasters, in addition to preventing and controlling the losses caused by natural disasters, it is also necessary to repair the traditional buildings that have been lost due to

Figure 6.   Grape house in Turpan.

natural disasters, which can maintain the characteristic culture of the original folk houses (Tang & Tan 2021). Therefore, in the protection of traditional folk houses, we can reshape the previous appearance of the ruins by consulting ancient history books and observing the current daily life situation, and return to the embodiment of daily value by exploring the memorial value of the ruins in parallel with the two lines of commemorative narrative and daily narrative. By organizing architecture, scenes and cultural activities, reproducing local culture, show the landmark scenes and rituals it reproduces while displaying intangible cultural heritage, so as to create a suitable cultural space carrier for the protection of intangible cultural heritage.

4.2   Modern building energy saving

After observing the pairs of traditional dwellings in different regions, for the areas with different climate conditions, the innovation of energy conservation and emission reduction for the local buildings can be simulated based on the adaptation mode of the traditional dwellings to the local climate according to local conditions.

4.2.1   Building insulation

The construction of houses must fully consider resisting the summer heat environment and taking good building thermal insulation measures (Zhou et al. 2015). Therefore, new thermal insulation materials such as expanded polystyrene board, extruded polystyrene board, rubber powder polystyrene particle thermal insulation mortar and polymer thermal insulation mortar can be considered. The wall adopts the external insulation system of the external wall, which is to put the thermal insulation materials outside the external wall. The external insulation system outside the external wall has a strong energy-saving effect, which can reduce the phenomenon of thermal bridges. The thermal insulation effect and economic utility are considerable and are widely used in recent construction engineering construction (Yi 2021). Compared with other external wall insulation systems, the external wall insulation system can protect the main structure of the building, extend the service life of the building, and does not occupy the internal space, ease the wall temperature changes, and greatly reduce the risk of thermal stress caused by thermal stress. Reducing the indoor energy loss caused by temperature regulation by using high-quality thermal insulation materials is also a good way to save energy.

4.2.2   Water resources utilization in coastal areas

In Zhoushan, a city lacking freshwater resources, rainwater collection and purification measures can be added to local buildings. Permeable coating materials are used on the roof to simulate the tile drainage of traditional buildings. This is more beautiful than the traditional tile-roofed house, and the rainwater collection efficiency is higher. After the rainwater is collected, it shall be stored indoors or in the reservoir on the roof for further filtration and disinfection, and activated carbon or light ceramist shall be added to absorb other impurities in the water, so as to meet the use requirements of domestic water. By installing a rainwater collection and treatment system in residence, the collected rainwater can be fully purified for domestic use. In addition to alleviating the problem of insufficient domestic water, it can also alleviate the pressure of the short supply of local waterworks.

4.2.3 *Utilization of wind and solar energy resources in inland areas*

For Turpan, an area rich in wind and solar energy, solar panels can be installed on the roof of civil buildings. At the same time, because there are few natural disasters with great damage to houses and many high-rise buildings in Xinjiang, wind power generation devices can be installed on the roof of high-rise houses according to the wind power fan, and the additional electric energy can be collected into the power grid and provided to other regions.

4.2.4 *Modern electronic management*

With the increasing development of modern science and technology, the automation and integrated control of modern buildings have been gradually integrated into people's daily life, such as access control video, mobile terminal control and smart home (Tao 2017). The rational distribution of indoor water supply, water and electricity, the automatic control of indoor temperature and the cooperation of intelligent furniture will be the direction of scientific and technological innovation.

4.3 *MODERN building emission reduction*

4.3.1 *Sewage recycling*

In the process of using domestic water, insufficient use of water resources often occurs. For example, the hand washing table in the kitchen or toilet is usually only used for washing vegetables or hands. However, the water still valuable should be discharged, which not only does not make good use of water resources, but also aggravates the pressure of sewage treatment. Therefore, such valuable water resources can be treated and supplied to other aspects for use, such as transforming the indoor drainage pipeline so that the water in the toilet sink can flow into the toilet for flushing. As for the sewage from the kitchen pool, because it may contain sludge, oil stains, many microorganisms and other pollutants, membrane separation technology can be adopted for solid-liquid separation treatment. After the proper treatment, the filtered sludge can be used for microbial fermentation, and the sewage discharged from the treated liquid is less difficult to treat, so as to achieve the effect of emission reduction.

4.3.2 *Green energy saving*

With the development of modernization, the carbon emissions of both outdoor and indoor environments are much higher than those of ancient traditional houses (Ayan 2020; Li & Liang 2021). An environment with high carbon emissions aggravates the greenhouse effect and does great harm. Adding greening to modern buildings can effectively reduce carbon emissions. In addition, building greening can also reduce the energy consumed by the building to adjust the room temperature by providing shading for the building envelope, partially isolating the heat exchange with the external environment, increasing the thermal resistance and thermal insulation coefficient of the envelope, and regulating ventilation, so as to provide energy conservation for the building on the other hand.

5 CONCLUSION

The difference in the natural environment between coastal and inland areas leads to the different styles of traditional residence. However, the traditional coastal and inland dwellings show the wisdom of ancient ancestors and the cultural style of harmonious coexistence between ancient people and nature, which are worthy of learning from modern architecture.

By investigating the site selection and layout of typical residential houses in coastal and inland areas, it can be seen that stone window structures in Zhoushan or wall materials with good thermal insulation effect in Turpan can adapt to the local environment and provide a good living environment. The rainwater collection and water-saving system in Zhoushan and ancient buildings such as Karez built in Turpan have reasonably collected and utilized natural resources. The construction of these facilities does not produce too much energy consumption, and the materials required for construction are only the simplest building materials.

Under the trend of energy conservation and emission reduction, this paper puts forward the innovative concept of modern architectural design. For example, using higher quality and better thermal insulation materials, combined with green natural building plants, can improve indoor air quality and reduce carbon emissions. Vigorously developing wind energy and taking solar energy resource utilization measures can improve the efficiency of building energy conservation and emission reduction. Combining the natural environment and natural resources, inheriting the advantages of traditional houses and reasonably applying them to modern buildings can bring people a better living experience.

REFERENCES

Adilijiang Ayan. (2020) Protection and continuation of traditional residential buildings in Xinjiang. *Dalian University of Technology*, (2020).
Chunli Zhang, Linghui Gu, Yanli Shao.(2013) "Analysis of energy conservation and building energy conservation in Zhoushan." *Building energy saving*, 11,41(2013): 45–46 + 69.
Jinhua Yi. (2021) Application strategy of external wall insulation materials under the energy-saving vision of green building. *Bulk Cement*, 4 (2021): 9–11.
Jinsong Gao. (2020) "Analysis of construction green construction technology innovation and energy conservation and environmental protection measures." *Engineering Construction and Design*, 16 (2020): 153–154.
Liping Zhu. (2009) Analysis of the survival wisdom of Zhoushan traditional residential buildings. *Decoration*, 10 (2009): 131–132.
Qihang Tian. (2021) Take the residential houses of Turpan as an example. *House*, 24 (2021): 27–28.
Rui Tao. (2017) Discussion on the protection and development of Island Ancient Villages. *Tianjin University*, (2017).
Shikuang Tang, Guangyi Tan. (2021) Residential Inheritance and Settlement Protection—Chinese Residential architecture Looking back. *Southern Building*, 4 (2021): 112–117.
Tao Li, Rui Liang.(2021) Comparison of Summer Thermal Environment between Traditional and New Residential Houses in Kashgar. *Resources and Environment in Arid Areas*, 35, 5 (2021): 56–62.
Tiejun Zhou, Dachuan Wang, Jianwu Xiong. (2015) Form, function and cognition: the role of building greening in green buildings and eco-city. *Chinese Science: Technical Science*, 45, 9 (2015): 951–963.

Advances in Energy, Environment and
Chemical Engineering – Abdullah & Osman (Eds)
© 2023 The Author(s), ISBN: 978-1-032-36083-6

Evaluation and empirical study of urban energy low-carbon transition based on set pair analysis under the dual-carbon target

Jiachun Zhang*

Department of Economics and Management, North China Electric Power University, Baoding, China

ABSTRACT: In the context of the "carbon peak and carbon neutral" policy, the energy transition is the main path for governments and enterprises to improve the efficiency of carbon emission reduction. Combining China's national conditions and energy development strategic goals, a universally applicable urban energy low-carbon transition indicator system has been constructed from the three levels of energy strategy, system and goals. The AHP-anti-entropy weight method is used to determine the combined weight of the indicators, and the subjective and objective weighting methods are scientifically and rationally combined. An evaluation model is constructed based on set pair analysis, a benchmarking system is established through comparison and optimization, and horizontal comparison is performed. With the help of scatter diagrams and radar diagrams, the development level of energy low-carbon transition in each city is visually analyzed. The empirical analysis results show that the evaluation model has obvious integration characteristics of regional differentiation and target consistency and the intuitive characteristics of the evaluation results, which will help enterprises and local governments to find effective ways to reduce carbon emissions in line with their actual conditions. Policy implementation provides scientific criteria.

1 INTRODUCTION

Energy is the lifeblood of the national economy. Cities are the core areas of human social and economic activities. The urban area accounts for only 2% of the global area but consumes 78% of the world's energy, while generating more than 60% of the global total carbon emissions and more than 90% of the air pollution, which can be considered the most important place of carbon emissions. It is the most important position of low-carbon energy transformation.

The Energy Transition Index released by the World Economic Forum is based on the concept of the "energy triangle" model, covering two types of indicators (Zhang et al 2021). The World Economic Forum and Accenture jointly publish the Global Energy Architecture Performance Index to evaluate the degree of economic support, the degree of environmental impact reduction, and the degree of the energy security risk of the energy architecture (Luo et al. 2021; Zhang 2013). The World Energy Council's Energy Trilemma Index ranges from energy performance to relationship performance. The two major directions have established an evaluation system (Wang et al. 2021). Zhang Ning et al. (Zhang & Zhang 2020) proposed an energy transition indicator system that covers energy mix, energy efficiency, energy security and energy prices. Zhang Li et al. (Wang et al. 2019), from the aspects of industrial structure, energy production and consumption, and carbon emissions, explained the background of low-carbon transformation in Huainan City and analyzed the necessity, key areas and development directions, key technologies and methods, and infrastructure construction of low-carbon cities. Luo et al. (Jiang 2020) studied the impact of policy constraints on low-carbon transformation, and the problem of insufficient supply is analyzed by

---

*Corresponding Author: 3469280464@qq.com

taking natural gas as an example. Wang Yongzhen et al. (Xiao & Xu 2021) took Suzhou as an example and established a comprehensive evaluation system for the energy transformation of industrial cities based on the structural interpretation model and analytic hierarchy method. Zhang Yixin et al. (Liu et al. 2005) constructed the evaluation index system, the transformation level was divided through undesirable output SBM, and the evaluation model was empirically analyzed. Wang Linyu et al. (Deng 2006), from the perspective of energy change driving urban development, proposed to comprehensively reflect on the evaluation indicators of urban energy transformation in different dimensions, such as the effectiveness, trend and power of urban energy reform and constructed an evaluation system. The empirical analysis results show that the development level of the city has a strong correlation with the effectiveness of energy transformation. Jiang Dalin constructed an analytical framework for sustainable development and comprehensively compared the characteristics of China's mid- and long-term energy low-carbon transition in terms of technology maturity, industrial impact, cost-effectiveness, and policy demands.

This paper constructs a comprehensive assessment method. The combination of the AHP-anti-entropy weight method makes the index weight more reliable. The mathematical theory of fusion set analysis constructs an evaluation model based on the ideas of identity, difference and opposition, aims at benchmarking to seek differences, and obtains a comprehensive ranking.

2 CONSTRUCT COMPREHENSIVE EVALUATION INDICATORS

This paper builds an urban energy low-carbon transformation index evaluation system from three levels and six dimensions of strategy, system and goal. This paper has 18 secondary indicators. Due to space limitations, this article explains only the main indicators.

2.1 Low-carbon energy policy indicators

The continuous development of the energy low-carbon transition process will also have an impact on the policy system. This paper puts forward two representative indicators for the proportion of fiscal expenditure on energy conservation and environmental protection and key energy construction projects.

2.2 Energy supply indicators

This paper proposes indicators: the proportion of installed renewable energy power generation, the standard coal consumption of thermal power plants, and the per capita natural gas supply.

2.3 Low-carbon consumption indicators

The three indicators selected in this paper are the proportion of new energy vehicles, the proportion of electric energy to terminal energy consumption, and the proportion of natural gas to primary energy consumption.

2.4 Low-carbon environmental indicators

The low-carbon environmental indicators proposed in this paper include carbon emission intensity, the comprehensive utilization rate of industrial waste, treatment capacity of sewage plants and excellent rate of ambient air quality.

2.5 Low carbon efficiency indicators

This paper establishes three indicators for energy consumption per unit of GDP, per capita comprehensive energy consumption and industrial added value energy consumption.

2.6 Low-carbon development indicators

This paper will account for the proportion of energy conservation and environmental protection and new energy industry output value, the proportion of tertiary industry output value and the proportion of electricity and heating industry output value in the index system.

3 MULTI-OBJECT SET PAIR ANALYSIS MODEL

The degree of similarity and contrast can be expressed by formula (1).

$$\mu = \frac{S}{N} + \frac{F}{N}i + \frac{P}{N}j = a + bi + cj \tag{1}$$

Where N represents the total number of sets of pairs of elements; S represents the number of sets of identical elements; F represents the number of elements that are bounded by the difference between the same and opposite; P represents the number of opposing elements of the set pairs; N=S+F+P; i and j represent the corresponding coefficients, i is valued in the interval $[-1,1]$, and j takes -1; a, b, and c represent the same degree, the degree of difference, and the degree of opposition of the set pair, respectively, and satisfy a+b+c = 1.

3.1 Determine the correlation of set pair analysis

For the multi-object evaluation system, Q= {F, Z, W, B}, its meaning can be expressed as a formula (2).

$$\begin{cases} F = \{f_1, f_2, \cdots, f_m\} \\ Z = \{z_1, z_2, \cdots, z_n\} \\ W = \{w_1, w_2, \cdots, w_n\} \\ D = \{d_{11}, d_{12}, \cdots, d_{mn}\} \end{cases} \tag{2}$$

Where, f_m represents the mth object to be evaluated; z_n represents the nth indicator; W represents the weight; d_{mn} represents the mth and nth metric value of the evaluation object.

The optimal and worst levels constitute the comparison domain of the indicator [U, V], which can also be recorded as the comparison interval [uz, vz]. Evaluation metric d_{fz} satisfies the comparison interval [uz, vz] (3):

$$\frac{d_{fz}}{u_z + v_z}, \frac{d_{fz}^{-1}}{u_z^{-1} + v_z^{-1}} \in [0, 1] \tag{3}$$

Where d_{fz} represents the actual value of the z indicator of the evaluation object f.

Thus, for any evaluation object f, define the set pair [f, u] in the comparison interval [uz, vz]. The weighted degree of association is expressed by equation (4).

$$\mu \left(F_f, U \right) = \sum_{z=1}^{n} w_z a_{fz} + \sum_{z=1}^{n} w_z b_{fz} i + \sum_{z=1}^{n} w_z c_{fz} j \tag{4}$$

Where w_z represents the weight of each metric and satisfies all weights adding up to 1.

3.2 Determine the weight based on the AHP-anti-entropy weight method

AHP method focuses on the subjective weight of indicators. The anti-entropy method is an objective weight determination method. Compared with the traditional entropy weight method, which better reflects the difference of the indicators.

3.2.1 AHP determines subjective weights
The subjective weight is calculated according to the AHP principle.

3.2.2 The anti-entropy law determines objective weights

There are m evaluation objects and n indicators in the system. Based on the principle of the entropy weight method, the calculation of index weights is improved. The steps are as follows:1) Metrics preprocessing; 2) Determination of anti-entropy values and objective weights.

3.2.3 Combined weights

The combination weights are calculated.

3.3 Relative proximity calculates the ordering of the evaluated objects

The relative closeness r of the development and optimal level of the object f is defined as:

$$\begin{cases} r = \frac{a_f}{a_f + c_f} \\ a_f = \sum_{z=1}^{n} w_z a_{fz}, \quad c_f = \sum_{z=1}^{n} w_z c_{fz} \end{cases} \tag{5}$$

The greater the relative closeness r, the better the evaluation object.

4 STUDY ANALYSIS

4.1 Data sources

At present, China's urban energy transformation is still at a low level of development, and low carbon is a high-standard task, so this paper selects five cities with higher comprehensive strength as sample cities for horizontal comparison. According to the index system and evaluation model of this paper, five cities were selected to carry out the assessment of low-carbon energy transformation. The basic data of this paper is more, and the data is mainly derived from the statistical yearbook of 5 cities, the statistics of the official agency gazette, the data crawling of the public website and the research results of other scholars.

4.2 Calculation and analysis

4.2.1 Determine the metric portfolio weights

According to the AHP Method, six experts from the energy industry were invited to score.

$$C = \begin{bmatrix} 1 & 0.5483 & 0.1680 & 0.0487 & -0.0051 & 0.9551 \\ 0.5483 & 1 & 0.0684 & -0.2167 & 0.0691 & 0.5538 \\ 0.1680 & 0.0684 & 1 & 0.5076 & 0.7287 & 0.2048 \\ 0.0487 & -0.2167 & 0.5076 & 1 & 0.2015 & 0.1009 \\ -0.0051 & 0.0691 & 0.7287 & 0.2015 & 1 & -0.0042 \\ 0.9551 & 0.5538 & 0.2048 & 0.1009 & -0.0042 & 1 \end{bmatrix} \tag{6}$$

Expert five has the lowest similarity, so the score of expert five is excluded, and the average weight of the remaining five experts is used as the final subjective weight. The subjective and objective weights and combined weights of all secondary indicators are plotted into a line chart, which intuitively reflects that the combined weights meet their characteristic requirements and indicate that they are relatively reasonable, as shown in Figure 1.

4.2.2 Determine the same degree and the degree of the opposite.

Comparing the actual values of the secondary indicators of the five cities, the optimal level U and the worst level V are determined according to the economic and cost of the indicators. According to the set pair analysis algorithm and the definition of the same degree, degree of difference, and degree of opposition, the same degree and degree of opposition of all secondary indicators are calculated, as shown in Table 1:

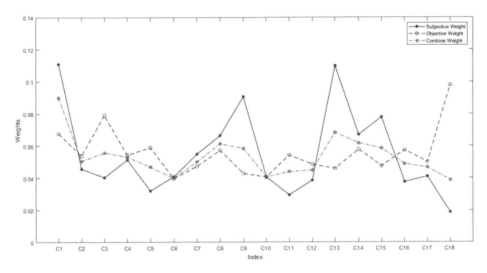

Figure 1.   Three weights of secondary indicators.

Table 1.   The degree of identity and opposition of the secondary indicators.

Shanghai i		Suzhou		Beijing		Tianjin		Shenzhen	
a_{fz}	c_{fz}	a_{fz}	c_{fz}	a_{fz}	c_{fz}	a_{fz}	c_{fz}	a_{fz}	c_{fz}
0.24	0.76	0.31	0.58	0.44	0.41	0.4	0.45	0.76	0.24
0.38	0.41	0.19	0.81	0.44	0.35	0.81	0.19	0.44	0.35
0.41	0.54	0.32	0.68	0.68	0.32	0.41	0.54	0.39	0.56
0.50	0.50	0.54	0.46	0.48	0.52	0.48	0.51	0.46	0.54
0.35	0.65	0.45	0.51	0.65	0.35	0.6	0.38	0.38	0.6
0.79	0.09	0.07	0.93	0.6	0.11	0.63	0.11	0.93	0.07
0.49	0.5	0.48	0.51	0.58	0.42	0.42	0.58	0.51	0.48
0.38	0.63	0.41	0.58	0.63	0.38	0.47	0.50	0.54	0.44
0.64	0.32	0.29	0.71	0.66	0.31	0.37	0.56	0.71	0.29
0.56	0.42	0.59	0.40	0.39	0.61	0.61	0.39	0.50	0.48
0.71	0.29	0.30	0.68	0.62	0.33	0.29	0.71	0.57	0.36
0.56	0.43	0.52	0.46	0.44	0.55	0.40	0.60	0.60	0.40
0.29	0.54	0.19	0.81	0.71	0.22	0.28	0.56	0.81	0.19
0.43	0.56	0.43	0.55	0.60	0.40	0.40	0.60	0.60	0.40
0.26	0.55	0.17	0.83	0.56	0.25	0.23	0.61	0.83	0.17
0.57	0.43	0.50	0.49	0.46	0.53	0.43	0.57	0.47	0.52
0.54	0.44	0.38	0.62	0.62	0.38	0.47	0.50	0.45	0.52
0.18	0.24	0.04	0.96	0.96	0.04	0.34	0.12	0.1	0.41

According to the set pair analysis algorithm, the degree of identity and opposition of all secondary indicators are calculated. The degree of set-pair connection of the five cities can be calculated. The degree of set-pair connection of the five cities can be calculated, as shown in Table 2:

4.2.3   *Relative proximity ranking and analysis*
The relative proximity degree r is calculated, and the results are as follows:

$$\begin{cases} r_{Shanghai} = 0.4763, r_{Suzhou} = 0.3485, r_{Beijing} = 0.6157 \\ r_{Tianjin} = 0.4748, r_{Shenzhen} = 0.6032 \end{cases}$$

Table 2. Set pair connection degree of five cities.

city	The degree of contact $\mu\{f,u\}$
Shanghai	$\mu\{f,u\}=(0.4223+0.1134i+0.4643j)$
Suzhou	$\mu\{f,u\}=(0.3280+0.0590i+0.6130j)$
Beijing	$\mu\{f,u\}=(0.5548+0.0990i+0.3462j)$
Tianjin	$\mu\{f,u\}=(0.4184+0.1189i+0.4627j)$
Shenzhen	$\mu\{f,u\}=(0.5526+0.0839i+0.3635j)$

The degrees of unity, difference and opposition of all indicators on radar charts are shown in Figure 2.

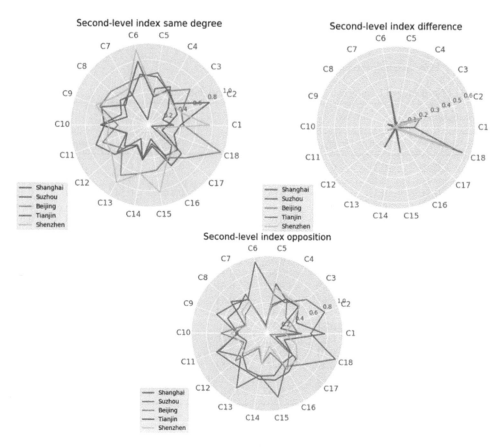

Figure 2. Degree of identity, difference, and opposition of the secondary indicators.

At the strategic level, Shenzhen ranks first. Tianjin ranks second. The cities at the bottom of the ranking should attach great importance to it and rationally plan and make adjustments.

At the energy system level, Beijing ranks first in both low-carbon energy supply and low-carbon consumption rankings. In terms of low-carbon energy supply, Beijing's new energy development and utilization have a good development. In the use of renewable energy and the low-carbon transformation of fossil energy, Shanghai, Shenzhen, Tianjin and Suzhou have their own advantages.

Shenzhen's per capita natural gas supply ranks fourth. The proportion of new energy vehicles in Suzhou is still very large compared with other cities.

At the energy target level, Shanghai ranks first, followed by Beijing and Shenzhen. In terms of energy development, Shanghai continues to lead in the proportion of energy conservation and environmental protection and new energy industry output value.

5   CONCLUSION

This paper constructed an index evaluation system for urban energy low-carbon transformation. The following conclusions are obtained:

1) The use of the set pair analysis method to comprehensively evaluate the index system, to the same evaluation cycle of the optimal value of urban energy low-carbon transformation, the scores of different cities at all levels of the energy low-carbon transformation can intuitively reflect the gap between the city itself and other cities.
2) The data is difficult to obtain, resulting in some more representative indicators not being used; follow-up research can continue to update the data to improve a more objective and comprehensive indicator system. In addition, the shortcomings of this paper are that the sample cities analyzed in the case study are all cities with a good level of urbanization development, which is limited by the high standards of urban low-carbon development and the difficulty of urban data collection and it is recommended that cities form a unified urban energy database platform as soon as possible to avoid the phenomenon of urban data "islands."
3) The empirical analysis results show that the economic strength of urban development is the basis of low-carbon energy transformation. So they can be ahead of economically underdeveloped cities in all aspects of low-carbon transformation and development of energy.

REFERENCES

Deng L Y. The Cross-Entropy Method: A Unified Approach to Combinatorial Optimization, Monte-Carlo Simulation, and Machine Learning[J]. *Technometrics*, 2006, 48(1): 147–148.

Guoliang Luo, Shanshan Liu, Xiaohui Yan et al. *Institutional constraints to China's low carbon transition: A case study of China's coal-to-gas program*, Structural Change and Economic Dynamics, Volume 57, 2021, Pages 121–135.

Jiang D L. Comprehensive comparative study on China's medium and long-term energy and low-carbon transition path[J]. *Coal Economic Research*, 2020, 40(11): 38–43.

Liu F C, Pan X F, Shi D G. Research on Regional Independent Innovation Capability Evaluation Based on Set Pair Analysis[J]. *China Soft Science*, 2005(11): 83–91+106.

Wang L Y, Guo L, Chen H, Zhou J W. Evaluation index and empirical research on urban energy reform [J]. *China Energy*, 2019, 41(04): 43–47.

Wang Y Z, Lin W, Li, Li C Y et al. Comprehensive evaluation of the energy transition of industrial cities: Taking Suzhou as an example[J]. *Global Energy Internet*, 2021, 4(02): 188–196.

Xiao X L and Xu Z C. Policy System Design of National Energy Transformation[J]. *E3S Web of Conferences*, 2021, 235: 01006-.

Zhang L, Sun S Q, Geng Y. Analysis of low-carbon transition of energy-based cities—Taking Huainan as an example[J]. *Northern Environment*, 2013, 25(12): 27–31.

Zhang N, Xue M M, Wu Z Y et al. Comparison and enlightenment of domestic and foreign energy transition[J]. *China Electric Power*, 2021, 54(02): 113–119+155.

Zhang Y X, Zhang J. The stages of low-carbon transformation of resource-based cities and their total factor productivity evaluation—Based on the SBM model of undesired output[J]. *Science and Technology Management Research*, 2020, 40(14): 76–83.

Advances in Energy, Environment and
Chemical Engineering – Abdullah & Osman (Eds)
© 2023 The Author(s), ISBN: 978-1-032-36083-6

Research on electricity market supervision system in the UK

Fan Zhang, Shijun Tian, Chen Lv* & Chenghui Tang
State Grid Energy Research Institute, Beijing, China

ABSTRACT: The electricity market in China is developing rapidly, and the scientific requirements for electricity supervision are very urgent. The British electricity market has developed
maturely, and the market supervision system has reference value for China. This paper studies
the British electricity market supervision system, including the relevant laws of electricity market supervision, the functions of electricity market supervision institutions, and the content of
electricity market supervision.

1 INTRODUCTION

China's electricity market is developing rapidly, and the market system structure has been initially formed. The market system structure covers regional and provincial levels in terms of space
and covers medium and long-term transactions and spot electricity transactions in terms of time
periods. In addition, the market architecture also includes various types of transactions, such as
electric energy transactions, ancillary services transactions, and renewable energy consumption
responsibility transactions. Therefore, the scientific requirements for power supervision are very
urgent. The British electricity market has developed maturely, and the market supervision system
has reference value for China. Therefore, this paper studies and analyzes the British electricity
market supervision system.

2 LAWS ON ELECTRICITY MARKET SUPERVISION IN THE UK

Electricity regulatory reform in the UK is recognized as a relatively successful and influential
reform of electricity regulatory legislation that has played an important role.

In 1989, the British revised "Electricity Law" passed the provisions on the privatization of power
enterprises in England, Wales and Scotland and began the first reform of the power industry. This
has established the legal basis and implementation procedures for the reform of the UK electricity
regulatory system. In the electricity market reform under the framework of the electricity law in
1989, the UK established the Electricity Regulatory Office as an independent regulatory agency. In
1989, the power law granted the supervisors of power regulatory agencies considerable legal power.
For example, after consultation with the minister of state responsible for the power industry, the
supervisors have the right to issue power generation, transmission, distribution and sale licenses
for enterprises. In 1990, the United Kingdom established a mandatory electricity trading system
in England and Wales. All the statutory operating rules of the electricity warehouse were clearly
defined in the "electricity warehouse and settlement rules" adopted by the Parliament.

The successful implementation of the power depot has accumulated rich experience: power quality and system security are guaranteed, new power producers are encouraged to enter the market,
and the introduction of power purchase side competition is allowed, etc. However, there are also

*Corresponding Author: Lvchen20130101@163.com

DOI 10.1201/9781003330165-11

some shortcomings in the operation of the power storage. In 2000, the United Kingdom canceled the mandatory power storage, implemented the new trading rules dominated by multilateral contracts, allowed the power contract to span from the day to the future, and reset the new power trading rules (NETTA). The goal is to realize the complete competition of the whole British power wholesale market. In 2004, the British Parliament passed the Energy Law, which mainly included the establishment of a unified British electricity trading and transmission system, measures to support the development of renewable energy, the appointment of a special administrator to maintain the operation of the network, and the parliament authorized the government to modify the current electrical regulations according to the decision of the Gas Electricity Market Office.

3 FUNCTIONS OF ELECTRICITY MARKET SUPERVISION INSTITUTIONS

3.1 *Gas and electricity markets authority*

In 1986, the Thatcher government passed the gas law to privatize the British natural gas company. At the same time, the bill partially lifted the regulation of natural gas supply and established a natural gas supply regulation market. Under the same background, the electric power law was passed in 1989, and the electric power supervision office was established to supervise the electric power market. In 2000, the utility law merged the gas supply office and the electricity regulatory office to form the Gas and Electricity Markets Authority. There are two main functions. One is to promote appropriate, effective competition, and the other is to supervise the electricity and natural gas markets. The Natural Gas and Electricity Market Committee is a non-ministerial government department responsible for the legislative body of Parliament. The management of the Market Committee comprises executive, non-executive members and a non-executive Chairman. Non-executive members are composed of experts in the fields of business, social policy, environmental protection, finance and European affairs. The chairman was appointed by the Minister of Commerce, Energy and Industry Strategy for five years. The committee's budget comes from annual licensing fees for energy companies. The operation of the committee maintains transparency by opening daily agendas to the public. The specific regulatory work of the Natural Gas and Electricity Market Committee is the responsibility of the Natural Gas and Electricity Market Office.

3.2 *Office of gas and electricity markets*

The Office of Gas and Electricity Markets (Ofgem) under the Natural Gas and Electricity Markets Bureau is responsible for supporting the specific supervision of the Markets Bureau. OFGEM includes several executives, non-executors and a non-executive chairman. Among them, the main work of non-executors is to analyze expert opinions and experiences on energy industries, social policies, and the environment. OFGEM decision makers have decision-making power on price and specific regulatory measures. The office has public relations departments, comprehensive functional departments, transmission supervision departments, distribution supervision departments, market supervision departments, sustainable development departments, E-service departments, etc. Each department undertakes different responsibilities. The main function of OFGEM is to regulate the UK electricity and gas markets, with the aim of protecting consumers' rights. The terms of reference of its supervision by the 1986 Natural Gas Law, 1989 Electric Power Law, 1998 Competition Law, 2000 Public Utility Law, 2002 Enterprise Law, 2004 Energy Law, 2008 Energy Law and other acts. Its main functions include supervising and preventing monopoly power in electricity and natural gas markets, so as to promote market competition and safeguard the interests of electricity and natural gas consumers; ensuring British energy supply, mainly by increasing the level of competition in the electricity and natural gas markets, encourage investment in the energy industry; adapting to climate change and promoting the sustainable development of energy economy, specifically helping power and natural gas enterprises to improve environment and efficiency, and providing help for energy consumption of low-income people. Power operation licenses are issued according to the authorization of the Department of Trade and Industry of the UK Department of Power Industry. The obligations are as follows: publicly modify the basis and content of the license; coordinate the relationship between power enterprises; handle user complaints, protect the interests of electricity consumers; investigate and deal with acts of unfair competition; publish the operation of the power industry and related information.

4 CONTENTS OF ELECTRICITY MARKET SUPERVISION

4.1 *Electricity market structure and market access supervision*

In terms of market structure, in order to promote competition in the electricity production market, regulators will ask the power companies suspected of monopolistic behavior to investigate MMC of the Commission on Monopolization and Merger to intervene in oligopoly. After the power restructuring in 1990, the two major power generation companies established in the power generation market - the State Power Corporation and the Power Production Company, which formed strong market forces and formed a dual monopoly market structure, and the price of electricity was basically determined by the two companies. In order to reduce the oligopoly power, the UK electricity regulatory authorities twice required the state power companies and electricity production companies to sell power generation assets in 1996 and 1999 to reduce their share in the power generation market. Since then, the share of power generation capacity owned by the State Power Corporation and the Power Production Corporation in the country's total power generation capacity has declined from 61.8 percent before sale to 29.5 percent after sale, significantly reducing the ability of the two companies to monopolize the market.

In terms of market access, on the one hand, the introduction of third-party access to the transmission network, that is, fair access to all market participants, provides a strong guarantee for the normal operation of the power grid and fair market transactions; the second is to gradually introduce the user's right of choice, gradually promote the process of user's free choice of power supply enterprises according to the load level, and stipulate the responsibilities and obligations of power supply enterprises to promote the competition in the power sales market.

4.2 *Price regulation*

Price regulation of monopoly enterprises is the core content of energy regulation. After industrial restructuring, the monopoly links mainly include transmission and distribution networks. The UK energy regulator (OFGEM) introduced a new price regulatory framework RIIO in 2010 to meet the reliability requirements of energy supply in the context of a low-carbon economy, using the RPI-X price regulatory model, the highest price limit model for monopoly enterprises.

1) RPI-X price regulation model. RPI-X price regulation model, namely the inflation rate (RPI) efficiency reduction coefficient (X), sets the price ceiling to encourage enterprises to reduce costs and improve efficiency. The efficiency coefficient is generally 2–5. As long as enterprises strive to improve their efficiency more than X level, you can get the corresponding benefits. As long as it does not exceed the maximum price limit, the pipeline enterprises can adjust the price in this period according to the price adjustment formula. The operation rule of RPI-X is to set the price standard for power grid enterprises in a specific regulatory cycle and adjust the price standard after the expiration of the regulatory period. Under the price control of the RPI-X model, the enterprise can adjust the annual price within the scope of price control. The ratio of increase and decrease is equal to the difference between the RPI (retail price index) of the previous year and the efficiency deduction coefficient X of the enterprise. The RPI-X price regulation model incentivizes pipeline companies to improve efficiency and reduce costs. In addition, the UK regularly reviews the price of power transmission and distribution networks and revises specific price formulation methods. In each revision, innovative methods will be introduced in combination with reality. For example, in the fourth price review of distribution

companies (DPCR4), the incentive compatibility menu contract method is used to set the capital expenditure limit of distribution companies.

2) RIIO price supervision model. In July 2010, OFGEM proposed a new RIIO regulatory model to deal with the future energy low-carbon development trend of how to meet the needs of energy consumers. The main content of the RIIO regulatory model is still to regulate the income of enterprises with monopoly market power in power transmission and distribution networks and natural gas pipeline network, in order to ensure that energy consumers pay reasonably and reflect the energy consumption price of the market efficiency. The meaning of RIIO refers to the fact that the income (Revenue) of power grid enterprises should be set with reference to incentive levels (Incentives), innovation (Innovation) and output (Outputs), namely, "income = incentive + innovation + output." RIIO model considers the reliability, safety, environmental friendliness, social responsibility and other service quality of power network and natural gas network enterprises, encourages enterprises to provide sustainable, long-term power or natural gas supply, and respond to consumer demand for certain technological innovations. The basic framework of the RIIO model includes: requiring natural gas or power network enterprises to provide specific business plans, including specific service matters and service currency value, market testing and other specific measures to determine the income of the enterprise; extending the review cycle of price control from 5 years (the requirements of RPI-X model) to 8 years, which is compatible with the technological innovation cycle and the life cycle of network assets. rewarding network enterprises for innovation and effective investment to stimulate investment; energy consumers and renewable energy developers (i.e., energy users) are given more options and voice to adjust the income during the price control period in the form of "customer satisfaction" by setting low carbon network, etc.

4.3 *Sustainable power grid supervision under RIIO model*

RIIO regulation model is applicable to the price regulation of power transmission and distribution networks and natural gas transmission and distribution networks. Therefore, the basic content of the RIIO regulation model is no longer restated here, and the meaning of the RIIO model in power grid regulation is concerned: the setting of power grid income pays more attention to the sustainable development of the economy and society. As an important platform for carbon emission reduction of the energy system and an important support force for sustainable development of the economy and society, the power grid has been highlighted. Pay more attention to the sustainable development of power grid enterprises, that is, to ensure that power grid investors and operators enjoy reasonable long-term returns and monetary value; it fully reflects the attention to the welfare of contemporary and the next generation of consumers, taking into account intergenerational equity; the key regulatory object of power grid under RIIO framework is output rather than cost. The so-called output includes grid reliability and security services, grid access conditions, universal services, environmental impact, consumer satisfaction and social responsibility. Therefore, compared with the RPI-X model, the power network supervision under the RIIO model pays more attention to the service output, such as paying more attention to the supervision of power supply service quality.

4.4 *Supervision of power supply service quality*

Supervision of power supply service quality focuses on the following aspects:

1) Power supply service quality: the number of users affected by power outages and the average outage time per 100 users in the power supply business area every year (power outages lasting more than 3 minutes are counted, and several outages caused by the same fault are extremely one outage statistics);

2) Telephone service quality: Evaluating through the third-party market survey, the quality of telephone service and the corresponding speed are mainly inspected in the telephone service inspection. The key sampling objects are the users who have recently called power outages, warranties or complaints to power suppliers. The telephone service quality of each power supplier

should be scored, and then certain fines should be imposed on enterprises below the standard according to the scoring results, while enterprises above the standard will receive rewards.

4.5 *Implementation of regulatory measures*

Ofgem has the power to enforce regulatory measures in the following aspects.

1) Supervise the implementation of licenses. Any enterprise that wants to produce power, run a power service network, or supply power should obtain the license provided by Ofgem. The licensed enterprises need to follow a series of provisions of the licensing requirements. If it is found that some enterprises have violated the relevant provisions, Ofgem can take the following coercive measures: issuing mandatory instructions to enterprises; the amount of economic punishment for enterprises can be as high as 10 % of their license turnover.
2) Supervise the enforcement of competition law. Ofgem's supervision of the level of energy market competition is mainly to increase the transparency of market competition, and reduce barriers to market competition and corporate behaviors that hinder competition. If an enterprise is in breach of competition law, the coercive measures that Ofgem can take include: issuing restraining orders for acts contrary to competition principles; economic punishment can be as high as 10 % of the global income of the enterprise.
3) Monitoring the implementation of consumer protection laws. Ofgem monitors whether energy companies comply with the UK and European consumer protection laws.

5 CONCLUSION

Based on the discussions presented above, the conclusions are obtained as below:

(1) Priority should be given to the enactment of regulations guaranteeing electricity regulation. A series of policy laws and regulations are issued by government departments or relevant agencies as codes of conduct, providing a legal basis for standardized supervision of electricity and a competitive environment for market participants.
(2) Independent power regulatory agencies should be set up specially. Electric power supervision departments are independent of other government departments and regulated subjects, with strong authority and neutrality. Electric power reform is carried out from top to bottom, the reform object is a powerful monopoly institution, and the independence and authority of government departments to carry out the reform is indispensable.
(3) To dynamically adjust the content and methods of power regulation. The supervision reform of the electric power industry should change with the macroeconomic situation. At different stages, electric power supervision has the task of adapting to the process of electric power reform.
(4) Emphasis should be placed on the capacity-building of regulators. Due to the complexity of the power system, the ability level of regulators must be adapted. Only when the regulatory level and talent allocation of regulators are in line with the development of the power industry can they play an effective role in regulation.

ACKNOWLEDGMENTS

This project is funded by the State Grid Corporation of China headquarters management science and technology project. Contract No.SGHBJY00NYJS2000020.

REFERENCES

James Murray. British electricity market reform: 10 questions to be answered [J]. *Low carbon world*, 2011 (05): 38–39.

Li Jixian, Xu Siyang. Enlightenment of British electricity market reform on the development of China's electricity market [J]. *Electricity*, 2021 (07): 1–4.

Liu Guanchi, Liu Zhigang. Operation Mechanism of British Electric Power Market and Its Enlightenment to China's Electric Power System Reform [J]. *Electronic World*, 2016 (16): 10–11.

Wen An, Huang Weifang, Liu Nian. Electricity Trading Balance Mechanism in British Electricity Market [J]. *Southern Power Grid Technology*, 2014, 8 (05): 1–5.

*Advances in Energy, Environment and
Chemical Engineering – Abdullah & Osman (Eds)
© 2023 The Author(s), ISBN: 978-1-032-36083-6*

Design and practice of geothermal cascade utilization model in Western Sichuan

Meilin Ye & Ting Ni*

College of Environment and Civil Engineering, Chengdu University of Technology, Chengdu, P.R. China

Lijun Liu

College of Energy, Chengdu University of Technology, Chengdu, P.R. China

Jiaxuan Li

Academic Affairs Office of Chengdu University of Technology, Chengdu, P.R. China

ABSTRACT: Geothermal resources in western Sichuan are extremely abundant, with total resources reaching 4.26×10^{16} kJ, but their utilization lacks reasonable planning, and the utilization amount and utilization efficiency need to be improved. According to domestic and international experience, the geothermal cascade utilization can effectively improve the utilization rate of geothermal resources. This paper designs two geothermal step utilization models for the western Sichuan plateau region. The model for winter is power generation → heating → water source heat pump → greenhouse → bath therapy → aquaculture. The model for summer is power generation → cooling → drying of agricultural products → bath therapy. Finally, the technical performance and environmental benefits of gradient utilization are verified by actual cases. The results show that the above two cascade utilization modes are technically feasible and have obvious energy-saving, and their application to actual projects is of great practical significance and long-term strategic significance to alleviate the energy pressure and achieve the goal of "double carbon" in Sichuan province.

1   INTRODUCTION

The plateau area in western Sichuan, with thin air and low oxygen content, does not burn conventional energy sufficiently and does not have high energy utilization. Therefore, there is a need to transform to new energy with clean and low carbon characteristics. Geothermal, as a zero-carbon, clean energy source, has the characteristics of stability and reliability (Sun 2014; You et al. 2007). According to expert statistics, the geothermal resources in the western Sichuan area are extremely rich, with total geothermal resources amounting to 4.26×10^{16} kJ (Fu 2009; Qu et al. 2019), which has great potential for development and utilization. However, at present, there is a gap in the level of utilization of geothermal resources in China compared with advanced foreign countries (Wang et al. 2014). Therefore, how to better develop geothermal energy to improve its utilization and efficiency plays an important role in relieving the pressure on energy supply and improving the ecological environment and is of great significance for China to achieve the goal of carbon neutral development by 2060.

Geothermal energy stepped utilization refers to the utilization of geothermal energy in steps from high to low according to the different temperatures of geothermal fluids (Pastor-Martinez et al. 2018; Rubio-Maya et al. 2015). According to domestic and international experience, it has been proved that the stepped utilization of geothermal resources can extract thermal energy from geothermal fluids in multiple stages, utilizes geothermal energy at multiple levels, expands the

---

*Corresponding Author: niting17@cdut.edu.cn

DOI 10.1201/9781003330165-12

overflow difference available for geothermal fluids, and realize the ideal recharge temperature (Feng 2014; Leveni et al. 2019). There are already many practical projects for geothermal energy gradient utilization in foreign countries. For example, a plant in Nevada, USA, has a geothermal energy cascade system for electricity generation and onion and garlic dehydration, which can dehydrate 34,019 kg of onions or 38,555 kg of garlic per day (Gordon 2004). A multi-stage stepped utilization system of geothermal water for circulating electricity, refrigeration, drying of chili and garlic, and hot springs has been built in Chiang Mai, Thailand (Amatyakul et al. 2016). In China, geothermal heating, combined with a heat pump heating cascade utilization system, was built in the six districts of Beiyuan homes in Beijing. This was the largest cascade utilization project in China at that time, using geothermal water as a resource combined with a water source heat pump to provide the heat source for buildings (Wu 2016). Although many researchers in China research geothermal energy terrace utilization, there are not many projects for practical application.

2 DESIGN OF THE CASCADE UTILIZATION MODE

2.1 *Overview of geothermal resources in Western Sichuan*

The western Sichuan region is rich in geothermal resources, and according to experts, the total geothermal resources in the western Sichuan region amount to 4.26×10^{16}kJ. Among them, more hot springs and spring groups are exposed in the western Sichuan plateau area (see Table 1). Through the survey, the utilization of geothermal resources in west Sichuan is mostly based on hot springs for leisure bathing, and the utilization is relatively single, lacking reasonable planning, resulting in the loss of much underground hot water for nothing (Li 2008; Luo et al. 2016; Wang et al. 2009). Secondly, in the plateau area of western Sichuan, the air is thin, and the oxygen content is low, so conventional energy does not burn sufficiently. Therefore, the western Sichuan region should give full consideration to the utilization of its geothermal resources and carry out reasonable planning to improve the amount and efficiency of geothermal resource utilization. In Table 1, the geothermal resources in the western Sichuan area are medium and high-temperature geothermal, and it is suitable to carry out geothermal cascade utilization and explore its potential for geothermal power generation (Sun 2019).

2.2 *Design of winter cascade utilization model*

According to the unique climate conditions in west Sichuan, geothermal power generation → heating → water source heat pump → greenhouse → bath therapy → aquaculture model can be used in winter (see Figure 1). The main reasons are: West Sichuan is located in the Tibetan Plateau, the winter is cold and long, the annual temperature is below 5°C for about 210 days, people's work and life is greatly affected, so there is a great demand for heating in winter in west Sichuan. In winter, the heat pump takes the geothermal drainage water of the geothermal heating system as the low-temperature heat source and extracts the heat from it to send to the high-temperature medium for heating, which further reduces the drainage water temperature and improves the available premium difference of the geothermal fluid (Li 2018). Geothermal is a compound resource, and using geothermal as the heat source for greenhouse greenhouses can effectively reduce greenhouse production costs and has good economic and environmental benefits. Hot spring water has a constant temperature, contains natural minerals, and is ideal for aquaculture and fish fry overwintering after water treatment (Tang et al. 2017). Hot spring physical therapy can be both fitness and cure and is a major tourism characteristic of the western Sichuan region. In addition, the application of geothermal in agriculture can not only effectively reduce production costs and enhance the competitiveness of modern agricultural products, but also help the construction of rural clean energy projects.

2.3 *Design of summer cascade utilization model*

In summer, power generation → cooling → drying of agricultural products → bath therapy geothermal cascade utilization model is adopted (see Figure 2). The high-temperature geothermal fluid extracted from the ground is pre-treated and first used to generate electricity. The generated geothermal return water can provide energy support for cooling units in cold storage and vegetable preservation warehouses, and can also be applied to production processes, air conditioning, etc. Immediately after that, the hot air of geothermal return water after cooling is sent to the drying channel through the fan for drying agricultural products, which is a tertiary utilization. After tertiary utilization, the geothermal water is treated and used for health physiotherapy, which is a quaternary utilization. The final tailwater is treated and recharged to the ground. Geothermal drying is conducive to creating special agricultural and sideline products and promoting "Sichuan" brand agricultural products to the whole country and the world. Secondly, summer is the peak season for tourism in western Sichuan, and hot spring bathing is a major selling point for local tourism development.

Table 1. Overview of geothermal resources in Western Sichuan.

| Geothermal resources | Number | Characteristics (hot water, flow) | Deep thermal storage temperature | Major geothermal zones | |
				Name	Fluid characteristics (temperature, flow rate)
Fresh water river (Ni et al. 2016; Qu et al. 2019; Zhang et al. 2021)	44	25.00–91.00°C 0.01–38.88L/s	93.49–150.00°C	Kangding-Yulin Palace	65.0–85.0°C, 0.50–2.78L/s
				Kangding-Zhonggu	65.0–83.0°C, 0.50–1.72L/s
				Danba-Dangling	41.0–78.0°C, 0.20–1.60L/s
Ganzi-Litang (Qu, et al. 2019; Zhang et al.)	42	25.00–96.00°C 0.11–24.05L/s	71.92–105.14°C	Gan Zi-Gan Inge	35.0–89.0°C, 0.25–5.72L/s
				Litang-Card ash	55.0–86.0°C, 8.5–19.5L/s
				Litang-Giza	41.0-81.0°C, 1.3–13.5L/s
Jinsha River (Ni et al. 2016; Qu et al. 2019; Yu et al. 2021)	37	25.00–98.00°C 0.02–30.50L/s	68.66–165.74°C	Batang-Tea luo	86.0–89.0°C, 0.98–4.10L/s
				Batang-Cuopu ditch	25.0–89.0°C, 0.15–5.69L/s

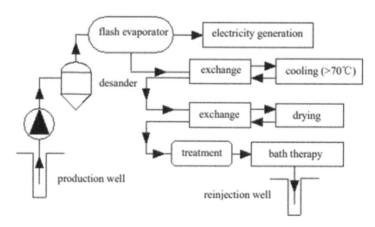

Figure 1. Geothermal winter cascade utilization mode in western Sichuan.

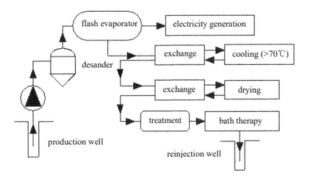

Figure 2.   Summer cascade utilization mode of geothermal energy in Western Sichuan.

3   CASE ANALYSIS

3.1   *Problem description*

A centralized heating project in the western Sichuan plateau area, with a heating area of about 100,000 m², is designed to use geothermal gradient utilization for heating. After heating load estimation, the peak system load is 4898kW; considering certain safety factors, the total design load of the system is about 5613kW. According to the field research on the hot springs around the project, the heat source is geothermal water from the hot springs about 10 km away from the project, and the temperature of the hot water at the mouth of the two hot springs is measured to be stable at about 80°C, and the flow rate of the water is 30m³/h and 40m³/h respectively. It is calculated that the heating power that can be provided by the hot springs after transportation is about 2963 kW. Considering the load demand of the heating project, the thermal power that the spa can provide cannot meet the heating demand, so the combination of heat pump units for combined heating is considered.

3.2   *Research methods*

3.2.1   *Construction of the cascade utilization system*
In this paper, taking into full consideration the local geothermal resources and combining them with the actual situation of the project, a geothermal cascade utilization system consisting of two systems, a geothermal heat exchange system, and a water source heat pump system, is established (see Figure 3).

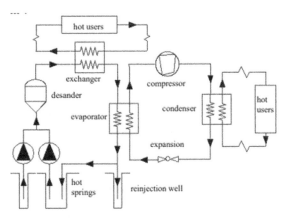

Figure 3.   Flow diagram of cascade utilization.

The hot water resource from the hot spring is diverted to the heat station, where the medium tap water is heated by heat exchange through a heat exchanger, and then the building is heated, which is the first stage. The second step is to introduce the geothermal tail water after heat exchange into the water source heat pump unit, input a small amount of electric energy, and continue to heat the building after heating. The geothermal tail water recharge temperature is 10°C. Among them, the heat pump unit with a heat production capacity of 2650 kW is selected for the system. The heating heat power provided by the second-stage utilization system composed of a geothermal heat exchange system and water source heat pump system is calculated to meet the heat load demand of the project (see Table 2).

Table 2. Main equipment and devices of the system.

Equipment	Number	Performance parameters	Equipment running power (kW)
Water source heat pump unit	1	Heating capacity: 2650 kW 55°C Residual heat: 32°C 70 m^3/h (including end heat exchanger, integrated hot water transmission and distribution system, ancillary equipment, control system)	417
Pick-up pump	2	Flow: 40 m^3/h Lift: 55 m	22
Heat Exchanger unit	1	Complete supply of equipment	
Heat storage tank	1	Volume: 150 m^3	

3.2.2 Technical performance evaluation method

Utilization rate and energy efficiency ratio are two relatively basic performance evaluation coefficients that can initially determine the technical feasibility of a gradient utilization system. The geothermal utilization rate is the ratio of the actual heat supplied by the geothermal to the heat available from the geothermal (Tang et al. 2017).

$$\eta = \frac{GC\,(t_1 - t_2)}{GC\,(t_1 - t_0)} = \frac{t_1 - t_2}{t_1 - t_0} \tag{1}$$

Where, η denotes geothermal energy utilization rate, %. G denotes geothermal water supply flow rate, kg/s. C denotes the specific heat of geothermal water, kJ/(kg-K). t_1 and t_2 denote geothermal water supply temperature and recharge temperature, °C. t_0 denotes the calculated minimum drainage temperature, °C.

Usually, the coefficient of performance (COP) is the ratio of heat production to effective input power when the heat pump is operated for heating under specified conditions (Tang et al. 2017), and sometimes it is also called the energy efficiency ratio according to its definition, which is also one of the coefficients for the performance evaluation of the geothermal energy terrace utilization system.

$$COP = \frac{Q}{W} \tag{2}$$

Where, COP denotes the energy efficiency ratio of geothermal energy heating. Q denotes the heat exchange of the heating part of geothermal energy utilization, kWh. W denotes the total energy consumption (electricity) of system operation, kWh.

3.2.3 Environmental benefit analysis method

In this paper, the annual energy saving and emission reduction of the system are calculated by the amount of standard coal saved. Based on this, the environmental benefit analysis of geothermal energy gradient utilization is carried out by borrowing the savings of treatment costs.

3.2.3.1 *Calculation of pollutant emission reduction*

The annual heat production of the geothermal energy gradient utilization system can be converted into standard coal. According to expert statistics, saving one ton of standard coal can reduce 2.493 tons of CO_2 emission, 0.075 tons of SO_2 emission, and 0.038 tons of NO_X emission; thus, we can get the reduction of pollutant emissions in one year of the operation of the system.

3.2.3.2 *Calculation of treatment cost saving*

According to the standard for saving pollutant treatment costs by using geothermal energy in Geologic Exploration Standard of Geothermal Resources (GB/T 11615-2010), the pollutant treatment cost can be saved by using geothermal energy to reduce 1kg of CO_2 by 0.1 yuan, 1kg of SO_2 by 1.1 yuan and 1kg of NO_X by 2.4 yuan. Therefore, the cost of CO_2, SO_2, and NO_X pollutants can be saved in one year of operation of this geothermal energy step utilization system.

4 RESULTS ANALYSIS

4.1 *Technical performance analysis*

Checking the design standard (JGJ26-2010), the average outdoor temperature of the heating period at the project site was obtained as -1.2°C. Therefore, from equation (1), it can be calculated that the energy utilization rate of this geothermal heating system can reach 86.2%. The energy consumption of this geothermal energy step utilization system is about 461kw at full load, and the heat production is 5613kw. Therefore, the theoretical energy efficiency ratio of this geothermal heating system can be calculated from equation (2) to be 13.49, which is higher than the theoretical energy efficiency ratio of the water source heat pump (COP = 7).

4.2 *Environmental benefit analysis*

In this project, the geothermal energy gradient utilization system can produce about 7.27×10^{10}kJ of heat per year (heating for five months in a year), which is converted into about 2480.59 tons of standard coal. According to expert statistics, saving 1 ton of standard coal can reduce 2.493 tons of CO_2 emission, 0.075 tons of SO_2 emission, and 0.038 tons of NO_X emission, which can get the reduction of pollutant emission in one year of this system operation (see Table 3).

Table 3. Reduction in pollutant emissions.

Name	CO_2	SO_2	NO_X
1t standard coal emissions (t)	2.493	0.075	0.038
Annual emission (t)	6184.11	186.04	94.26

According to the reduction of pollutant emission in Table 3 and the standard of saving pollutant treatment costs by using geothermal energy according to the specification (GB/T 11615–2010), we can get the saving cost of treating CO_2, SO_2, and NO_X pollutants in one year of operation of this geothermal energy gradient utilization system (see Table 4).

Table 4. Save on governance costs.

Name	CO_2	SO_2	NO_X
Expense standard (yuan/kg)	0.1	1.1	2.4
Cost saving (million yuan)	61.84	20.46	22.62
Summation (million yuan)		104.92	

The cascade utilization model was analyzed from two aspects: technical performance and environmental benefits. According to the results, the geothermal energy terrace utilization system is technically feasible, with obvious energy-saving and emission reduction effects, and has a certain promotion.

5   CONCLUSION

Western Sichuan is rich in geothermal resources, with a large number of hot springs exposed and good hydrothermal display, and it has a large potential for development and utilization. The geothermal resources in West Sichuan are medium and high-temperature geothermal resources, which are suitable for geothermal terrace utilization, and the potential of geothermal power generation can also be explored. This paper designs two geothermal utilization modes for winter and summer, taking into account the geothermal situation in western Sichuan. In winter, the geothermal power generation → heating → water source heat pump → greenhouse → physical therapy → aquaculture utilization mode is adopted; in summer, the power generation → cooling → drying of agricultural products → physical therapy utilization model is adopted.

Taking an actual project in the western Sichuan plateau as an example, the technical performance and environmental benefits of the cascade utilization are evaluated and analyzed. The results show that the energy utilization rate of the cascade utilization system is high, and the energy conservation and emission reduction effect is obvious. It is verified that the winter and summer cascade utilization model proposed in this paper has the significance of promotion in the western Sichuan region, and the application of this model to the actual project is of great practical significance and long-term strategic significance to alleviate the energy pressure and achieve the goal of "double carbon" in our province, which is worth promoting and applying.

ACKNOWLEDGMENT

This work was supported by the Sichuan Mineral Resources Research Center (Grant No. SCKCZY2020-ZC001).

LITERATURE

Amatyakul P, Boonchaisuk S, Rung-Arunwan T, et al. Exploring the shallow geothermal fluid reservoir of Fang geothermal system, Thailand via a 3-D magnetotelluric survey[J]. *Geothermics*, 2016, 64: 516–526.

E. Pastor-Martinez, C. Rubio-Maya, V. M. Ambriz-Díaz, J.M. Belman-Flores, J. J. Pacheco-Ibarra. Energetic and exergetic performance comparison of different polygeneration arrangements utilizing geothermal energy in cascade[J]. *Energy Conversion and Management*, 2018, 168

Feng G H, Fan W S, Sun Y D. Application analysis of deep geothermal cascade utilization in the field of building energy conservation in Xian County. *Construction Science and Technology*, 2014(02): 82~83.

Fu G H. *Research on hot spring type, causes and tourism development mode in Ganzi Prefecture*, Sichuan Province. Chengdu University of Technology, 2009.

Gordon R. *Empire energy*, Llc. GHC Bull 2004:11–3

Leveni M, Manfrida G, Cozzolino R, Mendecka B. Energy and exergy analysis of cold and power production from the geothermal reservoir of Torre Alfina. *Energy*. 2019; 180:807–818.

Li S. *Operation control and energy consumption optimization of water source energy pump system*. University Of Qingdao, 2018.

Li X Q, Hot spring characteristic and sustainable development in western Sichuan. *Journal of Anhui Agricultural Science*, 2008(16):6915–6917+6962.

Luo M, Ren R, Yuan W. Type, Distribution and genesis of geothermal resource in Sichuan. *Acta Geologica Sichuan*, 2016, 36(01):47–50+59.

Ni G Q, Wei Y T, Qu Z W, et al. Geothermal resource in Sichuan. *Acta Geologica Sichuan*, 2016, 36(2):239–242.

Qu Z W, Zhang H, Hu Y Z, et al. General situation and development regional division of geothermal resources in western Sichuan region. *Mineral Exploration*, 2019(5):1233–1242.

Rubio-Maya C, Ambríz Díaz V M, Pastor Martínez E, Belman-Flores J M. Cascade utilization of low and medium enthalpy geothermal resources - A review. *Renewable and Sustainable Energy Reviews*. 2015; 52: 689–716.

Sun D, Cao N, Liu X Z, Zhang Z P. Geothermal Resources and Development in Garzê Prefecture, Sichuan. *Acta Geologica Sichuan*, 2019, 39(01):133–138.

Sun J. Characteristics and sustainable development and utilization of geothermal resources. *Resources Economization & Environmental Protection*, 2014(04):11.

Tang Z W, Wang J F, Zhang H Y. *Geothermal energy utilization technology*. Beijing: Chemical Industry Press, 2017.1

Wang A M, Cao Z T, Shi G. Discussion on geothermal resources and engineering application in Tianjin. *District Heating*, 2014, 000(002):92–96.

Wang T, Kong C, Huang L H, et al. Discussion of the use of renewable energy in western Sichuan. *Refrigeration and Air Conditioning* (Sichuan), 2009, 23(04):109–113.

Wu Y L, Di Y H, Jiang H. Energy saving transformation of heating system in an agronomy garden in Beijing [J]. *Refrigeration and air conditioning* (Sichuan), 2016, 30 (06): 696–699 + 711.

You X C, Yao S Z, Yan S Q, et al. Utilization State and Protection Project in China Geothermal Area. *China Mining*, 2007, 16(6):1–3.

Yu C M, Zhang C, Yang Y, et al. Cascade utilization of geothermal resources in western Sichuan Province. *Natural Gas Exploration and Development*, 2021, 44(03):102–111.

Zhang J, Li W Y, Tang X C, et al. Geothermal data analysis at the high-temperature hydrothermal area in Western Sichuan. *Science China Earth Sciences*, 60: 1507–1521.

Zhang W, Wang G L, Zhao J Y, et al. Geochemical Characteristics of Medium-high Temperature Geothermal Fluids in West Sichuan and Their Geological Implications. *Geoscience*, 2021,35(01):188–198.

Advances in Energy, Environment and
Chemical Engineering – Abdullah & Osman (Eds)
© 2023 The Author(s), ISBN: 978-1-032-36083-6

Forest farm classification management strategy based on comprehensive carbon sink and economic benefit

Zhongyu Xiao*, Hongyu Cui*, Changguo Jia & Hongzhou Wang
Mathematical Experiment Center, School of Mathematics and Statistics, Beijing Institute of Technology, Beijing, China

ABSTRACT: With the increase in population and the shortage of resources, the concept of sustainable development is increasingly recognized. The management of forests around the world tends to pay more attention to the concept of sustainable development. However, at present, there are some problems in forest management, and the management and benefits of forests are the most attractive. Facing the present situation of forest management, we adopt the biomass inventory method of the relationship between biomass and stock and establish a carbon sink model to quantify the carbon sink capacity of trees and their products. On this basis, considering the life cycle of timber, long-term management of forest farms, and other factors, the entropy weight method evaluation model of comprehensive carbon sink and economic benefits of forest farm is given, and forest farm managers can formulate the optimal management strategy accordingly.

1 INTRODUCTION

Climate change poses a huge threat to life as we know it. To mitigate the effects of climate change, we need to take action to reduce the number of greenhouse gases in the atmosphere. Simply reducing greenhouse gas emissions is not enough. We need to work to increase our carbon dioxide stock, sequestering it from the atmosphere through the biosphere or mechanical means. This process is called carbon sequestration. The biosphere sequesters carbon dioxide in plants (especially large plants such as trees), soil, and water. Therefore, forests are integral to any climate change mitigation effort. (Siwa et al. 2016).

Globally, forest management strategies, including appropriate harvesting, benefit carbon sequestration. However, over-harvesting can limit carbon sequestration. Forest managers must find a balance between the value of forest products from logging and the value of allowing forests to continue to grow and sequester carbon. What's more, concerns are not limited to carbon sequestration and forest products. Forest managers often make forest management decisions based on the value of local forests. (Guang et al. 2011; Hou et al. 2012).

This paper will establish an evaluation model for tree carbon sequestration, evaluate the carbon sequestration effect of trees in the growth process, and compare the effects of natural death and felling for furniture and buildings on carbon reduction. On this basis, we give the optimal strategy model of forest farm management considering the two factors of carbon sequestration effect and wood value. At the end of this paper, we use the established model to evaluate the management strategies of the Saihanba Forest in China, the Tongass Forest in the United States, and the United Kingdom.

---

*Corresponding Authors: 1120193271@bit.edu.cn and cuihongyu200012@163.com

DOI 10.1201/9781003330165-13

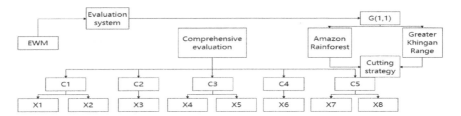

Figure 1.   Model frame.

2   MATERIALS AND METHODS

H-type main steel beams and steel channels, lightweight precast panels set upon the steel skeleton, shear keys connected to the main steel beams, and a post-pouring concrete layer.

2.1   *Assumption*

- The comprehensive value status of forests is defined as a quantitative assessment of indicators such as carbon absorption, economic benefits, potential carbon sequestration, biodiversity, and recreational and cultural value.
- Every tree grows according to the same rules. At the same time, the impact of natural disasters and other unexpected phenomena on the natural regeneration rate of trees is set to zero.
- The felling is done instantaneously (relative to the tree's growing period) and all at once.
- This paper only considers the carbon storage of forest standing trees and does not consider the relationship between the understory vegetation layer and volume (excluding the carbon storage of litter layer, shrub and grass layer, and soil layer).

2.2   *Model preparation*

1) Carbon Sink Value.

This paper defines the carbon sink value of forest resources as the product of forest carbon sink and the unit price of forest carbon sink. The forest carbon sink is the sum of the carbon content of various types of trees, and the unit price of the carbon sink is converted from the international carbon dioxide price. The higher the forest carbon sink value, the stronger the ability of the forest to regulate the natural environment. The importance of the forest can be reflected by the proportion of carbon sink value in the region's GDP X_2.

$$C_1 = \rho \cdot X_1 \tag{1}$$

$$X_1 = \sum B \cdot R_c \cdot S_c \cdot 10^{-6}, X_2 = \frac{C_1}{z} \tag{2}$$

In the formula, ρ is the price per ton of carbon, is the forest carbon sink, is the proportion of carbon sink value in the total production value, Sc is the area of each tree, B is the biomass in each tree, and is the carbon content of the forest component.

2) Economic Value.

In this paper, the economic benefit of forest is defined as the profit from the sale of forest products, which is the sum of the demand for each forest product and the corresponding profit. The higher the economic value, the greater the improvement it can bring to the lives of local residents.

$$C_2 = X_3 - v \tag{3}$$

$$X_3 = \sum n_i \cdot m_i \tag{4}$$

We divide forest products into four categories: wood, furniture, paper, and plywood. The price corresponding to each forest product is n_i (here, we calculate the price of the product that can be produced by one ton of wood), and m_i is the demand, which will be determined according to the composition of forest managers, and v is the cost of production.

3) Cultural Entertainment Value.

In this paper, tourism income X_4 is used to reflect the recreational value of the forest and per capita tourism consumption X_5 is used to reflect tourists' willingness to spend. The higher the per capita tourism consumption, the larger the potential consumption market in this area. The development of tourism can improve the utilization efficiency of forest resources and provide a decision-making basis for fully exploiting the value of forest culture and entertainment.

$$X_5 = \frac{X_4}{f} \tag{5}$$

In the formula, f represents the number of visitors per year, X_4 represents the annual economic income brought by the tourism industry, and X_5 represents the per capita tourism consumption level.

4) Potential Carbon Sequestration.

In this paper, the potential carbon sequestration is reflected by the total proportion (age group structure) X_6 of the young, middle-aged, and near-mature forests of various types of trees in the forest. The carbon rate is the fastest in the forest ecosystem, while the biomass of mature and over-mature forests basically stops growing, and their carbon absorption and release are basically balanced. (Zhong 2000) Therefore, the age composition of forests and the future carbon sequestration capacity of forests is closely related.

$$X_6 = \sum \frac{g_i + h_i + l_i}{S_c} \tag{6}$$

g_i, h_i, and l_i represent the areas of young, middle-aged, and near-mature forests of a tree respectively.

5) Biodiversity

In this paper, the biodiversity indicators are reflected by the regional biological species X_7 and the proportion of endangered species X_8. The biological species is the sum of the animals, plants, and fungi in the forest ecosystem. The more biological species, the higher the species richness and the stronger the self-regulation ability in the forest ecosystem. The proportion of endangered species can reflect the importance of the ecosystem to a certain extent.

Table 1. Vocabulary and symbols.

Symbols	Meaning
X_1	Forest carbon sink
X_2	Proportion of carbon sink value in GDP
X_3	Forest product revenue
X_4	Tourism economy
X_5	Per capita tourism consumption level
X_6	Age group structure
X_7	Biological species
X_8	Percentage of endangered species

2.3 *Evaluation method based on entropy weight method*

To avoid the influence of subjectivity on the quantitative results, we decide to use a data-driven weight calculation method—the entropy weight method (EWM), so as to establish a corresponding evaluation system to evaluate the pros and cons of the comprehensive conditions of each forest.

The entropy weight method is based on the principle of index variability; that is, the higher the variability, the higher the corresponding weight (Li & Li 2021). The steps to calculate the weight of each indicator are as follows:

a) Select n forests and the corresponding eight evaluation index samples, then xij represents the j th (j=1,2,...,8) of the i th forest (i=1,2,...,n) evaluation indicators.
b) Standardize the indicators. Since the indicators in this evaluation are all positive indicators, and the indicator values are all non-negative, the normalization method of column sum equal to 1 can be used for standardization directly.

$$z_{ij} = \frac{x_{ij} - x_{min}}{x_{max} - x_{min}} \tag{7}$$

zij is the variable obtained after normalization. x_{min} and x_{max} are the maximum and minimum values of each index in the evaluation system.

c) Calculate the proportion pij of the i th sample under the j th importance index.
d) Calculate the entropy value of the j th index.

$$e_j = -\frac{1}{\log(n)} \sum_{j=1}^{n} p_{ij} \cdot \log(p_{ij}) \tag{8}$$

e) Compute the information entropy redundancy.
f) Calculate the weight of each indicator.

3 RESULTS & DISCUSSION

3.1 *Practical application of the model*

The data of ten forest areas in four countries are calculated, including China, the United States, the United Kingdom, and Brazil. The weight of each secondary index of X1–X8 for the target layer is as follows:

Table 2. Indicator weights.

X_1	X_2	X_3	X_4
0.172	0.097	0.148	0.122
X_5	X_6	X_7	X_8
0.048	0.09	0.233	0.089

It can be seen that, since they are both forest ecosystems, the proportion of endangered species X_8 and the level of tourists' willingness to consume X_5 are roughly the same, with a small difference, a large entropy value, and a small entropy weight. However, the biological species X_7, carbon sink X_1 and economic value X_3 of different forests are affected by the forest area and main vegetation types, which results in the difference being large, the entropy value being small, and the entropy weight value being large.

The corresponding C1–C5 can be obtained from the corresponding secondary indicators, and it can be found that:

$$C_1 > C_5 > C_2 > C_3 > C_4$$

Carbon sink value and biodiversity are the main functions of forests; thus, the entropy weight is large. What follows are the economic effects, and finally, cultural entertainment value and potential carbon storage capacity. To a certain extent, it shows the effectiveness of the model.

Applying the above model to Saihanba National Forest Park and Greater Khingan Mountains in China, Tongass National Forest in the United States, Amazon Forest in Brazil, and Galloway Forest Park in Scotland, UK, the corresponding comprehensive scores are calculated as follows:

Table 3. Comprehensive evaluation table.

Forest name	Overview
Amazon rainforest	0.951
Saihanba National Forest Park	0.081069
Greater Khingan Mountains	0.354909
Galloway Forest Park, Scotland	0.0449
Tongass National Forest	0.23556

The comprehensive score of the forest should be positively correlated with its own forest size and has a certain relationship with its forest tree species. It can be seen from the evaluation results that the Amazon rainforest scores the highest, followed by the Greater Khingan Mountains and Tongass National Forest, which are basically consistent. We find that the area of the Amazon forest is about 84 times that of the Greater Khingan Mountains in China, but the overall score is not significantly different from that of the Greater Khingan Mountains, which means a large room remains for improvement. Therefore, this paper chooses to develop a rational and achievable management strategy for the Amazon forest, describe the relevant transition policies, and then discuss the practical impact of the implementation of the plan.

3.2 Forecasting based on time series models

Trees take 40 years or more to mature, and the corresponding carbon sequestration peaks at around 40 years and then gradually declines. Traditional forest management usually takes 2.5–3 times the rotation period as the length of the medium and long-term forest management planning cycle, which makes planning decisions achieve significant results (Liu 2013).

Therefore, this paper takes 120 years as the planning period. In these 120 years, the first 40 years are the first stage, the middle 40 years are the second stage, and the last 40 years are the third stage. Every ten years is a period of forest management adjustment, which is used to check, revise and guide the implementation plan.

The current mainstream forest management strategies are three modes: nursing-based management with tending thinning and rotation period as the cycle, continuous logging combined with selective cutting and tending to thin, and commercial-based management with large-area selective logging and clear-cutting (Ma et al. 2014). What's more, in a bid to make a comparison, we decide to add one more natural mode: the natural state of not resorting to logging. Based on the relationship between biomass and logging intensity 1.1, we obtain an estimate of the comprehensive score data 120 years after the implementation of various forest management strategies and compare it with the original management strategy. The results are shown in Figure 2.

$$B = k[(V - \alpha V_c)(1 + p)^n - V_c \cdot \frac{(1 + p)^n - (1 + p)}{p} - \beta V_c] + b \text{ (Zhang 2012)} \qquad (9)$$

V_n is the stock volume after n years (m^3), V is the actual stand volume (m^3), p is the stand growth rate, V_c is the quantitative annual cutting volume (m^3), α is the percentage of the number of trees felled before growth in each year to the annual harvest, and β is the percentage of annual cutting volume after forest growth to the annual harvest, satisfying $\alpha + \beta = 1$. Here we decide to use the compound interest of stand volume under the annual quantitative cutting of $\alpha = 0$ and $\beta = 1$. The formula, namely harvesting after growth, estimates the change of the stock volume and obtains the corresponding biomass according to the linear relationship (Fang et al. 1996).

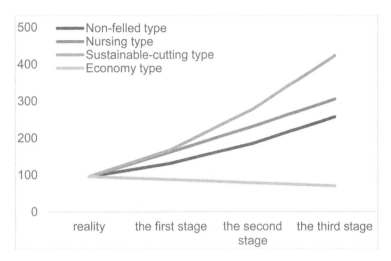

Figure 2.   Prediction of changes in comprehensive evaluation scores under different strategies.

It can be seen that different forest management strategies have different effects on the comprehensive scores of Amazon forest, among which the best effect is the sustainable forest management strategy combining selective cutting and tending to thin with rotation period as the cycle. (Bai et al. 2009).

3.3   *Policy making*

1) Logging strategy

During the mature period of trees, we can clearly find that their carbon content drops sharply, which is not suitable for carbon storage. Therefore, we convert mature trees into related forest products to better store carbon. In addition, the felled trees free up new space, which is conducive to the growth of new saplings. New saplings absorb carbon dioxide and continue a new cycle of carbon sinks, so felled trees also extend carbon sink life.

Considering the importance of the forest age structure, the specific nursing thinning strategy is to perform thinning on half of the forest area each time, with a thinning intensity of 20% and a thinning interval of 40 years. The results are expected to be shown in Table 4 after 120 years. The size of the forest will gradually recover, and the annual carbon dioxide absorption will increase from the original negative value to a positive increase.

Table 4.   Changes in carbon sinks.

Type \ Data	Sustainable-cutting type	Economy type
Reality (t)	84653546	84653546
The third stage (t)	92673355.62	62376297.05
CO2(t)	8019809.621	–22277248.95

From this, we can see that if the forest management continues in accordance with the original policy, the carbon absorption of the Amazon rainforest will be negative, indicating that the previous forest management policy has weakened the carbon absorption capacity of the Amazon rainforest. Instead, if we use the above-mentioned sustainable logging strategy, the net carbon dioxide absorption content is about 8 million tons, which is a significant improvement (Wu 2020).

2) Protection of forest

The Amazon rainforest is the most primitive tropical rainforest area in the world, and it is also the largest tropical rainforest area with the most species in the world, but human beings have not strengthened the protection of the Amazon rainforest.

At present, Brazil's deforestation of tropical rainforests has not only not weakened, but has intensified, which will not only affect Brazil's ecological environment, but also affect the global ecological carbon cycle. The Proceedings of the National Academy of Sciences has again published a "crisis report" on the Amazon rainforest. A recent severe drought and forest fires killed 2.5 billion trees and vines (plants) and turned one of the world's largest carbon sinks into one of the largest sources of pollution. (Xue & Tian 2021).

Frequent forest fires will not only have a direct impact on the climate of South America, such as reduced rainfall and drier climate, but also lead to drastic changes in forest structure and composition, followed by cascading changes in related forest composition. In the long run, the resulting carbon emissions may lead to more serious global warming. Therefore, it is very necessary to conduct regular forest inspections and do a good job of fire prevention. The specific measures are as follows:

1. Implement fire source management combined with dredging and blocking.
2. Strengthen the construction of forest aviation fire protection system.
3. Improve the duty system to ensure the smooth flow of information.
4. Strengthen the construction of a forest fire barrier system.
5. It is necessary to innovate the fire prevention supervision mechanism to eliminate fire hazards in a timely manner.

3) Industrial transformation

At present, the main sources of income for people in the Amazon rainforest are timber export and grain export, both of which depend on deforestation, which is also the reason why the current forest is over-deforestation, and the economic benefits have not been maximized.

Therefore, the government should encourage local residents to profit from by-products rather than deforestation, develop tourism and processing and manufacturing, make rational use of the local biological resources of the Amazon rainforest, and increase the value of cultural entertainment. What's more, we should establish a multi-industry chain integration development model.

Under our proposed management strategy, it is expected that by 2142, the rating of the Amazon rainforest will increase from 95.10 to 423.16, which is 6.07 times that of the Greater Khingan Mountains, an increase of 345%, and will gradually restore the previous forest scale and corresponding functional level.

Table 5. Composite score prediction.

Period — Forest	Reality	First stage	Second stage	Third stage
Amazon Rainforest	95.10	167.89	277.22	423.16
Greater Khingan Range	35.96	46.35	57.24	69.77

4 CONCLUSION AND OUTLOOK

In this article, we establish a mathematical model based on the entropy weight method, which can be used to evaluate the comprehensive status of each forest. Then, we predict the management status after the forest by time series model, comparing it with the scores of different new forest management policies, and get the optimal plan. This method is suitable for the comprehensive evaluation of forest scores. The main conclusions are as follows:

(1) Reasonable selective cutting reduces the density and biomass of trees in a short time, improves the light energy utilization rate and individual nutrition of trees, and reduces the competition among trees to a certain extent. What's more, it provides higher economic value and carbon sink, thus obtaining a higher comprehensive evaluation.

(2) Reasonable selective cutting makes smaller trees grow faster, leading to a rapid increase in productivity and biological growth rate. It can provide higher economic value and carbon sink, thus obtaining a higher comprehensive evaluation.

(3) In the long run, it is more beneficial to increase the carbon sequestration of forests than not cutting down trees, indicating that the ecological environment can be improved by artificial means.

(4) If we continue to operate according to the original policy and forest, the carbon absorption of the Amazon rainforest is negative, and the previous forest management policy has weakened the carbon absorption capacity of the Amazon rainforest. If we improve the cutting strategy with our method, the net carbon dioxide absorption content is about 8 million tons, which is significantly improved.

In the future, we will conduct more in-depth research and exploration and better improve the ecological environment through some artificial means. We will also apply it to different scenarios, such as the management of nature reserves.

REFERENCES

Bai Yanfeng, Jiang Chunqian, Zhang Shougong. (2009) Carbon Storage and Emission Reduction. *Acta Ecologica Sinica*, 29:399–405

Fang Jingyun, Liu Guohua, et al. (1996) Biomass and net production of forest vegetation in China. *Journal of Ecology*, 16:497–508

Guang Zheng, Gong Yuanbo, Chen Linwu, et al. Stability characteristics of soil organic carbon in shrub vegetation in arid valley areas of the upper reaches of the Minjiang River [J]. *Journal of Soil and Water Conservation*, 2011, 25(3): 209–214.

Hou Ping, Wang Hongtao, Zhang Hao, et al. Greenhouse Gas Emission Factors for China's Electric Power for Organization and Product Carbon Footprint [J]. *China Environmental Science*, 2012, 32(4): 961-967. DOI: 10.3969/ j.issn.1000-6923.2012.06.001.

Li Fang, Li Dongping. Combined Evaluation Model Based on Entropy Weight Method [J]. *Information Technology and Informatization*, 2021(9):148-150. DOI:10.3969/j.issn.1672-9528.2021.09.044.

Liu Jinwen. Analysis of forest maturity and rotation period [J]. *Heilongjiang Science and Technology Information*, 2013(17):283-283. DOI:10.3969/j.issn.1673-1328.2013.17.266.

Ma Zhengrui, Cheng Jimin, Hou Qingchun, et al. Study on the growth and carbon sequestration rate of typical forests of Larch and Pinus tabulaeformis in North China in Liupan Mountain [J]. *Journal of Northwest Forestry University*, 2014, 29(1): 8–14. DOI: 10.3969 /j.issn.1001-7461.2014.01.02.

Siwa·K·M, Tian Yu, Lei Lei, et al. Treatment and control of carbon dioxide, a greenhouse gas affecting the environment [J]. *Journal of Shenyang University of Chemical Technology*, 2016, 30(1): 90–96. DOI: 10.3969 /j.issn.2095-2198.2016.01.018.

Wu Mengyao. (2020) Establishment of Monitoring and Evaluation System for Forest Management Plan. *Electronic Journal of Master*, 04:1-50

Xue BeiBei, Tian Guo Shuang. (2021)Analysis of comprehensive benefits and influencing factors based on carbon sink wood compound management goal. *Journal of Nanjing Forestry University*(Natural Sciences Edition),45:205-212

Zhang Tiantian. *Study on biomass and carbon storage of larch plantation in North China* [D]. Beijing: Beijing Forestry University, 2012.

Zhong Zhihong. A preliminary study on the tending and transformation of broad-leaved secondary forest [J]. *Journal of Nanchang Water College*, 2000, 19(2): 54-56. DOI: 10.3969/j.issn.1006-4869.2000.02.014.

Advances in Energy, Environment and
Chemical Engineering – Abdullah & Osman (Eds)
© 2023 The Author(s), ISBN: 978-1-032-36083-6

Prediction research element analysis of coal quality based on multiple regression

Hongwei Liu*
Inner Mongolia Electric Power Research Institute, Inner Mongolia Autonomous Region, Hohhot, China

ABSTRACT: At present, the elemental analysis test cycle of coal-fired power plant boilers is long, the operation is complex, and the requirements for laboratory personnel are high. Most power generation enterprises can not normally carry out elemental analysis of coal. Taking the industrial analysis and element analysis of coal quality in western Inner Mongolia as the research object, different linear models are constructed using a multiple linear regression algorithm, and the general rules applicable to the element analysis and prediction of different coal types are found. The results show that the regression equations of carbon, hydrogen, oxygen, and coal quality obtained by multiple linear regression have good regression effects, and most of the regression errors of coal quality are within the acceptable range of engineering. The proposed method can provide a new idea for related professional and technical personnel to obtain coal element analysis and provide a method for online monitoring soft measurement of element analysis and online monitoring of economic indicators. It has a certain engineering application value.

1 INTRODUCTION

The industrial analysis indicators of coal quality include moisture, ash, volatile and fixed carbon. The test determines the moisture, ash, and volatile, while the fixed carbon is calculated by the subtraction method, which is not an independent variable. Elemental analysis of coal includes carbon, hydrogen, oxygen, nitrogen, and sulfur. The elemental analysis and industrial analysis of coal reflect the characteristics of coal quality from different aspects. At present, the test cycle of elemental analysis of coal-fired power plant boilers is long, the operation is complex, and the requirements for laboratory personnel are high. The vast majority of power generation enterprises cannot normally carry out elemental analysis of coal-fired power plants. In fact, almost all power generation enterprises can normally carry out industrial analysis of coal quality and accumulate a large amount of industrial analysis data. According to the GB10184-2015 specification for performance tests of utility boilers, the calculation of boiler performance, including the calculation of various losses and the calculation of boiler efficiency, requires element analysis data. Therefore, power generation enterprises cannot carry out an accurate evaluation of boiler efficiency. In order to solve this problem, the multiple linear regression analysis method is introduced to accurately calculate the analysis data of unknown basis elements by using the existing industrial analysis data of coal air drying basis (Liu 2005; Wu 2016).

2 INTRODUCTION TO MULTIPLE LINEAR REGRESSION ANALYSIS

2.1 *Multiple linear regression*

Usually, multiple independent variables affect a dependent variable. Multiple linear regression analysis refers to the establishment of a functional relationship between a dependent variable and

---

*Corresponding Author: 258422810@qq.com

DOI 10.1201/9781003330165-14

multiple independent variables and the use of sample data for analysis. Let the dependent variable be y, and the n independent variables affecting the dependent variable are $X_1, X_2, X_3, \ldots X_n$, assuming that the influence of each independent variable on the corresponding variable is linear, that is

$$Y = \beta_0 + \beta_1 X_1 + \beta_2 X_2 + \ldots \beta_n X_n + \varepsilon \tag{1}$$

It is called the overall regression model, $\beta_0, \beta_1, \beta_2, \ldots \beta_n$, is the regression coefficient, and is the random error.

3 ESTABLISHMENT OF MULTIPLE REGRESSION PREDICTION MODEL

There is a certain correlation between industrial analysis and elemental analysis of coal quality. Assuming that water, ash, volatile and fixed carbon of industrial analysis of coal quality are independent variables, x_1, x_2, x_3, x_4 respectively, and the indicators of elemental analysis include carbon, hydrogen, oxygen, nitrogen, and sulfur, which are set as dependent variables, are respectively, the following mathematical models are established:

$$
\begin{aligned}
&y_1, y_2, y_3, y_4, y_5 \\
&y_1 = \alpha_0 + \alpha_1 x_1 + \alpha_2 x_2 + \alpha_3 x_3 + \alpha_4 x_4 \\
&y_2 = \beta_0 + \beta_1 x_1 + \beta_2 x_2 + \beta_3 x_3 + \beta_4 x_4 \\
&y_3 = \delta_0 + \delta_1 x_1 + \delta_2 x_2 + \delta_3 x_3 + \delta_4 x_4 \\
&y_4 = \gamma_0 + \gamma_1 x_1 + \gamma_2 x_2 + \gamma_3 x_3 + \gamma_4 x_4 \\
&y_5 = \eta_0 + \eta_1 x_1 + \eta_2 x_2 + \eta_3 x_3 + \eta_4 x_4
\end{aligned}
\tag{2}
$$

$\alpha, \beta, \delta, \gamma_2, \eta$ are regression coefficients, respectively.

The air drying-based industrial analysis and received-based element analysis of all coal types in the sample library are from the test results of coal quality used by power plants in western Inner Mongolia. See Table 1 for specific data. The coal types include bituminous coal, anthracite, lean coal, lignite, and other coal types in western Inner Mongolia, which come from different mining areas and basically represent all coal types in western Inner Mongolia.

Table 1. Statistics of different coal quality analysis.

Variety of coal	Industrial analysis				Elemental analysis				
	Mad	Vad	Aad	Cad	Car	Har	Nar	St,ar	Oar
1	2.66	20.35	39.64	43.46	41.83	2.53	0.60	1.42	9.16
2	2.65	20.56	38.06	46.26	45.52	2.76	0.64	1.47	7.96
3	4.42	23.62	32.68	47.74	42.86	2.39	0.58	0.61	10.03
4	3.5	24.56	34.25	48.25	43.75	2.66	0.63	0.57	8.84
5	4.26	25.07	33.62	46.7	41.56	2.38	0.59	0.59	10.16
6	3.85	25.86	29.76	50.26	45.48	2.66	0.65	0.64	10.64
7	3.52	24.99	40	40.3	35.76	2.34	0.56	0.49	10.97
8	2.75	25.32	36.14	45.12	40.18	2.58	0.66	0.59	10.41
9	2.91	25.64	35	45.49	39.92	2.54	0.65	0.48	10.90
10	1.68	19.64	45.21	40.64	38.28	2.47	0.57	1.53	7.18
11	1.74	19.8	43.86	41.78	40.01	2.57	0.61	1.59	7.32
12	1.82	19.64	46.35	39.9	38.04	2.46	0.57	1.42	6.92
13	1.53	20.5	36.62	49.81	48.36	2.87	0.78	1.52	6.51
14	1.48	20.49	37.14	49.11	48.10	2.90	0.77	1.57	6.78
15	7.47	24.23	34.68	44.32	41.00	2.21	0.47	0.37	9.46

(continued)

101

Table 1. Continued.

Variety of coal	Industrial analysis				Elemental analysis				
	Mad	Vad	Aad	Cad	Car	Har	Nar	St,ar	Oar
16	7.02	23.48	37.09	42.78	38.19	2.15	0.46	0.37	8.72
17	7.88	25.63	31.66	46.06	41.65	2.40	0.52	0.51	9.60
18	7.56	24.06	34.44	44.58	38.58	2.07	0.47	0.34	8.74
19	8.72	26.62	31.95	44.46	39.75	2.16	0.51	0.38	10.24
20	3.66	24.14	29.5	53.97	48.96	2.82	0.58	0.94	7.33
21	4.28	26	22.76	60.67	54.89	2.99	0.59	0.80	6.74
22	3.04	26.89	27.5	57.04	50.47	2.76	0.53	0.73	6.97
23	8.44	29.32	20.09	55.22	46.68	2.77	0.52	1.47	8.97
24	10.67	28.14	21.53	51.98	45.27	2.67	0.51	1.03	9.58
25	19.86	26.72	17.56	47.46	45.66	2.63	0.52	0.81	10.59
26	1.18	18.71	43.2	44.07	42.01	2.63	0.61	1.55	6.22
27	1.2	18.29	44.49	42.92	41.53	2.59	0.60	1.45	6.38
28	1.26	18.82	42.64	44.94	43.65	2.66	0.64	1.49	6.05
29	1.14	18.27	45.2	42.06	40.16	2.57	0.57	1.17	6.76
30	1.18	18.22	43.99	42.98	41.01	2.56	0.59	1.32	6.84
31	1.26	18.66	42.32	45.42	43.42	2.63	0.62	1.34	5.91
32	1.22	18.42	43.36	44.46	42.66	2.58	0.62	1.43	5.87
33	1.68	19.6	38.84	48.53	46.74	2.72	0.62	1.01	6.20
34	1.3	18.68	41.66	45.9	43.52	2.58	0.62	1.11	6.24
35	1.38	19.54	40.24	47.62	45.96	2.74	0.64	1.28	5.71
36	1.26	18.98	40.5	47.18	45.72	2.73	0.63	1.45	5.90
37	0.66	20.66	40.11	48.77	46.78	3.05	0.66	1.19	5.12

The first 30 coal quality training samples in the sample library shall be used. The regression function in MATLAB software shall be regress, and the calling format shall be [b, bind, R, rint, s]=regress (y, x, 0.05). (Chen 2019; Li 2007) Assuming that water, ash, volatile and fixed carbon of industrial analysis of coal quality are independent variables, x_1, x_2, x_3, x_4 respectively, and the indicators of elemental analysis include carbon, hydrogen, oxygen, nitrogen, and sulfur, which are set as dependent variables, are respectively, the regression results are as follows:

$$y_1 = 37.0791 - 0.0508x_1 - 0.8720x_2 - 0.2814x_3 + 0.7752x_4 \tag{3}$$

$$y_2 = 6.5477 - 0.0556x_1 - 0.0695x_2 - 0.0586x_3 - 0.0124x_4 \tag{4}$$

$$y_3 = 74.6346 - 0.6988x_1 + 0.0921x_2 - 0.7435x_3 - 0.8364x_4 \tag{5}$$

$$y_4 = 4.3647 - 0.0463x_1 - 0.0331x_2 - 0.04x_3 - 0.0289x_4 \tag{6}$$

$$y_5 = 13.1402 - 0.0910x_1 - 0.2183x_2 - 0.1189x_3 - 0.0541x_4 \tag{7}$$

In matrix form is:

$$
\begin{pmatrix} C_{ar} \\ H_{ar} \\ O_{ar} \\ N_{ar} \\ S_{ar} \end{pmatrix} =
\begin{pmatrix}
37.0791 & -0.0508 & -0.8720 & -0.2814 & 0.7752 \\
6.5477 & -0.0566 & -0.0695 & -0.0586 & -0.0124 \\
74.6346 & -0.6988 & 0.0921 & -0.7435 & -0.8364 \\
4.3647 & -0.0463 & -0.0331 & -0.04 & -0.0289 \\
13.1402 & -0.0910 & -0.2183 & -0.1189 & -0.0541
\end{pmatrix}
\cdot
\begin{pmatrix} 1 \\ M_{ad} \\ V_{ad} \\ A_{ad} \\ C_{ad} \end{pmatrix}
\tag{8}
$$

The above correlation matrix realizes the regression from air drying-based industrial analysis to receiving-based element analysis for any coal type. With the increase of the sample library, the regression function regress can be called again to automatically update the regression matrix and establish a new regression matrix. Compared with the traditional fixed empirical formula, it has wider applicability.

4 EXAMPLE VERIFICATION

In order to verify the accuracy of the above regression model, the industrial analysis results of the last seven different coal qualities are brought into the above regression model for inspection, and the regression values are compared with the test values. The results are shown in Table 2. Equation (9) is used to calculate the relative deviation (Wu 2008).

$$\delta = \frac{\text{Estimat value}}{\text{Test value}} \times 100\% \tag{9}$$

Table 2. Comparison of element analysis test value and predicted value of test sample coal by regression analysis.

	Elemental analysis (%)														
	Car			Har			Nar			St,ar			Oar		
Variety of coal	Test value	Estimat value	Relative deviation	Test value	Estimat value	Relative deviation	Test value	Estimat value	Relative deviation	Test value	Estimat value	Relative deviation	Test value	Estimat value	Relative deviation
1	43.42	44.04	1.43	2.63	2.63	−0.10	0.62	0.6	0.42	1.34	1.46	8.53	5.91	6.01	1.86
2	42.66	43.21	1.29	2.58	2.60	0.94	0.62	0.61	0.85	1.43	1.44	0.53	5.87	6.05	3.07
3	46.74	46.59	0.32	2.71	2.71	0.13	0.61	0.62	1.82	1.01	1.46	44.87	6.20	5.79	6.53
4	43.52	44.58	2.42	2.59	2.66	2.94	0.62	0.63	1.45	1.11	1.50	35.86	6.24	6.08	2.54
5	45.96	45.56	0.88	2.74	2.66	2.87	0.64	0.61	5.39	1.28	1.38	8.13	5.71	5.72	0.13
6	45.72	45.64	0.19	2.73	2.69	1.27	0.62	0.63	1.26	1.45	1.51	4.16	5.90	5.93	0.45
7	46.78	45.55	2.64	3.05	2.62	14.18	0.66	0.58	12.33	1.20	1.16	3.05	5.12	5.46	6.64

Table 3. Test regression model statistics.

Element composition	Determination coefficient/R2	Statistic observations/F	Error variance estimation/S2
Car	0.9601	150.3703	0.8464
Har	0.9406	23.7543	0.0184
Nar	0.6188	10.1436	0.0025
St,ar	0.6217	10.2718	0.0941
Oar	0.9413	110.1756	0.1827

Judgment coefficient: the closer the value of [0,1] is to 1, the better the regression effect is; otherwise, the worse the regression effect is. The statistic observation value f is the overall significance test of the regression equation. The larger the value of f, the more significant the linear relationship between the independent variable and the dependent variable. It can be seen from Table 3 that the determination coefficients of carbon, hydrogen, and oxygen are close to 1, and the effect of the regression equation is excellent. The determination coefficient of nitrogen and sulfur is far less than 1, indicating that the regression effect is poor.

5 CONCLUSION

(1) Based on the results of industrial analysis and element analysis of different coal qualities in the west area of Inner Mongolia, the regression equations of element analysis and industrial analysis

were obtained using software to carry out multiple linear regression on 30 kinds of utility boiler coals. The regression equations were used to calculate the element analysis of 7 kinds of raw coals that were not involved in the fitting. The results showed that the correlation coefficients of carbon, hydrogen, oxygen, and regression equations were above 0.9, the regression effect was good, the fitting errors of most elements were within the acceptable range of engineering, and the fitting data were accurate. The correlation coefficient of the regression equation of nitrogen and sulfur is slightly low, the deviation between the fitting data and the actual results is large, and the accuracy is poor. Due to the small content of sulfur, it has little effect on the calorific value of coal, and the effect on boiler efficiency can also be ignored. Therefore, the regression model in this paper can be used to realize the application of boiler efficiency calculation.

(2) Compared with the online monitoring device for coal quality element analysis, this paper provides two methods with the advantages of accurate analysis, less investment, and good prediction effect compared with the commonly used empirical formula method. At the same time, it realizes the linear and nonlinear mapping from air drying-based industrial analysis to receiving-based element analysis and reduces the tedious conversion between benchmarks. It has certain engineering application value and provides a new idea for the technicians of power generation enterprises to carry out boiler performance evaluation and online monitoring of indicators.

ACKNOWLEDGMENTS

This work was financially supported by self-raised science and technology project of the Inner Mongolia Electric Power Research Institute branch of Inner Mongolia electric power (Group) Co., Ltd.(2021).

REFERENCES

Chen Xue(2019). Establishment of virtual coal quality database for online calculation of coal-fired boiler efficiency [J] *Journal of Shanghai University of Technology*, 41 (6): 546–551.

Li Taixing(2007). Research on general calculation model of coal quality element analysis based on MATLAB [J] *Boiler Technology*, 8 (5): 22–24.

Liufuguo(2005). Soft sensor real-time monitoring technology for elemental analysis and calorific value of coal-fired in utility boilers [J] *Chinese Journal of Electrical Engineering*, 5 (6): 139–145.

Wu Dehui (2008).A comprehensive prediction method for coal quality [J] *Thermal Power Generation*, 37 (1): 26–30.

Wu Xiaoyan (2016). Coal quality characteristic analysis and prediction model based on multiple linear regression [J], *China Science and Technology Information* (7): 17–19.

Advances in Energy, Environment and
Chemical Engineering – Abdullah & Osman (Eds)
© 2023 The Author(s), ISBN: 978-1-032-36083-6

Analytical model of user behavior based on carbon price demand response under carbon electricity market coupling

Huijuan Liu* & Jingqi Zhang
North China Electric Power University, Ltd. Beijing, China

Liang Sun & Xiaoliang Dong
Beijing Power Trading Center Co., Ltd. Beijing, China

Guoliang Luo & Dunnan Liu
North China Electric Power University, Ltd. Beijing, China

ABSTRACT: The national carbon emissions trading market aims to establish a market mechanism to reduce greenhouse gas emissions, and the power generation industry became the first pilot industry to participate in the national carbon emissions trading market in 2021. With the gradual advancement of carbon trading, there is a controversy in the market about what type of electricity market users need to bear the cost of excess emissions. Therefore, this paper proposes a two-layer optimization model based on the user utility formula of electricity consumption and the carbon emission intensity under the grid to promote the carbon emission reduction effect based on user benefits and grid carbon emission intensity.

1 INTRODUCTION

On March 15, 2015, the State Council of the CPC Central Committee issued "Several Opinions on Further Deepening the Reform of Electricity System" (No. 9), which aims to gradually introduce market competition in the areas of power generation, power sales, and incremental distribution, and a mature and perfect electricity market will promote the green and sustainable development of the industry. As a high-emission industry accounting for nearly 49% of the total carbon emissions in China, the power generation industry has become the first industry to be covered by the national carbon emission trading market in 2021. However, with the gradual advancement of the market, there has been controversy over what type of electricity market users need to bear the cost of excess emissions. In the electricity market, the power generation side is the direct emitter, but the electricity consumption side is the potential driver of carbon emissions. Without the active participation of the electricity side of the power market, it is difficult to effectively reduce carbon emissions by creating a carbon emissions trading market. Therefore, it is important to identify the market introduction sectors that can achieve the best emission reduction effect to build a carbon trading market.

2 CUSTOMER UTILITY FORMULA BASED ON ELECTRICITY CONSUMPTION

There are two ways to introduce demand response in a competitive electricity market: one is to increase the role of the demand side in the market based on price signals; the other is to incentivize

---

*Corresponding Author: liuhuijuan_2021@126.com

customers to respond promptly and curtail load when the system reliability is compromised by setting deterministic or time-varying policies. While the main demand response studies have focused on residential households, the model in this paper focuses on industrial and commercial end-users, who are the main market participants in the Chinese electricity spot market and the future carbon market. Based on profitability, industrial and commercial customers can also reduce their cooling, lighting, and production loads through demand response. Therefore, in order to establish the relationship between associated demand-side customers and carbon market-based demand response, it is important to first understand the utility equation when customers consume electricity.

Market users can be considered mutually independent decision makers in the electricity trading and carbon emissions trading markets. Since most industrial customers directly participate in the electricity spot market, we model the responses of market customers using a finite linear marginal utility function. Assume that there are N users in the market, and for user $n \in \{1,2,..., N\}$, his marginal effect can be expressed as

$$
\mathrm{MU}\left(P_n^t\right) = \begin{cases} \omega_n - \alpha_n P_n^t, 0 \leq P_n^t \leq \frac{\omega_n}{\alpha_n} \\ 0, P_n^t \geq \frac{\omega_n}{\alpha_n} \end{cases}
\tag{1}
$$

Where P_n^t is the total electricity used by the market demand side user n at time t (kWh). ω_n and α_n are predetermined parameters that ω_n represents the fixed marginal revenue per unit of electricity consumed by user n, and $\alpha_n P_n^t$ then represents the economic loss from the user's load change. The utility function of user n on the demand side of the market is the integral of the marginal utility in Equation (1). Therefore, the utility function of the market demand-side user n is

$$
\mathrm{U}\left(P_n^t\right) = \begin{cases} \omega_n P_n^t - \frac{\alpha_n}{2} P_n^{t2} + \varepsilon, 0 \leq P_n^t \leq \frac{\omega_n}{\alpha_n} \\ \frac{\omega_n^2}{2\alpha_n} + \varepsilon, P_n^t \geq \frac{\omega_n}{\alpha_n} \end{cases}
\tag{2}
$$

Since we assume that the average cost of electricity is the same for all industrial and commercial customers, for any customer $n \in \{1, 2, ..., N\}$. α_n can be simplified to $\omega_n = \alpha$. Therefore, the utility function of market demand side user n can be simplified as

$$
\mathrm{U}\left(P_n^t\right) = \begin{cases} \omega_n P_n^t - \frac{\alpha}{2} P_n^{t2} + \varepsilon, 0 \leq P_n^t \leq \frac{\omega_n}{\alpha} \\ \frac{\omega_n^2}{2\alpha} + \varepsilon, P_n^t \geq \frac{\omega_n}{\alpha} \end{cases}
\tag{3}
$$

For different ω_n, the utility functions all satisfy the non-decreasing property. When users consume more electricity, they accomplish more productive tasks and therefore receive more economic benefits. Moreover, in real production, more electricity consumption is always accompanied by the depreciation of production machines or other equipment, which leads to diminishing marginal utility of electricity use by users. And in general industrial or commercial sectors, depreciation is usually quadratic.

The regression results in Table 1 show the utility function parameters for different industry sectors. For industry-specific end-users, the ω_n is a predetermined parameter representing the marginal gross output of user n relative to the user's electricity consumption, and α_n is the marginal cost per unit of electricity used. Also, based on historical data, this paper assumes that an average of 36% of customer load is inelastic load $P_{n,\text{inelastic}}^t$ and the daily load of each customer cannot increase or decrease by more than 20%, i.e., the elastic load is in the interval $[P_{n,\text{min}}^t, P_{n,\text{max}}^t] = [80\%P_n^t, 120\%P_n^t]$. In addition, the experiment considers the electricity load of each node as a combination of user units based on the percentage of users in different industry categories in each node.

Table 1. Quadratic regression results for customer utility functions by sector.

Industry Type	$\omega_n/(\mathrm{Yuan/WWh})$	$\alpha_n/(\mathrm{Yuan/WWh}^2)$
Agro-processing industry	246312.85	−0.0258
Food manufacturing	316285.57	−0.0504
Garment industry	354303.05	−0.0318
Recreational products manufacturing	617064.15	−0.0503
Chemical industry	99428.45	−0.0032
Metal industry	35484.71	−0.0008
Automobile manufacturing	351619.29	−0.0170
Transportation industry	230125.96	−0.0218
Electronic manufacturing	194512.29	−0.0013
Instrument manufacturing industry	367284.05	−0.0707
Renewable resources industry	135373.19	−0.0564

3 CARBON EMISSION INTENSITY UNDER THE POWER GRID

From a spatial perspective, CEF in the grid refers to the virtual flow of carbon emissions between electricity production and consumption. In this paper, the CEF is applied to establish the link between the customer's electricity consumption and the generated carbon emissions. A mathematical model is used to quantify carbon emissions, where two parameters can be used to describe the distribution and movement of the CEF: the CEF rate and the carbon intensity. the CEF rate, which represents the "velocity" of the CEF, can be expressed as

$$\mathrm{R} = \frac{dF}{dt} \tag{4}$$

Where: F represents the total amount of carbon emissions flowing in or out, and t is the time required for the CEF to flow through all grid transmissions. Where the CEF rate is in tons of carbon dioxide per hour (t CO_2 e/h), carbon emission intensity ρ represents the density of the CEF relative to the active power flow. The carbon intensity is used to characterize the relationship between the CEF and the tide at the inflow and outflow nodes.

$$\rho = \frac{F}{G} = \frac{R}{P} \tag{5}$$

Where: G and ρ denote active electricity and active power, respectively. The unit of carbon intensity is tons of carbon dioxide per megawatt hour (t CO_2 e/ MW).

At present, it is common in the domestic power industry to calculate the nodal carbon intensity based on marginal carbon emissions, i.e., the total system carbon emissions E_0 under the existing power market supply and demand, and then calculate the total system carbon emissions E_1 when the electricity load of a node increases by 1 unit. After the two emissions calculations, the corresponding increased marginal carbon emissions (E_1 - E_0) is the nodal carbon intensity.

$$E_0 = \sum_{i=1}^{m} P_{gi} \cdot AD_{fuel} \cdot EF_{fuel} \tag{6}$$

Equation (6) explains the process of calculating the carbon emissions of the total power system. There are m units in the system, and P_{gi} represents the power of the first unit (MW). AD_{fuel} is the fuel consumption intensity of the unit, representing the amount of fuel to be burned per unit of the power output of the unit in tons per megawatt hour (t/MW). EF_{fuel} is the carbon intensity of the unit's fuel, representing the carbon dioxide emitted per unit of fuel burned in tons of carbon dioxide per ton (t CO_2 e/t).

4 TWO-LAYER OPTIMIZATION MODEL BASED ON CUSTOMER BENEFITS AND GRID CARBON INTENSITY

To analyze the behavior of users participating in the carbon and electricity trading markets, this paper establishes a research framework for carbon market-power market demand response based on two-layer optimization. As shown in Figure 1, the first level of optimization of the model represents the DR decision of the electricity market demand-side customers to maximize their revenue function, which introduces the arbitrage profit between the electricity market and the carbon market into the market user-side utility function; the second level of optimization calculates the optimal tide of grid operation based on the electricity demand of the electricity consumption side. The first level of optimization corresponds to the customer's demand response behavior, and the second level of optimization is related to the constraints on the distribution and system security conditions of the entire power system. In this paper, an iterative algorithm based on a two-tier optimization structure is used to discover the optimal power consumption dispatch to maximize the customer's benefit.

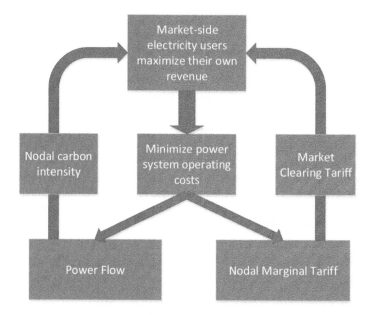

Figure 1.   Modeling framework for the bi-optimization of the integrated power system and emission trading market.

4.1   *The first layer of optimization: maximizing the revenue of market users*

In the model, the revenue equation of the customer considers the utility of the customer, the cost of electricity, and the environmental benefit from demand response, where the environmental benefit refers to the arbitrage profit of the electricity market and the carbon market. The arbitrage profit of the dual market arises when the customer reduces carbon emissions through demand response and brings the remaining carbon credits to the market for sale. For user n, the model defines the arbitrage profit function of the dual market A_n^t, at time, t∈ {1,2,...,T}. A_n^t can be expressed as

$$A_n^t = p_c \cdot (x_n^t - E_n^t \cdot p_n^t) \tag{7}$$

Where A_n^t represents the arbitrage profit of user n at time t, and p_c represents the market carbon price (unit: Yuan/t CO_2e). According to the rules of the existing carbon market, the carbon price in one day p_c can be regarded as fixed. where, for user n, the maximum carbon emission allowance at time t is x_n^t. We assume that the carbon emission allowances are apportioned to each time period with certain weights. When in a fixed time period, the user takes p_e^t (unit: yuan/kWh), the clearing electricity price consumes p_n^t kWh electricity, and the overall return formula is Payoff (p_n^t, E_n^t, p_e^t):

$$\text{Payoff}(p_n^t, E_n^t, p_e^t) = U\left(p_n^t\right) - p_e^t \cdot p_n^t + p_c \cdot (x_n^t - E_n^t \cdot p_n^t) \tag{8}$$

Therefore, when the user aims to maximize its revenue, the user's electricity consumption can be characterized as the solution to the following optimization problem

$$\max_{p_n^t} \sum_{T=1}^{T} [U\left(p_n^t\right) - p_e^t \cdot p_n^t + p_c \cdot (x_n^t - E_n^t \cdot p_n^t)] \tag{9}$$

$$s.t.$$

$$p_{n,\min}^t \leq p_n^t - p_{n,inelastic}^t \tag{10}$$

$$p_n^t - p_{n,inelastic}^t \leq p_{n,\max}^t \tag{11}$$

Where $p_{n,inelastic}^t$ represents the inelastic load at time t; Equation (9) and Equation (10) can be interpreted that this model considers the proportion of the elastic load of a single user to the total load, and excluding the fixed production load, the elastic load of the user at time t is in the interval $[p_{n,\min}^t, p_{n,\max}^t]$. We assume that the total load at each node under the grid is proportional to the electricity consumption of large power users, thus.

According to whether there is a mandatory compliance responsibility, China's carbon market is divided into a mandatory carbon market and a voluntary carbon market. Carbon emission allowance goods are only circulated in the mandatory type market, so this paper only considers the mandatory type carbon emission market at present.

4.2 Second layer of optimization: minimizing power system operating costs

In this paper, we use the Optimal Power Flow model, which is the simulated physical system clearing model.

$$\min_{P_g} F = \sum_{i=1}^{m} f_i(P_{gi}) \tag{12}$$

$$P_{gi\min} \leq P_{gi} \leq P_{gi\max}, i = 1, 2, ..., m \tag{13}$$

$$Q_{gi\min} \leq Q_{gi} \leq Q_{gi\max}, i = 1, 2, ..., m \tag{14}$$

Where: $P_{gi\max}\left(P_{gi\min}\right)$ is the maximum (small) active power of the first unit, Q_{gi} and $Q_{gi\max}\left(Q_{gi\min}\right)$ denote the reactive power and the maximum (small) reactive power of the ith unit. $P_g = (P_{g1}, P_{g2}, ..., P_{gm})$ is the power vector of the unit, and P_{gi} represents the power of the ith unit. $f_i(P_{gi})$ denotes the operating cost function of unit i, which is a quadratic function form, covering fixed running cost, variable O&M cost, startup cost, repair cost, and retirement cost F represents the total cost of the system m units. Equation (13), (14) represents the power constraint of the ith unit.

Where: $P_{gi\max}\left(P_{gi\min}\right)$ is the maximum (minimum) active power of unit i, and Q_{gi} and $Q_{gi\max}\left(Q_{gi\min}\right)$ represent the reactive power and the maximum (minimum) reactive power of unit i. $P_g = (P_{g1}, P_{g2}, ..., P_{gm}$ is the generation power vector of unit i, and P_{gi} represents the power of unit i. $f_i(P_{gi})$ represents the operating cost function of unit i. F represents the total cost of m units of the system. Equations (14) and (15) represent the power constraint of the i-th unit.

$P_l = P_{l1}, P_{l2}, \ldots, P_{ll}$ denote the transmission power vector of the system, and the transmission power of the ith line is P_{li}. The corresponding transmission constraint is

$$P_{li\,min} \leq P_{li} \leq P_{li\,max}, i = 1, 2, \ldots, l \tag{15}$$

where $P_{li\,max}(P_{li\,min})$ is the maximum (small) transmission power of the ith line.

For node, the voltage constraint is

$$V_{i\,min} \leq V_i \leq V_{i\,max}, i = 1, 2, \ldots, n \tag{16}$$

Here, the V_i denotes the voltage at node i, and $V_{i\,max}(V_{i\,min})$ is the upper (lower) limit of voltage that the corresponding node can withstand. When the system is operating, the transmission constraint of the line can cause transmission blockage, which in turn affects the stable operation. The safety constraints of the system are as follows.

$$g_{pi} = P_{gi} - P_{di} - \sum_{j=1}^{n} V_i V_j |Y_{ij}| \cos\left(\theta_i - \theta_j - \delta_{ij}\right) = 0 \tag{17}$$

$$g_{Qi} = Q_{gi} - Q_{di} - \sum_{j=1}^{n} V_i V_j |Y_{ij}| \sin\left(\theta_i - \theta_j - \delta_{ij}\right) = 0 \tag{18}$$

Equations (16) and (17) are the active and reactive power balance equations for node i, respectively.g_{pi} and g_{Qi} are the active and reactive power mismatch of the corresponding nodes, respectively. P_{di} and Q_{di} are the active and reactive loads of node i, and Y_{ij} is the conductance of the line from node i to j, and θ_i is the voltage V_i is the phase angle of δ_{ij} is the phase angle of the conductance Y_{ij} of the phase angle. Therefore, according to Eqs. (12)–(18), we can solve for the output of each unit in the system.

The CEF intensity of the node to which user n belongs requires solving the tide twice to calculate the carbon emission intensity of the whole node based on the marginal carbon emission. First, we calculate the total system carbon emission under the existing electricity market supply and demand and then calculate the total system carbon emission when one unit of electricity load is increased at one of the nodes, and the corresponding increased carbon emission is the carbon emission intensity of the node. To calculate the total system carbon emissions, we need to solve the tide to get the unit output, so the two tides here correspond to two total system carbon emissions calculations, and the specific calculation formula is shown in Equation (6).

5   CONCLUSION

Since the power side of the electricity market is often a potential driver of carbon emissions, this paper investigates the necessity of introducing user-side enterprises in the electricity market to the carbon market to better achieve the emission reduction goals of China's carbon emission market. In order to analyze the user behavior of bilateral users in the electricity market after participating in the carbon market, a two-layer optimization model of carbon market-electricity market demand response is proposed, which provides a theoretical basis for studying the impact of the existing carbon emission market on electricity prices in the electricity spot market as well as user behavior.

ACKNOWLEDGMENTS

This work was financially supported by the Science and Technology Project of State Grid Corporation of China (Research on Key Technologies of Power Market Design Serving the Construction of New Power System, 5100-202157292A-0-0-00).

REFERENCES

Chen H, Kang J N, Liao H, et al. Costs and potentials of energy conservation in China's coal-fired power industry: A bottom-up approach considering price uncertainties[J]. *Energy Policy*, 2017, 104: 23–32.

Chen Yi, Tian Chuan, Cao Ying, Liu Qiang, Zheng Xiao-Qi.Research on peaking carbon emissions of power sector in China and the emissions mitigation analysis[J]. *Climate Change Research*, 2020, 16(5): 632–640.

Jiang J, Xie D, Ye B, et al. Research on China's cap-and-trade carbon emission trading scheme: Overview and outlook[J]. *Applied Energy*, 2016, 178: 902–917.

Xu Guangda, Zhang Li, Liang Jun, et al. Estimation of Price Elasticity for Residential Electricity Demand Based on Electricity Consumption Features of Appliances[J]. *Automation of Electric Power Systems*, 2020, 44(13):48–55.

Xu Zhe, Huang Xiaoting, Chen Wei, et al. Power Quantity Division Model and Electricity Price Scheme Design Based on Transaction Curve [J]. *Southern Power System Technology*, 2020, 14(3):71–78.

Advances in Energy, Environment and
Chemical Engineering – Abdullah & Osman (Eds)
© 2023 The Author(s), ISBN: 978-1-032-36083-6

Main controlling factors of Xujiahe formation gas reservoir formation in Central Sichuan depression

Wenping Zhang, Xuncai Liu, Suotao Wang & Pan Zhang
No.2 Mud Logging Company, CNPC Bohai Drilling Engineering Company, Renqiu Hebei, China

Graciela Daniels*
Central Arizona College, Coolidge, AZ, USA

Shiela Kitchen
College of Arts and Science, University of New England, Armidale, Australia

Hanlie Cheng
Faculty of Contemporary Sciences and Technologies, South East European University, Ilindenska, Tetovo, Macedonia

ABSTRACT: Sichuan Basin is the largest gas-bearing basin in China. After decades of practice and exploration, Sichuan Basin has become the basin with the largest number of discovered gas fields in China. The current research shows that the complex geological processes in the geological history of the Xujiahe Formation in the deep part of the Central Sichuan depression make the trap types very complex, including stratigraphic traps, structural traps, and compound traps formed by the interaction of structural and stratigraphic elements. Based on the theory of petroleum geology and structural geology, the main controlling factors of the Xujiahe Formation gas reservoir in the Central Sichuan Basin are analyzed in this paper. The results show that the Xujiahe formation is dominated by clastic quartz sandstone, followed by lithic sandstone. Lithic quartz sandstone and lithic sandstone are the main rock types in the second member of the Xu2 Formation, and lithic quartz sandstone is the main rock type in the fourth member of the Xu4 Formation. The reservoir space mainly comprises primary intergranular and secondary dissolved pores. The favorable sedimentary environment and diagenetic epigenesis are the main controlling factors for reservoir development. The differential compaction on the palaeo-slope background can produce many micro-fractures and greatly improve the reservoir properties. This is of guiding significance to the exploration focus of the Xujiahe Formation, and the future exploration of the Xujiahe Formation should be reservoir and fracture prediction based on fine structure interpretation.

1 INTRODUCTION

Sichuan Basin is a large superimposed basin that has experienced a long Marine craton and Mesozoic-Cenozoic continental evolution from proterozoic to early Mesozoic. The Xujiahe Formation basin in the Upper Triassic is a foreland basin developed on the unified and stable cratonic basement. The basin is characterized by a small thrust and a large slope. The basin has experienced four evolution stages, with strong folding and deformation at the basin edge and well-developed thrust structures. The basin is mainly characterized by finishing and lifting. There is no significant structural deformation in the sedimentary strata, and a large area of gentle slope structure is developed, laying a foundation for forming a large area of lithologic gas reservoirs. The formation and

---

*Corresponding Author: graciela_daniels@stu.centralaz.edu

DOI 10.1201/9781003330165-16

evolution of the Sichuan Basin is not only restricted by the two global cycles but also controlled by the differences in basement rocks.

The north slope of the central Sichuan Paleo-uplift is located in the central-western part of the Sichuan Basin, and its main structure-sedimentary evolution process is closely related to Sichuan Basin. Therefore, its stratigraphic composition is roughly similar to that of the Sichuan Basin, but due to the local differential subsidence and the change of sedimentary environment, the strata in the study area have different characteristics. Jurassic and Cretaceous strata of Mesozoic are widely exposed on the surface, and Paleogene is distributed sporadically. The strata with the largest exposed area are mainly Jurassic strata. The formation, migration, and accumulation of oil and gas, as well as the final adjustment and reconstruction of oil and gas reservoirs, are largely controlled by the structural characteristics and evolution. Some researchers have found that deep faults greatly affect the tectonic deformation of sedimentary cap rocks, and the formation of flower-like structures is observed in the process of downward faulting through the basement and upward faulting to the shallow strata. In addition, the deep faults control the distribution of favorable deposits. On the other hand, they improve the physical properties of the Dachuan middle reservoir and communicate the source rocks and reservoirs.

The tectonic characteristics of the Xujiahe Formation basin are similar to the surface structure, and the structures of each section have a certain inheritance. The low-slow tectonic belt in the middle of Sichuan is located to the west of Mountain, east of Longquanshan fault, and reaches Daba Mountain and Micangshan frontier-belt in the north and Leshan-Yibin rhomboid-shaped fault area in the south. Structural belts are interrupted, and high and steep structures are not developed. The base belongs to a hard basement, with no big faults on the surface, a small dip Angle and small deformation strength of overlying sedimentary cover, weak structural fold strength, and small trap closure. Xujiahe Formation, as a transitional stratum from Marine to continental facies in Sichuan Basin, has a history of oil and gas exploration for many years and has made great achievements in the western Sichuan area, but there has been no great breakthrough in the central Sichuan area. Nonetheless, there has always been a large number of oil and gas shows, indicating that this area contains considerable oil and gas reserves. Xujiahe Formation is a continental clastic reservoir whose exploration and development technology are different from that of the Permian Marine carbonate reservoir in the Sichuan Basin. The strata of the Xujiahe Formation in the study area are flat and undeveloped. The strata of the Xujiahe Formation are from Marine facies to continental facies in the Sichuan Basin, which is mainly the product of continental deposits. There are two kinds of research results about sedimentary facies. One is recognition. It is delta front and plain facies. In addition, the Xujiahe Formation is considered to be the result of multistage fluvial facies oscillation superimposed deposition. Although there is a great difference between these two views, there is no dispute that the Xujiahe Formation reservoir is a typical continental lithologic stratigraphic reservoir. In addition, the Xujiahe river group is an important control and influence factor for the reservoir development and in the development of cracks. Areas of the development of tectonic activities are not closely related to the geomorphology and sedimentation, thus the river geomorphology of the local area has a vital significance in fracture development.

2 STRATIGRAPHIC CHARACTERISTICS AND SEDIMENTARY FACIES

2.1 *Formation characteristics*

During the deposition of the Xujiahe Formation, the Sichuan Basin gradually changed from Marine to continental deposition, so the Xujiahe Formation has unique characteristics. The lithologic assemblage and thickness of The Xujiahe Formation vary greatly from the west of Sichuan to the east of Sichuan, and there are great differences in stratification between the middle and the west of Sichuan. The thickness of The Xujiahe Formation varies greatly. According to the statistics on the drilling thickness of the Xujiahe Formation in the study area, the thickness of the Xujiahe Formation is between 480 m and 780 m, thick in the northwest and thinner in the southeast. The development status, lithology variation, thickness distribution, and plane distribution of the Xujiahe Formation are as follows:

(1) Paragraph 2. The second member of the Xujiahe Formation is dominated by grayish white, gray fine-grained to medium-grained sandstone, and coarse sandstone. The first member of the Xujiahe Formation and the third member of the Xujiahe Formation are delimited by black shale. Low natural gamma, medium to high resistance electrical characteristics, and easy to distinguish are its major characteristics. The lithology is fine and medium-grained feldspar quartz sandstone with a small amount of black-gray mudstone.

(2) Four paragraphs are required. The lithologic electricity of the Xu4 member is characterized by low natural gamma, low acoustic time difference, high density, and medium-high resistivity, which is easy to distinguish and identify. The thickness of four sections should be gradually thickened from northwest to southeast, with a thickness between 100m and 120m in most areas. The lithology is mainly light gray $\sim$ gray coarse-medium grain, medium grain, medium $\sim$ fine grain feldspar quartz sandstone, and lithic feldspar quartz sandstone, partially intercalated with thin layer shreccia.

(3) Six paragraphs are required. The upper part is mainly siltstone, the lower part is mainly gray black shale, and the bottom boundary is demarcated with the black shale of the fifth member of the Xujiahe Formation. The southern Sichuan area is usually divided into three layers. The upper and lower layers have similar lithology to the central Sichuan area, while the middle part is gray-black shale mixed with sandstone, and there are coal lines in some areas.

And quartz sandstone, such as the upper clip of dark gray mudstone and argillaceous siltstone.

2.2 *Sedimentary environment*

It is based on the observation and description of outcrop and core, including the characteristics of rock color, mineral composition, structure, sedimentary structure, sedimentary vertical evolution, profile structure, and sedimentary facies law. The Xujiahe Formation in the study area belongs to the delta-lacustrine sedimentary system. The well-preserved plant fossils, biodisturbance structures, and biological boreholes in members 1, 3, and 5 of The Xujiahe Formation indicate that the Xujiahe Formation in the study area is a littoral shallow lake sedimentary environment. In terms of maturity and grain size, the maturity of sandstones in the Xujiahe Formation in this area is good, the grain size is generally fine to medium, and the grinding is medium to good, reflecting the relatively stable delta-lacustrine sedimentary environment in this area. The sedimentary structures of the Xujiahe Formation in the study area are of various types, including wedge and trough cross-bedding and bottom scour structures reflecting river scour, as well as symmetrical ripple marks formed by shallow lake wave action. Therefore, from the point of view of sedimentary structure characteristics, the Xujiahe Formation was deposited in a delta-lacustrine environment.

2.3 *Reservoir characteristics*

Xujiahe Formation reservoir is mainly composed of Xu2, Xu4, and Xu6 members, and Xu1, Xu3, and Xu5 members are also developed, showing low porosity and low permeability on the whole. Reservoir pores are mainly residual primary intergranular pores and secondary dissolution pores. The reservoir is mainly controlled by sedimentation, diagenesis, and tectonics. The sand body of the Xujiahe Formation is widely developed. On the wide and gentle slope in the middle of the Sichuan Basin, the delta front sand body is widely developed in the low and gentle tectonic setting, which provides the foundation for the development of a large area of the high-quality reservoir. The second, fourth and sixth member reservoirs extend far in the plane, and the third and fifth member reservoirs also develop locally, forming the characteristics of multiple sets of reservoirs superposition on each other in the longitudinal direction and large area distribution in the plane. The maximum porosity of Xu6 member sandstone is 14.25%, with an average of 4.92%, mainly distributed in the range of 2%–13%, of which the porosity greater than 6.5% accounts for about

35%, with an average of 8.15%. The maximum permeability is 39.2 mD, with an average of 0.165 mD, mainly distributed between 0.002 mD and 1.0 mD. The average permeability in the reservoir and permeability section is 0.357 mD with porosity greater than 6.5%. The maximum porosity is 18.5%, with an average of 5.81%, mainly distributed in the range of 2%–12%, of which the porosity greater than 6.5% accounts for 42%, with an average of 8.96%.

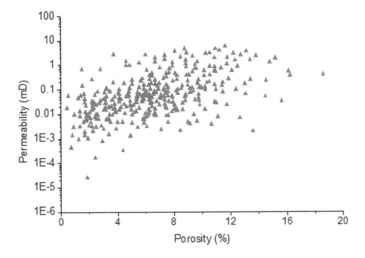

Figure 1.   Relation diagram of porosity and permeability of xu6 member.

By observing the thin sections of the cast sandstone core, it is found that most of the pores of the Xujiahe Formation in the Central Sichuan Basin are irregular in shape, and only a few are round to oval or long strips. The pore types of the reservoir are mainly residual primary intergranular pores, intergranular dissolved pores, and feldspar dissolved pores. A few of them are heterobasic pores and microfractures. Intergranular dissolution pore is the most important pore affecting reservoir physical property, and its development degree directly affects reservoir capacity.

3   MAIN CONTROLLING FACTORS OF GAS RESERVOIR FORMATION

3.1   *Tectonic factors*

Under the influence of multistage tectonic movement, Sichuan Basin experienced the multistage uplift-settlement-uplift process, in which the differential uplift-subsidence land-building movement was dominant in Sinian-Middle Triassic, and the compressional uplift and denudation were dominant after late Triassic. It is under the action of differential tectonic movement that the central Sichuan paleo-uplift emerges from nothing, develops, and finally forms its own unique petroleum geological characteristics. As a whole, the source rocks are rich in organic matter and have a high thermal evolution degree, the reservoirs are mainly Marine carbonate rocks with developed secondary pores and great lateral variation, and the cap rocks are distributed stably across the strata in the longitudinal direction.

3.2   *Hydrocarbon source factors*

Hydrocarbon source conditions are sufficient in central and western Sichuan Basin, and hydrocarbon generation intensity is more than 12×108 m^3/km$^2 \sim 100\times108$ m^3/km^2 in most areas. The source, reservoir, and cap are well matched, and natural gas is easy to migrate and accumulate. Exploration practice shows that most of the Xujiahe Formation gas reservoirs discovered so far are located in this area. Most of the source rocks of the Xujiahe Formation in the Upper Triassic entered the mature stage at the end of the Late Jurassic to generate oil and gas and entered the peak of hydrocarbon generation in the Cretaceous. In the process of thermal evolution, the thermal evolution degree of the source rock is mainly affected by the fluctuation of the overlying strata thickness. The thermal evolution degree is relatively high in the sag zone and relatively low in the uplift zone, indicating that the succession and development of regional uplift and sag is an important factor controlling the oil and gas evolution. The thermal evolution degree of the source rocks in the western, northern, and northeastern Sichuan Basin is relatively high and reaches high maturity stage, while the thermal evolution degree of the source rocks in the southern Sichuan Basin is relatively low.

3.3 *Accumulation characteristics*

Must reservoir longitudinal center of mainly distributed in the second paragraph, shall be four, six paragraphs, average on its longitudinal average porosity, permeability, reservoir thickness, high-quality reservoir thickness decreases gradually from bottom to top, development degree of the reservoir is presented from down to up variation trend gradually, the river with the Xujiahe group would require two, four and six sections of the sedimentary environment have a very close relationship. According to the relationship between sedimentary facies and reservoir physical properties in the study area, it is found that the favorable reservoir facies in the Upper Triassic are delta front underwater distributary channel sand body and mouth bar sand body with strong hydrodynamics, good sediment sorting, and high mineral maturity. In the underwater distributary channel development area, it has the characteristics of wide, deep, and high kinetic energy, forming large-scale and well-sorted coarse-grained sand bodies, and easy to form high-quality reservoirs. By comparing the structure and superposition patterns of different sedimentary microfacies in the study area, it is found that the underwater distributary channel sandstones have a lower content of argillaceous impurity and coarser grain size, so their porosity and permeability are better than other microfacies sandstones, indicating that the underwater distributary channel controls the distribution of high-quality sand bodies. Combined with drilling practice, many industrial gas Wells are located in the underwater distributary channel development area, which proves that favorable reservoir facies affect the distribution and formation of oil and gas reservoirs. The reservoir of the Xujiahe Formation in the study area is characterized by heterogeneity, low porosity, and low permeability, which prevents the horizontal migration of oil and gas in the sand body, resulting in the small size of favorable traps to capture oil and gas. Therefore, traps with high hydrocarbon enrichment degrees must have other ways to capture oil and gas. Within the study must be direct deposition of Yu Zhongsan upper center of erosion surface, its ray slope group strata in the indosinian period large area exposed to the surface erosion, erosion surface carbonate weathering leaching effect by atmospheric fresh water for a long time, solution pores, caves, and karst development has the good connectivity, can be a good hydrocarbon migration channel.

4 CONCLUSION

(1) Xujiahe Formation mainly comprises Xu2, Xu4, and Xu6 members. In terms of physical property distribution, the Xujiahe Formation sandstone has ultra-low porosity, ultra-low permeability $\sim$ low porosity, and low permeability on the whole, but the sandstone with better porosity and permeability in the longitudinal direction becomes an effective reservoir. Reservoir pores are mainly residual primary intergranular pores and secondary dissolution pores. The reservoir is mainly controlled by sedimentation, diagenesis, and tectonics.

(2) Based on the model of hydrocarbon accumulation and combined with the structure and distribution characteristics of oil and gas in the study area, the main controlling factors of hydrocarbon accumulation were summarized. Paleo-uplift is the orientation area of oil and gas migration and the most favorable place for oil and gas enrichment. Faults and fractures not only provide

favorable channels for oil and gas migration but also improve the permeability of reservoirs and create conditions for oil and gas accumulation. Relatively high porosity and high permeability sand bodies are the basis of lithologic reservoirs.

ACKNOWLEDGMENTS

This work was not supported by any funds. The authors would like to show sincere thanks to those techniques which have contributed to this research.

REFERENCES

Guo, X. Hu, D. Huang, R. Wei, Z. Duan, J. Wei, X. Miao, Z. (2020). Deep and ultra-deep natural gas exploration in the Sichuan Basin: Progress and prospect. *Natural Gas Industry B*, 7(5), 419–432.

Lei, X. Ma, S. Chen, W. Pang, C. Zeng, J. Jiang, B. (2013). A detailed view of the injection-induced seismicity in a natural gas reservoir in Zigong, southwestern Sichuan Basin, China. *Journal of Geophysical Research: Solid Earth*, 118(8), 4296–4311.

Tang, X. Jiang, Z. Jiang, S. Li, Z. Peng, Y. Xiao, D. Xing, F. (2018). Effects of organic matter and mineral compositions on pore structures of shales: A comparative study of lacustrine shale in Ordos Basin and Marine Shale in Sichuan Basin, China. *Energy Exploration & Exploitation*, 36(1), 28–42.

Zou, C. Yang, Z. Sun, Z. Zhao, S. Bai, Q. Liu, W. Yuan, H. Y. (2020). *"Exploring petroleum inside source kitchen": Shale oil and gas in Sichuan Basin*. Science China Earth Sciences, 63(7), 934–953.

Advances in Energy, Environment and
Chemical Engineering – Abdullah & Osman (Eds)
© 2023 The Author(s), ISBN: 978-1-032-36083-6

Combination and development prospect of intelligent waste classification technology and waste incineration power generation technology

Qimin Tian*
Beijing Forestry University, Beijing, China

ABSTRACT: Organic matter, combustible matter, and recyclable matter in Chinese municipal solid waste are increasing, and their utilization value is also increasing. Garbage classification before garbage disposal can reduce environmental pollution, realize resource reuse, and reduce carbon dioxide emissions. Garbage incineration for power generation is a promising way of garbage treatment, but the garbage must be sorted and collected before treatment to maximize the advantages of garbage incineration for power generation. In this paper, an intelligent garbage classification technology based on deep learning is designed to realize intelligent garbage sorting through professional annotation data and an intelligent classification algorithm, which saves social resources to a certain extent and lays a foundation for garbage power generation in the later stage. In addition, this paper also introduces the basic process of waste power generation technology and treatment methods of flue gas and leachate, and prospects for the development of waste incineration power generation technology.

1 INTRODUCTION

1.1 *The necessity and importance of garbage classification for garbage disposal*

As the primary link of the garbage treatment industry chain, garbage classification has an important impact on the reduction, recycling, and harmlessness of garbage treatment.

At present, the utilization value of garbage is increasing, but all the common disposal methods including sanitary landfill treatment, composting treatment, and garbage incineration power generation, cannot realize the maximum utilization of garbage value. Therefore, garbage classification before garbage disposal is very necessary.

If garbage classification is not carried out, mixed garbage collection is not conducive to the targeted treatment of different garbage and the reduction of garbage entering the terminal treatment facility (Zhuang 2020). On the one hand, valuable resources in the garbage are wasted; on the other hand, the cost of subsequent waste treatment increases, and the content of harmful substances after treatment also increases, which is more likely to cause environmental pollution. For example, If the battery and electronic products are burned together, there will be a lot of heavy metal pollution and secondary pollution problems through the exhaust gas (Jin 2021).

In general, garbage classification is conducive to reducing environmental pollution, realizing resource reuse, and reducing carbon dioxide emissions. Before garbage incineration, plastic, rubber, batteries, and other household garbage can be sorted out and recycled to reduce the harmful components in the smoke during incineration (Deng 2021). With the introduction of the "Ban on Waste", China officially banned the import of a variety of solid wastes including waste paper, which greatly reduced the quantity of fiber raw materials available to China's paper-making enterprises. Waste paper has become the strategic resource of the paper industry, and the waste pulp supports the development of China's paper industry to a large extent. If these raw materials are used for

---

*Corresponding Author: zstianqm@bjfu.edu.cn

DOI 10.1201/9781003330165-17

incineration power generation, it is a great waste of resources, and many industries in China cannot develop smoothly. In addition, research (Zou 2022) shows that after garbage classification, carbon emissions decrease significantly.

1.2 *Development and advantages and disadvantages of waste incineration power generation*

At present, landfill still accounts for the largest proportion of domestic garbage treatment in China, which occupies a large amount of land, leading to the phenomenon of garbage siege in many large and medium-sized cities in China (Cheng 2022). Although this method has relatively mature technology, simple operation and management, and low investment, it has a poor effect on volume reduction and capacity reduction, high processing difficulty, large occupation and is easy to produce secondary pollutants. Therefore, it is necessary to seek new technology to replace it. Among them, waste incineration power generation is a kind of waste treatment method with broad application prospects.

In recent years, China has paid more and more attention to carbon reduction, especially after the "dual carbon" policy was put forward. How to achieve the target on time or even in advance has become an important planning and target of local governments. Garbage incineration power generation is an important way to reduce carbon emissions. Its mechanism is to use incineration technology to replace landfills and other means and convert the heat generated by garbage incineration into usable electric energy to replace the fossil energy generated by coal and thermal power generation in the past. The generation of the same unit of electricity can release less carbon factor (Dong & Yue 2022) through the technology, thus achieving the effect of reducing greenhouse gas emissions.

However, waste incineration generally requires a minimum calorific value of waste above 3,360 KJ/Kg. When garbage classification is not properly carried out, the fuel must be added to assist combustion due to the low calorific value of mixed domestic garbage, which leads to an increase in operating costs. Moreover, the investment and operating costs of incineration treatment facilities are also high (Zhang et al. 2011), making large-scale applications difficult.

As there are great differences in the incineration process, main incineration pollutants, and control measures of different types of garbage (Ren 2021), the evaporation points of different kinds of heavy metals and their compounds are quite different, and the contents of domestic garbage are also different. Therefore, garbage should be sorted and collected before incineration for power generation. In this way, people can make the most of the advantages of waste incineration.

According to the research results (Liu et al. 2021), human toxicity potential and acidification potential are the main environmental influencing factors of waste incineration power plants, which together account for 72.8% of the overall environmental impact. Incineration is the main pollutant generation process, contributing 94.1% to the overall environmental impact. As an effective supplement to landfill power generation and coal-fired power generation, garbage incineration power generation has less impact on the environment, and it can bring significant energy saving and emission reduction benefits.

In addition, research results (Yao et al. 2019) show that the amount of domestic waste treated, the water content of domestic waste, installed capacity of biomass energy, average power line loss, and fossil fuel consumption in power generation are five important factors affecting GHG emission reduction potential.

2 INTELLIGENT GARBAGE CLASSIFICATION TECHNOLOGY BASED ON DEEP LEARNING

The flow chart of traditional secondary garbage sorting is shown in Figure 1, in which manual rough sorting, manual resorting, and selection are all heavily dependent on labor, with low sorting accuracy and many intermediate links, resulting in a waste of resources and certain risks.

Through professional annotation data and intelligent classification algorithms, intelligent garbage classification technology based on deep learning and visual recognition can replace manual work, improve accuracy, optimize paths, and realize intelligent sorting.

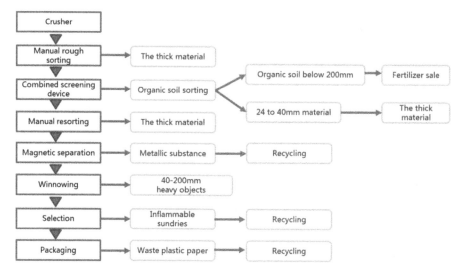

Figure 1.   Basic flow chart of traditional secondary garbage sorting.

2.1 *Intelligent garbage can and garbage classification one-stop technical solution*

An intelligent garbage can is specially designed for sorting and recycling recyclable garbage. It adopts Jetson Nano upper computer and STM32 lower computer as controllers, equipped with actuators such as steering gear, and uses photoelectric sensors to give full load alarm.

One-stop garbage sorting technology solution is personalized data with a system service package for secondary sorting enterprises to improve the efficiency and accuracy of secondary sorting. Its biggest advantage is that it provides a system with accurate identification technology. It can carry out machine learning and algorithm updating according to the changing requirements of garbage classification to connect with huge data, realize rapid migration of classification technology, and help garbage secondary sorting enterprises to quickly master classification technology. The interactive intelligent garbage classification system in one-stop technology solution is a visual and interactive intelligent garbage classification system platform for customers. Users can directly access the system and obtain massive intelligent garbage classification data, realizing the rapid migration and application of classification technology.

2.2 *Algorithms and training*

As the core of the intelligent garbage classification device, the YOLO algorithm is a real-time target detection algorithm with the advantages of fast detection speed and high accuracy. YOLOV3 algorithm is packaged in ROS (Robot operating system). After single garbage is dropped into the rotating tray through the drop port, the camera arranged at the best wide-angle position will transmit the picture in the drop port to the upper computer, which will monitor and identify the garbage input. After the upper computer confirms the garbage information, the ROS node is used to realize serial communication with the lower machine STM32, control the steering gear, make the rotating tray or rotating baffle move, block the fall of garbage, or swept into the corresponding garbage can, and then realize the effect of correct garbage classification.

2.3 Core technologies and components

The core technology lies in the upper computer Jetson Nano, the lower computer Stm32, Yolov3 algorithm, ROS, serial communication, PWM signal, and GPIO control. The basic idea is to deploy a deep learning algorithm for the upper computer to recognize images and send instructions through a serial port. The lower computer receives a signal from the serial port to control the steering gear rotation and prompt a full load alarm.

2.3.1 Upper computer and lower computer

The internal operation of our system mainly depends on the subscription and publishing of codes of the upper computer and the lower computer. The upper computer deploys the deep learning algorithm to recognize images and sends instructions through serial ports, while the lower computer receives serial ports. Among them, the upper computer mainly involves the code operation, and the control of the lower computer mainly involves the control principle and algorithm of the steering gear, GPIO control method, and serial communication.

2.3.2 Serial communication

With the method of asynchronous communication, the STM32 serial port 1 is used to communicate with the Jetson Nano. The Jetson Nano converts the picture information of the camera into instructions and sends it to the STM32, and the STM32 outputs the PWM wave to control the movement of the steering gear in the desired way.

2.3.3 Photosensor

Photoelectric sensors can sense when the trash can is fully loaded. In the process of each cycle monitoring, a full load test is carried out to ensure that garbage is not fully loaded before it can be put in. When the garbage is fully loaded, a full-load alarm will appear on the screen, reminding the user not to throw garbage in it. At the same time, the GMS module will send a message to the administrator to inform him to empty the garbage in time.

In addition, photoelectric sensors can assist garbage classification. According to the roughness of the surface of the object, the reflection of light on the surface of the object is different, so it can distinguish objects with different media such as metal and plastic.

2.4 Social meaning

Compared with artificial waste classification, the annual processing capacity of 300,000 tons using AI sorting can improve the efficiency by 40%, which greatly improves the efficiency and reduces the cost. In the new era, the internal sorting device of intelligent waste classification technology, which is characterized by high precision, high efficiency, and low time consumption, meets the urgent needs of the country for rapid processing of a large amount of stranded waste as well as effective and accurate recycling.

The intelligent terminal, which is applied in the collection process of efficient sorting, can actively complete further accurate sorting, and reduce the back-end link with the help of AI image recognition and robot system technology. It can also reasonably allocate resources to a material recycling company, waste power generation plant, and waste recycling disposal factory for management, which can save social resources to a certain extent.

3 WASTE INCINERATION POWER GENERATION TECHNOLOGY

3.1 The basic process of garbage incineration power generation

Garbage incineration power generation and conventional thermal power generation are the same in the process; their difference lies in that waste incineration power generation makes the use of burning garbage to release heat energy for power generation. In addition to a power station boiler, steam turbine, and generator, waste power generation equipment also includes a closed dump bin, waste incinerator, and other special equipment.

The system of a waste incineration power plant consists of a waste receiving, storage, and transportation system, incineration system, flue gas purification system, waste energy utilization system, residue treatment system, automatic control system, wastewater treatment system, input-output system, etc. Among them, the grate and combustion area have the functions of the combustion of garbage and primary air, controlling the amount of primary air, transporting the garbage from the feed zone to the slag notch evenly and slowly, controlling the import of secondary air, producing strong turbulent flow and so on, which are the important parts of waste incineration power generation.

3.2 Solution of common problems in the waste incineration process

In the process of garbage incineration, liquid slag may stick to the furnace wall and aggregate into a solid form and form slagging, which may cause large slag to fall on the grate to damage it. This problem can be solved by changing the distribution of primary air in the grate.

In addition, the main characteristic of the garbage incineration process is the unstable working condition. The stable combustion of garbage can be controlled by reasonable feeding and grate speed, matching primary air volume to ensure sufficient fuel-air volume and air pressure, matching primary air temperature and matching secondary air volume.

3.3 Flue gas treatment

The flue gas produced by waste incineration contains many harmful substances, with high gas humidity, complex composition, and toxic substances such as dioxin. Therefore, advanced technology, equipment and a stable flue gas purification process should be adopted.

Although the wet flue gas purification system has high performance, simple equipment, and high dust removal efficiency, the purification process produces a large sum of laundry wastewater, requiring many neutralizers and further treatment of the secondary products of desulfurization reaction. This process is more complex and has a higher input cost (Zhao 2020). Although the dry type has a low operation requirement and simple process, the removal efficiency is low. The semi-dry type makes full use of the waste heat of flue gas to evaporate the water in the slurry, and the reaction product is dry solid, so the concentration of $Ca(OH)_2$ required is low. What is more, the semi-dry type has high purification efficiency, which has been widely used.

3.4 Leachate treatment

Landfill leachate production is mainly affected by the composition, moisture, and storage days of incoming waste (Zhang 2016). The pollutant composition is complex, the concentration of organic pollutants and ammonia nitrogen is high, and the content of heavy metal ions and salts is high, so it needs to be treated according to national standards.

Research results (Guo et al. 2022) show that the main process of "anaerobic reactor + MBR system (two-stage A/O+ ultrafiltration membrane) + NF + RO + DTRO concentration reduction" can meet the requirements of leachate treatment. Therefore, the effluent can meet the standards for reuse, the concentrated liquid can be recovered for pulping, and zero discharge and full reuse of wastewater can be achieved.

3.5 Development prospect and outlook

China's garbage power generation business started late and is not perfect. But in the future, waste incineration power generation will gradually become the mainstream of waste treatment, while landfills will mainly be used for the residue after treatment and the final products that cannot be reused. During the "14th five-year plan", the waste incineration power generation industry will focus on "cost reduction and efficiency increase" and "pollution reduction and carbon reduction" under the influence of multiple factors such as subsidy "decline", waste classification, carbon emission reduction, and speed up to adapt to the new situation of green and low carbon (Zhao & Song et al. 2021).

Although many western countries are more advanced in garbage disposal technology and equipment, we should follow a path with Chinese characteristics. Garbage disposal technology and the creation of equipment should be innovated as soon as possible. Garbage classification, storage, and transportation, burning power generation, and late purification process ought to be optimized to further achieve emission peak and carbon neutrality.

4 CONCLUSION

Garbage should be sorted before incineration power generation. But traditional garbage classification is heavily dependent on labor, which causes a waste of resources and certain risks. To maximize the harmlessness, recycling, and reduction of waste, an intelligent waste classification technology was designed to promote the whole system of waste treatment and recycling. The author is the first researcher to first to combine the two technologies. In the future, researchers should further optimize the algorithm and design new materials for intelligent waste classification, and focus on the optimization of the waste incineration power generation process and the treatment of waste gas and waste residue.

REFERENCES

Cheng,H. (2022), "Development of waste incineration power generation industry in China", *Leather Manufacturing and Environmental Protection Technology,* Vol. 3 No. 01, pp. 152–154.

Deng, X. (2021) Incineration power generation can not ignore waste classification. *Anyang Daily.*

Dong, M. & Yue, M. (2022), "Legal issues and countermeasures of waste incineration for power generation in the context of ", *Xinjiang Social Science Forum,* No. 02, pp. 90–96+106.

Guo, Y., Li, R., Liang, Z., Ren, G. & Sun, S. (2022), "Zero Emission of leachate from waste incineration power plant", *Shandong Industrial Technology,* No. 02, pp. 37–42.

Jin, Q. (2021), "On the Advantages and Existing Problems of Garbage power generation", *China Equipment Engineering,* No. 13, pp. 220–221.

Liu, D., Wang, S., Xue, R., Gao, G. & Zhang, R. (2021), "Life cycle assessment of environmental impact on municipal solid waste incineration power generation", *Environmental Science and Pollution Research,* Vol. 28 No. 46, pp. 65435–65446.

Ren, T. (2021), "Waste incineration for power generation and environmental protection research", *Journal of Integrated Resources Utilization in China,* Vol. 39 No. 09, pp. 132–134+140.

Yao, X., Guo, Z., Liu, Y., Li, J., Feng, W., Lei, H. & Gao, Y. (2019), "Reduction potential of GHG emissions from municipal solid waste incineration for power generation in Beijing", *Journal of Cleaner Production,* Vol. 241118283.

Zhang, Y. (2016), "Review and prospect of domestic solid waste incineration treatment technology in China", *Environmental protection,* Vol. 44 No. 13, pp. 20–26.

Zhang, Y., Shang, X., Li, K., Zhang, C., Zhang, K. & Rong, H. (2011), "Present situation of municipal solid waste treatment technology and management countermeasures", *Journal of Ecology and Environment,* Vol. 2 No. 20, pp. 389–396.

Zhao, G. (2020), "Research and application of flue gas purification process of MSW incineration power generation", *Resource Conservation and Environmental Protection,* No. 6, pp. 128–129,131.

Zhao, H. & Song, Q. (2021), "Development prospect of Waste incineration power generation industry in China under the background of carbon peak and carbon neutralization", *Developmental Finance Research,* No. 04, pp. 11–20.

Zhuang, Y. (2020), "Present situation of municipal solid waste treatment technology and management countermeasures", *Engineering Research,* Vol. 13279–280.

Zou, X. (2022), "Study on the change of incineration plant carbon emission after waste classification", *Cleaning World ,* Vol. 38 No. 04, pp. 85–86+160.

Advances in Energy, Environment and
Chemical Engineering – Abdullah & Osman (Eds)
© 2023 The Author(s), ISBN: 978-1-032-36083-6

Research on risk identification and analysis of electricity market

Minfang Chen
Fujian Mianhuatan Hydropower Development Co., Ltd., Longyan, China

Zhanli Liu
Huadian Electric Power Research Institute Co., Ltd., Hangzhou, China

Jiongzhang Tang
Fujian Mianhuatan Hydropower Development Co., Ltd., Longyan, China

Faguo Wei
Huadian Shandong Energy Sales Co. Ltd., Jinan, China

Yeyang Zhu*, Jiguang Zhang & Baozhong Zhou
Huadian Electric Power Research Institute Co., Ltd., Hangzhou, China

ABSTRACT: The competition in the electricity market brings not only opportunities but also operational risks due to the volatility of electricity market prices and the uncertainty of demand. Identifying risk factors from the market and reasonably assessing risks has become the key to the sustainable operation and development of power generation companies and electricity sales companies. This paper proposes a risk prevention and control process mechanism for power generation enterprises for risk identification and analysis, sorts out the risk factors faced by power generation enterprises, and establishes four aspects of market risk, credit risk, operational risk, and compliance risk.

1 INTRODUCTION

Risks exist in the electricity market all the time. The main measures taken by market designers to avoid electricity price risks include: (1) introducing a competitive market structure, (2) introducing financial transactions, (3) introducing bilateral transactions, (4) designing a scientific pricing mechanism, appropriate market electricity, and electricity ratios.

Under the market environment, power generation enterprises are faced with many risk factors, and there is a certain mutual influence between the factors, which makes the risk management work more complicated. The study found that the factors that have a direct impact on the final risk factor of electricity revenue are electricity price control and adjustment policies, electricity price fluctuations, bidding strategies, and on-grid electricity. Controlling these risk factors can play a decisive role in effectively controlling the ultimate risk factor. The strongly related factors that influence each other include cost management and bidding strategy, production control, and production safety. Concentrating on controlling these two pairs of risk factors is beneficial to improving the efficiency of risk control and management. In terms of risk management and control, cost management and product safety management are mainly related to the production performance of the enterprise. Whether or not to participate in the spot market, it is necessary to improve the management level in this regard. Management and control of risks in the spot market are mainly aimed at the bidding

---

*Corresponding Author: zhuyeyangwork@163.com

DOI 10.1201/9781003330165-18

strategy, which can simultaneously alleviate the risk of electricity price fluctuation and power generation (Bjorgan et al. 1999).

1.1 *Contracts for difference*

Contracts for Difference (CFD) is a form of long-term bilateral power supply financial contract that avoids the risk of random electricity prices. It does not involve the physical execution of electricity generation and consumption, and only the difference between the settlement price and the contract price is used for cash settlement. For CFDs, market participants are usually placed under a unified regional real-time electricity price. The description is as follows. Assuming that there is a power plant on node i and a user who expects to trade Pi power on node j, the system's real-time electricity price in the market will be Pr. However, traders prefer to trade at the agreed electricity price Pc. This can be achieved indirectly by entering contracts for difference, which are defined as follows.

According to the contract for difference, the user pays the power plant $(Pc - Pr) \times Pi$, where Pc is the contract electricity price, Pi is the power traded in a certain period, and P is the market real-time electricity price. For the contract seller, when the real-time electricity price in the market is lower than the contract price, profit from CFDs can partially make up for the loss of selling electricity at a low price in the electricity spot market. For the contract buyer, when the real-time electricity price in the market is higher than the contract price, profits partially cover the cost of purchasing electricity at high prices in the electricity spot market. Therefore, CFDs have the function of locking the price for both parties to the transaction and avoiding the risk of market fluctuations (Huang et al. 2012).

1.2 *Electricity futures*

Futures refer to the trading method of financial contracts involving the sale of financial instruments or commodities for actual delivery in the future. Such contracts are called futures contracts. In addition to buyers and sellers, futures contracts can also be freely circulated in the futures market, but the transaction process must be carried out in the futures exchange. Compared with forward contracts, futures contracts are highly standardized, which greatly enhances the liquidity of futures contracts, and the financial nature of futures contracts makes futures trading, not for the actual power, but to gather huge social capital to discover the potential of power.

Electricity futures refer to electricity commodities that are traded at a specific price, delivered at a specific time in the future, completed within a specific period, and traded in the form of electricity futures contracts. For energy companies, the functions of power futures are mainly reflected in the following points.

(1) Risk aversion

The hedging function of futures trading can help electricity market participants avoid market risks that may be brought about by the high volatility of electricity prices. Hedging is the operation of using the futures market as a place to transfer price risks, using futures contracts as a substitute for purchasing commodities in the spot market in the future, and insuring future prices. During hedging, the operations performed on the spot and futures markets on the spot trading day have the same quantity and specifications but opposite directions (Bjerksund et al. 2010).

(2) Price discovery

Before the emergence of the futures market, spot market participants adjusted their trading strategies for production, consumption, and other links through real-time spot market prices. Therefore, spot market prices had a certain lag on the time scale. In addition, since spot trading participants are often directly related to the production and consumption of commodities, the scale of spot trading participants is limited, and the fragmented nature of spot trading makes spot market information asymmetric.

In contrast, the transaction price of the electricity futures market is the result of open competition by a large number of traders, which brings together the views of traders in all aspects of electricity generation, transmission, distribution, and use in the current situation trend of the electricity market.

1.3 *Financial transmission right*

Financial Transmission Right (FTR) is a financial contract used to avoid the risk of regional node electricity prices. It entitles the owner to be compensated for transmission congestion charges.

When the transmission network of the real-time market is blocked, it leads to different LMPs in different regions, that is, in the spot market, the electricity price of the generation node and the electricity consumption node are different, resulting in the difference between the electricity generation revenue on the generation side and the electricity purchase expenditure on the electricity consumption side. In terms of congestion cost, when network loss is not considered, the difference between the electricity purchase expenditure on the consumption side and the generation revenue on the generation side is the congestion surplus.

The financial transmission right settlement method is that a market entity holds the financial transmission right capacity Q_{ij} from i to j. In the spot market, the spot electricity prices of the two nodes are P_i and P_j respectively, then the financial transmission right settlement fee is $(P_i - P_j) \times Q_{ij}$. FTR has a clear direction, from node i to the FTR of node j. The price difference at the time of settlement is the price of node j minus the price of node i; if the price of node j is lower at the time of settlement, then the holder of FTR needs to pay extra. The fee corresponding to this part of the difference is the settlement of the responsible FTR. In addition, there is also option-type FTR, which is different from responsibility-type FTR in that if the electricity price of FTR pointing to the node is lower during settlement, the holder of option-type FTR can choose to give up the FTR. In other words, there is no need to pay part of the difference. FTR holders can still get this part of the blocking surplus (Vehviläinen & Keppo 2003).

Through the above analysis, the financial transmission right actually plays a role in fixing the transmission cost without considering the congestion cost. Corresponding to the function of CFDs and futures contracts to avoid future market price risks, FTR can provide energy companies with a channel to avoid congestion risks so that when there is a difference between the electricity price of the power outflow node and the power injection node, the transaction price of both parties is locked in one.

1.4 *Energy trading and risk management system*

The most common risk management tool used by energy companies for the electricity market trading business is the energy trading and risk management system (ETRM). Transactions are completed within a unified system with real-time reporting or with the assistance of a separate auditable data processing and reporting system, providing a holistic IT solution covering all aspects of energy trading and risk management.

The complete set of energy trading and risk management system (ETRM) generally includes the following functional modules.

Table 1. Energy trading and risk management system (ETRM) functional module.

Pre-trade support	Foreground/run	Middle ground	Background
Timing management	Trade	Transaction controls/restrictions	Documentation
Forecast (price, demand, and new energy)	Contract management	Credit risk	Confirm
Planning	Position management	Market risk	Settlement
Retail price	Fuel management	Compliance	Billing
	Unit combination	Market valuation and profit and loss	Inventory management
	Economic dispatch	Pressure test	Hedge accounting
	Emission	Margin management	Report
			Treasury management

2 RESEARCH ON RISK ASSESSMENT SYSTEMS AND PREVENTION AND CONTROL MECHANISM

2.1 *Risk identification*

The risk sources of power generation companies can be divided into market risk, operational risk, credit risk, and compliance risk.

Market risk refers to the risk of unanticipated potential losses in the process of business development between power generation companies and electricity sales companies due to changes in product prices and product liquidity in the electricity market. Market risks in the electricity market are mainly composed of three types of risks: price risk, supply risk, and liquidity risk.

(1) The generation of price risk is caused by the price changes of power products.
(2) The supply risk is caused by the difference between the actual physical power generation and the amount of medium and long-term contracts signed (Cancer Research UK 1975).
(3) Liquidity risk is mainly concentrated in the field of trading products in the electricity market. Usually, when the transaction party purchases, there is no corresponding trading product in the current market. The decomposition curve cannot fully meet the unit transaction requirements.

Credit risk, also known as default risk, refers to the risk of economic losses caused by counterparties failing to perform their contractual obligations. For the power generation side, its counterparty is a power sales company or a large power user in bilateral transactions, and a transaction clearing institution in centralized competitive transactions. For the electricity sales side, the counterparties of its wholesale transactions are power generation companies and transaction clearing institutions, and in retail transactions, they are the retail users of their agents.

Operational risk is the risk of loss resulting from inadequate or problematic internal procedures, people and systems, or external events. Compared with credit risk and market risk, operational risk originates from the business operations of the trading and marketing department and is an endogenous risk within the company's control. From empirical analysis, there is no relationship between a single operational risk factor and operational losses. There are clear, definable quantitative relationships. At the same time, operational risk covers a wide range and the situation is complex. It is difficult to cover all areas of operational risk with one method, and it is necessary to sort out the company's business processes separately.

Compliance risk refers to the failure of the company to comply with all applicable laws, administrative regulations, departmental rules and other normative documents, business rules, industry standards, codes of conduct, and professional ethics, which expose the company to legal sanctions, regulatory penalties, financial losses and uncertainty of reputation loss. Compliance risk may include the violation of a company's financial system and transaction regulations.

The following summarizes the main risk types and corresponding risk factors of electricity trading under the electricity spot market system.

2.2 *Risk analysis*

(1) Market risk
   a. Risk of changes in market trading rules. At present, the electricity market is constantly improving, and market trading rules are changing rapidly. Changes in settlement methods, assessment mechanisms, and quotation methods will have a great impact on power generation companies and their electricity sales companies. A deep understanding of market trading rules may help develop strategies to participate in the market. Trading rules are guidelines for participating in the market, and they have always tracked changes in trading rules.
   b. Market supervision risk. With the advancement of the electricity market, there are more and more players in the electricity market, and the electricity market supervision system is not yet perfect.

c. Market reform process risks. Since the No. 9 document, China's electricity market reform has been continuously advancing, and some substantial results have been achieved. However, throughout history, China has tried to carry out electricity market reform several times, but it has not been able to continue due to various reasons. If the process of the new round of electricity market reform slows down, it will affect the ability of power generation companies and electricity sales companies to adapt to the market.

d. Quotation strategy risk. Power generation enterprises will choose a certain quotation strategy when quoting, usually according to their conditions, combined with basic data such as weather, coal, and natural gas, they will also try to figure out the psychology of other power generation enterprises and guess the other party's quotation. Favorable quotation strategies, such as speculation and simulation are inherently uncertain, and the design of the internal decision-making mechanism of power generation enterprises will also affect the quotation strategy, which constitutes the risk of quotation decision-making (Krečar & Gubina 2020).

e. Risk of electricity price fluctuations. At present, many regional electricity markets in China are cleared according to the marginal electricity price, and the clearing price changes greatly, especially now that the spot market has begun to run continuously. There is great uncertainty, but this is also the most important factor that enterprises pay attention to and directly affects the income of enterprises.

f. Market supply risk. The electricity sold by time-sharing in medium and long-term contracts does not match the winning bid/power generation in the spot market, resulting in the need to buy/sell electricity with uncertain prices in the spot market; the electricity purchased by time-sharing in medium and long-term contracts does not match the actual electricity consumption of users, resulting in the need to purchase/sell electricity in the spot market with uncertain prices; the actual electricity consumption of the user is quite different from the forecast, which may cause the actual agency income to deviate from the forecast.

g. Market competition environment risk. There are more and more subjects participating in the electricity market, and the market environment is becoming more and more complex. To compete in market transactions, it is inevitable that some unfair competition methods will appear. For example, some electricity sales companies use intermediaries to obtain greater benefits. The uncontrolled problem of intermediaries results in cross-competition between customers. The price mechanism is affected by the distribution of interests, resulting in market chaos and, at the same time, leaving the hidden risk of the integrity of sales personnel. On the other hand, due to market share, the power generation enterprises have formed a group, lowered prices, or even formed an oligopoly.

(2) Credit risk

a. Risk of customer arrears. In the past power transactions, customers settled electricity bills through power grid companies, and power generation companies only faced one customer, the power supply company. The risk of power generation enterprises. For the situation where customers owe electricity fees, power generation enterprises do not have a complete set of mechanisms for power grid companies to collect electricity fees (Klessmann et al. 2008).

b. Customer performance risk. The risk of customer performance mainly refers to the fact that the customer cannot complete the power transaction according to the contract. The power generation company signs medium and long-term transactions with the customer and will arrange the company's unit start-stop plan and load arrangement according to the contract. The decision-making influence of the enterprise is also very large. At the same time, for electricity sales companies, in some regional markets, the assessment of electricity deviation is very serious.

(3) Operational risk

a. Market forecast risk. In the market, the supply relationship will determine the market price. Before participating in the market quotation, power generation companies usually have a market forecast. If the load forecast is high, the demand will be large, and the quotation may increase when the supply is in short supply; on the contrary, if the load forecast is low,

the demand will be small, and the supply will exceed the demand. Offers may be reduced. However, load and price forecasts also have many uncertain factors, so the forecast accuracy is not certain, so load and price forecasts also cause quotation decision risks.
b. Cost control risk. Power generation companies participate in the market to obtain benefits, and the quotation is based on cost estimates. Only when the total cost of the power generation enterprise is lower, the power plant will have more choices in the price declaration and can take the advantage of the price. For thermal power companies, fuel costs account for a high proportion, domestic coal prices fluctuate wildly according to market conditions, and fuel costs will also fluctuate greatly, which will affect companies' bidding decisions in the power market.
c. The risk of miscalculation of quotation. Due to human and tool reasons, errors in the market quotation process result in unreasonable market quotation decisions or power generation capacity calculations, potentially affecting the company's earnings.
d. Risks caused by the pressure and complexity of electricity market transactions. The trading patterns and rules of the electricity spot market are quite different from the current whole-sale electricity market based on physical delivery. These will lead to low work efficiency of traders, increase the difficulty of power generation companies in making quotation deci-sions, reduce the possibility of winning the bid and the room for profit, and may even cause more power generation and more money.
e. Risk of other human error in operation. No mid-to-long-term bilateral contracts were signed, and contracts were signed without authorization; contract terms were invalid due to human errors; management confusion was caused by the change of the paper version of the contract; information errors during the signing of the transaction authorization letter resulted in the invalidation of the power of attorney; due to human error, the company's key information (cost, quotation, finance, etc.) was leaked; due to human and tool reasons, the bilateral contract filing was not completed in the regulations; the calculation error led to the failure of the contract check.

(4) Compliance risk

Non-compliance of contract approval, management, and filing; compliance risk of cer-tificate reuse; whether the medium and long-term transaction business process meets the compliance requirements; whether the spot declaration business process meets the compli-ance requirements; whether the invoicing, receipt and payment management processes meet the compliance requirements.

3 CONCLUSION

Combined with the current situation of spot market construction and trial operation, this paper proposes a risk prevention and control process mechanism for power generation enterprises for risk identification and analysis, sorts out the risk factors faced by power generation enterprises, and establishes four aspects of market risk, credit risk, operational risk, and compliance risk. The risk assessment index system provides a reference and basis for risk management of power generation enterprises in actual operation.

REFERENCES

Bjerksund, P., Rasmussen, H., & Stensland, G. (2010). Valuation and risk management in the Norwegian electricity market. In *Energy, natural resources and environmental economics* (pp. 167–185). Springer, Berlin, Heidelberg.
Bjorgan, R., Liu, C. C., & Lawarree, J. (1999). Financial risk management in a competitive electricity market. *IEEE Transactions on power systems*, 14(4), 1285–1291.
Cancer Research UK. (1975) *Cancer statistics reports for UK.* http://www.cancerreseark.org/aboutcancer/statistics.

Huang, J., Xue, Y., Dong, Z. Y., & Wong, K. P. (2012). An efficient probabilistic assessment method for electricity market risk management. *IEEE Transactions on Power Systems*, 27(3), 1485–1493.

Klessmann, C., Nabe, C., & Burges, K. (2008). Pros and cons of exposing renewables to electricity market risks—A comparison of the market integration approaches in Germany, Spain, and the UK. *Energy Policy*, 36(10), 3646–3661.

Krečar, N., & Gubina, A. F. (2020). Risk mitigation in the electricity market driven by new renewable energy sources. *Wiley Interdisciplinary Reviews: Energy and Environment*, 9(1), e362.

Vehviläinen, I., & Keppo, J. (2003). Managing electricity market price risk. *European Journal of Operational Research*, 145(1), 136–147.

Advances in Energy, Environment and
Chemical Engineering – Abdullah & Osman (Eds)
© 2023 The Author(s), ISBN: 978-1-032-36083-6

Research and application of on-load voltage regulation transformer

Yu Zou*, Deyu Fu*, Xiaobing Chi*, Henglong Chen*, Zipei Guo* & Ming Wang*
Guangxi Power Grid Co., Ltd., Qinzhou Power Supply Bureau, China

ABSTRACT: In the process of power technology development, power transformer on-load voltage regulators have been widely used in power distribution systems. In recent years, on-load voltage regulator power transformers have been increasingly used in power systems. The on-load voltage regulator is widely used, which plays an important role in improving the voltage quality of the system and improving the reliability of the power supply. This paper briefly analyzes some on-load voltage regulation methods of power transformers, focuses on several new types of on-load voltage regulation transformers, and draws some conclusions. But what is frustrating is that there are still many off-load voltage regulating transformers operating in the power grid, and these off-load voltage regulating transformers are in urgent need of on-load voltage regulating transformers. To solve this problem, this paper introduces some traditional on-load voltage regulation methods, some new on-load voltage regulation methods, other innovative methods, and two common methods of on-site on-load voltage regulation transformation, which analyzes and compares the characteristics of these main schemes. The problems that should be paid attention to in the transformation work are also pointed out in this paper.

1 INTRODUCTION

Recently, on-load voltage regulation has gradually increased, which is also the result of higher and higher requirements for power supply quality. Although a lot of on-load voltage regulation transformers are installed and used in the power system, they have not played their role 100%. Here we introduce the role of on-load voltage regulation. The definition of a power transformer on-load voltage regulation technology is a transformer that can adjust the transformation ratio under load conditions. In the power system, with the improvement of users' requirements for the quality of grid voltage, the use of on-load voltage regulation technology in transformers is feedback to users' requirements, and it is also a new trend. The two most important factors on the on-load voltage regulator are the on-load tap-changer and motor mechanism and control. The most prone to failure is the electric mechanism, and the common failure is the switch linkage. All the faults and the many components of the motor mechanism are inseparably related to the complexity of wiring and the difficulty of debugging. Therefore, improving the design ideas of the transformer system and simplifying the complexity is our primary task. However, according to previous experience, there are still many off-load voltage regulating transformers still running in the power grid. Transforming this part of the transformer into an on-load voltage regulating transformer has a positive effect on improving the voltage quality of the system.

2 THE TRADITIONAL ON-LOAD VOLTAGE REGULATION METHOD

In a traditional transformer, the on-load voltage regulator is a mechanical tap-changer. Actions such as the driving gear of the mechanical switch can easily cause operation accidents. The occurrence

---

*Corresponding Authors: 27106124@qq.com; 1822059752@qq.com; 398155038@qq.com; 269407111@qq.com; 359088602@qq.com and wang_m.qzg@gx.csg.cn

DOI 10.1201/9781003330165-19

of these operation accidents will weaken the reliability of the transformer and bring certain safety hazards to the work. In addition, when the mechanical switch operates, an arc will be formed, and a certain arc will cause chronic ablation of the mechanical switch contacts. Another problem that cannot be ignored is that the generated arc will cause the transformer to deteriorate in oil quality, and then make the transformer in the transformer. The insulation capacity of the winding is weakened, resulting in the occurrence of a phase-to-phase short circuit or turn-to-turn short circuit. Because the response time of the mechanical switch action is generally about five seconds, which takes a long time, the transformer that has applied the on-load voltage regulation technology in the traditional sense can only be used for voltage regulation in a stable state. According to some research data, transformer accidents caused by tap-changers and faults account for 10% or even 20% of the total transformer accidents every year, while 500 kV transformers have the failure rate of the load switch as high as more than 20%, and accidents and failures are frequent.

3   A NEW METHOD OF ON-LOAD VOLTAGE REGULATION

Due to various problems in the traditional power transformer on-load voltage regulation technology, the new power voltage regulator on-load voltage regulation technology came into being. At this stage, the research and application of the new on-load voltage regulation technology are mainly reflected in the following aspects. For the improvement and perfection of the traditional mechanical on-load voltage regulation technology, the electronic switch circuit is added based on not changing the conventional mechanical on-load tap-changer, thus forming an improved mechanical voltage regulating transformer. In addition to maintaining the traditional selector, switch, motor mechanism, and other structures, transition resistance and thyristor are also added. This technology can greatly enhance the safety and stability of the converter, thus avoiding safety accidents to the greatest extent occur. Another new type of on-load voltage regulation technology for power transformers is the thyristor switching type voltage regulating transformer. This kind of transformer mainly uses a thyristor as the connection switch to realize the conversion process, so the quality and performance requirements of the thyristor are relatively high, and the cost is high. However, the on-load voltage regulation technology has good self-checking and fault alarm functions, so compared with other technologies, the safety and stability are higher, and the running speed is also faster with a good development prospect. Due to the various deficiencies and defects of traditional mechanical switches, various countries have actively researched new types of on-load voltage regulators, which can be divided into mechanical improvement type, auxiliary coil type, and three types of electronic switches.

3.1   *Mechanically improved on-load voltage regulation technology*

This type of transformer is composed of a traditional transformer plus a switching electronic circuit. Its tap changer only uses a small number of thyristors and a transition resistor. The mechanical switch and the electronic switch cooperate to limit the generation of arcs during operation. The advantage of this type of device is that there is no time control loop, and the capacity requirement of the thyristor is low. The loss of control of the thyristor will not damage the tap of the transformer, but the operation speed is not fast.

3.2   *On-load voltage regulation using auxiliary winding*

Control the conduction angle of the thyristor and superimpose the adjustable voltage on the transformer, so that the three-phase transformer and the step-up transformer are connected, and the reverse-connected thyristor is used as the connection medium. If the trigger of the thyristor is delayed, the short-circuit switch set in advance can prevent the rise. The voltage transformer is an open circuit. If the triggering of the thyristor is not delayed, it will trigger when the zero-crossing occurs, and the voltage will be loaded on the load in phase. In addition, the on-load voltage regulation method based on the auxiliary coil is used to realize arc-free operation. The specific method is to use a coil that can adjust the rated voltage and couple it into the transformer. In addition, a certain auxiliary voltage is added to the transformer to ensure that the superimposed voltage is in the same phase as the original voltage.

3.3 *Electronic switch-type on-load voltage regulation technology*

Due to the development and progress of electronic power technology, the power consumption of the thyristor can be reduced as much as possible by appropriately selecting the trigger time.

3.4 *Comparative analysis of the above three new voltage regulation technologies*

The mechanical improvement type on-load voltage regulation is characterized by high economy and low harmonic content, but the speed is slow, the arc is limited, and the operation of the transformer will not be affected when the thyristor is out of control; the electronic power switch type on-load voltage regulation is characterized by fast speed. The influence of the thyristor out of control is large, the harmonic content is low, but the price is high; the auxiliary coil type on-load voltage regulation is characterized by more harmonics, and the arc is not limited.

4 A NEW METHOD OF ON-LOAD VOLTAGE REGULATION

4.1 *Wearing shoes improve proposals*

The improvement method of wearing shoes is mainly to open the neutral point of the high-voltage three-phase coil of the main transformer, connect the voltage regulating coil of the compensator in series, and then connect the low-voltage side of the main transformer and the excitation coil of the compensating transformer in parallel. The on-load voltage regulation was successfully realized.

4.2 *Backpack improve proposals*

A backpack is a more economical and applicable transformation method when the non-excitation voltage regulation range of the transformer can meet the needs of power supply voltage fluctuations in the region. That is to remove the tap lead on the original off-excitation tap changer, remove the switch, install a jumper or linear on-load voltage regulator, and lead the original tap to the on-load voltage regulator.

5 COMMON METHODS FOR ON-SITE VOLTAGE REGULATION RETROFIT

5.1 *Direct substitution*

Using the original lead of the transformer to install the on-load voltage regulating switch, directly replace the original no-load three-phase or three no-load single-phase tap-changers with a five-speed three-phase composite on-load tap-changer to realize on-load voltage regulation. A composite tap-changer can be selected, which is small and easy to install. For the no-load tap-changer that was originally a neutral point voltage regulation, the on-load tap-changer with the same voltage regulation series and neutral point voltage regulation can be directly replaced. This method makes the stage voltage high, the circulating current is large in the tap change, the recovery voltage between the contacts is high, and the life of the on-load tap-changer is relatively short.

5.2 *Indirect compensation method*

The idea comes from the single main transformer of UHV and large capacity. To solve the three-phase connection voltage regulation, reduce the insulation level of the voltage regulation device, and reduce the transportation weight, the independent connection three-phase high-voltage neutral point voltage regulation is designed and produced. The device divides the main transformer and the on-load voltage regulation part into two structurally and uses them in series with each other electrically. A suitable location near the main transformer can facilitate the installation of the voltage regulating transformer, the voltage regulator, and the three-phase neutral point lead wire bracket. The advantage is that the renovation does not require a factory, so the construction period is greatly shortened; the second is that the number and range of voltage regulation stages are much larger than before the renovation, and when the operating voltage is high and the voltage regulating transformer is being tapped, it can compensate for the capacity passed by the transformer. When the voltage is low and the negative tap is running, the capacity of the main transformer should be offset; once a fault occurs in the on-load tap-changer or the compensation transformer, the lead of the entire set of on-load voltage regulation and compensation transformer can be disconnected and withdrawn from operation, and the three main transformers can be restored. The phase-neutral connection and the grounding protection device continue to be put into operation in the mode of no-load voltage regulation to supply power to the system.

6 PROBLEMS AND TROUBLESHOOTING IN OPERATION OF ON-LOAD TRANSFORMER

6.1 *Overhaul and test regularly*

The vacuum on-load switch can be disassembled for operation. Generally, after the switch is connected 2,000 times, it should be comprehensively inspected and repaired. It is recommended that the contents of the mailbox be sampled to ensure that its power-frequency withstand voltage does not fall below 25 kV. if the transformer passes the test, it does not require replacement, but if the transformer fails the test, replacement of the insulating oil is required.

The electric operating mechanism requires that they should be able to ensure that they are in a good maintenance and maintenance state for a long time. The gas protection and anti-riot devices equipped with the on-load gas tap-opening backbone and the phase sequence gas protection all need to be improved for a long time. Whenever a protection device is not functioning properly, it is important to check and identify the cause as soon as possible. During use, if the oil level increases or changes, people must disassemble the accelerator to open a switch and ensure that there is no oil leakage. It is necessary to disassemble and open the throttle and switch the throttle switch and seal it strictly to prevent leakage or oil leakage during daily use. If the engine sees a particularly high temperature or full fuel level after the car starts, it means that the transformer and the on-load tap start control switch have changed the tank and oil. There is a need to pay close attention to how to keep the oil level of the transformer of a motor, especially a box motor, above the tap starter motor switch in the actual usage process so that the oil level of the motor can be strictly controlled. The motor oil level in the box of the motor start switch is too deep to enter or affect the insulating oil quality of the motor.

For the vacuum on-load tap-changer tube, the vacuum switching element should be insulated according to the cycle, and the withstand voltage test should be carried out. High and low voltage work should be done before specific connections. Otherwise, problems such as power transmission will easily occur. By performing switch maintenance and lubrication, the connection between the power failure and the short board on the ground should be done, and the indoor temperature should not be lower than minus 5°C during the specific work process.

6.2 *Comply with the regulations and standardize the operation*

Since the on-load voltage regulating transformer cannot change the reactive power demand balance transfer of the system to avoid other problems in the power grid, this system should be able to fully integrate itself, optimize the configuration to ensure its capacity, and avoid problems caused by the on-load voltage regulating transformer, which could lead to large-scale power outages. Under normal circumstances, it is necessary to do the on-load tap-changer and automatic control mode. In the process of operation, it is necessary to make gear shift adjustments, and the second shift must be realized according to the corresponding requirements. At the same time, separate operation and step-by-step pressure regulation must be realized to ensure that the gear shift is performed after 1 minute after the pressure adjustment. For replacement, it is necessary to adjust two OLTC transformers running in parallel to find 85% of the transformer's rated current. At the same time, it is not allowed to use a single transformer continuously for too long, and it must be replaced in the same transformer, and then the corresponding operation is carried out. After 1 minute of conversion, it is necessary to check the changes in current and voltage. To prevent maloperation from causing other problems, the boost operation should be done to make it clear that the current one with less current will pass. The effective method is to improve and prevent its circulation from being too large. The step-down operation is the opposite of itself. When the operation is completed, the current size and distribution of the two parallel transformers should be checked to adjust the voltage. Also, the settings must be kept coordinated to ensure they work together. Through the parallel connection between the on-load voltage regulator transformer and the no-load voltage regulator transformer, the decomposition of the on-load voltage regulator transformer should be decomposed with the no-load voltage regulator transformer as much as possible to better control the large-capacity transformer and improve its reliability. To prevent and reduce its over-vibration voltage, it should try not to connect separately, and at the same time, try to control the no-load transformer as much as possible to prevent other problems.

6.3 *Correct setting and operate stably*

Setting the on-load tap changer's control parameters correctly is essential to ensuring its normal operation. It is generally the responsibility of the on-load tap-changer controller to determine the bus voltage, compare it with the reference bus voltage, and then issue a corresponding command to the bus motor switch to ensure that the bus motor runs in the desired direction. This allows the busbar voltage to be automatically regulated and transformed to ensure that the voltage of the transformer busbar is close to the voltage of the transformer on-load. Thus, the voltage, upper limit, lower limit, undervoltage, overcurrent, overload, and compensation resistance of the control can be correctly controlled before automatic voltage regulation as required. The correctness of the selection and setting of these control parameters directly affects the overall quality of the various power supply voltages. What is particularly important is the selection of the reference voltage value, and the corresponding control requirements must be done well. Judging from the development situation of the year, its control requirements are relatively obvious. Generally, it cannot be controlled and set according to 0.4 kV. In the specific development process, it needs to be set according to the actual situation to ensure its practicability. Sampling and measurement, as well as other factors, should be considered to ensure that the supply voltage meets the appropriate standard.

7 CONCLUSION

Through the above introduction and analysis and comparison, we believe that a large amount of on-load should be used in the power grid, which can improve the reactive power compensation and voltage regulation, and adjust the on-load compensation capacity of the main transformer in time, which can improve the capacitor investment rate. In the design of the main circuit of the on-load voltage regulating transformer switch, the problems of heat dissipation, protection, and triggering should be considered as much as possible to ensure the reliability of the device. The voltage regulating tap-changer ensures the quality of the power supply and truly achieves the safety of the power grid. As a reactive power compensation device, the power capacitor is used, and its reactive power output and operation are economical and high-quality. In addition, the cost requirements should be considered when using this technology. At present, there is no systematic research on the on-load voltage regulation technology of new power transformers in China. It will be possible to improve on-load voltage regulation technology if we can learn from some advanced research results from abroad, and then adapt them to the actual situation of local technology and economy. By doing this, the working performance of the power transformer will be greatly improved. In this regard, the above transformation methods have their unique features and are also suitable for the transformation of various types of transformers. It is an effective on-load voltage regulation transformation of power transformers to ensure the quality of power grid operation and improve the voltage qualification rate.

REFERENCES

Chen Song, Research on On-load Automatic Voltage Regulation of Lighting Transformer [J]. *Technology and Innovation*, 2019(23):122–123.

Kang Xueping, Adjust the gear position of the on-load voltage regulating transformer to generate circulating current to complete the differential load test [J]. *Modern salt chemical industry*, 2019, 46(05):67–69.

Li Xiaoshuang, Research on Mechanical Fault Diagnosis of Tap-changer of On-load Voltage Regulation Transformer [D]. *North China Electric Power University*, 2021.

Wei Zhenxiang, *Key technology and application of on-load voltage regulation distribution transformer based on planar structure switch*, Henan Province, Beijing Boruilai Intelligent Technology Zhoukou Co. ltd., 2021-04-18.

Xu Peng, Cao Yarong, Tang Ye, Zhang Kewei, Improvement of DC Resistance Test Method for Large On-load Voltage Regulation Transformer, [J]. *Transformer*, 2020, 57(02):70–73.

Advances in Energy, Environment and
Chemical Engineering – Abdullah & Osman (Eds)
© 2023 The Author(s), ISBN: 978-1-032-36083-6

Research on the evolutionary game of port emission reduction based on prospect theory

Xuan Zhang* & Yan Ping Meng
Shanghai Maritime University, Shanghai, China

ABSTRACT: In the context of green port construction, the interaction mechanism between local government and port enterprises in the port emission reduction problem is studied. A two-party evolutionary game model with the introduction of prospect theory is constructed and sensitivity analysis of key parameters is carried out. The results show that the initial probability affects the time for the system to reach stability; that different carbon tax regimes have less influence on local governments and significant influence on port enterprises; that the high cost of port enterprises has a decisive influence on the strategy choice of port enterprises; that the reputation loss of local governments has a decisive influence on the strategy choice of governments; that the higher the probability of ENGO discovery is, the faster the system reaches stability; and that the loss aversion coefficient and risk attitude influence the strategy choice of game parties.

1 INTRODUCTION

China is the world's largest port throughput, and many environmental problems have also been brought about by economic growth. Among them, the issue of port emission reduction has always been a hot topic at home and abroad, and port carbon emissions mainly come from the daily operation of port machinery and the pollution emissions generated by port operation activities such as ship berthing. For example, when moored in a port, ships use diesel-assisted engines to generate electricity, unload, and load activities, emitting large amounts of greenhouse gases, sulfur dioxide, and other harmful gases, which are released into the atmosphere, causing an increase in the concentration of carbon dioxide in the air, breaking the ecosystem cycle near the port. At the same time, because port environmental pollution has a spillover effect and a wide range of influences, if no active measures are taken to control pollution, it will bring environmental pollution to the cities near the port, and the health of urban residents will also be threatened. Therefore, port environmental governance is an urgent problem to be solved.

Port carbon emissions mainly come from the production and operation activities of the port and the entry and exit of ships. Solving the problem of port carbon oxide pollution is essential to reduce port pollution emissions and promote the construction of green ports. Meng et al. (2020) proposed the transformation of green smart ports; Sehwa et al. (2019) proposed that "green development" is the focus of sustainable research on ports from a qualitative point of view, encouraging the green operation of ports; Xia (2021) proposed the use of shore power facilities to build an evolutionary game model of the government and two different port companies to alleviate port pollution. Xiao et al. (2021) have established a three-way evolutionary game model composed of government regulators, blue carbon trading platforms, and news media, allowing the public to participate in the blue carbon governance of marine ranches and reflect the role of the public in the supervision. Boa (2020) has introduced third-party testing agencies in the construction of a two-way game

---

*Corresponding Author: 3462033890@qq.com

between governments and shipping lines on emissions reductions. Tan (2020) used the prospect value function to construct the perceived value construction and evolution game of gains and losses between environmental pollution enterprises and the surrounding masses and portrayed the cognition and decision-making laws of participants who are closer to reality. Based on prospect theory and evolutionary game, Sheng (2021) first established an evolutionary game model between local governments and polluting enterprises in the Taihu Lake Basin. According to the above findings, most scholars mainly study how port companies, governments, shipping companies, and other stakeholders can achieve equilibrium using game theory methods in the context of emission reduction policies. Although other indirect stakeholders are mentioned in these studies, most third-party testing agencies either use the media and the public as third parties or rarely address the role that ENGO can play in reducing emissions at ports. Secondly, many scholars choose to use the evolutionary game method to study the dynamic game between stakeholders, but the traditional evolutionary game method lacks the construction of the subjective cognitive level of the participants, and some scholars combine the prospect theory with the evolutionary game to introduce the decision maker's attitude to risk in the prospect theory so that the decision maker's perception effect of profit and loss has an impact on the decision. This makes the analysis of behavioral decisions of subjects and objects more realistic.

2 MODEL ASSUMPTIONS AND ESTABLISHMENTS

In the port emission reduction supervision system, the local government and port enterprises aim to maximize their respective interests, and the choice of behavior decision-making is risky. For the government, there is a need to worry about whether revenue can compensate for the financial pressure caused by the cost. Government fines need to be considered for port companies. This study uses game theory to analyze the behavioral decisions of the two, but the classical evolutionary game method still lacks the construction of the subjective cognitive level of the participants, especially the subjective perception of gains and losses by the local government and port enterprises, so the prospect theory is introduced.

According to the prospect theory proposed by Kahneman and Tversky (1979), the value function model is described as follows:

$$V(\Delta u) = \begin{cases} (\Delta u)^n & \Delta u > 0 \\ -\lambda(-\Delta u)^n & \Delta u < 0 \end{cases} \tag{1}$$

where Δu represents the actual benefits of the decision maker and Δu_0 is a reference point for the decision maker. $\lambda(\lambda \geq 1)$ is the loss avoidance coefficient. $n \in (0, 1)$ is the marginal sensitivity of decision makers to benefits; the greater the value, the more stakeholders prefer risk.

Without considering other external factors, it is assumed that the participating groups of port emission reduction groups are local governments and port enterprises, of which the local government's strategy is "supervision" and "not supervision" and the port enterprise's strategy is "emission reduction" and "non-emission reduction". Both local governments and port companies have limited rationality.

Under the total amount control system, the local government sets a certain limit on the carbon emissions of the port. Assuming the standard amount of pollution discharge s; the output of port enterprises is q; port enterprises do not reduce the amount of pollutant discharge s_1; the amount of sewage discharged by port enterprises to reduce emissions is s_2.

Suppose that regardless of the local government's choice of regulation, when port companies choose to reduce emissions, the local government can reap the benefits R. When the local government chooses to regulate, the local government needs to pay the cost of supervision C. When the local government chooses to supervise and the port enterprise does not reduce emissions, the local government punishes the port enterprise for exceeding the limit of pollutant discharge $t_1(s_1 - s)q$,

138

where t_1 is a carbon tax stipulated by the local government. When local governments choose not to regulate, they bring credibility loss to the government W.

Suppose that when port enterprises choose to reduce emissions, they need to invest in emission reduction costs, and it is secondary to the emission reduction. The emission reduction coefficient is assumed k, and the cost of reducing emissions for port enterprises is $k(s_1 - s_2)^2$.

Suppose that ENGO, as a social organization, can supervise and evaluate the pollutant discharge behavior of port enterprises, assuming that when the government department adopts non-supervision, ENGO finds that when the carbon emissions of port enterprises exceed the standard, it will report the violations of port enterprises to the relevant departments ∂; port companies will be punished accordingly $\partial t_2(s_1 - s)q$, where t_2 is a carbon tax punished by the local government for port enterprises that exceed emissions after the port enterprise is reported.

Based on the above assumptions, the perceived benefit matrix of local governments and port enterprises is shown in Table 1.

Table 1.   Revenue matrix

	Port	
	---	---
Government	Emission reduction y	Non-emission reduction $1 - y$
Supervision x	$R - C$ $-k(s_1 - s_2)^2$	$-C + V(t_1(s_1 - s)q)$ $V(-t_1(s_1 - s)q)$
Not supervision $1 - x$	$-\lambda(W)^n$ $-k(s_1 - s_2)^2$	$V(\partial t_2(s_1 - s)q) - \lambda(W)^n$ $V(-\partial t_2(s_1 - s)q)$

In Table 1, we examine the degree of risk appetite and loss avoidance of the local government and the port in the game, construct a game model based on the prospect theory and add the value function to the profit and loss of the game between the local government and the port for analysis, and establish the game gain perception matrix based on the value function. In the revenue matrix, $-\lambda(W)^n$ represents the perceived utility of the local government's reputational loss, which is the government's psychological loss; $V(-\partial t_2(s_1 - s)q)$, $V(-t_1(s_1 - s)q)$ is the perceived utility of fines for port enterprises, and it is the psychological loss of port enterprises; $V(t_1(s_1 - s)q)$, $V(\partial t_2(s_1 - s)q)$ is the perceived utility of fines collected by the government, which is the psychological benefit of the government.

The calculations allow local governments and port companies to replicate dynamic equations $F(x), F(y)$:

$$F(x) = (-1 + x)x(C + V(\partial t_2(s_1 - s)q) - \lambda(W)^n - (R - V(t_2(s_1 - s)q))y$$
$$+ V(t_1(s_1 - s)q)(-1 + y))$$
$$F(y) = (k(s_1 - s_2)^2 + V(-\partial t_2(s_1 - s)q) + V(-t_1(s_1 - s)q)x$$
$$- V(-\partial t_2(s_1 - s)q)x(-1 + y)y \tag{2}$$

According to the basic assumptions of evolutionary game theory, the participants are not completely rational but bounded rationality. This means that they cannot choose the best strategy at first, and they change their strategy over time. When participants go through multiple games without changing their strategy, the replicated dynamic system is stable. An evolutionary stabilization strategy (ESS) is the combination of strategies used by all participants. To get the steady state and ESS of the replicated dynamic system, $F(x) = 0$ and $F(y) = 0$.

By solving the equations, the equilibrium points $(0,0)$, $(0,1)$, $(1,0)$, $(1,1)$, and a mixed strategy point of the game system can be obtained ($\frac{-k(s_1-s_2)^2+V(-\partial t_2(s_1-s)q)}{-V(-t_1(s_1-s)q)+V(-\partial t_2(s_1-s)q)}$, $\frac{C-V(t_1(s_1-s)q)+V(\partial t_2(s_1-s)q)-\lambda(W)^n}{R-V(t_1(s_1-s)q)+V(\partial t_2(s_1-s)q)}$).

3 STABILITY ANALYSIS OF EQUILIBRIUM POINTS

The equilibrium point obtained by the replication dynamic equation is the local asymptotic stability point in the process of population evolution, not necessarily the evolutionary stability strategy point (ESS) of the system, according to the judgment method proposed by Friedman, the stability analysis of the equilibrium point of the system can be obtained by analyzing the local stability analysis of the Jacobian matrix of the system, for which the Jacobian matrix of the replicated dynamic equation is:

$$\begin{bmatrix} (-1+2x)(C+V_2-\lambda(W)^n+V_1(-1+y) \\ -(R+V_2)y) & -(R-V_1+V_2)(-1+x)x \\ (V_3-V_4)(-1+y)y & k(S_1-S_2)^2+V_4+V_3x-V_4x)(-1+2y) \end{bmatrix} \quad (3)$$

Based on Lyapunov's theory of system stability (Vadali 2015). According to this theory, the eigenvalues of the corresponding matrix of the system help to judge the stability of the system. If all the eigenvalues of the matrix are negative, the system is stable; if all the eigenvalues of the matrix are non-positive, and the eigenvalues equal to 0 have no multiple roots, it is determined that the system is stable in Lyapunov; otherwise, the system is unstable. Calculate the determinant and trace of the Jacobian matrix corresponding to the five equilibrium points, as shown in Table 2.

Table 2. Determinants and traces of a Jacobian matrix of five equilibrium points.

Point	λ_2	λ_2
E_1 $(0,0)$	$-C+V_1-V_2+\lambda(W)^n$	$-k(S_1-S_2)^2+V_4$
E_2 $(0,1)$	$-C+R+\lambda(W)^n$	$k(S_1-S_2)^2+V_4$
E_3 $(1,0)$	$C-V_1+V_2-\lambda(W)^n$	$-k(S_1-S_2)^2-V_3$
E_4 $(1,1)$	$C-R-\lambda(W)^n$	$k(S_1-S_2)^2+V_3$
E_5 (x^*, y^*)	Φ	0

Note: $\Phi=-\frac{(k(S_1-S_2)^2+V_3)(k(S_1-S_2)^2+V_4)(C+V_1+V_2-W)(-C+R+W)}{(R+V_1+V_2)(V_3-V_4)}$, $V_1=V(t_1(S_1-S)q)=(t_1(S_1-S)q)^n$, $V_2=V(\partial t_2(S_1-S)q)=(\partial t_2(S_1-S)q)^n$, $V_3=V(-t_1(S_1-S)q)=-\lambda(t_1(S_1-S)q)^n$, $V_4=V(-\partial t_2(S_1-S)q)=-\lambda(\partial t_2(S_1-S)q)^n$

For the evolutionary game model constructed by port emission reduction, the stability analysis is carried out, because the mixed strategy point (x^*, y^*) is Tr = 0, so the point is not an equilibrium point, so the remaining four points discuss the conditions for the existence of evolutionary stability strategy, and the following conclusions are obtained.

According to Table 2, it is found that E_1 and E_3, E_2 and E_4 are opposites λ_1, E_1 and E_2, E_3 and E_4 are opposites λ_2. In other words, when E_1 is a stable point, other points cannot become stable points and vice versa. By comparing the characteristic values of these four points, E_4 works best. In this case, local government supervision can reach port emission reduction. However, considering the participation of third parties, the local government may not supervise (the situation of E_3), so this study will discuss this state and analyze how to promote the system to reach an optimal state.

When E_3 is the final stabilization point, all its eigenvalues must be negative. Therefore, the model needs to be added to the inequality groups as a constraint.

$$\begin{cases} C-R>\lambda(W)^n \\ k(S_1-S_2)^2-\lambda(\partial t_2(S_1-S)q)^n<0 \end{cases} \quad (4)$$

4 NUMERICAL SIMULATIONS

This study mainly considers the role of the carbon tax system on port emission reductions in the context of prospect theory and considers the impact of ENGO's participation in port emission reductions. According to the above analysis, this article chooses to simulate the changes in both when port enterprises reduce emissions and the government does not supervise. This section is therefore satisfied E_3. The effects of t_1 t_2, W, k, ∂ and n on the behavioral strategies of local governments and port enterprises were analyzed. A replication dynamic system could be impacted by the initial probability of a party selecting a different strategy.

(1) Carbon tax t_1 and t_2.

To reflect the impact of the carbon tax on the replication dynamic system, we considered the impact of the carbon tax change on the participants, using the carbon tax and the initial probability as the amount of change, and simulated the system.

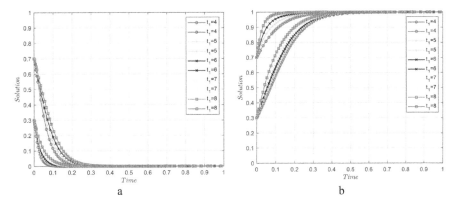

Figure 1.   Sensitivity analysis of t_1.

From Figure 1, it can be found that the strategy choice of the local government and port enterprises is (no regulation and emission reduction) regardless of the initial probability. From Figure 1a, it can be found that when the initial probability is low, the system reaches a stable state faster than when the initial probability is high, and as t_1 keeps getting larger, the government tends to stabilize slower. It can be seen from Figure 1b that regardless of the high or low initial probability. There is little impact on the port enterprises, which will choose the reduction strategy regardless of the initial probability and the system reaches a steady state quickly.

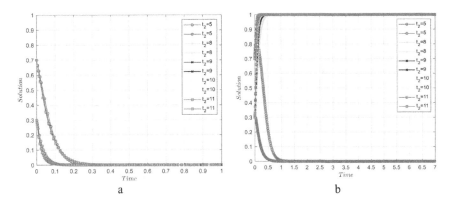

Figure 2.   Sensitivity analysis of t_2.

From Figure 2, it can be seen that the strategy choice of the local government and the port companies is (no regulation or abatement) regardless of the initial probability. From Figure 2a, we can find that the system reaches stability faster when the initial probability is low than when the initial probability is high, and the larger t_2 is, the faster the government reaches stability. According to Figure 2b, it can be found that regardless of the initial probability of high or low, when t_2 is small, port enterprises will choose not to reduce emissions, and as the value of the carbon tax increases, port enterprises change their strategy and choose to reduce emissions, and the system eventually tends to stabilize to 1.

(2) Emission reduction coefficient

Using the emission reduction factor as a variable, the evolution of port behavior is simulated, as shown in Figure 3.

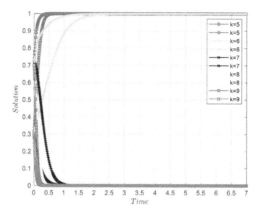

Figure 3.   Sensitivity analysis of emission reduction k.

As can be seen from Figure 3, k has a greater impact on port behavior decisions. Regardless of the probability when the cost coefficient is small, port enterprises will tend to reduce emissions at this time. When the emission reduction coefficient increases, port enterprises will have to invest a greater amount of money in emission reduction. After exceeding the cost budget of port enterprises, port enterprises tend to "not reduce emissions" currently. The faster the value of the port increases, the more port enterprises are inclined to adopt a "no emission reduction" strategy.

(3) Reputation of the local government

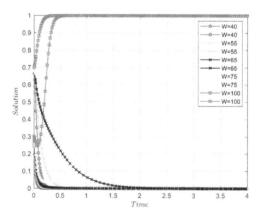

Figure 4.   Sensitivity analysis of local government reputation W.

From Figure 4, it can be seen that W has a greater impact on the behavior strategy of local governments. Regardless of the initial probability, the local government's strategy will shift from "no regulation" to "regulation". At the same time, when W increases to a larger size, the local government changes its strategy because W is a prospect value, which is affected by the local government's risk attitude coefficient and loss avoidance coefficient, and the local government's judgment of reputation loss has psychological factors.

(4) Probability of discovery

To illustrate the impact of ENGO's involvement in port emissions reductions, we suppose that ENGO finds that the probability of ports not reducing emissions is ∂. We take it as a variable and simulate the evolution of the system. The results are shown in Figure 5.

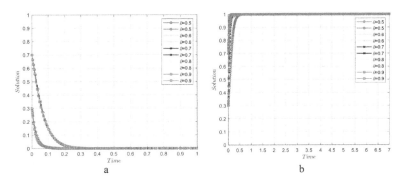

a   b

Figure 5.   Sensitivity analysis of ∂

From Figure 5, it can be seen that the stable state of the local government and port enterprises is (unregulated, and reduction in emissions). From Figure 5a, it can be seen that regardless of the initial probability, the stable state of the local government's strategy tends to be 0, and the stable state of the system tends to be faster when the initial probability is low than when the initial probability is high. As can be seen from Figure 4b, although overall ∂ does not have a greater impact on the strategy of port enterprises, it can still be seen from the figure that regardless of the initial probability, as ∂ continues to grow, the system tends to stabilize at 1.

(6) Risk attitude coefficient sensitivity analysis

It can be seen from the prospect theory that it is the risk attitude coefficient of the decision-making subject, and its value size represents the degree of risk preference of the subject participating in the decision-making. The impact of n on the behavior of governments and port companies are shown in Figures 6 and 7.

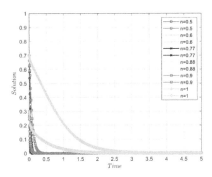

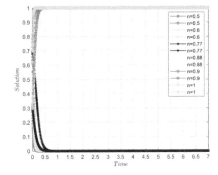

Figure 6. Sensitivity analysis of government n.   Figure 7. Sensitivity analysis of port n.

As can be seen from Figure 6, the local government eventually tends to be stable at 0 regardless of the initial probability of the local government. As n continues to increase, the local government reaches a stable state at a slower rate. It can be seen from the figure that as n increases, the speed of local government tends to stabilize and becomes slower, indicating that when the risk preference for local government is small, even when the risk preference reaches a large amount, local government still will not change its strategy. If people wish to improve the local government's risk preference, they may increase taxes or take other measures to encourage them to change the current strategy. It can also be seen that when n is equal to 1, the curve has a more obvious difference compared with other curves.

From Figure 7, we can see that n has a greater influence on the pollution of port enterprises. As can be seen from the figure, regardless of the probability of high or low, the steady state tends to zero at the beginning, but as n increases, the steady state tends to 1 and the time to reach stability becomes shorter. The port enterprise at first did not choose to reduce emissions, but when the risk attitude n increases, the greater the marginal progression of the perceived value of the port on profit and loss the port enterprise will face at this time. The port enterprise will then choose to "reduce emissions". In the case where n is equal to 1, the curve has a very different trend than other curves.

5   CONCLUSIONS

In this paper, the game behavior between local government regulation and port enterprise emission reduction is studied using evolutionary games to explore the cost of port enterprise regulation, local government reputation loss, the probability of carbon tax and ENGO discovery, members' risk attitude, and loss avoidance coefficient for analysis, and the following conclusions are drawn.

(1) Different initial probabilities have an impact on the time for the system to reach the steady state, and variables such as initial probabilities have a joint impact on the system. The time to reach stability varies for systems with different initial probabilities, and the joint influence of initial probabilities and variables can make the participants change their strategies.
(2) For local governments, the magnitude of the carbon tax value and the probability of discovery will only affect the trend of the local government's strategy choice. Changes in the value can make a difference in the length of time it takes for a local government to reach a steady state. Reputation loss has a decisive impact on the government. The larger the loss of reputation, the more the local government will choose to reduce emissions.
(3) For port enterprises, the carbon tax value t_2 has a strong influence on their decision-making, and as t_2 increases, port enterprises tend to reduce emissions. The reduction coefficient is decisive for port enterprises' decision-making, and the larger the reduction cost is, the more port enterprises tend to "not reduce emissions". Second, the probability of ENGO discovery only affects the stable trend of port enterprises' behavioral decisions. When the probability of discovery becomes larger, it can promote port enterprises to reach a stable state of "abatement" more quickly.
(4) The risk loss aversion factor has an impact on port enterprises, and all port enterprises will choose to "reduce emissions". This indicates that high loss-sensitive enterprises will choose to "reduce emissions". The loss aversion factor has a decisive influence on the government, and the high loss-sensitive government will choose to regulate. Risk attitude has a great influence on the behavioral decision of port enterprises. When the risk attitude coefficient becomes large, port enterprises will choose the stabilization strategy of "abatement". Risk attitude does not have a large impact on the government, but only affects the speed of stabilization of the local government, the less risk attitude, the faster the government tends to stabilize.

REFERENCES

B. Meng, H. Kuang, E. Niu, J. Li, and Z. Li. (2020). "Research on the transformation path of the green intelligent port: outlining the perspective of the evolutionary game "Government-Port-Third-Party organization, *sustainablity*, vol. 12, no. 19, p. 8072.

Bao Jiang, Xiaoqiong Wang, Hailiang Xue, Jian Li, Yu Gong. (2020). An evolutionary game model analysis on emission control areas in China, *Marine Policy*, Volume 118,104010.

Juqin Shen, Xin Gao, Weijun He, Fuhua Sun, Zhaofang Zhang, Yang Kong, Zhongchi Wan, Xin Zhang, Zhichao Li, Jingzhe Wang, Xiuping Lai. (2021). Prospect theory in an evolutionary game: Construction of watershed ecological compensation system in Taihu Lake Basin, *Journal of Cleaner Production*, Volume 291, 125929

Kahneman D Tversky A. (1979). Prospect theory: an analysis of decision under risk [J] *Econometrica* 47 (2): 263–292.

Qin Zhou, Kum Fai Yuen. (2021). Low-sulfur fuel consumption: Marine policy implications based on game theory, *Marine Policy*, Volume 124, 104304.

Sehwa Lim, Stephen Pettit. (2019). Wessam Abouarghoub, Anthony Beresford, Port sustainability and performance: A systematic literature review, Transportation Research Part D: Transport and Environment, Volume 72, Pages 47–64, ISSN 1361–9209.

Tan, D. Q., Xu, Hao. (2020). An evolutionary equilibrium analysis of environmental pollution group events based on prospect theory [J]. Operations Research and Management, 29 (05): 161–170.

Vadali, S. R., Kim, E. S. (2015). Feedback control of tethered satellites using Lyapunov stability theory. J. Guid. Contr. Dynam. 14 (4), 729e735.

Xiaole Wan, Shuwen Xiao, Qianqian Li, Yuanwei Du. (2021). Evolutionary policy of trading of blue carbon produced by marine ranching with media participation and government supervision, Marine Policy, Volume 124, 104302.

Xiaoyao Zhao, Lin Liu, Zhongjie Di, Lang Xu. (2021). Subsidy or punishment: An analysis of evolutionary.game on implementing shore-side electricity, Regional Studies in Marine Science, Volume 48, 102010.

*Advances in Energy, Environment and
Chemical Engineering – Abdullah & Osman (Eds)
© 2023 The Author(s), ISBN: 978-1-032-36083-6*

Analysis of governance modes of coal mining subsidence in China

Yong Chang
China Three Gorges Corporation, Beijing, China
Hohai University, Nanjing, China

Yaoping Bei, Binqing Yuan*, Zhousheng Cao, Dongmei Cao & Shuangquan Zhu
China Three Gorges Corporation, Beijing, China
China Three Gorges Renewables (Group) Corporation, Beijing, China

ABSTRACT: The large area of land subsidence caused by coal mining has a strong impact on the living of local people and the ecological environment. The efficient strategy to govern the coal mining subsidence area according to their characteristics is the key to local sustainable development. Currently, there are lots of successful governing cases, but there is still a lack of a summary of relevant governance methods and experiences. Based on the previous research and experiences, we analyze three potential factors that may influence the selection of governance modes of the coal mining subsidence. Then, five typical governance modes and their application conditions are further summarized. Based on the spatial distribution of the coal mining area, regional geographical environment, and economic development level, the coal mining area in China is divided into five different regions and the optimal governance model for each region is proposed. These results can serve as good guidance for the governance planning of the coal mining subsidence area in China.

1 INTRODUCTION

China is a big energy-consuming country. Under the energy structure of "rich coal, lean oil, and little gas" in my country, coal resources, as the main basic energy, have made outstanding contributions to the rapid development of the national economy in the past. According to the *China Statistical Yearbook 2020*, Chinese annual coal production increased from 1.080 billion tons in 1990 to 3.698 billion tons in 2018. Although the country has vigorously developed clean energy in recent years, the proportion of coal in consumption energy has gradually dropped from a high of 72.5% in 2007 to 59% in 2018. However, based on the natural endowments and strategic security attributes of coal, it is still difficult for China to change the coal-dominated energy structure for a long time in the future (Wang, 2018). The development of coal resources has supported the rapid development of the Chinese economy in recent years. Meanwhile, large-scale coal mining has also seriously damaged the ecological environment of coal mining areas (Duan 2020; Jia et al. 2020; Lei & Bian 2014; Liu & Gu 2006), induced a series of social problems, and seriously hindered regional social and economic development.

Compared with other countries in the world, the geological mining conditions of coal resources in China are relatively poor, most of which are well-mined. And the mining volume accounts for 95% of the total coal resources (Zhang & Zhang 2006). Long-term well mining will destroy the stress balance of the underground rock formation, causing deformation, displacement, and slump of the surrounding rock. Finally, the deformation or displacement will gradually extend to the surface to form a large-scale ground subsidence area, causing serious damage to the groundwater resources,

---

*Corresponding Author: 1918192610@qq.com

DOI 10.1201/9781003330165-21

surface soil, and vegetation in the area. In the 1980 s and 1990 s, China began to pay attention to the governance of coal mining subsidence areas, from simple land reclamation in the early stage to ecological environment restoration, and then gradually developed into a comprehensive governance method that considers the sustainable development of society, economy, and ecological environment (Liu & Lu 2009). Although China has carried out a lot of research and successful cases on the restoration and treatment of subsidence areas, most of the cases are limited to a certain area. For complex projects, selecting an appropriate development and governance model according to local conditions is the key to the sustainable development of the region. If the governance model is blindly copied, the governance effect is often counterproductive. In this regard, this paper analyzes and summarizes the potential factors affecting the selection of the coal mining subsidence management mode, and the existing coal mining subsidence management mode. The application of these conditions must be based on a thorough investigation of the relevant theories and practices of coal mining subsidence area control at the earliest possible stage. And on this basis, the applicable subsidence area governance mode is put forward according to the specific characteristics of different coal mining areas in my country.

2 DEVELOPMENT AND MANAGEMENT MODE OF COAL MINING SUBSIDENCE AREA

2.1 *Potential factors affecting the choice of governance model*

The selection of governance model in the coal mining subsidence area is affected by various factors. Bian (1999) proposed that the zoning affecting the reclamation conditions of the mining area should be dominated by natural geographical conditions. Meanwhile, the dry and wet conditions, topographic features, coal seam occurrence conditions, and development intensity of the mining area should be considered. Li (2019) pointed out that in addition to considering the natural geographical conditions, the comprehensive management of the subsidence area should also consider the social and economic factors, including the regional population, land use, and economic development of the mining area. In addition, the degree of ground subsidence in the coal mining area will also have a significant impact on local topography and land use patterns (Pang et al. 2018). Based on a comprehensive analysis, the selection of the governance model for the coal mining subsidence area should mainly consider three factors: natural geographical conditions, economic conditions, and local collapse degree in the subsidence area.

(1) Natural geographical conditions

The physical and geographical conditions are the basic conditions to be considered in the treatment of subsidence areas, including climatic conditions and topography. The climatic conditions of the subsidence area determine the abundance of rainfall and dry and wet conditions in the area. With abundant rainfall, the ecological environment in the humid area is easy to restore, and the area is easy to grow commercial crops. Therefore, the subsidence area should focus on farmland reclamation or industrial development. It is difficult to restore the ecological environment in arid and semi-arid areas. And land subsidence can easily lead to vegetation decline and surface desertification, further deteriorating the originally fragile ecological environment. Topographical conditions also affect the choice of governance mode for subsidence areas. Plain or hilly areas are suitable for land reclamation, while mountainous areas are only suitable for forestry or tourism development.

(2) Regional economic conditions

Regional economic conditions are important supporting conditions for the management of subsidence areas. On one hand, the economic conditions of the subsidence area determine the economic value of the land in the subsidence area. The stronger the economic conditions, the higher the industrial development and utilization value of the subsidence area. While the areas with poor economic conditions may only need to focus on simple ecological environment restoration. On the other hand, the development of tourism resources in the subsidence area requires a large amount of initial investment or later maintenance costs, and requires the area to have strong economic support capabilities.

(3) Degree of surface subsidence

The degree of surface subsidence is another important factor affecting the management of subsidence areas, which can be generally divided into three categories according to the degree of collapse (Pang et al. 2018): mild collapse, moderate collapse, and severe collapse as shown in Table 1. Mild subsidence generally has no stagnant water. After land remediation or reclamation, it can be used for crop planting, ecological environment restoration, or photovoltaic and wind power generation (green energy development). Moderate subsidence or severe subsidence is not suitable for planting crops due to the large slope. According to whether there is surface water in the subsidence area, there are certain differences in the governance model. When there is stagnant water in the subsidence area, the treatment of the subsidence area is more focused on water-related treatment, such as fish farming or ecological landscape construction. While the development and utilization value of the subsidence area in the area without water accumulation is relatively small, and the focus is on ecological environment restoration.

Table 1. Classification of subsidence degree of coal mining subsidence area.

Type	Degree of collapse
Mild collapse	Within 1.5 meters, no stagnant water
Moderate collapse	Between 1.5 and 3.0 meters; seasonal stagnant water or perennial water depth between 1.5 and 2.5 meters
Severe collapse	More than 3 meters; annual water accumulation of more than 2.5 meters

2.2 Management mode and applicable conditions of the subsidence area

There have been many successful cases of subsidence control and development in different regions of China. However, due to the differences in natural and geographical conditions, economic development levels, and the degree of subsidence of mining areas in different regions, coal mining subsidence areas are managed in various ways in different regions. At present, there is no unified treatment system for subsidence areas. By summarizing and analyzing the governance achievements and experiences of typical coal mining subsidence areas at home and abroad, the main governance modes include the following strategies (see Table 2).

(1) Agricultural governance mode

It is the most common governance model for subsidence areas, also known as the land reclamation mode. This model restores the land in the subsidence area to a state that can be re-cultivated and develops the agricultural economy by leveling and draining the subsidence area, digging deep and shallow padding, or filling it with coal gangue and fly ash (Li 2014). This model is simple and economical with low treatment cost, which can effectively supplement the cultivated land area in the subsidence area and alleviate the contradiction between man and land in the mining area.

(2) Breeding industry development mode

This model makes full use of many water resources in the subsidence area. It excavates and builds dams, or shapes and dredges, reforms the water intake and drainage systems in the long-standing water area in the subsidence area, and plans a large area of aquaculture waters (Li & Zhang 2019; Wang 2018). Fish and shrimp are cultivated in large-area deep water areas. Poultry such as ducks and geese are cultivated in shallow water areas. Economic crops (such as lotus root, watermelon, water chestnut, etc.) are cultivated to promote the increase of economic effects in the subsidence area. This governance model does not require a large amount of economic cost investment. It is suitable for areas with humid conditions and abundant rainfall, and plain and hilly areas with moderate or severe subsidence.

(3) Ecological governance mode

This refers to the planting of trees on the restored land, which plays the role of ecological restoration in subsidence areas, preventing sand and dust, and purifying the air [10]. The afforestation method is simple to operate, but the economic effect is not significant. This mode is mainly used in weak economic areas with low land value, uncultivated mountainous areas, or arid and semi-arid areas with scarce rainfall.

(4) Tourism development mode

Based on ecological restoration in the subsidence area, this mode includes measures to comprehensively transform the waters, mountains, or slopes in the subsidence area. And the water system dredge and the water quality purify. Moreover, it builds natural landscapes, plank roads, viewing towers, and basic service facilities through engineering construction. Various ornamental plants are cultivated, and ecological wetlands or mine parks with beautiful landscapes and a combination of water and land around the subsidence area are built (Ding et al. 2020; Li et al. 2020; Zhang 2019). The development of tourism resources in the subsidence area requires a lot of economic investment in the early stage and later operation and maintenance, so this model requires strong economic support conditions in the region.

(5) Green energy development model

This model has gradually developed with the continuous investment in green solar energy and wind energy by the state in recent years. It refers to the further construction of photovoltaic or wind energy power stations based on ecological restoration in the subsidence area, making full use of the land resources in the subsidence area. This model requires that the governance area has abundant solar or wind energy resources, and the land value is relatively cheap to reduce the cost of solar or wind energy generation. This model can also be combined with the aquaculture management model to build floating photovoltaic power stations on the water surface of moderate-to-severe subsidence water areas (Liang et al. 2019). For example, the floating photovoltaic power station built by *Three Gorges New Energy Co., Ltd.* locates in the Panji coal mining subsidence area in Huainan, Anhui Province. The research results in this area show that the appropriate photovoltaic panel coverage (about 50%) on the collapsed water body will not harm the water body's ecological environment, but it can prevent the water body from eutrophication to a certain extent. In addition, fish and shrimp can be cultivated under the floating photovoltaic power station to achieve "fishing and light complementation", making full use of the idle water surface, and reducing the cost of photovoltaic power generation. With China's long-term plan for carbon neutrality and carbon peaking, this subsidence area governance model may be used more frequently.

3   ZONING OF COAL MINING SUBSIDENCE CONTROL IN CHINA

3.1   *Distribution and mining status of coal resources in China*

China's coal reserves are abundant, but the spatial distribution is extremely uneven (China Energy Resources Report 2016), showing a spatial pattern of more in the north and less in the south, and more in the west and less in the east as shown in Figure 1. Coal reserves in the area north of Kunlun-Qinling-Dabie Mountains account for about 90% of national reserves. And it is concentrated in the area between Taihang Mountain and Helan Mountain, which is also the main coal mining area in North China. The coal resources in southern China are mainly concentrated in Yunnan, Guizhou, and Sichuan provinces, accounting for 90% of the total coal reserves in southern China. According to the *National Mineral Resources Planning 2016-2020*, China focuses on building 14 large-scale coal bases (Ding et al. 2020) as shown in Figure 1, namely Shendong, Jinbei, Jinzhong, Jindong, Mengdong (Northeast), Yungui, Henan, Luxi, Lianghuai, Huanglong, Jizhong, Ningdong, northern Shaanxi and Xinjiang bases. These coal bases are distributed in different provinces and regions in China, with coal mining rights accounting for 84.67%. The coal mines in production account for 76.31%. The actual coal production capacity of the coal base accounts for 89.09% of China's total

Table 2. Governance mode of coal mining subsidence area and its applicable conditions.

Governance mode	Climatic conditions	Topography	Economic conditions	Degree of surface subsidence
Agricultural governance mode	Rainy and humid	Plains or hills	–	Mild collapse
Breeding industry development model	Rainy and humid	Plains	–	Moderate collapse
Ecological governance model	Arid and semi-arid areas	Mountainous areas	Weak economic areas	–
Tourism development mode	–	–	Medium to strong economic regions	Moderate or severe subsidence (for eco-agricultural tourism, mild subsidence areas are applicable)
Green energy development model	Abundant solar and wind energy resources	Plains or hills	–	Mild collapse (floating photovoltaic power plants can be built on the water surface of moderately or severely collapsed water areas)

Note: "–" means no requirement for this condition.

actual production capacity, of which the main production areas are Inner Mongolia, Shanxi, and Shaanxi. In general, the distribution of these large coal bases can represent the main coal mining areas in China and the areas where coal mining subsidence exists or may occur in the future.

3.2 *The division of coal mining subsidence control area*

Since the main coal bases in China are distributed in different provinces, the natural geographical conditions of the regions where the coal bases are located and the economic conditions of each province are very different, so the optimal governance mode of coal mining subsidence areas in different regions should also have certain differences. According to the applicable conditions of different subsidence area governance modes in Section 2.2, the optimal governance mode planning of coal mining subsidence areas in different regions of China can be considered from two aspects: regional characteristics and local characteristics. On the one hand, regional planning mainly considers the climatic conditions, topography, and economic development level of the coal mining subsidence area. The optimal governance and development mode should be selected according to conditions such as urban and rural economic differences.

From a regional perspective, China is divided into different regions according to the differences in climatic conditions, topography, and economic development levels. The differences in climatic conditions across the country are mainly divided into humid and semi-humid regions and arid and semi-arid regions according to the demarcation line of 400 mm rainfall over the years, as shown in Figure 2(a). The topographical differences are mainly divided according to the three major steps in China. The east of the Daxinganling-Taihang Mountains-Wushan-Xuefeng Mountains is mainly distributed with hills, low mountains, and plains interlaced (Zhao et al. 1995) with flat terrain, as shown in Figure 2(b). The west area is mainly mountains and plateaus or basins. Except for some large basins, the terrain fluctuates greatly. The economic conditions of each province in the country refer to the average ranking of the gross domestic product (GDP) of each provincial administrative region from 2011 to 2020. The provinces whose 10-year average GDP in the ranking is smaller than Yunnan Province are defined as weak economic regions, as shown in Figure 2(c). Due to the

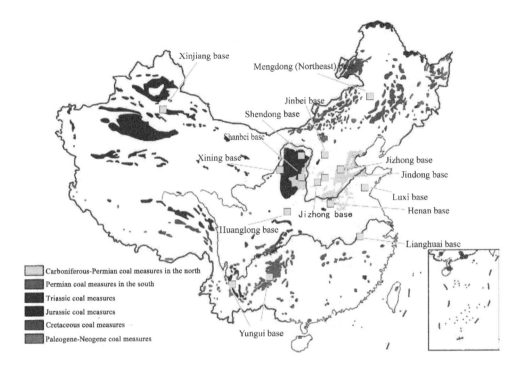

Figure 1.   Distribution of coal resources in China [18] (Source: China Energy Resources Report).

small area of the municipality directly under the Central Government, although the average GDP ranking of Tianjin is also behind Yunnan Province, its low GDP is mainly related to the area of the municipality directly under the Central Government. Finally, the coal mining subsidence area in China can be divided into 8 different areas, as shown in Figure 2(d), but three of the areas (V, VI, and VII) are only sporadically distributed and there is no concentrated coal mining area in the area. Therefore, there are 6 zones (see Table 3) in total. For the treatment of a specific coal mining subsidence area in each area, the degree of subsidence of the local subsidence area and whether it is in the periphery of the city or the township is further considered.

(1) Area I mainly includes Lianghuai Base, Luxi Base, Henan Base, and Jizhong Base. These coal bases are in humid or semi-humid areas, with relatively abundant rainfall and flat terrain. The economic development level of each mining area is relatively good. Therefore, the governance of coal mining subsidence areas should give priority to agricultural governance, aquaculture, or tourism development. Part of the idle land can also adopt the green energy development model. When dealing with a specific subsidence area, if coal mining subsidence mainly occurs in the inner or suburban areas of cities with strong economic development, the tourism development mode should be firstly considered because the area has strong economic conditions and urban greening construction needs. In urban and rural areas, the economic level is relatively weak. In areas with weak ground subsidence, the agricultural governance mode should be considered first. Aquaculture should be developed in areas with moderate to high ground subsidence. In some areas with large subsidence waters, water surface photovoltaic power plants can be introduced. The green energy development model can be used to realize "fishing and light complementation" and further improve the value of land development and utilization in the subsidence area.

(2) Area II contains most of the area of the eastern Mongolian base. The area is in a humid or semi-humid area, with relatively abundant rainfall and flat terrain, but relatively weak economic conditions. The management of coal mining subsidence areas should give priority to

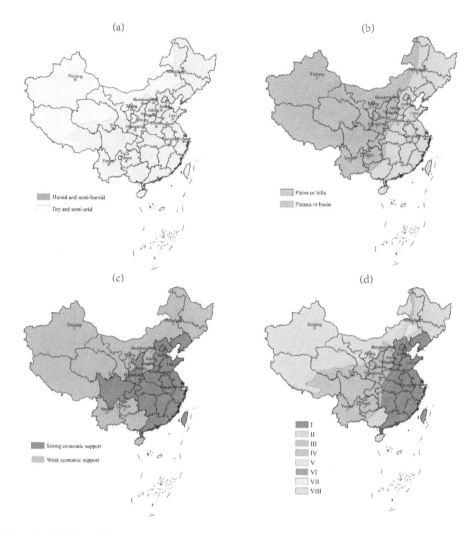

Figure 2. Division of development and governance of coal mining subsidence areas in China (I – humid and semi-humid + plains or hills + strong economic support; II – humid and semi-humid + plains or hills + weak economic support; III – humid and semi-humid + plateau or basin + Strong economic support; IV – humid and semi-humid + plateau or basin + weak economic support; V – arid and semi-arid + plain or hilly + strong economic support; VI – arid and semi-arid + plateau or basin + strong economic support; VII – arid and semi-arid arid + plains or hills + weak economic support; VIII – arid and semi-arid + plateau or basin + weak economic support).

agricultural management, aquaculture development mode, or ecological management mode. If the coal mining subsidence area has a moderate or higher degree of subsidence and is rich in water resources, priority should be given to the aquaculture development mode or ecological management mode. While areas with less subsidence may give priority to agricultural management mode.

(3) Area III mainly includes the Huanglong base and northern Shaanxi base. It is a humid and semi-humid area with good economic conditions, but the terrain is mainly plateau mountains or basins. The ecological governance mode or tourism development mode should be given priority to the governance of coal mining subsidence areas in this area. In addition, most of the areas are rich in wind and solar energy resources, hence green energy development models can be adopted in areas with flat terrain and less collapse.

(4) Area IV includes the Yunnan-Guizhou, Jinzhong, Jindong, and Jinbei bases. This area is a humid and semi-humid area with relatively weak economic conditions. The terrain is dominated by plateau mountains or basins. The governance of coal mining subsidence areas should be dominated by ecological governance or green energy development. For some cities with a strong economic level in the region, the tourism development model can also be considered.

(5) Area VIII only includes the Xinjiang base. This area is an arid and semi-arid area with relatively weak economic conditions. The terrain is dominated by plateau mountains or basins. The ecological governance mode should be given priority in the governance of coal mining subsidence areas. And the areas with less collapse degree can adopt the green energy development mode.

It should be noted that the division of the economy of the provinces in this paper is mainly based on the GDP ranking. The division is subject to a certain degree of subjectivity which means it cannot fully and accurately reflect the actual economic level of each province. With the development of China's economy, the economic development speed of each province is also inconsistent, and the economic level of each province and city will also change significantly in the future, which may lead to certain changes in the final regional division.

4 CONCLUSION

The development and management of coal mining subsidence areas are restricted by the natural geographical conditions, regional economic conditions, and the degree of ground subsidence in the subsidence areas. The measure to develop and control these subsidence areas according to local conditions is related to local social stability and sustainable economic development. The climatic conditions, topography, and economic level of major coal mining areas in China are significantly different. It leads to those different modes that must be considered in the development and management of different coal mining subsidence areas. Based on summarizing the development and governance mode and applicable conditions of coal mining subsidence areas, the governance of coal mining subsidence areas in China is preliminarily divided into four categories according to the distribution of major coal mining areas in China, the natural geographical conditions, and economic development levels of different regions. According to the specific location and degree of collapse of the subsidence area, the optimal development and governance mode is given. The results can guide the regional development and governance planning of coal mining subsidence areas in China.

ACKNOWLEDGMENT

This work was supported by the Science Project launched by China Three Gorges Corporation (Grant No. 34007172).

AUTHOR INFORMATION

Chang Yong (born July 1987), male, Ph. D., is an associate researcher at the School of Earth Science and Engineering, Hohai University. He was born in Jingmen, Hubei, and mainly engaged in the research of hydrogeology and engineering geology.

REFERENCES

Bian Z. F. Study on zoning coal mining areas by considering land reclamation conditions in China [J]. *Journal of China University of Mining & Technology*, 1999, 28 (3): 37–42.

Ding G. F., Lv Z. F., Cao J. C., et al. Analysis on status quo of development and utilization of large-scale coal bases in China [J]. *China Energy and Environmental Protection*, 2020, 42 (11): 107–110, 120.

Ding S. H., He J. J., Cheng C., et al. Ecological restoration path in Huaibei coal mining subsidence area: a case analysis of Nanhu coal mining subsidence area [J]. *Anhui Agricultural Science Bulletin*, 2020, 26 (7): 139–140, 150.

Duan Q. B. On the impact of coal mining on the surrounding environment [J]. *Inner Mongolia Coal Economy*, 2020, 306 (13): 151–152.

Jia Z. Q., Lv X. Q., Hu S. H., et al. Discussion on the influence of coal resource development on the ecological environment in Gansu Province [J]. *Energy Environmental Protection*, 2020, 34 (5): 104–108.

Lei S. G., Bian Z. F. Research progress on the environmental impact from underground coal mining in the arid western area of China [J]. *Acta Ecologica Sinica*, 2014, 34 (11): 2837–2843.

Li C., Liu M. B., Wu J. Y., et al. Research progress of the ecological restoration in subsidence areas [J]. *Journal of Anhui Agricultural Sciences*, 2020, 48 (7): 35–37, 73.

Li J. L., Zhang J. L. Heze City scientifically plans fishery governance in coal mining subsidence areas with the concept of green development [J]. *Guide to Getting Rich from Fisheries*, 2019, 520 (16): 25–28.

Li J. M., Yu J. H., Zhang W. Z. Spatial distribution and governance of coal-mine subsidence in China [J]. *Journal of Natural Resources*, 2019, 34 (4): 867–880.

Li S. Z. Present Status and Outlook on Land Damage and Reclamation Technology of Mining Subsidence Area in China [J]. *Coal Science and Technology*, 2014, 42 (1): 93–97.

Liang T., Wu J. L., Mi W. J., et al. The application of floating photovoltaic power generation system in coal mining [J]. *Construction & Design for Project*, 2019, 419 (21): 142–145.

Liu F., Lu L. Progress in the study of ecological restoration of coal mining in subsidence areas [J]. *Journal of Natural Resources*, 2009, 24 (4): 612–620.

Liu Y. B., Gu R. X. Study on the environmental problems and countermeasures of coal mine-based cities in our country [J]. *Journal of Zaozhuang University*, 2006,23 (2): 95–99.

One Hundred Achievements of China Geological Survey [M], Geological Press, 2016.

Pang J., Song X. H., Zhou M. J. Comprehensive Management Model for Coal Mining Subsidence Area: A Case Study of Heze City [J]. *Journal of Shandong University of Finance and Economics*, 2018, 30 (4): 109–119.

Wang H. N. New perspectives on coal resources distribution pattern and its fundamental function in China [J]. *Coal Geology of China*, 2018, 30 (7): 5–9.

Wang Q. G. *An analysis of innovative development paths for comprehensive management of coal mining subsidence areas* [C]. Proceedings of the 2018 Chinese Society for Environmental Sciences Science and Technology Annual Conference (Volume 1), 2018: 789–792.

Zhang H. X. Discussion on the development mode and countermeasures of wetland ecotourism in coal mining subsidence area: a case study of Xuzhou Jiuli Lake wetland park [J]. *Tourism Overview* (Second Half Month), 2019, 291 (6): 178–179.

Zhang S. Q., Zhang Y. Z. Overview of our country's Coal Resources, Production, and Environment [J]. *Environmental Protection*, 2006, 7 (13): 53–57.

Zhao J, Chen Y. W., Han Y. F. *Chinese Physical Geography* [M]. Beijing: Higher Education Press, 1995.

Advances in Energy, Environment and
Chemical Engineering – Abdullah & Osman (Eds)
© 2023 The Author(s), ISBN: 978-1-032-36083-6

Dynamic response analysis of pipe string under perforating detonation transient impact load

Pengfei Sang*
*Engineering Technology Research Institute of PetroChina Southwest Oil and Gas Field Company,
China*

Youcheng Zheng
PetroChina Southwest Oil and Gas Field Company, China

Yu Fan, Yufei Li, Linfeng Lu & Lin Zhang
*Engineering Technology Research Institute of PetroChina Southwest Oil and Gas Field Company,
China*

ABSTRACT: For the study of the effect of the shot hole blast, the current focus on the casing and cement ring damage or strength changes, the shot hole blast transient impact load on the shot hole section of the pipe column is less involved in the study. In particular, the dynamic response process and safety of the tubular column under impact loading and narrow, complex wellbore boundary conditions are still lacking in China. In view of this, a dynamic response model of the pipe column of the perforation section under the transient impact load of the perforation hole was established based on finite element simulation software, and the strength analysis of the pipe column of the perforation hole section was carried out. The response law of vibration displacement, velocity, acceleration, and stress to transient impact load was obtained for the perforation section, and the effect of length and wall thickness of the perforation section on the strength safety of the column was investigated. The example analysis shows that the column of the perforation tube is subjected to alternating cycles of tensile and compressive loads in the axial direction. Periodic changes in the displacement, velocity, and acceleration of the tube column from time to time. The farther away from the packer (the closer to the source of the shot hole explosion), the greater the speed and acceleration amplitude of the tube column vibration, the closer to the packer, the smaller the amplitude of the tube column vibration displacement, and the greater the equivalent force of the tube column. The length or wall thickness of the shot section column increases, the smaller the equivalent force of the column, the higher the security of the column strength. The research results can provide a reference for the optimization of perforating parameters and the design of string structure. The damage to the string caused by the transient impact load of perforating detonation can be reduced to ensure the safety of perforating operation.

1   INTRODUCTION

As E&P continues to move toward ultra-deep, high-temperature, high-pressure wells, hole-shot completion operations are becoming increasingly complex. The current perforating technology is constantly pursuing large aperture and high-density perforation, which will increase the total explosion energy of the perforation hole, causing strong fluctuations in the perforation fluid and violent vibration of the perforation section pipe column. Oil and gas play in the western onshore

---

*Corresponding Author: 798272782@qq.com

DOI 10.1201/9781003330165-22   155

fields in China and the Fort Worth Basin, Arkoma Basin, and the Gulf of Mexico in the United States have also encountered shot-hole accidents. To understand the impact of perforating detonation load on perforating string and downhole tool safety, engineers and technicians have conducted theoretical and experimental studies. Among them, Schlumberger has developed the shot-hole tubular column optimization design technology, which optimizes the shot-hole parameters and shot-hole tubular column configuration by predicting the shot-hole blast load and tubular column strength safety to achieve the purpose of preventing damage to the shot-hole tubular column and downhole tools. Following the idea of Schlumberger, the literature took a high-temperature and high-pressure ultra-deep well perforation completion tubular column as an example and analyzed the relationship between the structure of the body of the well and the length of the tubular column in the perforation section and the stress intensity of the tubular column.

The effect of the shot hole on the strength of the pipe column mainly lies in the vibration of the pipe column caused by the shot hole blast, that is, the response of the pipe column to the shot hole impact load, due to the complexity of the problem, at present mainly relies on finite element numerical analysis and experiments to study. The literature analyzed the relationship between the longitudinal vibration displacement and velocity of the shot pipe column with time under the action of axial impact load. In literature, Pulchra perforating engineering software was used to carry out implicit and explicit dynamic analysis of perforating string, and it was believed that the perforating string at the lower end face of the packer was constrained and had the largest force during perforating. In the literature, a structural dynamic model of the tubular column was established, and the dynamic response of the column under the action of the shot-hole blast wave was analyzed by ANSYS software, and the curves of radial displacement, contact reaction force, frictional resistance and axial force of the column were obtained with the depth of the well. The literature obtained the output characteristics and loading law of the shot-hole blast load in the shot-hole section of the pipe column using experiments and dynamics simulation and conducted a study of the dynamic response law of the pipe column. In the literature, based on the blast shock wave theory, the transient dynamics analysis software DYTRAN was used to simulate the blast shock wave transmission process of the shot-hole bomb and determine the law of axial displacement of the shot-hole tube column with time. In the literature, ground tests on the pressure and acceleration response at the end of the shot-hole column under the effect of blast impact were conducted, and it was found that the shot-hole blast products formed repeated impact effects on the end face of the column, resulting in axial and radial vibrations of the shot-hole gun and the shot-hole column.

According to the existing research base, the author applied ANSYS finite element analysis software to establish the finite element model of the pipe column in the shot hole section. Transient response and stress intensity analysis of the pipe column in the shot hole section. The response law of the vibration displacement, velocity, acceleration, and equivalent force of the pipe column in the shot hole section to the impact load of the shot hole is obtained. It also investigates the influence of the length and wall thickness of the shot section column on the stress intensity of the column and provides a reference for the optimal configuration of the shot section column.

2 FINITE ELEMENT MODELING OF THE SHOT SECTION PIPE COLUMN

The impact load acting on the tube column by the shot-hole blast is a function of time, and the loading time is very short, generally within tens to hundreds of milliseconds. To analyze the transient dynamic response of the pipe column under rapidly changing impact loads, the transient dynamics analysis module in ANSYS Workbench was applied to perform finite element analysis of the pipe column in the perforation hole section. As shown in Figure 1, a 20-m section of the perforation hole below the packer was intercepted on the 88.9 mm × 9.52 mm P110 tubing transmission perforation column as the study object. The yield strength of the oil pipe material is 758 MPa, the modulus of elasticity is 206 GPa, the Poisson's ratio is 0.3, and the density is 7.85 g/cm^3. As shown in Figure 2, an eight-node hexahedral solid cell is used to mesh the pipe column by applying the mapping mesh division method, and the 20-m pipe column is divided into 24,160 cells with 169,200 nodes.

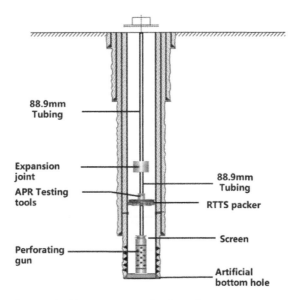

Figure 1. Schematic diagram of tubing transport perforating string.

The upper end of the pipe column of the perforation section cannot move due to the positioning effect of the packer, and the upper end of the pipe column of the perforation section is restrained by the fixed restraint condition. Shot hole blast impact load on the column of the action time is very short, and sudden changes in pressure are very large, it is difficult to use theoretical methods to get its accurate value. The wellbore pressure distribution during the perforation process is collected by a downhole perforation pressure monitor, and the data is extracted as the impact load on the tubular column, and the impact load is applied to the bottom end of the tubular column. The simplified shot-hole impact load-time curve is shown in Figure 3 with a peak pressure of 400 MPa and a loading time of 30 ms. The impact load shown in Figure 3 is applied to the finite element analysis model of the perforation section pipe column shown in Figure 2, and the dynamic response finite element analysis of the perforation section pipe column is carried out. The vibration displacement, velocity, acceleration, and equivalent force of the shot-hole section can be obtained as a function of time, according to which the dynamic response of the shot-hole section to the shot-hole impact load can be understood.

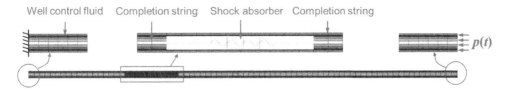

Figure 2. Finite element model of string dynamic response.

3 TRANSIENT RESPONSE ANALYSIS OF PIPE COLUMN IN SHOT HOLE SECTION

3.1 *Vibration displacement response of the perforated string*

Figure 4 shows the curves of vibration displacement of the pipe column at 5 m, 10 m, and 20 m from the packer with time. The vibrational displacement of the tubular column in the perforation

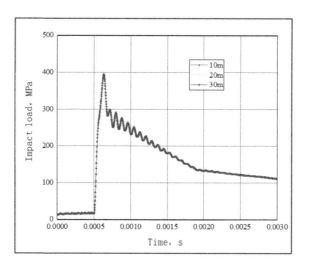

Figure 3. Perforating impact load-time curve.

section at different depths of the well varies periodically from time to time similar to a sine curve, and the periods are the same, all about 16 ms. The peak vibration displacement of the pipe column at 20 m, 10 m, and 5 m from the packer is 21 mm, 13 mm, and 8 mm, respectively, indicating that the closer to the packer, the smaller the vibration displacement of the pipe column.

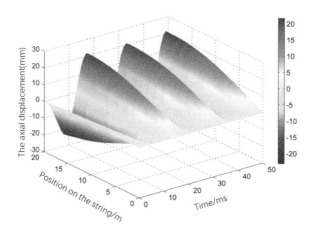

Figure 4. Variation curve of vibration displacement of the tubular column at different depths.

3.2 *Vibration velocity response of the perforated string*

The curves of vibration velocity of the pipe column at 5 m, 10 m, and 20 m from the packer as a function of time are shown in Figure 5. It can be seen that the vibration velocity of the tubular column varies periodically with time at different depths of the well within the same period. The peak vibration velocities of the pipe column at 5 m, 10 m, and 20 m from the packer are 4 m/s, 6 m/s, and 9 m/s, respectively. In other words, the farther away from the packer (closer to the source of the shot hole blast), the greater the amplitude of the vibration speed of the column.

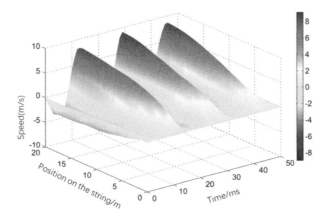

Figure 5. Vibration velocity variation curve for tubing columns at different well depths and perforation sections.

3.3 Vibration acceleration response of the perforated string

Figure 6 shows the curves of vibration acceleration of the pipe column with time at 5 m, 10 m, and 20 m from the packer. It can be seen that the vibration acceleration of the tubular column also varies periodically with time at different well depths within the same period. At a peak vibration acceleration of 3500 m/s^2 (about 350 grams), 2200 m/s^2, and 1600 m/s^2, the farther the pipe column is from the packer (closer to the shot hole blast source), the greater the vibration acceleration.

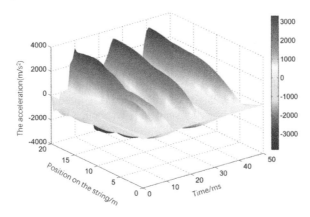

Figure 6. Different acceleration curves of tubular column vibration at different well depths and perforation sections.

3.4 Equivalent stress response of the perforated string

The equivalent force clouds of the pipe column at the packer at 6 ms and 10 ms are shown in Figures 7 and 8, respectively. It can be seen that the wave has not yet reached the packer at 6 ms, the equivalent force of the pipe column at the packer is small, and the maximum equivalent force is about 182 MPa. With the propagation of the shot-hole shock wave, the equivalent force of the pipe column at the packer gradually increases and reaches 441 MPa at 15 ms. In addition, it can be seen that the equivalent forces are not uniformly distributed on the cross-section of the pipe column, and the maximum equivalent forces always appear on the outer wall surface of the pipe column.

Figure 9 shows the curves of the maximum equivalent force of the pipe column at the packer and the full shot hole section as a function of time. It can be seen that the stress in the pipe column

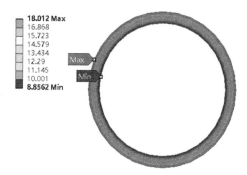

Figure 7. Equivalent force cloud of the pipe column at the packer at 6 ms.

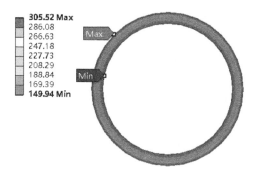

Figure 8. Equivalent force cloud of the pipe column at the packer at 15 ms.

at the packer at the initial moment is very small, indicating that the shot-hole shock wave has not yet reached the packer. After 4 ms, there is a sudden increase in equivalent force at the packer, indicating that the stress wave has been transmitted. After about 8 ms, the equivalent force of the pipe column at the packer almost coincides with the maximum equivalent force curve of the whole section of the pipe column, indicating that the equivalent force of the pipe column at the packer is the maximum.

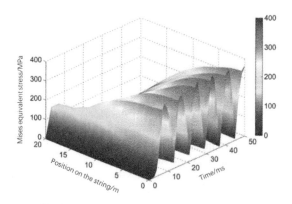

Figure 9. The curve of the maximum equivalent force of pipe column with time.

4 STRENGTH AND SAFETY ANALYSIS OF PERFORATED STRING

The results of the transient response analysis of the shot-hole section of the pipe column show that: under the action of the shot-hole impact load, the shot-hole section of the pipe column will occur axial vibration response, generating response load and response stress. If the maximum response equivalent force exceeds the yield strength of the tubular column material, it will cause the column to fail. Using the aforementioned finite element analysis model and method, the shot-hole impact load amplitudes of 100, 150, 200, 250, 300, and 350 MPa are assumed, respectively. The effect of the length and wall thickness of the shot on the maximum equivalent force is investigated and the results are shown in Figure 10. The maximum equivalent force is linearly related to the amplitude of impact load; the longer the pipe column in the shot hole section, the smaller the maximum equivalent force. Therefore, it is better to set a certain length of vibration damping oil pipe between the perforation gun and the packer when using the packer pipe column perforation process. The thicker the wall thickness of the pipe column, the smaller the maximum equivalent force and the safer the pipe column. Therefore, increasing the wall thickness can also improve the strength and safety of the perforation section of the pipe column.

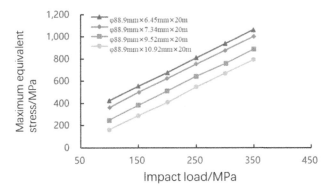

Figure 10. The maximum equivalent force of pipe column with different wall thicknesses under different amplitude impact loads.

Table 1. Maximum equivalent stress of perforated string with different lengths under different amplitude impact loads.

The length of the string (m)	Impact load (MPa)					
	100	150	200	250	300	350
20	248	398	498	677	798	857
30	217	369	469	649	769	826
40	203	357	456	637	758	812
50	195	351	448	629	752	805

5 CONCLUSIONS

The closer to the packer, the smaller the vibration displacement of the pipe column; at the beginning of perforation detonation, the vibration displacement of the string is very small, with the extension of time, the vibration displacement of the points on the column gradually increased, the column is subject to compression and tensile impact load alternately.

Vibration velocity and acceleration of perforating string vary periodically with time, and the period remains constant. The farther away from the packer (the closer to the source of the blast), the greater the vibration speed and vibration acceleration amplitude. The peak acceleration of vibration caused by the shot-hole impact can be hundreds of times the acceleration of gravity, resulting in a violent dynamic load.

Under the action of the shot-hole shock wave, the equivalent force of the pipe column at the packer is the largest, and the further away from the packer, the smaller the peak equivalent force. As a result, the pipe column of the perforation section is often deformed by plastic bending close to the packer or causes damage to the central pipe of the packer.

Under the same perforating impact load, the longer the perforating string, the thicker the string wall, and the smaller the maximum equivalent stress. Therefore, when using the packer column perforation process, it is better to set a certain length of vibration-damping oil pipe between the perforation gun and packer, and if necessary, a thick-walled oil pipe can be used to improve the strength and safety of the perforation section column.

REFERENCES

Baumann C, Williams H. Perforating high-pressure deepwater wells in the gulf of mexico [C]// SPE Annual Technical Conference and Exhibition. Denver: SPE, 2011.

Behrmann, L. A.; Li, J. L.; Venkitaraman, A.; Li, H. Borehole dynamics during underbalanced perforating [A]. Proceedings of the 1997 European Formation Damage Conference [C], 1997.

Birkhoff, Garrett; MacDougall, Duncan P.; Pugh, Emerson M.. Explosives Wtih Lined Cavties [J]. *Journal of Applied Physics*, 1948, Vol. 19 (6): 563–582.

Brink, DE; Poulter, TC; Caldwell, BM. Measurement of the Dynamic Characteristics of Perforating Shaped Charges by the Use of Ultra High-Speed Photographic Techniques [J]. *Journal of the American Society of Mining and Metallurgical Engineers*, 1960, Vol. 219 (11): 257–263.

Godfrey, WK; Methven, NE. Casing Damage Caused BY Jet Perforating [J]. *Journal of Petroleum Technology*, 1970, Vol. 22: 1054.

Groves, TK. Discussion of a Simple Treatment of Detonation Waves [J]. *Journal of Petroleum Technology*, 1967, Vol. 19 (9).

Hajianmaleki M, Daily J S. Critical-Buckling-Load Assessment of Drillstrings in Different Wellbores by Use of the Explicit Finite-Element Method [J]. *SPE Drilling & Completion*, 2014, 29 (2): 256–264.

Henderson S W, Navarette M. Shock Wave Boundary Effects on Jet Perforator Penetration in Stressed Berea Targets. *Society of Petroleum Engineers*, 1990.

Hill R J, Jarvie D M, Zumberge J, et al. Oil and gas geochemistry and petroleum systems of the fort worth basin [J]. *Aapg Bulletin*, 2007, 91 (4): 445–473.

Huang W, Gao D, Wei S, et al. Boundary Conditions: A Key Factor in Tubular-String Buckling [J]. *Spe Journal*, 2015, 20 (6).

S. C. Oliphant; R. Floyd. A Study of Some Factors Affecting Gun Perforating [J]. *Transactions of the AIME*, 1947, Vol. 170: 225–242.

Tariq, Syed M.. Analyses and Applications of Pressure, Flow Rate, and Temperature Measurements During a Perforating Run [J]. *SPE Production Engineering*, 1991, Vol. 6 (1): 83–92.

Yildiz, Turhan; Ozkan, Erdal. Pressure-Transient Analysis for Perforated Wells [J]. *SPE Journal*, 1999, Vol. 4 (2): 167–176.

Advances in Energy, Environment and Chemical Engineering – Abdullah & Osman (Eds)
© 2023 The Author(s), ISBN: 978-1-032-36083-6

Research on the countermeasures for the interactive development of Xiamen port and Longyan land port under the background of the carbon peak and carbon neutrality strategy

Xiaoqing Sun*
Chengyi University College, Jimei University, Xiamen, China

ABSTRACT: China proposes to strive to achieve a carbon peak and carbon neutral, which is the internal requirement for implementing the new development concept, building a new development pattern, and promoting high-quality development. It is also a major strategic decision made by the CPC Central Committee to coordinate the domestic and international situations. As a specific inland area with cargo distribution functions like those of the coast, the land port is a logistics center with port service functions such as customs declaration, inspection application, and issuance of bill of lading, which has helped in the forward movement of port functions and will effectively realize the function of cargo distribution. The paper analyzes the development of Longyan land port and the current situation along with the existing problems of goods imported and exported through Xiamen port. This puts forward the countermeasures to promote the coordinated development of Xiamen port and Longyan land port under the background of the double carbon strategy.

1 INTRODUCTION

With the successful implementation of the national western development strategy, the cascade transfer of coastal industries and the policy of expanding domestic demand, the initiative of inland regions to connect with the coast, and the rising demand for logistics channels, Xiamen Port actively plans "hinterland strategy" projects such as sea rail intermodal transport, accelerates the construction and operation of Longyan land port and achieves certain results.

Longyan land port is a regional logistics center along the border of Fujian, Guangdong, and Jiangxi with port functions, bonded warehousing, logistics distribution, commodity display, comprehensive services, and trade, a window for Longyan's opening to the outside world, and a platform for direct import and export logistics services. The land port has 9 major service functions, including direct import and export, on-site disassembly and assembly service, container return point, container management, and repair service, bonded warehousing, information system, docking of Xiamen container terminal, urban distribution, direct sales display of imported goods, and cross-border e-commerce. In 2021, the complete container volume of Longyan land port is anticipated to be 4,401 TEUs per year.

2 ANALYSIS OF THE CURRENT SITUATION OF LONGYAN LAND PORT CARGO IMPORT AND EXPORT THROUGH XIAMEN PORT

The main major customer groups of Longyan land port are *Zijin Copper Industry Co., Ltd.* in the mineral industry and wood and bamboo industry, and *Fujian Zhangping Mucun Forest Products Co., Ltd.* These two import and export containers rank first and second in the city. At the same time,

---

*Corresponding Author: sxq714@126.com

Longyan has formed key industrial clusters such as non-ferrous metals, machinery manufacturing, energy and fine chemicals, agricultural products processing, tobacco, building materials, and textiles, and is actively cultivating and developing emerging industries such as optoelectronics and new materials, and modern logistics. The container generation in the main hinterland of Longyan land port is about 39,401TEU. Longyan's goods flow to Xiamen port and Shenzhen port, of which 90% of the goods flow to Xiamen port.

The road transportation distances between Longyan land port and Xiamen, Shenzhen, and Quanzhou are about 150–270 km, 480–570 km, and 240 km, respectively. Xiamen has certain advantages, and Xiamen also has certain government subsidy policies. Xiamen still has certain advantages in the export environment.

Table 1. Comparison of the export environment of goods from Longyan land port to Xiamen, Shenzhen, and Quanzhou.

Summary of various costs starting from Longyan	Xiamen	Shenzhen	Quanzhou
Road transportation distance (km)	150–270	480–570	240
Road transportation time (hours)	4–7	12–15	7
Highway transportation cost (RMB)	2100–3780	5280–6270	3840
Stay time of trailer in the land port area (hours)	2	4	2
Government subsidy policy	Entering the land port, 1 US dollar, 3 RMB points, capped at RMB 3,000 per container	None	Entering the land port, 1 US dollar, 3 RMB points, capped at RMB 3,000 per container

3 MAIN PROBLEMS RESTRICTING THE INTERACTIVE DEVELOPMENT OF XIAMEN PORT AND LONGYAN LAND PORT

3.1 The location advantage of the land port

The location of the warehouse in the land port plays a great role in the company. Location selection has a very important relationship with transportation cost, competitiveness, and operation efficiency. Currently, the main wholesalers in Longyan urban area are concentrated around the West Fujian trading city, the warehouses are mainly located in the waste plant area of Longzhou Industrial Park, and the existing carload freight gathering areas are distributed near the West Fujian trading city. Longzhou Industrial Park is less than 4 km away from Minxi trading city, while Longyan land port warehouse is nearly 12 km away from Minxi trading city and close to carload freight gathering area, which is not conducive to customers' warehouse picking up and shipping to the surrounding counties. In terms of price, the Longyan land port warehouse is a standardized logistics warehouse, with an average storage price of RMB 15 per m^2 per month, while the general storage price in Longzhou Industrial Park and other places is 10 yuan per m^2 per month. For goods with low storage conditions, customers prefer to choose storage with a lower lease price.

3.2 Infrastructure conditions

In the face of fierce market competition, the export-oriented economic development in inland areas is still in a process of gradual development. In addition, in the face of fierce competition from other surrounding ports and other modes of transportation, the construction of the Longyan land port and railway port operation area is a systematic project, involving a wide range of coordinated departments and links, which still needs long-term huge multi-channel resources to support the operation. In addition, the large investment and long return period of the Longyan land port construction project are also objective problems.

3.3 *The overall customs clearance environment of the port*

At present, the customs clearance supervision of Xiamen port and Longyan land port lacks flexibility, and the investment in supervision labor and technical equipment is still insufficient, which directly affects the efficiency of cargo customs clearance. On the other hand, the overall customs clearance of Xiamen Port generally lacks a unified and coordinated management mechanism, and the one-stop port customs clearance function of "one declaration, one inspection, and one release" has yet to be fully realized.

3.4 *Informatization construction*

Xiamen port and Longyan land port have different informatization structures, and the cost of informatization construction is also very different. As a result, the implementation of information interaction platforms and other businesses in land ports cannot be carried out synchronously, which may cause some land ports to lose their development opportunities, which is not in line with the requirements of land port network construction. How to increase the investment scientifically, reasonably, and effectively in information construction needs further demonstration.

3.5 *The function of the land port needs to be further improved and innovated*

The "regional customs clearance" mode of the land port has certain restrictions on the qualifications of enterprises, while there are few enterprises in the inland region that meet the relevant qualifications, and the effective play of the service function of the land port is limited. In addition, the enthusiasm of shipping, agency, and transportation enterprises to participate in the land port business is not high, and the role of Longyan land port as an inland logistics node has not been fully played. The function of the Longyan land port needs to be further innovated and improved. It needs to be further studied to release administrative authority, break the shackles of the system, and increase bonus incentives.?

4 COUNTERMEASURES TO PROMOTE THE INTERACTIVE DEVELOPMENT OF XIAMEN PORT AND LONGYAN LAND PORT UNDER THE BACKGROUND OF THE CARBON PEAK AND CARBON NEUTRALITY STRATEGY

4.1 *Accelerate the construction of cargo passage between Xiamen port and Longyan land port*

We will further deepen and improve the planning of modern collection and distribution platforms and network systems, and speed up the connection between Xiamen port and Longyan railway lines, the construction of coastal freight railways, and coastal intercity railways. We will deepen the construction of Longyan land port as an extension and assembly center of Xiamen port in the inland hinterland, constantly strengthen its role as a "material base" and an important "feeding port" of Xiamen port, and help consolidate and expand the status of Xiamen port as a "home port". We also rely on the demand for round-trip cargo transportation in the inland economic hinterland of Longyan, make full use of the rich liner route resources between Xiamen port and major ports in Europe, North America, and ASEAN, and strengthen the construction of sea rail combined transport channel and high sea combined transport channel between Xiamen port and Longyan to make the goods between Longyan transit through Xiamen port.

4.2 *Improve the infrastructure construction of Xiamen port*

Promote the expansion project of a special railway line for port dredging in the Haicang port area and the development and construction of Zhangzhou Zhaoyin and Gulei railway branch lines, speed up the construction of large freight yards and container handling stations of Qianchang railway, build a land port through sea rail intermodal transport operation platform, and extend the railway channel to the end of Xiamen port's collection and distribution system to increase the railway collection and distribution volume and improve the railway freight channel of Xiamen port. By promoting the construction of "two depth" projects i.e., deep-water channel and deep-water berth, we will strengthen the service capacity of the port to undertake large container ships, improve the status of the international container trunk port, and raise the port's throughput, berthing capacity, and handling capacity to a new level to increase the flight density of the trunk liner routes of Xiamen port, and make some originally restricted custom clearance and shipping schedule. It often fails to catch the shipping schedule, so the local source of goods that leads to the loss of this part of the supply to other places can catch the shipping schedule. Through communication with the customs, the corresponding scheme shall be customized according to the needs of customers to try to ensure the timely declaration and release of goods. Based on direct customs clearance, combined with the regional customs clearance mode, the temporary urgent orders of some factories are directly pulled back to the Xiamen wharf after the goods are sealed with an electronic customs lock on the counter, and the loading of the wharf is obtained in time.

4.3 *Expand and strengthen the multimodal transport business between Longyan land port and Xiamen port*

Based on the construction of the railway branch line of the Haicang bonded port area, the container freight yard and loading and unloading area shall be mainly reconstructed according to the requirements of sea rail intermodal transport, the loading and unloading equipment shall be improved, and the container freight base land port direct sea rail intermodal transport operation platform suitable for the rapid operation of direct trains shall be built, so that the containers can be directly sent from the ship to the train, and the short barge and loading and unloading operations shall be reduced. At the same time, in the future improvement process of the Longyan "land port", a special railway line will be introduced for the direct movements of the train to the "land port" warehouse, and realize direct delivery with the trains. Finally, the efficiency of railway container train transportation organization will be improved, and a seamless connection between the highway and railway intermodal transportation will be achieved. Strengthen the management of freight organization, wagon organization, operation organization, and other aspects, optimize the operation scheme of container trains, and form a "direct road port mode". Actively coordinate and strive for greater support from Nanchang railway company for Xiamen port sea rail intermodal transport, cultivate and expand container sea rail intermodal transport business entities, and develop five scheduled sea rail intermodal trains serving the inland hinterland, including domestic and international five scheduled trains, focusing on Yingxiang Xiamen line, Ganlong Xiamen line, to further reduce railway freight rates.

4.4 *Encourage shipping companies to set up container return points at land ports*

The import and export business in the Longyan land port area is unbalanced, with fewer imports and more export, resulting in the reluctance of shipping companies to set up container return points in the Longyan land port in the early stage, and the process of market operation needs to be cultivated. It is suggested that a subsidy of not less than RMB 50 per TEU should be given to the shipping company that sets up a container return point in Longyan land port to store containers for the use of import and export enterprises. If the shipping company sets up an independent legal person subsidiary in the land port, an additional subsidy of not less than RMB 10 per TEU should be given. Further, freight forwarders, trailers, and other enterprises are encouraged to enter the Longyan land port, thereby transforming it into a real port, a platform for container management, full container transfer, handover, custody, LCL regrouping, cargo solicitation, transportation, trade, customs clearance, and other services to expand and strengthen the economies of scale of Longyan land port, thus attracting more sources of goods back to Xiamen port.

4.5 *Actively expand the business scope of the Longyan land port*

Through institutional innovation and trade facilitation, we will further increase the added value of trade in goods, promote the upgrading of trade in goods, strengthen the import function, vigorously develop the exhibition trade of imported goods and bonded futures delivery, and encourage enterprises in Longyan land port to carry out international and domestic trade as a whole and introduce more enterprises and commodities to carry out exhibition trade business in the land port, promote the formation of import trading centers for more commodities. Longyan land port should carry out multi-channel business promotions, provide customized services for customers, visit customers regularly in a planned way, hold recommendation meetings for some key development areas, and give publicity to medium and large customers in the region.

4.6 *Take multiple measures to attract the return of local goods*

Longyan land port business should expand its business based on the basic functions of the original port and bonded port, develop a variety of business forms and professional market businesses, promote the regional professional market influence, and let customers have a sticky effect on the land port. Limited by customs clearance and shipping schedule, some local cargo sources often fail to catch up with the shipping schedule, which leads to the loss of this part of cargo sources to other places. This phenomenon can be achieved by focusing on improving the flight density of Xiamen Port trunk liner routes and enhancing customs inspection communication to strive for the return of local cargo sources.

5 CONCLUSION

The land port is the bridgehead of the coastal economy extending inland and the core resource and key endowment of the inland economic development. Strengthening the joint development of Xiamen port and Longyan land port will promote the port function to drive the development of industrial clusters, thus promoting the development of the channel economy and becoming the development engine of the hub economy.

ACKNOWLEDGMENT

The project was financially supported by the Educational Research Project for Young and Middle-aged Teachers in Fujian Province (Grant No. JAT201033). And authors are very grateful for the assistance or encouragement from colleagues.

REFERENCES

Liu Qigang. Optimization of Service Measure Configuration Scheme for Railway Whole Logistics [J]. *China Railway Science*, 2017, 38(5):138–144.

Weixia Yang. Study on Countermeasure of the Joint Development of Manufacturing Industry and Land Port Logistics in Xi'an[C]//.Proceedings of 2017 6th EEM International Conference on Education Science and Social Science (EEM-ESSS 2017), 2017:48–53.

Wu Tiefeng, Zhu Xiaoning. Research on Program of Container Sea-Rail Intermodal Development [J]. *Journal of Beijing Jiaotong University* (Social Sciences Edition), 2011, 10(2): 27–32.

Xiaoqing Sun,Binlan Luo. A Study on the Development of Container Sea-rail Intermodal Transport of Xiamen Port in the Context of National Logistics Hub[C]//.Proceedings of 2020 6th International Conference on Energy Materials and Environment Engineering(ICEMEE 2020)(VOL.1), 2020:279–284.

Zhao Xin,Ma Xiaowei,Chen Boyang,Shang Yuping,Song Malin. Challenges toward carbon neutrality in China: Strategies and countermeasures[J]. *Resources, Conservation & Recycling*, 2022, 176.

Advances in Energy, Environment and
Chemical Engineering – Abdullah & Osman (Eds)
© 2023 The Author(s), ISBN: 978-1-032-36083-6

Research on islanding detection method of MMC-MG system with improved voltage phase shift

Xinggui Wang, Wanwan Dong*, Hailiang Wang & Jinjing Zhang
College of Electrical and Information Engineering, Lanzhou University of Technology, Lanzhou, China

ABSTRACT: To reduce the impact of unplanned islanding on the microgrids, reduce the outage range and improve the reliability of power supply, this paper proposes an islanding detection method for the MMC series structure microgrid system (MMC-MG). Firstly, the system topology and the control strategy of the inverter link are introduced; then, the variation of the frequency of the system in grid-connected operation and grid-connected to islanding under virtual synchronous generator control are analyzed. On this basis, an islanding detection method combining frequency and phase to improve voltage phase shift is proposed. Simulations show that the improved method can effectively detect the occurrence of unplanned islanding, reduce the probability of islanding detection misses, and ensure the smooth transition to islanded operation of the system.

1   INTRODUCTION

Microgrid operation from grid-connected to islanding can be divided into two cases: planned islanding and unplanned islanding (Zhang et al. 2019). The latter generally occurs in the case of grid faults, and if the state is not detected in time, it not only affects the simultaneous closing of the microgrids but also endangers the lives of maintenance personnel (Sun et al. 2016; Zhang et al. 2020). Therefore, the detection of unplanned islanding is an important aspect of microgrid control.

Islanding detection methods are divided into three categories: communication-based detection methods, inverter-based active detection methods, and passive detection methods (Abd-Elkader et al. 2018; Dutta et al. 2021; Wang & Huang 2020). Research by Xiao et al. (2022) proposed a hybrid islanding detection method combining adaptive feedback disturbance detection method and passive detection method to reduce the blind area of islanding detection. Yang et al. (2016) and Ma et al. (2021) have improved the frequency shift method by using the frequency difference between two cycles to determine the direction of application of the interference signal to improve the speed of detection. In addition, methods such as wavelet transform, random forest classifier, and neural network are also widely used in island detection (Allan & Morsi 2021; Dutta et al. 2021; Tao et al. 2017).

Voltage phase shift is used to determine whether unplanned islanding occurs by detecting the voltage phase difference between the two ends of the common coupling point (PCC), which has the advantages of being a simple algorithm, is easy to implement, and does not affect power quality (Chen et al. 2012). For the MMC-MG system controlled by a virtual synchronous generator (VSG), when unplanned islanding occurs, its voltage and frequency will only exceed the normal fluctuation range when there is a serious imbalance between active and reactive power. If the above method is used directly, there will be a large detection blind area. Therefore, this paper proposes an islanding detection method with improved voltage phase shift by combining frequency and phase. After detecting the islanding, the unplanned islanding is converted into active islanding by power regulation to guarantee system stability.

---

*Corresponding Author: 1570018138@qq.com

DOI 10.1201/9781003330165-24

2 SYSTEM TOPOLOGY AND INVERTER LINK CONTROL

2.1 *System topology*

The topology of the MMC-MG system is shown in Figure 1. The system is based on the MMC structure, where photovoltaic micro sources, wind micro sources, and hybrid energy storage systems are connected in parallel and then connected to the capacitor ends of the MMC half-bridge submodule to form a Generation Module (GM). Each bridge arm is composed of N generation modules and a reactor L.

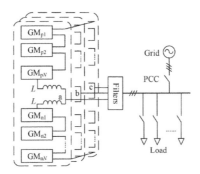

Figure 1.   MMC-MG system topology.

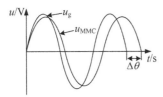

Figure 2.   Phase deviation of the voltage at both ends of the PCC.

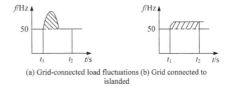

(a) Grid-connected load fluctuations (b) Grid connected to islanded

Figure 3.   Frequency change of the system under different operation modes. (a) Grid-connected load fluctuations (b) Grid connected to islanded.

2.2 *Inverter link control*

The system inverter link control mainly contains virtual synchronous generator control, voltage, current double closed-loop control, and carrier phase shift pulse width modulation. Among them, the active-frequency expression of the VSG is shown as follows

$$J\frac{\mathrm{d}\omega}{\mathrm{d}t} = \frac{P_\mathrm{m}}{\omega_\mathrm{n}} - \frac{P_\mathrm{e}}{\omega_\mathrm{n}} - D(\omega - \omega_\mathrm{n}) \tag{1}$$

$$P_\mathrm{m} = P_\mathrm{ref} + k_\omega(\omega_\mathrm{n} - \omega) \tag{2}$$

$$\theta_\mathrm{ref} = \int \omega \mathrm{d}t \tag{3}$$

where J is the rotational inertia; P_m and P_e are the mechanical power and electromagnetic power, respectively; D is the damping factor; ω is the actual angular frequency of the VSG output; ω_n is the rated angular frequency; θ_ref is the reference voltage phase angle; P_ref is the active power reference; k_ω is the modulation factor.

The reactive-voltage control equation of VSG can be expressed as

$$U_\mathrm{ref} = U_0 + k_\mathrm{q}(Q_\mathrm{ref} - Q) \tag{4}$$

where U_{ref} is the reference value of the VSG output voltage; U_0 is the nominal voltage; k_q is the reactive power regulation factor; Q_{ref} and Q are the references and actual values of reactive power.

3 IMPROVED ISLANDING DETECTION METHOD FOR VOLTAGE PHASE SHIFT

The excess or deficit power emitted by the system will change according to the active-frequency and reactive-voltage characteristics of the VSG, which can be expressed as

$$\begin{cases} \omega = \omega_n + \frac{P_{\text{ref}} - P_L}{k_\omega} = \omega_n + \frac{\Delta P}{k_\omega} \\ U_{\text{ref}} = U_0 + k_q(Q_{\text{ref}} - Q_L) = U_0 + k_q \Delta Q \end{cases} \tag{5}$$

Based on Equation (5), the change of active and reactive power will cause the system output voltage amplitude and frequency to change. Therefore, it can be determined whether the islanding phenomenon occurs in the system according to the rate of change of voltage and frequency. Figure 2 shows a diagram of the phase deviation of the voltage at both ends of the PCC when unplanned islanding occurs.

When the system operates grid-connected, the inverter link output voltage is consistent with the grid voltage, which can be expressed as

$$u_g = U_g \sin(\omega_g t + \theta_g) \tag{6}$$

where U_g, θ_g, and ω_g are the amplitude, phase, and angular frequency of the grid voltage.

When unplanned islanding occurs in the system, the inverter link output voltage can be expressed as

$$u_{\text{MMC}} = U_{\text{MMC}} \sin(\omega_{\text{MMC}} t + \theta_{\text{MMC}}) \tag{7}$$

where U_{MMC}, θ_{MMC}, and ω_{MMC} are the amplitude, phase, and angular frequency of the inverter link output voltage during island operation.

From Equations (6) and (7), the phase difference between the output voltage of the inverter link of the system before and after PCC disconnection can be expressed as

$$\Delta\theta = \frac{180}{\pi} \int |\omega_g - \omega_{\text{MMC}}| dt + |\theta_g - \theta_{\text{MMC}}| \tag{8}$$

The moment the system goes from grid-connected to an islanding equation: $\theta_g = \theta_{\text{MMC}}$. Therefore, the phase difference is generated by the accumulation of the difference between ω_{MMC} and ω_g for a long time. At this time, Equation (8) can be modified as

$$\Delta\theta = \frac{180}{\pi} \int |\omega_g - \omega_{\text{MMC}}| dt = 2\pi \cdot \frac{180}{\pi} \int |f_g - f_{\text{MMC}}| dt \tag{9}$$

When the microgrids are subjected to a small disturbance, they cause a change in grid frequency. This change acts on Equation (9) for a long time and can cause $\Delta\theta$ to exceed the threshold value, triggering an islanding error. Therefore, a phase shift within Δt is used for islanding detection, and Equation (9) is rewritten as

$$\Delta\theta = 360 \int_{t-\Delta t}^{t} |f_g - f_{\text{MMC}}| dt \tag{10}$$

Conventional voltage phase shifting is achieved by detecting the phase difference between the two ends of the PCC and then comparing it with the phase shift threshold $\Delta\theta_{th}$. If $\Delta\theta > \Delta\theta_{th}$, the system sends an islanding signal; if $\Delta\theta < \Delta\theta_{th}$, no islanding signal is sent. If the threshold value is selected too small, the island detection occurs by mistake, which will reduce the reliability of the detection. If the threshold value is selected too large, the detection time increases and the detection blind area also increases, reducing the effectiveness of island detection.

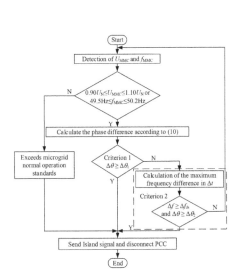

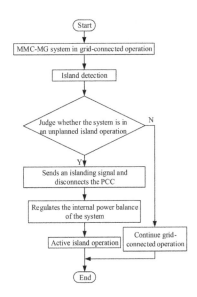

Figure 4.　Flow chart of island detection.

Figure 5.　Flow chart of system grid connected to island operation.

Figure 3 shows the frequency change of the system under different operation modes. The frequency change caused by the load will return to the stable value sooner and the frequency fluctuates more. After grid-connected to the islanded operation of the system, the frequency may rise or fall due to $P_{ref} \neq P_L$, and for a longer period. Therefore, the combination of frequency change and the phase shift is used as the judgment condition for island detection. It can effectively eliminate the influence of short-term frequency changes caused by load on the accuracy of the island detection algorithm, and reduce the probability of misjudgment of the algorithm while ensuring the sensitivity of island detection.

Figure 4 shows the flowchart of the islanding detection algorithm with an improved voltage phase shift. In the figure, $\Delta\theta_1$ is set larger, which is suitable for the case where the difference between P_{ref} and P_L is large. $\Delta\theta_2$ is set smaller, which is suitable for the case where the difference between P_{ref} and P_L is small. The specific steps are as follows:

(1) Detecting the output voltage and frequency of the inverter link of the system. If the amplitude exceeds $0.90\ U_N \leq U_{MMC} \leq 1.10\ U_N$ and 49.5 Hz $\leq f_{MMC} \leq 50.2$ Hz. An islanding signal is issued and the PCC is disconnected.

(2) If the amplitude does not exceed the range given in step (1), then criterion 1 is activated. Calculate the phase difference according to Equation (10). If $\Delta\theta \geq \Delta\theta_1$, the system sends an islanding signal; otherwise, go to step (3).

(3) Activation criterion 2. Only when the frequency change throughout the integration interval satisfies $\Delta f \leq \Delta f_{th}$ and $\Delta\theta \geq \Delta\theta_2$, the system sends out an island signal; otherwise, return to step (1).

4　SYSTEM GRID CONNECTION TO ISLAND SWITCHING PROCESS IN CASE OF UNPLANNED ISLANDING

In general, when unplanned islanding occurs in a system, its internal power is unbalanced. When the power generated by the microgrids is not enough to support the load demand, the system should give priority to supplying the important loads and achieve unplanned islanding to planned islanding operation. Figure 5 shows the flow chart of the changes from grid-connected control to the islanded control when unplanned islanding occurs.

The steps are as follows:

(1) Put in islanding detection when the system is in grid-connected operation.
(2) Determe whether unplanned islanding of the system has occurred.
(3) If unplanned islanding occurs, the system sends an islanding signal and disconnects the PCC. Adjust the internal power balance of the system to active island operation; otherwise, continue grid-connected operation.

5   SIMULATION ANALYSIS

To verify the effect of islanding detection under different operating conditions, the system is initially a grid-connected operation and unplanned islanding occurs at 1 s. The simulation parameters are shown in Table 1.

Table 1.   System simulation model parameters.

Parameters	Value	Parameters	Value	Parameters	Value
GM rated voltage (V)	160	Filter inductors (mH)	1	Rotational inertia (kg • m^2)	0.3
Sub-module capacitance (μF)	4400	Filter capacitor (μF)	30	Damping factor (N·s·m^{-1})	10
Number of bridge arm sub-modules	4	Grid-rated voltage (V)	380	Active power reference value (kW)	20
Bridge arm reactance (mH)	0.5	Grid-rated frequency (Hz)	50	Reactive power Reference value (var)	

5.1   Simulation of the unimproved island detection method

5.1.1   Active power mismatch of 1.5% and 10%
Load active power $P_L = 20.3$ kW and $P_L = 18$ kW, reactive power $Q_L = 0$ var, and active power mismatch of 1.5% and 10%. During simulation, $\Delta\theta_1 = 5°$ and 8°, and $\Delta t = 0.5$ s. Figure 6 shows the effect of load power variation on islanding detection. When $P_L = 20.3$ kW, the unimproved islanding detection algorithm fails to detect the islanding signal at both $\Delta\theta_1 = 5°$ and 8°, resulting in missed islanding signal detection. And with $P_L = 18$ kW, the system output frequency is stable at 50.07 Hz. As time goes by, the phase difference between the voltages at the two ends of the PCC becomes larger and larger. When $\Delta\theta_1 = 8°$, the system islanding detection time is 0.28 s; when $\Delta\theta_1 = 5°$, the system islanding detection time is 0.168 s.

5.1.2   Load power variation during grid-connected operation
Initially, $P_L = 19$ kW, with a sudden addition of 5 kW at t = 1 s to simulate the small disturbance to the system during grid-connected operation to verify the accuracy of islanding detection. Figure 7 shows the islanding detection waveform for load fluctuation during grid connection. When $\Delta\theta_1 = 8°$, the system does not send an islanding detection signal; when $\Delta\theta_1 = 5°$, the system sends an islanding signal at 1.22 s, but the system is in grid-connected operation and the islanding detection is a malfunction.

5.2   Simulation of the improved island detection method

5.2.1   Active power mismatch of 1.5% and 10%
The improved island detection method consists of a combination of criterion 1 and criterion 2, where $\Delta\theta_1 = 8°$ and $\Delta\theta_2 = 3°$. Figure 8 shows the islanding detection waveform of the system after

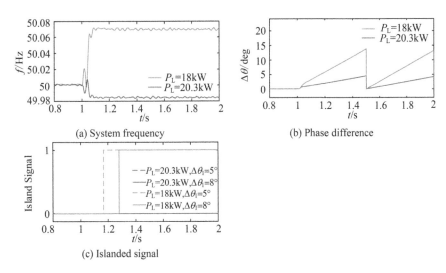

Figure 6.　Effects of load power variation on islanding detection.

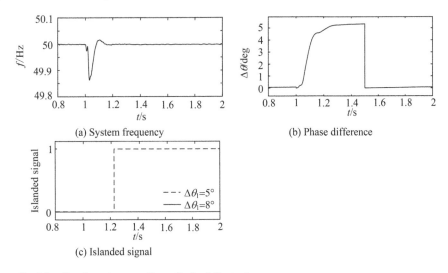

Figure 7.　Islanding detection waveforms for load fluctuation.

the improved method. Active power mismatch of 1.5%, the improved method sends an islanding signal at 1.4 s. At a system active power mismatch of 10%, the improved method sends an islanding signal at 1.09 s, the detection time was reduced by 0.19 s compared to Figure 6(c). Simulations show that adding criterion 2 reduces the probability of missing an island detection and can speed up islanding detection when the active power mismatch is small.

5.2.2　*Load power variation during grid-connected operation*
Figure 9 shows the islanding detection waveform of the improved method during load fluctuations. Although the phase shift exceeds $\Delta\theta_2$, the system frequency fluctuates more than 0.1 Hz in a fixed period, so the islanding detection does not send out a signal.

Here, we set $P_L = 25$ kW. Figure 10 shows the waveform diagram of the system grid-connected to islanded under unplanned islanding. Unplanned islanding occurs in the system at 1 s. However, the power generated by the microgrids is insufficient, causing the system frequency to drop, the

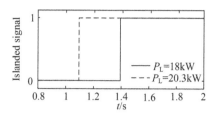

Figure 8.   Waveforms of island detection after improved method.

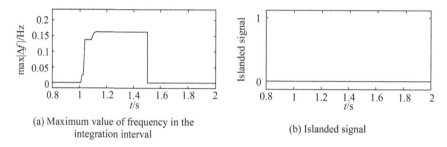

(a) Maximum value of frequency in the
integration interval

(b) Islanded signal

Figure 9.   Islanding detection waveforms of the improved method during load fluctuations.

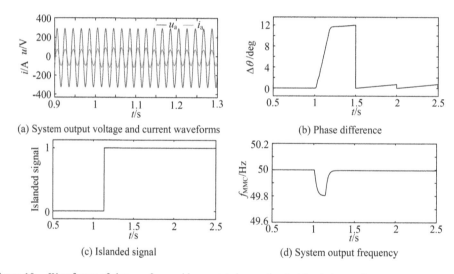

(a) System output voltage and current waveforms

(b) Phase difference

(c) Islanded signal

(d) System output frequency

Figure 10.   Waveforms of changes from grid-connected operation to islanded operation.

system sends out an islanding signal at 1.135 s. Then, to remove the excess load to ensure the system power supply and demand balance, the system frequency changes to 50 Hz. In this way, the conversion from grid-connected to islanded systems in unplanned islanding situations is completed.

6   CONCLUSIONS

This paper analyzes the variation of the frequency of the MMC-MG system under virtual synchronous generator control in different operating states, an islanding detection method with improved voltage phase shift by combining frequency and phase is proposed. The results show that the improved method can effectively avoid the false islanding detection caused by load disturbance during the grid-connected operation of the system. The improved method is effective in detecting the occurrence of islanding, regardless of whether the power match is large or small. The system can smoothly transition to islanded operation after unplanned islanding occurs, without affecting the stability of system voltage and frequency.

ACKNOWLEDGMENT

This work was supported by the National Natural Science Foundation of China (Grant No. 51967011).

REFERENCES

Abd-Elkader A G, Saleh S M, Eiteba M B M. (2018) A passive islanding detection strategy for multi-distributed generations. *International Journal of Electrical Power & Energy Systems*, 99: 146–155.

Allan O A, Morsi W G. (2021) A new passive islanding detection approach using wavelets and deep learning for grid-connected photovoltaic systems. *Electric Power Systems Research*, 199: 107437.

Chen Shaojie, Qian Suxiang, Xiong Yuansheng, et al. (2012) Islanging detection based on phase jump and the voltage perturbation combination [J]. *Power Electronics*, 46 (06): 7–9.

Dutta S, Olla S, Sadhu P K. (2021) A secured, reliable and accurate unplanned island detection method in a renewable energy based microgrid. *Engineering Science and Technology, an International Journal*, 24 (5): 1102–1115.

Ma L, Guo X, Wei L. (2021) *An improved islanding detection algorithm based on AFDPF*. E3S Web of Conferences, 257 (5): 02049.

Sun Zhenao, Yang Zilong, Wang Yibo, et al. (2016) A Hybrid Islanding Detection Method for Distributed Multi-inverter Systems. *Proceedings of the CSEE*, 36 (13): 3590–3597+3378.

Tao Weiqing, Wang Leqin, Gu Zhixia, et al. (2017) Application of SVD and neural network in islanding detection [J]. *Power System Protection and Control*, 45 (02): 28–34.

Wang Kangzhong, Huang Chun. (2020) Novel Islanding Detection Method Based on AFD with No NDZ and Low Distortion. *Proceedings of the CSU-EPSA*, 32 (03): 59–65.

Xiao Xin, Hu Zheng, Li Peng. (2022) Islanding Detection Based on Rate of Change of Frequency in Power Grids. *Control Engineering of China*, 29 (02): 315–321.

Yang Huidong, Wu Lang, Li Xinru, et al. (2016) An improved islanding detection method based on alternate active frequency drift. *Power System Protection and Control*, 44 (16): 50–55.

Zhang Liming, Hou Meiyi, Zhu Guofang, et al. (2019) Seamless transfer strategy of operation mode for microgrid based on collaborative control of voltage and current. *Automation of Electric Power Systems*, 43 (05): 129–135+158.

Zhang Mingrui, Wang Junkai, Wang Jiaying, et al. (2020) Study on hybrid islanding detection and operation mode transition of microgrid. *Power System Protection and Control*, 48 (02): 1–8.

Advances in Energy, Environment and
Chemical Engineering – Abdullah & Osman (Eds)
© 2023 The Author(s), ISBN: 978-1-032-36083-6

Application of microservices in power trading platforms

Ning Yang*
NARI Group Corporation (State Grid Electric Power Research Institute), Nanjing, China
Beijing Kedong Electric Power Control System Co., Ltd. Beijing, China

Yongwei Liu
State Grid Hunan Electric Power Co., Ltd. Changsha, China

Wentao Lv, Chuncheng Gao, Shiqiang Zheng & Qian Zhang
NARI Group Corporation (State Grid Electric Power Research Institute), Nanjing, China
Beijing Kedong Electric Power Control System Co., Ltd. Beijing, China

ABSTRACT: With the acceleration of China's power system reform, the power trading market is booming, the volume of market elements such as market subjects and trading varieties is growing rapidly, the frequency of transactions is becoming more and more real-time, and the complexity of transaction clearing and settlement is increasing, therefore, introducing microservices architecture in the power trading platform, using microservices technology architecture to transform the power trading system, enhance the load performance and scalability of the system, and explore the application of microservices in the power platform.

1 INTRODUCTION BACKGROUND

In recent years, with the gradual acceleration of the process of electricity marketization, the main body of the electricity market is developing to diversification, the number is increasing day by day, new electricity trading business is emerging, and information technology is also developing continuously. At the same time, the spot market and auxiliary service market are gradually opened, distributed trading, renewable energy quotas and other emerging trading varieties are also joining the market, and the form of trading organization and trading varieties are constantly enriched, but the development of each provincial power market has its characteristics, and the power trading rules change frequently, the traditional single structure of the power trading platform has been difficult to meet the needs of users. To cope with the above status quo, the power trading platform as the technical carrier of power trading business needs to have the ability to rapid response to business needs. Therefore, it is necessary to realize distributed architecture transformation based on microservices to cope with the deepening of power system reform and deepen the support of the existing system for business.

The power trading platform is mainly for power generation enterprises, power users, and power sales companies, which are market players carrying out power trading business. The platform is built on the State Grid cloud platform with microservice architecture, and the business middle platform is built to support the national power trading business.

---

*Corresponding Author: vitasll@126.com

DOI 10.1201/9781003330165-25

2 MICROSERVICES

2.1 *Introduction to microservices*

Microservices refer to the division of a single application into a set of small services, which coordinate and cooperate with each other to provide ultimate value to users. The microservice framework is a service that consists of many small services, each running in a separate process, and through lightweight communication mechanisms (such as RPC), complete the entire application communication, and vertical division for the business, for automation independent deployment, to ensure a minimum of centralized management of a service.

Microservices architecture is an emerging software architecture that refers to a loosely coupled, service-oriented architecture with a certain bounded context. Each service has its processing and lightweight communication mechanisms and can be deployed on single or multiple servers. Microservice architectures create applications around business domain components that can be developed, managed, and iterated independently.

The core idea of microservices is a way to implement a single application in the field of application development using a series of tiny services. The purpose of microservices is to effectively split the application, splitting a complete application into several service functions for agile development, rapid deployment, and service reuse.

One of the biggest advantages of microservice applications is that microservices can handle functional bottlenecks by scaling component instances. This allows for more efficient use of computing resources, and the system only needs to deploy computing resources for additional components, rather than deploying a full application. The result is that more resources are available for other tasks. Another benefit of microservice applications is that with applications designed with microservices, developers can update a single component of the application without affecting other parts, which can speed up development and support continuous improvement and refinement of the application.

2.2 *Microservice architecture*

Microservice architecture generally includes load balancing, service gateway, local cache, service management layer, microservice layer, remote cache, data access layer, message queue layer, etc.

(1) Load balancing is responsible for extending the bandwidth of network devices and servers, increasing throughput, enhancing network data processing capabilities, and improving network flexibility and availability.

(2) The service gateway is responsible for authentication of access clients, anti-replay and anti-data tampering, service authentication of function calls, desensitization of response data, traffic, and concurrency control, etc.

(3) Local caching is responsible for reducing the frequency of service calls and prompts access speed. Local caching generally uses automatic expiration and allows for some data latency in business scenarios.

(4) The service management layer is responsible for the registration, discovery, and governance of microservices.

(5) The microservice layer includes atomic services for each business, realizing various types of business such as adding, deleting, checking, and computing processing.

(6) Remote caching is a layer of distributed caching in front of the access DB, reducing the number of DB interactions and increasing the TPS (throughput) of the system.

(7) The data access layer is responsible for adding, deleting, and checking database tables, and the data volume of a single table is too large to do the processing of the data in separate libraries and tables.

(8) The message queue layer is responsible for decoupling dependencies between services, and asynchronous calls can be performed by way of MQ.

3 MICROSERVICE TRANSFORMATION OF POWER TRADING SYSTEM

3.1 *Overall program*

Based on the many advantages of microservice architecture, this paper proposes a technical solution to transform the power trading system with microservices. Figure 1 depicts the architecture of the power trading system applying microservices.

(1) Load balancing. The front-end of the system uses load technology to ensure that the traffic can be distributed to the back-end services in a balanced manner. In addition to the hardware-based F5, NGINX and LVS are added as software implementations of load balancing.

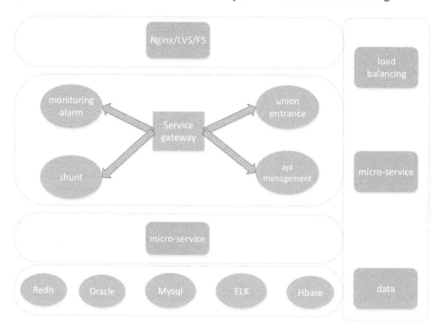

Figure 1.   Electricity transaction system architecture based on microservice.

(2) API service gateway includes service gateway, monitoring and alarming, unified entrance, triage and permission, and API management, and provides a unified interface between the system and users, with several service governance functions such as triage, monitoring, and alarming in addition to user-user interaction functions.
(3) Servitization and RPC use the thrift-based RPC framework to abstract business functions such as market management, user registration, transaction organization, and market settlement into independent services, using zookeeper-based service registration and discovery, making the server side and the caller completely decoupled.
(4) The data layer supports not only traditional relational databases, but also in-memory databases and distributed databases, and abstracts various databases into a unified data service to be transparent to users.

3.2 *Basic principles of microservice transformation*

The basic requirements for microservice transformation of the system are to ensure the rationality of microservice splitting and microservice design, the original business process, and data structure is not affected as much as possible, in addition, the rationality and stability of microservice interfaces need to be ensured.

How to slice and dice services is the core issue of implementing microservices. To ensure that the implementation of microservices can achieve the desired goals, the following principles need to be followed.

(1) Maintain a single business. At the beginning of the power reform, business and demand will keep changing. To maintain the stability of the microservice architecture, the business functions of microservices should be kept single, and a microservice only does one thing. In this way, no matter how the business changes, these basic services can maintain stability.

(2) It should be ensured that any individual service is testable. A service that is not testable will not guarantee effective and stable business.

(3) Gradual and orderly progress. The power trading system has more functional modules and complex business, the implementation of microservices cannot be achieved overnight and must follow the principle of gradual progress, from coarse to fine, from easy to difficult, and gradually switch the system functions to ensure the stable operation of the whole system.

4 KEY TECHNOLOGY POINTS

The process of implementing microservice transformation involves many key technical points such as RPC communication, load balancing, service registration, discovery, etc. The following is a description of these key technical points.

4.1 RPC key communication technology protocols

RESTful and RPC are two common communication methods used in microservices. RESTful (representational state transfer) is strictly a web design pattern that specifies that the client communicates with the server via an HTTP protocol. RPC (remote procedure call) is a computer communication protocol. The protocol allows a program running on one computer to call a subroutine on another computer without the programmer having to additionally program the interaction.

RPC and RESTful have their advantages in practice. RESTful is based on HTTP, easy to debug, implement, and cross the security system, with lower development costs. The RPC method is also used in the practice of microservices in major Internet companies.

In practice, this paper uses the Thrift framework as a means of communication, with the following features.

(1) Cross-language, with Thrift IDL defining the format of the interaction data and providing tools to translate the code into different language implementations for invocation.

(2) High performance. The Thrift framework itself provides the thrift binary protocol in small size and occupies less traffic. In a highly concurrent environment, it can save bandwidth consumption, while the Thrift framework can provide synchronous and asynchronous protocols, which can be adjusted according to the results of the stress test.

(3) Wide applications. Because the Thrift framework technology is mature and easy to operate and maintain, it has a wide range of applications in Taobao, Facebook, and other Internet companies, as well as in other famous open-source projects, including Scribe and HBase.

4.2 Service registration and discovery mechanism

In the implementation of a microservice system, RPC client and server find each other through service registration and service discovery, the server-side server is registered in the service management system through "service registration", and the client side finds the server side registered in the service management system through "service discovery". The server of the service provider is registered in the service management system through "service registration", and the client discovers the service provider registered in the management system through "service discovery" (as shown in Figure 2) so that the client can be notified of changes when the server is added or taken offline.

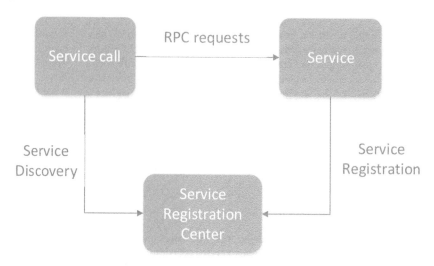

Figure 2.   Service registration and service discovery.

Zookeeper is an open source, highly available distributed orchestration service that provides naming management, configuration services, service registration, service discovery, distributed queues, and other functions for distributed systems. 1) Permanent nodes, which are created and will not be deleted unless actively operated. 2) Temporary nodes, which have a lifetime of one session and will be deleted once the session is over.

Zookeeper has the following features.

(1) Zookeeper cluster ensures high availability through leader election and data replication, and each node has a backup of stored data.
(2) The sequential nature of Zookeeper receiving updates on the same client.
(3) Eventual consistency. The data backups of each node of the cluster will be consistent in the end.

Based on the above features of Zookeeper, service registration and service discovery can be done using Zookeeper. The service provider and the caller agree on a directory on Zookeeper, such as "/root" and "/service", to be used for service registration and service discovery. When the service provider joins the cluster, it creates a temporary node under the node, set IP, port, and other information to the node, and the caller listens to the "/root" and "/service" directory. When the caller listens to the creation of a node in the directory, it gets the IP, port, and other server information from the node and brings the discovered server information to the list of available calls for the service. When a node is deleted, it means that the service provider and Zookeeper have lost the session, which means that the service provider has lost its service capability due to downtime, so the client needs to remove the server corresponding to the node from the list of services to achieve failover.

4.3   *Load balancing mechanism*

In a microservice architecture, to prevent individual servers from crashing due to overload, there is a need to distribute traffic more evenly, and this technique is called load balancing. There are many techniques to implement load balancing, both software and hardware, depending on the application scenario, different load balancing techniques can be selected.

The power trading system is exposed externally as an HTTP interface, and Nginx can be chosen as a means of load balancing. Nginx is running on top of the network layer 7 protocol, which is easy to implement corresponding load balancing policies through domain names, URI structures, etc. Load balancing policies can also be added to the service discovery. If the service is stateless, a simple random approach usually works well. If the service is stateful and the client-server interaction depends on the information from the previous interaction, a consistent hash can be used to maintain the relationship between the client and the service.

4.4 *Parameter optimization*

In engineering practice, parameter optimization is a very important aspect in the implementation of microservices, involving the setting of parameters such as the number of connections, timeout times, number of threads, etc. Nowadays, the means of optimizing experimental parameters in an automated way are usually used, and such means can be summarized in the following steps.

(1) Set 1 parameter interval.
(2) Set N parameter detection points.
(3) Experimentation on N probe sites.
(4) The probe point with the best statistical results.

For some special parameters, the theory can be used for optimization. Here, we take the number of threads as an example. Equation (1) is more widely accepted, where T is the number of threads and N is the number of cores.

$$T = N + 1 \qquad (1)$$

Equation (1) does not consider specific practical situations, such as the gradual increase in IO and the gradual decrease in CPU processing time despite the implementation of microservices. Therefore Equation (1) can be modified in the following way.

$$T = (t_w + t_p)/t_p \times N \qquad (2)$$

where t_w is the thread wait time; t_p is the thread operation time. The formula reflects that the higher the CPU utilization, the fewer the number of threads should be set, too many threads are set, but the performance degradation caused by thread switching, the more IO, the more CPU waiting time, the number of threads should be increased to provide CPU utilization.

5 PRACTICAL EFFECTS

To verify the performance of microservices, stress tests were conducted on the original monolithic architecture solution and the solution after the implementation of the microservices transformation. The test environment was 10 servers, and 4 of them were deployed as independent parallel microservices for the power trading module in the microservices transformation scheme, and Table 1 and Table 2 show the stress test results.

5.1 *Stress test*

Table 1 shows that the TPS (processing per second) of the monolithic architecture solution gradually saturates as the number of pressure test threads increases, and the CPU usage increases as the TPS increases. Table 2 shows that the effect of the modified microservice solution does not improve significantly, or even decreases slightly when adding one IO request at a lower concurrency level. The advantages of microservices gradually emerge when the pressure test threads exceed 70, with TPS, average response time, and CPU utilization all outperforming the monolithic solution. As the degree of concurrency increases, the advantages of the microservice solution become more obvious. When the number of tested threads reaches 150, the monolithic solution gradually reaches saturation and the throughput decreases significantly, while the throughput of the microservice architecture increases further, which shows that the microservice solution is more adaptable to highly concurrent business scenarios.

Table 1. Data of single structure system for stress testing.

Number of processes	Number of processes per second	Average response time (ms)	CPU utilization (%)
50	5076	4.16	21
70	7329	13.77	43
100	8035	40.6	62
150	7625	134.9	77

Table 2. Data of microservice system for stress testing.

Number of processes	Number of processes per second	Average response time (ms)	CPU utilization (%)
50	5013	4.16	20
70	8766	12.01	37
100	9035	17.82	53
150	12308	22.07	64

5.2 *Experimental analysis*

From the stress test, the larger the business volume, the more obvious the performance of microservices. The transaction management module is the most load-intensive place, and microservice slicing for it makes the performance optimization more obvious. At the same time, after slicing out the services, it can more easily cope with the actual needs of changing trading business and increasing trading varieties. In addition, as the system failure of the monolithic structure can easily lead to the overall system unavailability, this study also shows that the system undergoes microservice transformation, which will gradually stabilize with the service framework, thus effectively reducing the overall system failure rate.

6 CONCLUSION

This paper analyzes the disadvantages of poor system scalability, difficult maintenance, and slow iteration speed of monolithic structures, and discusses the general technical scheme of transforming the architecture of power trading systems based on microservices, as well as the key technologies applied. It is also demonstrated through experiments that the implementation of microservice transformation will effectively improve the concurrency performance of the system, and the robustness and version iteration speed of the system will be improved.

The implementation of microservice transformation can improve system performance, scalability, and maintainability, but it also increases the complexity of the system and puts forward higher requirements for both development and operation, and maintenance. In terms of registration, management, and monitoring of microservices, it will also increase the investment in system construction. Therefore, when it comes to whether an information system needs to be fully transformed with microservices, it needs to be evaluated and verified accordingly according to its business characteristics.

ACKNOWLEDGMENT

This work was supported by the Science and Technology Foundation of SGCC— "Key Technologies for Operation Optimization of New Generation Power Trading Platform" (Grant No. 5400-202040230A-0-0-00).

REFERENCES

Feng Xinyang, Shen Jianjing. REST and RPC: Comparative analysis of two web services architectural style [J]. *Journal of Chinese Mini-Micro Computer Systems*, 2010, 7 (7): 1393–1395 (in Chinese).

Kong Xiangrui, Li Peng, Yan Zheng, et al. Stress testing on electricity market with retail transactions opened [J]. *Power System Technology*, 2016, 40 (11): 3279–3286 (in Chinese).

Slee M, Agarwal A, Kwiatkowski M. Thrift: scalable cross-language services implementation [J]. *Facebook White Paper*, 2007 (5): 8.

Wang Wei, Yu Lihua. RPCI: RPC for internet applications [J]. *Computer Engineering And Applications*, 2013, 49 (21): 106–110 (in Chinese).

Zhang Xian, Zheng Yaxian, Geng Jian, et al. Design of direct trading platform for electricity users and power generation enterprises supporting whole business procedure [J]. *Automation of Electric Power Systems*, 2016, 40 (3): 122–128 (in Chinese).

Advances in Energy, Environment and
Chemical Engineering – Abdullah & Osman (Eds)
© 2023 The Author(s), ISBN: 978-1-032-36083-6

The evolutionary game between government and shipping company under the background of shore power

Yulin Zhao* & Yanping Meng
Shanghai Maritime University, Shanghai, China

ABSTRACT: To solve the "shore-side hot and ship-side cold" problem in the promotion process of shore power, this paper constructed a two-party evolutionary game model between the local government and the shipping company, and revealed the evolution law of the two parties' behavior strategy, and simulated the key factors in the model by using MATLAB. The simulation results show that, under the influence of income, cost, and other factors, the reduction of regulatory costs and the increase of fines will make the evolution rate of shipping companies' behavior faster, but the selection strategy will change with the cycle, and will not reach a stable equilibrium state, and reducing the extra cost brought by using shore power can effectively increase the willingness of shipping companies to use shore power. Based on this, the basic suggestions for accelerating the application and popularization of shore power are put forward.

1   INTRODUCTION

Ports are important transportation channels for international goods trade, and the port economy is booming in the context of the global economy. However, due to the increase in shipping activities, emissions of a large number of harmful substances surge greenhouse gases, and ports and shipping activities have also become the main activities of energy consumption and pollution sources. The most important reason for the pollution of the ship in port is to use the auxiliary engine to provide power for ship work. Shore power refers to berthed ships turning off auxiliary engines and using clean energy from the port to power onboard systems and meet the power needs of all onboard facilities such as lighting and unloading facilities. According to the literature, shore power has been proved to be an effective solution for port emission reduction.

To promote the adoption of shore power, the Chinese government provides considerable subsidies to ports and ships for equipment installation and shore power facility operation. From the current policy perspective, the government is still encouraging ports to strengthen their construction, but the focus has shifted to urging normal use of shore power. In September 2021, the Ministry of Communications of China revised the "Management Measures for Ports and Ships using shore power", adding the content that ships will face fines if they fail to use shore power in accordance with regulations, indicating that in the process of promoting shore power in the future, the punishment of ships may be inevitable.

At present, research on the promotion of shore power mainly focuses on the environmental benefits and construction of shore power, and few studies focus on urging the normal use of shore power. This paper studies the synergistic effect between local government and shipping companies in the implementation of shore power abstracts the influencing factors of strategic choice between local government and shipping companies from the perspective of bounded rationality and constructs an evolutionary game analysis model. In addition, the paper uses MATLAB simulation

---

*Corresponding Author: 2437600694@qq.com

DOI 10.1201/9781003330165-26

to analyze the influence of different strategies on the income of both parties and focuses on the influence of government supervision costs, fines, and additional costs of using shore power by shipping companies on the decision-making behavior of both parties, providing a reference for standardizing the behavior of shipping companies and promoting the application process of shore power.

2 MODEL DESCRIPTION

From the current situation, although the price of shore power is lower than that of fuel, shipping companies are not willing to use shore power because shore power equipment needs daily maintenance, and operators need to connect shore power during use. Therefore, the government will urge shipping companies to use shore power through penalties. This paper considers the shore power application promotion system consisting of local government and shipping companies. In the system, the shipping company chooses whether to use shore power according to regulations, and the local government penalizes the shipping company for not using shore power according to regulations. In this process, the government chooses to regulate the behavior of shipping companies and pay the cost of regulation, or it chooses not to. Shipping companies choose to abide by regulations and pay shore power costs, and other additional costs associated with it, such as maintenance costs of shore power equipment and time costs for connecting shore power facilities or choose not to use shore power, pay fuel costs, and pay fines in case of government supervision. When the shipping company does not use shore power, the government needs to pay the environmental treatment cost of pollution caused by auxiliary power generation. When the shipping company uses shore power, the government will gain the environmental benefits brought by the use of shore power. This paper adopts the evolutionary game model, establishes the benefit payment matrix, and uses the Jacobian matrix to calculate the evolutionary stability strategy and the corresponding conditions of stakeholders. Finally, numerical simulation is carried out to illustrate the influence of related parameters on the evolutionary game among stakeholders.

3 MODEL

3.1 *Basic assumptions and model parameters*

Based on the analysis of the game relationship between the stakeholders of the local government and the shipping company, the following hypotheses are proposed.

(1) The local government's strategy selection space is {regulation, no regulation} and the corresponding probability is {x, 1-x}. The shipping company's strategy selection space is {compliance, violation} and the corresponding probability is {y, 1-y}.

(2) The economic parameters set in the model are as follows.
I refers to the environmental benefits achieved by the Government when the shipping company complies with regulations and uses shore power. C_g is the environmental treatment cost paid by the government when the shipping company does not use shore power. D_0 refers to the initial revenue of the shipping company. C_f is the fuel cost paid by the shipping company. C_{sp} is the shore power cost paid by the shipping company. C_{ex} refers to the maintenance, time, and other additional costs incurred by the shipping company when using shore power. F_s is the government fine imposed on a shipping company for not using shore power as required. b is the cost of regulation paid by the government.

3.2 *Analysis of income matrix and income function*

Based on the above assumptions, the mixed strategy game matrix between the local government and the shipping company can be obtained, as shown in Table 1.

Table 1. Local government and shipping company game revenue matrix.

Game player		The shipping company	
		S_1 strategy y	S_2 strategy $1-y$
The local government	G_1 strategy x	$[I-b, D_0 - C_{sp} - C_{extra}]$	$[-C_g + F_s - b, D_0 - C_f - F_s]$
	G_2 strategy $1-x$	$[I, D_0 - C_{sp} - C_{ex}]$	$[-C_g, D_0 - C_f]$

(1) The expected returns of the local government under G_1 and G_2 strategies are as follows:

$$U_{g1} = y(I-b) + (1-y)(-C_g + F_s - b) \tag{1}$$
$$U_{g2} = yI + (1-y)(-C_g) \tag{2}$$

Then the average expected revenue of the local government can be described as:

$$\overline{U}_g = xU_{g1} + (1-x)U_{g2} \tag{3}$$

Thus, the replication dynamic equation of the probability of local government choosing regulatory strategy can be deduced as:

$$F(x) = x(U_{g1} - \overline{U}_g) = x(1-x)(F_s(1-y) - b) \tag{4}$$

(2) The expected revenue of the shipping company using S_1 and S_2 are as follows:

$$U_{s1} = x(D_0 - C_{sp} - C_{extra}) + (1-x)(D_0 - C_{sp} - C_{ex}) \tag{5}$$
$$U_{s2} = x(D_0 - C_f - F_s) + (1-x)(D_0 - C_f) \tag{6}$$

Then the average expected revenue of the shipping company can be described as:

$$\overline{U}_s = yU_{s1} + (1-y)U_{s2} \tag{7}$$

Therefore, the replication dynamics equation of the probability that shipping companies choose to use shore power strategy as required can be deduced as:

$$F(y) = y(U_{s1} - \overline{U}_s) = y(1-y)(C_f - C_{sp} - C_{ex} + xF_s) \tag{8}$$

Therefore, the two-dimensional replication dynamic system formed by the government and the shipping company is composed of Equations (4) and (8)

4 STABILITY AND SIMULATION ANALYSIS OF GAME MODEL

4.1 Calculation of evolutionary equilibrium points

The dynamic system of replication reflects the strategy adjustment process of the players of bounded rationality through learning and imitation. To find the equilibrium point when the two players reach the stable state, we set the dynamic equation as $F(x) = 0, F(y) = 0$. There are four equilibrium points $E_1(0,0), E_2(0,1), E_3(1,0), E_4(1,1)$. In addition, when $C_{sp} + C_{extra} \leq C_f + F_s, 0 \leq (F_s - b)/F_s \leq 1$, $E_5(x^*, y^*)$ may also be equilibrium points, where $x^* = (C_{sp} + C_{extra} - C_f)/F_s, y^* = (F_s - b)/F_s$.

4.2 *Evolutionary stability analysis*

According to Friedman's method, if the determinant of the Jacobian is greater than zero and the trace is less than zero, then the system is stable and the equilibrium point evolves a stable strategy (ESS). Otherwise, the system is unstable and the equilibrium point is the saddle point. According to Equations (4) and (8), the Jacobian matrix of the system can be described as:

$$
J = \begin{pmatrix} \dfrac{dF(x)}{dx} & \dfrac{dF(x)}{dy} \\[2mm] \dfrac{dF(y)}{dx} & \dfrac{dF(y)}{dy} \end{pmatrix} = \begin{pmatrix} (1-2x)(F_s(1-y)-b) & -x(1-x)F_s \\[2mm] y(1-y)F_s & (1-2y)(C_f - C_{sp} - C_{ex} + xF_s) \end{pmatrix}
$$

(9)

Table 2. Stability analysis of equilibrium points.

Points	$Det(J)$	$Trc(J)$	Stability	Stability condition
$(0,0)$	$(b-F_s)(C_{sp}+C_{ex}-C_f)$	$C_f + F_s - (C_{sp}+C_{ex}+b)$	Condition (1)	$b > F_s$, $C_{sp}+C_{ex} > C_f$
$(0,1)$	$-b(C_{sp}+C_{ex}-C_f)$	$C_{sp}+C_{ex}-C_f-b$	Condition (2)	$C_{sp}+C_{ex} < C_f$
$(1,0)$	$-(b-F_s)(C_{sp}+C_{ex}-C_f-F_s)$	$-(C_{sp}+C_{ex}) +C_f+b$	Condition (3)	$b < F_s$, $C_{sp}+C_{ex} > C_f+F_s$
$(1,1)$	$b(C_{sp}+C_{ex}-C_f-F_s)$	$C_{sp}+C_{ex}+ b-C_f-F_s$	Condition (4)	$C_{sp}+C_{ex} > C_f+F_s$ $C_{sp}+C_{ex}+b < C_f+F_s$
(x^*,y^*)	$\dfrac{b(C_{sp}+C_{ex}-C_f)(b-F_s)(C_{sp}+C_{ex}-C_f-F_s)}{F_s^2}$	0	Unstable	–

The five equilibrium points were substituted into the Jacobian matrix, and the determinant values and traces were shown in Table 2. And only if $\det(J) > 0$ and $trc(J) < 0$, the equilibrium point is the evolutionarily stable strategy of the system. Since trc (J) of $E_5(x^*,y^*)$ is identical to 0, the two conditions in Condition (4) are mutually exclusive. Therefore, only equilibrium points $E_1(0,0), E_2(0,1), E_3(1,0)$ are studied, and the stability strategy of the evolutionary game is discussed in three situations.

(1) Scenario 1: If the parameters meet Condition (1), that is, if the government's regulatory cost is greater than the fine benefit, and the sum of shore power cost and additional cost of using shore power is greater than the fuel cost then the government tends to choose "no regulation" as the stability strategy, and the ship company's stability strategy is "violation". $E_1(0,0)$ is the gradual stability point; {no regulation, violation} is the evolutionary stability strategy, that is, neither party participates in the process of promoting the use of shore power.
(2) Scenario 2: If the parameters meet Condition (2), that is, if the sum of shore power cost and additional cost of using shore power is less than fuel cost, then $E_2(0,1)$ is the gradual stability point; {no regulation, compliance} is the evolutionarily stable strategy of the system, that is, the shipping company will use shore power without government intervention.
(3) Scenario 3: If the parameters meet Condition (3), that is, if the government's regulatory cost is less than the penalty benefit, the sum of shore power cost and additional cost of using shore power is greater than the sum of fuel cost and penalty cost. Therefore, the government tends to choose "regulation" as the stabilization strategy, while the shipping company will choose a "violation" strategy to achieve stabilization. $E_3(1,0)$ is the gradual stability point of the system, that is, in the application and promotion of onshore power, the government actively supervises, but shipping companies do not have the driving force to use onshore power.

5 NUMERICAL SIMULATION AND PARAMETER ANALYSIS

We use MATLAB to simulate the evolution process and take (0, 0) as the starting point to discuss the optimal path from the least ideal state to the ideal state. Parameter values are mainly based on two aspects: (1) literature and actual research and (2) the equality balance principle. In the simulation, the willingness of local government and shipping companies to use shore power is decreasing successively, so the initial probability of government supervision and shipping company compliance is 0.6 and 0.4, respectively.

(1) The impact of regulatory cost (*b*) on system evolution

Other parameters are kept unchanged to reduce the government's regulatory costs, and six simulations were conducted. The simulation results of the evolutionary game are shown in Figure 1. As can be seen from Figure 1a, the change of regulatory cost *b* does not change the violation strategy of the shipping company, but only the regulatory probability of the government. The reduction of the regulatory cost *b* may promote the evolution of the local government from a "no regulation" strategy to a "regulation" strategy.

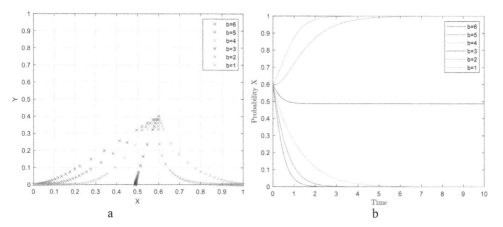

Figure 1.   Parametric analysis of the regulatory cost (*b*).

As can be seen from Figure 1b, when $b = 3$, the probability of the government choosing the "regulation" strategy will not trend to 0 or 1, indicating that it cannot reach a stable state. When $b < 3$, the probability of the government choosing the "regulation" strategy tends to be 1, and the smaller the value of *b*, the faster the convergence rate is. When $b > 3$, the probability of the government choosing the "regulation" strategy tends to be 0, and the greater the value of *b*, the faster the convergence rate.

This means for the local governments to take more action to let the shipping company choose compliant to use shore power strategy so that on the one hand can make the government save the environmental pollution caused by the shipping company to use fuel management costs. On the other hand, this can also make the local government environmental benefits brought by the shipping company to use shore power. However, the supervision cost of the local government should be controlled within a reasonable range. Only in this way can the local government choose the "regulation" strategy.

(2) The impact of fines F_s on system evolution

Ensure that other parameters remain unchanged, and increase the government's fine, and six simulations were conducted. The simulation results of the evolutionary game are shown in Figure 2, showing that changes in the government's penalty for shipping company violations may also change the system from a stable state of {no regulation, violation} to an unstable state.

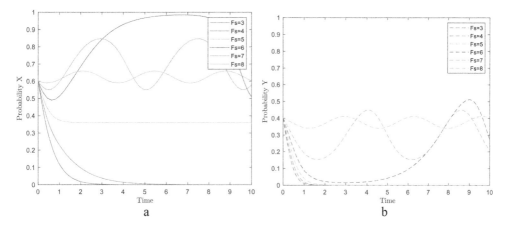

Figure 2.　Parametric analysis of fines (F_s).

As can be seen from Figure 2a, when $F_s = 5$, the probability of the government choosing the "regulation" strategy will not trend to 0 or 1, that is, it cannot reach a stable state. When $F_s < 5$, the probability of the government choosing the "regulation" strategy will trend to 1, and the smaller the value of b, the faster the convergence rate will be. When $F_s > 5$, if the fine continues to increase, the government's game curve becomes shorter and the amplitude becomes smaller, indicating that although increasing the fine will speed up the local government's choice of "regulation" strategy, it cannot reach the ideal state.

As can be seen from Figure 2b that when $F_s \leq 5$, the probability of a shipping company choosing a "compliance" strategy tends to be 0, and the smaller the penalty (F_s) is, the faster the convergence rate is. When $F_s > 5$, if the fine continues to increase, the game curve of the shipping company becomes shorter and the amplitude becomes smaller, indicating that although the increase in fine will accelerate the speed of the "compliance" strategy of the shipping company, it cannot reach the ideal state.

This shows that for the local government, increasing fines cannot make the enterprise reach a stable state, but keep it in a state of the game with the local government. Therefore, the government's single fine policy cannot completely drive the shipping company from the initial strategy of violation to abide by the rules.

(3) The impact of the extra cost (C_{ex}) of shore power on system evolution.

Other parameters remain unchanged to reduce the extra cost of using shore power for the shipping company, and six simulations were conducted. As shown in Figure 3a, the additional costs associated with changing the use of shore power by shipping companies may shift the system from a state of {regulated, non-regulated} to a state of {unregulated, non-regulated}.

As can be seen from Figure 3b, when $C_{ex} = 2$, the probability of the shipping company choosing a "compliance" strategy does not tend to 0 or 1, that is, it cannot reach a stable state. When $C_{ex} < 2$, the probability of the shipping company choosing the "compliance" strategy tends to be 1, and the smaller the extra cost caused by the use of shore power is, the faster the convergence rate is. When $C_{ex} > 2$, the probability of the shipping company choosing the "compliance" strategy tends to 0, and the greater the extra cost brought by the use of shore power, the faster the convergence rate.

For shipping companies, if the extra cost brought by the use of shore power is low enough that the sum of the extra cost and shore power cost is less than the fuel cost, even if the local government chooses the strategy of "no regulation", shipping companies will change from the "violation" strategy to the direction of "compliance". Therefore, the government can subsidize the extra cost of using shore power for shipping companies, such as giving priority to berthing ships using shore power to reduce their time cost, and promoting the technical standardization

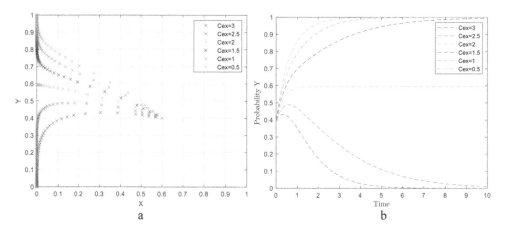

Figure 3. Parametric analysis of extra cost (C_{ex}).

of shore power equipment to improve the operational efficiency of ships accessing shore power to increase the willingness of shipping companies to use shore power.

6 CONCLUSION AND ENLIGHTENMENT

In view of the problem of "shore-side heat and ship-side cold" in the process of shore power application and promotion, this paper uses evolutionary game theory to construct a two-party evolutionary game model between local government and shipping company, analyzes the influencing factors of strategy choice of each participant in the process of shore power promotion, and uses the influencing factors to regulate the behavior choice of game subjects. In this paper, the simulation starts from the unsatisfactory evolutionary stability point and further analyzes the key factors that improve the willingness of the shipping company to use shore power and promote its compliance. The main conclusions are as follows.

(1) For the government, when the cost of supervision is high, the government will be less active in choosing a "regulation" strategy. When the cost of supervision is reduced to a certain extent, the government will be more active in choosing a "regulation" strategy. Until the cost of supervision reaches a value, the government will choose a "regulation" strategy. This shows that the cost of supervision of local government should be controlled within a reasonable range. Only in this way can the government choose the strategy of "regulation".

(2) When the government imposes a fixed penalty on the shipping company for violations, the increase of the penalty accelerates the evolution rate of the behavior of the participants, but the selection strategy will change with the cycle and cannot reach a stable equilibrium state.

(3) For shipping companies, if the extra cost brought by the use of shore power is low enough that the sum of the extra cost and shore power cost is less than the fuel cost, even if the local government chooses the strategy of "no regulation", shipping companies will change from the "violation" strategy to the direction of "compliance ". Therefore, the promotion of the use of shore power should not only rely on the increase of fines but also need to reduce the additional costs brought by the use of shore power to improve the willingness of shipping companies to use shore power.

Based on the above research conclusions, the following suggestions are put forward to further promote the application process of shore power:

(1) We should improve the supervision and punishment measures for the use of shore power. The shipping company's illegal behavior of not using shore power should be punished following relevant regulations. Change the fixed penalty strategy, form effective dynamic supervision and punishment mechanism from various aspects, and improve the effectiveness of supervision and punishment in the application and promotion process of onshore electricity to reduce the violations of shipping companies and improve the environmental benefits of port cities.

(2) We should also strengthen the support for the promotion of shore power technology, promote the technical standardization of shore power equipment to improve the operation efficiency of ships accessing shore power, and give priority to the berthing of ships using shore power to reduce their time cost, thereby improving the willingness of shipping companies to use shore power.

REFERENCES

Ballini, F. & Bozzo, R. (2015). Air pollution from ships in ports: The socio-economic benefit of cold-ironing technology. Res. *Transport. Bus. Manage.* 17, 92–98.

Chen, J. et al. (2019). The relationship between the development of global maritime fleets and GHG emission from shipping. *J. Environ. Manag.* 242, 31–39.

Lu, H. & Huang, L. (2021) Optimization of Shore Power Deployment in Green Ports Considering Government Subsidies. *Sustainability*, 13, 1640.

Sciberras, E. A et al. (2016). Cold ironing and onshore generation for airborne emission reductions in ports [J]. *Proceedings of the Institution of Mechanical Engineers, Part M: Journal of Engineering for the Maritime Environment*, 2016, 230: 67–82.

Winkel R et al. (2016). Shore side electricity in Europe: potential and environmental benefits. *Energy Policy.* 88, 584–93.

Yin, M et al. (2020) Policy implementation barriers and economic analysis of shore power promotion in China [J]. *Transportation Research Part D: Transport and Environment*, 87: 102506.

Zis, T et al. (2014). Evaluation of cold ironing and speed reduction policies to reduce ship emissions near and at ports, *Marit. Econ. Logist.* 16 (4), 371–398.

Advances in Energy, Environment and
Chemical Engineering – Abdullah & Osman (Eds)
© 2023 The Author(s), ISBN: 978-1-032-36083-6

Evaluation of development quality of natural gas hydrate

Qiannan Yu
College of Mechanical and Electrical Engineering, Guangdong University of Petrochemical Technology, Maoming, China

Han Zhang
Guangdong Huanqiu Guangye Engineering Co., Ltd, Guangzhou, China

Zhijing Chen, Kun Zhang* & Ning Li
College of Mechanical and Electrical Engineering, Guangdong University of Petrochemical Technology, Maoming, China

ABSTRACT: Natural gas hydrate has the potential to be the most promising alternative energy source for oil and gas. Evaluation of development quality is a key issue for the development of natural gas hydrate reservoirs which has typical characteristics of high risk, high difficulty, and high input. The factor set is established based on the main controlling factors, and the scoring criteria of the main controlling factors are defined. The weighted set is established by the comprehensive weight calculated by the analytic hierarchy process, and then a fuzzy comprehensive evaluation of the development quality of the natural gas hydrate reservoir is further completed. This method is used to evaluate the development quality of Messoyakha and other typical natural gas hydrate reservoirs, and the applicability and feasibility of the method are verified. The method of development quality evaluation may provide theoretical basis and technical support for the efficient development of natural gas hydrate reservoirs.

1 INTRODUCTION

Natural gas hydrate with huge reserves and high combustion value is the most promising alternative energy source for oil and gas (Collett 2002). Development of natural gas hydrate has typical characteristics of high risk, high difficulty, and high input, and evaluation of development quality in technical and economic aspects is a key issue for natural gas hydrate reservoirs, as the development programme design is risky and limited by burial conditions, development technology and engineering facilities (Sahu 2020; Zhao 2020). At present, research on the development quality of natural gas hydrate is still in the initial stage. The applicability of evaluation methods widely used in conventional oil and gas development such as evaluation expert investigation, tree decision-making, grey correlation analysis, comprehensive scoring, and artificial neural network method is limited. The method of development quality evaluation based on fuzzy comprehensive evaluation has good applicability and may provide theoretical basis and technical support for the efficient development of natural gas hydrate reservoirs.

2 METHODOLOGY

Compared with evaluation methods commonly used for complex systems, the fuzzy comprehensive evaluation method is preferred for the evaluation of the development quality of natural gas hydrate

---

*Corresponding Author: zhangkun@gdupt.edu.cn

DOI 10.1201/9781003330165-27

reservoirs. The fuzzy comprehensive evaluation method can judge an object based on the comprehensive study of multiple relevant factors and turn the qualitative evaluation into a quantitative evaluation based on the affiliation theory of fuzzy mathematics.

2.1 Set of factors

A set of factors refers to the set of key factors that influence the judgment of objects, and a set of factors for the evaluation of the development quality of gas hydrate reservoir are defined as below.

$$Z = \{z_1, z_2, \cdots, z_n\} \tag{1}$$

Where Z = factor set, z_i = key factors, i=1,2,...,n.

2.2 Set of weights

Influencing degree of factor z_i varies greatly, and the impact weights of each influencing factor x_i are characterized. The weights of each relevant factor are assigned as a fuzzy subset X of the set of weights on the set of factors.

$$X = \{x_1, x_2, \cdots, x_n\} \tag{2}$$

Where X = weight set, x = evaluation factor, i=1,2,...,n.

In practice, it is generally required that x_i satisfy normalization and non-negative conditions.

$$\sum x_i = 1 \tag{3}$$

$$x_i \geq 0 \tag{4}$$

Where i=1,2,...,n.

X value in weight set is an objective quantification of subjective judgments using dimensionless processing, weighing the influence degree of parameters represented by influencing factor, which can be calculated by hierarchical analysis depending on the degree of the proportion of each factor in the quality evaluation.

2.3 Set of evaluation

Based on the qualitative evaluation of the development quality of the gas hydrate reservoir, a numerical interval for scoring the development quality is set, corresponding to different evaluation results. The evaluation set is defined as below.

$$Y = \{y_1, y_2, \cdots, y_m\} \tag{5}$$

Where Y = evaluation set, y = evaluation factor, i = 1,2,...,n.

Assuming that qualitative evaluation of the development quality of gas hydrate reservoir has four grades of excellent, good, moderate, and poor, corresponding to the evaluation results of Level I, Level II, Level III, and Level IV. The scoring intervals correspond to the four grades as shown in Table 1.

2.4 Evaluation results

Factor in the set of factors refers to the single-factor evaluation result in a fuzzy subset on X, further building a single-factor evaluation vector for the factor set.

$$R_i = \{r_{i1}, r_{i2}, \cdots, r_{im}\} \tag{6}$$

Where R_i = single-factor evaluation vector, i = 1,2,...,n.

Table 1. Scoring intervals of development quality evaluation of natural gas hydrate reservoir.

Level	Evaluation grades	Score
Level I	Excellent	>80
Level II	Good	75-80
Level III	Medium	70-75
Level IV	Poor	<70

Multiple single-factor evaluations are characterized by means of affiliations on intervals divided by sensitivity parameters to form the single-factor evaluation matrix.

$$R = \begin{bmatrix} R_1 \\ R_2 \\ \vdots \\ R_n \end{bmatrix} = \begin{bmatrix} r_{11} & r_{12} & \cdots & r_{1m} \\ r_{21} & r_{22} & \cdots & r_{2m} \\ \vdots & & & \vdots \\ r_{n1} & r_{n2} & \cdots & r_{nm} \end{bmatrix} \tag{7}$$

The composite judgment of a particular judgment object is calculated from the weight set X and the single-factor judgment matrix R as the evaluation set.

$$Q = X \cdot R \tag{8}$$

Using the weighted average comprehensive evaluation model, the operator pair $(\bullet, +)$ is selected, and the evaluation result is obtained. q_j value is the comprehensive evaluation result:

$$q_j = \sum_{i=1}^{n} x_i r_{ij} \tag{9}$$

Where j = 1,2,…,m.

The comprehensive evaluation results can be compared with the dividing intervals in Table 2 to determine the current development quality level of natural gas hydrate reservoirs.

3 MAIN CONTROLLING FACTORS

There are many factors affecting the development quality of natural gas hydrate reservoirs, and the degree of influence of different factors varies greatly. The establishment of factor sets in the evaluation process is too complicated, and the calculation of weight sets is difficult, resulting in limited accuracy of the evaluation results of the development quality of natural gas hydrate reservoirs. The main controlling factors of gas hydrate reservoir development quality should be clarified, and then the weights of each factor in the set of factors can be calculated by hierarchical analysis.

3.1 *Fundamental property parameters*

The development quality of natural gas hydrate deposits should be evaluated from different perspectives such as deposit resources, gas production potential, engineering implementation difficulty, and safety technology requirements (Li 2016; Max 2005; Yang 2019).

The factors influencing the development quality of natural gas hydrate reservoirs are mainly engineering-geological parameters, reservoir property parameters, and fluid properties parameters in three categories with a total of 28 parameters.

Table 2. Basic property parameters of natural gas hydrate reservoir.

Category	Parameter
Engineering geological parameters (EGP)	Reservoir burial depth D_r
	Depth of Covered water D_w
	Reservoir inclination A_r
	Distance from the coast L_r
	The amount of reservoir resources R_r
Reservoir property parameters (RPP)	Original formation temperature T_i
	The temperature difference between the formation and bottom of the well T_{gp}
	Original formation pressure p_i
	The pressure difference between the formation and bottom of the well p_{gp}
	Raw water saturation S_{wi}
	Original hydrate saturation S_{hi}
	specific heat of rocks C_r
	Thermal conductivity of rock K_r
	Formation permeability K
	Formation rock porosity $\varnothing_0$
Fluid property parameters (FPP)	Air-water capillary force p_c
	Specific heat of Methane C_{vg}
	Specific heat of Water C_w
	Specific heat of Hydrate C_h
	Water density ρ_w
	Hydrate density ρ_h
	Water viscosity μ_w
	Methane viscosity μ_g
	Thermal conductivity of Methane K_g
	Thermal conductivity of water K_w
	Thermal conductivity of hydrate K_h
	Balance pressure p_{eq}
	Constant decomposition rate k_d

3.2 Main controlling factors

The main controlling factors are influencing factors affecting significance. Sensitivity S_i can be characterized as the degree of distortion of the objective function due to a change in the variable of interest.

$$S_i = \frac{\Delta \overline{F}_i / F_o}{|w_i|} \tag{10}$$

$$F_o = \int_0^T R_o V_1, V_2, \cdots V_n, t dt \tag{11}$$

$$\Delta F_i = \int_0^T [R_i V_1, V_2, \cdots V_n, t - R_o V_1, V_2, \cdots V_n, t] dt \tag{12}$$

$$w_i = \frac{V_i - V_{io}}{V_i} \tag{13}$$

Where S_i = sensitivity, F = function of V, R = correlation coefficient, V = variable in development process, n = number of V, T = time, w_i = offset factor.

Parameter sensitivity analysis is based on numerical simulation, and economic evaluation results are based on some typical gas hydrate reservoirs.

According to the sensitivity, the above parameters are sorted, and the parameters with the greatest sensitivity are selected, including reservoir burial depth D_r, original formation temperature

Table 3. Sensitivity analysis results of parameters of natural gas hydrate development.

Category	Parameter	Sensitivity
Engineering geological parameters (EGP)	D_r	1.73
	D_w	6.12×10^{-2}
	A_r	7.85×10^{-4}
	L_r	8.91×10^{-3}
	R_r	9.79×10^{-2}
Reservoir property parameters (RPP)	T_i	10.35
	T_{gp}	3.09×10^{-2}
	P_i	1.03
	P_{gp}	5.08×10^{-2}
	S_{wi}	4.65×10^{-2}
	S_{hi}	5.31
	C_r	7.63×10^{-3}
	K_r	2.55
	K	9.01×10^{-2}
	$\varnothing_0$	0.97
Fluid property parameters (FPP)	p_c	2.72×10^{-2}
	C_{vg}	5.12×10^{-2}
	C_w	9.65×10^{-3}
	C_h	3.88
	ρ_w	4.07×10^{-2}
	ρ_h	1.56×10^{-3}
	μ_w	1.29×10^{-2}
	μ_g	2.75×10^{-2}
	K_g	3.23×10^{-5}
	K_w	9.68×10^{-5}
	K_h	2.24×10^{-4}
	P_{eq}	7.24
	k_d	1.08

T_i, original formation pressure P_i, original hydrate saturation S_{hi}, formation of rock thermal conductivity K_r, formation of rock porosity $\varnothing_0$, hydrate specific heat C_h, hydrate phase equilibrium pressure P_{eq} and hydrate decomposition rate constant k_d, as the main controlling factors in the factor concentration of natural gas hydrate reservoir development quality evaluation.

3.3 Scoring criteria

Referring to the conventional gas field development experience and gas hydrate indoor research results, the different factors were assigned scores to determine the scoring basis for the main controlling factors.

4 WEIGHT OF MAIN CONTROLLING FACTORS

Using the analytic hierarchy process to clarify the weight of each factor in the factor set, it is needed to first establish a judgment matrix of relative importance among factors, solve the judgment matrix eigenvectors, and then calculate the weights of the relative importance of each factor within the hierarchy, and weight the calculation to obtain the comprehensive weight.

Table 4. The score of main controlling factors for natural gas hydrate reservoir development.

Score	60	70	80	90	100
D_r	>900	600-900	300-600	150-300	<150
P_i	<1	1-5	5-10	10-15	>15
T_i	<5	5-10	10-20	20-30	>30
S_{hi}	<0.1	0.1-0.3	0.3-0.5	0.5-0.7	>0.7
K_r	<0.2	0.2-0.4	0.4-1	1-2	>2
$\varnothing_0$	<0.25	0.25-0.35	0.35-0.45	0.45-0.55	>0.55
C_h	>2100	1900-2000	1800-1900	1700-1800	<1700
P_{eq}	>10	8-10	6-8	4-6	<4
k_d	<10-16	10^{-16}-10^{-15}	10^{-15}-10^{-14}	10^{-14}-10^{-13}	>10^{-13}

4.1 Hierarchy of main controlling factors

The main controlling factors for the development quality of natural gas hydrate reservoirs have a clear hierarchical structure.

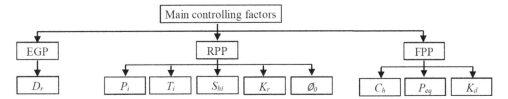

Figure 1. Hierarchical structure chart of main controlling factors in development quality evaluation of natural gas hydrate reservoir.

In the first level of the hierarchy, there are three factors: engineering-geological parameters(EGP), reservoir properties parameters(RPP) and fluid properties parameters(FPP). There is D_r in the second level of EGP, there are five factors P_i T_i, S_{hi}, K_r and $\varnothing_0$ in the second level of RPP, and there are three factors C_h, P_{eq} and k_d in the second level of FPP.

4.2 Intra-level weight

The 1-9 scale method uses scale value to compare the importance of different elements, and the larger the scale value, the more significant the importance of two elements compared to each other. A judgment matrix is established, the scores of main controlling factors are assigned based on the feature assignment method, and the weight values of each parameter are calculated.

The calculation of the judgment matrix W, the weight coefficient X_i is obtained by calculating the maximum feature root of the judgment matrix W and the corresponding feature vector of the judgment matrix.

$$W = \begin{bmatrix} C_{11} & C_{12} & \cdots & C_{1n} \\ C_{21} & C_{22} & \cdots & C_{2n} \\ \vdots & & & \vdots \\ C_{n1} & C_{n2} & \cdots & C_{nn} \end{bmatrix} \tag{14}$$

Where C_{ij} = scale value of the factor i compared to the factor j.

$$C_{ii} = 1 \tag{15}$$

$$C_{ji} = 1/C_{ij} \tag{16}$$

Where i,j=1,2,...,n.

When testing the consistency of the judgment matrix, the ratio of the consistency index (CI) and the stochastic consistency index(RI) is called the tested coefficient (CR). When CR is less than 0.1, the judgment matrix passes the consistency test, otherwise, the judgment matrix is readjusted until it passes the consistency test.

4.3 Weight of the first level

In the first level, the degree of three types of parameters changes continuously during the development process. The judgment matrix of the first level is established, and weights are calculated.

Table 5. Judgment matrix of the first level.

Parameter	EGP	RPP	FPP	Weight
EGP	1	0.2	0.2	0.0909
RPP	5	1	1	0.4545
FPP	5	1	1	0.4545

4.4 Weight of the second level

The judgment matrix of the second level of reservoir properties parameters (RPP) and fluid properties parameters (FPP) is established separately, and weights are calculated.

Table 6. Judgment matrix of reservoir property parameters.

Parameter	T_i	P_i	S_{hi}	K_r	$\varnothing_0$	Weight
T_i	1	7	2	5	7	0.5065
P_i	0.143	1	0.333	2	1	0.0197
S_{hi}	0.5	3	1	3	3	0.2470
K_r	0.2	0.5	0.333	1	2	0.0854
$\varnothing_0$	0.143	1	0.333	0.5	1	0.0695

Table 7. Judgment matrix of fluid property parameters.

Parameter	C_h	P_{eq}	k_d	Weight
C_h	1	0.5	3	0.3090
P_{eq}	2	1	5	0.5816
k_d	0.3333	0.2	1	0.1095

Calculation results are checked against evaluation results and the weight of each main controlling factor to meet consistency requirements.

4.5 Set of weights

The comprehensive weights of the main controlling factors for the development of natural gas hydrate reservoirs are calculated based on the weight of the first level and the weight of the second level.

Table 8. Comprehensive weight of main controlling factors for the development of natural gas hydrate reservoirs.

	EGP		RPP			FPP			
Parameter	D_r	T_i	P_i	S_{hi}	K_r	$\varnothing_0$	C_h	P_{eq}	k_d
Weight	0.0909	0.2302	0.0417	0.1122	0.0388	0.0316	0.1404	0.2643	0.0498

Synthetic weight set X= [0.0909,0.2302,0.0417,0.1122,0.0388,0.0316,0.1404,0.2643,0.0498].

5 DEVELOPMENT QUALITY EVALUATION OF TYPICAL NATURAL GAS HYDRATE RESERVOIRS

Exploration and production test work were conducted in Messoyakha, Mallik, Alaska, Nankai Trough, and Shenhu Sea Area, basic data and test production gas information were obtained. Evaluation of the development quality of typical gas hydrate reservoirs using the evaluation method in this paper is used to further explore the applicability of the evaluation method.

Table 9. Main controlling parameters for development quality evaluation of Messoyakha gas hydrate reservoir.

	EGP		RPP			FPP			
Natural gas hydrate reservoirs	D_r	T_i	P_i	S_{hi}	K_r	$\varnothing_0$	C_h	P_{eq}	k_d
Messoyakha	750	9.1	7.8	0.45	0.56	0.25	1800	6.8	2.9×10^{-14}
Mallik	950	9.8	11.2	0.4	0.89	0.35	2000	10.2	3.7×10^{-14}
Alaska	650	9.5	7.5	0.4	0.72	0.27	1800	6.6	2.5×10^{-14}
Nankai Trough	300	10.6	13.5	0.35	1.37	0.35	2000	10.0	1.2×10^{-13}
Shenhu Sea Area	150	10	8.9	0.3	1.58	0.38	2000	8.5	9.8×10^{-12}

Messoyakha's one-factor judgment matrix R=[70,80,80,80,80,70,80,80,80], Mallik's one-factor judgment matrix R=[60,80,70,80,80,80,70,60,80], Alaska's one-factor judgment matrix R=[70,80,70,80,80,70,80,80,80], Nankai Trough's one-factor judgment matrix R=[90,90,80,80,90, 80,70,70,90], Shenhu Sea Area 's one-factor judgment matrix R=[90,90,70,80,90,80,70,70,100].

The quality evaluation results are calculated by using a weighted average type of composite judging model.

Table 10. Evaluation results of a single factor of typical gas hydrate reservoirs.

Natural gas hydrate reservoirs	EGP	RPP				FPP				Evaluation results Q	Development quality level
	D_r	T_i	P_i	S_{hi}	K_r	$\varnothing_0$	C_h	P_{eq}	k_d		
Messoyakha	70	80	80	80	80	70	80	80	80	78.77	II
Mallik	60	80	70	80	80	80	70	60	80	69.93	III
Alaska	70	80	70	80	80	70	80	80	80	76.95	II
Nankai Trough	90	90	80	80	90	80	70	70	90	80.04	I
Shenhu Sea Area	90	90	70	80	90	80	70	70	100	81.63	I

Moridis analyzed the development quality of the Messoyakha gas hydrate reservoir and rated it as good but average in terms of economics in his study (Moridis 2008). The evaluation results were in the middle to upper range of typical gas hydrate reservoirs, and roughly the same result as obtained by the evaluation method in this paper. The development quality ratings of natural gas hydrate reservoirs in Messoyakha, Mallik, Alaska, Nankai Trough, and Shenhu Sea Area are Level II, Level III, Level II, Level I, and Level I respectively. The evaluation results are generally consistent with the test and production data in relevant studies and reports. Therefore, the quality evaluation method established in this paper for natural gas hydrate reservoirs development is fundamentally applicable and generalizable.

6 CONCLUSION

A fuzzy integrated evaluation method based on the analytic hierarchy process is proposed to carry out the development quality evaluation of natural gas hydrate reservoirs. Sensitivity analysis was performed on basic property parameters of natural gas hydrate reservoirs, and nine parameters with the greatest sensitivity were selected as main controlling factors, and the scoring criteria of main controlling factors were clarified. The comprehensive weight values of the main controlling factors were obtained by the analytic hierarchy process, and a comprehensive weight set was established. A fuzzy integrated evaluation model was developed using basic property parameters and development data of Messoyakha and other typical natural gas hydrate reservoirs to evaluate the development quality, and the applicability and feasibility of the development quality evaluation method proposed in this paper were verified.

ACKNOWLEDGMENTS

This project has been carried out under the framework of the Maoming Municipal Science and Technology Project (2020513), and the study was financially supported by the Foundation for Young Talents in Higher Education of Guangdong, China (2019KQNCX084).

REFERENCES

Collett, T. S. (2002). The energy resource potential of natural gas hydrates. *AAPG Bulletin*, 86(6), 1971–1992.

Li, X. S., Xu, C. G., Zhang, Y.(2016). Investigation into gas production from natural gas hydrate: A review. *Applied Energy*, 172, 286–322.

Max, M. D., Johnson, A. H., & Dillon, W. P. (2005). *Economic geology of natural gas hydrate* (Vol. 9). Springer Science & Business Media.

Moridis, G. J., Collett, T. S., Boswell, R. (2008). Toward production from gas hydrates: assessment of resources, technology, and potential. *In SPE unconventional reservoirs conference*. OnePetro.

Sahu, C., Kumar, R., & Sangwai, J. S. (2020). A comprehensive review on exploration and drilling techniques for natural gas hydrate reservoirs. *Energy & Fuels*, 34(5), 11813–11839.

Yang, L., Liu, Y., Zhang, H. (2019). The status of exploitation techniques of natural gas hydrate. *Chinese Journal of Chemical Engineering*, 27(4), 2133–2147.

Zhao, J., Liu, Y., Guo, X. (2020). Gas production behavior from hydrate-bearing fine natural sediments through optimized step-wise depressurization. *Applied Energy*, 260, 114275.

Advances in Energy, Environment and
Chemical Engineering – Abdullah & Osman (Eds)
© 2023 The Author(s), ISBN: 978-1-032-36083-6

Distributed sewage pretreatment and exhaust device in the drainage system

Ling Shi, Xinyue Wang, Lishan Rong*, Man Lu, Jiyang Wang, Kun Wang & Yilong Luo
School of Civil Engineering, University of South China, Hengyang, China

ABSTRACT: "Distributed sewage pretreatment and exhaust device in drainage system" is a new drainage system based on the principle of sedimentation tank and connector. The two major devices of anti-clogging and exhaust are installed on the inspection wells at the nodes with flat terrain and large flow in the urban drainage pipeline, and the principle of sedimentation tank and connector is used to deal with the problem of sewer pipeline blockage. In addition, the automatic exhaust is realized by using natural energy such as solar energy and wind energy through the exhaust device. The principle of the device is clear, simple, and feasible, and it can be used for a long time only after fixed-point inspection for sludge cleaning and pipeline maintenance.

1 INTRODUCTION

With the development of society, the scale of cities is expanding day by day, and the requirements for drainage are getting higher and higher, whether in life or in industry. The urban drainage system has the functions of sewage and rainwater collection, transportation, and distribution, and is an important infrastructure for urban construction. The smoothness of the drainage system is not only related to urban flood control and drainage, but also affects the urban landscape and environmental sanitation, and restricts the level of urban development. It directly affects the daily life and physical and mental health of the general public. After the long-term operation of the pipeline, sludge and domestic waste accumulate, making the diameter of the pipeline gradually decrease, which affects the drainage speed and causes blockage of the pipeline (Li 2012), and has a great impact on the normal operation of the city, even causing serious urban flooding in the flood season. At the same time, toxic gases such as biogas will also be produced in the underground pipeline. Biogas, as the main gas causing sewer safety accidents, has two main impacts. One is the poisoning incident caused by toxic and harmful gases such as H_2S when the staff enters the sewers for maintenance; the other is the explosion caused by the flammable and explosive gases such as CH_4 when they meet the ignition source (Su 2021). At present, there have been many gas explosion incidents in sewer pipes in various parts of my country, and the safety problem should not be underestimated. This study has important practical significance for solving the problems of drainage pipe network blockage and pipeline dangerous gas emissions and provides a reference for solving urban waterlogging and building a sponge city.

At present, the traditional dredging methods include the winch dredging method, high-pressure water jet dredging method, and water scouring dredging method (Li 2017), which have the characteristics of low cleaning cost, fast speed, high cleaning rate, and no environmental pollution. The process is water-intensive and requires a lot of human assistance. Once there is too much silt that cannot be flushed or the diameter of the pipeline is limited, manual cleaning is required. At this time, the pipeline is filled with toxic gases and liquids, which will threaten the safety of the

---

*Corresponding Author: 148351409@qq.com

dredging workers. Therefore, the drainage pipe network system needs a device that can effectively solve the blockage of the pipe and avoid manual exposure to harmful gases.

In recent years, scholars at home and abroad have done a lot of research on pipeline blockage and gas emissions. For instance, Li Aimin et al (2019) developed a dredging robot for sewer pipes, which realized the robot walking and dredging. Zhang Dong (2014) and others designed an intelligent cleaning device, which uses ultrasonic sensors, cameras, and robotic arms to monitor and clean up the conditions in the pipeline. Gu Peiyue (2015) and others designed a sewer blockage early warning device based on SI1000. The above-mentioned dredging and gas warning devices can be used for reference, but at present, there is no device that fundamentally prevents siltation and combines dredging and exhausting, and the principle is complicated and the operation is difficult, requiring professionals to operate. Installation costs are also higher.

Therefore, based on the existing research, we will focus on solving the problem of pipeline blockage and gas emission and design a new drainage system through improvement and innovation. At present, this new type of distributed drainage system is aimed at the nodes with flat terrain and large flow in urban drainage pipelines. It adopts the node pretreatment method and uses the principle of sedimentation tank and connector to deal with the problem of sewer pipeline blockage. According to the literature research of Liu Zhichang (2011), the anti-clogging device adopts the sonar detection method for pipeline deposits. The sensor head is submerged, and the sonar detection can analyze the shape of the sediment and its variation range. The data is accurate and reasonable. Desilting is carried out by the silt pumping truck at a fixed point, which effectively avoids the situation of pipeline blockage. Automatic exhaust is realized by using natural energy such as solar energy and wind energy through the exhaust device, and the odor and combustible gas are discharged from high places. The device can solve the problems of pipeline blockage and gas discharge to a certain extent.

2 DEVICES

The distributed sewage pretreatment and exhaust device in the drainage system includes an anti-clogging device and an exhaust device. The sediment is settled by the principle of sedimentation tank and communication device, and then the suction truck is connected with the suction pipe, the valve on the upper part of the suction pipe is manually opened, and finally, the suction truck is started for suction and dredging. The harmful gas in the pipeline drives the fan blades to pump the gas to the outside under the action of wind energy. At the same time, under the action of radiant energy, sunlight penetrates through the outer glass and irradiates the inner black glass. The temperature of the air section wrapped by it increases, the gas expands, and the pressure drops, which promotes the directional flow of the gas in the pipeline and realizes the directional and safe discharge of harmful gases. The device diagram is as follows.

2.1 *Anti-blocking device*

The anti-clogging device is mainly composed of a sedimentation tank, a fixed suction pipe, and an intercepting grid, which can intercept and temporarily store some of the easily blocked objects in the sewage, so as to reduce the possibility of clogging unknown pipe sections of the sewer. There will be easily blocked objects in urban underground pipelines. In order to centrally clean the easily blocked objects to prevent them from blocking the pipeline, the existence of the sedimentation tank can temporarily store the easily blocked objects in the pipeline, which is convenient for cleaning the sediment in the pipeline in the future. To a certain extent, it saves manpower and material resources and is easy to clean.

2.1.1 *Sedimentation tank*
This structure combines the principle of the connector at the sedimentation tank, that is, when the sedimentation tank is filled with sewage, the liquid level in the two connected pipe sections is

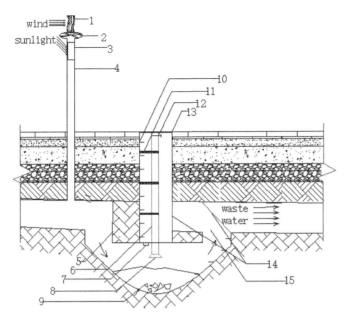

Figure 1.   Device structure. 1- Vertical fan blade 2- Horizontal suction fan 3- Black body 4- Exhaust pipe 5- Water inlet 6- Fixed suction pipe 7- Check valve 8- Bell mouth 9- Sediment 10- Step 11- Fixed bracket 12- Valve 13-Manhole cover 14-Folding blocking grid 15-Water outlet.

horizontal; When the upstream sewage flows through the inspection well, it first flows downward from the water inlet. When the pool is full of sewage, it overflows through the pool outlet and flows to the downstream drain. By increasing the path of sewage flowing through the inspection well, the sedimentation percentage of the easily blocked objects is increased; at the same time, due to the large space in the pool, the horizontal velocity of the water flow decreases rapidly after entering the pool under the condition of stable flow, which is conducive to the easy blockage in the pool. Settlement in the vertical direction.

The structure diagram of the sedimentation tank is as follows:

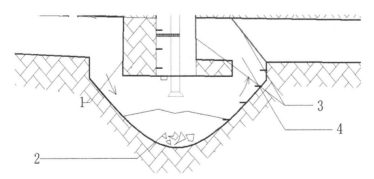

Figure 2.   Sedimentation tank. (1-water inlet 2-sediment 3-folding blocking grid 4-water outlet 5-step).

2.1.2   *Suction tube*
The suction pipe leads from the inspection well to the sedimentation tank, which is easy to operate, takes less time to use, and has an ideal effect on the cleaning of the sedimentation tank.

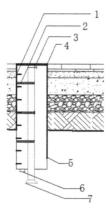

Figure 3.　Suction tube. (1-step 2-fixed bracket 3-valve 4-well cover 5-blocking grid vertical section 6-check valve 7-bell mouth).

The structure diagram of the suction pipe is as follows:

The suction pipe is mainly composed of a normally closed valve on the upper part, a straight pipe, and a bell mouth on the lower part; the bell mouth mainly increases the suction area during operation, so that the easily blocked objects can enter the suction pipe more easily. The municipal management department regularly sends a suction truck to clear the easily blocked objects; before the suction truck starts to work, it is first necessary to connect the suction truck with the suction pipe, manually open the valve on the upper part of the suction pipe, and then start the suction truck Carry out suction. When the suction truck is full or the easy blockages in the pool are cleared, the suction truck stops working and closes the valve.

2.2　Exhaust

The device is composed of fan blades, two-color vacuum glass tubes, etc., and its power comes from solar energy and wind energy. Its main function is to promote air circulation in the exhaust pipe, discharge harmful gases, reduce pipe corrosion and protect personal safety.

Considering various factors, the device is designed to be about 5m high, which can be placed in the park as a sign of scenic spots, and lightning protection lightning terminals are installed around the device.

The device structure diagram is as follows:

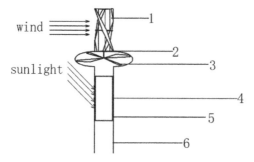

Figure 4.　Exhaust device. (1-vertical fan blade 2-drive shaft 3-horizontal exhaust fan 4-black body 5-transparent glass 6-exhaust pipe).

2.2.1 *Fan blade*

The fan blade consists of a wind receiving blade in the vertical direction and a suction blade in the horizontal direction. Under the action of wind energy, the fan blade rotates and drives the fan blade to rotate through the gear at the end, so as to pump the gas in the pipeline to the outside.

2.2.2 *Double layer vacuum glass tube*

The double-layer vacuum glass tube is divided into an outer layer and an inner layer, the outer layer is transparent glass, and the inner layer is black glass. The black glass adopts RLHY-black body radiation energy production paint. Under the action of radiant energy, sunlight penetrates the outer glass to the inner black glass, the black glass absorbs heat and heats up, and through heat transfer, the temperature of the wrapped air section rises. High, the gas expands and the pressure drops, which promotes the directional flow of the gas in the pipeline.

3 CONTROL METHOD

The nodes with flat terrain and large flow in the urban drainage pipelines according to the drainage pipe network census and the data of the sewage treatment company, and install the device on the basis of the inspection well.

A vertical section interception grid is set at the connection between the inspection well and the downstream pipeline, and an oblique section interception grid is set at the connection between the water outlet channel and the downstream pipeline. The objects flowing down the sewage are intercepted by the intercepting grid, and the filtered sewage is finally discharged through the downstream pipeline. When the upstream sewage flows through the inspection well, it will first flow into the sedimentation tank from the water inlet. At the same time, blockages such as suspended solids and garbage in the sewage settle vertically at the bottom of the sedimentation tank. The sewage from the upstream pipeline enters the sedimentation tank through the water inlet. When passing through the water outlet, the sewage that is not settled in time can be intercepted by the inclined section grid.

In the third step, when the sonar detects that the blockage in the sedimentation tank reaches the standard, the municipal department will send a suction truck to clear the blockage. Before the suction truck starts to work, it is first necessary to connect the suction truck with the suction pipe, manually open the valve on the upper part of the suction pipe, and then start the suction truck for suction. After completion, the suction truck stops working and closes the valve.

In the fourth step, the device can be placed in the park as an attraction sign, and air terminations are installed around the device. The fan blade is designed to be composed of a wind receiving blade in the vertical direction and a suction blade in the horizontal direction. At the same time, the inner layer of black glass in the double-layer vacuum glass tube adopts RLHY-black body radiation production capacity paint. Under the action of solar energy and wind energy, the air in the pipe is promoted to flow in a directional manner, and the odor and combustible gas generated in the underground pipe are continuously extracted and concentrated at high places.

4 ECONOMIC ANALYSIS

After data analysis, the main cost of the device is about 35,000 yuan, about 30,000 yuan for sedimentation tanks and inspection wells, and about 5,000 yuan for exhaust pipes, suction pipes, blocking grids, and black body paint. And it only needs to be selectively installed on the main high-flow nodes. At the same time, each system costs about 10,000 yuan per year for pipeline maintenance and parts replacement. According to data analysis, the processing cycle in the area is about once a month, and the labor costs for fixed-point silt extraction and maintenance are about 5,000 yuan per cycle per system. The cost per cycle of manual dredging and cleaning of traditional decentralized pipelines in the same area exceeds 10,000 yuan. In contrast, the periodic cleaning of the device can save more than 50% of labor costs. The device runs stably, safely, and reliably, without consuming a lot of manpower and energy, and it can be used for a long time only by silt extraction and maintenance based on the test results in the later stage, resulting in stable economic benefits.

5 CONCLUSION

Compared with the prior art, the device provides a set of equipment that integrates anti-blocking and exhausting. The overall structure of the device is simple, and the principles of the communication device, sedimentation tank, solar energy, and wind energy used are simple and feasible, and easy to operate. In addition, the node preconditioning method is adopted, which reduces the treatment burden of the sewage treatment plant. The device does not require a large-scale transformation of the drainage pipe network, thereby reducing the financial pressure on the government. The device research of this project solves the problem of pipeline blockage and gas emission to a certain extent, promotes the green and sustainable development of the city, and conforms to the concept of sponge city development.

REFERENCES

Gu, P.Y., He, W.C., Zhou, Y.W. (2015) Sewer blockage warning device based on SI1000 [J]. *Internet of Things Technology*, 5(03):19–20+23.

Li, A., Yu, H., Li, Y.P., etc. (2019) Design of pipeline dredging robot [J]. *Light Industry Science and Technology*, 35(03):64–65.

Li, H.F. (2012) Research on Hydraulic Desilting Technology of Large Diameter Drainage Pipeline[D]. Chongqing University, 2–3.

Li, J.L., Jin, B.C., Yang, Z.Y. (2017) A Brief Discussion on the Desilting Technology of Municipal Drainage Pipes [J]. *Shanxi Architecture*, 43(27):104–105.

Liu, Z.C. (2011) *Research on sediment deposition and control technology of combined drainage pipelines*[D]. Hunan University.

Su, S., Yang, X.Y., Chen, S., Peng, S.N. (2021) Research on the explosion of biogas in urban sewers [J]. *City Gas*, (S1):153–157.

Zhang, D., Jiang, Y.K., Zheng, Z., etc. (2014) Intelligent sewer cleaning device [J]. *Internet of Things Technology*, 4(04): 13–14.

Advances in Energy, Environment and
Chemical Engineering – Abdullah & Osman (Eds)
© 2023 The Author(s), ISBN: 978-1-032-36083-6

Analysis of China's energy flow diagram in 2018

Quanhong Yuan*, Ning Zhang*, Yafen Xu & Lixing Ma
Guangdong University of Science & Technology, Dongguan, China

ABSTRACT: Nowadays China has entered a critical period of comprehensive green transformation of economic and social development. At present, scholars mostly use the Sankey diagram to draw energy flow diagrams, but there are many lines and many detailed data that are difficult to view. In this paper, according to the energy statistics, China's energy flow diagram in 2018 is drawn by simplified symbols, and the supply side, processing and conversion links, and consumption side of energy were analyzed. According to fossil energy consumption, the reasons for the increase in energy consumption were analyzed. The results show that in 2018, the total primary energy production in China increased by 8.6% compared with that in 2016, and the self-sufficiency rate of China's energy supply was 78.6%, China's total energy consumption in 2018 increased by 7.5% compared with 2016. Finally, some suggestions for energy conservation and energy transformation are put forward.

1 INTRODUCTION

Since the reform and opening up, China's economy has developed rapidly for a long time and made great achievements. But at the same time, China's energy consumption and carbon dioxide emissions have also increased rapidly. Before 2002, China's GDP (gross national product, the same below), energy consumption, and carbon dioxide emissions all grew steadily, and since then, the growth rate has all accelerated, with GDP increasing at first and then slowly, carbon dioxide emissions continuing to grow rapidly, and energy consumption growing steadily.

In order to grasp the new situation, new problems, and new trends in China's energy field in recent years, so as to better control the total amount of fossil energy, save energy resources and accelerate green development. It is necessary to compile China's energy flow diagram and carbon flow diagram. The energy flow diagram, also known as the energy system network diagram or system energy flow and energy efficiency diagram, is a tool to visually analyze the energy supply and demand balance and effective utilization degree of a given energy system. It can intuitively and vividly show the general situation of energy statistics of a region or country, so it has been widely used in the world. For example, Lawrence Livermore National Laboratory in the United States has published the energy flow diagram of the United States every year since 1972; Since 1974, the British Ministry of Trade and Industry has published the British Energy flow diagram every year (Shi 2014).

Chinese scholars' research in this field can be divided into two stages. In the early stage, we mainly paid attention to the study of an energy flow diagram. Since 1985, Tsinghua University Institute of Nuclear Energy first drew the energy flow diagram of Beijing. Since then, Li Zheng and Xie Shichen have drawn the energy flow diagram of China from 2003 to 2010 (Cao 2012; Li 2006). In recent years, scholars have begun to pay attention to the research of energy flow diagrams and carbon flow diagrams at the same time. Since Hu Xiulian, World Wide Fund for Nature (WWF) and Natural Resources Conservation Association (NRDC) first published the energy flow diagram

---

*Corresponding Authors: 3301954919@qq.com and 1565617037@qq.com

and coal flow diagram of China in 2012 (Hu 2012), Fan Jingli and Zhang Hao have drawn the energy flow diagram and carbon flow diagram of China from 2015 to 2017 (Fan 2018; Yu 2019; Zhang 2018) respectively. In addition, some scholars have studied provincial energy flow diagrams (Dong 2020; Yi 2010; Yang 2013; Zhang 2011).

2 ANALYSIS OF CHINA'S ENERGY FLOW DIAGRAM IN 2018

This paper refers to the guidelines of the international energy statistics system and the practices of related literature, appropriately simplifies China's Energy Balance Table -2018 (Energy Statistics Department of National Bureau of Statistics 2020), and combines the import (export) volume with the refueling volume of domestic (foreign) aircraft and ships overseas (domestic) into the total import (export) volume. Secondly, coal washing, coking, and coal gasification are combined into coal products, and gasoline, diesel, solvent oil, and liquefied petroleum gas are combined into oil products.

At present, scholars mostly use the Sankey diagram to draw energy flow diagrams. However, due to the diverse links of the energy system from the production end to the consumption end, there are many lines of the Sankey diagram, and many detailed data are difficult to view. Therefore, in this paper, simplified symbols are used to draw the energy flow diagram, vertical and horizontal unidirectional arrow lines are used to indicate the flow direction, and the number of each tributary is marked so that the whole diagram is clear at a glance. China's energy flow in 2018 is shown in Figure 1.

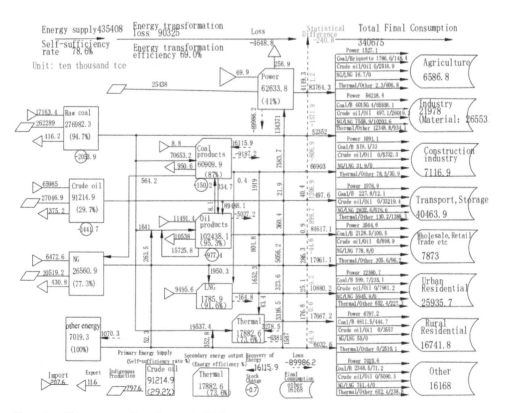

Figure 1. China's energy flow diagram in 2018.

2.1 *The pressure on the supply side continues to increase*

The total production of primary energy in China in 2018 was 34.2×10^8tce (electrothermal equivalent calculation method, the same below), which increased by 8.6% compared with 31.5×10^8tce in 2016. Among them, raw coal, crude oil, natural gas (NG), primary electricity (water, nuclear power, wind power, photovoltaic power generation, etc.), and other energy sources account for 76.6%, 7.9%, 6%, 7.4% and 2.1% of the total primary energy production respectively. Compared with 2016, the main change in primary energy production structure is that domestic crude oil production continues to decline by 1.2%, while natural gas increases by 2.2%, primary electricity increases by 0.7%, and other changes are small. In terms of import and export, China's net energy import increased further in 2018, in which the total energy import (including the refueling of domestic aircraft and ships abroad, the same below) totaled 11.07×10^8tce, the total energy export totaled 1.3×10^8tce, and the net import was $9.77 \times 10^8$8tce. The net import of primary fossil energy is 8.84×10^8tce, and the net import of secondary fossil energy is 0.95×10^8tce. Table 1 shows China's energy security in 2018, in which the self-sufficiency rate of LNG, crude oil, and natural gas is low. Compared with 2016, the export volume of electric power increased by 19.6%, and the imports of liquefied natural gas, oil products, natural gas, crude oil, and raw coal increased by 107.4%, 45.6%, 30.9%, 21.5%, and 9.5% respectively, and the self-sufficiency rates of primary fossil energy and secondary fossil energy were 78.5% and 95.2% respectively, and the imports of fossil energy continued to increase. All kinds of energy stocks increased by 46.23 million tons of standard coal, the total energy supply was 4.35 billion tons of standard coal, and the self-sufficiency rate of energy supply was 78.6%. The problem of energy security was not optimistic.

2.2 *The overall efficiency of energy processing and conversion is basically unchanged*

From 2016 to 2018, the overall efficiency of China's energy processing and conversion was basically flat, at 69.0%. Among them, thermal power and heating increased by 0.5% and 1.1% respectively, but coal products and oil refining decreased by 0.5% and 2.4% respectively, and the liquefaction efficiency of natural gas was basically the same.

In 2018, the loss of China's energy processing process was 110989.4×10^4tce, among which the loss of thermal power generation, power grid transmission and distribution, coal products, oil products, natural gas, and liquefied natural gas converted into standard coal was 89986.2×10^4tce, 4119.3×10^4tce, 9187×10^4tce, and $5,027 \times 10^4$tce respectively.

2.3 *China's energy consumption side continues to grow*

In 2018 China's total final energy consumption was 34.1×10^8tce, an increase of 6.9% over 2016. Among them, the energy consumption of the primary, secondary, and tertiary industries accounted for 1.9%, 58.5%, and 18.8% of the final energy consumption, respectively, which increased by 0.3%, 5.5%, and 7.4% compared with that of 2016. The energy consumption of the secondary and tertiary industries increased greatly, and the pressure on energy conservation and emission reduction became prominent.

In 2018 the terminal consumption of raw coal and coal products in China was 5.2×10^8tce and 6.7×10^8tce, respectively, which decreased by 14.2% and -0.8% compared with 2016. The processing amount of raw coal increased by 7.5%, among which coal-to-oil increased the most, reaching 153.9%, coal-to-gas increased by 52.9%, heat supply increased by 18.9%, and thermal power generation increased by 11.5%. In 2018, 73% of raw coal consumption was used in industry, and it is promising for the industrial sector to speed up "electricity instead of coal".

In 2018 the total consumption of crude oil and oil products in China was 497.6×104tce and $84,617 \times 10^4$ tce, which increased by −44.7% and 8.9% respectively compared with 2016. Among them, the transportation, warehousing, and postal services account for 4.1% of the total oil consumption, the industrial sector accounts for 30.3%, and the residents' living consumption accounts for 11.8%.

In 2018 the terminal consumption of natural gas and liquefied natural gas in China was 17961×10^4 tce and 10880×10^4 tce, respectively, which increased by 14.9% and 101.3% compared with 2016. Among them, the natural gas consumption of industry, construction, residents, agriculture, transportation and wholesale, retail, accommodation, and catering increased by 53.9%, 25.2%, 20.6%, 16.8%, 11.6%, and 10.1% respectively.

In 2018 China's total electricity consumption was 8.38×10^8 tce, which increased by 17% compared with 2016. Among them, transportation, wholesale, retail, accommodation and catering, construction, residential life, industry, and agricultural natural gas consumption increased by 28.5%, 24.8%, 22.4%, 19.4%, 14.3%, and 13.8% respectively. In 2018 China's per capita electricity consumption was 6482.1kW·h, 27.6% more than the world average.

In 2018, China's energy transportation loss was $4,648.8 \times 10^4$ tce, and the balance difference was -240.8×10^4 standard coal, among which the difference (supply-consumption) of raw coal, coal products, and natural gas was negative, and the statistics need to be improved. The loss of natural gas is 2.42 million tons of standard coal. Natural gas leakage not only wastes energy, but also its main component methane has an impact coefficient of 23 times that of CO_2, which will seriously aggravate the greenhouse effect.

3 CONCLUSIONS

In this paper, the energy flow diagram is drawn with simplified symbols, the main conclusions can be summarized as follows:

(1) In 2018 China's primary energy production and energy consumption continued to increase, with the import growth rate of liquefied natural gas, oil products, natural gas, crude oil, and raw coal at the top. The dependence of fossil energy on foreign countries continued to increase, and the problem of oil safety remained significant.
(2) In terms of energy processing and conversion, the production of coal products, oil products, and electricity in China continued to increase in 2018, and the power generation of renewable energy increased rapidly, but the overall efficiency of energy processing and conversion remained basically unchanged.
(3) In 2018 China's total energy consumption was 4.356 billion tons, with the primary, secondary, and tertiary industries increasing by 0.3%, 5.5%, and 7.4% respectively compared with 2016. The energy consumption of the secondary and tertiary industries increased greatly, and the pressure on energy conservation and emission reduction was high.

Facing the shortcomings of fossil energy, China must accelerate the green transformation of energy. Control coal resources, Accelerate the research and application of clean coal technology, eliminate backward production capacity, shut down small coal stoves and eliminate loose coal; Ensure oil safety, speed up bio-oil substitution, vigorously develop new energy vehicles, increase natural gas exploitation, speed up pipeline construction, increase reserve base and popularize natural gas consumption; Accelerate the "two alternatives" and vigorously develop renewable energy such as primary electricity and hydrogen.

ACKNOWLEDGMENT

I would like to show gratitude to the Dongguan Science and Technology Commissioner Project Fund (20201800500602) for its support of the study.

REFERENCES

Bai Mei. The direction and path of high-quality development of the power industry during the 14th Five-Year Plan period. *Price Theory and Practice*, 2020 (08): 4–10,44.

BP Statistical Review of World Energy 2020.https//www.bp.com/en/global/corporate/energy-economics/statistical-review-of-world-energy.htmlJune.2020.

Cao Huaishu, Liao Hua, Wei Yiming. Analysis of China's energy flow in 2010. *China Energy*, 2012,34 (04): 29–31+25.

Dong Donglin, Zhang Yiyan, Gang Lin. Shandong energy balance analysis and energy demand forecast based on the energy flow diagram. *Coal Economic Research*, 2020,40 (02): 31–38.

Energy Statistics Department of National Bureau of Statistics, *China Energy Statistics Yearbook* 2019. Beijing: China Statistics Press, October 2020.

Fan Jingli, Wei Shijie, Zhang Xian. Analysis of China's energy flow and carbon flow in 2015. *journal of Beijing institute of technology* (Social Science Edition), 2018,20 (04): 40–45.

Hu Xiulian. Energy Flow Map and Coal Flow Map of China in 2012 [EB/OL].www.wwfchina.org/.

Li Zheng, Fu Feng, Ma Linwei. China's energy flow diagram based on the energy balance table. *China Energy*, 2006,28 (9): 5–10.

National Bureau of Statistics, China Statistical Yearbook 2019. Beijing: China Statistics Press, September 2019.

Shi-Chen Xie. The driving forces of China's energy use from 1992 to 2010: An empirical study of input-output and structural decomposition analysis[J]. *Energy Policy*, 2014,73.

Yang Lei, Xu Jun. Analysis of Guangdong's energy supply security based on an energy flow diagram. *Ecological Economy*, 2013 (05): 86–91.

Yi Jingwei, Zhao Daiqing, Cai Guotian. Energy Flow Diagram and Energy Balance Analysis of Guangdong Province. *China and Foreign Energy*, 2010,15 (04): 95–101.

Yu Pengwei, Zhang Hao, Wei Shijie, Qi Zirui. Analysis of China's energy flow and carbon flow in 2017. *Coal Economic Research*, 2019,39 (10): 15–22.

Zhang Hao, Fan Jingli, Wang Hang, Zhang Xian. Analysis of China's energy flow and carbon flow in 2016. *China Coal*, 2018,44 (12): 15–19+50.

Zhang Ming. Shandong energy consumption analysis based on the energy flow diagram. *china population resources and environment*, 2011,21 (07): 46–50.

Advances in Energy, Environment and
Chemical Engineering – Abdullah & Osman (Eds)
© 2023 The Author(s), ISBN: 978-1-032-36083-6

Evaluation and optimization of sand control methods under different reservoir conditions in Bayan oilfield

Xuemin Bai, Gaofeng Li, Zhen Qin*, Changqing Ma, Jianning Wang, Chong Li, Chunsheng Zhang, Fei Zhao & Taoning Zhang
Engineering Technology Research Institute of Huabei Oil Company, Renqiu City, Hebei province, China

Wei Li
Downhole Services Branch of CNPC Bohai Drilling Engineering Company Limited, Renqiu City, Hebei province, China

ABSTRACT: In view of the characteristics of strong reservoir water sensitivity and low reservoir compaction in the Bayan oilfield, considering the changes in production conditions and the influence of later gas injection and water injection measures on sand production, the static and dynamic sand production conditions and laws of the reservoir are obtained through experimental and numerical simulation research. On the basis of sand production prediction, the sand control methods under different reservoir conditions are evaluated and optimized, and the sand control methods are recommended. It provides key support for the implementation of sand control technology in gas injection and water injection production.

1 INTRODUCTION

Jihua x, Linhua x, and Xinghua x oilfields are the main oil-producing areas of the Bayan oilfield. They are characterized by strong reservoir water sensitivity and low reservoir compaction. They are typical loose sandstone reservoirs. Sand production occurs during oil testing and production. In the early stage, sand production was initially avoided by reasonably controlling the production differential pressure and working system. However, the change of later production conditions and the implementation of water injection production measures are susceptible to leading to sand production problems.

Based on the test and analysis results of the sand production law, the adaptability evaluation and productivity evaluation comparison of different sand control methods of Jihua x, Linhua x, and Xinghua x are carried out, and reasonable sand control strategies and methods are recommended for different well conditions. Its essence is to match the applicable conditions of different sand control methods with the specific conditions of specific oil and gas wells. The reasonable well completion method is the one with the highest comprehensive matching degree. The basic idea and process of sand control mode evaluation and optimization in this paper are shown in Figure 1: (a) Collect geological production data; (b) Sand production prediction and sand production risk assessment; (c) Technical adaptability evaluation of well completion mode; (d) Comparative evaluation of well completion methods and productivity; (e) Comprehensive evaluation and decision-making (Wang et al. 2015).

---

*Corresponding Author: cyy_qinz@petrochina.com.cn

    DOI 10.1201/9781003330165-30

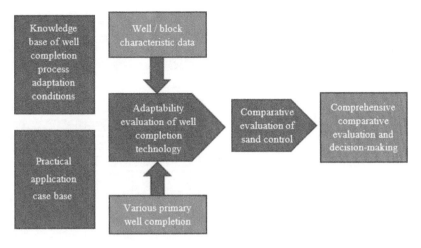

Figure 1.   Evaluation and optimization of sand control methods.

2   BASIC IDEAS FOR OPTIMIZING SAND CONTROL METHODS

Choosing a reasonable sand control method is the primary condition for successful sand control. At present, there are many kinds of sand control methods and processes, each with its own characteristics and adaptability, and the sand control effects are also different. For a specific sand-producing reservoir or oil and gas well, the evaluation and optimization of the sand control method is an important job, and it is also one of the core contents of comprehensive decision-making of oil and gas well sand control. The optimization of the sand control method is to evaluate the adaptability of various sand control processes to specific oil and gas reservoirs, blocks, or well numbers according to the prediction, or actual sand production of oil and gas wells, considering rock physical properties, fluid physical properties, production mode, and working system. According to the adaptability knowledge base of various sand control processes and combined with expert experience, the type or mode of the sand control process is determined (Dong et al. 2016).

The basic idea of sand control method evaluation and optimization is shown in Figure 2. Firstly, the technical adaptability of sand control technology is evaluated according to the adaptive condition knowledge base of sand control technology, the actual sand control application case base, and the geological and production data of target oil and gas wells or reservoirs. The methods mainly include the direct experience comparison method, experimental chart method, artificial neural network method, and comprehensive fuzzy evaluation method. For sand control methods with good technical adaptability, it is needed to continue to evaluate and compare the productivity after sand control, and finally, comprehensively select the most appropriate type of sand control process.

3   EVALUATION AND OPTIMIZATION METHOD AND PROCESS OF SAND CONTROL MODE

3.1   *Selection of evaluation and optimization methods of sand control methods*

This section adopts the comprehensive fuzzy evaluation method, and the process is shown in Figure 3. Many factors are considered in the selection of sand control methods, and the adaptive limits of many factors are difficult to determine, which makes it very difficult to optimize sand control methods. At present, the selection of sand control methods is mainly determined by experience, which inevitably has some one-sidedness and limitations. Comprehensive fuzzy evaluation is a mathematical method widely used in the fields of method optimization and pattern recognition. In recent

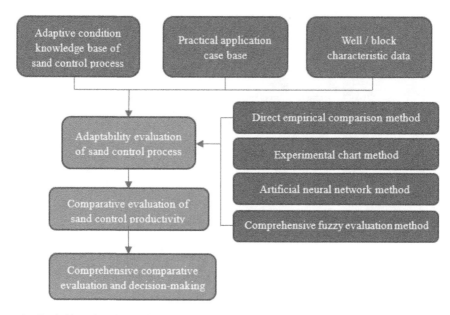

Figure 2.   Basic idea of sand control method evaluation and optimization.

years, it has been gradually applied to the evaluation and optimization of sand control schemes. Using the principle of fuzzy mathematics, the comprehensive evaluation and optimization of various well completion methods considering various factors are carried out, and the comprehensive adaptability evaluation index is calculated (Jing et al. 2021; Tian et al. 2021; Zhou 2017).

Figure 3.   Evaluation process of comprehensive fuzzy evaluation method.

3.2   *Comparative evaluation of the productivity of different sand control methods*

Productivity evaluation is to set typical parameters of different sand control methods and use the productivity evaluation model to predict the completion productivity ratio. After the technical evaluation of the sand control completion mode is completed, productivity evaluation and comparison are carried out to optimize the reasonable completion mode with good technical adaptability and high productivity.

Different completion/sand control methods cause different flow resistance areas at the bottom of the well, forming different additional flow pressure drops and skin coefficients, but they can be divided into two categories: the one-way circular pipe flow model and the radial flow model. The completion productivity ratio (PR) is used to measure the productivity under different sand control methods, that is, take the open hole completion method as the comparison reference, and calculate the ratio of the productivity under different completion methods to the productivity under the same conditions.

4 EVALUATION AND OPTIMIZATION RESULTS OF SAND CONTROL METHODS IN THE TARGET WORK AREA

4.1 *Jihua x block*

The reservoir profile is 1800-2800m deep, the well type is vertical well and directional well, and the well deviation angle is 3.4-21.8 degrees; the medium value of formation sand particle size shows the phenomenon of coarse and fine differentiation, and the depth is lower than < 2000m, and the medium value of reservoir is lower than 0.05-0.1mm; the median value of reservoir with depth > 2500m is 0.2-0.35mm, which is generally uneven formation sand as shown in Figure 4.

The law of sand production is that the low value of the critical production differential pressure of sand production in each layer is mainly about 0.5-5Mpa as shown in Figure 5, with an average of 2.14Mpa. Among the three blocks, the risk and severity of sand production are the highest.

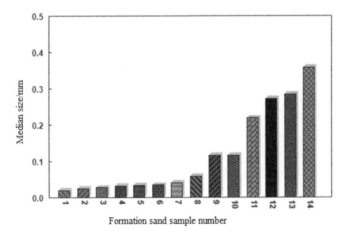

Figure 4.   Median grain size of formation sand samples.

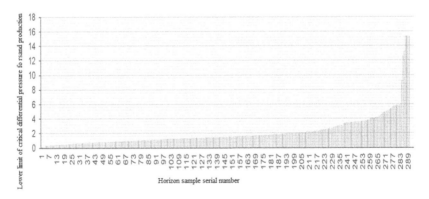

Figure 5.   Critical production pressure difference of sand production.

According to the knowledge base of the adaptive conditions of different sand control methods and the basic data of well Jihua x-1, the comprehensive fuzzy evaluation method is used to evaluate the technical adaptability of well completion methods, as shown in Figure 6. It is concluded that considering the difficulty of squeeze filling of new wells, for the well layer with the median sand particle size<0.1mm, the screen pipe circulating gravel filling and independent mechanical screen pipe sand control methods are recommended. If there is a demand for increasing production, fracturing and filling can also be considered; for the medium coarse sand well layer with the median sand particle size>0.2mm, in order to reduce the cost, the recommended sequence is independent mechanical screen pipe and screen pipe circulating gravel filling.

	Well completion mode	Total additional skin	Capacity ratio
A	Conventional fracturing and filling sand control	7.6832	1.0000
B	Sand control by squeezing gravel outside the pipe	25.3474	0.3218
C	Chemical sand consolidation and control	1.2823	0.8916
D	Mechanical sieve tube sand control	7.6832	1.0000
E	Mechanical sieve tube circulating gravel filling	33.6887	0.3844

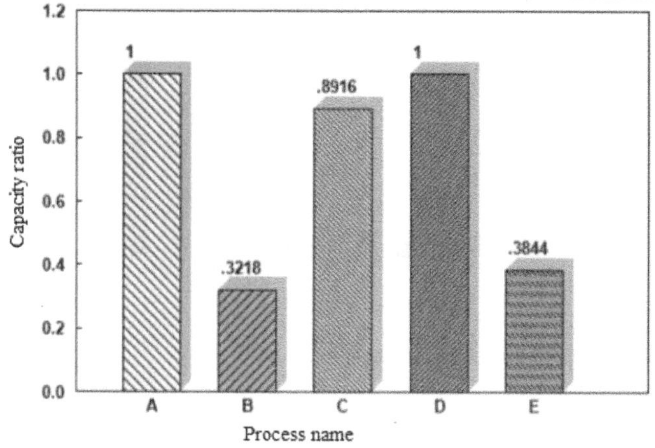

Figure 6.   Productivity comparison of sand control methods.

4.2 *Xinghua x block*

The reservoir profile is 4200-4800m deep, the well type is vertical, and the well deviation angle is 0.6-1.6 degrees; the median grain size of formation sand is 0.18-0.2mm, and the uniformity coefficient is 2.17-2.82, belonging to uniform sand.

The sand production law is that the low value of the critical production differential pressure for sand production in each layer is mainly about 2-6Mpa, and the sand production risk and severity are in the middle among the three blocks. Figure 7 is the cumulative distribution curve of formation sand in Xinghua x block.

According to the knowledge base of different sand control methods and the basic data of Xinghua x, the comprehensive fuzzy evaluation method is used to evaluate the technical adaptability of well completion methods, draw the conclusions as shown in Figure 8, and recommend the sand control methods of independent mechanical screen pipe and screen pipe circulating gravel packing in turn.

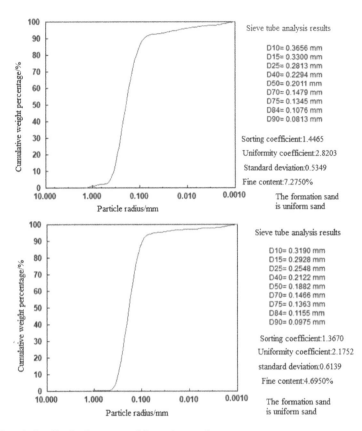

Figure 7.   Cumulative distribution curve of formation sand.

	Sand control type	Technical evaluation index
D	Mechanical sieve tube sand control	0.6203
E	Mechanical sieve tube circulating gravel filling	0.6069
C	Chemical sand consolidation and control	0.4919
A	Conventional fracturing and filling sand control	0.4462
B	Sand control by squeezing gravel outside the pipe	0.3497

Figure 8.   Recommended sand control methods.

4.3   *Linhua x block*

The reservoir is generally 3300-5400m deep, and the well type is mainly directional well, with a deviation angle of 33-36 degrees; the median grain size of formation sand is 0.05-0.05mm, and the uniformity coefficient is 22.5, belonging to extremely uneven sand for the Figure 9.

217

The sand production law as shown in Figure 10 is that the low value of the critical production differential pressure for sand production in each layer is mainly about 4-20Mpa, and the sand production risk and severity are the lowest in the three blocks.

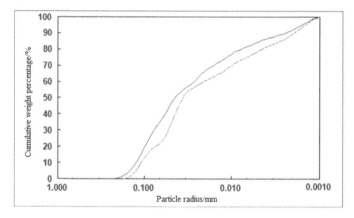

Figure 9. Cumulative distribution curve of formation sand.

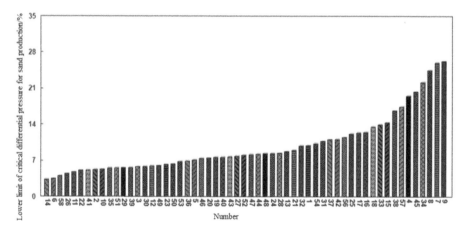

Figure 10. Critical pressure difference in sand production.

According to the actual situation of the block, the sand control scheme recommends no sand control in the early stage, and the sand control method after water breakthrough in the later stage recommends screen pipe circulating filling or independent screen pipe.

5 CONCLUSION

The low value of critical production pressure difference of sand production in each layer of Jihua x block is mainly about 0.5-5 Mpa, with an average of 2.14 Mpa. Among the three blocks, the risk and severity of sand production are the highest. Recommended sand control schemes: (1) considering the difficulty of squeezing and filling in new wells, for the well layer with the median sand particle size<0.1mm, it is recommended to use screen pipe circulating gravel filling and independent mechanical screen pipe sand control; (2) If there is a demand for increasing production, fracturing and filling sand control can also be considered.

The low value of critical production pressure difference of sand production in each layer of the Xinghua x block is mainly about 2-6 Mpa, and the risk and severity of sand production are in the middle among the three blocks; the sand control methods of independent mechanical screen pipe and screen pipe circulating gravel packing are recommended in turn.

The low value of critical production pressure difference of sand production in each layer of Linhua x block is mainly about 4-20 Mpa. Among the three blocks, the risk and severity of sand production are the lowest. The sand control scheme recommends no sand control in the early stage, and the sand control method after water breakthrough in the later stage recommends screen pipe circulating filling or independent screen pipe.

The sand control method optimized in this paper can be used for sand control optimization. It is suggested to consider the complex conditions such as reservoir heterogeneity in the target work area, carry out micro sand production process simulation and optimization in the middle and later stage, and guide the implementation of sand control in the middle and later stage.

REFERENCES

Dong C.Y, Zhang Q.H, Gao K.G, Li X.B, Dong S.X. (2016) Study on sand control screen media deformation and sand retention precision[J]. *China Petroleum Machinery*, 44(10): 97–102.

Jing T, Hou H.T, Teng X.W, Li J.X, Duan Z.G. (2021) Development and application of high efficiency and safety sand control filter pipe[J]. *Petrochemical Industry Application*, 40(7).

Tian B, Wei Y.S, Wang H.L, Xing H.X, Deng. H. (2021) Research and prospect of sand control history of unconsolidated sandstone reservoir in the eastern South China Sea[J]. *Petrochemical Industry Application*, 40(2).

Wang D.J, Dong C.Y, Peng Y.D, Luo Q.M, Fu J.J, Peng J.W. (2015) Experimental study on evaluation and selection of mechanical sand control screens for the cretaceous reservoir in Chunguang Oilfield [J]. *Science Technology and Engineering*, 15(25): 120–124.

Zhou Y. (2017) Development strategy research on thin-interbedded sand-bearing heavy oil reservoirs in Wa83 Block[J]. *Special Oil & Gas Reservoirs*, 24(4): 83–87.

Advances in Energy, Environment and
Chemical Engineering – Abdullah & Osman (Eds)
© 2023 The Author(s), ISBN: 978-1-032-36083-6

Prediction of heavy oil productivity based on modified steam-assisted gravity drainage technology in double horizontal wells

Xingzhou Chen*

Liaohe Oilfield Exploration and Development Research Institute, Panjin, Liaoning, China

ABSTRACT: China is rich in heavy oil and gas resources in middle-shallow areas. With the continuous large-scale exploitation of oil and gas resources, the traditional oil production technology has been unable to meet the current production demand. Based on the investigation and summary of previous studies, taking the Paleogene typical well group of Liaohe Oilfield as an example, using the combination of numerical simulation and reservoir engineering, this paper establishes the theoretical simulation of steam-assisted gravity drainage of double horizontal wells, divides the production stages of steam-assisted gravity drainage, and modifies the previous productivity prediction model of steam-assisted gravity drainage of double horizontal wells. Among them, the numerical simulation results show that taking the daily oil production as the standard, the steam-assisted gravity drainage of double horizontal wells can be divided into three stages, namely, increased production, stable production, and reduced production. The corresponding steam cavity development can also be divided into three stages: the formation of the oil drainage slope of the steam cavity, the vertical expansion and horizontal expansion of the steam cavity, and the expansion of the steam cavity to the lower part of the steam injection well. The prediction results show that the expansion angle of the steam chamber has an obvious influence on the oil discharge rate of the steam chamber in the stable production stage, and the accuracy of the modified productivity model is higher. The modified steam-assisted gravity drainage technology of double horizontal wells established in the experiment has a certain theoretical guiding significance for heavy oil recovery and productivity prediction.

1   INTRODUCTION

China's heavy oil resources are widely distributed with a resource reserve of about 300×10^8t. At present, with the large-scale exploitation of middle-shallow oil and gas resources, the traditional thermal oil recovery technology has great limitations. In order to further improve the recovery of the heavy oil reservoir, it is very necessary to convert new production methods. At present, the development of heavy oil mainly includes four technologies: steam-assisted gravity drainage (SAGD), gas injection solvent extraction of heavy oil, gravity-assisted reservoir burning, and horizontal fracture-assisted steam oil displacement. Among them, SAGD technology has been widely used abroad. Its applicable standards are as follows (Table 1).

At present, SAGD mainly consists of two well spacing methods: SAGD of double horizontal well and SAGD of combination with vertical well and horizontal well. Among them, the SAGD method of double horizontal wells is to drill a pair of parallel horizontal wells near the bottom of the reservoir. The upper horizontal well is used for steam injection and the lower horizontal well is used for oil production (Figure 1). According to the principles of the above methods, most scholars put forward the capacity prediction model of SAGD. However, these models have some

---

*Corresponding Author: chenxz01@petrochina.com.cn

DOI 10.1201/9781003330165-31

Table 1. SAGD applicable standards.

Index	Standard	Unit
Burial depth	<1000	m
Porosity	>20	%
Permeability	>0.5	mD
Core intersection	>15	m
Oil saturation	>50	%
Viscosity	$>10^4$	mPa·s

shortcomings. First, the stage division of SAGD is not clear. Previous experimental models failed to combine the expansion of the steam chamber and the transformation of the saturation field with production indexes such as daily oil production and water cut. Second, previous experiments failed to effectively study the productivity prediction model in different stages under the double horizontal well and did not effectively predict the daily oil production in different stages. Through numerical simulation, this paper divides the production stage of SAGD of double horizontal wells. By studying the oil drainage mechanism of SAGD and combining it with the theoretical derivation of the Darcy equation and the shape of the steam cavity front, the productivity model of different production stages of the double horizontal well is modified, and the measured results are used for test and analysis.

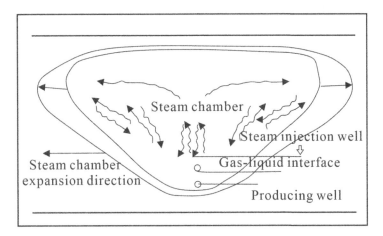

Figure 1. Schematic diagram of the SAGD mechanism.

2 BACKGROUND

D-84 block of Liaohe Oilfield is located on the west slope of the Western Sag of Liaohe depression. The study area is relatively complete, and a NW trending fault is developed in the northeast. The oil-bearing area of the D-84 block is about 6.2 km², and three sets of oil-bearing strata are developed in Paleogene vertically. Among them, the crude oil of the Xinglongtai reservoir is super heavy oil, the crude oil density is 0.975 g/cm³, the reservoir temperature is 44.6°C, and the formation pressure is 7.5 MPa. Under the condition of reservoir temperature, the viscosity of degassed crude oil is about 15×10^4mPa·s, but the crude oil viscosity is very sensitive to temperature change.

3 METHODS

The numerical simulation theoretical model takes the Paleogene typical well group of Liaohe Oilfield as an example. The grid division adopts 20m*40m*25m, in which the x-axis direction is 20m*5m, the y-axis direction is 40m*10m, and the z-axis direction is 10*5+15*2=80m. The production operation steps are: steam huff and puff first. The upper horizontal well is huff and puff for 6 cycles, 200 days a week, and the gas injection is 6000m³. Then huff and puff from the lower horizontal well for 6 cycles, and the gas injection volume is 5000m³. The upper and lower horizontal wells huff and puff for 6 cycles at the same time, and the steam injection volume is 5500m³. Then turn to the SAGD opportunity. after 18 cycles of steam huff and puff, the thermal connection is formed between horizontal wells. The temperature between injection and production wells increases from 40°C to 200°C, and the pressure decreases to 3.1MPa. Finally, it needs to turn to SAGD production control. Steam injection in an upper horizontal well, steam injection volume of a single well is 400m³/day. The daily liquid production of a horizontal well is controlled at 480 m³/day. The production injection ratio of steam huff and puff and steam-assisted gravity drainage production process is controlled to 1.2.

4 RESULTS AND DISCUSSIONS

4.1 *Simulation results*

Taking the daily oil production as the standard, the headless water SAGD can be divided into three stages: increase production stage is 140 days, stable production stage is 5260 days, and decline production stage is 1600 days (Figure 2).

Taking the water cut as the index, it can also be divided into three stages: water cut decreasing stage, water cut stable stage, and water cut rising stage (Figure 3).

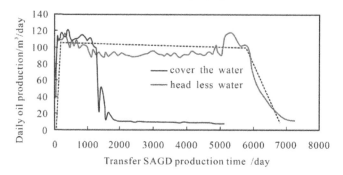

Figure 2. Variation characteristics of daily oil production.

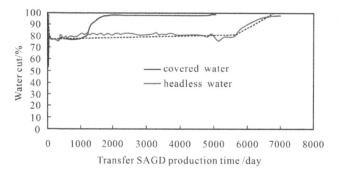

Figure 3. Variation characteristics of water cut.

Each production stage corresponds to a different stage of steam chamber development. The oil drainage slope is formed in the increase production stage, the steam chamber expands vertically and horizontally in the stable production stage, and the steam chamber expands to the lower part of the steam injection well in the decline production stage.

4.2 *Production capacity prediction model and results*

According to the expansion process of the steam chamber of SAGD in double horizontal wells and the corresponding division of the production process, the prediction formula of oil drainage production in different stages of SAGD is deduced.

Increasing production stage

$$q_L = 2\sqrt{\frac{\beta K g \alpha \varphi \Delta S_o h}{m v_s}}$$

Stable production stage

$$q_L = 0.0658\theta\sqrt{\frac{K g \alpha \varphi \Delta S_o h}{m v_s}}$$

Decline production stage

$$q_L = 1.2248\sqrt{\frac{K g \alpha \varphi \Delta S_o h}{m v_s}} - 0.8165 \frac{t^2 K g \alpha}{W^2 m v_s h \varphi \Delta S_o}\sqrt{\frac{K g \alpha h \varphi \Delta S_o}{m v_s}}$$

K is permeability, mD; g is gravitational acceleration, m/s^2; ΔS_o is oil saturation difference, %; Φ is porosity, %; α is thermal conductivity; h is reservoir height, m; m is constant; v_s is the oil viscosity, m/s; t is time, s; W is well spacing, m.

Due to the short time of increasing the production stage, the key to SAGD productivity prediction is a stable and declining production stage. The range of expansion angle in numerical simulation is 22°~27°, which is taken as 20° after correction. The prediction results are shown in Figure 4, and the modified capacity model is more in line with the numerical simulation results.

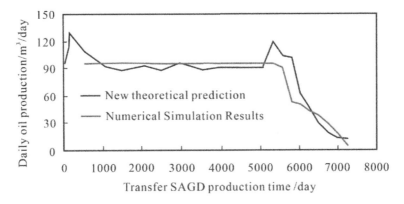

Figure 4. Prediction results of modified SAGD model.

5 CONCLUSION

Through the above methods and the results and discussion, this paper draws the following conclusions.

Based on the numerical simulation of steam-assisted gravity drainage technology in double horizontal wells, it is divided into three stages of heavy oil injection and production: increasing production stage, stable production stage, and declining production stage. The corresponding development degree of the steam chamber can be divided into three stages. The oil drainage slope is formed, the steam chamber expands vertically and horizontally, and the steam chamber extends to the lower part of the steam injection well. On this basis, the modified production capacity model is established, and the prediction results are consistent with the numerical simulation results.

ACKNOWLEDGMENTS

This work was financially supported by the National Natural Science Foundation of China and Major Special Projects of PetroChina.

REFERENCES

Akin, S., Bagci., Suat. (2001). A laboratory study of single-well steam-assisted gravity drainage process. *Journal of Petroleum Science & Engineering*, 32(1), 23–33.

Butler, R.M. (1987) Rise of Interfering Steam Chambers. *Journal of Canadian Petroleum Technology*, 87: 03–07.

Butler, R.M., Mcnab, G.S., Lo, H.Y. (2010). Theoretical studies on the gravity drainage of heavy oil during in-situ steam heating. *Canadian Journal of Chemical Engineering*, 59(4), 455–460.

Chen, Q., Gerritsen, M.G., Kovscek, A.R. (2007). Effects of reservoir heterogeneities on the steam-assisted gravity-drainage process. *SPE Reservoir Evaluation & Engineering*, 11(05): 921–932.

Fernandes, G.M.D., Diniz, A.A.R., Araújo, E.A., Rodrigues, M.A.F., Gurgel, A.R., Dutra, T.V. (2017). Economic analysis of oil production by applying steam-assisted gravity drainage (sagd) to reservoirs from the Potiguar basin. *Energy Sources Part B Economics Planning & Policy*, 12(5): 428–433.

Griffin, P.J., Trofimenkoff, P.N. (1986). Laboratory studies of the steam-assisted gravity drainage process. *Aostra J. of Research*, 2(4): 197–203.

Ma, Z., Leung, J.Y., Zanon, S., Dzurman, P. (2015). Practical implementation of knowledge-based approaches for steam-assisted gravity drainage production analysis. *Expert Systems with Applications an International Journal*, 42(21): 7326–7343.

*Advances in Energy, Environment and
Chemical Engineering – Abdullah & Osman (Eds)
© 2023 The Author(s), ISBN: 978-1-032-36083-6*

Research on incremental distribution network operation mode for source network load storage interaction

Youzhong Miao & Feng Zhao
State Grid Jibei Electric Power Co., Ltd., Beijing, China

Hui Li*, Xinjian Chen, Guangbiao Wang, Dongming Zheng, Shaobo Chai & Xu Han
Tianjin Tianda Qiushi Electric Power High Technology Co., Ltd., Tianjin, China

ABSTRACT: In the market-oriented environment, the operation mode of the incremental distribution network will be more diversified. By analyzing the source network load storage interaction mechanism of an incremental distribution network, this paper puts forward the cascade utilization strategy of internal resources of an incremental distribution network for source network load storage interaction. At the same time, combined with the development status of China's power market, this paper puts forward the operation and investment strategy of incremental distribution network, including focusing on promoting cooperation with industrial investment companies, continuing cooperation with energy storage equipment manufacturers, and technical service providers, and carrying out key customer cooperation through the territorial resources and channels of municipal power grid companies / collective enterprises. Incremental distribution network operators should reasonably allocate resources to obtain the maximum market benefits under the condition of comprehensively considering various market participation conditions and benefits.

1 INTRODUCTION

Developing demand side response to enable power users to participate in the power market is an important research and practice direction in the construction of the power market. Although the development of demand response in China is late, there are many kinds of demand-side resources in China. Gaspari, lorenzoni, frías, et al (2017) divided uncontrollable generalized demand side response resources into two categories according to certain characteristics: the first category is completely uncontrollable resources, mainly the load on the user side, which respond mainly through electricity price signals; The second category is controllable resources, mainly including energy storage system and distributed power generation. Perdan and Azapagic (2011) described different types of generalized demand-side resources from the perspective of energy interconnection and studied the composition and classification of generalized demand-side resources from many aspects. (Zhao et al. (2021) made an in-depth study on the optimal operation of multi-modal microgrids from the aspects of generalized demand-side resources and their classification, response technology, collaborative optimization, and so on.

Demand side response is more and more widely used in China, and its related theories and application research are gradually increasing. However, with the development of power system reform and energy Internet, in order to ensure the stable operation of the power grid and improve energy utilization efficiency, how to build a set of power demand side resource value evaluation system for the power grid prior to decision-making is a key problem. In view of this, through the research on the characteristics and related theories of power demand-side resources, this paper explores

---

*Corresponding Author: 1718241359@qq.com

DOI 10.1201/9781003330165-32

its comprehensive value in power supply, power grid, users, and environment, and constructs a value evaluation system to fully reflect the comprehensive value of power demand-side resources. A decision was supported for the stable operation of the power grid.

2 INTERACTION MECHANISM OF SOURCE AND LOAD IN THE INCREMENTAL DISTRIBUTION NETWORK

The source load interaction of an incremental distribution network can be divided into two types: price demand response and incentive demand response. Among them, price-based demand response means that power companies adjust users' power consumption behavior by formulating special electricity prices, mainly because generally, users' power consumption behavior is inversely proportional to electricity price. When the electricity price increases, rational users will reduce their power consumption in this period; When the electricity price decreases, users will increase their electricity consumption during this period. In the current application, the price-based demand response can be further divided into as follows: time of use price (TOU), which combines the peak and trough of load with the time, and is divided into several periods. Each period corresponds to a power price. In order to reduce the peak and fill the valley, the high price is adopted in the peak stage and the low price is adopted in the trough stage. Peak electricity price (CPP), based on time of use electricity price, adopts higher electricity price to restrain users' peak electricity consumption. For example, the electricity price at peak load is about 5 times that at peak time, which makes users' electricity cost soar at peak time, so as to effectively reduce the power load at peak time. However, peak electricity prices are generally used only in extreme cases. Real-time tariff (RTP) refers to the electricity price changes with the passage of time and can reflect the marginal cost of power production in real-time, and its update cycle is shorter than a time-of-use tariff.

Incentive demand response means that power grid enterprises and power users specify the response time and capacity in advance, then sign relevant contracts and call directly when necessary. Incentive demand response mainly includes: direct load control (DLC), which means that when the power system is in short supply, the power grid enterprise controls the electrical equipment of the power users who have reached the demand response contract with the help of the remote control system, so as to adjust the power load to alleviate the supply shortage of the power system, and give corresponding compensation to the power users after the response is completed; Interruptible load (IL) refers to that in the peak period of load, the power grid enterprise cuts off the power supply of some electrical equipment with low requirements for power reliability according to the response contract signed in advance, so as to reduce the power load in the peak period and alleviate the pressure of safe and stable operation of the power grid; Demand side bidding (DSB) means that power users actively change their power consumption behavior to participate in demand response, and obtain corresponding subsidies and benefits by means of bidding. However, the demand side bidding mechanism is mainly open to large users, and small residential users cannot participate due to index restrictions; Emergency demand response (EDR) refers to the behavior that can respond to the sudden emergence of the power grid in a short time, the power grid company issues the response demand temporarily, and the power users voluntarily reduce their power load according to the load reduction requirements of the power grid company and can obtain corresponding subsidies afterward.

3 CASCADE UTILIZATION STRATEGY OF SOURCE LOAD INTERACTION RESOURCES IN THE INCREMENTAL DISTRIBUTION NETWORK

Different power demand-side resources have different effects when participating in power grid regulation. The power demand side resources with higher comprehensive value will bring greater benefits to the power grid when participating in power grid regulation. At the same time, the operation of the power grid will also be affected by different factors such as climate environment, human error, and equipment failure. Under different circumstances, the impact received by the power grid is different. Some minor impacts can calm down the recovery capacity of the power grid itself. When the power grid itself cannot be restored, the power demand side resources can be called for regulation. In order to improve the use efficiency of the power grid transferring power demand-side resources and prevent resource waste, a cascade utilization strategy based on power demand-side comprehensive value is proposed in this paper.

(1) Principle of cascade utilization of power demand side resources

According to the severity of the impact encountered in the operation of the power grid and the urgency of the recovery capacity required by the power grid, the power grid demand is divided into three levels: emergency demand, medium emergency demand, and ordinary demand. Call the power demand side resources with different value levels under different levels, that is, call the power demand side resources with the highest value under emergency demand; Call medium value power demand-side resources in a medium emergency; In the case of ordinary demand, call the power demand side resources of general value, so as to achieve the cascade utilization purpose of "high-value emergency use and low-value ordinary use".

(2) Cascade utilization method of power demand side resources

The comprehensive value of power demand side resources willing to participate in power grid regulation shall be evaluated in advance, the comprehensive value evaluation results of all power demand side resources shall be sorted, and the value grade shall be established. At the same time, combined with the needs of power grid enterprises and the operation status of the power grid, and following the principle of cascade utilization, sign transfer contracts with each power demand side resource provider, so as to transfer the most appropriate power demand-side resources to effectively solve the problems in power grid operation and quickly Effectively help the power grid restore stable operation. On the one hand, the cascade utilization strategy of power demand-side resources can help the power grid make full use of power demand-side resources; on the other hand, it can also effectively improve the enthusiasm of power demand-side resource providers to participate in demand response.

Table 1. Energy demand of different users.

Customer type		Energy needs			
	Electric energy	Thermal energy	Cold energy	Natural gas	
Resident user	$\checkmark$		Related to region and climate	/	$\checkmark$
Industrial users	$\checkmark$		Industry related	Industry related	
Business users	$\checkmark$		Related to region and climate	/	$\checkmark$
Public demand	Road and traffic	$\checkmark$	/	/	/
	communal facilities	$\checkmark$	Related to region and climate	/	/
	Green space and square	$\checkmark$	/	/	/

The park contains various types of energy users, namely energy demanders. In different types of parks, the types of energy demanders are different, which makes the energy demand of each park different. From the fundamental point of view of energy demand, the analysis of the actual demand of energy consumers is the basis of energy demand prediction. Based on the current situation of energy demand and the actual situation of energy users in China's incremental distribution parks, it can be obtained that the energy consumers in China's parks mainly include residential users, industrial users, commercial users, and public demand, among which the public demand includes road and traffic energy demand, public facilities energy demand, green space, and square energy demand.

4 OPERATION STRATEGY OF INCREMENTAL DISTRIBUTION NETWORK UNDER MARKET ENVIRONMENT

At present, in China's power trading market, some regions have issued corresponding rules for demand-side resources to participate in the demand response market and auxiliary service market, so as to encourage the construction of a competitive retail market. In Jiangsu, Ningxia, Jilin, Mengdong, Shanghai, and other places, relevant market construction pilot work to promote new energy consumption has been started. Therefore, a series of large-scale demand-side users such as virtual power plants, comprehensive energy parks, and electric vehicle charging column networks fully tap the market potential on the demand side by integrating resources, so as to achieve a win-win plan for enterprises and users.

Power demand-side users should reasonably allocate resources to obtain the maximum market benefits under the condition of comprehensively considering various market participation conditions and benefits. In addition to demand response and ancillary service market transactions, users can make full use of the flexibility of demand-side resources and the flexibility of supply-demand interaction to fully tap the market possibilities. Considering the power market-oriented to the flexible interaction between supply and demand, this paper summarizes six modes of demand side participation in the power market, and their characteristics, benefit sources, and capacity realization methods are shown in Table 1. It is worth noting that different modes of demand side resource participation need to consider the maturity of policy and technology, and select the market according to local conditions in order to obtain the maximum benefits.

Table 2. Participation of the demand side in electricity market transactions.

Pattern	Characteristic	Benefit source	Applicable conditions
Auxiliary service peak shaving	Easy to start and have a market foundation	Power plant allocation	Areas with insufficient peak shaving capacity and large peak valley difference
Demand response market	The compensation price is high and the income is considerable	Government subsidies	
Curve tracking of power abandonment in a low valley	Curve tracking	Abandonment price reduction and time of use price	In new energy-rich areas, users are more flexible and consume rich renewable energy
New energy and user substitution	Restrain volatility and reduce the deviation	New energy subsidies	
Distributed energy, energy storage, and flexible application	Time value, value-added service model	Time-sharing spread	Regions carrying out spot electricity trading
Medium and long-term power trading with curve considering spot	Building a flexible and interactive value system	Electricity value + electricity value	Future electricity market

228

(1) It is necessary to focus on promoting cooperation with industrial investment companies. Invested by new energy and other industries, power grid companies provide customer mining, project matching, engineering EPC, and asset custody operation services, optimize their project investment efficiency and capital liquidity, reduce construction costs, reduce post-investment operation costs and risks, and improve profitability.

(2) It is needed to continue to cooperate with energy storage equipment suppliers and technical service providers. Bring energy storage equipment suppliers and service providers into the product library of the shared service platform. Cooperate with energy storage equipment suppliers for supply chain optimization and integration, as well as joint product development and application. Energy storage equipment suppliers are responsible for equipment integrated production and after-sales service. The technical service provider is responsible for the operation and maintenance, operation optimization, value-added service development, and other services after construction.

(3) Efforts should be made to focus on cooperation with key customers such as local bus companies, taxis, and 5G facility operators through the territorial resources and channels of municipal power grid companies / collective enterprises. The Localization Company of un.com, the platform operator, carries out key customer project development, EPC Construction, and daily operation and maintenance services.

5 CONCLUSION

This paper studies the operation mode of incremental distribution networks under the interaction of source network load. In order to achieve the double carbon goal, power users will gradually enter the power market and participate in trading. Incremental distribution network aggregates multi-type resources such as user load, distribution network, and distributed generation, and has the natural interaction attribute of source network load. In the market-oriented trading environment, incremental distribution networks can obtain benefits by optimizing trading strategies and participating in multi-type market transactions. At the same time, it can give full play to the company's resource advantages, take operation and maintenance as the starting point, provide users with services such as the safe operation of distribution network and intelligent operation and maintenance of new energy stations, improve customers' professional energy consumption level and obtain reasonable operation and maintenance service fees.

ACKNOWLEDGMENTS

This work was financially supported by the Science and technology project of State Grid Jibei Electric Power Co., Ltd (Research on the strategy of mixed ownership participating in incremental distribution investment business).

REFERENCES

Gaspari M, Lorenzoni A, Frías P, et al. (2017) Integrated Energy Services for the industrial sector: an innovative model for sustainable electricity supply, *Utilities Policy*, (45): 118–127.

Perdan S, Azapagic A. (2011) Carbon trading: current schemes and future developments, *Energy Policy*, 39(10): 6040–6054.

Qian Jiaxin, Wu Jiahui, Yao Lei, et al. (2021) Comprehensive performance evaluation of a CCHP-PV-Wind system based on energy analysis and a multi-objective decision method, *Power System Protection and Control*, 49(2): 130–139 (in Chinese).

Tim B, Maria S, Edward M, Carly M, Matthew H, Jeff H, Sarah M.(2020) Business models and financial characteristics of community energy in the UK, *Nature Energy*, 5(2):63–72.

Zhao Pu, Zhou Man, Gao Jianyu. et al. (2021) Evaluation method for a park-level integrated energy system based on electric power substitution, *Electric Power*, 54(4): 130–139 (in Chinese).

Advances in Energy, Environment and
Chemical Engineering – Abdullah & Osman (Eds)
© 2023 The Author(s), ISBN: 978-1-032-36083-6

Experimental research on leakage and diffusion characteristics and leakage quantity prediction of underwater gas pipeline

Shaoxiong Wang*, Jing He* & Xueling Chang
Zhejiang Scientific Research Institute of Transport, Hangzhou, China

ABSTRACT: The leak experiment for underwater gas leakage was carried out by using a self-designed circular pipeline and air as an experimental medium. The diffusion characteristics of gas in water and the variation of flow rate as a function of different water depth, hole diameter, and leakage pressure were studied. The results show that the leakage rate has a linear relation, second order relation with the leakage pressure and diameter of the hole respectively, while it decreases with the increase of water depth. The change of the hole diameter has the greatest influence on the leakage rate, followed by the pressure, while the effect of water depth is the smallest. The nonlinear fitting of the experimental data is carried out to obtain the quantitative relationship between the release flow rate and the hole diameter, leakage pressure, and water depth. At the same time, by introducing coefficient α, a quantitative formula for predicting the leakage rate of underwater gas pipelines under the conditions of small hole diameter ($d \leq 20$ mm) and subsonic flow ($p \leq 90$ kPa) is obtained.

1 INTRODUCTION

In the process of natural gas transportation, the water environment will inevitably be encountered. Underwater gas pipelines are often subjected to various damages such as improper design and installation, corrosion, mechanical or material failure, and natural hazards, which increase the risk and leakage accidents (Liu et al. 2017). Compared with onshore pipelines, the environment in which the underwater gas pipeline is located is more complicated. Once leaks occur and bubble plumes are formed, not only fires, explosions, and water pollution may be caused, but also serious casualties and property losses (Li et al. 2018). Therefore, scientific understanding of the leakage and diffusion law of underwater gas pipelines has practical guiding significance for preventing potential leakage accidents and formulating risk assessment schemes.

At present, most of the theoretical studies on underwater gas pipeline leakage are based on the CFD model. Cloete et al. developed a coupled volume of fluid and discrete phase model to simulate the plume and free surface behavior resulting from a sub-sea gas pipe rupture (Cloete et al. 2009). Olsen et al. proposed a CFD model based on the Euler-Lagrange criterion to simulate the release of bubble plumes in open water, which took into account the inhibition of free surface on turbulence and the change of bubble size during the rise of bubble plumes (Olsen & Skjetne 2016). Tessarolo proposed a numerical model based on the Lagrange method to simulate the diffusion of oil and gas in the deep-water blowout and verified three entrainment relations through the model. He found that the JETLAG relation was the most consistent with the experimental results, but this model did not take into account the changes in the physical and chemical processes of oil and gas and the formation of hydrates under water (Tessarolo & Innocentini 2016). Wu K et al. developed a CFD model to describe the behavior of a subsea gas release and the subsequent rising gas plume, and compared the applicability of four different numerical methods in capturing the characteristics of rising bubble plume. The CFD results were compared with experimental data, and it is found that the large eddy simulation results have the highest consistency (Wu et al. 2017).

---

*Corresponding Authors: wsxupc@163.com and 853245516@qq.com

DOI 10.1201/9781003330165-33

There is no doubt that the above research provides good theoretical guidance for predicting plume dynamics. However, most of the established numerical models are based on many assumptions and have not been fully verified. The experimental research is relatively scarce and the focus is also on the diffusion of the gas in the water and neglects the leakage characteristics of the gas pipeline. Prediction formulas for the leakage rate of overhead pipelines have become mature, but few for underwater pipelines. Consequently, the purpose of this paper is to investigate the natural gas leakage for underwater pipelines through the designed circular pipeline. Finally, the quantitative relationship between the amount of natural gas leakage and the hole diameter, leakage pressure, and water depth is obtained by nonlinear fitting, which provides a theoretical and experimental basis for the risk assessment of underwater gas pipelines and related simulation research.

2 EXPERIMENTAL DEVICE AND METHOD

To investigate gas release and dispersion from underwater gas pipelines leak, an underwater gas pipeline leakage experimental system, which is shown in Figure 1, was designed and constructed. The total length of the pipeline is 251.5 m and the inner diameter is 42 mm. For safety reasons, the air was used as the experimental medium instead of natural gas. The main structure of the leakage device is shown in Figure 2. The water tank is made of high-pressure plexiglass. The leak branch is located above the center of the main pipe and is connected by a pneumatic valve that can be switched instantaneously. The end of the branch pipe is connected with a detachable leakage orifice. The main equipment and specifications of the whole experimental system are listed in Table 1.

By changing the diameter of the leaking holes (1mm, 2mm, 3mm), water depth (30cm, 60cm, 90cm), and leakage pressure (40kPa, 50kPa, 60kPa, and 70kPa), the variation of flow rate and pressure in the pipeline and the diffusion characteristics of the gas in the water are obtained.

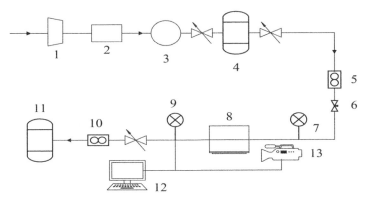

Figure 1.   Experimental system for leakage of underwater gas pipeline: 1-air compressor; 2-cold dryer; 3-filter; 4&11-buffer tank; 5&10-mass flow meter; 6-neddle valve; 7&9-pressure gauge; 8-water tank; 12-computer and 13-high speed camera

Table 1.   Main equipment and specifications of the experimental system.

Equipment	Type	Number	Parameters	Uncertainty
Air compressor	GA30CP-13	1	P=1.3MPa; Wt=61.5L/s; Pt=30kW	–
Pressure gauge	–	2	0-0.6MPa	0.5%
Mass flow meters	–	2	0.01–1m^3/h	0.2%
High-speed camera	FASTCAM SA-X2	1	Shooting speed = 5000 frames/s	–

Figure 2. Main structure of leakage device.

3 RESULTS AND DISCUSSIONS

3.1 *Analysis of leakage process*

Figure 3 shows the variation of flow rate with time before and after leakage (this figure is an example of the leakage with P=70kPa, d=3mm, and h=90cm). As can be seen from the figure, when no leakage occurs, the flow rate remained the same before and after the leakage point at about 16.59Nm³/h. At about 20 s, due to the small hole opening momentarily, the flow rate before the leakage point begins to rise, while the flow rate after the leakage point drops sharply. When the leakage lasts for about 60s, owing to the close of the small hole by the pneumatic valve, the flow rates in the pipeline return to the initial value.

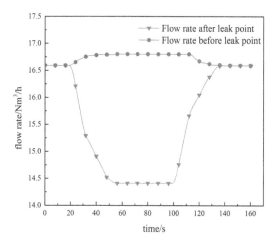

Figure 3. Curve of flow rate with time before and after the leak point.

Figure 4 is a schematic showing the amount of released gas as a function of the hole diameter at P = 70 kPa. As it can be seen, when the leakage pressure and the water depth are constant, the leakage amount gradually increases with the rise of the hole diameter and has an approximate power function relationship. This is because the larger the hole, the larger the flow area of the gas at the moment of the leak, and the larger the leakage amount. In Figure 5, the change of the leakage amount is shown for different pressure at h = 90 cm. It can be seen from the figure that when the hole diameter and water depth are fixed, the leakage amount increases with the pressure and shows a linear relationship. This is because the greater the pressure, the greater the initial momentum of gas rushing out of the leakage hole, and the greater the leakage amount. Figure 6 shows the amount of leakage as a function of water depth at d = 3 mm. As can be seen from the figure, when the pressure and the hole diameter are constant, the leakage amount decreases with the increase of the water depth. This is because as the water depth increases, the back pressure above the leakage hole accordingly increases. Taking the leakage for d=2mm, p=40kPa, and h=90cm as an example, when the hole diameter changes from 2mm to 3mm, the pressure rises from 40KPa to 60kPa and the water depth decreases from 90cm to 60cm, that is, the variation range of the three is 50%, the increase of leakage amount was 106%, 41%, and 5%, respectively. It can be seen that the change of the hole diameter has the greatest influence on the leakage amount, followed by the pressure, and the influence of the water depth is the smallest.

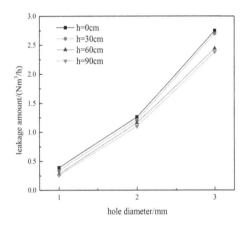

Figure 4.   Variation of leakage amount as a function of the hole diameter at P=70 kPa.

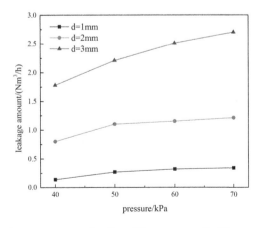

Figure 5.   Variation of leakage amount as a function of the pressure at h=90cm.

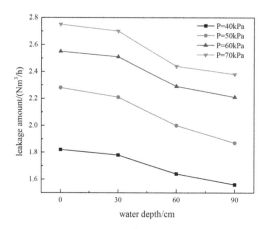

Figure 6. Variation of leakage amount as a function of the water depth at d=3mm.

3.2 Leakage flow rate prediction

According to the statistics of the European Gas Pipeline Incident Data Group, the probability of small hole leakage accounts for about 83% of the total pipeline leakage accidents, which is the main damage form to gas pipelines (Arnaldos et al. 1998). Since the diameter of the hole in the experiment is 1mm, 2mm, and 3mm, this paper mainly discusses the calculation of the leakage flow rate in the form of small-hole damage in the underwater pipeline.

Figure 7 is a schematic of the small hole model for gas pipeline leakage. Point 1 is the initial point, and the flow rate in the pipeline is Q. A small-hole leakage occurs at a distance of L from point 1. Point 3 is the small-hole leakage point, while point 2 is a point in the pipeline at the same vertical line as leakage point 3. For small hole leakage, the leakage flow rate is much smaller than the flow rate, so the pipe pressure is little affected by the leakage, that is, $P1 = P2 = P3$. In order to calculate the amount of natural gas, it is necessary to determine whether the flow of the gas is a sonic flow or a subsonic flow, which depends on the critical pressure ratio CPR (Lu et al. 2014) (Eq. 1).

$$CPR = \frac{P_c}{P_2} = \left(\frac{2}{k+1}\right)^{\frac{k}{k-1}}$$ (1)

Where, P_c is the critical pressure, Pa; P_2 is the absolute pressure of point 2, Pa; K is the isentropic index of natural gas.

When $p_a/p_3 > CPR$, the flow of gas at the leakage point is subsonic flow, which belongs to subcritical flow. The leakage amount can be calculated from Eq. (2).

$$Q = C_0 A p_2 \sqrt{\frac{M}{ZRT_2} \frac{2k}{k-1} \left[\left(\frac{p_a}{p_2}\right)^{\frac{2}{k}} - \left(\frac{p_a}{p_2}\right)^{\frac{k+1}{k}}\right]}$$ (2)

When $p_a/p_3 \leq CPR$, the flow of gas at the leakage point is sonic flow, which belongs to the critical flow. The leakage amount can be calculated from Eq. (3).

$$Q = C_0 A p_2 \sqrt{\frac{Mk}{ZRT_2}\left(\frac{2}{\gamma+1}\right)^{\frac{k+1}{k-1}}}$$ (3)

Where, Q is the leakage amount, Nm^3/h; p_a is atmospheric pressure, Pa; C_0 is the flow coefficient; A is the leakage area, m^2; T_2 is the gas temperature, K; R is the general gas constant, $J/(k\mathrm{mol} \cdot K)$; M is the molar mass of the gas, g/mol.

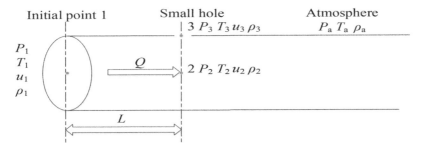

Figure 7. Schematic of the small hole model for gas pipeline leakage.

The formulas mentioned above are suitable for the case of small-hole leakage in the overhead pipeline. The main difference between the gas pipeline crossing the water and the overhead pipeline is that the environmental pressure of the underwater gas pipeline is larger than atmospheric pressure, so the environmental pressure on pipelines must be taken into account when calculating the leakage amount of underwater pipelines. Based on this, Dou et al revised Formula (2) and Formula (3) and proposed that environmental pressure P_e of the underwater pipeline should replace atmospheric pressure P_a of the overhead pipeline (Dou 2015). The formula is as follows (Eq. 4; Eq. 5).

$$p_e = p_a + p_w \tag{4}$$

$$p_w = \rho_w g h \tag{5}$$

Where P_e is the ambient pressure of the pipeline, Pa; p_a is the atmospheric pressure on the water surface, Pa; p_w is the pressure on the pipeline from the water, Pa; ρ_w is the density of water, kg/m³; h is the water depth; m.

However, when the underwater pipeline leaks, the situation is very complicated. Compared with the overhead pipeline, the pressure of the gas injected into the water will not be instantaneously reduced to the ambient pressure, but will fall in a certain pressure gradient and will not stabilize at ambient pressure. Therefore, the above formula cannot accurately calculate the leakage amount of underwater gas pipelines.

Ebrahimi-Moghadam conducted a numerical simulation to develop a few equations to estimate leakage from above-ground and buried urban natural gas pipelines by considering the impact of various parameters such as the pipeline and hole diameters (Amir Ebrahimi-Moghadam 2018). He fitted the simulation data and obtained the prediction formula for the buried pipeline (Eq. 6). However, he did not consider the effect of soil depth on the amount of leakage in the simulation.

$$Q = 0.44(1 + \beta^4)d_{orf}^2 P \tag{6}$$

Where, $\beta = \frac{d_{orf}}{D_p}$ is the ratio of the hole diameter to the pipe diameter.

Through the above analysis of the factors affecting the amount of leakage in the underwater gas pipeline, it can be found that the hole diameter, leakage pressure, and water depth all play a decisive role. Based on the result, the volumetric flow rate of leaked gas has a power function, linear relation with the diameter of the hole and leakage pressure respectively while it decreases with the increase of water depth. This is consistent with the variation of the buried pipeline leakage simulated by Ebrahimi-Moghadam. Therefore, this paper fits the experimental data and obtains the prediction formula for the leakage flow rate of the underwater gas pipeline (Eq. 7).

$$Q = 0.0261 \left[(h + 750.9)^{-0.951} + 0.167 \right] \cdot d^2 \cdot P \tag{7}$$

Figure 8 is the error distribution of the leakage amount calculated by the fitting formula and experiment. Compared with the experimental data, the relative error range calculated by the fitting formula is −22%∼+26%, and the average relative error is 9.27%, which indicates that the fitting formula is suitable for the leakage amount prediction of the underwater gas pipeline.

However, Eq. 7 is only applicable to this experimental system. In order to obtain a quantitative relationship for the leakage amount prediction of underwater gas pipelines under the conditions of small hole leakage (d≤20 mm) and subsonic flow (P≤90 kPa), compared with the mature theoretical model for calculating gas leakage of the overhead pipeline, Eq. 7 is modified by introducing coefficient α.

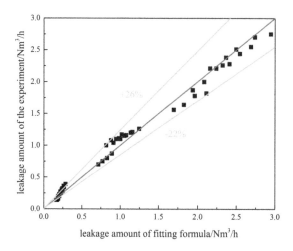

Figure 8. The relative error distribution between the fitting formula and experiments.

When $p_a/p_3 = CPR$, P_3 equals about 90 kPa at this time. The maximum leakage pressure of the experimental system is 70 kPa, which is less than 90 kPa. Therefore, the flow of the gas in the experimental system is subsonic. Eq. 2 can be used to calculate the gas flow rate when the water depth is 0 cm. Comparing the theoretical value of the empirical formula with the experimental value under the condition of this experiment (h=0 cm), the following formula is obtained by introducing coefficient α (Eq. 8)

$$Q_T = \alpha Q_e \tag{8}$$

Where Q_T is the gas leakage amount calculated by the theoretical model, Q_e is the gas leakage amount measured by experiment.

The quantitative relationship between the coefficient α and the hole diameter and leakage pressure is obtained by fitting (Eq. 9):

$$\alpha = 2.225 d^{0.195} \cdot (P + 20.012)^{-0.325} \tag{9}$$

Combined with Eq. 7. Eq. 8 and Eq. 9, the prediction formulas for underwater gas pipelines with a pressure range of 0-90 kPa and hole diameter of 1-20 mm are obtained (Eq. 10).

$$Q = 0.0261\alpha\left[(h + 750.9)^{-0.951} + 0.167\right] \cdot d^2 \cdot P \tag{10}$$

Taking the hole diameter of 10-20 mm and leakage pressure of 40-90 kPa as examples, the prediction formula Eq. 10 is validated by comparing the theoretical formula Eq. 2, and its relative error is shown in Table 2. According to the table, the relative error range of the predicted formula is −16.745%∼14.314% and the average relative error is 7.569%, which meets the accuracy requirement of leakage amount calculation in engineering applications.

Table 2. Leakage prediction results and error analysis.

Leakage hole diameter/mm	Leakage pressure/kPa	Leakage amount /Nm3/h		Relative error %
		Predictive value	Theoretical value	
10	40	20.269	18.614	−8.165
	50	24.099	24.132	−0.139
	60	27.690	27.691	−6.341
	70	31.093	34.865	−10.821
	80	34.338	40.009	−14.172
	90	37.452	44.985	−16.745
15	40	49.358	44.882	9.973
	50	58.683	54.297	8.078
	60	67.430	66.522	1.365
	70	75.714	78.447	−3.483
	80	83.618	90.019	−7.111
	90	91.201	101.217	−9.895
20	40	92.811	84.456	9.893
	50	110.346	96.529	14.314
	60	126.792	118.261	7.214
	70	142.370	139.461	2.086
	80	157.232	160.035	1.751
	90	171.491	179.941	4.696

4 CONCLUSION

A self-designed experimental circular pipeline is used to investigate the underwater gas leakage and diffusion characteristics under different operating conditions. The specific conclusions are as follows.

(1) When leakage occurs, the flow rate before the leakage point increases, while the flow rate after the leakage point decreases sharply. The flow rate of leaked gas has a power function, linear relation with the diameter of the hole and leakage pressure respectively while it decreases with the increase of water depth. The change of the hole diameter has the greatest impact on the leakage, followed by the leakage pressure, and the influence of water depth is the smallest.
(2) The quantitative relationship between the leakage amount of the underwater gas pipeline and the hole diameter, leakage pressure and water depth is obtained by nonlinear fitting. By introducing the coefficient α, the quantitative relationship for predicting the leakage amount of underwater gas pipeline under the condition of small hole leakage ($d \leq 20$ mm) and subsonic flow ($p \leq 90$ kPa) is obtained, which has a reference value for the leakage prediction of the actual pipeline after leakage accident.

ACKNOWLEDGMENTS

This work was supported by the National Key Research and Development Project (Project Number:2016YFC0802104) and the 2021 science and technology plan project of the Zhejiang Provincial Department of transportation (Project Number: 2021020).

REFERENCES

Amir Ebrahimi-Moghadam. CFD analysis of natural gas emission from damaged pipelines: Correlation development for leakage estimation [J]. *Journal of Cleaner Production*, 2018.

Arnaldos J, Casal J, Montiel H, et al. Design of a computer tool for the evaluation of the consequences of accidental natural gas releases in distribution pipes [J]. *Journal of Loss Prevention in the Process Industries*, 1998, 11(2):135-148.

Cloete S, Olsen J E, Skjetne P. CFD modeling of the plume and free surface behavior resulting from a sub-sea gas release [J]. *Applied Ocean Research*, 2009, 31(3):220–225.

DOU Z Y. *Analysis of leakage and diffusion rule and consequence of natural gas pipeline in water* [D]. Southwest Petroleum University, 2015.

K. Wu, S. Cunningham, S. Sivandran. Modelling subsea gas releases and resulting gas plumes using Computational Fluid Dynamics [J]. *Journal of Loss Prevention in the Process Industries*, 2017, 49(03):411–417.

Li X H, Chen G M, Zhang R. Simulation and assessment of underwater gas release and dispersion from subsea gas pipelines leak [J]. *Process Safety and Environmental Protection*, 2018, 14(05):17–22.

Liu C W, Li Y X, Fang L P, et al. Leakage monitoring research and design for natural gas pipelines based on dynamic pressure waves [J]. *Journal of Process Control*, 2017, 50:66–76.

Lu L, Zhang X, Yan Y, et al. Theoretical Analysis of Natural-Gas Leakage in Urban Medium-pressure Pipelines[J]. *Journal of Environment & Human*, 2014, 1(2): 71–86.

Olsen J E, Skjetne P. Modelling of underwater bubble plumes and gas dissolution with an Eulerian-Lagrangian CFD model [J]. *Applied Ocean Research*, 2016, 59:193–200.

Tessarolo L F, Innocentini V. Evaluation of entrainment formulations for liquid/gas plumes from underwater blowouts [J]. *Journal of Geophysical Research-Oceans*, 2016, 121(7):5530–5366.

*Advances in Energy, Environment and
Chemical Engineering – Abdullah & Osman (Eds)*
© 2023 The Author(s), ISBN: 978-1-032-36083-6

Study on comprehensive evaluation indexes of power tower resource sharing by using ISM method

Jie Ma*, Chengjie Wang*, Peng Wu & Haifeng Zheng
State Grid Energy Research Institute Co., Ltd., Beijing, China

Xiangjun Meng
State Grid Shandong Electric Power Company, Jinan, China

Zhenkun Wang
Economic&Technology Research Institute, State Grid Shandong Electric Power Company, Jinan, China

ABSTRACT: 5G is the key technology to promote the development of China's digital economy and realize carbon peaking as well as carbon neutralization. Power tower resource sharing can effectively reduce the development cost of the 5G network by relying on widely distributed power basic resources. However, power tower resource sharing is still in the early stage of development, it is necessary to build a systematic and scientific comprehensive evaluation indexes system to effectively screen the project promotion and development mode. Using the methods of literature reading and expert consultation, this paper constructs the primary indicators of power tower resource sharing from economy, society and environment dimensions. After that, this paper uses the ISM method to optimize the comprehensive evaluation indexes, and finally obtains 12 key indexes for power tower resource sharing comprehensive evaluation.

1 INTRODUCTION

Under the background of China's high-quality economic development and the goals of carbon peak as well as carbon neutralization, the development of the digital economy will play an important role in improving the efficiency of all factor production and promoting the low-carbon transformation. As a new generation of communication means, 5G is an important technical basis for China to realize the great leap forward development of the digital economy. However, the 5G network still faces practical problems such as high construction costs and long-time cycles.

In recent years, in order to further promote the construction of the 5G network, China has put forward the combination of 5G development and the sharing economy. Meanwhile, China has successively issued a series of policy documents aiming to reduce relevant construction costs and time costs by guiding the efficient reuse of resources.

As the concrete embodiment of sharing economy in the electric power field, power tower resource sharing can effectively improve the utilization efficiency of social resources, and help the rapid development of the national 5g as well as the digital economy.

However, power tower resource sharing is currently in the exploratory stage, and only a few works of literature put forward relevant evaluation indexes. Nevertheless, several pieces of literature propose the evaluation indexes for resource sharing in other areas. A performance evaluation method for provincial government information resource sharing platform based on big data was proposed by Jun Li (Li et al. 2021). Ning Ning utilized a fuzzy comprehensive evaluation method to evaluate

---

*Corresponding Authors: majiesxyq@163.com and wangchengjie@sgeri.sgcc.com.cn

the high-quality resource-sharing courses (Ning et al. 2018). A set of evaluation indexes of national scientific data sharing platform for population and health was proposed by Zanmei Li (Li et al. 2018).

In order to fill this gap, comprehensive evaluation indexes applied to power tower resource sharing are proposed in this paper, which systematically considers the impact of economic, social, and environmental aspects.

2 METHODOLOGY AND DATA

2.1 *ISM method*

The first step is to select the preliminary index according to project characteristics, which should follow the principles of hierarchy, comprehensiveness, independence, and operability.

The second step is to analyze the correlation between any two evaluation indexes. The relevant position in the matrix should be marked as 1 if they are related to each other. Otherwise, the value should be marked as 0. In this way, the adjacency n-matrix A_1 can be obtained as shown in eq.1, where n is the number of indicators.

$$A_1 = \begin{bmatrix} 1 & 1 & ... & 0 \\ 1 & 1 & ... & 0 \\ ... & ... & ... & ... \\ 0 & 0 & ... & 1 \end{bmatrix} \tag{1}$$

The third step is to establish an indicator impact diagram. The A_1 should be processed first to obtain A_2 by employing eq.2, then the matrix A_2 should be rearranged from front to back according to the number of "1" in each row. After that, each index can be distributed hierarchically according to the indexes framed by each maximum unit submatrix. In this way, the influence relationship between indicators at different levels can be determined.

$$(A_1 + I)^{k-1} \neq (A_1 + I)^k = (A_1 + I)^{k+1} = A_2 \tag{2}$$

Where I indicates the unit matrix, and k means the number of iterations.

The fourth step is to select the top and bottom indicators as the achievement index group and driving index group, which are also the final output indicators.

2.2 *Data description*

This paper applied the ISM method to construct and optimize the evaluation index system of power tower resource sharing. The methods of literature research and expert consultation are employed to carry out the preliminary evaluation indicators of economic, social, and environmental benefits of power tower resource sharing.

The preliminary indexes are shown in Table 1.

Table 1. Preliminary evaluation indicators of power tower resource sharing.

One-level index	Two-level index	Number
Economic benefit evaluation index	The annual income of the project	A1
	Primary cost input	A2
	Life cycle cost	A3
	Payback period	A4

(continued)

240

Table 1. Continued.

One-level index	Two-level index	Number
	Internal rate of return	A5
	net present value	A6
	Return on investment	A7
	Investment profit rate	A8
	Present value rate of investment	A9
Social benefit evaluation index	System energy supply reliability	B1
	Average failure rate of equipment	B2
	Sustainability of cooperation model	B3
	Contract performance capability	B4
	Ease of equipment maintenance	B5
	Personal risk of equipment maintenance	B6
	Impact on power grid equipment	B7
	Maintenance misoperation risk	B8
	Resource reuse rate	B9
	Total cost savings	B10
	Influence of city appearance	B11
Environmental benefit evaluation index	CO_2 emission reduction	C1
	SO_2 emission reduction	C2
	NOx emission reduction	C3
	Land resource saving area	C4

3 RESULTS AND DISCUSSION

3.1 *Economic benefit index screening*

According to the results conducted in Section 2.2, there are 9 preliminary economic benefit evaluation indexes, and their relationship can be represented by using the adjacency matrix as shown in Figure 1.

	A1	A2	A3	A4	A5	A6	A7	A8	A9
A1	1	0	0	1	0	1	1	0	0
A2	0	1	1	1	0	1	0	0	1
A3	0	0	1	0	0	0	1	0	0
A4	0	0	0	1	0	0	0	0	0
A5	0	0	0	0	1	1	0	0	0
A6	0	0	0	0	1	1	0	0	0
A7	0	0	0	0	0	0	1	1	0
A8	0	0	0	0	0	0	1	1	0
A9	0	0	0	0	0	0	0	0	1

Figure 1. The adjacency matrix of preliminary economic benefit evaluation indexes.

By using MATLAB software and the method introduced in Section 2.1, the reachability matrix of the economic benefit evaluation indexes in the power tower resource sharing project can be obtained as shown in Figure 2.

	A1	A2	A3	A4	A5	A6	A7	A8	A9
A1	1	0	0	1	1	1	1	1	1
A2	0	1	1	1	1	1	1	1	1
A3	0	0	1	0	0	0	1	1	0
A4	0	0	0	1	0	0	0	0	0
A5	0	0	0	0	1	1	0	0	1
A6	0	0	0	0	1	1	0	0	1
A7	0	0	0	0	0	0	1	1	0
A8	0	0	0	0	0	0	1	1	0
A9	0	0	0	0	0	0	0	0	1

Figure 2.   Reachability matrix of economic benefits comprehensive evaluation indexes.

It can be seen that A9, A4, and A7 constitute the highest level of the system. Meanwhile, A1 and A2 constitute the lowest level. It should be noted that A5 and A6 are exactly the same in the matrix, which means the correlation between them is much higher than that with other indicators. In order to further simplify the evaluation index system, A6 will be eliminated and only A5 will be retained in the final index screening. Due to the same reason, A7 is eliminated and only A8 will be retained.

In this way, the matrix can be rearranged as shown in Figure 3.

	A9	A4	A8	A3	A5	A1	A2
A9	1	0	0	0	0	0	0
A4	0	1	0	0	0	0	0
A8	0	0	1	0	0	0	0
A3	0	0	1	1	0	0	0
A5	1	0	0	0	1	0	0
A1	1	1	1	0	1	1	0
A2	1	1	1	1	1	0	1

Figure 3.   Rearranged matrix of economic benefits comprehensive evaluation indexes.

According to the reachability matrix, the hierarchical structure of different indexes can be drawn in Figure 4. It can be seen that A1 and A2 are critical to the project, which are drivers of A4 and

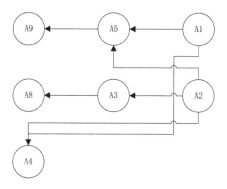

Figure 4.   Hierarchical structure of economic benefits comprehensive evaluation indexes.

242

A5, while A2 is a separate driver of A3. At the same time, A3 and A5 are the driving factors of A8 and A9, respectively. In other words, A3 and A5 are only used as intermediate variables in the system, and their evaluation effect can be completely replaced by the other indicators. Therefore, in the final optimization, only A1, A2, A4, a8, and A9 indicators are retained.

3.2 Social benefit index screening

According to the results conducted in Section 2.2, there are 11 preliminary social benefit evaluation indexes, and their relationship can be represented by using the adjacency matrix as shown in Figure 5.

	B1	B2	B3	B4	B5	B6	B7	B8	B9	B10	B11
B1	1	0	0	0	0	0	0	0	0	0	0
B2	1	1	0	0	0	1	1	1	0	0	0
B3	0	0	1	0	0	0	0	0	1	1	0
B4	1	0	0	1	0	1	0	1	0	0	0
B5	1	0	0	0	1	1	0	1	0	0	0
B6	0	0	0	0	0	1	0	0	0	0	0
B7	1	0	0	0	0	0	1	0	0	0	0
B8	0	0	0	0	0	1	0	1	0	0	0
B9	0	1	0	1	1	0	1	0	1	0	0
B10	0	0	0	0	0	0	0	0	0	1	0
B11	0	0	0	0	0	0	0	0	0	0	1

Figure 5. The adjacency matrix of preliminary social benefit evaluation indexes.

By using the same method shown in Section 3.1 the reachability matrix and the rearranged matrix can be conducted as shown in Figure 6 and Figure 7.

It can be seen that B1, B6, B10, and B11 constitute the highest level of the system. Meanwhile, only B3 constitutes the lowest level. As the same analysis method of economic benefit evaluation indexes, B7, B8, B4, B5, B2, and B9 are intermediate variables, although they are not at the same level. Therefore, only B1, B6, B10, B11and B3 will be restrained.

	B1	B2	B3	B4	B5	B6	B7	B8	B9	B10	B11
B1	1	0	0	0	0	0	0	0	0	0	0
B2	1	1	0	0	0	1	1	1	0	0	0
B3	1	1	1	1	1	1	1	1	1	1	0
B4	1	0	0	1	0	1	0	1	0	0	0
B5	1	0	0	0	1	1	0	1	0	0	0
B6	0	0	0	0	0	1	0	0	0	0	0
B7	1	0	0	0	0	0	1	0	0	0	0
B8	0	0	0	0	0	1	0	1	0	0	0
B9	1	1	0	1	1	1	1	1	1	0	0
B10	0	0	0	0	0	0	0	0	0	1	0
B11	0	0	0	0	0	0	0	0	0	0	1

Figure 6. Reachability matrix of economic benefits comprehensive evaluation indexes.

3.3 Environmental benefit index screening

It can be seen that the environmental benefit indexes can be divided into two categories: pollutant emission indexes and resource conservation indexes. Since the emissions of various pollutants

	B1	B6	B10	B11	B7	B8	B4	B5	B2	B9	B3
B1	1	0	0	0	0	0	0	0	0	0	0
B6	0	1	0	0	0	0	0	0	0	0	0
B10	0	0	1	0	0	0	0	0	0	0	0
B11	0	0	0	1	0	0	0	0	0	0	0
B7	1	0	0	0	1	0	0	0	0	0	0
B8	0	1	0	0	0	1	0	0	0	0	0
B4	1	1	0	0	0	1	1	0	0	0	0
B5	1	1	0	0	0	1	0	1	0	0	0
B2	1	1	0	0	1	1	0	0	1	0	0
B9	1	1	0	0	1	1	1	1	1	1	0
B3	1	1	1	0	1	1	1	1	1	1	1

Figure 7. Rearranged matrix of social benefits comprehensive evaluation indexes.

considered in the power tower sharing all come from the processing and manufacturing of steel, cement, and copper in the equipment transformation as well as the new equipment production, the main difference only comes from the different emission coefficients, which are closely related to each other in the changing trend. Therefore, only CO_2 emission reductions are retained in the emission indexes.

3.4 *The optimized comprehensive evaluation index*

According to the above screening process, an optimized comprehensive benefit evaluation index database of power tower resource sharing projects can be formed, which is shown in Table 2:

Table 2. The optimized evaluation indicators of power tower resource sharing.

One-level index	Two-level index	Number
Economic benefit evaluation index	Annual income of the project	A1
	Primary cost input	A2
	Payback period	A4
	Investment profit rate	A8
	Present value rate of investment	A9
Social benefit evaluation index	System energy supply reliability	B1
	Sustainability of cooperation model	B3
	Personal risk of equipment maintenance	B6
	Total cost savings	B10
	Influence of city appearance	B11
Environmental benefit evaluation index	CO2 emission reduction	C1
	Land resource saving area	C4

4 CONCLUSION

Based on the results and discussions presented above, the conclusions are obtained as below:

(1) It is shown that the annual income of the project and primary cost input are the driving indexes of power tower resource-sharing economic benefit, while the present value rates of investment,

payback period, and investment profit rate are the achievement indexes. The other indexes can be treated as intermediate variables which are discarded.

(2) It is shown that only the sustainability of cooperation model is the driving index of power tower resource sharing social benefit, while system energy supply reliability, personal risk of equipment maintenance, total cost savings, and influence of city appearance are the achievement indexes.

(3) The CO_2 emission reduction and land resource saving area are both driving and achievement indexes of power tower sharing environmental benefit, therefore both of them should be restrained. In addition, since SO_2 and NOx emissions are highly correlated with CO_2 emissions, both these two indexes are discarded finally.

ACKNOWLEDGMENTS

Thanks to the State Grid Corporation Science and Technology Project "Research on the Value Mining Technology, Business Model and Benefits Evaluation Method of Power Tower Resources" (Contract No. 1400-202157213A-0-0-00).

REFERENCES

Jun Li, Aihua Zheng, Weiwei Liu, Kun Ma (2021) *Performance Evaluation Method of Provincial Administrative Information Resource Sharing Platforms Based on Big Data.*, 35(4):381–386.

Ning Ning, Junqun Wu, Xueyan Jing, Yanhua Hao, Lili Chen (2018) Application of fuzzy comprehensive evaluation method in the evaluation of top-quality resource-sharing courses. *China Medical Education Technology.*, 32(2): 169–171.

Zanmei Li, Zhibin Hu, Haixia Sun, Junlian Li (2018) Research on Evaluation Index for Resources in the National Scientific Data Sharing Platform for Population and Health. *China Digital Medicine.*, 13(3): 02–04.

*Advances in Energy, Environment and
Chemical Engineering – Abdullah & Osman (Eds)
© 2023 The Author(s), ISBN: 978-1-032-36083-6*

Construction of evaluation index system under the coupling effect of the carbon market and power market

Yue Zhao*, Sijie Liu*, Guobing Wu, Yang Bai & Qiuna Cai
Power Dispatching and Control Center of Guangdong Power Grid Co., Ltd., Guangzhou, Guangdong Province, China

Pengcheng Zha* & Xinyu Meng
Beijing Tsintergy Technology Co., Ltd., Haidian District, Beijing, China

ABSTRACT: With the deepening of the power system reform, the proportion of power market-oriented transactions is increasing. At the same time, the national carbon market only includes the power industry in the early stage of construction. Aiming at the current situation of high coupling between the carbon market and the power market, this paper first summarizes the necessity of a comprehensive evaluation of the carbon power market from the background and purpose; then this paper analyzes the overall requirements and basic principles for the construction of carbon market and power market; based on the stability of the market operation, the effective transmission of prices and the promotion of emission reduction effects, this paper considers the fairness, operability, and problem orientation of indicators to build a complete evaluation index system for the carbon-power market; finally, the characteristics of the carbon market and power market in different construction stages are analyzed, and the evaluation focus of different stages is expounded. This article provides a quantifiable evaluation index system for the coordinated operation of the carbon market and the power market, which can effectively promote the construction of the carbon market and the power market.

1 INTRODUCTION

With the proposal and advancement of the dual carbon goals, the national unified carbon market was officially launched in July 2021. As of December 31, 2021, the cumulative transaction volume of carbon emission allowances in the national carbon market reached 179 million tons. In the initial stage of the establishment of the carbon market, the "National Carbon Emissions Trading Market Construction Plan (Power Generation Industry)" clarified that the carbon market is a policy tool to control greenhouse gas emissions.

At the same time, the first batch of national electricity spot pilots have carried out long-term settlement trials many times, the construction of the second batch of pilots has been put on the agenda, and the construction of the electricity spot market has entered the deep end. With the proposal of a new power system, the construction of the power market still faces many challenges in terms of adaptability and effectiveness.

At present, most of the existing studies only focus on the construction of the evaluation system of a certain market, and the research on the evaluation system of the carbon market focuses on the analysis, demonstration, and evaluation methods of the effectiveness of the carbon market. Some studies have constructed an index system based on the historical carbon trading data of pilot carbon markets in 7 provinces and cities in my country, and analyzed the effectiveness of the pilot

---

*Corresponding Authors: zhaoyue165@126.com; liusijie@gd.csg.cn and zhapc@tsintergy.com

                                              DOI 10.1201/9781003330165-35

carbon markets (Wu et al. 2021); some studies have measured and evaluated the effectiveness and development of China's pilot carbon markets (Lv et al. 2019).

Regarding the evaluation of the power market, researchers at home and abroad have conducted in-depth research on the evaluation of the power market, and have achieved certain results. Some studies have expounded and analyzed the principles and construction methods of the power market evaluation index system (Liu et al. 2005). According to the characteristics of China's power system and the development focus of power market construction, some studies have constructed a complete power market adaptability assessment framework from four aspects: effectiveness, fairness, security, and scalability of the power market. And based on the evaluation method combining fuzzy comprehensive evaluation and analytic hierarchy process, the comprehensive quantitative and effective evaluation of the power market based on the index system is realized (Yan et al. 2018). In addition, based on the typical power market model, some studies have established an indicator system and evaluation method for the performance evaluation of the power market including five aspects (Li & Tan 2008). Some studies combine multifractal theory with the power market and then use electricity price to evaluate the risk value of the power market (Liu et al. 2013). Some studies have introduced the evaluation system adopted in the European power market and clarified its evaluation system framework, including three aspects of the competition field index and the market sustainable development index (Li 2010; Wen et al. 2009).

To sum up, most of the existing research focuses on the evaluation of the carbon market or the power market, and the evaluation index system has not been established for the market construction and operation effect under the coexistence of the two markets. The development of the carbon market has externalized the cost of carbon emissions, which will directly impact the transaction price and competition pattern of the power market. At the same time, the price of the power market will affect the structure of the power supply, which will have a certain impact on the development of the carbon market. The coordinated operation of the market will assist in the orderly promotion of the transformation of the power system to a clean and efficient power system.

Based on the above analysis, this paper proposes a set of evaluation index systems suitable for the current construction stage of the carbon-power market, which provides a quantifiable evaluation method for the coordinated operation of the carbon market and the power market.

2 BACKGROUND AND PURPOSE OF CARBON-POWER MARKET EVALUATION

The purpose of constructing the carbon-power market evaluation index system is that the two markets have a deep coupling background and consistent construction goals. In addition, the evaluation results can provide a basis for subsequent policy formulation.

2.1 *Carbon-power market evaluation background*

The carbon market is policy-driven, and its construction purpose is to promote efficient emission reduction and achieve high-quality dual-carbon goals; the power market is demand-driven, and the purpose of its establishment is to efficiently allocate electricity resources through market-oriented means to achieve electricity supply and demand. balance. In terms of market operation, in the future, with the gradual realization of carbon peaking and carbon neutrality goals, the continuous development of carbon sequestration, technological emission reduction, and other means, carbon emissions will gradually decrease, and the scale of the carbon market will gradually shrink; With the deepening of reform, the proportion of power market-oriented transactions has been increasing, and the space requirements of the power market have gradually expanded.

The carbon-power market deeply integrates the trading products, management institutions, participants, market mechanisms, and other elements of the power market and the carbon market. On the power generation side, the cost of power generation and the cost of carbon emission together form the price of carbon-electricity products. Through dynamic price adjustment, the market competitiveness of clean energy is continuously improved, and clean substitution is promoted; In the related transaction mechanism of the energy industry, energy-consuming enterprises automatically bear the cost of carbon emissions when purchasing energy, forming a price advantage of clean electric energy over fossil energy, and encouraging the development of electric energy substitution and electrification on the energy-consuming side.

The carbon-power market is oriented to promote the development of high-efficiency units and clean energy. It can integrate the relatively scattered climate and energy governance mechanisms and participants to achieve efficient coordination of goals, paths, and resources, effectively solving the current existence of separate operations of the two markets. To promote the implementation of high-efficiency emission reduction plans and paths, stimulate the active emission reduction power of the whole society, and promote the optimal allocation of high-quality power generation resources.

2.2 *Carbon-power market evaluation objectives*

Considering that the carbon market and power market are still in the critical construction period, scientifically summarize the existing construction and operation status, find out the key issues hindering the construction and operation efficiency of the carbon-power market, and guide the subsequent policy adjustments and key issues of the carbon market and power market. Indicator adjustment is an important application of carbon-power market evaluation, which includes the following three aspects.

(1) Reasonable setting of the total quota

The national carbon market total control target should be "moderately tight", and at the same time consider objective factors such as the rigid growth of electricity under economic development, the difficulty of changing the power structure dominated by coal power in the short term, regional economic development and uneven distribution of energy resources. Through the evaluation of the carbon-power market, objectively reflect the current quota setting and provide a basis for the next step to formulate reasonable carbon quotas.

(2) Sustainable development of the market

As the main body participating in the power market and the carbon market at the same time, the excessive carbon quota setting will greatly increase the power generation cost of thermal power, significantly reduce its competitiveness in the power market, and make it difficult for thermal power companies to survive; at the same time, High-energy-consuming enterprises not only bear the emission cost in the carbon market but also need to accept the carbon emission cost transmitted from the power generation side. The transitional quota setting is not conducive to the survival of high-energy-consuming users in the power market. Therefore, through this evaluation index system, it can directly reflect the survival of entities participating in both the carbon market and the power market, and ensure the sustainable and long-term operation of the market.

(3) Unification of clean development goals

The construction goals of the carbon market and the power market are unified, to further promote the realization of clean and efficient energy development. Through the carbon-power market evaluation index to analyze the emission reduction effect of the carbon market and the power market, and based on the principle of goal orientation, analyze whether the market operation effect is unified with the target construction, and form a closed loop of the market from purpose-result-feedback-revising.

3 CARBON-POWER MARKET EVALUATION INDEX SYSTEM

According to the overall requirements and basic principles of the construction of the carbon market and power market, starting from the stable operation of the market, the effective transmission of prices, and the promotion of emission reduction effects, and considering the fairness, operability, and problem orientation of indicators, the proposed carbon -The indicator body of power market evaluation is as follows.

3.1 Carbon-power market relevance evaluation index

The carbon-power market correlation index mainly reflects the degree of correlation between the carbon-power market. The final manifestations of the carbon market and the power market are the carbon price and the electricity price respectively. By setting the relevant index and time lag index of carbon allowance transaction price and power market transaction price, we pay attention to whether there is a correlation between carbon and electricity price fluctuations and analyze the price transmission process of the carbon-power market.

Through the analysis of the correlation between carbon and power market, the influence degree of a carbon price on the fluctuation of electricity price can be obtained, which can provide a reference for the direction of power market construction from the perspective of electricity price stability or timely fluctuation. At the same time, we can evaluate the impact of the carbon market on the power industry from the perspective of the cycle and the magnitude of the impact of carbon prices on electricity prices. Provide a reference for the subsequent inclusion of more industries in the carbon market. The specific evaluation indicators of carbon-power market correlation are as follows.

(1) Carbon and power price-related indices

The carbon electricity price correlation index refers to whether there is a causal relationship between the carbon allowance transaction price and the power market clearing price in time series. The indicator reflects whether the carbon market and the power market have price correlations.

The specific calculation method is to obtain the clearing price of the power market and the transaction price of the carbon market daily within the same calculation period to form a daily price curve and calculate the relevant index of the carbon electricity price through the correlation function. The calculation formula is:

$$A1 = f(l_c, l_e)$$

Among them, A1 is the relevant index of carbon electricity price, and l_c and l_e are the daily change curves of carbon price and electricity price, respectively. A1 is generally in the range of 0 to 1. When A1>0.75, it is considered that there is a strong correlation between carbon electricity prices, and it is generally believed that there is an obvious causal relationship, that is, the increase in carbon costs has a high probability of causing an increase in the clearing price of the power market.

(2) Carbon electricity price conduction time lag index

The transmission time lag of carbon electricity price refers to the time when electricity price fluctuation lags behind carbon price fluctuation on the same time scale. The time lag index reflects when the current carbon price fluctuation will affect the power market clearing price in the later period. Generally, the sliding window method is used to calculate, and the calculation formula is:

$$A2 = f(l_c, l_e, t)$$

A2 is the carbon-electricity price conduction time lag index, t is the window step size of the sliding window, and f is the sliding window model function for calculating the lag time. The smaller the A2, the shorter the lag time, indicating that the power generation enterprises are more sensitive to carbon price fluctuations, the carbon electricity price transmission is smooth, and the carbon price can better play the role of the price signal.

(3) Carbon electricity price conductivity

Carbon electricity price transmission rate is defined as power market price change/carbon market price change, reflecting the degree of transmission from a carbon price to electricity price. A higher conductivity indicates that the two markets are deeply coupled, and most changes in carbon prices will be transferred to electricity prices. A lower conductivity indicates that the two markets are less coupled, and carbon prices and electricity prices are relatively independent. In addition, the carbon electricity price conductivity can also reflect the proportion of carbon cost occupied by electricity cost.

The calculation method is:

$$A3 = \Delta p / \Delta c$$

Among them, A3 is the carbon electricity price conductivity, Δp and Δc are the power market price change and the carbon market price change, respectively. When A3=10%, it means that 10% of the carbon cost change will be transferred to the power market price change, and when A3 is close to 1, indicating that the carbon electricity price transmission is smooth.

3.2 Carbon-power market stability evaluation indicators

The carbon-power market stability indicator focuses on the initial stage of the carbon market construction, and only includes the two market fluctuation degrees in the case of the power industry.

(1) Changes in the income of thermal power units

The change in the income of thermal power units is defined as the difference between the income of thermal power units in the current evaluation period and the income of thermal power units in the previous evaluation period.

The calculation formula is:

$$B1 = B_i - B_{i-1}$$

Among them, B1 is the change in the income of thermal power units, Bi and Bi-1 are the income of thermal power units in the ith evaluation period and the income of thermal power units in the (i-1)th evaluation period, respectively. Among them, the carbon cost should be included in the cost of thermal power generation when calculating the income of thermal power units.

(2) Changes in settlement prices on the user side

The change in the settlement price on the user side refers to the change in the settlement price of users in the power market before and after the carbon market is launched. The smaller the change, the higher the stability of the carbon-power market.

The calculation formula is:

$$B2 = C_i - C_{i-1}$$

Among them, B2 is the change of the settlement price on the user side, and Ci and Ci-1 are the settlement price of the user side in the ith evaluation period and the settlement price of the user side in the (i-1)th evaluation period, respectively.

3.3 Evaluation indicators of emission reduction effect

The power industry is one of the industries with large carbon emissions in my country, which is why the national unified carbon market was first included in the power industry. The scope of the industries to be included will continue to be expanded in the future, and industries with high carbon emissions such as metallurgy will also be gradually included in the country. To unify the carbon market, the current carbon market uses the industry benchmark method to issue carbon allowances. In addition, my country's current carbon market uses free allowances to issue. It is necessary to evaluate the emission reduction effect of the power industry, which is beneficial to provide a reference for the allocation of carbon allowances, the way of issuance, and the construction of power market mechanisms.

(1) Changes in the proportion of power generation of coal-fired enterprises

The change in the proportion of power generation of coal-fired enterprises mainly reflects the impact of the carbon market on the current power supply structure. By comparing the changes in the proportion of power generation of coal-fired enterprises in a certain period after the operation of the carbon market, it shows that under the background of increasing carbon costs, the overall Changes in the power structure, as well as changes in the power generation and income levels of coal companies.

The calculation formula is:

$$C1 = \alpha_i - \alpha_{i-1}$$

Among them, C1 is the change in the proportion of power generation of coal-fired enterprises, αi and α_{i-1} are the proportion of power generation of coal-fired enterprises in the ith evaluation period, and the proportion of power generation of coal enterprises in the (i-1)th evaluation period, respectively.

(2) Changes in carbon emissions per kilowatt-hour

Changes in carbon emissions per kilowatt-hour reflect the changing trend of the power industry's emission reduction effect. By comparing the changes in carbon emissions per kilowatt-hour after the carbon market operates within a certain period, it reflects the current total carbon quota setting, quota allocation, and the level of power marketization. The effect of emission reduction in the power industry.

The calculation formula is:

$$C2 = D_i - D_{i-1}$$

$$D_i = M_i / P_i$$

Among them, C2 is the change of kWh carbon emissions, and Di and Di are the kWh carbon emissions in the i-th evaluation period and the (i-1)-th evaluation period, respectively. Mi is the carbon emissions of the power industry in the i-th evaluation cycle. Data acquisition can be calculated based on the carbon verification data of carbon-emitting companies and the carbon emissions data of various types of power sources. Pi is the total power generation of the power industry in the i-th evaluation period.

4   PHASED ASSESSMENT OF CARBON-POWER MARKET

After determining the evaluation index system under the coupling effect of the carbon market and the power market, considering that in the different construction stages of the carbon market and the power market, the focus of market construction and operation indicators is different, and it is necessary to combine the carbon market activity, transaction volume Evaluate the current construction stage of the carbon market and the power market and reasonably consider the phased focus of the carbon-power market evaluation.

In the early stage of carbon market construction, the carbon market only included the power industry, and the marketization of the power industry was relatively low, and it was difficult to transmit carbon prices through electricity prices. Therefore, it is necessary to focus on the correlation and stability of the carbon-power market, especially the carbon-electricity price transmission process. The transmission process of carbon price depends on the power market, and a low degree of conduction is not conducive to the healthy development of the power industry and energy security, nor is it conducive to the continuous operation and effect of the carbon market. Only by realizing the smooth transmission of carbon electricity prices can the cost pressure of coal-fired power companies be reduced, and at the same time, price signals can be transmitted to end-users, which is conducive to energy conservation and emission reduction for the end-users.

In the late stage of carbon market construction, the carbon-power market has formed a relatively mature and stable price transmission system, and the carbon-power market evaluation focuses on the emission reduction effect of the power industry. By monitoring the changes in the power generation ratio of coal enterprises in the power industry, as well as indicators such as carbon emissions per kilowatt-hour, combined with the country's "dual carbon" goals and plans, starting from carbon market mechanisms such as the total amount of carbon allowances and distribution methods, by properly adjusting the market mechanism to encourage power generation enterprises to actively reduce carbon emissions through technological innovation and market transactions.

5   CONCLUSION

Under the background that the national carbon market is only included in the power industry, it is necessary to analyze the current construction of the carbon-power market with scientific and comprehensive evaluation indicators, to provide a reference for the adjustment of the market mechanism. Relevant studies at home and abroad focus on the single market evaluation system and indicators of the power market or carbon market but have not analyzed the market construction and operation level from the perspective of carbon-power market coupling.

There is a deep coupling relationship between the carbon market and the power market, including the overlap between the carbon market and the power market in terms of market scope, emission reduction promotion mechanism, and market entities. Therefore, the construction of the carbon-power market needs to be promoted simultaneously. This paper constructs the carbon-power market evaluation index system from the three dimensions of carbon-power market correlation, stability, and emission reduction effect, corresponding to the key tasks of carbon-power market construction in different periods. With the continuous advancement of power market reform, the accumulation of market operation data, and the adjustment of carbon market-related mechanisms, the carbon-power market evaluation indicators and calculation methods proposed in this paper can be further improved according to the market construction level and operation. In addition, the method proposed in this paper can also provide a reference for the setting of allowances and allocation methods when the carbon market is subsequently incorporated into other industries.

ACKNOWLEDGMENTS

This work was financially supported by the Science & Technology Project of China Southern Power Grid Cor-poration (036000KK52200041).

REFERENCES

Li Bo. Evaluation of market power in Nordic electricity market(2010). In: *Electric Power*, 43(12): 74–77.

Li Hanfang, Tan Zhongfu (2008). Construction research on index system of electricity market performance evaluation. In: *Electric Power*, 41(6): 29–32.

Liu Dun, Chen Xueqing, He Guangyu (2005) General Principle and Constitution Process of Evaluating Indices System for Electricity Market. In: *Automation of Electric Power*, 29(23): 2–14.

Liu Weijia, Shang Jincheng, Zhou Wenwei(2013). Evaluation of value-at-risk in electricity markets based on multifractal theory. In: *Automation of Electric Power Systems*, 2013, 37(7): 48–54

Lv Jingye, Cao Ming, Li Penglin (2019) An Empirical Analysis of the Effectiveness of China's Carbon Market. In: *Ecological Economy*, 35(7): 13–18.

Wen Fushuan, Wang Qin, Liu Min(2009). An evaluation system for the European electricity market. In: *Proceedings of the CSU-EPSA*,21(3): 23–31.

Wu W G, Zhu Y L, Gu G T. Determinants of the effectiveness of China's pilot carbon market (2021). In: *Resources Science*, 43(10): 2119–2129.

Yan Yu, Lin Jikeng, Hou Yanqiu (2018) Construction of the Framework for Adaptability Evaluation of Electricity Market. In: *Electric Power*, 51(12): 149–157.

Advances in Energy, Environment and Chemical Engineering – Abdullah & Osman (Eds)
© 2023 The Author(s), ISBN: 978-1-032-36083-6

Research on energy saving and carbon reduction technology for coal-fired power plant pulverizing system

Xiu Liu*
Huadian Electric Power Research Institute Co., Ltd., Hangzhou, China

Deyun Zhang & Yizhang Hu
Huadian Power International Co., Ltd., Tianjin Development Area Branch, Tianjin, China

Zhian Ma, Fang Dong & Zhansheng Shi
Huadian Electric Power Research Institute Co., Ltd., Hangzhou, China

ABSTRACT: By analyzing and researching the working principles and equipment characteristics of several typical coal-fired power plant pulverizing systems, the key to energy consumption in the operation of the pulverizing system is found, the potential for energy saving and carbon reduction in the pulverizing system is excavated, an energy saving optimization plan is proposed, and several typical cases are listed. The results show that the energy-saving optimization and transformation of the pulverizing system can reduce energy consumption, improve unit economy and safety, and reduce CO_2 emissions.

1 INTRODUCTION

The dual carbon goals put forward that the issue of how high-emission enterprises can achieve energy conservation and carbon reduction, and ultimately achieve the carbon neutrality goal, has aroused extensive thinking. To achieve energy saving and carbon reduction in coal-fired power plants, it is necessary to fully tap the performance potential of each equipment, especially the pulverizing system, whose power consumption accounts for about 20% of the power consumption of the plant. The operating parameters of the pulverizing system should be optimized to reduce energy consumption is of great significance for improving boiler efficiency, energy saving, and carbon reduction. There are various types of milling systems and complex equipment, so the key parts of energy saving and carbon reduction are also different. Huang Kui (Huang 2016) proposed graded high-chromium steel balls and high-chromium liner technology to reduce the energy consumption of steel ball mills. Wang Shiqiang et al (Wang et al. 2021) optimized the overall flow field of the medium-speed coal mill to improve its performance of the coal mill. Dong Fang et al. (Dong et al. 2021) reduced the energy consumption of the direct-blown pulverizing system of the double-inlet and double-outlet coal mill by optimizing the separator and pipe network system. Ai Dagao (Ai 2016) studied the effect of fine powder separator on the performance of milling system. Zhang Hailong et al. (2020) described the primary air in the pulverizing system and the boiler combustion system. How to fully tap the potential of energy saving and carbon reduction through the upgrading of milling system equipment and optimization of operating parameters still needs further analysis and research.

---

*Corresponding Author: xiu-liu@chder.com

DOI 10.1201/9781003330165-36

2 WORKING PRINCIPLE AND MAIN TYPES OF MILLING SYSTEM

The pulverizing system is an important part of the boiler side of the coal-fired power plant and is responsible for providing qualified pulverized coal for boiler combustion. The pulverized coal is screened by the separator, and the qualified pulverized coal is stored in the powder silo or directly sent to the boiler for combustion. The pulverizing system mainly includes direct blowing pulverizing system and an intermediate storage silo pulverizing system. There are several types of primary fan pulverizing systems, medium-speed grinding positive pressure direct blowing cold primary fan pulverizing system, and the intermediate storage silo pulverizing system is mainly a low-speed ball mill intermediate storage silo pulverizing system (including Hot primary air powder and exhaust deliver powder).

3 INFLUENCING FACTORS AND OPTIMIZATION SCHEME OF ENERGY SAVING IN PULVERIZING SYSTEM

3.1 *Steel ball mill*

According to experience and theoretical analysis, the weight of the lining plate and the steel ball is the main reason that affects the running current of the ball mill, and the material of the lining plate and the steel ball affects the service life of the equipment. At present, the energy-saving scheme of the steel ball mill mainly adopts the technical route of optimizing the gradation of high-chromium steel balls and the corresponding wave parabolic double-valley curved high-chromium liner. Improve the carrying capacity of the liner, improve the grinding efficiency of the coal mill, thereby reducing power consumption, and the improvement of the material can prolong the service life of the equipment.

3.2 *Medium speed coal mill*

The internal flow field of the medium-speed coal mill can be divided into three areas: the separator part, the inner cavity of the coal mill, and the nozzle ring area. The flow fields of the three parts can affect the efficiency of the medium-speed coal mill. The optimization of the separator can improve the separation efficiency and volume strength, improve the separation treatment while ensuring the fineness of pulverized coal, and can be flexibly adjusted to adapt to different coal qualities. The optimization of the flow field in the coal mill cavity can improve the conveying output and grinding output of the coal mill. The optimization of the lower flow field can reduce the leakage of primary air and improve the primary separation capacity. By optimizing the flow field of the medium-speed coal mill, the pulverizing capacity is improved, the fineness of pulverized coal can be adjusted, the adaptability to coal types is improved, and the power consumption of pulverizing is reduced.

3.3 *Meal separator*

At present, the coarse powder separator mainly adopts radial coarse powder separator and biaxial coarse powder separator. Biaxial coarse powder separators are mostly used in intermediate storage silo-type milling systems. The performance parameters of the coarse powder separator have a great influence on the performance of the pulverizing system. For example, the performance of the coarse powder separator is poor, and there are often problems such as large system resistance, a large amount of powder returning, and large fineness of coal powder, which increases the energy consumption of pulverizing and reduces coal consumption. The unqualified powder will affect the combustion of the boiler. At the same time, problems such as the baffle plate of the coarse powder separator and the wear of the center cylinder are also found in the actual operation, which reduces the performance of the coarse powder separator. By optimizing and transforming the coarse powder separator, the resistance of the system can be reduced, the cycle rate can be reduced, the energy consumption of pulverizing can be reduced, and the quality of pulverized coal can be improved.

3.4 *Fine powder separator*

The fine powder separator is mainly used in the intermediate storage silo-type pulverizing system, and the centrifugal fine powder separator is generally used, then sent to the boiler for combustion. The poor performance of the fine powder separator will cause the phenomenon of too much powder in the tertiary air, which will cause the center of the furnace flame to move up, the wall temperature of the superheater to be too high, the carbon content of the fly ash to increase, and the high powder content of the tertiary air will also lead to powder discharge. The impeller of the fan is seriously worn, which increases the power consumption of the powder discharge fan. Therefore, improving the performance of the fine powder separator can match the combustion stability, reduce the coal consumption, and also reduce the energy consumption of the power exhaust fan and reduce the maintenance workload.

3.5 *Powder feeding tube*

The diameter of the powder feeding pipe, the number of elbows, and the arrangement will affect the resistance of the milling system, thereby affecting the processing of the milling system. In severe cases, when the primary air pressure reaches the limit, the output of the pulverizing system is still insufficient, and the energy consumption of pulverizing is high. At the same time, because the resistance coefficients of each pipeline are inconsistent, there may be deviations in the different outlet air volumes and powder volumes of the same coal mill, which affects the boiler. burning effect. At present, the method of arranging and adjusting shrinkage holes is generally used for leveling. Although this method has the effect of leveling the resistance, it will further increase the resistance of the pipe network. Therefore, in order to increase the output, it is necessary to reduce the resistance of the pipe network system. The methods of increasing the diameter of the powder feeding pipe, optimizing the pipe arrangement, optimizing the number of elbows, and the radius of curvature can be adopted.

3.6 *Primary wind*

The low operating efficiency of the primary fan will lead to an insufficient output of the pulverizing system, the output of the coal mill cannot reach the maximum, and the current of the primary fan is too large. By optimizing and improving the efficiency of the primary fan, the output of the pulverizing system can be increased while reducing the fineness of pulverized coal. The hot primary air duct mostly adopts the form of a T-shaped tee, which leads to uneven primary air pressure at the inlet of each coal mill, and the output of coal mills with too low air pressure is low. The phenomenon of powder pipe blocking will seriously affect the combustion of the boiler. Therefore, optimizing the hot primary air duct will reduce the deviation of the inlet air pressure of each coal mill, and improve the performance of the pulverizing system and the safety and economy of the boiler.

4 ENERGY SAVING CASE

4.1 *Double-in and double-out ball mill direct-blowing pulverizing system*

The boiler of a power plant unit is a $2 \times 600MW$ supercritical variable pressure DC Bunsen boiler, the pulverizing system is a double-in and double-out pulverizer, and the designed coal is lean coal. As the actual coal type deviates from the design coal type, the power plant has a low output of the pulverizing system and high plant power consumption. At full load, all 6 coal mills must be put into operation, and it is impossible to achieve 5 operations and 1 backup. Therefore, it is urgent to conduct in-depth research and analysis on the pulverizing system of the power plant to improve the economic benefits of the power plant.

After analysis, the pulverizing system of the plant mainly has problems such as unbalanced resistance of the pipe network, small diameter of the pulverized coal pipe, and poor performance of the separator. By adjusting the resistance of the pipe network and optimizing the separator for the pulverizing system, the output of the pulverizing system has been improved, the fineness of the pulverized coal is significantly reduced, and the overall performance index of the pulverizing system has been significantly improved. The average maximum output of the pulverizing system before the transformation was 44.3t/h, and the maximum output after the transformation was 53.34t/h, an increase of about 20%, the average pulverizing unit consumption was reduced by 4.73, and the coal fineness was reduced by 2.25%–8.34%. The burnout degree and boiler efficiency of inferior coal provide effective technical support, and the annual cost saving is about 3.91 million yuan.

4.2 Low-speed ball mill intermediate storage bin type pulverizing system

The 2×300MW heating unit in the first phase of a power plant is an intermediate storage silo type pulverizing system, and each furnace is equipped with 4 steel ball mills. The actual operation found that, to a certain extent, the boiler pulverizing system has problems such as the low output of the coal mill, heavy wear at the outlet of the coal mill, and high cycle rate of the coarse powder separator. The performance test and optimization adjustment of the 2D milling system were carried out. The results are shown in Table 1. Even after the optimization and adjustment of the milling system, the above problems have not been solved well. In order to promote energy saving and consumption reduction, the two pulverizing systems of 2A and 2D were renovated, mainly replacing the whole coarse powder separator with a new type of high-efficiency coarse powder separator, and upgrading the original fine powder separator into a high-efficiency fine powder separator. After the transformation, the output of the pulverizing system increases by about 20%, and the pulverizing unit consumption decreases by 3-4.35kwh/t; under the condition that the fineness R90 of the pulverized coal powder in the pulverized bunker is not much different, the pressure drop of the fine powder separator after the transformation decreases by 821-1954Pa; the separation efficiency is increased by 10%, the pulverized coal uniformity index of the powder bin and the coarse powder separator is significantly increased; the heat loss of solid incomplete combustion is further reduced.

Table 1. Operating parameters of the coarse and fine powder separator.

Parameter	Before the test	After trial adjustment	Separator Technical Specification
Output (maximum) t/h	48	53	31.1 (The maximum is 41.5)
Coal fineness R90 %	8.64	7.12	8
Coal fineness R200 %	0.04	0.2	–
pulverized coal uniformity index	1.45	1.43	>1.1
Meal Separator Efficiency %	34.19	57.95	75
Cycle magnification	4.22	2.5	1.7–2.0
Coarse Separator Pressure Drop Pa	1600	1217	<780
Fines Separator Efficiency %	90.97	90.10	90
Fines Separator Pressure Drop Pa	1300	1510	<1000
Milling unit consumption kWh/t	29.13	26.59	–

4.3 Retrofit of fine powder separator in storage pulverizing system in steel ball mill

The #1 boiler of the first phase of a thermal power project adopts a steel ball mill intermediate storage pulverizing system. Each boiler is equipped with four DTM380/720 steel ball mills and GXF3500 fine powder separators. The efficiency of the fine powder separator is low. The test data shows that the efficiency of the fine powder separator is 83.7% when the maximum air volume of the powder discharge machine is high, and the efficiency is far lower than the design value of 90%, which causes the current of the exhaust fan to increase. The impeller wears seriously, and the tertiary air contains a large amount of powder, which seriously affects the combustion efficiency of the boiler and the nitrogen oxide discharge at the boiler outlet.

Through the transformation of the pulverizing system, the combustion of the boiler is stabilized, the amount of desuperheating water is reduced, and the exhaust gas temperature of the boiler is reduced by 2-4 degrees. According to the previous comparison, the running time of a single pulverizing system is reduced by about 2 hours per day; the current of the pulverizer is reduced by about 3.5A under the same air volume; Calculated at 1.8 million tons of coal, the single-machine modification of the pulverizing system of two fine powder separators can reduce the annual power consumption of the plant by 960,000 kWh, and increase the on-grid electricity revenue by 400,000 yuan per year. Through the transformation of the equipment, a single unit reduces the coal consumption for the power supply by about 1.87g/kWh and saves about 4,000 tons of standard coal per year.

5 CONCLUSION

In our country, the problems of an energy shortage, energy saving and emission reduction, environmental protection, and physical health are becoming increasingly prominent, especially the proposal of the dual carbon goal, the importance of energy saving, and carbon reduction in coal-fired power plants is particularly significant. By optimizing and transforming the pulverizing system, the performance of the separator can be improved, the fineness of pulverized coal can be adjusted, the adaptability of coal types can be improved, the uniformity index of pulverized coal can be improved, the flow resistance of the system can be reduced, the combustion of the boiler can be stabilized, and the combustion loss and NO_x emissions, improve equipment safety and stability, reduce coal consumption for power generation, achieve significant energy saving and carbon reduction effects, and considerable economic and social benefits.

ACKNOWLEDGMENTS

This work was financially supported by the China Huadian Co. Key Program (CHDKJ-2021-01-18) and the Key R&D Program of CHD (CHDKJ19-01-88).

REFERENCES

Ai Daogao. (2016). Retrofit and Effect Analysis of Fine Powder Separator on 300MW Middle Storage Pulverizing System Boiler. Technology and Enterprise (09), 195.

Dong Fang, Ma Zhian, Peng Li, Shi Zhansheng, Liu Guanqing & Li Zonghui. (2021). Optimizing and reformation of direct blowing pulverizing system with double-inlet and double-outlet ball mill. China Powder Science and Technology (02),82–86.

Huang Kui. (2016). Application of gradation high chrome steel ball and high chromium liner technology in energy saving transform of ball mill. Hunan Electric Poser (06), 90–91.

Wang Shiqiang, Tian Hang & He Lanqing. (2021). Optimization and Retrofit for Flow Field of a ZGM113 Medium-speed Coal Mill. Power Equipment (04), 292–296.

Zhang Hailong, Yang Guangrui, Jing Xinjing, Xing Leqiang, Liu Chao & Dang Lijun. (2020). Research on energy-saving optimization for medium-speed mill pulverizing systems. Huadian Technology (06), 25–30.

Advances in Energy, Environment and Chemical Engineering – Abdullah & Osman (Eds)
© 2023 The Author(s), ISBN: 978-1-032-36083-6

Prediction of NO_x emission concentration based on data from multiple coal-fired boilers

Lin Cong, Xigang Yang, Yong Zhang* & Baosheng Jin
School of Energy and Environment, Southeast University, Nanjing, China

ABSTRACT: In order to solve the prediction problem of NO_x emission concentration in coal-fired power plants, four forecasting models are established. Through the correlation analysis of three types of parameters (boiler structure parameters, coal quality parameters, operating parameters) and NO_x emission concentration, the correlation degree between various influencing factors and NO_x emission concentration can be obtained. The parameters with a high correlation with NO_x emission concentration are selected as the input of the model to reduce the complexity of the model and prevent the model from overfitting. The four prediction models based on the grid search method are established. The results showed that among the four prediction models, the Random Forest and Bagging of prediction models have the best prediction effects. From the four evaluation indicators, the prediction effects of the two prediction models are not much different, but the Bagging prediction model has an over-fitting problem, therefore, the Random Forest prediction model is the most suitable for NO_x emission concentration prediction. When boiler structure parameters and coal quality parameters are taken as the input of the forecasting models, the forecasting models have a better prediction effect and higher prediction accuracy, and the prediction models have better generalization performance. By predicting the results, you can find that the random foresting prediction model can better predict NO_x emission concentration through coal quality parameters and boiler structure parameters.

1 INTRODUCTION

With the rapid development of the economy and the improvement of people's requirements for the comfort of life, the demand for electricity is increasing year by year. In the era of mainly thermal power generation, the amount of coal consumed by thermal power plants has also increased year by year. But coal-fired power plants will produce a large amount of nitrogen oxides from boilers combustion which will cause a series of environmental problems (Yuan 2020), such as acid rain and photochemical smog. The emission of NO_x can cause extremely serious damage to the environment and cause economic losses to people. In order to resolve the serious pollution issues, the country has released stringent emission standards for air pollutants. The NO_x emission limit for coal-fired Industrial boilers is 200–300 mg·Nm^{-3} (9% O_2) and 100 mg·Nm^{-3} (6% O_2) for the thermal power plants (National Standards) of the People's Republic of China 2014) (Zhu 2021). With the promulgation of a series of national policies and regulations in China, finding more effective ways to optimize the combustion process and reduce NO_x emissions has become a concern for many Scholars (Yuan 2020). Therefore, reducing the emission concentration of NO_x has become the top priority of the power plant's tasks.

---

*Corresponding Author: zyong@seu.edu.cn

DOI 10.1201/9781003330165-37

Some of the techniques that have been proposed to alleviate the negative impacts of NOx emissions include selective catalytic and noncatalytic reduction, employing low NO_x burners, excess air reduction, adsorption, and absorption, and flue gas recirculation (Mostafa 2019). Furthermore, boiler combustion optimization has received special attention as a way to minimize NOx emissions in coal-based power plants (Chen 2021). At present, the combustion operation of the domestic power stations is mainly adjusted by the operation personnel based on their own engineering experience. However, due to the complexity of boiler combustion, there are many factors that affect the NO_x emission concentration. Therefore, only relying on engineering experience to adjust combustion can often not achieve ideal results (Liang 2021). With the maturity of data storage technology and the development of artificial intelligence technology, data-driven modeling methods have become more and more common. Compared with the mechanism model, the data model is a black box model. It does not need to pay attention to the internal structure of the system and can be used to solve the modeling problems of the mechanism with complex or even unknown. Artificial intelligence technology has been widely used in process modeling because of its good fitting ability for nonlinear functions. Many scholars began to apply statistical methods to boiler modeling, such as Linear Regression (LR) and Support Vector Regression (SVR) have achieved good results (Wang 2021; Yuan 2020). By establishing an accurate boiler combustion system model and then using an intelligent optimization algorithm to optimize the parameters of the model, the goal of reducing the emission concentration of NO_x is finally achieved.

Hong et al. (Hong 2021) established a power station boiler combustion control model by using RBF/BP neural network coupled fuzzy rules, which effectively reduced NO_x emission concentration and improved power generation efficiency to a certain extent. Liang et al. established the NOx emission prediction model by using the weighted least squares support vector machine algorithm. Then the improved multiverse optimization algorithm was used to optimize the model parameters. Compared with the unoptimized model, the prediction accuracy was improved. Xie et al. (Xie 2021) used the bidirectional gating cyclic neural network to predict NO_x emission concentration. Compared with the traditional GRU neural network, Bi-GRU neural network made full use of the time sequence information before and after the sample data and reduced the error. Jin et al. (Jin 2021) established the forecasting model of combustion thermal efficiency and NO_x emission concentration by using the Online Sequential Extreme Learning Machine (OS - RELM) algorithm, which can better reflect the real operation state of the boiler and achieve the optimal balance between high efficiency and low pollution of boiler combustion. In addition, Elman network (Chen 2021; Li 2018; Peng 2016), LSSVM (Lv 2013; Li 2013) and LSTM neural network (Jongchol 2021; Li 2013; Qin 2019) are also used for NO_x emission prediction.

At present, most of the forecasting methods using machine learning and intelligent optimization algorithms use the same unit operating data to predict the NO_x emission concentration or NO_x emission amount. This prediction method ignores the influence of boiler structure parameters on NO_x emission concentration or NO_x emissions. Wang et al. (2002) discussed and analyzed the NO_x emissions of two W-flame furnaces in terms of operation, coal type, and equipment structure. They found that the furnace structure parameters have a certain influence on NO_x emissions. Therefore, the boiler structure parameters also have a certain influence on the NO_x emission concentration. In this paper, based on the operating data of multiple coal-fired units, variable selection is made for multiple factors such as boiler structure parameters and coal quality parameters, and multiple machine learning algorithms are used to establish multiple NO_x emission concentration prediction models. Then, the grid search algorithm is used to optimize the parameters of the four established NO_x emission concentration prediction models, and the model with the best prediction effect is selected to achieve the purpose of reducing the NO_x emission concentration of the boiler. Through the prediction model, the appropriate coal type can be selected according to the boiler structural parameters to achieve the purpose of reducing NO_x concentration.

2 MODELING PRINCIPLE AND RESEARCH OBJECT

2.1 *Modeling principle*

Machine learning is divided into unsupervised learning and supervised learning. Four supervised learning algorithms in machine learning are selected to predict the NO_x emission concentration.

When establishing the forecasting models, the main factors affecting NO_x emission concentration (boiler structural parameters, coal quality parameters, and operating parameters) were taken as the input samples of modeling, and NO_x emission concentration was taken as the output samples of modeling. The structure of the prediction model for the NO_x emission concentration of the boiler is shown in Figure 1.

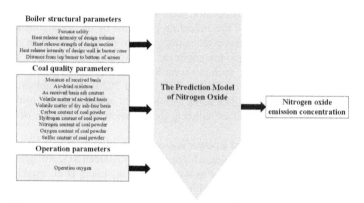

Figure 1.   The forecasting model structure of boiler NO_x emission concentration.

2.2 *Modeling principle*

The research object is a number of hedged coal-fired units above 600MW. The research institute provides the boiler structure parameters, coal quality parameters, and operating parameters of multiple coal-fired units. The concrete content of these three parameters is shown in Table 1.

Table 1.   The concrete content of the three parameters.

Parameter	Concrete parameter	Unit
Boiler structural parameters	Furnace capacity	m^3
	Heat release intensity of design volume	kW/m^3
	Heat release strength of design section	MW/m^2
	Heat release intensity of design wall in burner zone	MW/m^3
	Distance from the top burner to the bottom of the screen	m
Coal quality parameters	The moisture of the received basis	%
	Air-dried moisture	%
	As received basis ash content	%
	Volatile matter on air-dried basis	%
	Volatile matter of dry ash-free basis	%
	The carbon content of coal powder	%
	Hydrogen content of coal power	%
	Nitrogen content of coal powder	%
	The oxygen content of coal powder	%
	The sulfur content of coal powder	%
Operation parameters	Operation oxygen	%

3 CORRELATION ANALYSIS

Correlation analysis is defined as the analysis of two or more correlated variable elements, so as to measure the degree of correlation between two variable factors (Luo 2018). The correlation coefficient is a statistical indicator that reflects the closeness of the relationship between variables. There are three main correlation coefficients: the Pearson correlation coefficient and the Spearman correlation coefficient.

3.1 *Pearson correlation coefficient*

The Pearson correlation coefficient is the most commonly used method when studying the correlation between two variables. It needs to meet the following two conditions:
(1) It must be assumed that the data are obtained in pairs from a normal distribution.
(2) The data is equidistant at least within the logical range.
Assuming that there are two variables X and Y, the formula for solving the Pearson correlation coefficient is as follows:

$$\rho_{XY} = \frac{cov(X,Y)}{\sigma(X)\sigma(Y)} == \frac{E((X-\mu_X)(Y-\mu_Y))}{\sigma(X)\sigma(Y)} = \frac{E(XY)-E(X)E(Y)}{\sqrt{E(X^2)-E^2(X)} \cdot \sqrt{E(Y^2)-E^2(Y)}} \quad (1)$$

where *cov* represents the covariance; σ represents the standard deviation; E represents the expectation; μ_X and μ_Y represent the average values of X and Y respectively.

3.2 *Spearman correlation coefficient*

If the two variables do not meet the two conditions of the Pearson correlation coefficient, the Spearson correlation coefficient can be used to replace the Pearson correlation coefficient.
The variables X and Y are sorted in order of rank from small to large. And if the data is repeated, the average value of their order is taken. The Spearson correlation coefficient can be calculated by the following formula:

$$r_s = 1 - 6\left[\frac{\sum_{i=1}^{N}(R_i - Q_i)^2}{n(n^2-1)}\right] \quad (2)$$

where R_i and Q_i are the rank of X and Y respectively.
The correlation analysis method used in this paper is to calculate the Spearson correlation coefficient between each variable and the NO_x emission concentration.
According to the data provided, the Spearman correlation coefficient between them is calculated and presented in the form of a heat map (Figure 2). The scale on the right side of Figure 2 shows the colors of different correlation coefficients corresponding to different shades. According to Figure 2, we can see the degree of correlation between different factors and the concentration of NO_x emissions.
According to Figure 2, it can be seen that the NO_x emission concentration is negatively correlated with furnace capacity, moisture of received basis, air-dried moisture, volatile matter of air-dried basis, volatile matter of dry ash-free basis, the hydrogen content of coal power, and oxygen content of coal powder. The correlation coefficients between these factors and NO_x emission concentration are −0.29, −0.46, −0.43, −0.62, −0.47, −0.5, −0.49 respectively. The NO_x emission concentration is positively correlated with heat release intensity of design volume, heat release strength of design section, heat release intensity of design wall in burner zone, distance from top burner to bottom of the screen, as received basis ash content, the carbon content of coal powder, the nitrogen content of coal powder, the sulfur content of coal powder, operation oxygen. The correlation coefficients between these factors and NO_x emission concentration are 0.27, 0.37, 0.39, 0.42, 0.58, 0.07, 0.14, 0.5, 0.28, respectively. Therefore, it can be understood that the increase of furnace capacity,

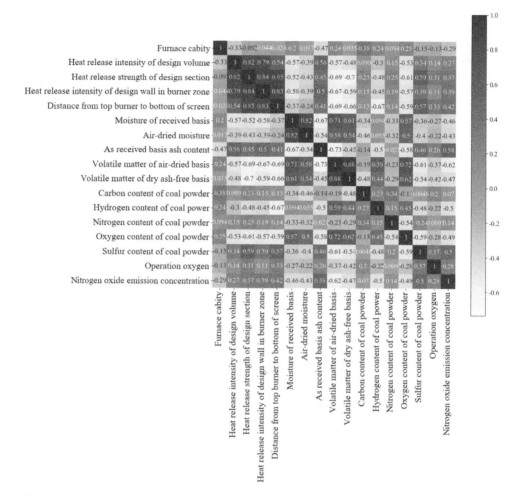

Figure 2. Correlation coefficient between different factors and NO$_x$ emission concentration.

moisture of received basis, air-dried moisture, volatile matter of air-dried basis, volatile matter of dry ash-free basis, hydrogen content of coal power, and oxygen content of coal powder, can cause the reduction of NO$_x$ emission concentration. With heat release intensity of design volume, heat release strength of design section, heat release intensity of design wall in burner zone, distance from top burner to bottom of the screen, as received basis ash content, the carbon content of coal powder, the nitrogen content of coal powder, the sulfur content of coal powder, operation oxygen, NO$_x$ emission concentration will increase.

In coal quality parameters, most factors are highly correlated with NO$_x$ emission concentration. Among them, the volatile matter of air-dried basis of coal powder of the highest degree in relation to the concentration of NO$_x$ emission, the correlation coefficient can reach -0.62, which is a moderately negative correlation. And the carbon content of coal powder is least related to the concentration of NO$_x$ emission, the correlation coefficient of only 0.07, belongs to the weak correlation. In the boiler structure parameters, the distance from the top burner to the bottom of the screen has the highest correlation with NO$_x$ emission concentration, with a moderate positive correlation coefficient of 0.42, while the heat release intensity of design volume has the lowest correlation with NO$_x$

emission concentration, with two correlation coefficients of only 0.27, which is a weak correlation. Among the operating parameters, the amount of operation oxygen is weakly correlated with the concentration of NO_x emissions concentration. On the whole, the correlation degree between coal quality parameters and NO_x emission concentration is higher than the correlation degree between boiler structure parameters and NO_x emission concentration and the correlation degree between operating parameters and NO_x emission concentration.

4 NO_x EMISSION CONCENTRATION PREDICTION METHODS

4.1 *Random forest algorithm*

The random forest algorithm consists of an ensemble of decision trees used for regression or classification tasks (Mourad 2020). The process of constructing a model using the random forest algorithm is: first of all, using the bootstrap method to randomly select n samples from the original training set and constructing n decision trees. And then, assuming that there are m features in the training sample data, then the best feature is selected for each split and each tree keeps splitting like this until all the training examples of the node belong to the same category. Then they maximized the growth of each decision tree without any pruning. Finally, the generated multiple classification trees are formed into a random forest, and the random forest classifier is used to classify and regress the new data. For classification problems, the final classification result is determined by the voting of multiple tree classifiers; for regression problems, the final prediction result is determined by the average of the predicted values of multiple trees.

4.2 *Bagging algorithm*

Bagging is an ensemble method conceived to improve the stability and accuracy of machine learning algorithms adapted to classification and regression. It is capable to reduce the variance, resists over-fitting, added to its ability to be used with any basic Algorithm (Mourad 2020). The mechanism of the Bagging algorithm is for us to sample T sampling sets containing m samples. Then a learner is trained based on each sample training, and then the learners are combined. The voting method is used for classification tasks, and the average method is used for regression tasks. In this paper, the base learner of the Bagging algorithm uses a decision tree.

4.3 *KNN algorithm*

KNN algorithm is also known as the K-nearest neighbor algorithm. Predictions are made for a new instance (x) by searching through the entire training data set for the K most similar instances (the neighbors) and summarizing the output variable for those K instances (Chipindu 2020). When it is necessary to predict an unknown sample, the prediction result is determined by the K neighbors closest to the sample. KNN can be used for classification problems as well as regression problems. When predicting a classification problem, the largest number of categories (or the most weighted) among the K neighbors is used as the prediction result; when predicting a regression problem, the average (or weighted average) of the K neighbors is used as the prediction result.

4.4 *XGBoost algorithm*

The extreme gradient advancement decision tree is an integrated machine learning algorithm based on decision trees. It uses the gradient ascent framework. And it is improved on the basis of the GBDT algorithm. It is suitable for classification and regression problems. XGBoost algorithm integrates a number of weak classifiers to form a strong classifier. The basic principle is to add different trees to the model in turn and generate different forms of tree models by splitting features.

5 CASE ANALYSIS

5.1 Sample selection

In this paper, a number of coal-fired boilers of different grade specifications were taken as the research object, and a group of stable operation data was collected from each unit. A total of 180 groups of data were collected. Due to the lack of some parameters, the data with null values were deleted. Therefore, 145 sets of data are retained, and these 145 sets of data are used as the total sample. The sample data were divided into training samples and test samples according to the ratio of 7:3. The training samples were used to train the forecasting models, and the test samples were used to test the generalization ability of the models.

Each group of sample data includes 17 dimensions. Normally, multidimensional data may contain tens or hundreds of attributes, where several attributes in the dataset may be irrelevant to the pattern classification in machine learning (Srivihok 2015). In order to reduce the complexity of the models and avoid over-fitting of the forecasting model, the input variables of the model should be reduced. It can be seen from Figure 2 that the correlation between some factors and NO_x emission concentration is not high. In order to reduce the complexity of the model and avoid over-fitting the forecasting models, 10 factors with a high correlation with NO_x emission concentration are selected as the input variables of the models, including heat release intensity of design wall in burner zone, distance from top burner to bottom of the screen, moisture of received basis, air-dried moisture, as received basis ash content, volatile matter of air-dried basis, volatile matter of dry ash-free basis, hydrogen content of coal power, oxygen content of coal powder, and sulfur content of coal powder. The dimension of each set of data was reduced from 17 to 11, and the 11th dimension was the NO_x emission concentration of the boiler, 11th dimension for boiler NO_x emission concentration, specific data are shown in Table 2.

Table 2.   Parameters of each unit.

Unit	1	2	3	……	145
Heat release intensity of Design wall in burner zone	1.637	1.660	1.660	……	1.388
Distance from top burner to bottom of screen	21.5190	27.2223	27.2223	……	19.4440
Moisture of received basis	22.70	9.80	7.60	……	24.20
Air-dried moisture	9.96	4.28	1.52	……	11.00
As received basis ash content	24.14	26.88	24.67	……	5.07
Volatile matter on air-dried basis	28.781080	18.062020	15.462390	……	40.061920
Volatile matter of dry ash-free basis	46.48	26.88	21.42	……	48.24
Hydrogen content of coal power	2.64	2.94	2.79	……	3.46
Oxygen content of coal powder	9.42	4.38	5.02	……	12.96
Sulfur content of coal powder	0.62	0.70	0.68	……	0.70
NOx emission concentration	227.03	367.00	519.50	……	300.10

5.2 Data normalization processing

Due to the different dimensions of the parameters of the samples collected by the research institute, if the data are directly input into the four algorithms for training, the training effect of machine learning is bound to be affected. Therefore, the normalization method should be adopted to normalize data to [-1,1].

5.3 Evaluation index

In order to further compare the prediction effect of the four forecasting models, coefficient of determination, mean absolute error, mean absolute percentage error, and root mean square error are used as evaluation indexes to evaluate the prediction effect of the models. R2, MAE, MAPE, and MSE are defined as follows (Tang 2019; 2020; Yuan 2020):

(1) R2 (coefficient of determination): the numerator represents the sum of the squares of the true value and the predicted value, and its calculation formula is as follows:

$$R^2 = 1 - \frac{\sum_i \left(\hat{y}_i - y_i\right)^2}{\sum_i \left(\bar{y}_i - y_i\right)^2} \tag{3}$$

(2) MAE (mean absolute error): its calculation formula is as follows:

$$MAE = \frac{1}{n}\sum_{i=1}^{n}|\bar{y}_i - y_i| \tag{4}$$

(3) MAPE (average absolute percentage error): its calculation formula is as follows:

$$MAPE = \frac{100\%}{n}\sum_{i=1}^{n}\left|\frac{\hat{y}_i - y_i}{y_i}\right| \tag{5}$$

(4) MSE (mean square error): its calculation formula is as follows:

$$MSE = \frac{1}{n}\sum_{i=1}^{n}\left(\hat{y}_i - y_i\right)^2 \tag{6}$$

where n is the number of test data; y_i is the actual value; $\hat{y}_i$ is the predicted value.

5.4 *Analysis of prediction results*

Based on the grid search method, the optimal parameters of the four models are obtained, and the 10 relevant variables selected after the correlation analysis are used as the input of the four NO_x emission concentration prediction models. The training set is used to train the four forecasting models, and the test set is used to evaluate the prediction accuracy and generalization of the four forecasting models.

In order to compare and analyze the accuracy of the four prediction models, we illustrate the prediction effects of the four prediction models from the relative prediction errors and four evaluation indicators, so as to select the NO_x emission concentration prediction model with the best prediction effect.

According to Figure 3, when we use the training set data to train the four models, the prediction results of the four prediction models obtained are roughly similar. The prediction results of the four models follow the trend of NO_x emission concentration to a certain extent. In the four NO_x emission concentration prediction models, the fitting effect of the Random Forest prediction model and Bagging prediction model is better, while the fitting effect of the KNN prediction model and XGBoost prediction model is poor. According to Figure 4, it can be seen that when we use the test set data to test the four prediction models, the Random Forest prediction model has the best follow-up effect, followed by the Bagging prediction model. But, the XGBoost prediction model has many points with poor prediction effects.

According to Figure 5 and Figure 6, the four evaluation indicators (R^2, MAPE, MAE, and MSE) are used as the evaluation criteria. From the perspective of the coefficient of determination, the coefficients of determination of the Random Forest prediction model for the training set and test set are 0.87441 and 0.702701, respectively, and the coefficients of determination for the Bagging prediction model are 0.91665 and 0.679907, respectively, which are higher than KNN prediction

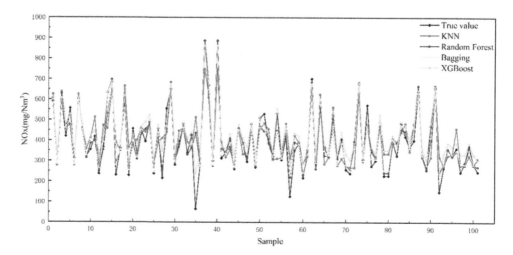

Figure 3. The prediction results of different algorithms (training set).

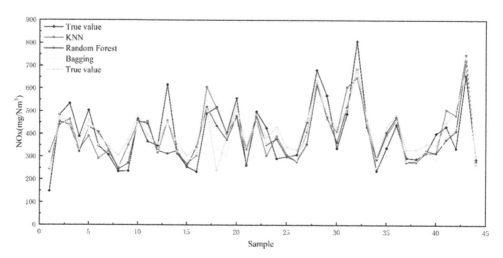

Figure 4. The prediction results of different algorithms (test set).

model and XGBoost prediction model (the training set is 0.682854 and 0.858807 respectively, and the test set is 0.668982 and 0.655892 respectively). From the training set, the coefficient of determination of the Bagging prediction model is the highest, and from the test set, the Random Forest prediction model has the highest coefficient of determination. From the perspective of MAPE, the training set and test set of the Random Forest prediction model are 14.00144 and 13.85841, respectively, and the training set and test set of the Bagging prediction model are 11.47145 and 14.23021, respectively, which are lower than KNN prediction model and XGBoost prediction model. From the perspective of MAE, the training set and test set of the Random Forest prediction model are 32.54633 and 54.39937, respectively, and the training set and test set of the Bagging prediction model are 28.03753 and 56.96863, respectively, which are lower than KNN prediction model and XGBoost prediction model. From the perspective of MSE, the training set and test set of the Random Forest prediction model are 53.6573 and 74.40745, respectively, and the training set and test set of the Bagging prediction model are 43.70851 and 77.20721, respectively, which are lower than KNN prediction model and XGBoost prediction model. From the four evaluation indicators,

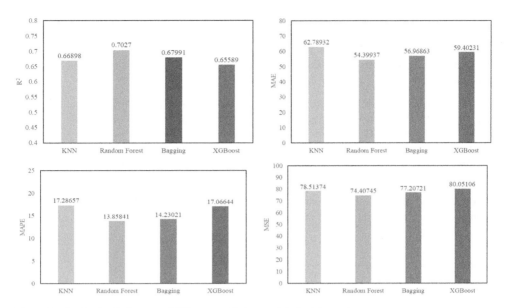

Figure 5. Evaluation indicators of different prediction models (training set).

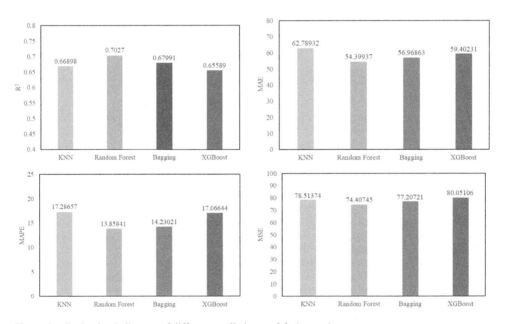

Figure 6. Evaluation indicators of different prediction models (test set).

the prediction effects of the Random Forest prediction model and the Bagging prediction model are roughly similar, and the results are not much different. The KNN prediction model and XGBoost prediction model are inferior to the Random Forest prediction model and Bagging prediction model from the training set and test set.

6 CONCLUSION

In this paper, a variety of algorithms are used to predict NO_x emission concentration, and a variety of evaluation indexes are used to evaluate multiple prediction models, so as to obtain the following conclusions:

(1) According to the correlation analysis, it can be seen that the influence of coal quality parameters on the NO_x emission concentration is greater than the influence of the boiler structure parameters on the NO_x emission concentration.
(2) Compared with the Bagging prediction model, the KNN prediction model, and the XGBoost prediction model, the random forest prediction model has high prediction accuracy, small prediction errors, and better generalization. Therefore, the prediction effect is better.

Appropriate boiler operation parameters are selected for different boiler structures, which is a problem that needs to be improved in the future.

ACKNOWLEDGMENTS

Over the course of my research and writing this paper, I would like to express my thanks to all those who have helped me.

I would like to express my gratitude to all those who helped me during the writing of this thesis. I am particularly indebted to Dr. Zhang Yong who gave me kind encouragement and useful instruction all throughout my writing.

Sincere gratitude should also go to all my learned Professors and warm-hearted teachers who have greatly helped me in my study as well as in my life.

REFERENCES

Chen XiaoJuan, Zhang Haiyang, Xing Xiaoxue, Qin Hongwu, Wójcik Waldemar T. Modeling NO_x Emissions with an Intelligent Combinatorial Algorithm[J]. *Mathematical Problems in Engineering*, 2021,2021:

Guoqiang Li, Peifeng Niu, Weiping Zhang, Yongchao Liu. Model NO_x emissions by least squares support vector machine with tuning based on ameliorated teaching–learning-based optimization[J]. *Chemometrics and Intelligent Laboratory Systems*, 2013,126:

Guoqiang Li, Peifeng Niu, Weiping Zhang, Yongchao Liu. Model NO_x emissions by least squares support vector machine with tuning based on ameliorated teaching–learning-based optimization[J]. *Chemometrics and Intelligent Laboratory Systems*, 2013,126:

HONG Chang-shao, HUANG Jun, GUAN Ying-yuan, MA Xiao-qian. Combustion Control of Power Station Boiler by Coupling BP / RBF Neural Network and Fuzzy Rules [J]. *Journal of Engineering for Thermal Energy and Power*, 2021,36(04):142–148.

Hui Wang, Guobao Zhang, Yongming Huang, Yongchun Zhang. *Study on boiler's comprehensive benefits optimization based on PSO optimized XGBoost algorithm*[J]. E3S Web of Conferences, 2021,261:

Jin Peng. Research on Modeling of Combustion System based on OS-ELM Algorithm [J]. *Electric Tool*, 2021(02):14–16.

Jongchol Kim. *An Air Pollution Prediction Scheme Using Long Short Term Memory Neural Network Model*[J]. E3S Web of Conferences,2021,257:

Liang Tao, Jin Yunjie, Jiang Wen, Liu Zihao. Optimization of NO_x emissions from coal-fired boilers based on improved MVO and WLSSVM [J/OL]. *China Measurement & Testing Technology*:1–8[2021-09-26].

Luo Changchun. Fault Diagnosis of Thermal Power Equipment Based on Parameter Correlation Analysis [J]. *Power Equipment*, 2018,32(06):429–433.

Mourad Azhari, Abdallah Abarda, Altaf Alaoui,Badia Ettaki, Jamal Zerouaoui. Detection of Pulsar Candidates using Bagging Method[J]. *Procedia Computer Science*, 2020,170:

Peng Tan, Ji Xia, Cheng Zhang, Qingyan Fang, Gang Chen. Modeling and reduction of NO X emissions for a 700 MW coal-fired boiler with the advanced machine learning method[J]. *Energy*, 2016,94:

Rattanawadee Panthong, Anongnart Srivihok. Wrapper Feature Subset Selection for Dimension Reduction Based on Ensemble Learning Algorithm[J]. *Procedia Computer Science*, 2015,72:

Seyed Mostafa Safdarnejad, Jake F. Tuttle, Kody M. Powell. Dynamic modeling and optimization of a coal-fired utility boiler to forecast and minimize NO_x and CO emissions simultaneously[J]. *Computers and Chemical Engineering*, 2019,124:

W. Mupangwa, L. Chipindu, I. Nyagumbo, S. Mkuhlani, G. Sisito. Evaluating machine learning algorithms for predicting maize yield under conservation agriculture in Eastern and Southern Africa[J]. *SN Applied Sciences*, 2020,2(406):

Wang Chunchang,Zhou Hongguang.Influence of boiler structural parameters on NO_x emission[J]. *Thermal Power Generation*, 2002(01):11–13+37-0.

XiaLi, 12, Peifeng Niu, Jianping Liu, Qing Liu. Improved Teaching-Learning-Based Optimization Algorithm for Modeling NO_x Emissions of a Boiler[J]. *Computer Modeling in Engineering & Sciences*, 2018,117(1):

XIE Ruibiao,LI Xinli,WANG Yingnan,YANG Guotian. NO_x emission prediction of coal-fired power plants based on PSO and Bi-GRU [J/OL]. *Thermal Power Generation*, 2021(10):87–94[2021-09-26]

You Lv, Jizhen Liu, Tingting Yang, Deliang Zeng. A novel least squares support vector machine ensemble model for NO x emission prediction of a coal-fired boiler[J]. *Energy*, 2013,55:

Yuan Zhaowei, Meng Lei, Gu Xiaobing, Bai Yuyong, Cui Huanmin, Jiang Chengyu. Prediction of NO_x emissions for coal-fired power plants with stacked-generalization ensemble method[J]. *Fuel*, 2020(prepublish):

Yuan Zhaowei, Meng Lei, Gu Xiaobing, Bai Yuyong, Cui Huanmin, Jiang Chengyu. Prediction of NO_x emissions for coal-fired power plants with stacked-generalization ensemble method[J]. Fuel,2020(prepublish):

Z. Tang, Y. Li, and A. Kusiak, "A deep learning model for measuring the oxygen content of boiler flue gas," *IEEE Access*, vol. 8, pp. 12268–12278, 2020.

Zepeng Qin, Chen Cen, Xu Guo. Prediction of Air Quality Based on KNN-LSTM[J]. *Journal of Physics: Conference Series*, 2019,1237(4):

Zhenhao Tang, Xiaoyan Wu, Shengxian Cao. Adaptive Nonlinear Model Predictive Control of the Combustion Efficiency under the NO_x Emissions and Load Constraints[J]. *Energies*,2019,12(9):

Zhu Guangqing, Gong Yanhao, Niu Yanqing, Wang Shuai, Lei Yu, Hui Shi'en. Study on NO_x emissions during the coupling process of preheating-combustion of pulverized coal with the multi-air staging[J]. *Journal of Cleaner Production*, 2021,292(prepublish):

Advances in Energy, Environment and
Chemical Engineering – Abdullah & Osman (Eds)
© 2023 The Author(s), ISBN: 978-1-032-36083-6

A green certificate allocating model considering carbon flow tracking and consumer carbon emission level

Zhengang Xie, Yixin Guo, Xiangyuan Lyv, Anqi Chen & Chun Xiao*
State Grid of Shanxi Electric Power Company Marketing Service Center, Taiyuan, China

ABSTRACT: To better manage the demand side of the power grid, this paper proposes a two-stage model for allocating green responsibility certificates for electricity consumption that considers carbon flow tracking and consumers' carbon emission rating to motivate consumers to take better responsibility for carbon reduction. The responsibility for carbon emission reduction is shared among the responsible parties in a common and differentiated manner. The simulation results show that the method can effectively assess the carbon emission level of consumers and achieve the effective allocation of green responsibility certificates, reducing the equivalent carbon emission status of high carbon consumers.

1   INTRODUCTION

Due to the excessive consumption of fossil energy, global temperatures are rising and carbon dioxide (CO_2) content has reached 77% (Kang et al. 2009). The power industry has become the main body of carbon dioxide emissions and should bear the heavy burden of carbon emissions reduction (Hu 2021). Clean power and low-carbon power are important development directions for the future power system (Zhou 2018).

Zhou Tianrui and other scholars proposed the concept of carbon emission flows such as power system carbon emission flows and nodal carbon potentials (Zhou et al 2012), and established a model of power system carbon emission flows without considering network losses and its calculation method (Zhou et al 2012); Feng and Yang (2016) improved the carbon emission flow theory and proposed a carbon emission flow calculation method applicable to the system with network losses. Promoting carbon emission reductions requires synergies with carbon trading. Chinese scholars have researched carbon markets and carbon trading (Wang et al 2019) and have developed clearing models to promote clean energy consumption and low carbon system operation. China has formulated a series of supportive policies to promote the rapid and steady development of renewable energy, including a renewable energy quota and a green certificate trading mechanism (An 2017, Dong & Shi 2019). Luo, Gong, Jiang, et al (2015) reduced the carbon intensity of system nodes by allocating green certificates based on carbon flow tracking, but it does not consider carbon emissions along with carbon intensity.

In order to actively guide electricity consumers to reduce carbon emissions, this paper builds a two-stage model for allocating green responsibility certificates for electricity consumption that considers carbon flow tracking and carbon emission rating.

2   CARBON FLOW TRACKING MODEL

A carbon flow is a virtual carbon network flow that is dependent on the power flow to secure tributary currents in the power system (Zhou et al 2012). Figure 1 reveals the flow mechanism of carbon emissions flows that are generated from the production of power plants, enter the grid and

---

*Corresponding Author: chengycn@hotmail.com

DOI 10.1201/9781003330165-38

flow through the grid with the system currents and finally flow to the load nodes on the consumer side.

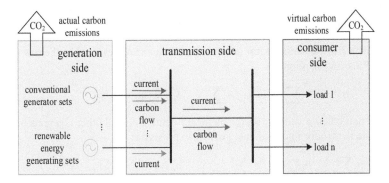

Figure 1. Carbon emission flow processes with energy flows.

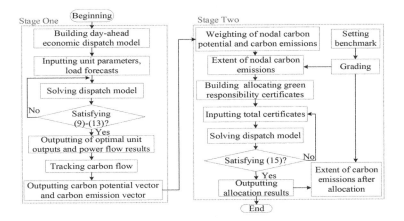

Figure 2. Two-stage green certificate allocation flow chart.

2.1 Nodal carbon potential

The nodal carbon potential is the equivalent carbon emission value on the generation side caused by a node consuming a unit of power (Luo et al. 2015), typically in $kgCO_2/(kW \cdot h)$. To calculate the nodal carbon potential, the proportional sharing principle, an important assumption in tidal analysis, is used as the basic assumption. The nodal carbon potential is calculated as follows:

$$e_{i,t} = \frac{\sum_{k \in \Omega_i^G} P_{Gk,t} \cdot e_{Gk} + \sum_{j \in \Omega_i^N} S_{ji,t} \cdot \rho_{ji,t}}{\sum_{k \in \Omega_i^G} P_{Gk,t} + \sum_{j \in \Omega_i^N} S_{ji,t}} \tag{1}$$

Where $e_{i,t}$ is the carbon potential of node i; Ω_i^G is the set of generating units with active currents injected into node i; Ω_i^N is the set of nodes with active currents injected into node i; e_{Gk} is the carbon intensity of generating unit k, $P_{Gk,t}$ is the active power of generating unit k at moment t at injected node i; $S_{ij,t}$ is the active currents flowing from node i to node j at moment t; $\rho_{ji,t}$ is the carbon flow density of branch $j - i$ at moment t.

2.2 Calculation method for carbon flow tracking

The carbon potential and emissions of each load node of the system can be obtained through carbon flow tracking (Zhou et al 2012), which in turn guides the consumer in reducing carbon emissions. The active power passed by a system node is equal to the sum of the active power injected into all upflow branches and the active power injected into the units connected to that node, also the node active flux matrix P_N, the denominator in Equation (1) can be represented by the row i and column i element of P_N as in Equation (2).

$$\sum_{k\in\Omega_i^G} P_{Gk,t} + \sum_{j\in\Omega_i^N} S_{ji,t} = (\eta_N^{(i)})^T P_N \eta_N^{(i)} \tag{2}$$

where $\eta_N^{(i)}$ is a unit column vector with the element i being 1 and all other elements being 0.

Similarly, the numerator in equation (1) can be expressed as

$$\sum_{k\in\Omega_i^G} P_{Gk,t} \cdot e_{Gk} + \sum_{j\in\Omega_i^N} S_{ji,t} \cdot \rho_{ji,t} = (\eta_N^{(i)})^T (P_G^T E_G + P_B^T E_N) \tag{3}$$

where P_G is the system unit injection distribution matrix, E_G is the unit carbon potential vector, P_B is the branch power flow distribution matrix, and E_N is the system node carbon potential distribution vector.

The above equation is expanded into matrix form and collapsed to give the system node carbon potential distribution vector as:

$$E_N = (P_N - P_B^T)^{-1} P_G^T E_G \tag{4}$$

The nodal carbon emission distribution vector for the system is

$$C_N = P_L E_N \Delta t \tag{5}$$

where P_L is the load distribution matrix and Δt is the time scale.

3 RATING OF THE EXTENT OF CARBON EMISSIONS OF CONSUMERS

In this paper, the average node carbon potential and total carbon emissions during the dispatch time are taken for a comprehensive evaluation, i.e., the extent of carbon emissions. In order to simplify the model for implementation, this paper is based on the linear weighted summation method, using weighting factors to convert multiple indicators into a single indicator problem. As the carbon potential and carbon emissions are of different magnitudes, they need to be averaged separately, so the extent of carbon emissions can be expressed as:

$$
\begin{aligned}
f &= w_1(C_{Li}/C_{av}) + w_2(E_{Li}/E_{av}) \\
C_{av} &== \sum_{i=1}^{N} C_{Li}/N \\
E_{av} &= \sum_{i=1}^{N} C_{Li}/\sum_{i=1}^{N} P_{Li}/\Delta t
\end{aligned}
\tag{6}
$$

where f is the extent of carbon emissions; N is the number of consumer nodes; w_1 and w_2 are the weights of the two evaluation components, $w_1 + w_2 = 1$, respectively; C_{Li} and E_{Li} are the carbon emissions and average carbon potential of node i, respectively; C_{av} and E_{av} are the average carbon emissions and carbon potential of the system, respectively; P_{Li} is the electricity consumption of node i.

The extent of carbon emissions benchmark value K is defined as

$$K = [w_1(C_{av}/C_{av}) + w_2(E_{av}/E_{av})] \times (1 - \beta) = 1 - \beta \tag{7}$$

Where β is the system reduction factor, the value of which is influenced by local conditions and industry conditions and is specified by the government, and is taken as 0.05 in this paper.

According to the carbon emission extent benchmark value can be divided into five levels (Luo et al. 2015), as shown in Table 1. α is the extent of carbon emissions of consumers for the rating standard segmentation factor.

Table 1. Rating criteria for the extent of carbon emissions on the consumer side.

Ratings	Zoning features	Grading Criteria
A	Green zoning	$f < (1 - 2\alpha)K$
B	Low Emission Zones	$(1 - 2\alpha)K \le f \le (1 - \alpha)K$
C	Standard Emission Zones	$(1 - \alpha)K < f \le (1 + \alpha)K$
D	High Emission Zones	$(1 + \alpha)K < f \le (1 + 2\alpha)K$
E	Very high emission zoning	$f > (1 + 2\alpha)K$

4 TWO-STAGE GREEN RESPONSIBILITY CERTIFICATE ALLOCATION MODEL

4.1 *Stage one day-ahead economic optimization scheduling model*

In the first stage, the optimal economic dispatch of the power system is carried out to determine the optimal output of the units and to track the carbon flow according to the power flow distribution, so as to obtain the nodal carbon potential and carbon emissions of the consumer. The objective function of the Stage One dispatching model is to minimize the generation cost of the units.

$$\min F_1 = \sum_{t=1}^{T} \left[\sum_{i=1}^{n} \left(a_i P_{e,it}^2 + b_i P_{e,it} + c_i \right) + \sum_{i=1}^{m} \beta_i P_{g,it} + \alpha P_{w,t} \right] \tag{8}$$

where F_1 is the system cost of generation; n is the total number of thermal units and m is the total number of gas units; a_i, b_i and c_i are the generation cost factors for thermal units; α is the generation cost factor for wind units; β_i is the generation cost factor for gas unit i; $P_{e,it}$ and $P_{g,it}$ are the generation power of thermal units and gas units respectively at time t; $P_{w,t}$ is the amount of wind power consumed at time t.

Binding conditions:

(1) Load balance constraint

$$P_{lt} = \sum_{i=1}^{n} P_{e,it} + \sum_{i=1}^{m} P_{g,it} + P_{w,t} \tag{9}$$

where P_{lt} is the electrical load demand for time period t.

(2) Thermal power unit constraints

$$\begin{aligned} P_{e,i}^{\min} \le P_{e,it} \le P_{e,i}^{\max} \\ r_{e,id}^{\max} \le P_{e,it} - P_{e,it-1} \le r_{e,iu}^{\max} \end{aligned} \tag{10}$$

Where $P_{e,i}^{\min}$ and $P_{e,i}^{\max}$ are the maximum and minimum output limits for unit i respectively; $r_{e,id}^{\max}$ and $r_{e,iu}^{\max}$ are the maximum downward and maximum upward ramp rates for unit i respectively.

(3) Gas unit output

$$\begin{aligned} P_{g,i}^{\min} \le P_{g,it} \le P_{g,i}^{\max} \\ r_{g,id}^{\max} \le P_{g,it} - P_{g,it-1} \le r_{g,iu}^{\max} \end{aligned} \tag{11}$$

Where $P_{g,i}^{\min}$ and $P_{g,i}^{\max}$ are the upper and lower limits of the output of gas unit i respectively; $r_{g,id}^{\max}$ and $r_{g,iu}^{\max}$ are the maximum downward and upward ramp rates of gas unit i respectively.

(4) Wind turbine output

$$P_{w,t} \le P_{w,t}^p \tag{12}$$

where $P_{w,t}$ is the predicted value for the wind turbine in time period t.

(5) DC current constraint

$$
\begin{aligned}
P_{ij,t} &= (\theta_{i,t} - \theta_{j,t})/x_{ij} \\
-P_{ij}^{\max} &\le P_{ij,t} \le P_{ij}^{\max} \\
-\theta_i^{\max} &\le \theta_{i,t} \le \theta_i^{\max} \\
\theta_{ref} &= 0
\end{aligned}
\tag{13}
$$

Where $P_{ij,t}$ is the active current between node i and node j; $\theta_{i,t}$ and $\theta_{j,t}$ are the nodal voltage phase angles of node i and node j *respectively*; x_{ij} is the reactance of line $i - j$; $P_{ij}^{\max}$ is the upper capacity limit of line $i - j$; $\theta_i^{\max}$ is the maximum voltage phase angle of node i; θ_{ref} is the balanced nodal phase angle.

4.2 *Stage two green responsibility certificate allocation model*

Taking the extent of carbon emission of system consumers as known, a green responsibility certificate allocation model is built to solve the problem of allocating green power certificates on the consumer side. In order to satisfy the principle of fairness, the minimum gap between consumers with different carbon emission extent levels across the network is set as the objective function of the model.

$$
\begin{aligned}
\min F_2 &= \sqrt{\frac{1}{N}\sum_{i=1}^{N}(f_i' - \mu)^2} \\
f_i' &= w_1[(C_i - G_{ci}E_{av})/C_{av}] + w_2[(C_i - G_{ci}E_{av})/P_i/\Delta t/E_{av}] \\
\mu &= w_1[(\sum_{i=1}^{N} C_i - G_P E_{av})/I/C_{av}] + w_2[(\sum_{i=1}^{N} C_i - G_P E_{av})/\sum_{i=1}^{N} P_i/\Delta t/E_{av}]
\end{aligned}
\tag{14}
$$

where F_2 is the difference in carbon emissions of consumers across the network; f_i' is the carbon emissions of consumer i after allocating green responsibility certificates; μ is the average carbon emissions across the network after allocating green responsibility certificates; G_{ci} is the number of green responsibility certificates by consumer i; G_P is the total allocated green responsibility certificates.

The constraint is a quantitative constraint on the number of certificates assigned.

$$
\begin{aligned}
G_{ci} &\ge 0 \\
\sum_{i=1}^{N} G_{ci} &\le G_P
\end{aligned}
\tag{15}
$$

The Stage Two model is solved using the CPLEX solver and the flowchart for the solution of the model is shown in Figure 2.

5 EXAMPLE ANALYSIS

5.1 *Database*

In this paper, the IEEE30 nodes system is used for arithmetic analysis. The four thermal units are connected to #1, #5, #8, and #13, the wind turbine to #11, and the gas turbine to #2. The carbon potential and carbon emission weights are set to 0.5. Figure 3 shows the electricity load demand and the forecasted wind power output during the dispatch cycle, which is 24h. The scheduling interval, $\Delta t = 1h$, $\alpha = 0.25$, and $G_P = 1200$.

5.2 *Analysis of the results of the one-stage model*

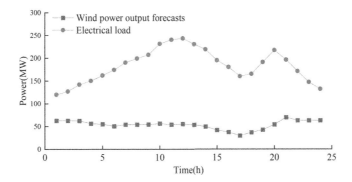

Figure 3.   Electric load and wind power output forecasts.

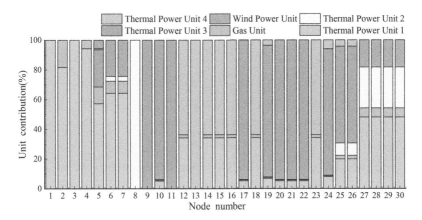

Figure 4.   Unit contribution of each node during0-1 period.

As can be seen from Figure 4, nodes #1, #3, #8, #9, #11, and #13 all have only output contributions from one unit, the reason being that the above nodes are all direct access nodes to the generating units or nodes closer to the generating units; thermal unit 1 provides unit output to the vast majority of the nodes, the reason being that the maximum output of this unit is higher than the other units, and the unit has a large ramp rate and can respond quickly to load changes.

According to Figure 5, the fact is that #5, the majority of the load, is supplied by thermal and gas-fired units, resulting in high average carbon potential at nodes such as #14, #15, and #16. In contrast, nodes #20, #21, and #22, which are supplied by wind turbines, not only have low carbon emissions but also have low carbon potentials. Therefore, both carbon emissions and carbon potential should be considered when assessing the level of carbon emissions of consumers.

5.3 *Analysis of the results of the two-stage model*

Table 2 reflects the carbon emission effect of the integrated carbon emissions and carbon potential. The results of the carbon emission extent rating can provide targeted guidance for carbon emission reduction activities on the consumer side, encouraging consumers with a rating of D or E to participate more in carbon emission reduction activities.

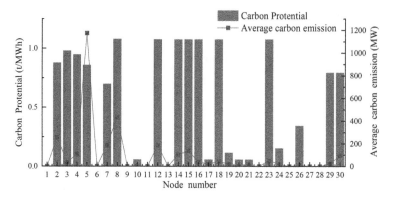

Figure 5. Average carbon potential and carbon emissions at nodes.

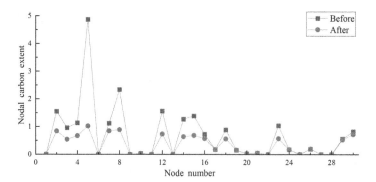

Figure 6. Comparison of the extent of nodal carbon emissions before and after certificate allocation.

Table 2. Consumer-side carbon emission level rating.

Node	Extent of carbon emissions	Ratings	Green responsibility certificate	Node	Extent of carbon emissions	Ratings	Green responsibility certificate	Node	Extent of carbon emissions	Ratings	Green responsibility certificate
1	\	\	\	11	\	\	\	21	0.050025	A	0
2	1.553204	E	208	12	1.562003	E	174	22	\	\	\
3	0.95259	C	28	13	\	\	\	23	1.035672	C	40
4	1.1304536	C	76	14	1.265866	D	90	24	0.175445	A	0
5	4.868373	E	1640	15	1.384321	D	122	25	\	\	\
6	\	\	\	16	0.721233	C	14	26	0.195298	A	0
7	1.1151815	C	81	17	0.173914	A	0	27	\	\	\
8	2.336105	E	476	18	0.88374	C	28	28	\	\	\
9	\	\	\	19	0.15601	A	0	29	0.560218	B	2
10	0.0321116	A	0	20	0.026599	A	0	30	0.821629	C	21

According to Figure 6, the carbon emission extent of the remaining nodes decreased in different ranges, especially #5, and #8. This indicates that green responsibility certificates can guide consumers to use clean electricity, help promote emission reduction behavior on the consumer side and assist the grid in demand-side management.

6 CONCLUSION

This paper proposes a two-stage green responsibility certificate allocation model, which guides consumers to take active responsibility for carbon emissions and to guide the grid in demand-side management.

(1) The carbon flow tracking model can fully reflect the proportional contribution of units at load nodes and realize the traceability of carbon emissions at nodes. The rating of the extent of carbon emissions of consumers comprehensively reflects the differences in different consumers.

(2) The green responsibility certificate allocation model can reasonably allocate green responsibility certificates according to the extent of carbon emissions of consumers and satisfy the common and differential responsibility of carbon emission reduction of all network consumers.

ACKNOWLEDGMENTS

This work was financially supported by the Science and Technology Project of State Grid of Shanxi Electric Power Company (520531220003).

REFERENCES

An Xue-Na, Zhang Shao-Hua, Li Xue. Equilibrium analysis of oligopoly power market considering green certificate trading[J]. *Power System Automation*, 2017,41(09): 84–89.

Dong FG, Shi L. Design and simulation of renewable energy quota system and green certificate trading mechanism[J]. *Power System Automation*, 2019,43(12): 113–121.

Feng Xin, Yang Jun. Improvement and refinement of carbon emission flow theory considering network losses[J]. *Power Automation Equipment*, 2016,36(05): 81–86.

Hu Angang. China's goal to reach the carbon peak by 2030 and the main ways to achieve it[J]. *Journal of Beijing University of Technology* (Social Science Edition), 2021,21(03): 1–15.

Kang, Chongqing, Chen, Qixin, Xia, Qing. Research perspectives on low carbon power technologies[J]. *Power Grid Technology*, 2009,33(02): 1–7.

Luo Yifan, Gong Yu, Jiang Chuanwen, et al. Study on consumer-side carbon emission intensity rating and green power certificate allocation[J]. *Hydropower Energy Science*, 2015,33(10): 199–203.

Wang Hui, Chen Bobo, Zhao Wenhui, et al. Multi-entity bidding game of cross-provincial clean energy power trading with carbon trading synergy[J]. *Power Construction*, 2019,40(06): 95–104.

Zhou Dan. China's energy mix to gradually shift to clean power generation [J]. *China and Foreign Energy*, 2018,23(06): 38.

Zhou T R, Kang CQ, Xu Q Y, et al. A preliminary study on the theory of power system carbon emission flow analysis[J]. *Power System Automation*, 2012,36(07): 38–43.

Zhou Tianrui, Kang Chongqing, Xu Qianyao, et al. A preliminary study on the calculation method of carbon emission flow in power system [J]. *Automation of Electric Power System*, 2012,36(11): 44–49.

Advances in Energy, Environment and
Chemical Engineering – Abdullah & Osman (Eds)
© 2023 The Author(s), ISBN: 978-1-032-36083-6

Research on the development of traditional power enterprises under the background of new energy

Yihang Lyu* & Jingjing Kong*
Chengdu Neusoft University, Chengdu, China

ABSTRACT: In 2021, the global epidemic situation is complex and changeable. China's economic development is facing the triple pressure of shrinking demand, supply shock, and weakening expectations. The external environment is becoming more complex, severe, and uncertain. The energy and power industry is facing the arduous test of balancing multiple objectives of ensuring supply, adjusting structure, and stabilizing growth. From cases in recent years, operating conditions of A company, this paper mainly focused on operating income, operating income, core competitive advantages, such as to carry out the analysis, and found that the traditional energy companies still give priority to coal or gas power generation, and new green energy proportion is very small. It is proposed that the company can rely on national policy resources such as carbon peak and carbon neutrality to build a first-class green and low-carbon listed electric power company. The company shall strengthen the business of coal power, gas power, and biological power generation, vigorously develop new energy, promote the integrated development of new power generation methods, and build an ecological civilization power generation enterprise.

1 INTRODUCTION

In 2021, the global epidemic situation is complex and changeable. China's economic development is facing the triple pressure of shrinking demand, supply shock, and weakening expectations. The external environment is becoming more complex, severe, and uncertain. The energy and power industry is facing the arduous test of balancing multiple objectives of ensuring supply, adjusting structure, and stabilizing growth. According to the statistical data of the national power industry in 2021 released by the national energy administration, the power consumption of China's whole society in 2021 was 8.31 trillion kWh, with a year-on-year increase of 10.3%. The significant growth of power consumption was mainly affected by the continuous recovery and development of the domestic economy, the low base in the same period of the previous period, the rapid growth of foreign trade exports, and other factors. Among them, the power consumption of the primary industry was 102.3 billion kWh, a year-on-year increase of 16.4%; the electricity consumption of the secondary industry was 5.61 trillion kWh, a year-on-year increase of 9.1%; the electricity consumption of the tertiary industry was 1.42 trillion kWh, a year-on-year increase of 17.8%; the domestic electricity consumption of urban and rural residents was 1.17 trillion kWh, a year-on-year increase of 7.3%. With the in-depth promotion of carbon peaking and carbon neutralization, China has comprehensively promoted the large-scale and high-quality development of wind power and solar power generation, accelerated the growth rate of new energy installed capacity, and continuously increased the proportion of total installed power generation capacity in China.

By the end of 2021, China's installed capacity of wind power was 328 million kW, a year-on-year increase of 16.6%; the installed capacity of solar power generation was 306 million kW, a

---

*Corresponding Authors: 4761420@qq.com and 774518325@qq.com

DOI 10.1201/9781003330165-39

year-on-year increase of 20.9%; the installed capacity of coal-fired power was 1.11 billion kW, a year-on-year increase of 2.8%. In addition, the wind power generation capacity was 652.6 billion kWh, a year-on-year increase of 40.5%; solar power generation was 325.9 billion kWh, a year-on-year increase of 25.1%; coal power generation was 5.03 trillion kWh, an increase of 8.6% year-on-year, accounting for 60.0% of the full caliber power generation, a decrease of 0.7% year-on-year. Combined with the installed capacity and power generation, coal power is still the main power supply in China, and it is also the basic power supply to ensure a safe and stable power supply in China. In 2021, affected by safety inspection, environmental protection supervision, a limited quota of imported coal, and other factors, the coal supply continued to be tight, the price of electric coal repeatedly hit record highs, and the overall operating performance of thermal power enterprises continued to be under pressure due to the rise of fuel costs. Traditional power enterprises are facing great challenges.

2 CASE ANALYSIS

2.1 Company profile

The company is mainly engaged in the investment, construction, operation, and management of power projects, and the production and sales of power. It belongs to the power and heat production and supply industry in the industry classification guidelines for listed companies issued by the CSRC. Since its establishment, the company has always adhered to the business purpose of "taking capital from the people, using capital for electricity and benefiting the public" and the business policy of "focusing on electricity and diversified development". It focuses on the main business of electric power, with a diversified power structure. It has a variety of energy projects such as large-scale coal-fired power generation, natural gas power generation, wind power generation, and hydropower generation, and provides users with reliable and clean energy through the power grid company. By the end of the reporting period, the company had a controllable installed capacity of 29.9426 million kW, including 28.2292 million kW for holding and 1.7134 million kW for equity participation. Among them, the installed capacity of coal-fired power generation holdings is 20.55 million kW, accounting for 72.8%; the installed capacity of gas power holding is 5.472 million kW, accounting for 19.4%; the holding installed capacity of wind power, hydropower, biomass, and other renewable energy power generation is 2.2072 million kW, accounting for 7.8%. In addition, the company's entrusted installed capacity is 8.854 million kW, and the above controllable installed capacity and entrusted installed capacity total 38.7966 million kW.

2.2 Data analysis

In 2021, benefiting from the sustained and steady recovery of China's economy, the power demand of the whole society in Guangdong Province increased faster than expected, reaching 786.663 billion kWh, a year-on-year increase of 13.58%. In 2021, Guangdong received 189.385 billion kWh of Western power, a year-on-year decrease of 7.97%; in terms of installed capacity, the newly put into operation nuclear power and thermal power units in the province have a capacity of 5.79 million KW, and the newly added unified commissioning installed capacity accounts for 3.65% of the unified commissioning capacity. Affected by the reduction of power transmission in the west, the shortage of gas generators caused by high natural gas prices, and the simultaneous increase of peak shaving demand caused by the sharp growth of renewable energy in Guangdong Province, the utilization hours of coal-fired power generation in Guangdong Province increased significantly. During the reporting period, the on-grid power of the company's coal machinery was 85.519 billion kWh, a year-on-year increase of 51.43%; the on-grid power of gas turbine was 16.627 billion kWh, a year-on-year increase of 42.27%; the on-grid power of wind power was 1.742 billion kWh, a year-on-year increase of 41.97%. From 2016 to 2021, the company's operating revenue increased year by year, from 22.68 billion yuan in 2016 to 44.17 billion yuan in 2021. The specific data are shown in Figure 1:

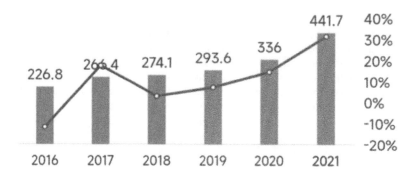

Figure 1.   Chart of revenue in recent years.

Table 1.   Composition of electricity sales revenue.

	business income	Operating costs	Gross profit margin	Increase or decrease in operating income	Increase or decrease in operating costs	Increase or decrease in gross profit ratio
Electricity sales revenue	43,278,023,920	46,541,774,902	−7.54%	31.90%	7%	−27.14%
Coal fired power generation	35,283,584,475	39,459,402,833	−11.84%	44%	92.84%	−28.33%
Gas power generation	6,373,234,192	6,083,234,210	4.55%	−9.33%	17.72%	−21.93%
Wind power generation	1,078,426,815	570,951,289	47.06%	42.68%	73.98%	−9.52%
Biomass power generation	478,196,815	372,045,928	22.20%	5.84%	3.50%	1.76%
Hydropower	64,581,623	56,140,642	13.07%	−10.53%	−8.95%	−1.50%

98.66% of the company's operating income comes from electricity, steam sales, and labor services. It can be seen from the composition of the power sales business in Table 1 that the company's traditional coal-fired power generation and gas-fired power generation with relatively serious pollution are 35.3 billion yuan and 6.3 billion yuan respectively, totaling 41.6 billion yuan, accounting for 96.2% of the total power sales revenue of 43.2 billion yuan Other types of green energy, including wind power generation, biomass power generation and hydraulic power generation, account for less than 5%. It shows that green energy is still in an absolutely weak position in traditional energy enterprises.

2.3   Competitiveness analysis

2.3.1   The company is large in scale and rich in resources
The total controllable installed capacity and entrusted managed installed capacity of the case company are 38.7966 million KW, accounting for about 24.4% of the total installed capacity of Guangdong Province. It is the largest power-listed company in Guangdong Province. Taking advantage of its resources, technology, asset scale, and other advantages, the controlling shareholders of the company have always actively supported listed companies to become better, stronger, and bigger.

2.3.2 *Market competitive advantage is obvious.*

With high parameters, large capacity, high operation efficiency, low coal consumption, stable operation, and superior environmental protection performance, the company has strong market competitive advantages. In 2021, the company completed a total of 84.261 billion kWh of market electricity, the scale of electricity sales continued to rank first in the province, and the price of electricity sales was better than the average level of the province. The company gives full play to the three advantages of scale, brand, and service, and provides auxiliary value-added services such as peak shaving, frequency modulation, and standby for the power grid with the marketing service network all over the province, the technical accumulation, and comprehensive resources of deep cultivation of the power industry, as well as high-quality value-added services such as comprehensive energy conservation and power consumption consulting for users.

2.3.3 *Good financial position*

At present, the total assets of the case company reach 100 billion, and the cash flow of the stock business is relatively abundant, which provides good support for the sustainable development of the company. The company has a good asset-liability ratio and rich financing channels. It can make full use of internal and external financial resources and provide a strong financial guarantee for the production and operation of the company, the construction of key projects, and the rapid development of the new energy industry.

3 FUTURE DEVELOPMENT TRENDS AND ENTERPRISE DEVELOPMENT DIRECTION

At present, China's power generation industry continues to present a diversified competition pattern. The main generator units of the case company are concentrated in Guangdong Province, where there are many other generators and are greatly affected by the west to East power transmission. China's power supply structure is mainly thermal power generation. In recent years, with the proposal of the "double carbon" goal and the in-depth promotion of power supply-side reform and other policies, new and renewable energy such as wind power, photovoltaic, nuclear power, hydropower, and biomass power generation have developed rapidly, and thermal power has gradually changed from the main power supply to the basic power supply of peak shaving and frequency modulation. The central economic work conference in 2021 proposed that the gradual withdrawal of traditional energy should be based on the safe and reliable substitution of new energy. Based on the basic national conditions dominated by coal, we should pay attention to the clean and efficient utilization of coal and promote the optimal combination of coal and new energy. In the future, with the further promotion of energy-saving and consumption reduction transformation, flexibility transformation, and heating transformation of coal-fired units, coal-fired units with large capacity, high parameters, and advanced energy-saving will still be important power support. In addition, clean and efficient gas turbine generator sets are conducive to enhancing the power grid peak shaving capacity, safety, and reliability of the new power system with new energy as the main body, helping to build a clean, low-carbon, safe and efficient energy system, and have a certain development space. To sum up, in the future, the thermal power industry will mainly rely on the development of high-capacity, high parameter, advanced energy-saving coal power and accelerating the development of gas power to optimize the power supply structure, improve the quality and efficiency by improving the technical R & D strength, strive for electricity and electricity price by active marketing, and reduce costs by fine management; at the same time, the company will actively grasp the development trend of accelerating energy transformation under the objectives of "carbon peak" and "carbon neutralization", actively expand new energy project resources through self-construction and acquisition, and make every effort to promote the leapfrog development of new energy and promote the transformation of clean and low-carbon energy of the company.

In the future, the company can focus on energy production and supply, give consideration to comprehensive energy services, focus on the goal of carbon peak and carbon neutralization, based on Guangdong and facing the whole country, build a first-class green low-carbon power listed company, coordinate safety, and development, optimize and strengthen coal power, gas power, and biomass power generation, and vigorously develop new energy, energy storage, hydrogen energy, and land park development. Efforts should be made to fully promote the leapfrog development of new energy, grasp the window period of thermal power development, and speed up the development and construction of key projects; it is necessary to explore and carry out project distribution, promote the integrated development of various new power generation modes, and build an ecological and civilized power generation enterprise.

4 CONCLUSION

In 2021, the external environment will become more complex, severe, and uncertain. China's economic development will face the triple pressure of demand contraction, supply shock, and weakening expectations, and the energy and electric power industry will face the arduous test of balancing multiple goals of ensuring supply, adjusting structure, and stabilizing growth. Starting with the operating situation of company A in recent years, this paper focuses on the analysis of operating revenue, the composition of operating revenue, and core competitive advantages. It finds that traditional energy enterprises are still dominated by coal power generation or gas power generation, and the proportion of new green energy is very small. It proposes that the company can rely on national policy resources such as carbon peak and carbon neutrality to build A first-class green and low-carbon listed electric power company. We will strengthen the business of coal power, gas power, and biological power generation, vigorously develop new energy sources, integrate and develop new power generation methods, and build an ecological civilization power generation enterprise.

However, due to the limitations of the case company, it can not represent the whole traditional power enterprises in China. If we want to further study the development and transformation of the whole traditional power enterprises, we need to obtain more sample information and data.

ACKNOWLEDGMENTS

This research was supported by the 2022 teaching reform project of Chengdu Neusoft University.

REFERENCES

Gao Wei The production and sales of new energy vehicles continued the momentum of rapid growth, and the performance of leading enterprises in the first quarter was bright [n] *Economic information daily*, 2022-04-20 (003).

Li Zhen *Ecological business model of new energy vehicles and its impact on brand competitiveness* [D] Beijing Jiaotong University, 2021 DOI:10.26944/d.cnki.gbfju. 2021.000125.

Nie Yihua *Research on China's marketing strategy of GF company's new energy forklift battery* [D] Yunnan Normal University, 2021.

Shen Keli, Cheng Fan, Wu Qiong, Li Qifen, Chen Min. The electric power enterprise comprehensive energy service transformation think - in its Shanghai electric power, for example [J]. *Journal of Shanghai energy conservation*, 2022 (01): 9–15. DOI: 10.13770/j.carol carroll nki issn2095-705-x. 2022.01.003.

Zheng Houqing, CUI Weiping. Annual observations and reflections on the digital transformation of energy and electric power enterprises [J]. *China Power Enterprise Management*, 2022(07):74–75.

Zhou Yahong, Puyu Road, Chen Shiyi, Fang Fang Government support and the development of new industries – Taking new energy as an example [J] *Economic research*, 2015,50 (06): 147–161.

*Research and development of renewable
energy structure and performance*

*Advances in Energy, Environment and
Chemical Engineering – Abdullah & Osman (Eds)
© 2023 The Author(s), ISBN: 978-1-032-36083-6*

A mobile evaporation tank for radioactive wastewater

Yuankai Du*
School of Materials Science and Engineering, Wuhan Institute of Technology, Wuhan, China

Xinyi Xie
School of Optoelectronic Information and Energy Engineering, Wuhan Institute of Technology, Wuhan, China

ABSTRACT: The device relates to a mobile evaporation tank for radioactive wastewater, which belongs to environmental protection treatment equipment. Radioactive wastewater refers to the waste liquid containing radionuclides, which usually contains radioactive elements, nitrate, or hydrochloride, and may also have a large amount of acid and alkali. It has strong radiation, is difficult to handle, and has the tendency to scale. In order to meet the needs of production separation and subsequent environmental protection, nuclide heavy metals and acids need to be separated. In the traditional separation methods, the neutralization precipitation method, distillation method and ion exchange resin method are widely used. Still, because of the high acidity, the neutralization method will consume a lot of alkalies and produce a lot of secondary pollutants. Distillation with high acidity consumes a lot of energy and requires extremely high equipment. Moreover, due to the scaling of a large number of heavy nuclide metals, it affects heat transfer, has the risk of limited concentration explosion, and will also produce secondary pollutants produced by cleaning the equipment. The membrane method, especially reverse osmosis, membrane distillation, electrodialysis and other technologies, has become a relatively convenient and easy technology. However, the need for heating, pressurization and power consumption have caused problems such as complex equipment, the complex treatment caused by membrane pollution, low membrane life and increased secondary pollution. Evaporation pond technology has the advantages of convenient equipment, low investment and no operation cost. It has become an essential economic, and effective treatment in conditional areas.

1   INTRODUCTION

In the process of nuclear energy utilization, radioactive wastes are inevitably produced because the objects of production and operation are radioactive materials (Jeong & Choi 2021). It is estimated that during the operation of a million-kilowatt PWR nuclear power unit, about $3000m^3$ of radioactive waste liquid is produced annually, of which the low-discharge waste liquid accounts for about 90% (Phillip et al. 2022; Yan et al. 2021). Because of the large amount of waste liquid production, the potential environmental harm is great (Chen et al. 2016). Therefore, more technologies for the treatment of medium and low radioactive wastewater have been developed, including evaporation concentration, chemical precipitation, ion exchange, and natural evaporation pond technologies (Galeev et al. 2015; Hong 2019; Ji et al. 2019). More mature treatment technologies in other fields are also constantly introduced into the field of radioactive wastewater treatment, such as membrane treatment technology (Zhang & Liu 2020), biological treatment technology (Beji & Merci 2018), and improved application of the above technologies (Zhang et al. 2019).

---

*Corresponding Author: 1462429664@qq.com

2 THE TECHNICAL DEFECTS OF THE TRADITIONAL EVAPORATION POOL

Traditional evaporation ponds cause soil pollution, but in recent years, especially for radioactive wastewater treatment, mobile evaporation ponds have been adopted, which also leads to increased equipment costs and secondary pollution (Liu & Wang 2013). And the traditional evaporation pool technology or mobile evaporation pool technology has not solved the following difficulties (Eun et al. 2019; Ghasemipanah 2013; Yeszhanov et al. 2021).

2.1 *Large floor area*

Limited by the low evaporation efficiency, the general evaporation pool covers a large area, which limits the use of its place, with only great cheap land being used.

2.2 *Open space*

Open space produces the following problems: 1) Secondary pollution of the external air environment, such as the release of radioactive dust and volatilization of many pollutants due to the drying of radioactive materials. 2) Open type caused by birds into the problem, so that possible pollution and spread. 3) The problem of rain flooding due to the opening.

2.3 *Temperature and humidity are affected by the external environment*

Because the evaporation pool is one of the environments with the outside environment, there is no way to control the temperature and humidity affected by the outside environment, but also make its application by regional restrictions, can only be used for large evaporation, rainfall less area.

3 SPECIFIC SCHEME OF MOBILE EVAPORATION POND FOR RADIOACTIVE WASTEWATER

The device comprises water tank 5 and the above plastic greenhouse cover 1. The water tank is evenly distributed up and down two rows of pipes, the upper pipe is the upper exhaust pipe 3, and the lower pipe is the collecting pipe 7. Plastic greenhouse 1 also has a suction tube. In the so-called

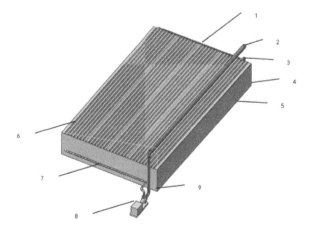

Figure 1.   Three-dimensional view of mobile evaporation tank for radioactive wastewater. 1. Plastic greenhouse cover, 2. Suction pipe, 3. Exhaust pipe, 4. Drain, 5. Sink, 6. Absorbent cloth, 7. Heating condensing tube, 8. Fan and 9. Salt outlet.

Figure 2.   Perforation drawing of exhaust and suction pipes.

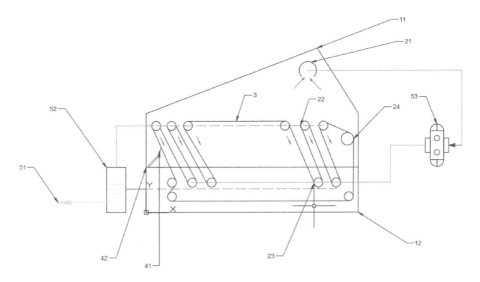

Figure 3.   Schematic diagram of a v-type desalting collector. 11. Plastic greenhouse cover, 12. Sink, 21. Suction pipe, 22. Exhaust pipe, 23. Heating condensing tube, 24. Transfer roller, 3. Absorbent cloth, 41. Desalination device, 42. Brine mouth, 51. Water outlet; 52. Exhaust water collector and 53. Fan.

plastic greenhouse cover 1, there is a long tube for suction tube 2; one end of the suction tube is sealed, one end of the opening is connected to the fan 8 inlet and suction tube 2 from the sealing end to the opening end, distributed from small to large air inlet. Fan 8 has its import connected to the suction pipe 2, exports connected to the heating condenser pipe; its effect is high at the top of the plastic greenhouse cover 1 [gas, after the suction 2, pressure into the condenser pipe heating through Fan 8. External heat condenser pipe 7 is immersed into liquid waste to be processed, thus realizing the condensation of the gas temperature inside the condenser pipe heating 7, and water flow. The exterior realizes the heating of the salt pond and raises the temperature of the tank. The two ends of the heating and condensing pipe 7 are connected in parallel to the two pipes to form the inlet and outlet. The inlet is connected to the discharge port of fan 8, while the outlet is connected to the exhaust pipe. The so-called exhaust pipe 3 is blocked at one end of the pipe, and the other end is the open end. From the sealed end to the open end, there are air inlet holes from small to large. The open end is connected directly to the exhaust outlet of the heating and condensing pipe 7 through a three-way aggregation in parallel. The absorbent cloth 6 is continuously wound on the exhaust pipe 3 and the heating condensing pipe 7 through a drive roller 10, which can roll freely. The so-called absorbent cloth 6 is wound and rolled in the sink, and the salt on the surface is scraped off through a V-shaped mouth. The scraped salt falls into the salt storage chamber and is discharged through the salt outlet. The so-called absorbent cloth material for radiation resistance, UV resistance, acid and alkali resistance and a hydrophilic polymer can be woven through the formation of at least one side with strong water absorption capacity and air contact area.

4 THE APPLICATION OF MOBILE EVAPORATION POND FOR RADIOACTIVE WASTEWATER WAS EXPLORED

4.1 *Experimental comparison of static moving evaporation tank*

The sink is made of PVDF material with a thickness of 10mm and a length, width, and height of 5m*2m*1.5m, respectively. The height of the plastic greenhouse is 2m, and ss316L is the support frame, covered with PI transparent plastic film material with a thickness of 0.1mm. The heating and condensing pipes, exhaust pipes and suction pipes are all ss316L, with a mirror surface and a thickness of 2 mm. The suction pipe is DN50, the heating and condensing pipe, and the exhaust pipe is DN20. The distance between the heating condensing pipe and the bottom of the tank are 800mm, and the distance between the exhaust pipe and the heating condensing pipe is 1800mm. The exhaust pipe is 80, with equal spacing of 42mm, and the heating condensing pipe is 80, with equal spacing of 42mm; absorbent cloth uses radiation-resistant PP non-woven fabric, adding absorbent fiber. Fan ventilation is 5000 m³/H. A 5% solution of sodium chloride was prepared, the height of water in the tank was 1000mm, and the speed of the driving roller was 50 rpm/min. When the surface of the absorbent cloth is dry, and salt has crystallized, the rotating roller is started to harvest the salt of the absorbent cloth. At the same time, the absorbent cloth will be adsorbed again on the crystalline salt, and then remain static.

Similarly, in comparison to the traditional evaporation pool, only the water tank is made of PVDF material. The thickness is 10mm, the size is 5m*2m*1.5m in length, width and height, respectively, the sodium chloride 5% solution is prepared, and the water volume in the tank is 1000mm in height.

A comparative experiment was conducted on the above two evaporation ponds in Tianjin Binhai New Area and Yumenguan, Gansu Province, from July to October. The average maximum temperature in Tianjin Binhai New Area is 25.5°C, the average minimum temperature is 18.5°C, the average monthly precipitation is 117.49mm, and the average wind speed is 7.78km/h. From July to October, the average maximum temperature in Yumenguan, Gansu province is 27.5°C, the average minimum temperature is 10.5°C, the average monthly precipitation is 0.16mm, and the average wind speed is 11.15km/h. Both evaporation tanks were placed in outdoor conditions, and compared evaporation was measured. The results are shown in Table 1 below. Traditional evaporation pools in dry areas develop a high evaporation surface, which is four times that of data in the Tianjin region, but for the new type of evaporation pool, arid and humid areas, its evaporation. The evaporation rate of the new evaporation pool is usually 10-50 times that of the traditional evaporation pool. Even in Tianjin, the evaporation volume reaches 0.023m/d in rainy July and August. The main reason is that the evaporation surface of the new evaporation tank reaches 640m², while the traditional evaporation tank only has 10m². It's 64 times larger than conventional evaporation. In addition, the temperature of the traditional evaporation pond varies greatly from day to night, but as a closed new evaporation pond, it has a better insulation effect.

4.2 *Experimental comparison of dynamic mobile evaporation tank*

In the first instance, the water level in the mobile evaporation pool and the traditional evaporation pool is reduced to 700mm. Then we start the drive roller, so that the speed of the absorbent cloth is 10 m/d. Nothing else has changed. After that, evaporation was calculated and listed in Table 1 below.

The comparison experiments of the two evaporation pools mentioned above were conducted in Tianjin Binhai New Area and Yumen, Gansu province, respectively, from July to October. The results are shown in Table 1 below. The evaporation rate can be further increased by 20-50% in both Tianjin Binhai New Area and Yumen, Gansu Province, in the dynamic mobile evaporation pool experiment. The main reason is that the evaporation surface of the dynamic moving evaporation pond reaches 736 m². The dynamic operation of the suction cloth can clean and scrape the salt deposition on the surface of the condensing heating tube, improve the heat transfer capacity and enhance the heating and condensing effect.

Table 1. Comparative experiment of old and new evaporation ponds.

		Tianjin Binhai New District		Gansu Province Yumen City	
	Experiment time	Jul. 10th to Aug. 10th	Sep. 10th to Oct. 10th	Jul. 10th to Aug. 10th	Sep. 10th to Oct. 10th
Plan A	Static mobile evaporation tank (evaporation m/D)	0.023	0.018	0.026	0.021
Plan B 0	Dynamic mobile evaporation tank (evaporation m/D)	0.032	0.022	0.035	0.028
Plan C	Conventional evaporation tank (evaporation m/D)	0.0005	0.00001	0.002	0.0015

5 CONCLUSION

The absorbent cloth designed by this device has a high water absorption capacity, increases the huge surface area, and accelerates the evaporation rate. The internal circulation of the gas improves the gas flow on the evaporation surface and the vapor pressure difference, which accelerates the evaporation rate. The sealed chamber and condensing heating pipe design raises the evaporation pool temperature and speeds up the evaporation rate. The timely separation of brine reduces the concentration of liquid in the evaporation tank, which also increases the evaporation rate. The invention also has the advantages of a small site, only using solar energy and a small amount of energy consumption, and has high evaporation efficiency. This product adopts closed space and enhanced treatment, so that the application of evaporation pond technology breaks through the restrictions of environment, region, and other factors.

REFERENCES

Beji T, Merci B. Development of a numerical model for liquid pool evaporation[J]. *Fire Safety Journal*, 2018, 102: 48–58.

Chen D, Zhao X, Li F, et al. Rejection of nuclides and silicon from boron-containing radioactive wastewater using reverse osmosis[J]. *Separation and Purification Technology*, 2016, 163: 92–99.

Eun H C, Jung J Y, Park S Y, et al. Removal and decomposition of impurities in wastewater from the HyBRID decontamination process of the primary system in a nuclear power plant[J]. *Journal of Nuclear Fuel Cycle and Waste Technology* (JNFCWT), 2019, 17(4): 429–435.

Galeev A D, Salin A A, Ponikarov S I. Numerical simulation of evaporation of volatile liquids[J]. *Journal of Loss Prevention in the Process Industries*, 2015, 38: 39–49.

Ghasemipanah K. Treatment of ion-exchange resins regeneration wastewater using reverse osmosis method for reuse[J]. *Desalination and Water Treatment*, 2013, 51(25–27): 5179–5183.

Hong S J, Wang E S, Park C W. Heat transfer characteristics of falling film and pool boiling evaporation in hybrid evaporator in vapor compression system[J]. *Applied Thermal Engineering*, 2019, 153: 426–432.

Jeong K S, Choi M S. Conceptual design of a container with drainage system for treating and transporting the radioactive wastes under water during decommissioning of nuclear facilities[J]. *Annals of Nuclear Energy*, 2021, 154: 108110.

Ji W T, Zhao E T, Zhao C Y, et al. Falling film evaporation and nucleate pool boiling heat transfer of R134a on the same enhanced tube[J]. *Applied Thermal Engineering*, 2019, 147: 113–121.

Liu H, Wang J. Treatment of radioactive wastewater using direct contact membrane distillation[J]. *Journal of hazardous materials*, 2013, 261: 307–315.

Phillip E, Khoo K S, Yusof M A W, et al. Mechanistic insights into the dynamics of radionuclides retention in evolved POFA-OPC and OPC barriers in radioactive waste disposal[J]. *Chemical Engineering Journal*, 2022, 437.

Yan Z, Jiang Y, Liu L, et al. Membrane Distillation for Wastewater Treatment: A Mini Review[J]. *Water*, 2021, 13(24): 3480.

Yeszhanov A B, Korolkov I V, Dosmagambetova S S, et al. Recent progress in the membrane distillation and impact of track-etched membranes[J]. *Polymers*, 2021, 13(15): 2520.

Zhang X, Gu P, Liu Y. Decontamination of radioactive wastewater: State of the art and challenges forward[J]. *Chemosphere*, 2019, 215: 543–553.

Zhang X, Liu Y. Nanomaterials for radioactive wastewater decontamination[J]. *Environmental Science: Nano*, 2020, 7(4): 1008–1040.

*Advances in Energy, Environment and
Chemical Engineering – Abdullah & Osman (Eds)*
© 2023 The Author(s), ISBN: 978-1-032-36083-6

Study on pore-throat distribution characteristics of Chang 8 reservoir in Huanxi area, Ordos Basin

Jinsheng Zhao*
School of Petroleum Engineering, Xi'an Shiyou University, Xi'an, Shaanxi, China

Jingyang Ma
No. 6 Oil Production Plant, Changqing Oilfield, Xi'an, Shaanxi, China

Du Chang & Yaocong Wang
Oil and Gas Technology Research Institute, PetroChina Changqing Oilfield Company, Xi'an, Shaanxi province, China
National Engineering Laboratory of Low-permeability Oil & Gas Exploration and Development, Xi'an, Shaanxi, China

Yingjun Ju
No. 6 Oil Production Plant, Changqing Oilfield, Xi'an, Shaanxi, China

ABSTRACT: For the tight sand reservoirs, the distribution characteristics of pore-throat directly affect the seepage law of oil and water in reservoirs, and then affect the productivity of oil wells. It is necessary to understand the pore-throat distribution characteristics. In this study, the pore-throat structure and distribution characteristics of the Chang 8 Reservoir were investigated by means of a scanning electron microscope, casting thin section, high-pressure mercury injection and nuclear magnetic resonance. The results show that the pore types of Chang 8 Reservoir in the study area are mainly intergranular dissolved pores, followed by intragranular dissolved pores, granular dissolved pores, and intergranular pores, micropores, and mold pores are the least. The throat types are mainly sheet, curved sheet, and bundle throat. The distribution curve of pore-throat radius is mainly bimodal with the characteristics of high left peak and low right peak, indicating high content of small pores and low content of large pores. Nine of the 10 rock samples are mainly micropores with an average of 64.41%, followed by small pores with an average of 21.87%, followed by medium pores with an average of 10.98%, and the smallest is large pores with an average of 2.74%.

1 INTRODUCTION

Ordos Basin, located in northwestern China, is the second largest inland sedimentary basin in China, with an area of about $37 \times 104 km^2$. It is an important oil-gas-bearing basin and oil-gas production base in China, with a huge amount of oil and gas resources (Yang 2012, Li et al. 2012). Yanchang Formation of Triassic is one of the favorable layers for oil accumulation, which has important research and economic value (Yang 2002). Chang 8 Reservoir in the Huanxi area is the main reservoir in this area, with an average permeability of $0.185 \times 10\text{-}3\mu m^2$. It is a typical tight reservoir with a complex pore-throat structure. To guide the next development, it is necessary to understand the pore-throat distribution characteristics.

The distribution characteristics of pore-throat directly affect the seepage law of oil and water in reservoirs and then affect the productivity of oil wells. The pore-throat structure and distribution characteristics have been studied by many methods, including scanning electron microscopy, cast

---

*Corresponding Author: jszhao@xsyu.edu.cn

DOI 10.1201/9781003330165-41

thin section, high-pressure mercury injection, and nuclear magnetic resonance. Scholars at home and abroad have carried out relevant studies on the distribution characteristics of pore-throat (Wang et al. 2021, 2019; Zhong et al. 2021). Zhong Hongli et al. studied the distribution characteristics of multi-scale microscopic pore-throat in a tight sandstone reservoir by combining high-pressure mercury injection and nuclear magnetic resonance technology (Zhong et al. 2021). Wang Junjie et al. studied the microscopic pore structure of reservoir by casting thin section, scanning electron microscope, and mercury injection test, especially characterized the pore throat size distribution characteristics of sandstone reservoir using high-pressure mercury injection and constant speed mercury injection (Wang et al. 2021). Wang Wei et al. analyzed and compared the pore throat types and distribution characteristics of Chang 6 and Chang 7 reservoirs in Jiyuan area of Ordos Basin using casting thin sections, scanning electron microscopy and constant rate mercury injection, and discussed the causes of their differences (Wang et al. 2019). Although many scholars have studied pore throat distribution characteristics of reservoir, the research methods are not diversified enough. In this paper, the pore-throat structure and pore-throat distribution characteristics of the target reservoir were characterized using the scanning electron microscope, casting thin section, high-pressure mercury injection, and nuclear magnetic resonance, which is helpful to guide the oil well development.

2 PORE-THROAT DISTRIBUTION CHARACTERISTICS

2.1 *Pore types*

According to the statistics of 17 cast thin sections and 286 scanning electron microscope diagrams, the pore types and pore characteristics of Chang 8 Reservoir in the Huanxi area are summarized in Table 1 and Figure 1. As Table 1 shows, among the cast thin sections of 16 core samples, 10 thin sections showed poor pore development, spotted distribution, and poor connectivity, accounting for 62.5%, with average face porosity of 5.04%, average pore diameter of 43.63 μm and average homogeneity coefficient of 0.33. The other 37.5% of the samples showed well-developed pores with relatively uniform distribution and good connectivity, with the corresponding average face porosity of 18.37%, average pore diameter of 76.02 μm and average homogeneity coefficient of 0.38. From Figure 1 and Figure 2, we can see that the pore types are mainly intergranular dissolved pores, followed by intragranular dissolved pores, granular dissolved pores, and intergranular pores, micropores, and mold pores are the least.

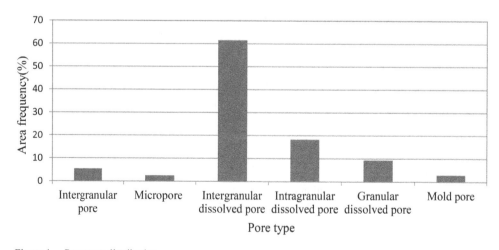

Figure 1. Pore type distribution.

Table 1. Statistical table of pore characteristics.

| Core number | Porosity character | | | | | | | |
	Face porosity, %	Ratio surface, μm^{-1}	Pore-throat ratio	Homogeneity coefficient	Pore diameter, μm	Shape factor	Coordination number	Sorting coefficient
M91-2	4.03	0.53	7.04	0.33	35.91	0.86	0.07	23.8
M91-3	3.89	0.52	6.56	0.27	43.68	0.87	0.04	37.88
M117-2	3.38	0.52	6.23	0.31	37.94	0.87	0.04	26.3
M117-3	2.78	0.68	9.84	0.23	13.15	0.91	0	9.18
M131-2	9.16	0.63	7.26	0.23	35.19	0.84	0.08	27.83
L82-2	5.36	0.6	5.82	0.37	53.56	0.84	0.06	37.27
L82-3	2.75	0.56	5.1	0.41	37.79	0.85	0.03	20.95
M166-3	6.38	0.49	4.93	0.4	90.18	0.82	0.17	61.99
M173-2	6.55	0.44	6.7	0.36	53	0.53	0.23	35.23
B68-2	6.16	0.6	8.97	0.34	35.94	0.82	0.05	26.8
B53-2	13.8	0.33	5.9	0.41	70.29	0.62	0.49	40.29
B71-3	17.59	0.52	5.61	0.25	82.43	0.76	0.32	75.18
H12-2	26.08	0.48	5.99	0.34	91.53	0.72	0.52	57.77
H41-3	21.75	0.33	6.13	0.36	78.5	0.47	0.89	48.68
L364-3	16.63	0.37	5.25	0.42	85.71	0.72	0.48	42.86
M131-3	14.34	0.53	6.87	0.41	47.66	0.76	0.18	28
Average	10.61	0.51	6.47	0.34	57.74	0.76	0.24	38.52

2.2 Throat type

There are five types of throats in sandstone reservoirs, which are pore narrowing throat, necking throat, bundle throat, sheet, and curved sheet throat. According to the observation and statistics of cast thin sections and scanning electron microscopy (Figure 3), the throat types of Chang 8 Reservoir in the study area are mainly sheet, curved sheet, and bundle, among which curved sheet throat is highly developed, and a part of necking throat is also developed.

2.3 Characteristics of pore-throat variation

We can see from the characteristic parameters of the capillary curve of the tested samples in Table 2 that as the permeability of the three samples is close, there is no significant difference among the characteristic parameters. The average median pore throat radius is 0.029 μm, the average pore throat radius is 0.171 μm, and the average throat sorting coefficient is 2.228. The overall performance is characterized by a small pore throat and poor sorting. According to the classification standards for the pore structure of the Chang 8 member reservoir (Zhang et al. 2020), the pore-throat type is between micropore-throat type and small pore-throat type. It also can be seen from Figure 4 that there is no obvious platform end in the capillary pressure curves in Chang 8 Reservoir of the study area, reflecting poor pore-throat sorting.

2.4 Characteristics of pore-throat distribution

The T_2 spectrums of 10 rock samples were tested by NMR test. By corresponding pore-throat radius distribution curve of mercury injection test and T_2 spectrum distribution curve of NMR test, the pore-throat radius r and T_2 corresponding to the peak value were found, and the conversion coefficient C=0.03 was obtained according to C=r/T_2. Based on the conversion coefficient C, T_2 spectrum curves of each rock sample are converted into pore-throat radius distribution curves, as shown in Figure 5. According to the value of transverse relaxation time T_2, the pores of the cores are

a. 53, 2886.15m, Chang 8
Intragranular dissolved pores

b. B53, 2886.15m, Chang 8
Intragranular dissolved pores

C. H41, 2682.15m, Chang 8
Intergranular dissolved pores

d. H41, 2682.15m, Chang 8
Intergranular dissolved pores

Figure 2. Intergranular dissolved pores and intragranular dissolved pores.

Figure 3. Different throat types.

classified into four groups, i.e., micropores (<1 ms), small pores (1-10 ms), medium pores (10-100 ms), and large pores (>100 ms) (Zhao et al. 2021). The true pore size of micropores, small pores, medium pores, and large pores is 0-0.03, 0.03-0.3, 0.3-3, and >3 μm respectively. The proportion of pore throats of different sizes in each rock sample is summarized in Table 3.

We can find from Figure 5 that the distribution curve of the pore-throat radius of Chang 8 Reservoir in the study area is mainly bimodal, which is also typical of tight sandstone. The 10 samples all show the characteristics of high left peak and low right peak, indicating high content of small pores and low content of large pores. As can be seen from Table 3, except for the uniform pore throat size distribution in core sample No.6, the other nine rock samples showed the largest

Table 2. Characteristic parameters of capillary curve.

Core Sample	Permeability $(10^{-3}\mu m^2)$	Porosity (%)	Displacement pressure (MPa)	Pore throat median radius (μm)	Average pore throat radius (μm)	Sorting coefficient of throats	Maximum mercury saturation (%)	Mercury withdrawal efficiency (%)
M91-2	0.103	9.7	0.66	0.038	0.182	2.296	76.25	27.49
B71-3	0.133	9.8	0.66	0.028	0.19	2.283	76.83	32.97
B68-2	0.112	10.9	0.66	0.022	0.142	2.104	77.98	44.06
Average	0.116	10.1	0.66	0.029	0.171	2.228	77.02	34.84

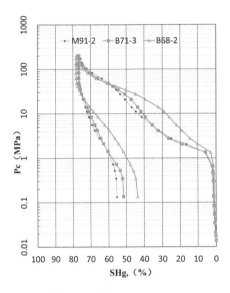

Figure 4. Capillary pressure curves of three samples.

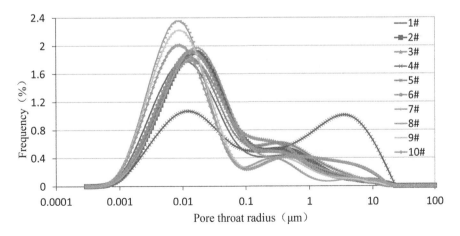

Figure 5. The distribution curve of pore throat of core samples.

Table 3. Pore throat distribution of core samples.

| Core sample | Pore throat distribution (%) | | | |
	Micropore <0.03 μm	Small pore 0.03–0.3 μm	Medium pore 0.3-3 μm	Macropore >3 μm
1#	59.90	27.30	10.39	2.40
2#	74.96	14.76	8.54	1.74
3#	77.98	14.14	5.87	2.01
4#	62.64	26.23	10.06	2.29
5#	67.68	13.35	13.09	5.88
6#	36.61	19.03	24.50	19.86
7#	58.50	24.87	14.41	2.21
8#	60.70	20.57	13.16	5.57
9#	59.02	27.07	11.82	2.09
10#	58.28	28.54	11.44	1.74

proportion of micropores with an average of 64.41%, followed by small pores with an average of 21.87%, followed by medium pores with an average of 10.98%, and the smallest is large pores with an average of 2.74%. For a tight sandstone reservoir, the pore throat in the reservoir is mostly micropores, which leads to the low movable fluid porosity and the lower natural productivity of the reservoir. This is also the main reason why the tight reservoir must be fractured for reservoir transformation before production.

3 CONCLUSIONS

In this study, the pore-throat structure and distribution characteristics of the Chang 8 Reservoir were investigated using the scanning electron microscope, casting thin section, high-pressure mercury injection, and nuclear magnetic resonance, the detailed conclusions are as follows:

(1) The pore types of Chang 8 Reservoir in the study area are mainly intergranular dissolved pores, followed by intragranular dissolved pores and granular dissolved pores, and intergranular pores, micropores, and mold pores are the least. The throat types are mainly sheet, curved, sheet and bundle throat.
(2) The pore-throat type is between the micropore-throat type and the small pore-throat type. There is no obvious platform end in the capillary pressure curves in Chang 8 Reservoir of the study area, reflecting poor pore-throat sorting.
(3) The distribution curve of the pore-throat radius of Chang 8 Reservoir in the study area is mainly bimodal with the characteristics of high left peak and low right peak, indicating high content of small pores and low content of large pores. Nine of the ten rock samples are mainly micropores with an average of 64.41%, followed by small pores with an average of 21.87%, medium pores with an average of 10.98%, and large pores with an average of 2.74%.

ACKNOWLEDGMENTS

This research is supported by the Science and Technology Plan Project of Shaanxi Province (No. 2021JM-411), the National Natural Science Foundation of China (No.52174031), and the Youth Innovation Team of Shaanxi Universities.

REFERENCES

Jinsheng Zhao, Pengfei Wang, Haien Yang, et al. (2021) Experimental Investigation of the CO_2 Huff and Puff Effect in Low-Permeability Sandstones with NMR. *ACS Omega*, 6 (24), 15601–15607.

Li Wei, Liu Luofu, Wang Yanru, et al. (2012) Application of logging data in calculating mud shale thickness. *China Petroleum Exploration*, 17(5): 32–35, 70.

Wang Junjie, Wu Shenghe, Xiao Shuming, et al. (2021) Distribution characteristics of micro pore throat size of turbidite sandstone reservoir in middle sub-member of 3rd member, Shahejie Formation in Dongying Depression. *Journal of China University of Petroleum* (Edition of Natural Science), 45(4): 12–21.

Wang Wei, Zhu Yushuang, Yu Caili, et al. (2019) Pore size distribution of tight sandstone reservoir and their differential origin in Ordos Basin. *Natural Gas Geoscience*, 30(10):1439–1450.

Yang Junjie. (2002) *Tectonic evolution and oil-gas reservoirs distribution in Ordos Basin*. Beijing: Petroleum Industry Press.

Yang Youyun. (2012) Characteristics of the depositional systems and sequence evolution of the Yanchang Formation in the southern Ordos Basin. *Geological Bulletin of China*, 24(4): 369–372.

Zhang Xianglong, Liu Yicang, Zhang Pan, et al. *Reservoir Classification Evaluation and Prediction in Terms of Sedimentary Microfacies: A Case Study from Chang 8 Section in Huanxi–Pengyang South Section, Ordos Basin* [J]. https://doi.org/10.14027/j.issn.1000-0550.2020.076.

Zhong Hongli, Zhang Fengqi, Zhao Zhenyu, Wei chi, Liu Yang. (2021) Micro-scale pore-throat distributions in tight sandstone reservoirs and their constrain to movable fluid. *Petroleum Geology & Experiment*, 43(1):77–85.

Advances in Energy, Environment and
Chemical Engineering – Abdullah & Osman (Eds)
© 2023 The Author(s), ISBN: 978-1-032-36083-6

Discussion on streamline numerical simulation of water flooding development in low-permeability sandstone reservoir

Haili You*

No. 4 Oil Production Plant of Daqing Oilfield Company Ltd, Daqing, Heilongjiang, China

ABSTRACT: Low-permeability sandstone reservoirs are widely distributed in China. In specific production, if the liquid flow path and injection production relationship cannot be fully understood, the injection production well pattern cannot be reasonably improved, and the production effect will not be ideal. Based on this, this paper simulates and analyzes the streamline numerical simulation of water flooding development in the production of this kind of reservoir. The numerical results show that the optimal injection production ratio is 1.0, the optimal well spacing is 150 m, and the lower the starting pressure, the better the production effect.

1 INTRODUCTION

Reasonable injection production ratio and well spacing are the necessary conditions for good results in oilfield development. There are many methods to determine the optimal injection production ratio and the optimal well spacing, including the field test method, historical experience method, etc. Based on other methods, this paper uses a numerical simulation method to determine the optimal injection production ratio, the optimal well spacing and other oilfield development parameters. In the process of using the streamline method to simulate water flooding development of low-permeability sandstone reservoir, we first need to understand the streamline numerical simulation method, and then analyze its mathematical model, including model hypothesis, basic calculus equation, seepage equation, auxiliary equation of seepage method model, and analysis of initial and boundary conditions of seepage model. Finally, the streamline numerical simulation analysis is carried out, including the impact of the well pattern on its development, the impact of injection production ratio on its development, the impact of well spacing on its development and the impact of start-up pressure on its development. In this way, the influence of various factors on water flooding development of low-permeability sandstone reservoirs can be clarified, so as to provide a scientific reference for subsequent development.

2 OVERVIEW OF THE STREAMLINE NUMERICAL SIMULATION METHOD

The so-called streamline numerical simulation method is to transform the original problem of solving the saturation field based on a two-dimensional network or three-dimensional network into a one-dimensional problem of solving the saturation along the streamline (Yu 2021). In this way, it can not only significantly reduce the calculation in the solution, but also further save the operation time and play a significant advantage in processing high-precision geological models.

---

*Corresponding Author: dqyouhaili@petrochina.com.cn

DOI 10.1201/9781003330165-42

3 ANALYSIS ON STREAMLINE MATHEMATICAL MODEL OF WATER FLOODING DEVELOPMENT IN LOW-PERMEABILITY SANDSTONE RESERVOIR

3.1 Model assumptions

Before the streamline mathematical model of water flooding development in low-permeability sandstone reservoir is established, we must make scientific assumptions about the specific reservoir and fluid conditions. The following are the model assumptions in this simulation: 1) the influence of temperature is not considered; 2) fluid viscosity and compressibility are not considered; 3) it is assumed that the fluid flow in the medium follows the characteristics of non-Darcy seepage; 4) it is assumed that there are only two fluids in porous media: water and oil; 5) The effects of starting pressure, capillary force, and gravity on the fluid in the medium are considered.

3.2 Basic calculus equation

When the fluid flows through the formation, it follows the law of conservation of mass and energy, which is consistent with the continuity equation and non-Darcy motion equation. The oil-water two-phase continuous seepage equation is:

$$\frac{\partial(\phi\rho_o S_o)}{\partial t} + \nabla \cdot (\rho_o \bar{v}_o) = q_o \tag{1}$$

$$\frac{\partial(\phi\rho_w S_w)}{\partial t} + \nabla \cdot (\rho_w \bar{v}_w) = q_w \tag{2}$$

where ϕ represents the rock porosity in the reservoir; ρ_o and ρ_w represent the density of oil phase and water phase; S_o and S_w represent the oil content saturation and water content saturation in the reservoir; q_o and q_w represent the mass flow during the injection of oil-phase and water-phase fluid(where the injection is positive and output is negative) (Song 2021).

In the simulation of nonlinear seepage law, the quasi-starting pressure gradient model is to extend the seepage curve artificially, replace the curve with a straight one, and express the nonlinear seepage law of fluid in porous media, so as to realize the scientific reflection of the problems in starting pressure gradient. The seepage equation is as follows:

$$\bar{v}_o = \frac{KK_{ro}}{u_o}[\nabla \cdot (p_o - \rho_o gD) - G_o] \tag{3}$$

$$\bar{v}_w = \frac{KK_{rw}}{u_w}[\nabla \cdot (p_w - \rho_w gD) - G_w] \tag{4}$$

where K represents the absolute permeability of the reservoir; K_{ro} and K_{rw} represent the relative permeability of oil-phase and water-phase fluids; u_o and u_w represent the viscosity of oil phase and water phase fluid; p_o and p_w represent the pressure of oil phase and water phase fluid; g represents the acceleration of gravity; D represents the depth calculated from a datum plane; G_o and G_w represent the starting pressure gradient of the oil phase and water phase (Chen 2021).

3.3 Seepage equation

$$\nabla'\left\{\frac{KK_{ro}}{u_o}[\nabla \cdot (p_o - \rho_o gD) - G_o]\right\} + q_o = \frac{\partial(\phi\rho_o S_o)}{\partial t} \tag{5}$$

$$\nabla'\left\{\frac{KK_{rw}}{u_w}[\nabla \cdot (p_w - \rho_w gD) - G_w]\right\} + q_w = \frac{\partial(\phi\rho_w S_w)}{\partial t} \tag{6}$$

3.4 Streamline method model auxiliary equation

When oil-water two-phase flows in the formation, its law is consistent with the following auxiliary equations.

The first is the saturation equation:

$$S_o + S_w = 1 \tag{7}$$

The second is the relative permeability equation:

$$K_{ro} = f(S_o) \tag{8}$$

$$K_{rw} = f(S_w) \tag{9}$$

The third is the capillary pressure equation:

$$P_c = P_o + P_w \tag{10}$$

where P_c represents the capillary pressure of oil-water two-phase, and its unit is mPa.

3.5 Initial and boundary conditions of the seepage model

In this model, the initial conditions are expressed as follows:

$$p(x,y,z)|_{r=0} = p_o(x,y,z) \tag{11}$$

$$S(x,y,z)|_{r=0} = S_o(x,y,z) \tag{12}$$

The internal boundary conditions can be divided according to the constant liquid production and constant oil production:

$$Q_o = q_{oconst} \tag{13}$$

$$Q_l = q_{lconst} \tag{14}$$

The outer boundary condition is:

$$\frac{\partial p}{\partial n}|_G = 0 \tag{15}$$

where n represents the outer normal direction of the outer boundary G of the model.

4 STREAMLINE NUMERICAL SIMULATION ANALYSIS OF WATER FLOODING DEVELOPMENT IN LOW-PERMEABILITY SANDSTONE RESERVOIR

4.1 Simulation analysis of the influence of well pattern on its development

In the development of the low-permeability sandstone reservoir studied this time, the oil wells are set as vertical wells. This analysis is mainly carried out through five points, inverse seven points, inverse nine points and row well pattern. The simulation analysis shows that the well pattern density will affect the water flooding development of the low-permeability sandstone reservoir. The greater the density is, the faster the oil production rate is. However, with the passage of time, the water content in the produced oil will continue to rise. Therefore, in the later stage of production, the formation energy can be supplemented by well pattern density adjustment or other measures to reduce the water content in the oil (Zhang 2021).

4.2 Simulation analysis on the influence of injection production ratio on its development

In this simulation, the well pattern is set as the inverse nine points, and the injection production ratio is set as 0.8, 1.0 and 1.2, respectively. The simulation analysis shows that the injection production ratio will affect the water flooding development of low-permeability sandstone reservoirs. The larger the injection production ratio is, the faster the oil production rate is. However, with the passage of time, the water content in the produced oil will continue to rise. Further analysis shows that when the injection production ratio is 1.0, the formation and fluid parameters can achieve the best effect, so 1.0 can be used as the optimal injection production ratio in the reservoir.

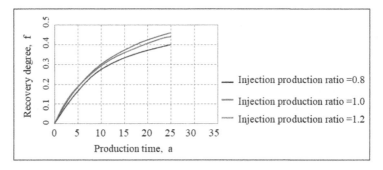

Figure 1.   Oil recovery degree changes with time at different injection production ratios.

4.3 Simulation analysis of the influence of well spacing on development

In this simulation, the well pattern is set as the inverse nine points, the injection production ratio is set as 1.2, and the well spacing is set as 90m, 120m, 150m, 180m and 210m, respectively. Through simulation analysis, it can be seen that in the water flooding development of low-permeability sandstone reservoir, the well spacing will have an impact on it. The larger the well spacing is, the slower the oil production rate is. With the passage of time, the degree of influence gradually decreases. The smaller the well spacing is, the greater the water content in the oil rises, and the rising speed is very fast. Through further analysis, it is found that when the well spacing is 150m, the recovery effect is the best, and the rise of water content is the slowest. Therefore, 150m can be used as the optimal well spacing of the reservoir.

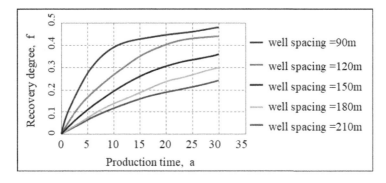

Figure 2.   Oil recovery degree changes with time under different well spacing.

4.4 Simulation analysis of the influence of start-up pressure on its development

In this simulation, the well pattern is set as the inverse nine points, the injection production ratio is set as 1.2, the well spacing is set as 150m, and the start-up pressure is set as 0.015 mPa/m and 0.03 mPa/m, respectively. Through simulation analysis, it is found that the start-up pressure will affect it. The greater the gradient of start-up pressure is, the stronger the nonlinear characteristics of seepage are, the lower the seepage capacity in the reservoir is, and the worse the reservoir development effect is (Guan 2020). Therefore, in the process of water flooding development of low-permeability sandstone reservoir, petroleum enterprises should reduce the start-up pressure gradient as much as possible according to the actual situation, so as to ensure the production effect.

5 CONCLUSION

Based on the results and discussions presented above, the conclusions are obtained as below:

(1) When water flooding development technology is used to exploit low-permeability sandstone reservoirs, in order to achieve a good development effect, petroleum enterprises and relevant technicians must analyze oil recovery degree and water content by means of a streamline numerical model, so as to determine the influence of various factors on development effect.
(2) Then, they should take the specific simulation analysis results as an effective basis, reasonably plan the production mode of this kind of reservoir, and ensure the degree of production and reduce the water cut through the control of various technical parameters. This will have a very far-reaching significance for reservoir exploitation in the form of low-permeability sandstone and the good operation and development of the oil industry.

REFERENCES

Guowei Zhang. (2021) *Study on Vector Characteristics and Optimal Matching of Water Injection Development in Heterogeneous Sandstone Reservoir*. D. China University of Geosciences, 12–13.
Liangye Song. (2021) *Experimental Study on Water Flooding in Low-permeability reservoir*. D. Northeast Petroleum University, 31–33.
Taotao Chen. (2021) *Numerical Simulation of the Effect of Low-frequency Pulse Wave on Porosity and Permeability Parameters of Low-permeability Reservoir*. D. Xi'an Shiyou University, 18–19.
Yun Guan. (2020) *Study on Optimization of Unstable Water Injection Mode in Low-permeability Reservoir*. D. China University of Petroleum (Beijing), 28–29.
Zhao Yu. (2021) Analysis on Development Effect of Low-permeability Reservoir in Xiaermen Oilfield. *J. Inner Mongolia Petrochemical Industry*, (08): 122–124.

*Advances in Energy, Environment and
Chemical Engineering – Abdullah & Osman (Eds)
© 2023 The Author(s), ISBN: 978-1-032-36083-6*

Optimization of multi-branch well development scheme of Liu3 member in Wushi oilfield

Yanjun Li & Lei Ma*
CNOOC Zhanjiang Branch Company, Zhanjiang, Guangdong, China

Songtao Ren
Engineering Technology Zhanjiang Branch of CNOOC Energy Development Co., Ltd., Zhanjiang, China

Zhujun Li & Longbo Sun
CNOOC Zhanjiang Branch Company, Zhanjiang, Guangdong, China

Junlong Yang
Engineering Technology Zhanjiang Branch of CNOOC Energy Development Co., Ltd., Zhanjiang, China

Jing Xu
CNOOC Zhanjiang Branch Company, Zhanjiang, Guangdong, China

Jinsheng Zhao*
School of Petroleum Engineering, Xi'an Shiyou University, Xi'an, Shaanxi, China

ABSTRACT: Liu3 member in Wushi oilfield is a low-permeability reservoir with complicated characteristics of faults and multiple layers vertically, leading to a lower natural production capacity, so it is necessary to adopt appropriate production increasing measures to improve well productivity. In this study, the well type and development scheme of multi-branch wells are systematically optimized through the numerical simulation method. The results show that horizontal well development can obtain higher initial fluid production but lower ultimate recovery due to limited controlled reserves. Inclined and V-shaped wells have better control of reserves, resulting in higher ultimate recovery. Increasing the number of fishbone laterals and multi-bottom laterals can improve the recovery rate. Still, considering drilling costs, the single bottom V-shaped well was the best, followed by the 10-branch fishbone well. For the Liu3 member of well area 9 in Wushi block, the water injection development scheme of double bottom well +V shaped well combination is the best, and the recovery degree of 20 years is 24.8%.

1 INTRODUCTION

Wushi block is located in the eastern part of Wushi Sag in the northern Wan Basin of the South China Sea. It is an important crude oil-producing area in the west of the South China Sea and has wide development prospects. However, the Wushi block is mainly a fault block lithologic reservoir with complex geological reservoir characteristics and mainly a low-grade reservoir. Although the reserves are considerable, it is difficult to develop, and economic development is poor. The reservoir longitudinal connectivity is poor, the single sand body of the thin interbedded reservoir is thin, and the sand bearing rate is low. The understanding of sand body connectivity is not perfect yet. The reservoir of Liu3 member is a typical low permeability reservoir with poor physical properties,

---

*Corresponding Author: malei1@cnooc.com.cn and jszhao@xsyu.edu.cn

permeability less than 60mD, and porosity less than 20%. Conventional well is difficult to release the production. Compared to horizontal wells, multi-lateral wells can communicate multiple zones longitudinally to maximize the release of well production.

Since the 1980s, multi-lateral wells have been widely used in oil production because of higher production rates than single-lateral wells, better access to the reservoir, lower drilling costs, and lower demand for offshore platform space (Aubert 1998; El-Sayed et al. 2001; Elyasi 2016; Joshi 1988; Lim et al. 1998; Salas et al. 1996, Shi et al. 2019; Song et al. 2019). Due to the complex reservoir characteristics of the Liu3 member in the Wushi block, thin interbedded and well-developed faults, it is necessary to optimize the well type and development plan of multi-branch Wells. In this paper, taking the Liu3 member of well area 9 in the Wushi block as the research object, the numerical simulation method is adopted to systematically optimize the well type and development scheme of multi-branch wells.

2 OPTIMIZATION OF MULTI-BRANCH WELL DEVELOPMENT SCHEME

2.1 *Overview of the research region*

Figure 1 is the plane structural map of the study area. It can be seen from Figure 1 that the reservoir in the study area is complicated by faults and distributed in multiple layers vertically. According to the connectivity of sand bodies, the reservoir is divided into three layers, 1, 2, and 3, from the top-down, with a reserve of 1,3731 million tons, which is a small reservoir. Faults are distributed in the south and north of the reservoir.

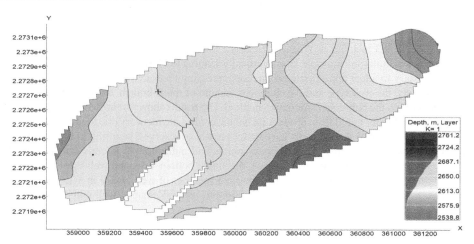

Figure 1. The plane structural map of the study area.

2.2 *Optimization of well types*

The mechanism model was constructed with the no.1 oil formation as the prototype. The grid size was $51 \times 40 \times 41$, the grid size was $30m \times 30m \times 0.976m$, and the model size was $1200m \times 1500m \times 40m$. The physical properties of each layer were the same as that of no.1 oil formation. Horizontal well, inclined well, V-shaped well, and fishbone well are optimized. The schematic diagram of various well types is summarized in Figure 2.

Ten well types were simulated to predict recovery over 20 years, with a focus on recovery per km drilling footage, taking into account costs. The letters on the horizontal axis indicate various well types, where H is a horizontal well, D is an inclined well, V is a V-shaped well, B2 is a

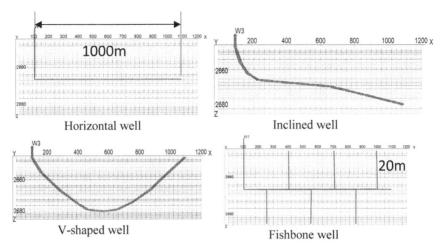

Figure 2. Schematic diagram of various well types.

fishbone well with 2-branch, B4 is a fishbone well with four branches, B6 is a fishbone well with six branches, B10 is a fishbone well with ten branches, B14 is a fishbone well with 14 branches, DB is dual-branch horizontal well and TB is a three-branch horizontal well. The predicted results are shown in Figures 3-5.

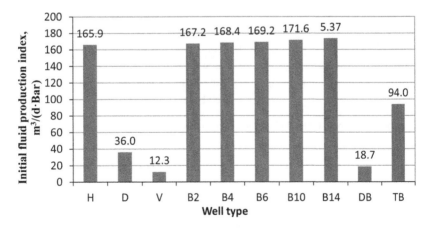

Figure 3. Recovery rate of different well types.

As can be seen from Figures 3-5, the initial production index of horizontal Wells is high, but the ultimate recovery factor is low. This is because horizontal wells usually pass through high-permeability layers, but reservoirs are separated, so the control effect of horizontal wells on reserves is poor, and the ultimate recovery factor is low. Inclined and V-shaped wells have a lower initial production index but better control of reserves, resulting in higher ultimate recovery. For fishbone wells, the ability to cross multiple zones resulted in a higher initial production index and eventual recovery. Considering drilling costs, the single bottom V-shaped well was the best, followed by the 10-branch fishbone well.

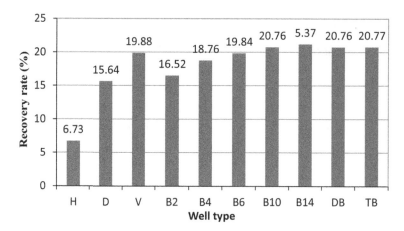

Figure 4. Recovery rate of different well types.

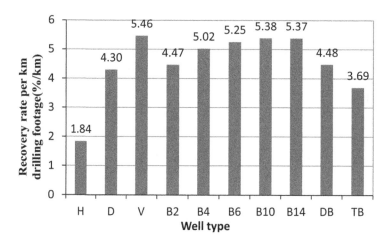

Figure 5. Recovery rate per 1000m drilling footage of different well types.

2.3 *Optimization of development mode*

The study area is an approximate parallelogram of 2km×1km, and its interior is complicated by three faults. In the section, the horizontal wells pass through three thin layers with good physical properties. There are seven development schemes designed in this study. Scheme I is a depletion development mode with nine V-shaped single bottom horizontal wells, which are evenly arranged in three oil formation groups. Scheme II is a depletion development mode with 9 V-shaped bottom horizontal wells, in which six V-shaped double-bottom horizontal wells are arranged in oil formation groups 1, 2, and three V-shaped single-bottom horizontal wells are arranged in oil formation group 3. Scheme III is obtained by acidizing the wells in Scheme II with a skin factor of -1.5. Scheme IV is a depletion development mode with the combination of a dual-bottom well and a V-shaped well, and on the basis of Scheme I, 2 to 12 short radius fishbone branches are drilled. Schemes V, VI and VII convert W19, W22, and W25 into water injection wells, respectively, when the pressure of the depleted oil group drops to 14MPa on the basis of schemes I, II and IV. Production and injection allocation data for each well is summarized in Table 1. Seven development schemes are simulated for 20 years. The predicted results are shown in Figure 6–7.

Table 1. Production and injection allocation data for each well.

Well name	Fluid production (m³/d)	Water injection rate (m³/d)
W18	21	/
W19	21	26
W20	21	/
W21	49	/
W22	49	61
W23	49	/
W24	48	/
W25	48	60
W26	48	/

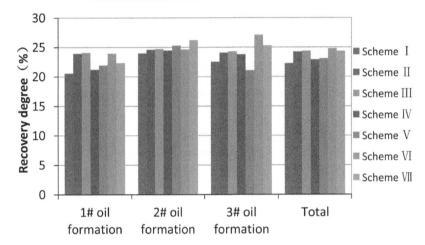

Figure 6. Recovery degree of various development schemes in different oil formations.

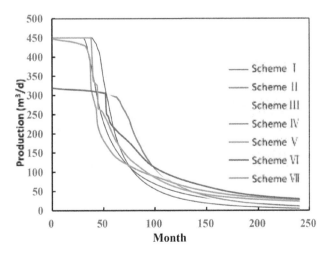

Figure 7. Monthly oil production of various development schemes.

As can be seen from Figure 6, due to the different physical properties, structure, and oil-water layer distribution of each oil reservoir group, the suitable development scheme for every oil reservoir group is different, resulting in different recovery degrees of seven development schemes in each oil reservoir group after 20 years of simulated production. It can be seen from the simulation results that for the No. 1 oil reservoir group, the recovery degree of development plan I III and VI is the highest; for the No. 2 oil reservoir group, the recovery degree of development plan VII is the highest; for the No. 3 oil reservoir group, the recovery degree of development plan VI is the highest. Overall, development plan 6 had the highest recovery rate of 24.8%. In addition, as can be seen from Figure 7, the production of development option 6 was also at a high level throughout the simulation.

3   CONCLUSIONS

In this study, the well type and development scheme of multi-branch wells are systematically optimized through the numerical simulation method. The detailed conclusions are summarized as follows:

(1) For thinly interbedded reservoirs, horizontal well development can obtain higher initial fluid production, but lower ultimate recovery due to limited controlled reserves. Inclined and V-shaped wells have better control of reserves, resulting in higher ultimate recovery.
(2) Increasing the number of fishbone laterals and multi-bottom laterals can improve recovery rate, but considering drilling costs, the single bottom V-shaped well was the best, followed by the 10-branch fishbone well.
(3) For the Liu3 member of well area 9 in Wushi block, the water injection development scheme of double bottom well +V shaped well combination is the best, and the recovery degree of 20 years is 24.8%.

ACKNOWLEDGMENTS

This research is supported by the Science and Technology Plan Project of Shaanxi Province (No. 2021JM-411), the National Natural Science Foundation of China (No.52174031), and the Youth Innovation Team of Shaanxi Universities.

REFERENCES

Aubert W G (1998) Variations in multi-lateral well design and execution in the Prudhoe Bay Unit[R]. *SPE* 39388-MS.
El-Sayed A A H, Ai-Awad M N J, Al-Blehed M S (2001) An economic model for assessing the feasibility of multi-lateral wells [J]. *Journal of King Saud University: Engineering Sciences* 13(1):153–176.
Elyasi S (2016) Assessment and evaluation of degree of multi-lateral well's performance for determination of their role in oil recovery at a fractured reservoir in Iran [J] *Egyptian Journal of Petroleum* 25(1): 1–14.
Joshi S D (1988) Augmentation of well productivity with slant and horizontal wells [J]. *JPT* 40(6):729–739.
Lim B, Xi L, Quan C, (1998) The design considerations of a multi-lateral well[R]. *SPE* 48845-MS.
Salas J R Clifford P J Jenkins D P (1996) multi-lateral well performance prediction[R]. *SPE* 35711.
Shi Y, Song X And Li J (2019) Numerical investigation on heat extraction performance of a multi-lateral-well enhanced geothermal system with a discrete fracture network[J]. *Fuel*, 244: 207–226.
Song X, Zhang C, Shi Y, (2019) Production performance of oil shale in-situ conversion with multi-lateral wells[J]. *Energy*, 189:116145.

*Advances in Energy, Environment and
Chemical Engineering – Abdullah & Osman (Eds)*
© 2023 The Author(s), ISBN: 978-1-032-36083-6

Effect of nodules on electrolyte flow and Cu^{2+} concentration distribution in copper electrolytic refining

Zerui Wang, Yi Meng, Chun Li, Jun Tie* & Rentao Zhao
School of Mechanical and Materials Engineering, North China University of Technology, P.R. China

ABSTRACT: In the copper electrolytic refining process, the mass transfer process between cathodes and anodes is mainly achieved through the electrolyte flow. Under normal conditions, the electrolyte is in a stable flow state, and the Cu^{2+} ions are evenly distributed. However, if nodules are formed on the cathode surface, the electrolyte will flow upward, and the flow rate will gradually increase as the nodules grow, which may lead to more impurity particles floating in the electrolyte. The growth of nodules will reduce the Cu^{2+} concentration below the nodules and increase the Cu^{2+} concentration near the anode opposite to the nodules, thus forming a serious concentration difference polarization.

1 INTRODUCTION

In the copper electrolytic refining process, the flow of electrolytes is the primary way of mass transfer of Cu^{2+}, making the ion distribution between electrode plates more uniform (Andersen 1983).

Mutschke (2010) studied the natural convection and its contribution due to changes in copper ion concentration caused by electrode reactions on the anode surface and cathode surface. Wang (Hongdan 2019; Zhou 2016) investigated the flow variation caused by different inlet and outlet positions and explored the optimal outlet and inlet configurations of the electrolyte, providing guidance for the optimization of the electrolyzer design.

These previous studies were conducted under normal electrolytic conditions without consideration of nodules, and the problem of nodules is a widespread concern in industrial production (Adachi 2018; Meng 2022; Nakai 2018). However, nodules will change the electrolyte flow and lead to uneven ion concentration distribution in the electrolyte (Kalliomaki 2021). Therefore, this paper used the computer simulation method to investigate the effects of nodules with different heights on the flow pattern of electrolytes and the distribution of Cu^{2+} concentration.

2 MODELING

2.1 Control equations

The cathodic reduction of copper is carried out in a CuSO$_4$-H$_2$SO$_4$-H$_2$O solution. Assuming that all electrolytes are ionized completely and satisfy the condition of electrical neutrality, the Nernst-Planck equation (1) describes the mass transfer process in the electrolyte.

$$N_i = -D_i \nabla c_i - z_i v_i F c_i \nabla E_l + \boldsymbol{u} c_i \tag{1}$$

---

*Corresponding Author: tiejun67@263.net

where D_i is the diffusion rate of ions, z_i is the charge of ions, v_i is the mobility of ions, and c_i is the concentration of ions.

The velocity $\boldsymbol{u}$ in equation (1) represents the flow velocity of the electrolyte, which is the natural convection velocity. The natural convection is formed by the concentration gradient, which is caused by the decrease in copper ions near the cathode surface and the increase in copper ion concentration near the anode surface. The natural convection velocity $\boldsymbol{u}$ can be determined by the Navier-Stokes equation:

$$\rho \left(\frac{\partial \boldsymbol{u}}{\partial t} + \boldsymbol{u} \cdot \nabla \boldsymbol{u} \right) = -\nabla p + \nabla \cdot \left(\mu \left(\nabla \boldsymbol{u} + (\nabla \cdot \boldsymbol{u})^T \right) - \frac{2}{3} \mu \left(\nabla \cdot \boldsymbol{u} \right) \boldsymbol{I} \right) + F \tag{2}$$

where p is the fluid pressure, ρ is the fluid density, and μ is the fluid dynamic viscosity.

The reduction reaction rate on the cathode surface is controlled by the Bulter-Volmer equation.

2.2 *Model building*

In this study, the three-dimensional cathodic and anodic reactions are modeled using the electrode surface interface in the "triple current distribution - electrically neutral (tcd)," the electrolyte flow is modeled using the "laminar flow" module, and the simulation of cathode surface growth is modeled with the "Deformation Geometry" module. The geometry model and the mesh profile are shown in Figure 1, which is a geometry of $60 \times 60 \times 25$mm, with 10mm, 15mm, 20mm, and 23mm height nodules at the center of the cathode, respectively.

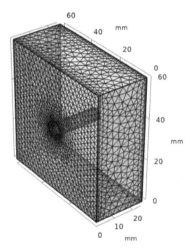

Figure 1.   Three-dimensional geometric model and mesh profile.

3   RESULTS

3.1 *Effect of nodules on electrolyte flow*

The flow of electrolytes is observed in three central cross-sections, xy, xz and yz, and the corresponding flow diagrams are given in Figure 2. As can be seen in the figure:

(1) When there is no nodule, the electrolyte flows downward in the middle part between the electrodes, upward near the cathode, and downward near the anode.
(2) When there is a nodule, the flow pattern of electrolytes is disturbed. The nodule separates the electrolyte between electrodes into two circulation areas, and the electrolyte in the middle of the electrode plates flows upward.

(3) As the height of the nodule increases, the flow rate increases, the upward flow area of the electrolyte also increases, and the upward flow area is closer to the anode plate, which may bring more anode sludge into the electrolyte.

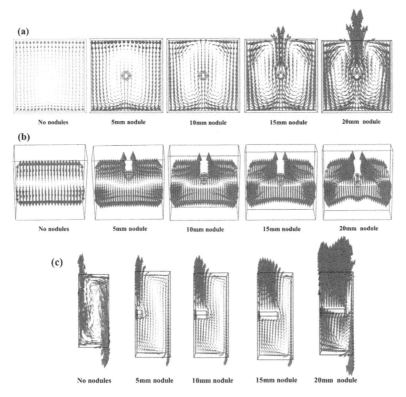

Figure 2.   Schematic diagram of the flow pattern of electrolyte (a) yz section (b) xy section (c) xz section.

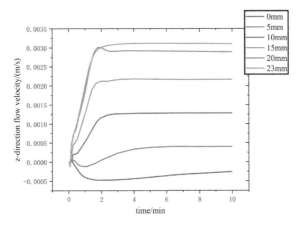

Figure 3.   Upward flow velocity distribution of electrolyte at each nodule height.

Figure 3 shows the flow velocity distribution in the vertical direction on the nodule side. The flow velocity in the middle of the electrolyte is stable at -5×10^{-4} m/s when there is no nodule, which means that the electrolyte is flowing downward and the flow velocity is slow. With the growth of the nodule, the flow direction of electrolyte changes upward, and the flow speed increases gradually with the increase of nodule height. When the nodule height of the nodule is 24 mm, the flow speed of the electrolyte reaches 3×10^{-3}m/s, which is significantly higher than the speed without the nodule.

3.2 *Effect of nodules on concentration*

The presence of nodules affects the mass transfer process in the electrolyte. Figure 4 shows the distribution of Cu^{2+} concentration in the presence of different nodules. From Figure 4(a), it can be seen that the Cu^{2+} concentration is 625 mol/m^3 when there are no nodules. When there is a nodule, the ion concentration below the nodule is reduced by 5-10 mol/m^3. With the increase of the nodule height, the ion concentration on the surface of the anode relative to the nodule gradually increases, and the ion concentration on the surface of the anode reaches 950 mol/m^3 when the nodule height is 23 mm.

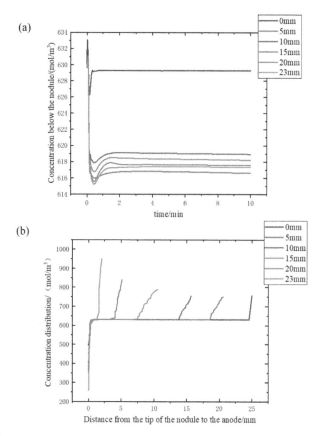

Figure 4.   (a) Cu^{2+} concentration distribution at the bottom of each nodule (b) Cu^{2+} concentration distribution from the surface of each nodule to the anode surface.

4   CONCLUSION

The effects of the nodules with different heights on the electrolyte flow and Cu^{2+} distribution have been investigated. It was found that the presence of a nodule would change the original flow of the electrolyte, forming a loop flow on both sides of the nodule, changing the flow direction of the middle part of the electrolyte from downward to upward. With the increase of the nodule height, the upward flow area on both sides of the nodule would increase, and the speed would increase, which might bring more anode impurity particles into the electrolyte.

The growth of the nodule will impede the mass transfer process on the cathode surface below the nodule, resulting in a localized decrease of Cu^{2+} concentration on the cathode surface and an increase of Cu^{2+} concentration on the corresponding anode surface, which will significantly affect the anode current density distribution.

ACKNOWLEDGMENTS

The authors would like to acknowledge support from the National Natural Science Foundation of China (21978004).

REFERENCES

Adachi K. (2018) FEM Simulation of Nodulation in Copper Electro-refining. *Rare Metal Technology* 2018:215–222.

Andersen TN. (1983) Nodulation of electrodeposited copper due to suspended particulate. *Journal of Applied Electrochemistry*, 13: 429–438.

Hongdan W. (2019) Effect of jet flow between electrodes on power consumption and the apparent density of electrolytic copper powders. *Powder Technology*, 343:607–612.

Kalliomaki T. (2021) Industrial validation of conductivity and viscosity models for copper electrolysis processes. *Minerals Engineering*, 171:107069.

Meng Y. (2022) Cathode current variation and nodulation morphology of copper electrolytic short circuit. *Chinese Journal of Nonferrous Metals*, 32:262–270.

Mutschke G. (2010) On the origin of horizontal counter-rotating electrolyte flow during copper magnetoelectrolysis. *Electrochimica Acta*, 55:1543–1547.

Nakai Y. (2018) Experimental Modeling of Nodulation in Copper Electrorefining. *Rare Metal Technology* 2018:319–323.

Zhou P. (2016) Evaluation of flow behavior in copper electro-refining cell with different inlet arrangements. *Science Direct*, 27:2282–2290.

Advances in Energy, Environment and
Chemical Engineering – Abdullah & Osman (Eds)
© 2023 The Author(s), ISBN: 978-1-032-36083-6

Simulation of zinc electrolyzer based on equivalent circuit model

Zeyue Yin
School of Mechanical and Materials Engineering, North China University of Technology, Beijing, China

Dexiu Zhang, Juhong Zhao, Feifei Ma & Duoxiang Xi
Northwest Lead-Zinc Smelter, Baiyin Nonferrous Group Co.Ltd, Baiyin, Gansu, China

Jun Tie* & Rentao Zhao
School of Mechanical and Materials Engineering, North China University of Technology, Beijing, China

ABSTRACT: In the process of zinc electrolytic deposition, the anode current distribution and the tank voltage of the electrolyzer are extremely important production parameters. Maintaining the normal current distribution and cell voltage is of great significance to energy-saving and the improvement of zinc quality. In this paper, based on the theory of the equivalent circuit model, combined with the mechanism of electrochemical reaction, the equivalent circuit simulation model of zinc electrolyzers has been established using the Matlab/Simulink simulation software. The current and voltage changes under two different operations of zinc discharge and anode cleaning are simulated by changing input parameters and circuit structure, providing guidance for optimizing process parameters and production operations.

1 INTRODUCTION

Due to its superior corrosion resistance and mechanical properties, zinc is one of the most important irreplaceable raw materials in the construction and development of the national economy. Currently, more than 80% of zinc worldwide is produced by the wet metallurgy process. Zinc electrolytic deposition, as the most important link in the wet process of zinc refining, directly determines the product quality and economic efficiency.

An electrolyzer is the carrier of zinc electrowinning production. Its electrode current and cell voltage contain important information about the production changes in the electrolytic deposition process, which has positive guiding significance for the diagnosis of cell condition (Wiechmann et al. 2006, Zeng et al. 2020). In this paper, the equivalent circuit simulation method is used to simulate the circuit structure and electrochemical reaction of the electrolyzer. The circuit model is built by Matlab/Simulink simulation software, and the equivalent circuit simulation model of the electrolyzer in line with the industrial production situation is established. Using the model, the effects of the two most common operations in production, cathode zinc discharge and anode cleaning, on the current distribution of electrolyzer and cell voltage are studied.

2 RESEARCH METHODS AND RESULTS

2.1 *Electrolyzer circuit simulation mode*

At present, zinc plants mainly use the Walker type conductive row structure for electrolyzer circuit connection; that is, lead-silver alloy plate anode and calendered aluminum plate cathode are hung in

---

*Corresponding Author: tiejun67@263.net

DOI 10.1201/9781003330165-45

parallel and staggered in the electrolyzer with direct current, and the inner circuit of the cell shows the circuit relationship of anode-anode parallel connection, cathode-cathode parallel connection, and anode-cathode series connection. The DC current flows from the anode conductive busbar, through the conductive row, anode plate, zinc sulfate mixture, and cathode plate, and out of the cathode conductive busbar.

In the circuit, metal conductors such as anode internal resistance, cathode internal resistance, and conductive row can be regarded as fixed-resistance; the voltage drop caused by the theoretical decomposition of voltage and overvoltage can be equivalent to nonlinear resistance by the current-voltage relationship. Among them, the cathode overvoltage can be neglected because the proportion is too small and is not considered in the circuit. It also includes the variable resistance, such as anode slime resistance, and its resistance value will change with the simulation time. Figure 1 is a schematic diagram of the multi-circuit equivalent circuit of some zinc electrolyzers.

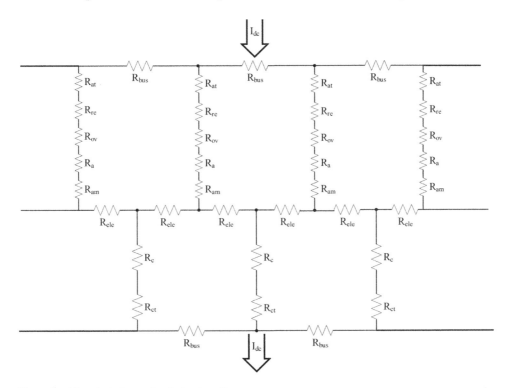

Figure 1.   Zinc electrolyzer circuit structure diagram.

The meaning of each symbol is shown in Table 1.

The decomposition voltage of the electrochemical reaction is calculated using the Nernst equation shown in equation (1).

$$E = E^0 - \frac{RT}{ZF} ln \frac{a_p}{a_R} \tag{1}$$

where R is the molar gas constant; F is the Faraday constant; Z is the stoichiometric coefficient; a_p and a_R are the activities of each reactant and product in the electrolysis reaction equation, respectively; E^0 is the standard state electric potential.

The overvoltage of the electrode reaction kinetics is calculated using the Tafel equation.

$$\eta = a + b * lgi_a \tag{2}$$

Table 1. Identifier meaning.

Symbols	Physical meaning
R_{bus}	Conductive row resistance
R_{at}	Anode contact resistance
R_{re}	Reaction resistance
R_{ov}	Overvoltage resistance
R_a	Anode internal resistance
R_{am}	Anode mud resistance
R_{ele}	Electrolyte resistance
R_c	Cathode internal resistance
R_{ct}	Cathode contact resistance
I_{dc}	DC Current

Among them, i_a represents the current density through the anode, A/cm^2; a represents the Tafel constant; b represents the Tafel slope.

For the metal conductor resistance, contact resistance, and electrolyte resistance, the measurement data and research results of literature (Tang 2004; Wu et al. 2003; Zhong 2008) are used to calculate this part.

In this paper, a zinc electrolyzer with 51 anodes and 50 cathodes is simulated, and the input parameters of some models are shown in Table 2.

Table 2. Partial model input parameters for zinc electrolyzer.

DC current	Pole spacing	Temperature	Zn^{2+}	H$_2$SO$_4$	Electrode working surface
33000A	30mm	40°C	52g/L	134g/L	6000cm^2

2.2 Simulation of the zinc extraction operation

Figure 2 shows the variation of the tank voltage of the three cells during the simulated cathode zinc discharge process. This is due to the presence of the anode slime resistance module in the model, which simulates the continuous deposition of anode slime during production. In the case of tank 02, the tank voltage rises instantaneously to 3.918 V at 150 s, which is 0.6-0.7 V higher than that of normal electrowinning production, equivalent to about 20% of normal electrowinning production. It can also be seen from the graph that the voltage of tank 01 and tank 03 remain relatively stable, indicating that the operation of zinc discharge in tank 02 does not have a significant impact on the tank voltage of the upstream and downstream electrolyzers.

For electrolyzer 02 in the zinc discharge operation, the anode currents of the first and last anodes of the electrolyzer, A01, and A51, show a large variation of 200-260A, which is equivalent to 40%-50% of the normal electrolysis. In contrast, the rest of the anode currents do not vary by more than 10%, as shown in Figure 3.

2.3 Simulation of anode cleaning

Figure 4 depicts the change of tank voltage with the number of cleaned anode plates in the process of anode cleaning. The linearity of the fitted curve of voltage (U) - number of cleaned anodes (x) is very high, so it is considered that the relationship between tank voltage and the number of cleaned anodes in the process of anode cleaning can be seen as a linear change, as the anode cleaning proceeds, the tank voltage decreases with the increase of the number of cleaned anode plates, and finally the tank voltage decreases to 3.172V, which is 47mv less than that before cleaning.

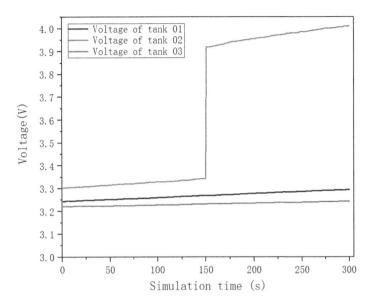

Figure 2. Variation of tank voltage during simulation of zinc discharge operation.

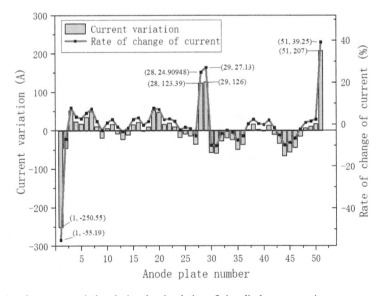

Figure 3. Anode current variation during the simulation of zinc discharge operation.

Assuming a current efficiency of 90%, the electricity consumption of zinc precipitation per ton of cathode could be reduced by 40kW·h /t-1after cleaning the anodes, which brings significant economic benefits.

Figure 5 gives a graph of the anode current change and rate of change of the electrolyzer after proposing the anode plates A25 and A27. From the graph, it can be seen that after proposing the anode plates A25 and A27, the current of each anode in the electrolyzer shows an increasing trend, and the anode plates A24, A26, and A28 in adjacent positions have the most obvious change in current value, with a change of 35A-70A, accounting for 7%-12% of the initial current.

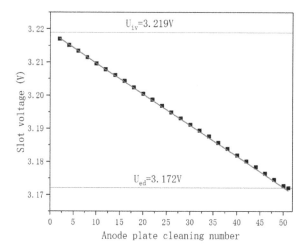

Figure 4.   Anode plate cleaning number versus tank voltage.

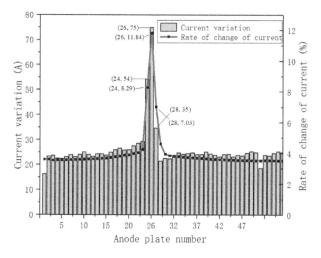

Figure 5.   Anode plate cleaning number versus current change.

In the process of anode cleaning, the resistance of the anode plate decreases after the cleaning work is completed, and the anode current is slightly increased, while other parts of the anode that are not cleaned share the decrease of the anode current. However, the anode current distribution is almost unchanged (within 1%) after the scraping work is completed in the whole cell, so it is believed that the anode current distribution is greatly affected by the operation of removing the anode plate, while the scraping work does not have an excessive effect on the current distribution.

3   CONCLUSIONS

An equivalent circuit simulation model of zinc electrolyzers has been established to study the influence of different operations on the current distribution and tank voltage, and the following conclusions can be drawn:

(1) During the zinc extraction operation of the electrolytic cell, the cell voltage will increase abnormally by 0.6-0.7V when half of the cathodes are drawn out. The current of the anodes located

at both ends is most heavily affected by this operation, while the upstream and downstream cell voltages remain stable. Therefore, the first and last anode of cells should be checked more often to prevent breakage.

(2) During the anode cleaning operation, the cell voltage decreases linearly as the number of cleaned anode groups increases; the effect of withdrawing the anode plate to be cleaned on current reaches the greatest for the adjacent anodes; the overall anode current distribution does not change significantly while the tank voltage decreases after the entire tank is cleaned. Therefore, the cleaning cycle of anode mud should be reduced to save power consumption.

ACKNOWLEDGMENTS

The authors of this paper would like to acknowledge support from the National Natural Science Foundation of China (21978004).

REFERENCES

Tang S C. (2004) Comprehensive production of reduced direct current consumption with zinc electrolysis process. *Hunan Nonferrous Metals*.

Wiechmann E P, Vidal G A, Pagliero A J. (2006) Current-source connection of electrolytic cell electrodes: an improvement for electrowinning and electrorefining. *IEEE Transactions on Industry Applications*, 42:851–855.

Wu R, Oliazadeh M, Alfantazi A M. (2003) Electrical conductivity and density of NiSO4/H2SO4solutions in the range of modern nickel electrorefining and electrowinning electrolytes. 33:1043–1047.

Zeng Q Y, Li C, Meng Y, et al. (2020) Analysis of interelectrode short-circuits current in industrial copper electrorefining cells. *Measurement*, 164: 108015

Zhong S P. (2008) Electrochemical behavior of Pb-Ag-Bi alloys as anode in zinc electrowinning. *Chinese Journal of Process Engineering*.

Advances in Energy, Environment and
Chemical Engineering – Abdullah & Osman (Eds)
© 2023 The Author(s), ISBN: 978-1-032-36083-6

Research on implementation guidelines for carbon neutrality in automobile industry

Nan Wen & Tongzhu Zhang*
China Automotive Technology & Research Center Co., Ltd., China

ABSTRACT: With the proposal of China's 3060 double carbon target, all industries begin to pay attention to and implement carbon emission management. As a pillar industry of the national economy in China, material acquisition, production, use, and scrapping of raw materials for automobiles and spare parts produce a large amount of carbon emissions. It is of great significance for the automobile industry to actively explore the implementation methods and paths of carbon neutrality, carry out energy conservation and emission reduction as soon as possible and achieve carbon neutrality, which is of great significance for realizing the national carbon neutrality goal in 2060. In this paper, the entities, subject matters, implementation paths, quantification methods, emission reduction, carbon offset, carbon declaration, and carbon information disclosure in the automobile industry are studied and analyzed in order to guide the implementation and certification of carbon neutrality at the organization and product level and so on in the automobile industry.

1 INTRODUCTION

Carbon neutrality refers to achieving a state of net-zero carbon dioxide emissions. This can be achieved by offsetting the carbon dioxide generated directly or indirectly by countries, regions, enterprises, groups, or individuals within a certain period by means of tree planting solutions, energy saving, emission reduction, et. At present, the demonstration of carbon neutrality is not considered a component of corporate legal responsibility, but more of the autonomous responses from enterprises to fulfill their social responsibility. The demonstration of carbon neutrality is subject to strict specifications and cannot be treated randomly.

The automobile industry has a long industrial chain with a wide range of coverage; either the production or the use of automobiles throughout the entire life cycle will generate greenhouse gases. Although the country has not clearly stated the time and requirements for achieving carbon neutrality for specific industries, regions, companies, and products, the time node for achieving carbon neutrality by 2060 at the national level has been given. The automobile industry has spearheaded energy conservation and emission reduction to achieve carbon neutrality, reflecting its social responsibility as the pillar industry of the national economy. At present, international auto companies, including BMW, Volkswagen, Mercedes-Benz, Bosch, Nissan, and other auto and spare parts companies, have successively demonstrated commitment to carbon neutrality or fulfilled their commitments. However, details such as who should demonstrate the commitment, what the carbon-neutral subject is, and how to implement, prove and demonstrate a commitment to carbon neutrality still remain unclear, and not many auto companies have demonstrated the commitment to carbon neutrality. Companies that have demonstrated commitments to carbon neutrality also hold the mentality of responding first and then planning specific plans. Since the demonstration of carbon neutrality commitment is vaguely defined and lacks uniform standards, it is difficult for

*Corresponding Author: zhangtongzhu@catarc.ac.cn

                                          DOI 10.1201/9781003330165-46

outsiders to understand the strategies most companies have adopted to achieve carbon neutrality, let alone how well those strategies work. Therefore, it is urgent to clarify the implementation of carbon neutrality in the automotive industry as soon as possible.

2 NATIONAL AND INTERNATIONAL STANDARDS FOR CARBON NEUTRALITY

PAS2060 published by the British Standards Institution (BSI): The *Specification for the Demonstration of Carbon Neutrality* published in 2010 is the first standard in the world to propose a carbon neutrality demonstration. In 2014, a new version of PAS2060 was released: "Specification for the Demonstration of Carbon Neutrality" (PAS2060,2014). The standard puts forward relevant standardized requirements such as the certification process of carbon neutrality, the determination and verification of carbon emissions of the subject matter, carbon footprint quantification, carbon neutrality commitment, carbon emission reduction, carbon offset, demonstration of carbon neutrality commitment, and maintaining carbon neutrality status. The Carbon Neutrality Working Group of the Greenhouse Gas Management Sub-Standard Committee of the International Organization for Standardization ISO/TC207/SC7 Environmental Management Standards Committee has launched the research on the carbon neutral ISO 14068 standard, which is expected to be completed and released in 2023. The standard is still in the draft stage, and discussions focused on the scope of the standard, definitions of core terms, emission reduction requirements, and exchange of carbon neutrality information. In 2019, the Ministry of Ecology and Environment released the *Implementation Guidelines for Carbon Neutrality of Large-Scale Event (Trial)* (Implementation Guidelines for Carbon Neutrality of Large-Scale Event,2019), which stipulated the basic requirements and principles, carbon neutrality process, commitment and evaluation of large-scale events. In 2021, Beijing issued the local standard DB11/T 1861-2021 *Implementation Guidelines for Carbon Neutrality of Enterprises and Institutions* (DB11/T 1861,2021), which stipulated the implementation process, preparation, implementation, evaluation and commitment demonstration of carbon neutrality for enterprises and institutions in Beijing.

At present, China has not issued the implementation guidelines for carbon neutrality in the auto industry. Most auto companies are in the wait-and-see stage to implement carbon neutrality. Therefore, it is of great significance to research the implementation guidelines of carbon neutrality as soon as possible, thus providing guidance for the automobile industry to implement carbon footprint accounting, commitment demonstration, carbon emissions disclosure, carbon emission reduction, carbon offset, and carbon certification, and to speed up the development process of carbon neutrality in the automobile industry.

3 ANALYSIS OF THE MAIN BODY AND SUBJECT MATTER OF CARBON NEUTRALITY IMPLEMENTATION IN THE AUTOMOBILE INDUSTRY

(1) Main Body of Carbon Neutrality Implementation. The main body of carbon neutrality implementation is mainly independent entities that seek to achieve carbon neutrality, such as countries, communities, organizations, companies, divisions, departments, households, and individuals. The main body of the implementation of carbon neutrality in the automobile industry mainly includes entities such as automobile enterprises, parts and components enterprises, and recycling enterprises. If the goal of carbon neutrality is required to be decomposed within the enterprise, we can decompose the main body into entities with independent operating forms such as departments, branches, subsidiaries, and some sites within the automobile enterprise. These entities should use the equity ratio approach or the control approach to define which GHG emissions should be accounted for.

(2) Subject Matters of Carbon Neutrality Implementation. The subject matter of carbon neutral implementation mainly includes carbon neutrality in all aspects of any object selected and defined by the implementing entity, including activities, products, services, buildings, projects

and major developments, towns, events, etc. For the automobile industry, the Subject Matters of carbon neutrality implementation include the automobile enterprises, their branches, sub-brands, departments, buildings, and other subordinate parts of the automobile enterprises. It can also include the products of automobile enterprises, such as auto & spare parts, reused parts, re-manufactured products, echelon utilization products, and recycled products. If auto companies also provide related services, such as R&D services, logistics services, maintenance services, auto leasing services, auto financing services, etc., these services can also be considered as the subject matters for the implementation of carbon neutrality in the automobile industry. If the automotive industry organizes large-scale events, such as large automotive conferences, and exhibitions, such large-scale events can also be considered the subject matter of carbon neutrality in the automobile industry. After the subject matter is determined, the subject matter and the adopted carbon accounting methodology shall remain unchanged during the whole process of carbon neutrality implementation.

4 IMPLEMENTATION PROCESS OF CARBON NEUTRALITY IN THE AUTOMOBILE INDUSTRY

To achieve carbon neutrality for the companies, products, services, etc., auto companies should comply with the relevant rules of carbon neutrality standards, and demonstrate and achieve their carbon neutral commitment by determining the subject matters, accounting for the carbon footprint, developing a management plan, demonstration of carbon neutrality commitment, reducing the carbon footprint, re-quantifying the remaining carbon footprint, and implementing carbon offsets. In general, the basic process of implementing carbon-neutral demonstrations in the automobile industry is shown in Figure 1 below (Yao Tingting Chen Zeyong, 2011).

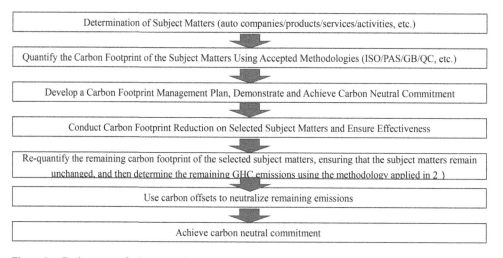

Figure 1. Basic process for implementing a carbon neutral commitment in the automobile industry.

Since it is difficult to achieve carbon neutrality only by reducing direct carbon emissions, the method of carbon offset should also be adopted to balance the carbon emissions. However, carbon neutrality cannot be achieved entirely by carbon offsets, and actions to reduce carbon emissions of the subject matter become essential. Under normal circumstances, the carbon footprint quantification of the target should be carried out in the first application cycle (period 1).

In the early stage of the second application cycle (period 2), enterprises shall demonstrate commitments to carbon neutrality, and formulate concrete plans for the implementation of carbon emission reduction. At the end of the second application cycle (period 2), enterprises shall offset the remaining carbon emissions, and achieve the carbon neutral commitment.

Considering the situation that a newly established enterprise wishes to demonstrate the carbon neutrality commitment at the end of the first application cycle, or the enterprise wishes to issue a prior demonstration at the end of the first application cycle by tracing the historical emission reduction of the previous several years. To facilitate the advancement of carbon neutrality, auto companies are allowed to achieve carbon neutrality entirely through carbon offsets at the end of the first application cycle (period 1). In addition, auto companies cannot achieve carbon neutrality only through carbon offsets in the subsequent application cycles, and must clearly stipulate the absolute value of carbon emissions or reduction in the carbon emissions intensity of the subject matter.

The implementation of carbon neutrality by auto companies can be implemented on a continuous rolling basis. After achieving carbon neutrality in the first application cycle, auto companies can commit to the next application cycle and achieve the carbon neutral commitment of the next application cycle after reducing emissions and implementing carbon offsets. The implementation of each application cycle can be done on a rolling basis, as shown in Figure 2 below [PAS2060,2014].

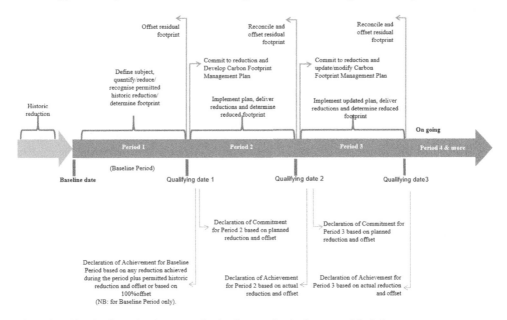

Figure 2. Circular flow of carbon neutrality implementation in the automobile industry.

5 QUANTITATIVE ACCOUNTING METHODS FOR CARBON EMISSIONS OF SUBJECT MATTERS IN THE AUTOMOBILE INDUSTRY

Since different subject matters have different carbon emission sources, accounting boundaries, accounting results uses, carbon emission reduction plans and carbon emission reduction technical solutions. The automobile industry should use different accounting methods to quantify the carbon emissions for different subject matters. The methodology used to quantify and account for GHG emissions should be implemented in a certain order of priority. The methodological sequence defined in PAS 2060 is "International Standard (ISO) -> Internationally recognized standard (e.g. regional or national standard) -> Recognized industry standard or trade method", while the carbon accounting methodological priorities specified in DB11/T 1186-2021 is "Relevant standards for carbon dioxide accounting and reporting requirements in Beijing -> Guidelines for greenhouse gas accounting methods and reporting for industry and enterprises issued by the state -> Internationally recognized or commonly used relevant greenhouse gas quantification standards". Before an auto company decides which set of methodologies should be adopted or specified to quantify the carbon footprint of the target, it is necessary to confirm whether there are certain regulations that have relevant provisions for the carbon accounting of the subject matters in the automobile industry.

Relevant standards need to be formulated for the quantification and accounting of carbon emissions at the organizational level of automobile companies, and the relevant standards are currently in the process of formulation and have not yet been released. As such, the quantification of carbon emissions at the organizational level of auto companies can refer to ISO 14064-1, GHG Protocol for Corporate Accounting, GB/T 32150, and other standards to carry out the accounting of carbon emissions at the corporate organizational level (ISO 14064-1, 2018) (The GHG Protocol,2021) (GB/T 32150, 2015). Quantitative accounting of carbon emissions at the automotive product level requires the formulation of carbon footprint quantification accounting standards for auto and spare parts. Since relevant standards have not yet been released, enterprises can refer to ISO 14067, GHG Protocol for Product Life Cycle Accounting and Reporting Standard, PAS 2050, and other standards accounting for the carbon footprint throughout the product life cycle (ISO 14067, 2018) (The GHG Protocol, 2021) (PAS 2050,2011). Quantitative accounting of carbon emissions at the project level can be carried out with reference to the currently issued standards such as ISO 14064-2, GHG Protocol for Project Accounting, and GB/T 33760-2017(ISO 14064-2, 2019) (The GHG Protocol for Project Accounting,2021) (GB/T 33760, 2017). For the quantification of carbon emissions in large-scale events carried out by automobile companies, please refer to the "Implementation Guidelines for Carbon Neutrality of Large-Scale Event (Trial)" issued by the Ministry of Ecology and Environment in 2019.

For different subject matters in the automotive industry, the quantification and accounting methods of carbon emissions used and the scope of carbon emissions for accounting are also different. The main carbon emission accounting methodologies for different subject matters in the automobile industry are listed in Table 1 below:

6 CARBON EMISSION REDUCTION AND CARBON OFFSET IN THE AUTOMOTIVE INDUSTRY

(1) Carbon Emission Reduction. Auto companies shall adopt the best-fit approaches to reduce greenhouse gas emissions by combining their practical conditions to ensure the realization of the emission reduction targets determined in the plan. Auto companies should avoid completely adopting the method of outsourcing resources to achieve carbon neutrality. The reduction of greenhouse gas emissions is a necessary condition for carbon neutrality. Auto companies should implement management plans for their carbon footprints and quantify their emissions reductions. The scope of carbon emissions that should be accounted for and the way to reduce emissions are also different for different subject matters. For carbon emission reduction at the organizational level, auto companies can focus on reducing carbon emissions from the use of renewable energy, the improvement of production energy efficiency, and the capture and storage of direct carbon emissions. For the carbon footprint reduction of auto and spare parts, auto companies can reduce carbon emissions by controlling the production process of the entire industry chain, such as using clean energy, low-carbon materials, and recycled materials in production process. Besides, auto companies can also reduce emissions by increasing the recycling of materials, improving the energy efficiency of vehicles, and using alternative low-carbon fuels, renewable electricity, and carbon-neutral fuels. For the emission reduction of large-scale events in the automobile industry, auto companies can also reduce carbon emissions from various carbon emission sources by using renewable clean energy, avoiding waste, and recycling materials according to the carbon emission sources within the scope of accounting.

Table 1. Scope of carbon emission accounting for different subject matters in the automobile industry.

Subject Matters	Scope	Emission Type	Emission Source
Organizational Level	Scope 1	Direct energy emissions (greenhouse gas emissions from the combustion of fossil fuels (natural gas, gasoline, diesel, coal, etc.)), direct carbon emissions from production processes such as chemical reactions, direct fugitive emissions	Greenhouse gas emissions, also known as direct emissions, stemming from sources of direct, ancillary, or ancillary production systems owned or controlled by an entity.
	Scope 2	Indirect energy emissions (electricity, chilled water, hot water, compressed air, etc.), outsourced steam, others, etc.	The power station, four major crafts, other workshops, others (living, office environment, etc.)
Whole vehicle	Entire Life Cycle	Acquisition of raw materials (including the use of recycled materials), vehicle production and use, fossil energy consumption in the fuel cycle (production and use), outsourced electricity, etc.	Direct and indirect emissions from raw material production, production of auto and spare parts, vehicle use, and re-production.
Spare Parts	Life Cycle/ Partial Carbon Footprint	Acquisition of raw materials for auto parts, production and use of spare parts, fossil energy consumption in the reproduction stage, outsourced electricity, etc.	Raw material acquisition, transportation, production and use of spare parts, equipment used in the re-production stage, etc.
Auto Service	Service Period	Fossil energy consumption, outsourced electricity, etc., vary by service type	Energy-consuming equipment, vehicles, etc. used to provide services.
Large-Scale Events	Whole Process	Fossil fuel consumption, outsourced electricity, transportation and travel, accommodation and catering, conference supplies, and waste disposal.	Transportation, catering equipment, etc. used before, during, and after the event.

Auto companies should formulate plans for carbon emission reduction solutions, including but not limited to technical solutions and quantities of energy-saving measures, time and scope of implementation, required funds and sources, and target emission reductions, and develop emission reduction strategies regarding the replacement rate improvement of renewable energy and the provision of alternative carbon-containing raw materials, including the type and quantity of renewable energy and alternative raw materials, time and scope of substitution, the required funds and sources, the target emission reductions, etc.

Carbon Offset. Auto companies can use carbon quotas in the local or national carbon emission trading market to offset the remaining carbon emissions that exceed their own quotas at the organizational level. Besides, they can also use carbon credits to offset the remaining carbon emissions, and the items involved in the offset should meet the criteria of additionality, permanence, leakage, and double counting. Carbon emissions at the level of auto products, auto services, auto activities, etc. can be offset by means of carbon credits. Auto companies can purchase the carbon credits from nationally certified projects, and give priority to the Forest Carbon Sequestration Projects (FCSPs) and China Certified Emission Reduction (CCER) projects, to purchase the approved certified emission reduction in the Carbon Generalized System of Preferences (carbon GSP). Enterprises can also use their self-developed projects to offset the certified emission reductions generated by self-developed emission reduction projects outside the accounting boundary of the subject matters or offset the certified emission carbon-sink generated by their own carbon-sink projects outside the accounting boundaries of subject matters. After the carbon offset is completed, the auto company should keep proper records for the documents proving the carbon offset process, including the type of carbon emissions offset, the amount of offset, the type of offset and the projects involved, the number and type of carbon credits used, the time period in which the credits were generated, and the cancellation instructions.

7 SPECIFICATION FOR DECLARATIONS AND DISCLOSURE OF CARBON NEUTRALITY

(1) Declarations of Carbon Neutrality. There are three main ways of carbon neutral declarations: "declarations of commitment to carbon neutrality" "declarations of achievement of carbon neutrality", and "unified declarations of achievement and commitment". The declarations of Carbon Neutrality need to be verified, and there are three verification methods including "Independent thirdparty certification", "other party validation" and "self-validation". The description of carbon neutral declarations differs from different verification methods.

"Declarations of commitment to carbon neutrality" refer to companies that have not yet achieved carbon neutrality and make a commitment to achieving carbon neutrality in the future. At this stage, auto companies who give the commitment are required to determine the subject matters of carbon neutrality, and develop corresponding plans for carbon footprint management, describing how to achieve carbon neutrality with respect to the subject matters. "Declarations of achievement of carbon neutrality" require entities to make carbon footprint reductions and offset remaining carbon emissions for selected subject matters, ensuring that carbon neutrality has been achieved. It applies only to the scope and cycle that has been validated, and further validation is required if it needs to be extended to future cycles. "Unified declarations of achievement and commitment" are applicable to entities that desire to declare the achievement of carbon neutrality while fulfilling their commitment to carbon neutrality in the future. The description of carbon neutral declarations differs from different verification methods. Please refer to Table 2 below.

(2) Carbon Disclosure. Carbon information disclosure refers to the information disclosure of the greenhouse enterprise generated by the enterprise in the process of production and operation, so as to provide investors with environmental protection information about the enterprise. After the carbon neutral declaration is approved, companies can disclose carbon neutral information, and promote it to consumers on websites, social responsibility reports, environmental development reports, media, publicity materials, and product labels. At present, most of the carbon neutrality actions carried out by domestic enterprises are process control and tracking management at the level of their own enterprises and products, resulting in a weak willingness to disclose carbon information and a lack of sufficient public supervision. The non-disclosure of carbon emission information has become an important factor restricting the development of carbon neutrality. The main content of corporate carbon emission information disclosure is carbon emission data. If these data are provided by the company based on self-validation, it will lead to low credibility and may also be questioned by the media and the public. Therefore, it is recommended that companies conduct carbon-neutral information disclosure through third-party verification or other party validation.

8 CONCLUSION

Under the pressure of China's pledge to be carbon neutral by 2060, all industries are taking action to achieve carbon neutrality as soon as possible within their capabilities. This is critical to achieving national carbon neutrality goals. To achieve carbon neutrality, enterprises should clarify the implementation entities of various types of carbon neutrality, determine the emission reduction responsibilities of various entities, and form a worldwide mutual recognition and industry-uniform accounting and verification methodology for the subject matters of carbon neutrality defined by various entities. Then, it is attempted to promote the realization of carbon neutrality from micro to macro in all walks of life through systematic accounting and statistics of carbon emission data of various subject matters, development of various carbon emission reduction technologies, and the employment of economic methods such as carbon trading and carbon credit. We could move

Table 2. Declaration and verification methods of carbon neutrality in the automobile industry.

Declarations and Verification Methods of Carbon Neutrality		Description	Duration
Declarations of commitment	self-validation	Self-declared: [XX AUTO Company] will achieve carbon neutrality of [Subject Matter: Company/Product] that meets [XXX Carbon Neutrality Standard] within [Qualification Date] from [Base Date].	Up to one year
	Third-party Verification	Certified by [XX Institution], [XX AUTO company] will achieve carbon neutrality of [Subject Matter: Enterprise/Product] that meets [XX carbon neutrality standard] within [Qualification Date] from [Base Date]	
	other party validation	Declared by [XX Agency], [XX AUTO company] will achieve carbon neutrality of [Subject Matter: Enterprise/Product] that meets [XX carbon neutrality standard] within [Qualification Date] from [Base Date]	
Declarations of achievement	self-validation	Self-declared: [XX AUTO Company] has achieved carbon neutrality of [Subject Matter: Company/Product] that meets [XXX Carbon Neutrality Standard] within [Qualification Date] from [Base Date].	Permanently valid and cannot be extended to subsequent cycles.
	Third-party Verification	Certified by [XX Institution], [XX AUTO company] has achieved carbon neutrality of [Subject Matter: Enterprise/Product] that meets [XX carbon neutrality standard] within [Qualification Date] from [Base Date]	
	Other party validation	Declared by [XX Agency], [XX AUTO company] has achieved carbon neutrality of [Subject Matter: Enterprise/Product] that meets [XX carbon neutrality standard] within [Qualification Date] from [Base Date]	
Unified declarations of achievement and commitment	self-validation	Self-declared: [XX AUTO Company] has achieved carbon neutrality of [Subject Matter: Company/Product] that meets [XXX Carbon Neutrality Standard] within [Qualification Date] from [Base Date] and promises to maintain it until [Next Qualification Date]	declarations of achievement are permanently valid, and Declarations of commitment are valid for one year.
	Third-party Verification	Certified by [XX Institution], [XX AUTO company] has achieved carbon neutrality of [Subject Matter: Enterprise/Product] that meets [XX carbon neutrality standard] within [Qualification Date] from [Base Date], and promises to maintain it until [Next Qualification Date]	
	other party validation	Declared by [XX Agency], [XX AUTO company] has achieved carbon neutrality of [Subject Matter: Enterprise/Product] that meets [XX carbon neutrality standard] within [Qualification Date] from [Base Date], and promises to maintain it until [Next Qualification Date]	

much closer to the goal of realizing carbon neutrality at the national level only if individuals strive to achieve carbon neutrality.

There are many types of enterprises in the automobile industry, including enterprises of passenger cars, commercial vehicles, etc., as well as enterprises that produce spare parts such as power batteries, engines, and transmissions, and enterprises that are engaged in recycling and dismantling, reuse and remanufacturing of automobile products. Auto companies have a wide variety of products and rich layers, so the implementation entities and subject matters of carbon neutrality are also diverse. Since there is neither a broad consensus on carbon emission accounting and control, nor a unified carbon accounting standard methodology and carbon verification and certification system, the automobile industry does not have enough carbon emission data, resulting in difficulties in providing necessary data support for the subsequent development of low-carbon economy and carbon neutrality. In order to facilitate the early achievement of carbon neutrality among domestic auto companies, auto products, and relevant entities of the entire upstream and downstream industry chain, automotive enterprises should conduct research on implementation guidelines for carbon neutrality, and conduct continuous exploration, implementation, supplementation and improvement regarding systems for carbon emission management, carbon emission accounting, carbon data accumulation, carbon emission reduction, carbon verification, carbon offset, carbon certification, carbon neutral demonstration, carbon information disclosure, etc.

REFERENCES

Announcement No. 19, 2019 of the Ministry of Ecology and Environment, *"Implementation Guidelines for Carbon Neutrality of Large-Scale Event (Trial)"*

DB11/T 1861-2021*"Implementation Guidelines for Carbon Neutrality of Enterprises and Institutions"*

GB/T 33760-2017*"Technical specification at the project level for assessment of greenhouse gas emission reduction—General requirements"* [s] Beijing: China Standards Press, 2017

GB/T 32150-2015 *"General Guideline of the Greenhouse Gas Emissions Accounting and Reporting for Industrial Enterprises"* [s] Beijing: China Standards Press, 2015

ISO 14064-2:2019 Greenhouse gases — Part 2: Specification with guidance at the project level for quantification, monitoring, and reporting of greenhouse gas emission reductions or removal enhancements

ISO 14064-1:2018 Greenhouse gases — Part 1: Specification with guidance at the organization level for quantification and reporting of greenhouse gas emissions and removals

ISO 14067-2018 Greenhouse gases—Carbon footprint of products — Requirements and guidelines for quantification

PAS 2060-2014 Specification for the demonstration of carbon neutrality

PAS 2050:2011 Specification for the assessment of the life cycle greenhouse gas emissions of goods and services

WRI, WBCSD. The GHG Protocol: Corporate Accounting and Reporting Standard

WRI, WBCSD. The GHG Protocol: Product Life Cycle Accounting and Reporting Standard

WRI, WBCSD. The GHG Protocol for Project Accounting.

Yao Tingting, Chen Zeyong. Analysis of the International Standard for Carbon Neutrality [J] *Electron Mass.* 2011(01):59-61

Advances in Energy, Environment and
Chemical Engineering – Abdullah & Osman (Eds)
© 2023 The Author(s), ISBN: 978-1-032-36083-6

Practical application of converting food waste into energy

Yuan Yao*
China Urban Construction Design & Research Institute Co. Ltd., Beijing, China

Sibo Wang
CUCDE Environmental Technology Co. Ltd., Beijing, China

Jingwen Liu
China Urban Construction Design & Research Institute Co. Ltd., Beijing, China

Jie Wang
Chongqing Sanfeng Sci&Tech Co. Ltd, Chongqing, China

Canwei Zhou
China Urban Construction Design & Research Institute Co. Ltd., Beijing, China

ABSTRACT: Transforming food waste into energy products can not only solve the shortage of fossil fuels, but also solve the problem of environmental pollution. At present, the main food waste treatment processes are landfill, incineration, anaerobic fermentation, aerobic composting, direct drying as feed, hydrolysis, and microbial treatment technology. Through the analysis of the existing food waste treatment technology route, the anaerobic fermentation technology is relatively advanced and reliable. A practical application of anaerobic digestion of food waste was discussed in the current study. It is analyzed the methane production amount is 7565 m^3/d with the process scale is 200 t/d, and the total annual on-grid power generation capacity of this project is 243.19×10^4 kWh/a. The practical application of converting food waste into energy via anaerobic digestion is a promising technical route.

1 INTRODUCTION

The impact of food waste on the environment and society is becoming increasingly serious. A large number of food waste mixed with domestic waste in the terminal treatment facilities could significantly impact the normal operation of terminal treatment (Karmee, 2016).

Food waste, the largest proportion of municipal solid waste, mainly refers to food residues and waste edible oils and fats produced in food processing, catering services, collective meals, and other activities other than the daily life of residents. Its composition is complex. It is a mixture of oil, water, peel, vegetables, rice noodles, fish, meat, bones, waste tableware, plastic, paper towels, and other substances.

Food waste has a high moisture content, and moisture accounts for 80 to 90% of the total weight of the waste. The content of organic matter is high. Under high-temperature conditions, it is easy to rot and deteriorate, producing a foul smell. Salt and oil content is much higher than that of other organic waste.

Food waste has a high content of organic matter, and the collection and transfer process leakage will pollute the air, soil, and water sources, seriously interfering with people's normal life. Most

---

*Corresponding Author: 15201443594@139.com

DOI 10.1201/9781003330165-47

parts of the country are accustomed to feeding food waste directly to livestock and poultry as cheap feed. However, in addition to containing metals, toothpicks and plastics, and other sharp objects, urban food waste will hurt the digestive tract of livestock and poultry, but also contains a large number of pathogenic microorganisms, parasites, and their eggs, which are easy to cause zoonotic diseases after feeding livestock and poultry.

In addition, there is a food chain danger in the direct feeding of livestock and poultry by food waste: firstly, the biotoxins produced by pathogenic microorganisms are enriched in the livestock and poultry and then transferred to the human body through the food chain; secondly, the food waste contains a large amount of homologous protein of the fed livestock and poultry, which has major safety risks, and the Ministry of Agriculture promulgated several measures for expressly prohibits the use of animal-derived feed to feed ruminants.

Meanwhile, food waste is rich in organic nutrients (Wen, Wang, and Clercq, 2015). After rational disposal, it is an important source for the production of organic fertilizers and bioenergy (Li, 2016, Zhang, Su, Baeyens, and Tan, 2014), and is a high-value biomass resource. The research on the resource treatment technology of food waste has been deepening, and the resource of food waste is the future trend of domestic food waste treatment.

At present, the main food waste treatment processes are landfill, incineration, anaerobic fermentation, aerobic composting, direct drying as feed, hydrolysis, and microbial treatment technology (Karmee and Lin, 2014, Pöschl, Ward and Owende, 2010). The more advanced food waste treatment technology in foreign countries is mainly distributed in European countries, and the food waste treatment technology in South Korea and Japan in Asian countries is also more advanced. Due to China's eating habits, the food waste composition is quite different from that of European and American countries and is closer to Japan and South Korea. Therefore, foreign treatment technology is unsuitable for Chinese food waste treatment.

Before the pilot cities were implemented, Xining, Ningbo, Beijing, Shanghai, and other cities carried out the harmless and resource utilization of food waste, which provided a good experience for the promotion of food waste treatment technology in China.

2 DEVELOPMENT OF FOOD WASTE TREATMENT TECHNOLOGY

Food waste treatment technology can be divided into four stages of development.

2.1 *First stage*

In 2010, the state implemented the previous treatment technology of the pilot cities for the resource utilization and harmless treatment of food waste, which was classified as the first stage of treatment technology.

The core of the treatment technology is to treat food waste as a raw material that can produce high added value. Fertilizer technology is used in Xining, Ningbo, Beijing, and other cities, and the fertilizer produced is used for crops, fruit trees, aquaculture and livestock, poultry breeding, etc.; swill oil extraction is mainly used to make "crude oil," and biodiesel is made in some cities.

2.2 *Second stage*

From 2010 to 2012, with the possibility of potential threats to people's health posed by food waste fertilizer or feed, the Ministry of Agriculture partially withdrew the fertilizer and feed production and sales licenses issued by food waste treatment enterprises, and food waste treatment enterprises faced a wide range of technological transformation. In 2010, the state implemented the work of the first batch of pilot cities for food waste, providing better policy support for developing treatment technology. Therefore, 2011 to 2012 is the second stage of food waste treatment technology development.

The core of the treatment technology is that anaerobic fermentation develops rapidly in food waste, and anaerobic fermentation of food waste produces biogas, a typical example of resource utilization. The representative city is Chongqing.

2.3 Third stage

In the process of anaerobic fermentation operation, the anaerobic treatment unit faces more technical problems in material control, reaction temperature, homogenization reaction, and equipment manufacturing. With the promulgation and implementation of the "Technical Specification for The Treatment of Food Waste" (CJJ 184-2012), quantitative restriction indicators have been made for the technical parameters of anaerobic fermentation, and the "damp heat" technology has gradually developed and applied. After 2012, it is the third stage of food waste treatment technology development.

The core of the treatment technology is to change the characteristics of the dry anaerobic fermentation material of the second stage of the material, change the solid content of the anaerobic fermentation material through the "damp heat" technology of the previous stage, and use the anaerobic liquid phase material to solve many difficult problems in the anaerobic fermentation. Representative cities are Suzhou, Lijiang, and so on.

2.4 Fourth stage

On December 21, 2016, President Xi chaired the 14th meeting of the Central Financial and Economic Leading Group and delivered an important speech proposing to universally implement the garbage classification system in China. Therefore, the treatment and disposal of organic waste have entered a new stage.

In March 2017, the National Development and Reform Commission and the Ministry of Housing and Urban-Rural Development issued the "Implementation Plan for the Classification System of Domestic Waste," which requires the implementation of compulsory classification of domestic garbage in 46 cities and the recycling rate of municipal solid waste will reach more than 35% by the end of 2020. Vegetables, melons, and fruits produced in the wholesale market of agricultural products, carrion, minced meat, eggshells, livestock, and poultry products offal, etc. It encourages the use of perishable waste to produce industrial grease, biodiesel, feed additives, soil conditioners, biogas, etc., or joint disposal with straw, feces, sludge, etc. Cities that have carried out pilot projects for food waste treatment should promote full regional coverage based on stable operation. Cities that have not yet built food waste treatment facilities may temporarily not require residents to separately classify "wet waste" of food waste. Severely crack down on and prevent the production and circulation of "gutter oil." It is strictly forbidden to use urban household garbage directly as fertilizer.

Under the new situation, organic waste treatment has entered a new stage of "diversified collaborative treatment." On the one hand, its technical characteristics are that catering waste, food waste, feces, sludge, straw, and other materials are passed through different pretreatment processes and then enter the anaerobic fermentation system for collaborative disposal, and biogas is generated for use. On the other hand, the construction of environmental protection industrial parks that coordinate food waste treatment facilities with harmless treatment facilities for domestic waste, construction waste treatment facilities, and sewage treatment facilities are a major development trend today.

3 TECHNICAL ROUTES OF FOOD WASTE TREATMENT

According to the statistical analysis of the technical routes of the food waste treatment projects that have been built, under construction, and are under construction in China, and combined with the relevant provisions of the "Technical Specifications for the Treatment of Food Waste"

(CJJ 184-2012), the technical route of food waste treatment can be summarized as "pretreatment + harmless treatment + resource utilization." Among them, "pretreatment" includes sorting, crushing, dehydration, oil-water separation, etc. "Harmless treatment" includes a process for obtaining feed and fertilizer raw materials by adding high-temperature strains or using anaerobic fermentation techniques. "Product resource utilization" includes the post-treatment of feed and fertilizer raw materials or the resource utilization of biogas and biogas residue.

Overall analysis, the current domestic food waste treatment technology route includes three categories: The first is the technical route of dry feed as the treatment target; the second is the technical route of aerobic fermentation of biological feed or fertilizer as the treatment target; the third is the technical route of using anaerobic biogas for resource utilization.

The treatment concept of "harmless treatment + product resource utilization" is the same, but the technical routes used are different. The following is an analysis of three processing technology routes.

3.1 The technical route for dry feed

This technical route is to dehydrate, dry, and sterilize the food waste to form animal feed with a moisture content of less than 15%, and its dehydration process is divided into high-temperature dehydration, fermentation dehydration, frying, and dehydration.

This treatment process is the original food waste treatment process, and its advantages are simple to process, easy to operate, and low operating costs. However, the animal protein in its feed may cause a potential, uncertain risk of disease, i.e., homologous contamination, when eaten by the same species of animals. In addition, the salt content of this feed fluctuates greatly, and when the salt content is high, it will have a certain impact on the growth of livestock and poultry. Therefore, the quality and safety of the products of this feed are not easy to control, and there are irreversible safety hazards. Therefore, this process cannot meet the requirements of the state for the harmlessness of food waste and cannot be used as a recommended process.

3.2 The technical route aimed at aerobic fermentation to produce biological feed or fertilizer

This technical route represents Beijing Gaoantun Food Waste Treatment Plant, Chengdu Shuangliu Food Waste Treatment Center, Guangzhou, and Nanjing food waste treatment centers. This technology is characterized by the addition of microbial strains, and the products after high-temperature treatment are organic feed and organic fertilizer.

From food waste to feed is the shortest path of material resources, and the loss of resources is also the smallest, so the feed of food waste is a higher degree of resource utilization. However, the state currently has strict policies and technical regulations on the feed of food waste because there may be a potential and uncertain risk of spreading diseases. The Ministry of Agriculture promulgated the Administrative Measures for the Safety and Health of Feed Products of Animal Origin in 2004 and the Measures for the Administration of Production Licenses for Feed and Feed Additives in 2004 and explicitly prohibited the use of animal-derived feed to feed ruminants CJJ 184-2012 specifies strict processes for the use of "feed treatment."

The use of high temperatures to make organic fertilizers also has certain problems in the use of fertilizers. Salt is found in food waste, and salt is also found in direct fertilizer. Long-term application in water-scarce areas may cause salt accumulation and damage the soil ecological environment. In addition, due to China's agricultural fertilization habits, chemical fertilizers dominate, and the amount of organic fertilizer produced after the treatment of food waste is not ideal.

3.3 The technical route of anaerobic biogas production for resource utilization

This technical route is to use the anaerobic fermentation of food waste and the biogas produced for energy recycling: biogas is used as fuel to provide heat for the normal operation of the system;

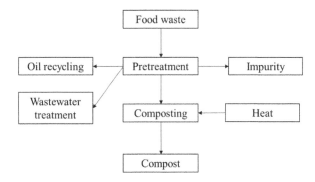

Figure 1.   Scheme of the technical route for compost.

biogas is purified as on-board natural gas; biogas is used for the use of electricity in the cogeneration supply system, heat or transported to off-site users.

Currently, more than 80% of the food waste treatment facilities are in operation, under construction, or are planning to be built using this technical route. From the perspective of resource utilization, biogas is used as biomass energy in production and life, reducing the consumption of external energy, which is in line with the national circular economy industrial policy. From the aspect of handling waste, anaerobic digestion is one of the effective ways to "reduce, harmless, and recycle" food waste. From the aspect of technical application, anaerobic technology for food waste treatment, through learning from foreign treatment technology, after several years of exploration, has gradually formed a "localization" food waste anaerobic treatment technology. According to the current direction of technological development, the anaerobic treatment technology of food waste in China will gradually become a trend.

The anaerobic fermentation technology is relatively advanced and reliable through the analysis of the existing food waste treatment technology route. It is in line with the national industrial policy and development direction, and there is no potential safety hazard of feed and fertilizer; the product is biogas, and the use of biogas provides energy for the treatment system; the product can be sold smoothly, which can ensure the long-term sustainable treatment of food waste; the number of successful domestic application cases has gradually increased; suitable for large-scale continuous production; the secondary environmental pollution is small; it is easy to control; the site selection is relatively easy; the investment is moderate. Moreover, anaerobic fermentation technology embodies a variety of concepts such as energy conservation and environmental protection, circular economy, etc., which can achieve the coordination and unification of environmental, social, and economic benefits and is of great significance to the sustainable development of the environment and economy. It is mainly reflected in the following aspects: 1. The biogas produced after anaerobic fermentation is clean energy. 2. After anaerobic fermentation of solid substances, it can produce high-quality organic fertilizers or soil amendments. 3. In the process of transforming organic matter into methane, "reduction and recycling" have been realized. 4. The biogas produced by anaerobic fermentation can be used for biogas boilers, biogas power generation, or natural gas production, reducing greenhouse gas emissions. 5. The kitchen waste moisture content is high, and using anaerobic fermentation treatment requires almost no need to adjust the moisture content, saving water consumption. 6. Oil and fat recovery to produce "crude oil" to recover energy and resources as much as possible.

Based on the above analysis, the technical route of kitchen waste treatment is proposed to choose the "technical route with anaerobic biogas production for resource utilization as the treatment target." Including pretreatment system (including grease recovery and utilization system), anaerobic treatment system, anaerobic product treatment and utilization system (biogas utilization system, biogas sludge, and sludge treatment or utilization system and biogas sludge treatment system), deodorization system".

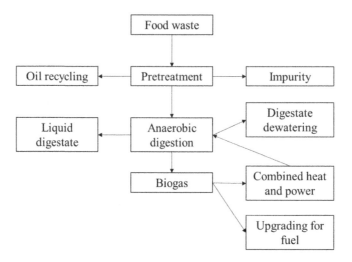

Figure 2. Scheme of the technical route for energy.

4 PRACTICAL APPLICATION OF ANAEROBIC DIGESTION OF FOOD WASTE

4.1 *Amount of biogas produced*

The practical project uses a wet, medium-temperature anaerobic system to produce biogas. According to the calculation of the amount of biogas in the sewage treatment anaerobic system, usually converted according to the inlet water COD, the theoretical methane production is 1 g COD to produce about 0.35 L of methane. The COD of the liquid phase material entering the wet medium temperature anaerobic section is about 100,000~150,000 mg/L according to the experience data that has been operated in China, and the average value of this project is 125,000 mg/L. After passing through this system, the removal rate is about 90%; that is, the removal amount is 112500 mg/L. According to the process flow, when the process scale is 200 t/d, the liquid phase feed volume of the treatment system is calculated as 192.13 m^3/d.

Therefore, the methane production amount Q=112500×192.13×0.35/1000 m^3/d=7565 m^3/d.

The methane content of biogas produced is about 40 to 60%, and the amount of biogas produced is 13754 m^3/d according to 55%, and about 16.75 t/d is calculated according to the specific gravity of mixing.

4.2 *Amount of biogas consumption*

The total amount of steam required for the food waste treatment process is 23.56 t/d. According to the generator set parameters, the waste heat boiler can produce about 500 kg of steam per hour and 12.0 t of steam per day. The total amount of steam required for the steam boiler is 11.56 t/d, which is 2.78×10^7 kJ/d.

According to the operating parameters of the biogas generator set of the project, the hot water waste heat power of the biogas generator set is about 600.0 kW, which can meet the summer hot water heating and equipment insulation needs of the project, and the additional heating required by the biogas boiler in winter is 286.1 kW, which is 2.47×10^7 kJ/d.

Referring to other similar craft food waste projects in the whole country, the calorific value of biogas generated by anaerobic digestion of this project is calculated according to 21500 kJ/m^3, and the thermal efficiency of steam boilers is 90%.

It is calculated that the digest gas consumption of the steam boiler is 1436 m^3/d in summer and 2714 m^3/d in winter.

4.3 *Amount of biogas power generation*

The energy balance of a biogas generator varies depending on the type of generator and the operating conditions. According to the experience data of domestic biogas power generation operation, the calorific value of biogas in this project is about 20,000~23,000 kJ/m^3, and the biogas power generation capacity is about 2.0 kWh/m^3.

(1) Power generation during the non-heating season

According to the above data parameters, the amount of biogas that can be used for biogas power generation in the non-heating season of the second phase project is: 13754-1436 = 12318 m^3/d;

The annual non-heating season time of this facility is 245 days, and the annual power generation capacity of the non-heating season under the normal working conditions of the project is:

(12318 m^3/d×2.0 kWh/m^3×245 d/a) = 603.6×10^4 kW·h.

The installed capacity required for non-heating season biogas generator sets is 1026.5 kW.

(2) Heating season power generation

According to the above data parameters, the amount of biogas that can be used for biogas power generation in the heating season of the project is 13754-2714=11040 m^3/d;

The annual heating season of the practical project is 120 days, and the annual heating season power generation under the normal working conditions of this project is:

(11040 m^3/d×2.0 kWh/m^3×120 d/a) = 265.0×10^4 kWh

The installed capacity required for the heating season biogas generator set is 920.0 kW.

Considering the fluctuation of biogas production and methane content in biogas, and considering the possibility of generating electricity from all biogas, the total installed capacity of this project is determined to be 1560 kW. A 1560 kW generator set was selected. The total annual power generation of the project is 868.5 ×10^4 kWh.

4.4 *Amount of biogas power generation on the Internet*

According to the above calculations, the daily power generation capacity of the generating unit in the heating season of this project is 22080.7 kWh, and the daily power generation capacity of the generator set in the non-heating season is 24635.6 kWh; the daily power consumption of this project is 17133 kWh. Therefore, the daily internet power available in this project during the heating season is 4948.7 kWh, and the daily internet available during the non-heating season is 7502.6 kWh.

The total annual on-grid power generation capacity of this project is 243.19×10^4 kWh/a.

Table 1. Energy information of the anaerobic digestion plant.

Processes	Data
Biogas generation	13754 m^3/d
Biogas consumption	2714 m^3/d
Biogas power generation	868.5×10^4 kWh
Biogas power generation on the Internet	7502.6 kWh
Total amount of on-grid power generation capacity	243×10^4 kWh/a

5 CONCLUSION

The anaerobic fermentation technology is relatively advanced and reliable through the analysis of the existing food waste treatment routes. For a practical application of anaerobic digestion of food waste, the methane production amount is 7565 m^3/d. The amount of biogas produced is 13754 m^3/d with a methane percentage of 55%, and about 16.75 t/d is calculated according to the specific gravity of mixing. It is calculated that the digest gas consumption of the steam boiler is 1436 m^3/d in summer and 2714 m^3/d in winter. The total annual on-grid power generation capacity of this project is 243.19$\times$10^4 kWh/a.

REFERENCES

Karmee S and Lin K 2014 *Theor. Chem. Acc.* **2** 4
Karmee S, 2016 *Renew Sustain. Energy Rev.* **53** 945
Li Y 2016 *Procedia Environ. Sci.*, **31** 40
Pöschl M, Ward S and Owende P, 2010 *Appl. Energy.* **87** 3305
Wen Z, Wang Y and Clercq D 2015 *J. Clean.* **118** 88
Zhang C, Su H, Baeyens J and Tan T 2014 *Renew. Sustain. Energy Rev.* **38** 383

*Advances in Energy, Environment and
Chemical Engineering – Abdullah & Osman (Eds)
© 2023 The Author(s), ISBN: 978-1-032-36083-6*

Evaluation of favorable areas for hydrocarbon accumulation in W oilfield based on geographic information system

Heng Wang*

PetroChina Liaohe Oilfield Company, Panjin, Liaoning, China

ABSTRACT: W oilfield is rich in oil/gas resources. The oilfield is in a stage of high-level oil/gas exploration and development, but oil/gas production has decreased year by year. In order to achieve the goal of increasing reserves and improving the proven rate of oil/gas resources, it is urgent to reevaluate favorable areas for oil/gas accumulation. Based on this, taking the Paleogene Shahejie Formation of H oilfield as an example, combined with seismic, logging and analysis, and test data, this paper uses the fuzzy set theory and the data superposition and data fusion method of GIS on the ArcGIS platform to evaluate the dominant factors of hydrocarbon accumulation in the study area and predict the distribution of favorable areas of hydrocarbon accumulation. The results show that the source rock conditions of Paleogene Shahejie Formation in the W oilfield are rich, with a total organic carbon content of 0.01% ~2.31%, asphalt A content of 0.0141%~0.3167%, and total hydrocarbon content of $151\mu g/g$~$2221\mu g/g$. The sedimentary facies are transitional from Delta to fluvial facies. The reservoir lithology is mainly gray and light gray argillaceous coarse siltstone and calcareous fine sandstone, with an average porosity of 11.7% and an average permeability of 25mD. The average thickness of regional caprock in the development area is about 25m. Faults are developed, most of which are oil-source faults, which can be used as a good channel for vertical migration of oil/gas. The corresponding membership functions are set for four different variables, and the relevant data are fused by Gamma fuzzy integration method. It is concluded that the favorable hydrocarbon accumulation area of Shahejie Formation in the W oilfield is about 115 km^2, and the secondary favorable reservoir forming area is about 42 km^2. The prediction results help guide the next geological exploration direction of the oilfield.

1   INTRODUCTION

The essence of the prediction and evaluation of favorable areas for hydrocarbon accumulation is to comprehensively consider the control of source rocks, reservoirs, caprocks, faults, and other factors on oil/gas, find out the coincidence position of the favorable area in each factor, and finally obtain the formation and distribution of favorable target areas. Although this method can quickly form decision-making objectives for geological exploration results, it has some shortcomings. The first is the accurate positioning of spatial position. Different maps are formed in the geological thinking of different interpreters. Therefore, it is difficult to unify in the same coordinate system and can only form a general direction. Secondly, there is uncertainty in different map data. For example, when classifying favorable reservoirs, although different levels have different understandings, its scope will be determined once defined. This is an either this or that division method, and the result may differ from the actual condition (Naghadehi et al. 2014). Different experts and scholars have put forward different solutions to this problem, and the fuzzy logic method is one of the important methods (Khaleghi et al. 2013; Lisa et al. 2012). In the process of mineral resources exploration,

---

*Corresponding Author: wangheng2@petrochina.com.cn

 DOI 10.1201/9781003330165-48

both subjective data and objective data are formed. The fuzzy logic method can better integrate these two kinds of data (Libor 2008; Warick et al. 2003).

W oilfield is an inherited anticline structural belt developed under the background of buried hill, which mainly develops two groups of fault systems: Northwest and northeast. Many sets of reservoirs, such as the Shahejie formation, Dongying Formation, Guantao Formation, and Minghuazhen Formation, were developed in the Paleogene of the region, which provided a favorable place for hydrocarbon accumulation. The buried depth of oil reservoirs in the W oilfield is 1000m~4300m. The oil layer is vertically distributed in multiple oil-bearing series and overlapped and connected horizontally, thus forming the current scale. Currently, the W oilfield is in the high-level stage of development as a whole. The annual newly added proved reserves are far from meeting the needs of stable production, and the resource reserves are seriously insufficient. Based on this, this paper uses ArcGIS 10.2 platform and combines GIS with fuzzy logic theory to comprehensively evaluate and predict the favorable areas of hydrocarbon accumulation in the study area. The prediction results are expected to provide a theoretical basis for the next exploration and development of the oilfield

2 MATERIALS AND METHODS

2.1 *Materials*

The data required for evaluating favorable areas for hydrocarbon accumulation include the geochemical characteristics of source rocks, sedimentary facies and reservoir thickness, caprock thickness, fault distribution, oil/gas reservoirs distribution, etc.

2.2 *Methods*

The membership function is the basis of fuzzy logic decision-making. After data preprocessing, the membership degree of each grid of each graph can be calculated, so each graph can be regarded as a fuzzy set (Ikonomopoulos et al. 1993; Milenova 1997). Currently, the commonly used membership functions include Gaussian, increasing, and decreasing functions (Figure 1).

The basic principle of data fusion using GIS and fuzzy logic method is to generate fuzzy membership degree from the spatial data related to generation, storage, and cover according to the method of fuzzy data, so that each map can be regarded as a fuzzy set, and then use the AND, OR, SUM, PRODUCT or GAMMA operation method of the fuzzy set to combine all fuzzy sets in the same area to obtain the fuzzy set of favorable area (Ping et al. 1991), and finally divide the fuzzy set, and divide the target area into different levels. The processing flow is shown in Figure 2.

3 RESULTS AND DISCUSSION

W oilfield has abundant oil source conditions. W oilfield develops two sets of source rocks of Es_3 member and Es_1 member. The kerogen type is mainly the type I-II, with high hydrocarbon generation potential. Among them, the source rock of Es_3 member is dark mudstone of deep lake facies, with a total organic carbon content of 0.28%~2.31% and asphalt A content of 0.04%~0.3167%, total hydrocarbon content of $587\mu g/g$~$2221\mu g/g$. The source rock of Sha-1 member is also lacustrine dark mudstone, with a total organic carbon content of 0.1%~0.87% and asphalt A content of 0.0141%~0.1246%, total hydrocarbon content of $151\mu g/g$ ~$676\mu g/g$. The sedimentary facies from Es_3 member to Dongying Formation in W oilfield gradually transition from Delta facies to fluvial facies. The upper member of Es_3 is dominated by Delta, shallow lake, and deep to semi-deep lake deposits. The Delta is generally distributed in the southwest-northeast direction. The second member of the Shahejie Formation is dominated by delta deposits, and shore shallow lake deposits are locally developed. The Delta is distributed in the southwest-northeast direction. The lower member of Es_1 is dominated by delta and shore shallow lake sediments, with clastic rock

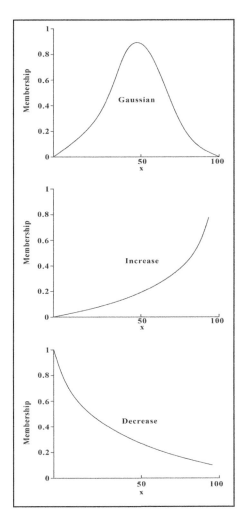

Figure 1.　Membership function model.

beach bar developed, and the beach bar trend is southwest and northeast. The upper member of the Es₁ member is mainly deposited by the meandering river and delta plain, and the river trend is generally SW-NE. The Dongying Formation is dominated by meandering river deposits, and the river trend is generally SW-NE. Under the influence of sedimentary facies control and late diagenesis, the thickness of the reservoir is 10m~110m, with an average of 85m. The lithology is mainly gray and light gray argillaceous coarse siltstone and calcareous fine sandstone. The porosity is 8%~17.0%, with an average of 11.7%, and the permeability is 10mD~38.1mD, with an average of 25mD. Regional mudstone caprock is developed in the study area, with a thickness of 10m~40m and an average of 25m. Faults are developed on the plane. The Western fault strikes northeast and tends to Southeast, with a fault distance of about 50m and an extension of about 14km. The fault is secondary to several small faults in the north, northeast, and southeast directions. The central fault strikes northeast and tends to the northwest, with a fault distance of about 200m and an extension of more than 20km.

The degree of membership is the basis of fuzzy decision-making. After data preprocessing, the membership degree of each grid of each graph can be calculated, so each graph can be regarded as

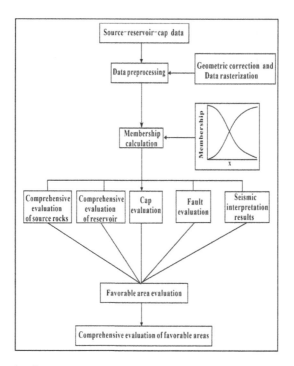

Figure 2.   Data processing flow.

a fuzzy set. Several types of membership functions widely used at present have been summarized above. Different membership functions are selected for different elements. If the larger the original value is, the more unfavorable it is to the favorable target, and the decreasing membership function is adopted. If the smaller the original value is, the more favorable the target is, and the increasing membership function is adopted. If the favorable target increases first and then decreases with the increase of the original value, the Gaussian membership function is used. In the above, we selected four factors, source rock, reservoir, caprock, and fault, to evaluate the favorable area of hydrocarbon accumulation. According to the characteristics of these four elements and their relationship with the distribution of oil/gas reservoirs, different types of membership functions can be selected for different elements. For example, the increased large membership function can be selected for hydrocarbon expulsion intensity of source rock and caprock thickness. The Gaussian membership function is selected for the dominant sedimentary facies, and the decrease membership function is selected for the distance from the fault (Table 1). AND is the most rigorous evaluation method; that is, the best results can be achieved when all elements are the best simultaneously. OR is the most relaxed evaluation method; that is, one of the elements is the best. In the selection of gas-bearing target areas, source rocks, reservoirs, caprocks, and fracture density all have an important impact on the results. Therefore, OR operation is not feasible. Still, if AND operator is adopted, some regions of interest may be omitted due to too strict and considering the uncertainty of the original data. The result of SUM operation is larger than that of OR, and the result of PRODUCT operation is smaller than that of AND operation, while GAMMA combines the situation of SUM and PRODUCT, ånd the results of different values are different. Based on this consideration, this paper adopts the GAMMA operation method. According to the above membership function, the fuzzy sets of four different elements can be obtained, and the GAMMA operator is used for fuzzy fusion. The larger the calculated membership value is, the more favorable the target area is. As shown in Figure 3, the assignment range of reservoir distribution probability under the control of a single element is from 0 to 1. When different elements are superimposed, the distribution probability of oil/gas control of each functional element needs to be weighted average to obtain the membership degree of favorable hydrocarbon accumulation areas. The membership degree greater than 0.4 is selected as the favorable area for hydrocarbon accumulation. Those with membership degrees greater than 0.2 and less than 0.4 are selected as secondary favorable areas for hydrocarbon accumulation. The results show that the favorable areas in the study area are mainly concentrated in the middle of the exploration area, with an area of about 115 km². The secondary favorable areas are distributed around the exploration area, with an area of about 42 km².

Table 1.   Selection of membership functions of different elements.

Order number	Controlling factors	Membership function
1	Source rock	Increase
2	Sedimentary facies	Gaussian
3	Cap	Increase
4	Falut	Decrease

The prediction results show that the range of favorable areas for oil and gas exploration in the W oilfield has increased, and the area of favorable areas has increased by 35% compared with the previous exploration range. The data fusion method based on fuzzy logic overcomes the binary logic phenomenon of traditional methods. Using this method is very positive for W oilfield to expand the exploration scope of favorable areas for hydrocarbon accumulation in Paleogene strata. For high-level oil/gas exploration areas, this method effectively improves the potential reserves of oil and gas. However, the research also has some shortcomings and needs to be improved. One is the definition of membership functions of different variables. The membership function is different, the membership degree is also different, and the result must have an impact. Secondly, the synthesis method of fuzzy sets also has a great impact on the results: on the one hand, the results of different

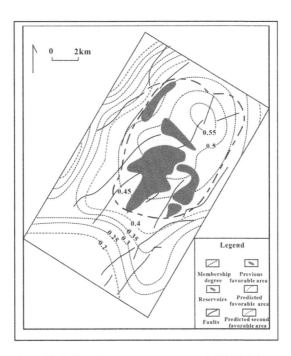

Figure 3.   Prediction of favorable hydrocarbon accumulation areas in W Oilfield.

synthesis methods (fuzzy addition, fuzzy and, fuzzy product, fuzzy algebraic sum, and gamma) are different; on the other hand, even if the same synthesis method, such as GAMMA synthesis method, has different values, its results are also different. Currently, there is no fixed standard for the determination of membership function or the selection of fuzzy set synthesis method, resulting in the uncertainty of new results. Determining the membership function and a more reasonable synthesis method of the fuzzy set needs to be combined with practice.

4 CONCLUSIONS

By collecting the existing geological, geophysical, and other relevant data of W oilfield, prepro-cessing them and integrating them into the ArcGIS platform, and then using the spatial analysis function of ArcGIS, especially the superposition analysis function, to integrate the results of source rock data, sedimentary facies data, cap rock data and fracture analysis, the area of favorable oil and gas accumulation area of W oilfield is calculated, and the following two points are obtained. (1) GIS has a powerful spatial analysis function. In particular, the combination of GIS and fuzzy logic can give full play to the powerful graphic decision-making function of GIS and combine the characteristics of geologists' fuzzy thinking. The reasoning is more reasonable, and the result aligns with reality. (2) W oilfield is rich in Paleogene oil/gas resources and has a wide exploration range in favorable areas. Among them, the favorable hydrocarbon accumulation area is mainly located in the middle of the exploration area, with an area of about 115 km^2. The secondary favor-able hydrocarbon accumulation area is mainly located around the exploration area, with an area of about 42 km^2. The area of favorable areas has increased by 35% compared with the previous exploration range. The combination of GIS and fuzzy algorithm has certain guiding significance for the exploration direction of oil fields with high exploration degrees.

ACKNOWLEDGMENT

This work was financially supported by the National Natural Science Foundation of China and Major science and technology projects of CNPC.

REFERENCES

Ikonomopoulos, A., Tsoukalas, L.H., Uhrig, R.E. (1993) Integration of neural networks with fuzzy reasoning for measuring operational parameters in a nuclear reactor[J]. *Nuclear Technology*, 104:1(1):1–12.
Khaleghi, B., Khamis, A., Karray, F.O. (2013). Multisensor data fusion: A review of the state-of-the-art[J]. *Information Fusion*. 14(1):28–44.
Libor, B. (2008) On the difference between traditional and deductive fuzzy logic[J]. *Fuzzy Sets and Systems*. 159(10):1153–1164.
Lisa, B., Raul, Z., Alejandro, E. (2012) Geographic information system–based fuzzy-logic analysis for petroleum exploration with a case study of northern South America[J]. *AAPG Bulletin*. 96(11):2121–2142.
Milenova, B. (1997) Fuzzy and neural approaches in engineering: By Lefteri H. Tsoukalas and Robert E. Uhrig, John Wiley & Sons, New York: 1997, 587 pp, ISBN 0-471-16003-2[J]. *Neural Networks*, 10(9):1740–1741.
Naghadehi, K.M., Hezarkhani, A., Honarpazhouh, J. (2014) Integration multisource data for mineral exploration using fuzzy logic, case study: Taknar deposit of Iran[J]. *Arabian Journal of Geosciences*. 7(8):3227–3241.
Ping, A., Moon, W.M., Rencz, A. (1991) Application of fuzzy set theory to integrated mineral exploration[J], *Canadian journal of exploration geophysics*, 27(1):1–11.
Warick, B., David, G., Tamas, G. (2003) Use of fuzzy membership input layers to combine deductive geolog-ical knowledge and empirical data in a neural network method for mineral-potential mapping[J]. *Natural Resources Research*. 12(3):183–200.

Advances in Energy, Environment and
Chemical Engineering – Abdullah & Osman (Eds)
© 2023 The Author(s), ISBN: 978-1-032-36083-6

Research on the characteristics of water plant sludge and its application in sewage treatment

Yiyang Liu & Shoubin Zhang*
University of Jinan, Jinan, P.R. China

Jingying Chen
Shandong Jinnuo Construction Project Management Co., Ltd, Qingdao, P.R. China

Shikai Zhao
Shandong Industry Ceramics Research and Design Institute, Zibo, P.R. China

Wenhai Jiao
Jinan Municipal Engineering Design & Research Institute (Group) Co., Ltd, Jinan, P.R. China

ABSTRACT: In the process of water treatment, a large amount of sludge will be generated in the water treatment plant. Resource utilization of water plant sludge is an effective way to treat it. In this paper, the production process, composition, and physical and chemical properties of water plant sludge are introduced in detail. At the same time, combined with the research on the resource utilization of sludge from water supply plants in sewage treatment in recent years, the research progress of water plant sludge in terms of adsorbents, constructed wetland substrates, and coagulants are systematically summarized. The study can provide a reference in application in sewage treatment.

1 INTRODUCTION

After the water is purified by the water supply plant, it will produce a large amount of water supply plant sludge while producing drinking water that meets the water quality standards. The sludge of the water supply plant contains a large amount of aluminum and iron elements, and the internal pore size is rich, which has a good adsorption capacity for pollutants and has obvious advantages in sewage treatment (Sotero-Santos et al. 2007). In recent years, domestic and foreign research on the application of sludge in water plants in sewage treatment has gradually increased, and a lot of results have been achieved. This paper summarizes the different utilization methods of sludge in sewage treatment in water supply plants and provides new ideas for its popularization and application in sewage treatment (Zhou &Haynes 2011).

2 SOURCES AND PROPERTIES OF WATER PLANTS SLUDGE

2.1 *Sources of water plant sludge*

Most of the water supply plants use surface water as raw water for treatment, and the main processes are coagulation, sedimentation, filtration, and disinfection. Among them, the coagulant can

---

*Corresponding Author: zhangshoubin1980@vip.163.com

destabilize the tiny suspended solids and colloidal particles in the water through electrical neutralization, adsorption bridging and net sweeping, etc., and aggregate each other into larger flocs, which are then precipitated and removed under the action of gravity. A small number of flocs with poor sedimentation performance are retained and removed by the filter material layer in the filter tank. The sludge produced by the sludge discharge in the sedimentation tank and the backwash water in the filter tank after collection and treatment is the water plant sludge. Due to the difference in raw water quality, the types and dosages of chemicals used in tap water treatment are different, and the composition and output of water plant sludge are also different. The statistical results of sludge production in some water plants in China are shown in Table 1 (Wang & He 2002).

Table 1. Statistics of sludge production in some water supply plants in China.

Water plant	Water production ($10^4 m^3$/d)	Dry sludge yield (t/d)	Dosage (mg/L) Aluminum salt	Iron salts
Beijing No. 9 Water Plant	150	39	1	1
Shanghai Minxing Water Plant	7.2	12	3.8	
Shijiazhuang Runshi Water Plant	30	12		
Shenzhen Meilin Water Plant	60	50	1.6	
Guangzhou Xizhou Water Plant	50	74	3.3	9.6
Baoding Second Water Plant	26	12		
Fuzhou Water Plant	60	42.6	19	

Note: "–" in the table denotes no statistical data

In the actual process, the filter washing wastewater after sedimentation in the drainage tank enters the concentration tank together with the sludge in the sludge drainage tank, and the water content of the sludge is reduced by gravity concentration in the concentration tank. In order to make the sludge easier to dewater, sludge pretreatment must be carried out before dewatering to reduce the specific resistance of the sludge. The conventional sludge treatment process in the water supply plant is shown in Figure 1 (Cheng 2016).

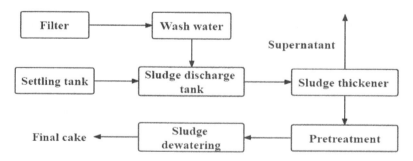

Figure 1. Conventional treatment process of water plant sludge.

2.2 Physicochemical properties of water plant sludge

The material form of sludge in water supply plants is generally brownish-gray viscous flocs and contains odor. Table 2 shows the nutrients in the conventional water plant sludge (Environmental Protection Agency 2006).

Table 2. Nutrient components in water plants sludge.

Nutrients	Range	Nutrients	Range
Total solids content /%	8.1~81.0	Organic nitrogen /%	0.752(0.399)
Volatile solid content /%	9.3~29.1	Ammonia nitrogen /%	0.016(0.016)
Electrical conductivity(μs/cm)	563.8(530.2)	Nitrate nitrogen	0.003(0.003)
pH	7.0~8.8	Total phosphorus /%	0.228(0.248)
Kjeldahl nitrogen /%	0.495(0.256)	Total potassium /%	0.225(0.317)

The main components of water supply plant sludge include soil particles, metal hydroxides, humus, and a small number of algae and bacteria, mainly inorganic components, and less organic components (Guan et al. 2005; Leader et al. 2008). The main inorganic ions are Al, Fe, Ca, and Mg, of which a large amount of Al and Fe are mainly introduced into the sludge flocs by the coagulant added by the water plant along with the sedimentation. The main element content and phase composition of the water plant sludge are shown in Figure 2 (Chen 2017).

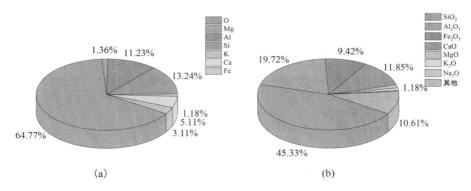

Figure 2. Main element content and phase composition of water plant sludge.

3 APPLICATION OF WATER PLANT SLUDGE IN SEWAGE TREATMENT

3.1 *As adsorbent for sewage treatment*

Zhao tested the aluminum sludge after drying in the Irish waterworks, and the results showed that aluminum accounted for 16.4% to 43% of the dry weight of the sludge, and its existence was mostly amorphous. Amorphous aluminum ions have larger porosity and specific surface area than crystalline aluminum ions (Zhao et al. 2010). Hovsepyan et al. used the sludge of a water supply plant as the adsorbent for mercury in their experiments (Hovsepyan & Bonzongo 2009). The maximum adsorption capacity of mercury measured by the Langmuir model was 79 mg/g, which was in line with the first-level model of adsorption kinetics. The study showed that the sludge of the water supply plant could effectively remove mercury from water. Makris et al. conducted experiments using sludge from a water supply plant as an adsorbent for perchlorate (2004). The study showed that the maximum removal rate of perchlorate was 76% after 24 h, and finally, perchlorate was degraded to chloride.

The water plant sludge contains a large amount of Ca^{2+}, Fe^{3+}, Al^{3+}, and other ions. These ions can combine with PO_4^{3-} when hydrolyzed to form a precipitate on the surface of the sludge, which leads to the adsorption and removal of phosphorus. Zifu Li et al. studied the phosphorus adsorption capacity of different types of sludge from five water plants in different cities (Li et al. 2013).

The phosphorus adsorption test showed that all types of water supply sludge had good adsorption capacity for phosphorus. However, the sludge of the water supply plant often contains a large number of particles and organic matter, which makes the effective phosphorus removal adsorption sites on the sludge of the water supply plant less, which affects its adsorption capacity. Therefore, it is often necessary to properly modify the water plant sludge to change its physical and chemical properties, increase the effective adsorption sites, and improve the adsorption performance of phosphorus. Currently, the common modification methods include thermal modification, acid modification, and modification by adding metal elements.

Li Yibing et al. used sludge from a water supply plant to treat phosphorus-containing wastewater (Li et al. 2018). The study showed that under the optimal experimental conditions, the phosphorus removal rate was 90.93%, and the adsorption capacity was 0.60 mg/g. The adsorption kinetics conformed to the Lagergren pseudo-second-order kinetic model. Wang Xin et al. studied the removal effect and mechanism of phosphorus by using sludge from a water supply plant loaded with iron compounds as adsorption material (Wang et al. 2016). The results show that the phosphorus removal rate is inversely proportional to the initial concentration of phosphorus solution. Due to the saturation of the adsorption sites of the adsorbent, the phosphorus adsorption curve gradually tends to be stable. The test results show that the conditions for the complete removal of phosphorus in the solution are that the phosphorus concentration is less than or equal to 10 mg/L, and the dosage is 2.5 g/L. Niu Yafen et al. carried out adsorption experiments on acid-modified and thermally-modified water plant sludge (Niu 2019). The results show that the adsorption capacity can reach 3.79 mg/g using 1 mol/L sulfuric acid to modify the water plant sludge. The sludge adsorption capacity of the water supply plant after high-temperature treatment at 900°C can reach 6.22mg/g. The adsorption process conformed to the pseudo-second-order kinetic equation, which was more favorable for the adsorption of phosphorus under acidic conditions.

3.2 *As constructed wetland substrate*

Using water supply plant sludge as the substrate of constructed wetlands is low-cost, not only can remove COD and SS, but also has a good removal effect on nitrogen and phosphorus in sewage.

Yuansheng Hu et al. used water plant sludge as a constructed wetland matrix filler to treat sewage (Hu et al. 2014). When the nitrogen load is $46.7g \cdot Nm^{-2} \cdot d^{-1}$, the removal rate of total nitrogen can reach 90% after 280d. It shows that the sludge from the water supply plant is used as the filler matrix of the constructed wetland and has a good removal effect on nitrogen in the sewage. Cao Jiashun et al. compared water supply plant sludge with common filter sand and explored its phosphorus removal performance through static adsorption experiments and dynamic phosphorus removal experiments (Cao et al. 2015). Studies have shown that the phosphorus absorption capacity of water supply plant sludge is much higher than that of sand, and the saturated phosphorus adsorption capacity can reach 1.273 mg/g. Whether used alone or in combination with other filter media, water supply sludge can significantly improve the phosphorus removal capacity of wetlands and prolong the service life of wetlands.

Gu Pengfei used the ceramsite fired from the sludge of the water supply plant for tidal flow constructed wetlands, aerated biological filters, and submerged flow constructed wetlands to treat rural sewage and micro-polluted river water (Gu 2017). The processing results are shown in Table 3. It can be seen from the treatment results that the ceramsite fired from the sludge of the water supply plant has a good treatment effect on nitrogen, phosphorus, and COD in the sewage.

3.3 *As coagulant for sewage treatment*

Basibuyuk et al. used iron-containing waterworks sludge as a coagulant to treat vegetable oil wastewater (Basibuyuk & Kalat 2004). Studies have shown that sludge and alum have good removal effects on grease, COD, and suspended solids. The combined use of 12.5mg/l ferric chloride and 1000mg/l sludge can achieve a 99% removal rate of oil and total suspended solids and an 83% COD removal rate. Siriprapha et al. used sludge from a water supply plant as a coagulant to treat

Table 3. The effect of water plant sludge ceramsite as filler on sewage treatment.

Treatment process	Sewage type	NH3-H Removal rate	TN Removal rate	TP Removal rate	COD Removal rate
Tidal flow constructed wetland	Rural sewage	72.1±7.46%	61.01±6.75%	69.77±9.24%	–
Biological aerated filter	Slightly polluted river water	96.32%	37.70%	–	80%
Subsurface flow constructed wetland	Rural sewage	26.33%	39.35%	51.81%	47.40%

surfactant wastewater (Siriprapha et al. 2011). Studies have shown that the combined use of water plant sludge and alum can remove 80% of the surfactant in wastewater, far exceeding the effect of alum treatment alone.

4 CONCLUSION

The special structure and properties of the water plant sludge make it possible to use it in various fields, especially in the field of water environment treatment. However, most of the practical application of sludge in water supply plants is still in the laboratory stage, the practical application of engineering is limited, and the future research space is very large. In short, the resource utilization of sludge will become the mainstream direction of sludge disposal in the future.

ACKNOWLEDGMENTS

This work was financially supported by the Key R&D Program of Shandong Province (2017GSF17105).

REFERENCES

Basibuyuk M, Kalat D G. (2004) The use of water works sludge for the treatment of vegetable oil refinery industry wastewater. *Environ Technol*, 25(3):373–380.
Environmental Protection Agency. Office of research and Development. (2006) *Technology transfer handbook: management of water treatment plant residuals*. 36(4): 1334–1336.
Guan X H, Chen G H, Shang C. (2005) Reuse of water treatment works sludge to enhance particulate pollutant removal from sewage. *Water Research*, 39(15): 0-3440.
Hovsepyan A, Bonzongo J C J. (2009) Aluminum drinking water treatment residuals (Al-WTRs) as sorbent for mercury: Implications for soil remediation. *Journal of Hazardous Materials*, 164(1): 73–80.
Jiashun Cao, Wenjie Zhu, Chao Li, et al. (2015) The Research of Phosphate Removal by Using Water Treatment Residuals as Filter Materials in Constructed Wetlands. *Environmental Science and Technology*, 28(05): 16–20.
Leader J W, Dunne E J, Reddy K R. (2008) Phosphorus sorbing materials: Sorption dynamics and physicochemical characteristics. *Journal of Environmental Quality*, 37(1): 174–181.
Makris K C, El-Shall H, Harris W G, et al. (2004) Intraparticle phosphorus diffusion in a drinking water treatment residual at room temperature. *Journal of Colloid & Interface Science*, 277(2): 417–423.
Pengfei Gu. (2017) *Research on New Type of Water Plant Sludge Carrier Application in Sewage Treatment*. Lanzhou University of Technology.
Qinhua Wang, Junlan He. (2002) Determination of sludge yield in water purification plant and selection of related parameters. *China Water & Wastewater*. 08:64–66.

Siriprapha J, Sinchai K, Suwapee T, et al.(2011) Evaluation of reusing alum sludge for the coagulation of industrial wastewater containing mixed anionic surfactants. *J Environ Sci*, 23(4): 587–594.

Sotero-Santos R B, Rocha O, Povinelli J. (2007) Toxicity of ferric chloride sludge to aquatic organisms. *Chemosphere*, 68(4): 628–636.

Xin Wang, Xiaoyu Ma, Wen Zhou, et al. (2016) Water supply sludge loaded iron compound on behavioral effect and functional mechanism of phosphorus removal. *Chinese Journal of Environmental Engineering*, 10(10):5420–5428.

XueLi Cheng. (2016) *Utilization of sludge to water resources*. Xi'an University of Architecture and Technology.

Yafen Niu. (2019) *Study on adsorption of phosphorus in water using waterworks sludge and the synthesized sludge ceramsite*. North China University of Water Resources and Electric Power.

Yang Chen. (2017) *Study on Preparation Ceramsite in Water Supply Sludge and its Application of Phosphorus Remove*. Shandong University.

Yibing Li, Ruiqi Hu, Yanping Zhang, et al.(2018) Research on the adsorption capability of aluminum-containing sludge for the phosphorus-containing wastewater in waterworks. *Industrial Water Treatment*, 38(05):30–34.

Yuansheng Hu, Yaqian Zhao, Anna Rymszewicz.(2014) Robust biological nitrogen removal by creating multiple tides in a single bed tidal flow constructed wetland. *Science of The Total Environment*,470–471,1197–1204.

Zhao Y Q, Babatunde A O, Hu Y S, et al. (2010) A Two-Prong Approach of Beneficial Reuse of Alum Sludge in Engineered Wetland: First Experience from Ireland. *Waste and Biomass Valorization*, 1(2): 227–234.

Zhou Y F, Haynes R J. (2011) Removal of Pb(II), Cr(III) and Cr(VI) from Aqueous Solutions Using Alum-Derived Water Treatment Sludge. *Water Air Soil Pollut*, 215(1-4):631–643.

Zifu Li, Nan Jiang, Fengfu Wu, et al. (2013) Experimental investigation of phosphorus adsorption capacity of the waterworks sludges from five cities in China. *Ecological Engineering*, 53: 165–172.

Advances in Energy, Environment and
Chemical Engineering – Abdullah & Osman (Eds)
© 2023 The Author(s), ISBN: 978-1-032-36083-6

Research on road performance and dynamic mechanical properties of rubber/SBS composite modified asphalt mixture

Pengcheng Sun
Zhejiang Institute of Transportation Science, Hangzhou, China
Zhejiang Key Laboratory of Inspection and Maintenance Technology for Highway and Bridge, Hangzhou,
China

Chunmei Pan
Zhejiang Communications Investment Expressway Operation Management Co, Ltd, Hangzhou, China

Fengxia Chi*
Zhejiang Institute of Transportation Science, Hangzhou, China
Zhejiang Key Laboratory of Inspection and Maintenance Technology for Highway and Bridge, Hangzhou,
China

Zheng Li
Zhejiang Communications Investment Expressway Operation Management Co, Ltd, Hangzhou, China

Chenchen Zhang & Yihan Sun
Zhejiang Institute of Transportation Science, Hangzhou, China
Zhejiang Key Laboratory of Inspection and Maintenance Technology for Highway and Bridge, Hangzhou,
China

ABSTRACT: Asphalt pavement is prone to various diseases under the influence of external factors for a long period of time, and rubber/SBS composite modified asphalt can effectively improve the quality of the mixture and prolong the service life of the road. Through the designed gradation of SMA-13, the result is compared with SBS-modified asphalt from the perspective of road performance and dynamic mechanics. The experimental results show that the rubber/SBS composite modified asphalt displayed several physical properties such as having good high-temperature performance, high viscosity, and being easy to store. Other properties like the high-temperature deformation resistance and water damage resistance achieved a good result when the oil-stone ratio reached 6.6%. The viscoelastic properties and high-temperature deformation resistance of the mixture were evaluated from the perspective of dynamic mechanics. The research results provide a theoretical reference for improving the durability of the pavement and the quality of road use, as well as the popularization and application of rubber/SBS composite modified asphalt.

1 INTRODUCTION

The general solution to reduce road surface diseases such as low durability, high prone to rutting, and high-water damage is the asphalt modification method (Tang et al. 2020), and the SBS composite modified method is currently the most common method implemented. Compared with the traditional modified asphalt, the SBS modified asphalt can significantly improve surface rutting, water damage, and other road surface-related disease. It can improve the road surface durability and is one of the best asphalt materials.

---

*Corresponding Author: 811315060@qq.com

The utilization of the asphalt rubber modification to improve the properties of high mixture materials has been the priority of road engineering research (Chen et al. 2018). Chen et al. (2016) prepared rubber-modified asphalt by pretreatment with rubber powder furfural extraction oil, which significantly improved the high- and low-temperature performance and aging resistance of the mixture. Huang et al. (2001) discovered the absorption of oil increases the viscosity of asphalt when analyzing the swelling phenomenon of rubber powder, which has a significant role in the temperature-sensitivity performance of asphalt materials and the durability of the mixture. Yang et al. (2009) concluded that rubber asphalt can still be in the liquid-solid phase under high-temperature conditions, so the ratio of liquid and solid phase properties can be adjusted by controlling the swelling of rubber crumbs (Liu et al. 2010), thus improving the performance of asphalt mixes.

The effect of the single modification of rubber asphalt is more often studied and is directed at evaluating the performance of the asphalt itself. In this paper, a comparative study of the road performance of two modified asphalts is carried out by designing SMA-13 asphalt mixes. At the same time, its importance for improving pavement durability is evaluated from a dynamic mechanical perspective (Ji et al. 2018).

2 MATERIALS

2.1 Asphalt

The rubber/SBS composite modified asphalt and SBS modified asphalt used in this paper are from a bitumen company from Jiang Su province, China. According to the test specifications for asphalt materials, the measured performance of the two modified asphalts is shown in Table 1.

Table 1. Asphalt performance index.

Experimental objectives	Rubber/SBS composite modified asphalt	SBS modified asphalt
Softening point/°C	90.0	76.3
Cone penetration/0.1mm	48.0	58.5
5°C ductility/cm	31.1	34.6
60°C dynamic viscosity/10^4Pa·s	23.1	5.8
Segregation/(48h,163°CC)	1.6	3.4
PG grade	94-28	78-16

Based on Table 1, the rubber/SBS composite modified asphalt has a wide range of temperature domains and applicable temperature range, with a softening point of 90 C. At the same time, its 60 C dynamic viscosity is also relatively high, which displayed a better resistance to shear deformation. The level of asphalt viscosity directly affects the adhesion effect between it and the stone, which produces a thin film on the stone, and the greater the thickness, the better the adhesion (Joseph et al. 2021). At the same time, asphalt viscosity can effectively improve the resistance of the mixture to water damage (Wang et al. 2017).

2.2 Aggregate

In this paper, the coarse aggregate selected is basalt, the fine aggregate selected is limestone, and the main index of each material is shown in Table 2.

2.3 Fiber

Fiber can improve the practicality of the high mixture of composite materials, and this paper used plant fiber (external 0.3%) with the index requirements shown in Table 3.

Table 2. Aggregate technical index.

Class	Technical Specification	Test result	Technical requirements	Experimental method
Coarse aggregate	Los Angeles wear value/%	10.3	≤ 22	T0317
	Apparent specific gravity/$(g \cdot cm^{-3})$	2.933	≥ 2.6	T0304
	Adhesivity/grade	5	≥ 4	T0616
	Water absorption%	1.7	≤ 2.0	T0307
Fine aggregate	Soil content/%	1.8	≤ 3	T0333
	Sand content/%	71	≥ 60	T0334
Mineral fines	Hydrophilic coefficient	0.58	≤ 1.0	T0353
	Plasticity index/%	3.3	≤ 4	T0351

Table 3. Technical index of lignin fiber.

Technology index	Test result	Technical requirement
Ash content/%	19.8	18 ± 5 No volatiles
pH value	7.2	7.5 ± 1.0
Oil absorption rate/%	5.6	≥ 5 times its own weight
Water content/%	1.3	<5.0

3 MIX DESIGN

3.1 Asphalt mixture grading determination

Based on the sieving results, three gradation levels of coarse, medium, and fine are designed, and the gradation design parameters are shown in Table 4.

Table 4. SMA-13 gradation design.

Sieve size/mm	Gradation type				
	Upper limit	Lower limit	Gradation 1	Gradation 2	Gradation 3
16	100	100	100.0	100.0	100.0
13.2	100	90	90.1	90.3	90.6
9.5	75	50	65.0	65.9	66.9
4.75	34	20	28.9	30.6	32.4
2.36	26	15	20.0	21.8	23.5
1.18	24	14	17.3	18.6	19.8
0.6	20	12	14.9	15.7	16.5
0.3	16	10	12.6	13.1	13.5
0.15	15	9	11.3	11.5	11.7
0.075	12	8	10.3	10.4	10.4

3.2 Determining the optimal oil-to-stone ratio

In the process of determining the optimal oil to stone ratio, the initial value of the oil to stone ratio was set to be 6.6%, and 6.3% for the SBS modified asphalt. With 0.3% as the design step, the Marshall volume indexes were evaluated under the conditions of 5.7%, 6.0%, 6.3%, 6.6% and 6.9% of oil to stone ratio, and the grade 2 was determined as the design grade based on pass experiences.

Table 5. Proportion of ore and material of SMA-13 asphalt mixture.

Mixture type	Proportion of minerals/%				
	9.5-16mm	4.75-9.5mm	0-2.36mm	mineral fines	Fiber
SMA-13	35	44	11	10	0.3

4 ROAD PERFORMANCE EVALUATION

This section compares the road performance of the two asphalt mixtures based on the aspects of performance index requirements and performance effects evaluated.

4.1 High-temperature performance

The experiment will be conducted under two different temperature settings (60 C and 70 C), with a wheel pressure of 0.7 MPa. The specimen forming size is set according to the JTG E20-2011 specification, and the test results are shown in Figure 1.

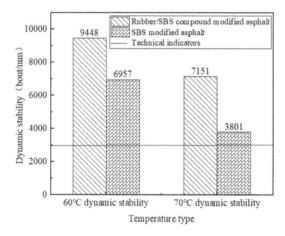

Figure 1. Dynamic stability of asphalt mixtures.

The results from Figure 1 show that the test results under both temperature conditions meet the index requirements. The dynamic stability of rubber/SBS composite modified asphalt is 9448 times/mm at 60 C, which is 35.8% compared to SBS modified asphalt, which indicates that the rubber/SBS modified asphalt is the better option to meet the road performance requirements. The test results at 70 C displayed an even higher improvement rate of 88.1%. The dynamic stability decreases when the test temperature increases. At 70 C, the value of the dynamic stability of the rubber/SBS composite materials is 7151 times/mm, which is 24.3% lower compared to the result at 60 C. Even so, the rubber/SBS composite modified asphalt mixture at high temperatures is not prone to plastic deformation and shows better temperature stability.

4.2 Low-temperature performance

The test temperature for the low-temperature trabecular test was -10 C during the test, the loading rate was 50mm/min, and the specimen size was set according to the test specification JTG E20-2011. The test results are shown in Figure 2.

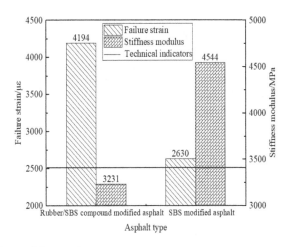

Figure 2.   Asphalt mixture failure strain.

In Figure 2, there is a large difference in the breaking strain between the two, with a difference of 1564 $\mu\varepsilon$. The breaking strain of the rubber/SBS composite modified asphalt mixture is $4194\mu\varepsilon$, which is 59.5% higher. The main reason for this phenomenon is that the rubber particles are more elastic. The elastic strain energy is large at low-temperature conditions, and in the process of compound modification with SBS, the spatial grid structure formed by the asphalt binder plays an important role in the low-temperature stability of the mixture. The stiffness modulus reflects the ability of the mix to resist deformation within the elastic limit, and SBS-modified asphalt mixes exhibit strong brittle fracture at -10 degrees, with small deformation, large modulus, and poor low-temperature crack resistance.

4.3 *Water stability*

This section will evaluate the water stability of asphalt mixtures based on the aspect of immersion damage and freeze-thaw condition. The test conditions are based on the specification of JTG E20-2011, and the result is shown in Figure 3.

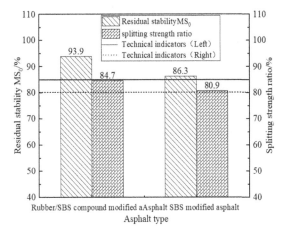

Figure 3.   Asphalt mixture water stability.

Based on the result from Figure 3, the rubber/SBS composite modified asphalt mixture displayed better water stability, with a residual stability MS_0 of 93.9%. In contrast, SBS-modified asphalt only achieved a result of 86.3%. The water resistance of SBS is low, which is not suitable for regions with a high intensity of precipitation. The rubber/SBS composite modified asphalt is mixed with rubber powder particles, which plays a good filling effect in the skeleton structure of the mixture embedded, making it difficult for rainwater to enter the internal structure of the mixture, and the resistance to water damage is enhanced.

5 THE RESEARCH ON THE DYNAMIC MECHANICAL PROPERTIES

This section evaluated the changes in viscoelastic properties of the mixture from the aspect of dynamic modulus, combined with the phase angle and the dynamic modulus master curve.

5.1 *Dynamic modulus and phase angle*

The experiment is based on the specification of JTG E30-2011 under six frequencies which are $-10°C$, $5°C$, $20°C$, $30°C$ and $40°C$. Figure 4 and Figure 5 displayed the experimental result of the dynamic modulus and phase angle.

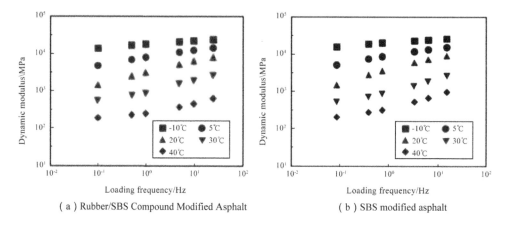

(a) Rubber/SBS Compound Modified Asphalt   (b) SBS modified asphalt

Figure 4.   Dynamic modulus of asphalt mixtures.

Based on Figure 4, when the loading frequency increases, the dynamic modulus displays linear displacement from low to high, with the rate decreasing gradually. It varies negatively with temperature; the higher the temperature, the smaller the modulus. Under low-temperature conditions, the dynamic modulus of SBS-modified asphalt is higher due to its poor elasticity, while the opposite is true under high-temperature conditions. As can be seen from Figure 5, in the high-temperature high-frequency state, the phase angle of rubber/SBS composite modified asphalt is larger. When the asphalt changes from the elastic state into a viscous state, the bonding force can be improved while strengthening the effect of deformation resistance at the same time, so that the mixture is not prone to elastic deformation.

5.2 *Dynamic modulus master curve*

Combined with the experimental results from section 5.1, the Sigmoid function model is used to describe the dynamic modulus master curve. According to the time-temperature equivalence principle, the master curve of the dynamic modulus of asphalt mixture is constructed with 20°C as the reference temperature, as shown in Figure 6.

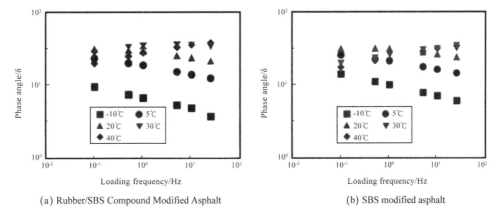

(a) Rubber/SBS Compound Modified Asphalt   (b) SBS modified asphalt

Figure 5.   Rubber/SBS composite modified asphalt mixture.

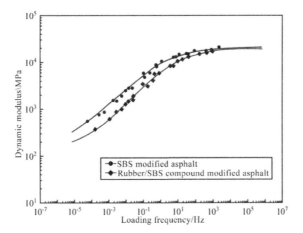

Figure 6.   Main dynamic modulus curves of asphalt mixtures.

Based on the result from Figure 6, the dynamic modulus of the mixes all changes positively with the shrinkage frequency, which indicates the enhanced deformation resistance of the mixes. When the mixture is in the low-frequency range, the better elastic recovery makes the dynamic modulus value grow linearly. In the high-frequency range, the dynamic modulus of the rubber/SBS composite modified asphalt exceeds the SBS modified asphalt, with an overall tendency of low moderate, which indicates the strengthening of the intermix embedding and the mixture skeleton structure has a greater impact on the mixture performance.

6   CONCLUSION

In conclusion, this paper studied the performance of rubber/SBS composite modified asphalt, designed the SMA-13 gradation, evaluated the road performance in terms of resistance to high and low temperature, water damage, and other indicators, and evaluated the resistance to defor-mation and changes in viscoelastic properties from a dynamic mechanical point of view. The main conclusions are as follows:

(1) The characteristics of rubber/SBS composite modified asphalt with high viscosity make it form a good coating effect with the stone material, which increases the adhesion and also enhances the shear deformation resistance of the mixture

(2) Combined with the designed SMA-13 mix, with its unique dense skeleton embedded structure makes, the coarse and fine aggregates are fully combined, significantly improving the overall strength of the mix. In addition, the rubber/SBS composite modified asphalt mixture can effectively reduce the permanent deformation of the pavement due to vehicle loading, and its low-temperature cracking and water damage resistance are also better than SBS modified asphalt, significantly improving the quality of road use.

(3) Dynamic mechanical properties show that the viscoelastic properties of rubber/SBS composite modified asphalt mixes have better elastic recovery ability at low temperatures and are less prone to cracking; at high temperatures. They are less prone to plastic deformation due to their good temperature sensitivity, which enhances the bonding effect of the mix and improves the quality of road use.

ACKNOWLEDGMENTS

This work was supported by the Science and Technology Planning Project of the Zhejiang Provincial Department of Transportation (2021013) and the 2021 Institute Support Project (ZK202101).

REFERENCES

Chen Meng, Zhang Tao. Research Progress of Modified Asphalt using Waste Rubber Powder in Road Engineering [J]. *Shanxi Communications Technology*, 2016(02): 45–49.

Chen Ruikao, Wei Xueli, Song Ling, et al. Road performance of asphalt mixture mixed with quantitative anti-rutting agent [J]. *Science technology and engineering*, 2018, 18(36): 120–124.

Huang Peng, Lv Weimin, Zhang Fuqing, et al. Research on Performance and Technology of Rubber Powder Modified Asphalt Mixture [J]. *China Journal of Highway and Transport*, 2001(S1): 6–9.

Ji Zezhong, Liu Jiawei, Xu Kai. Research on Properties of Rubber /SBS Composite Modified Asphalt and its Mixture [J]. *New building materials*, 2018, 45(04): 124–128+132.

Liu Jiamin, Cui Yingming. Application of rubber in modification of road asphalt [J]. *Shanxi architecture*, 2010, 36(14): 150–152.

Podolsky Joseph H., Chen Conglin, Buss Ashley F, Williams R. Christopher, Cochran Eric W. Effect of bio-derived/chemical additives on HMA and WMA compaction and dynamic modulus performance[J]. *International Journal of Pavement Engineering*, 2021, 22(5).

Tang Wei, Zhan He, Wang Yuying, et al. Research on application of MP-5 Thin Coating in Cement Concrete Bridge Deck "White to Black" [J]. *Journal of dalian jiaotong university*, 2020, 41(03): 87–92.

Wang Huifeng. *Research on Material Design and Technical Performance of High Friction Thin Layer Coating* [D]. Chang'an University, 2017.

Yang Renfeng, Dang Yanbing, Li Aiguo. Research on Quality Evaluation Index of Rubber Asphalt [J]. *Highway*, 2009(06): 174–178.

Advances in Energy, Environment and
Chemical Engineering – Abdullah & Osman (Eds)
© 2023 The Author(s), ISBN: 978-1-032-36083-6

Comparative analysis on mechanical behavior of temporary brackets for three kinds of girder bridges

Jing Ji
Department of Architectural Engineering, Qiqihar Institute of Engineering, Qiqihar, Heilongjiang, China
College of Civil and Architectural Engineering, Northeast Petroleum University, Daqing, Heilongjiang, China

Jinjin He*
College of Civil and Architectural Engineering, Northeast Petroleum University, Daqing, Heilongjiang, China

Ming Xu & Meihui Zhong
Department of Architectural Engineering, Qiqihar Institute of Engineering, Qiqihar, Heilongjiang, China

Liangqin Jiang
College of Civil and Architectural Engineering, Northeast Petroleum University, Daqing, Heilongjiang, China

ABSTRACT: Due to the large span of the steel box girder, temporary brackets must be set at the segment connections during installation; thereby, a safe and feasible bracket scheme needs to be designed. In order to ensure the safe and reliable construction of the whole steel box girder hoisting process, by referring to engineering example, the mechanical behavior of temporary brackets for three kinds of girder bridges is analyzed. Based on MIDAS/Civil software, the strength, stiffness, and stability of the temporary brackets for the three steel box girder forms in the construction process are analyzed, respectively. The results show that the strength, stiffness, and stability of the temporary bracket structures for three kinds of steel box girder bridges all meet the construction requirements, and the structures are safe and reasonable. It can provide one reference value for the construction of a temporary bracket for the steel box girder.

1 INTRODUCTION

With the rapid development of highway construction in our country, the number of steel box girders was constantly increasing, especially in super-long sea-crossing bridges and hub-interconnecting three-dimensional viaducts. Steel box girders were favored and paid attention by more and more bridge engineers. Due to their advantages of a lightweight, good torsional resistance, fast construction speed, and reduced impact on traffic, the steel box girders were favored and paid attention by more and more bridge engineers (Li 2014; Ji et al. 2020; Zhang 2012). Steel tube temporary bracket as a kind of bracket structure was often used in the construction of steel box girders. Especially in the hoisting and construction process of large steel box girder bridges, temporary brackets were required to support each block, and then welded into a whole to complete the final construction process. The appearance of finite element analysis software effectively predicts the structure rupture and collapse in the construction process, which provides a strong theoretical basis for avoiding the occurrence of these phenomena. Ji Jing et al. (Ji et al. 2020, 2021; Yue 2021) have studied the

---

*Corresponding Author: hejin09122022@163.com

finite element analysis and simulation of structures. In recent years, with the development of the steel box beam, the scope of application has been wider and wider. People gradually realized that the lack of carrying capacity of the temporary bracket structure would cause engineering accidents, so safety calculation research on the temporary bracket structure has been carried out.

Huang Peng (Hou et al. 2018) took a cable-stayed bridge as the background, and four space finite element bracket models on the general finite element software were built in this article. The structural characteristics were compared and analyzed, as well as the theoretical basis for the bridge construction was provided.

The research object of this paper is the temporary bracket for the steel box girder of a bridge spanning the main line. In order to ensure the safety of the construction stage, finite element modeling is carried out for the three types of brackets of the temporary bracket construction structure, and the strength, stiffness, and stability are checked and analyzed, respectively, which provides reliable theoretical support for the actual construction operation.

2 PROJECT OVERVIEW

2.1 Steel box girder overview

The steel box girder adopted two kinds of three-dimensional crossing bridges in this bidding section, including the separated three-dimensional crossing bridge and the inter-connecting three-dimensional crossing bridge. According to the design data of this tender section, there are eight steel box girder bridges under the main line in this tender section. The width of the bridge deck adopts four forms: 8.5m, 10m, 12m, and 16.5m, and these cross sections are arranged as shown in Table 1. The upper structure adopts two types of span combination structure, including 30+48+30 m and 35+60+35 m, and the lower structure adopts a column bridge pier, vase pier, and rib platform.

Table 1. The cross-sectional arrangement.

The width of the bridge (m)	The cross-sectional arrangement
8.5	(0.75+7.0+0.75)m
10	(0.75+8.5+0.75)m
12	(0.75+12+0.75)m
16.5	(0.75+16.5+0.75)m

2.2 Bracket structure

The temporary bracket is supported by a $\phi325\times8$mm spiral welded tube, and the steel tubes are connected by $\phi114\times5$mm and $\phi180$mm$\times5$mm. H400$\times$400$\times$20$\times$20mm section steel or double-spliced I30a is placed on the steel tube, a steel tube adjustment tube is placed on the section steel to support the steel box girder section, and the foundation adopts a reinforced concrete structure.

The form of box girder is divided into three types: single-box single-chamber, single-box double-chamber, and single-box three-chamber, so for different forms of the steel box girder, temporary brackets with different structural forms should be used for construction. According to the calculation and statistics of different forms of steel box girders, three forms of brackets are proposed. Among the three forms, the ones with the largest load are the Lalin River Separation Bridge (ZJ1), Lanling Interchange Ramp Bridge (ZJ2), and Changsheng Village Separation Bridge (ZJ3). The three types of bracket structures are shown in Figure 1 as follows.

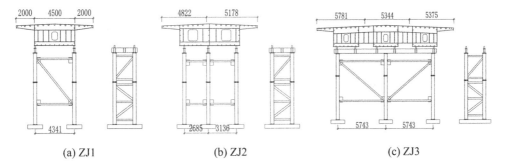

(a) ZJ1	(b) ZJ2	(c) ZJ3

Figure 1. Three types of stent structures.

3 FINITE ELEMENT MODEL

3.1 *Model establishment*

In recent years, finite element analysis software has been applied to various fields and has strong practicability in solving a wide range of problems. Midas/Civil software is used to simulate and analyze the whole process of the temporary bracket construction structure. The beam elements are used to simulate a temporary bracket. The steel tube brackets, distribution beams, and column top beams are rigidly connected, the height of the model is taken as 6.4m, and the distance between each group of steel brackets is taken as 3m-5m.

According to the bracket design, Midas Civil finite element analysis software is used for calculation, and the overall structure model is established. The arrangement of the structural brackets supporting the steel box girder is shown in Figure 2, respectively.

(a) ZJ1	(b) ZJ2	(c) ZJ3

Figure 2. Computational model.

3.2 *Load cases*

Nodal loads in the downward direction (in units of t to simulate pressure, as shown in Figure 3) are applied to the model, and nodal loads in the horizontal direction (in units of t to simulate wind loads, as shown in Figure 3) are applied to the overall model.

4 RESULTS ANALYSIS

To guarantee the accuracy of the calculation results, the bracket calculation is carried out with the main beam bracket system under the most unfavorable working condition. The temporary bracket

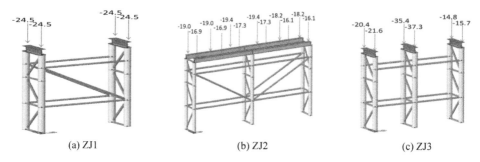

(a) ZJ1   (b) ZJ2   (c) ZJ3

Figure 3.   Load model (t).

structure adopts the Q235B steel grade, and its chemical composition and mechanical properties should meet the relevant requirements of GB/T 1591-2008.

The cast-in-place bracket structure adopts the limit state design method based on the probability theory and is designed with the design expression of the partial coefficient.

Table 2.   Specific parameters of the material.

Material model	E/MPa	γ/kg/m^3	ξ/10^{-6}	f/MPa	f_v/MPa	f_{ce}/MPa
Q235	206000	7850	12	215	125	325
Q345				310	180	400

Note: E is the elastic modulus; γ is the capacity; ξ is the coefficient of linear expansion; f is the design value of tensile, compressive, and flexural resistance; f_v is the shear design value; f_{ce} is the design value of end face pressure.

4.1   Bracket strength check

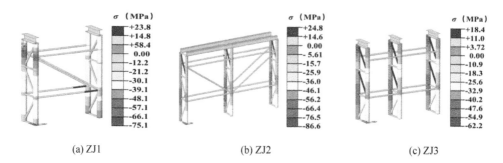

(a) ZJ1   (b) ZJ2   (c) ZJ3

Figure 4.   Maximum combined stress of stent.

As shown in Figures 4(a)-4(c), the maximum combined stress of the Lalin River Separation Bridge bracket is equal to 75.1 MPa, that of the Lanling Interchange Ramp Bridge bracket is equal to 62.2MPa, and that of the Changsheng Village Separation Bridge bracket is equal to 62.2MPa. According to the standard, the maximum combined stress of the bracket cannot exceed 215 MPa, and the strength of the three brackets can meet the construction requirements.

4.2   Bracket stiffness check calculation

As shown in Figures 5(a)-5(c), the maximum compression deformation of the steel tube of the Lalin River Separation Bridge is equal to 1.2mm, that of the steel tube of the Lanling Interchange

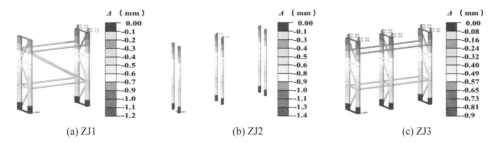

| (a) ZJ1 | (b) ZJ2 | (c) ZJ3 |

Figure 5.   Compression deformation of steel tube.

Ramp Bridge is equal to 1.4mm, and that of the steel tube of Changsheng Village Separation Bridge is equal to 0.9mm. According to the standard, the maximum compression deformation of the steel tube cannot exceed 6.0mm, and the stiffness of the three brackets can meet the construction requirements.

4.3   Stability check of the bracket

In order to calculate the overall stability of the bracket, the buckling analysis on the calculation model was carried out, critical analysis was carried out for the load variables wind load and construction and crowd load, and the number of modes is 1. The critical load deformations of the calculation model buckling coefficients as shown in table 3.

Table 3.   First-order buckling coefficient of the bracket.

Bracket name	First-order buckling coefficient
ZJ1	28.6
ZJ2	28.4
ZJ3	24.0

The first-order buckling mode is the overall buckling form; the minimum critical load factor is 24, which meets the stability requirement.

Due to the eccentric load of the temporary bracket, the buckling analysis should be carried out according to the eccentric load; in order to calculate the overall stability of the bracket, the buckling analysis is carried out on the calculation model. The self-weight of the temporary bracket is set as an invariant beam load, the eccentric beam weight is set as a variable load for critical analysis, and the number of modes is 3. The results show that the first-order buckling mode is an overall buckling form; the minimum critical load factor is 31.96, which meets the stability requirement.

4.4   Basic verification

The maximum reaction force is 420kN, which is calculated according to 669.4kN due to less than 669.4kN.

The foundation of strip foundation adopts a reinforced concrete structure with a size of 1400×3000×300mm. The size and reinforcement arrangement as shown in Figure 6.

The maximum bending moment of the strip foundation is designed according to the maximum support reaction force shown in the reaction force diagram. The maximum reaction force is 669.4kN. In the calculation, the steel tube is used as a support, and the ground is subjected to a uniform load on the foundation, so the bending moment of the strip foundation is equal to 150.05kN·m, which meets the verification requirements.

361

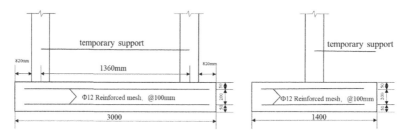

Figure 6. Schematic diagram of foundation.

5 CONCLUSION

Based on Midas/Civil software, the mechanical behavior of temporary brackets for three kinds of girder bridges is carried out in this paper. The strength, stiffness, and stability of the temporary brackets for the three steel box girder in the construction process are obtained. The influence of different bracket forms on the construction of girder bridges on site is studied. The conclusions can be drawn as follows.

(1) The structural strength, stiffness, and stability can meet the construction requirements. Based on the finite element analysis, the safety and reliability of the construction of the temporary bracket structure of the steel box girder spanning the bridge on the main line are ensured.
(2) It has great guiding significance for on-site construction and provides a rich experience for the bracket construction of similar projects in the future.

ACKNOWLEDGMENTS

The authors are grateful for the financial support from the Natural Science Foundation of Heilongjiang Province (Grant No. LH2020E018); Opening Fund for Key Laboratory of The Ministry of Education for Structural Disaster and Control of Harbin Institute of Technology (Grant No.HITCE201908); Scientific Research Fund of Institute of Engineering Mechanics, China Earthquake Administration (Grant No.2020D07) and Northeast Petroleum University Guided Innovation Fund (Grant No. 2020YDL-02).

REFERENCES

Hou H.N., Zhang X.W., Song, J.X. (2018) Design and analysis of temporary support of steel box girder. *J. Henan Univ. Urban Constr.* 27(5): 1–7.
Ji, J., Song, Z.F., Zhang, Y.F., Yu, D.Y., Jiang, L.Q., Liu, Y.C. (2020) Axial compressive properties of short columns of solid-web GFRP pipe-steel-steel-concrete combination. *J. Northeast Petroleum Univ.*, 44(3): 107–118+11-12.
Ji, J., Yang, M., Xu, Z., Jiang, L., Song, H. (2021) Experimental study of H-shaped honeycombed stub columns with rectangular concrete-filled steel tube flanges subjected to axial load. *Adv. Civ. Eng.*, 2021(11): 1–18.
Ji, J., Yu, D.Y., Jiang, L.Q., Liu, Y.C., Yang, M.M., Song, H.Y. (2020) Research on the axial compressive bearing capacity of solid-web double-steel-tube special-strength concrete composite short columns. *Build. Struct*, 50(5): 120–129.
Li Q.M. (2014) S-shaped viaduct concrete box girder replaced by steel box girder design. *World Bridge.*, 42(3): 80–84.
Yue, C.G. (2021) Analysis on the construction technology of steel box girder hoisting of urban bridges. *World Bridge.*, 40(1): 37–41.
Zhang X.Y. (2012) Stress analysis of flat steel box girder jacking construction of suspension viaducts. *Eng. Constr.*, 35(4): 805–807.

Advances in Energy, Environment and
Chemical Engineering – Abdullah & Osman (Eds)
© 2023 The Author(s), ISBN: 978-1-032-36083-6

Research and application of 10 kV intelligent transformer based on on-load regulation technology in voltage management

Yu Zou, Yifeng Su, Yinguan Song, Xingsong Chen, Yan Zeng & Zhenjun Xie
Guangxi Power Grid Co. Ltd Qinzhou Power Supply Bureau, China

ABSTRACT: At present, the voltage of residential users in China should be stable between 198 V and 235.4 V. However, due to many factors causing the abnormal voltage in the power distribution field, heavy management work, and large investments, the voltage of residential users in China often does not meet the national standard's requirements for normal voltage. This project is mainly to partially control the voltage abnormality in the power distribution field, that is, to control the output voltage of the distribution transformer, so that the output voltage of the distribution transformer in the distribution station area can reach GB/T 12325-2008 "Power Quality," the voltage fluctuation range specified in the standard of "Power Supply Voltage Deviation."

1 INTRODUCTION

Currently, the main measure to control the output voltage of distribution transformers in China is to adjust the voltage from higher-level transmission lines or substations, to achieve the purpose of the normal input voltage on the high-voltage side of distribution transformers. However, due to the higher purchase cost of the equipment with higher voltage levels, lower regulation accuracy, and poorer self-regulation capability, it is more economical, more reliable, and accurate to conduct voltage governance in the field of power distribution with lower voltage levels. China's main network transformers have gradually adopted the on-load voltage regulation technology in recent years, but the distribution network field still uses outdated non-excitation voltage regulation technology. The voltage regulation range of this technology is small, only ±5%. The risk factor is high, the work efficiency is low, and the voltage regulation requirements of the distribution transformer cannot be met, so the output voltage of the transformer is still uncontrollable. Because of the above problems, the abnormal input voltage of the high-voltage side of the distribution transformer caused by higher-level transmission lines or substations can be effectively managed by improving the voltage regulation capability of the distribution transformer. And the governance accuracy is higher, the method is more effective, and a more accurate governance effect can be achieved.

2 THE PAST RESEARCH SITUATION OF THE PROBLEMS TO BE SOLVED BY THIS PROJECT AND THE EXISTING TECHNICAL BOTTLENECKS

Due to the problem of unqualified voltage in the distribution station area, although China Southern Power Grid Corporation has the project research experience to control the voltage at the end of the line, that is, the user side, the process structure of the on-load tap-changer of the voltage regulating transformer and the transformer oil tank topics such as optimization, intelligent controller and remote data intelligent monitoring system have not yet carried out in-depth research on systems similar to this project. No related research has been done on the cost control of the new on-load tap-changer, the voltage regulating transformer manufacturing process, and the programming logic and principle of the intelligent controller, which are mainly studied in this project.

DOI 10.1201/9781003330165-52

At present, the main technical bottlenecks of on-load voltage regulation distribution transformers are:

2.1 *The voltage regulation amplitude is small, and the large-scale on-load voltage regulating transformer needs to improve the transformer coil and the overall oil tank structure.*

2.2 *The manufacturing cost of transformers is relatively high. According to relevant market statistics, the manufacturing cost of on-load voltage regulating transformers is more than 25% higher than that of conventional on-load voltage regulating transformers of the same model and capacity.*

2.3 *The cost of the on-load voltage regulating switch is relatively high. At present, most of the on-load voltage regulating switches of domestic on-load voltage regulating transformer manufacturers adopt the form of outsourcing, and the core technology of the switch is still in the hands of the relevant manufacturers.*

2.4 *The defects of the switch intelligent controller are serious. At present, the intelligent controller that controls the action of the on-load voltage regulating transformer voltage regulating switch has various logics, and the programming principles are diversified. Frequent action or no action when receiving a signal of sudden voltage change.*

Because of the above technical bottlenecks and obstacles in the commercialization and application of on-load voltage regulator transformers, the research and development of low-cost and superior performance on-load voltage regulator transformers is very important for the early large-scale application of this product and greatly improving the overall power quality of residential users.

3 THE RESEARCH FOUNDATION AND CONDITIONS OF THE PROJECT UNIT

3.1 *Theoretical research basis*

3.1.1 *Research on aging of transformer insulating materials*
In the research and development process of oil-immersed transformers, in addition to the insulation and heat resistance design, it is also necessary to fully study and confirm the cooling method, material compatibility, and accessories selection. The transformer is closely related to the selected oil and paper system in these aspects, and the selection of different insulating materials will have different effects on it. The heat dissipation performance of the insulating oil will directly affect the cooling effect of the transformer, which in turn affects the thermal aging of the insulating paper.

The aging experiment of insulating paper and insulating oil has very high requirements for environmental control. It is necessary to avoid contact with air to prevent oxygen and moisture in the air from entering the oil-paper system and affecting the aging process. Therefore, it is necessary to design the aging container to ensure airtightness strictly. To more truly reflect the different thermal stresses that the oil-paper system bears during the actual operation of the transformer, a dual-temperature aging model is generally designed regarding the standard IEC 62332-1. The dual-temperature aging tank is shown in Figure 1 below. According to the standard, the material was tested at three temperature points and four aging times.

For insulating papers and oils aged at different time points, measurements of typical basic properties are required. And on this basis, we conducted the insulating oil DGA analysis.

It is generally believed that the life of transformer oil and paper insulation depends on the performance level of the insulating paper. When the DP value of the insulating paper drops to 200, it is considered that its life has reached the end; that is, the transformer has reached its operating life. The aging experiment will use the 200DP value as the criterion to study the life curve of the oil-paper system under different moisture contents and obtain its temperature index. Combined with the comparative study of basic performance, the insulation performance of the system is comprehensively judged, and its effectiveness in improving the overload resistance of the transformer is verified.

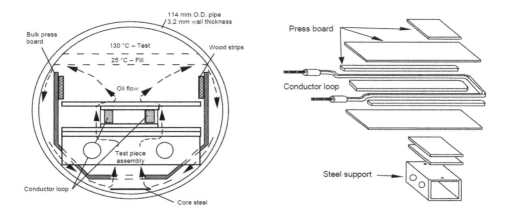

(a) Aging unit cross-section　　　　(b) Schematic diagram of transformer winding insulation simulation

Figure 1.　Schematic diagram of the design of the dual-temperature aging model.

3.1.2　*Selection of voltage regulation mode of on-load tap-changer*

Whether it is a vacuum on-load tap-changer or a non-vacuum on-load tap-changer, the transformer change can be adjusted by changing the miscellaneous books of the primary or secondary side windings, usually through linear voltage regulation, positive and negative voltage regulation, and coarse and fine adjustment. Three pressure regulation methods are realized.

There are three basic voltage regulation methods, each of which has its own voltage regulation characteristics and scope of application, and the insulation requirements are different. Coarse and fine voltage regulation and positive and negative voltage regulation are usually suitable for larger 220 kV transformers. However, the tap winding of linear voltage regulation has always been connected in series with the basic high-voltage winding. Compared with positive and negative regulation or coarse and fine regulation, there will be no problems such as potential connection and potential suspension of the tap winding. Moreover, there is no changeover selector device on the vacuum on-load tap-changer, the overall structure is relatively simple, and it is more suitable for 10 kV and 35 kV small transformers. Therefore, the voltage regulation mode of the vacuum on-load tap-changer developed in this project selects linear voltage regulation.

3.2　*Laboratory conditions*

3.2.1　*Transformer dual temperature aging test system*

The dual-temperature aging test system comprises a dual-temperature aging tank and an electrical control system for the aging tank. The dual temperature aging tank is composed of conductors, heaters, tanks, sensors, and pressure gauges. The electrical control system of the aging tank is composed of a voltage regulator, a system control cabinet, a large current generator, and a temperature monitor.

Features: The temperature monitoring range of the dual-temperature aging tank is 0–400°C, and the power of the heating tube at the oil temperature can reach 500 W. The aging tank is a combination of conductors and insulating materials in a certain proportion to simulate the actual operation of the transformer equipment. The temperature sensor embedded in the aging tank monitors the temperature of each part of the insulation system and realizes the effective control of its temperature through the temperature control unit.

3.2.2　*Electrical fast transient burst test device*

The electrical fast transient burst test device simulates and examines the performance of electronic and electrical equipment under the influence of such transient disturbances.

4 MAIN IMPLEMENTATION

4.1 *Main technical content*

4.1.1 *Research and development of on-load tap-changer*
The tap-changer to be developed in this scheme is a horizontal on-load vacuum tap-changer. The horizontal on-load tap-changer transmission device includes a gear motor located on the transmission device installation plate, a permanent magnet electromagnet, and a gear mechanism.

4.1.2 *Design of tap changer control circuit*
The tap changer has nine grades of voltage regulation selection terminals, which are respectively connected with nine coil taps on the transformer. When the controller detects a voltage failure on the high-voltage side, it sends an instruction to control the mechanism on the tap changer to realize the upshift and downshift of the coil voltage.

4.2 *Research and development of intelligent controller*

The intelligent controller is an important part of the wide-amplitude voltage regulating transformer. By collecting the voltage on the high-voltage side of the transformer and processing the data by the ARM chip, it controls the on-load tap-changer according to the predetermined logic to realize automatic voltage regulation. Signal sampling circuit mainly collects transformer temperature, oil pressure, current, voltage, and other signals; ARM chip is the core of intelligent controller, responsible for data processing, logic control, etc.; LCD screen provides Chinese display interface, which is convenient for parameter setting and query; WIFI GPRS module is a backup module, which is convenient for data collection and transmission when the transformer is connected to the grid. In actual operation, these two modules are not installed.

4.3 *Research and development of wide-amplitude voltage regulating transformer*

The on-load vacuum tap-changer is installed in the transformer oil tank. According to the structure and working characteristics of the on-load tap-changer, the ordinary S13 oil transformer needs to be transformed.

4.3.1 *Coil design*
The coil of the ordinary S13 oil transformer has only three taps, and the wide-amplitude voltage regulating transformer has nine tap positions, and the number of turns of each coil needs to be recalculated. The voltage regulation range of the tap changer is ±5%*4, and the stage voltage of each gear is 500 V.

4.3.2 *Design of the fuel tank*
The horizontal on-load switch is installed in the fuel tank, and the volume of the fuel tank needs to be increased synchronously. At the same time, it is necessary to consider the increase of the taps, the reliability of the tap fixing, and the improvement of the insulation performance.

5 MAIN TECHNICAL DIFFICULTIES AND SOLUTIONS

5.1 Research on heat dissipation and dielectric properties of small volume transformers. The horizontal on-load tap-changer is used in the wide-amplitude voltage regulating transformer. The volume of the oil tank is smaller than that of the cylindrical on-load tap-changer, and the layout of the box is more compact. The question is how to ensure the effective heat dissipation of the transformer in a limited space while analyzing hot spots and simulation, as well as its test environment.

Solution: Through the dual-temperature aging experiment, compare the performance change characteristics of different oil-paper materials, and use microscopic analysis and observation methods, combined with the existing S13 oil change research basis, to study the aging mechanism of the wide-amplitude voltage regulating transformer.

5.2 Anti-interference design of the electronic controller. Electronic components are easily disturbed by the environment. When the transformer is working, the coil and iron core generate a large electromagnetic field, which poses a severe test to the reliability of the signal acquisition and data transmission of the intelligent controller. The controller takes power from the output of the transformer, and the connection and withdrawal of the load on the line will generate pulse group, overvoltage, and other disturbances on the line, which puts forward higher requirements for the controller to resist electromagnetic interference.

Solution: The company has a rich technical accumulation in power electronics and has mature anti-interference design solutions for AC power circuits, sampling circuits, and RS485 communication circuits. The hardware design of the intelligent controller adopts mature and stable circuits as much as possible. The reliability evaluation in the design stage will be strengthened, and the electronic components will be left with sufficient margin; in the prototype verification stage, the company's EMC laboratory will conduct a thorough test and repeated verification in the company's EMC laboratory.

5.3 Handling of the contact bounce problem when the tap changer is switched rapidly. After the tap changer gear selection is completed, the switching between the two gear coils is completed within tens of milliseconds, and the three groups of main and auxiliary vacuum tubes connected to the three-phase transformer coils complete the transformation of six working states in a short time, and the switching speed is fast. It will cause the bounce of the vacuum tube contact. If the bounce is large, it will cause the contact to burn and reduce the mechanical and electrical life of the vacuum tube.

Solution: After the structure is designed, we use the Ansys simulation software to perform mechanical simulation; in the prototype test, we use a high-voltage switch characteristic tester and a decomposition switch tester to test the bounce of the contacts.

6 CONCLUSION

This project mainly deals with the problem of large voltage fluctuations in the grid-connected lines of small power stations in the Qinzhou network area. With a new wide-range voltage regulation pole number, through the design of horizontal vacuum on-load tap-changers, the development of embedded systems based on ARM chips, through a special algorithm to achieve automatic voltage adjustment. It guarantees customers' electricity consumption and provides ensures reliability for the safe and economical operation of the power grid. Through theoretical analysis, simulation calculation, and experimental verification, systematic research is carried out from the aspects of mechanism analysis, device research and development, monitoring analysis, etc., to form a set of governance methods for large voltage fluctuations, and finally carry out application practice in power supply companies in the Qinzhou Power Supply Bureau area. 1) When the voltage fluctuates greatly in the line, the intelligent wide-amplitude voltage regulating transformer automatically adjusts the output voltage of the transformer according to the preset voltage regulating control logic to improve the power quality of the user end. 2) The intelligent wide-amplitude regulating transformer monitors the oil temperature, oil pressure, and electrical parameters of the high and low voltage sides of the transformer in real-time and displays and inquires through the LCD screen, which facilitates the maintenance efficiency of the maintenance personnel of the power supply bureau. 3) When the input voltage of the high-voltage side of the distribution transformer fluctuates within the adjustable range, the output voltage meets the national standard. When the voltage deviation is greater than the adjustable range, the deviation of the output voltage of the transformer should be reduced as much as possible. 4) The voltage deviation in the Taiwan area has been greatly improved, the power quality has been improved, and the complaint rate from the residential users has been reduced.

REFERENCES

Bai Fengjun, *Industrialization of high-voltage and large-capacity intelligent power transformers*, Shandong Province, Shandong Dachi Electric Co. Ltd., 2017-06-26.

Gao Fei, Zhao Xiaoyu, Zhao Pengcheng, Research on the overall joint debugging and detection method of intelligent power transformer [J]. *Smart Grid* 2017, 5(03):282–287.

Huang Changjun, Residential Client "Low Voltage" Governance Mechanism [J]. *Chinese Power Enterprise Management*, 2019(03):64–65.

Li Chun, Qinwei, Li Sun, Research on field wiring and protective measures for test site of intelligent secondary components of power transformers [J]. *Electric Automation*, 2016, 38(06):51–53+61.

Liu Jun, *Design and application of intelligent transformer online monitoring system* [D]. Shenyang Agricultural University, 2018.

Shi Yuxi, Application Research of Smart Transformer in Smart Grid System [J]. *Technology and Market*, 2016, 23(12):233.

Yu Zhangli, *Design and application of intelligent transformer integrated online monitoring system* [D]. North China Electric Power University (Beijing), 2016.

Advances in Energy, Environment and Chemical Engineering – Abdullah & Osman (Eds)
© 2023 The Author(s), ISBN: 978-1-032-36083-6

Impacts of produced polymer on oilfield injecting waters quality during polymer flooding

Yan Guo*, Meimei Du, Lizhen Wang, Limin Liu, Mei Bin, Jing Wang & Zhen Qin
Engineering Technology Research Institute of Huabei Oilfield Company of Petro, Renqiu City, Hebei province, China

ABSTRACT: The effect of different concentrations and molecular weights of the produced polymer on the water quality standard of the extracted water was investigated using water quality analysis. Experimental results show that the produced polymers flocculate or encapsulate fine suspended matter in the water to form flocs increasing the median diameter of the particles in the injected water. The presence of an emulsified layer of polymeric effluent results in a lower oil content determination than the actual value, while the presence of the produced polymer also leads to an increase in the rate of injection water fouling. In addition, as polymer concentration increases, injection water quality tends to improve.

1 INTRODUCTION

In recent years, polymer flooding for enhancing oil recovery (EOR) has been used in the oil and gas industry, and while the benefits have been remarkable, many effluent-containing polymers have been generated. The direct discharge of polymer-containing effluent will cause environmental pollution, and it is not economically advantageous to treat it to meet the discharge standards, so injecting it back into the reservoir is the best way (Deng 2002; Lai 2020; Li 2021). Therefore, whether the quality of the polymeric effluent can meet the requirements of the re-injection water is the key issue to be faced when injecting it back into the reservoir.

Compared with primary and secondary oil recovery effluent, polymer flooding (PF) has a high degree of emulsification, a small particle size of emulsified oil droplets, strong electronegativity at the oil-water interface, and strong interfacial film strength and viscoelasticity, resulting in high stability of the oil-water emulsion and increased difficulty in oil-water separation. In addition, as the separated effluent contains polymers, the oil content of the effluent increases significantly and it is difficult to achieve oil-water separation through gravity separation, so a large amount of clear water agent needs to be added, but the flocculent formed after the combination of polymer and clear water agent can form viscous sludge with oil droplets, which is easy to block the pipeline and cause the consumption of clear water agent to increase (Li 2020, 2016; Lun 2018).

The low molecular mass and a high degree of hydrolysis of the produced polymers contained in the polymeric effluent cause many problems in the treatment and re-injection of the polymeric effluent. For example, the produced polymers in the effluent containing aggregates are not compatible with the water treatment chemicals, and the treatment effect becomes worse at low dosing levels or even has the opposite effect of increasing the suspended solids content of the extracted water (Wang 2012; Ye & Iop 2018). Some of the viscous products suspended in the extracted water enter the filter, resulting in serious blockage of the filter due to scaling and sand accumulation, accelerated contamination rate of the filter media, poor backwashing effect, and serious excess oil content and suspended solids content in the treated effluent (Zhao 2014; Zhong 2018).

---

*Corresponding Author: cyy_guoy@petrochina.com.cn

Most of the research on the water quality of polymeric wastewater has been focused on the comparison of the water quality of polymeric wastewater with that of ordinary wastewater, but the specific effects of the produced polymer on the water quality indicators of polymeric wastewater have not been reported. In this study, the effect of the produced polymer on the main water quality indicators such as suspended solids content, diameter median of suspended solids, oil content and scale forming ion concentration of the effluent was investigated in the laboratory. The results of the study are important references for the field regulation of the quality of effluent-containing aggregates and the development of water quality standards for injecting water quality.

2 EXPERIMENTAL SECTIONS

2.1 *Simulation of produced polymers*

The polymers with molecular weights of 25 million and 19 million were weighed and prepared into a 5,000 mg/L polymer solution with distilled water, degraded by UV light for 30 min and then thermally degraded at 60°C for 5 days. The basic parameters of the produced polymer are shown in Table 1.

Table 1. Basic parameters of the simulated produced polymer.

The relative molecular weight of polymers	25 million		19 million	
Polymer concentration (mg/L)	5000	100	5000	100
Five-day relative viscosity (mPa•s)	1689	11.8	1482	8.3

2.2 *Preparation of water sample*

The water quality analysis data from the oilfield-produced water without produced polymer are shown in Table 2. The simulated produced polymer, which has undergone photodegradation and thermal degradation, was added to the produced water without polymer at different concentrations to determine the changes in water quality before and after the addition of produced polymer.

Table 2. Water quality parameters for injecting water without polymer.

Water sample	Oil content	Suspended solid content	Median diameter	Ion concentration (mg/L)		
				Mg^{2+}	Ca^{2+}	HCO_3
Produced water	42.1 mg/L	14 mg/L	2.19 μm	7.2	81.6	1642.7

3 RESULTS AND DISCUSSION

3.1 *Effect of polymer concentration on extracted water quality*

3.1.1 *Constant suspended solids*

As can be seen from the results in Table 3, the addition of different concentrations of simulated produced polymers that have undergone photodegradation and thermal degradation to the non-polymerized output water resulted in little change in the suspended solids content because the polymer-containing water samples to be measured in this study were placed in a constant temperature water bath at 60°C for 30 min and the filter membrane was rinsed with deionized water at 60°C after filtration to remove the effect of the polymer on the determination of the suspended solids.

Table 3. Effect of produced polymer concentration on water quality indicators.

Polymer concentration (mg/L)	0	50	100	150
Suspended solid (mg/L)	13	14	18	14
Median diameter (μm)	2.19	4.54	5.67	7.95
Oil content (mg/L)	42.1	29.2	25.7	19.5
Mg^{2+}, Ca^{2+} (mg/L)	88.8	28.1	29.4	27.2
HCO_3^- (mg/L)	1642.7	1468.7	1531.8	1553.5

3.1.2 Increase in median diameter

The addition of the produced polymer increases the median diameter of the suspended solids, and the higher the concentration of the produced polymer, the greater the increase in the median particle size. This is mainly because the polymer molecules contain a variety of reactive groups which adsorb the solid suspended particles in the water, and through bridging, cause the suspended matter in the water to gather to form larger suspended flocs, these organic flocs are strong and can deform, and are the main substances causing clogging of the filter media in the field equipment and damage to the strata.

3.1.3 Reduced oil content

The addition of 50 mg/L, 100 mg/L and 150 mg/L of photodegraded and thermally degraded simulated produced polymer to the non-polymerized output water reduced the oil content of the injected water by 31%, 39%, and 54%, respectively, and the higher the concentration of the produced polymer, the greater the reduction in oil content. On the one hand, the presence of an emulsified layer in the conventional infrared light oil measurement method was found in the determination of the oil content of the polymeric effluent, resulting in the incomplete transfer of the oil in the emulsified layer to the cyclohexane, which led to low results; on the other hand, more than 90% of the oil beads in the polymeric effluent were $\leq$ 10 μm in size, which prevented the complete extraction of the oil contained in the water. The presence of the emulsified layer and the small particle size of the oil beads were due to the activation of the produced polymer.

3.1.4 Decrease in magnesium and calcium ion concentration

The addition of the produced polymer reduces the concentration of magnesium and calcium ions in the solution. The reduction in the concentration of scale-forming ions is partly due to the formation of calcium carbonate scale precipitation and partly due to chelation by the produced polymer, the ability of the produced polymer to chelate scale-forming ions increases with increasing concentration, and the test results show that the concentration of scale-forming ions does not decrease linearly with increasing polymer concentration, indicating that the scale-forming ions are mainly precipitated as calcium carbonate scale.

3.2 Influence of polymer relative molecular mass on wastewater quality

Table 4. Effect of relative molecular mass of polymer produced on water quality (polymer concentration: 100 mg/L).

	25 million	19 million
Suspended solid (mg/L)	18	15
Median diameter (μm)	5.67	6.01
Oil content (mg/L)	25.7	24.0

The relative molecular mass of the polymer is also an important factor affecting water quality. The experimental results are shown in Table 4. From the data in the table, it can be seen that when the polymer with higher relative molecular weight is added to the sewage, the corresponding solid suspension content and oil content are correspondingly higher.

4 CONCLUSIONS

(1) The addition of the produced polymer to the injection water increases the concentration from 0 mg/L to 150 mg/L, and the median diameter of the injected water increases, mainly through the bridging effect of the polymer active groups, which causes suspended matter of various particle sizes in the injected water to collect and form coarse flocs.
(2) The presence of an emulsified layer of polymeric effluent and the finer particle size of the polymeric effluent oil beads result in a lower oil content value than the actual value.
(3) Low concentrations of produced polymer will slightly accelerate calcium carbonate scaling in the injection water and reduce the concentration of scaling ions. Scaling is inhibited by slowing down the rate of scaling at > 100 mg/L of yield polymer.
(4) The higher the relative molecular weight of the produced polymer, the more significant the deterioration of the water quality of the effluent containing the polymer.

ACKNOWLEDGMENT

This study was financially supported by the Huabei Oilfield Science Research Project (Grant No. 2022-HB-E08). We are very grateful for the useful suggestions from the Engineering Technology Research Institute of Huabei Oilfield.

REFERENCES

Deng, S.B. 2002. Effects of alkaline/surfactant/polymer on stability of oil droplets in produced water from ASP flooding. Colloids and Surfaces a-Physicochemical and Engineering Aspects, 211(2-3), 275–284.
Lai, N.J. 2020. Polymer flooding in high-temperature and high-salinity heterogeneous reservoir by using diutan gum. Journal of Petroleum Science and Engineering, 188.
Li, A.C. 2021. Experimental analysis on water quality of produced sewage in polymer flooding oilfield. Fresenius Environmental Bulletin, 30(1), 349–357.
Li, L. 2020. Factors affecting the performance of forward osmosis treatment for oilfield produced water from surfactant-polymer flooding. Journal of Membrane Science, 118457.
Li, Z.H. 2016. Formation damage during alkaline-surfactant-polymer flooding in the Sanan-5 block of the Daqing Oilfield, China. Journal of Natural Gas Science and Engineering, 35, 826–835.
Lun, W.J. 2018. Research on Effects of Ions on the Determination of Suspended Solid Content in the Oilfield Injection Water. In 2nd International Workshop on Advances in Energy Science and Environment Engineering (AESEE), Vol. 1944 Zhuhai, PEOPLES R CHINA.
Wang, X.Y. 2012. Contribution of Main Pollutants in Oilfield Polymer-Flooding Wastewater to the Total Membrane Fouling Resistance. Separation Science and Technology, 47(11), 1617–1627.
Ye, B.J. 2018. Characteristic analysis and evaluation of chlorine dioxide mixed with oilfield sewage. In 4th International Conference on Environmental Science and Material Application (ESMA), Vol. 252 Xian, PEOPLES R CHINA.
Zhao, Q.S. 2014. Research of Water quality in Typical Low Permeability Oilfield Produced Water Treatment. In International Conference on Mechatronics Engineering and Computing Technology (ICMECT), Vol. 556-562 Shanghai, PEOPLES R CHINA, pp. 867–871.
Zhong, H.Y. 2018. Microflow Mechanism of Oil Displacement by Viscoelastic Hydrophobically Associating Water-Soluble Polymers in Enhanced Oil Recovery. Polymers, 10(6).

Advances in Energy, Environment and
Chemical Engineering – Abdullah & Osman (Eds)
© 2023 The Author(s), ISBN: 978-1-032-36083-6

Research on environmentally friendly scale inhibitors and their mechanism of action

Zhan Liu*
Institute of Energy Sources, Hebei Academy of Science, Shijiazhuang, Hebei, China
Hebei Engineering Research Center for Water Saving in Industry, Shijiazhuang, Hebei, China

Na Li
Hebei Sun-water Treatment Co., Ltd., Shijiazhuang, Hebei, China

Mei-fang Yan, Yu-hua Gao & Hai-hua Li
Institute of Energy Sources, Hebei Academy of Science, Shijiazhuang, Hebei, China
Hebei Engineering Research Center for Water Saving in Industry, Shijiazhuang, Hebei, China

ABSTRACT: Several new environmentally friendly water treatment agents that exist today are described in detail, and the results of research scholars are examined and studied, which provides a reference for future research. The purpose of this study is to obtain water treatment agents that will perform better, be more environmentally friendly, and have a broader range of applications, based on previous research.

1 INTRODUCTION

In China, industrial water consumption accounts for around 40% of total water consumption, while industrial cooling water accounts for 80% of industrial water consumption. Therefore, the recycling of industrial cooling water can greatly reduce water consumption and save limited water resources. Among the industrial water treatment chemicals, scale inhibitors are commonly used in large quantities, which are of great significance in preventing equipment scaling and reducing energy consumption, as well as extending the life of the equipment and ensuring safe operation of equipment.

The industrial circulating water treatment agents currently on the market are mainly phosphorus-based, which can eutrophicate the water bodies and they are very difficult to biodegrade. They are non-environmentally friendly products (Cheng et al. 2018; Geng & Tang 2015; Feng 2017; Migahed et al. 2016; Wei & Zhong 2015). Therefore, the development of a new type of scale inhibitor with excellent performance and environmentally friendly is the direction of water treatment agent development.

2 RESEARCH STATUS OF ENVIRONMENTALLY FRIENDLY WATER TREATMENT AGENTS

Three new environmentally friendly water treatment agents have been widely researched: polyaspartic acid (PASP), polyepoxysuccinic acid (PESA), and itaconic acid (IA) copolymer (Wang 2016).

---

*Corresponding Author: liuzhan1216@126.com

2.1 Polyaspartic acid (PASP)

PASP had been studied extensively by scholars at home and abroad since the 1990s. Due to the amino as well as carboxylic acid groups contained in its structure, it has good scale inhibition properties. PASP has the advantage of being phosphorus-free, non-toxic, biodegradable, and has the function of chelating, which has the effect of preventing scale generation (Wang 2015). In recent years, the synthesis and applications of polyaspartic acid have received worldwide attention (Chen et al. 2015; Koskan & Larry et al. 1993; Rahman & Gagnon 2014; Sangeetha et al. 2016).

The average molecular weight of polyaspartic acid is between 1,000 and 2,000. Some academics have polymerized it to obtain the product polyaspartic acid, which has a scale inhibition effect on both $Ca_3(PO_4)_2$ and $CaCO_3$, but the scale inhibition effect of the drug does not compare favorably with commercially available agents. The synthesis of polyaspartic acid proceeds as follows.

$$(1)$$

$$(2)$$

$$(3)$$

2.2 Polyepoxy succinic acid (PESA)

PESA was developed by an American company in the 1990s (Brown et al. 1991; Fuknmoto et al. 1990). PESA is phosphorus-free, nitrogen-free, and biodegradable. It is a promising environment-friendly scale inhibitor, which has become a hot spot for domestic and international research (Brown & McDowell 1991; Xiong et al. 1999). The preparation and performance of PESA and the synergistic effect of PESA with other agents have been studied in China. It can be concluded that PESA has a good scale and corrosion inhibition performance, and its biodegradation rate can reach 80% on the 28th day, which is an easily biodegradable water treatment agent.

R.D. Bush (Bush & Heinzman 1987) synthesized PESA in 1987 and obtained a mixture of epoxy succinic acid homopolymer and tartaric acid salt. In China, Xiong Rongchun and others conducted the earliest research on PESA. After synthesizing the intermediate product ESA, they finally made ESA polymerize under certain conditions to obtain PESA.

The synthesis reaction equation is as follows.

$$\text{(maleic anhydride)} + H_2O \xrightarrow{\text{Hydrolysis}} \text{(disodium maleate, ONa/ONa)} \xrightarrow{\text{Epoxidation}} \text{(epoxysuccinate, ONa/ONa)} \tag{4}$$

$$\text{(epoxysuccinate, ONa/ONa)} \xrightarrow{\text{Polymerisation}} H \!\!-\!\!\left[\!-O-\underset{\underset{NaOOC}{|}}{\overset{\overset{H}{|}}{C}}-\underset{\underset{COONa}{|}}{\overset{\overset{H}{|}}{C}}-\right]_n\!\!-OH \tag{5}$$

After reviewing a large amount of literature, it can be concluded that polyepoxy succinic acid has a certain scale inhibition effect on many types of scale, such as the most common calcium carbonate, calcium sulfate, etc. Moreover, polyepoxy succinic acid also has certain corrosion inhibition properties, but its scale inhibition and corrosion inhibition effects are not very outstanding, and these shortcomings have been restricting its development.

2.3 *Itaconic acid copolymers*

Itaconic acid (IA) copolymers are derived from the polymerization of its monomers. IA is a product of biological fermentation, its molecules do not contain phosphorus and nitrogen, are non-toxic, and are easily biodegradable, so it is widely used in water treatment agents. The presence of two -COOH groups in IA gives water treatment agents containing IA monomers good negative electrical properties and a certain complexing ability for scale-forming ions. The presence of -COOH makes the water treatment agent containing IA monomer have good negative electrical properties and scale inhibition ability (Fu 2014).

Itaconic acid homopolymers were synthesized by Zhang Yanhe (Zhang et al. 2006). They used itaconic acid as raw material and a compound molecular chain transfer agent to adjust the relative molecular mass in the presence of a redox-type initiator.

The synthesis equation is shown as follows.

$$x\,CH_2=\underset{\underset{COOH}{|}}{C}-CH_2-COOH \xrightarrow{\text{Homopolymerization}} \left[\!CH_2-\underset{\underset{COOH}{|}}{\overset{\overset{CH_2COOH}{|}}{C}}\!\right]_x \tag{6}$$

Scale and corrosion inhibition experiments show that when the mass concentration of itaconic acid homopolymer in the solution is 5 mg/L, the static scale inhibition rate can reach 100%. However, the corrosion inhibition performance needs to be improved, and the corrosion inhibition rate is extremely low when itaconic acid homopolymer is used alone at a low concentration. The degradation rate of itaconic acid homopolymer was 76.56% at 28 d. It has excellent biodegradability and is an environmentally friendly agent.

3 THE MAIN MECHANISM OF ACTION OF ENVIRONMENTALLY FRIENDLY SCALE INHIBITORS

While people are developing new environmentally friendly scale inhibitors, they have also conducted in-depth research on their scale inhibition mechanism. The following effects or synergies between them are mainly considered (Al-Roomi & Hussain 2016; Cheng et al. 2003; Duan & Ren 2015; Li et al. 2015; Shi et al. 2016; Wang et al. 2018; Yang 2018). After the scale inhibitor is added to the circulating water system, it will produce negatively charged anions. They chelate with the metal cations in the water (such as calcium ions, magnesium ions, etc.), and they can form more stable chelating compounds, thus making the limit concentration of metal cations in the aqueous solution to be increased, so that it plays a role in solubilization to a certain extent (Fu et al. 2012; Wang et al. 2017).

3.1 *Coagulation and dispersion mechanism*

The above-mentioned environmentally friendly scale inhibitors are polymer-based. They will produce ionization in the aqueous solution to generate negatively charged anions, which occur through physical and chemical adsorption and scale deposits in water microcrystal adsorption. The surface of the scale microcrystals with electric ions will not be able to coalesce together due to the mutual repulsion of the same charge. Consequently, not only disperses the scale particles in the water effectively but also improves the solubility of the ions in the water, thereby reducing the tendency for scale formation. The coalescence and dispersion effect of polymeric scale inhibitors is more significant than the chelation and solubilization effect. In the medium and high hardness water systems, coalescence and dispersion play an irreplaceable role (Wang et al. 2001).

3.2 *Mechanism of lattice distortion*

As the cooling water in the circulating cooling water system is continuously reused, the water in the system gradually evaporates and concentrates. The temperature of the water body increases, increasing the concentration of inorganic salt ions in the circulating water system. When the concentration of ions in the system reaches a critical concentration, they will precipitate out in the form of crystals. Without the participation of scale inhibitors, the scale will form dense and regular crystals according to the law of crystal growth. When scale inhibitors are added to circulating cooling water systems, the interaction between their molecules and inorganic salt ions disrupts the nucleation phase of scale crystals, which affects the normal growth of scale crystals, making them less regular and loose, and very easy to flow away with the water (You 2010).

3.3 *Low dose effect mechanism*

In polyaspartic acid, polyepoxysuccinic acid, and poly (epsilon-containing acid) water treatment agents contain a large number of carboxylic acid groups. The carboxylic acid groups have a high complexing capacity for scale-forming cations in the water column and can still exhibit excellent and exceptional scale inhibition even when the scale inhibitor dosage is relatively small. When scale inhibitors are added to a system, the number of metal ions stabilized in the aqueous solution is much greater than it would be if they were obtained by stoichiometry. The main reason for this is that when scale inhibitors are present in the circulating water system, the scale inhibitor will prevent the precipitation of scale crystals through a variety of different mechanisms, and the ions of the scale inhibitor and inorganic salts can interact with each other without stoichiometric ratios.

4 CONCLUSION

In summary, researchers have synthesized a variety of new environmentally friendly water treatment agents and studied their performance, the mechanism of action has also been initially explored.

The results show that the new scale inhibitors have certain scale inhibition performance, but there is also room for improvement.

In the future, we should further study the mechanism of scale inhibition of environmentally friendly water treatment agents, and conduct targeted research on the mechanism of action of different molecular groups. This can inspire us to select the effective group to modify the environmentally friendly scale inhibitor in a more targeted manner, hoping to obtain a new environmental-friendly scale inhibitor with better performance.

ACKNOWLEDGMENT

This work was supported by the Natural Science Foundation of Hebei Province, the Key Basic Research Project (Grant No. 18964005D), and the science and technology projects of the Hebei Academy of Sciences (Grant No. 21712).

REFERENCES

Al-Roomi Y M, Hussain K F. Potential kinetic model for scaling and scale inhibition mechanism [J]. *Desalination*, 2016, 393:186–195.

Brown J M, Mcdowell J F, et al. Methods of controlling scale formation in aquous systems [P].US5 062 962, 1991–11

Brown M J, McDowell J F. Method of controlling scale formation in aqueous systems:US5062962 [P], 1991.

Bush R D, Heinzman S W. Either hydroxypolycarboxylate detergency builders[P].US4 654 159, 1987–3.

Chen J, Xu L H, Han J, Su M, Wu Q. Synthesis of modified polyaspartic acid and evaluation of its scale inhibition and dispersion capacity [J]. *Desalination*, 2015, 358 (7):42–48.

Cheng Y S, Sun Y C, et al. Corrosion and Scale Resistance of Environment-friendly Water Treatment Agent [J]. *Henan Sciences*, 2018, 36(11):1715–1722.

Cheng Y Z, Zhai X H, Ge H H, et al. Scale inhibitory mechanism of scale inhibitor and its performance evaluation [J]. *East China Electric Power*2003 (07):14–18.

Duan J J, Ren C M. Study on the scale inhibition mechanism of organophosphonic acid scale inhibitors and the evaluation method of their high temperature resistance [J]. *Science & Technology Information* 2015, 13(02):104–105.

Feng Y L. Research Progress of Green Asparagine as Green Water Treatment Agent [J]. *Chemical Engineering Design Communications*, 2017, 43(10):1–2.

Fu C E, Zhou N M, Xie H, et al. Research progress in the studies on the structure-activity relationship for scale inhibitors in circulating cooling water systems [J]. *Industrial Water Treatment* 2012, 30(12): 30–37.

Fu H L. *Synthesis and performance study of environmentally friendly itaconic acid copolymers* [D]. Hebei University of Technology, 2014.

Fuknmoto Y, et al. *Water Treating agents for prevention of metal corrosion and scale generation*[P]. JP 04-166298, 1990-10.

Geng C Z, Tang M R. Progress on polyaspartic acid as an environmental friendly water treatment agent [J]. *Applied Chemical Industry*, 2015(7):1350–1353.

Koskan P, Larry, Meah, et al. *Production of high molecular weight polysuccinimide poly- succinimide and high molecular weight polyaspartic acid from maleicanhydride and ammonia* [P]. US5219952.1993

Li H, Yu S H, Yao Q Z, et al. Chemical control of struvite scale by a green inhibitor polyaspartic acid [J] *RSC Advances*, 2015, 5(111): 91601–91608.

Migahed M ARashwan S MKamel M M, et al. Synthesis, characterization of polyaspartic acid-glycine adduct and evaluation of their performance as scale and corrosion inhibitor in desalination water plants [J]. *Journal of Molecular Liquids* 2016224:849–858.

Rahman M S, Gagnon G A. Bench-scale evaluation of drinking water treatment parameterson iron particles and water quality [J]. *Water research*, 2014, 32 (48):137–147.

Sangeetha, Meenakshi, Sundaram. Corrosion inhibition of am inated hydroxylethyl cellulose on mild steel in acidic condition [J]. *Carbohydrate Polymers*, 2016, 150 (9):13–20.

Shi S C, Zhao X W, Wang Q, et al. Synthesis and evaluation of polyaspartic acid/fusfurylami-ne graft copolymer as scale and corrosion inhibitor [J]. *RSC Advances*, 2016, 6:102406–102412.

Wang R, Zhang Qi, Ding J, et al.. Research Progress on the Mechanism of Scale Inhibitors [J]. *Chemical Industry and Engineering* 2001, 18(2): 79–87.

Wang S Q. *Research on application and synthesis of polyaspartate inhibitor* [D]. Jilin University 2015.

Wang S Z. *preparation and performance research of no phosphorus efficient green scale inhibitor* [D]. Wuhan Institute of Technology, 2016.

Wang Y L, Li C Q, Gu L F, et al. Research Progress in Inhibition Mechanism and Performance Evaluation Method of Scale Inhibitor [J] *Guangdong Chemical Industry* 2018, 45(12):192–193.

Wang Y.W., Li A, Yang H. Effects of substitution degree and molecular weight carboxymethyl starch on its scale inhibition [J], *Desalination* 2017, (48): 60–69.

Wei J, Zhong H B. Modification Methods for Polyasparatic Acid [J]. *Guangdong Chemical Industry*, 2015, 42(12):118.

Xiong R C, Wei G, Zhou Di. Synthesis of green scale inhibitor polyepoxysuccinic acid [J]. *Industrial Water Treatment* 199919(3):11–13.

Yang X. *Synthesis of Heterocyclic and Sulfonic Acid Modified Polyaspartic Acid Composite Scale Inhibitors and Study of Synergistic Performance* [D]. Henan University 2018.

You X F. Analysis of the Magnetic Scale Control Mechanism [J]. *Journal of Education Institute of Taiyuan University* 2010, (4): 89–91.

Zhang Y H, Liu Z F, et al.. Research on synthesis and performance of itaconic acid homopolymer [J]. *Industrial Water Treatment*, 2006(08):21–22.

*Advances in Energy, Environment and
Chemical Engineering – Abdullah & Osman (Eds)
© 2023 The Author(s), ISBN: 978-1-032-36083-6*

Progress of resource technology of phosphorous wastewater

Zhaohe Dong*
Suzhou Foreign Language School, Suzhou, Jiangsu, China

ABSTRACT: The progress of chemical and physical methods for phosphorus removal from wastewater in recent years is reviewed, such as crystallization of magnesium ammonium phosphate, chemical coagulation and precipitation, adsorption, and so on. Due to the characteristics of high adaptability, high efficiency, and low cost, composite phosphorus removal methods and solid removal phosphorus technology are important development trends. To promote the development of phosphorus removal technology and the application of phosphorus recovery, it is necessary to improve relevant regulations and technical standards as soon as possible.

1   INTRODUCTION

Water eutrophication is an important factor of water pollution, among which the excessive discharge of phosphorus is one of the reasons. At the same time, phosphorus is an important non-renewable resource, and it is of great significance to develop phosphorus recycling (Golroudbary SRA et al. 2019). Johanna Grames et al (2019) studied the relationship between economic decision-making and phosphorus resource cycling and developed a general equilibrium model including the flow of matter (Figure 1).

From the source of phosphorus and phosphorus in the natural circulation process (Figure 2), the source of wastewater containing phosphorus is a major motivation, mainly including sewage and industrial wastewater, rainfall, snow, surface runoff, and livestock manure. The phosphorus pollution contribution is agricultural drainage, living drainage, and industrial drainage. Human activities are the main factor that causes excessive phosphorus content in water, and excessive phosphorus in water will lead to eutrophication of water, causing a variety of hazards.

Although some people believe that phosphorus recovery is not a sustainable solution in the long term, the research and application of phosphorus recycling have not stopped so far. Kalimuthu Senthilkumar et al. (2014) (Mollier et al. 2014) took France as an example to study the current situation and potential of phosphorus recovery in waste and believed that although industrial waste, domestic wastewater, and urban wastewater are treated and reused, a considerable proportion of phosphorus is still discharged into water or landfills, resulting in waste. EW Mohammad et al. (2018) (Mohammad et al. 2019) established a system kinetic model study that improving phosphorus recovery could reduce the phosphorus import level by 79% in the European Union.

Phosphate, used for phosphate fertilizer production in China, accounts for more than 85% of the total phosphate consumption in China while existing phosphate reserves are only enough to maintain for about 70 years (Alexander Maurer et al. 2018). Most phosphorus sources in China's inland water are crop planting. A total of 1.31 million tons of phosphorus were discharged from Chinese aquaculture wastewater, which can replace about 3 million tons of phosphate ore (in P_2O_5). These phosphorus elements can produce about 9.41 million tons of guano stone, thus reducing the phosphorus loss by 35% (Liu et al. 2020). The sludge produced by sewage treatment also contains

---

*Corresponding Author: dongzhaohe321418@outlook.com

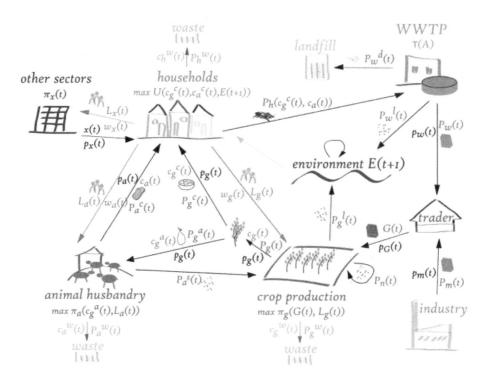

Figure 1. Outline of the model framework of phosphorus recycling. We denote with the letter "g" the grain field, "a" the animal husbandry, "h" the households, "w" the wastewater treatment plant (WWTP), "m" the mineral fertilizer from industry, and "$\times$" the other sectors. The subscripts of the consumption c, prices p, and phosphorus P variables describe the source of the flow and the superscript describes the destination.

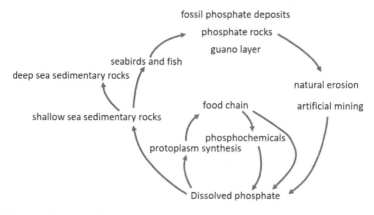

Figure 2. Schematic diagram of the circulation process of phosphorus in nature.

rich phosphorus resources, and the phosphorus in the sludge can be released into the water by certain technical means.

To recycle the phosphorus in the wastewater, a variety of phosphorus-containing wastewater treatment methods have been developed at home and abroad, mainly including physical, biological, and chemical methods. Among them, the chemical physical method includes the crystallization method, chemical coagulation precipitation method, adsorption method, micro-electrolysis method, membrane separation method, etc. Each has advantages and disadvantages (Wu et al. 2019; Yang 2020) (Table 1).

Table 1. Comparison of several chemical and physical phosphorus removal methods.

Methods for phosphorus removal	Advantages	Weaknesses	Application range
Chemical coagulation and precipitation method	Simple operation, stable effect, a wide range of application	The treatment cost is high, and the subsequent sludge treatment is complicated	High concentration of industrial wastewater
Adsorption method	Low cost, easy to operate, and environmentally friendly	Strong selectivity, small adsorption capacity	Small-scale use, specific wastewater use
Crystallization process	Wide range of use, environmentally friendly, high processing efficiency	Mainly affected by the operation, the processing results are unstable	High concentration of industrial wastewater
Microelectrolysis method	Low cost, simple operation, and covers a small floor area	The treatment effect is unstable, and the reaction medium is easy to fail	High concentration of industrial wastewater
Membrane separation process	Good treatment effect, small floor area, stable effect	High processing cost, strong selectivity	Specific industrial wastewater

This paper summarizes the progress of chemical physics of phosphorus removal in wastewater and discusses the effective way of phosphorus recycling.

2 THE CRYSTALLIZATION METHOD OF MAGNESIUM AMMONIUM PHOSPHATE

In the wastewater phosphorus removal method, the magnesium ammonium phosphate (MAP) crystallization method has the advantages of fast reaction, high efficiency, and recyclable resources, which has a good development expectation. The main presence form of phosphorus in the pH range in the presence of MAP crystals is HPO_4^{2-} or $H_2PO_4^{-}$, but not PO_4^{3-}, and protons form during MAP formation (Zheng et al. 2019).

$$Mg^{2+} + NH_4^{+} + HPO_4^{2-} + 6H_2O \rightarrow MgNH_4PO_4 \cdot 6H_2O + H^{+} \qquad (1)$$

$$Mg^{2+} + NH_4^{+} + H_2PO_4 + 6H_2O \rightarrow MgNH_4PO_4 \cdot 6H_2O + 2\,H^{+} \qquad (2)$$

Copper, zinc, aluminum, and iron ions in wastewater affect the linear growth rate of magnesium ammonium phosphate crystals, in which Cu^{2+} (0.2–0.5 mg/kg) can promote the linear growth of magnesium ammonium phosphate crystals, but its nucleation rate decreases. Zn^{2+} can promote the nucleation rate growth of magnesium ammonium phosphate, but its linear growth rate decreases; Al^{3+} grows better for magnesium ammonium phosphate crystals than Fe^{3+} (Hutnik et al. 2019, 2020).

Taking aquaculture wastewater as an example, as the total phosphorus (TP) content in low-concentration anaerobic biogas slurry wastewater is 138.3 mg/L, the optimal pH value of MAP synthesis is 10, and the optimal Mg: P: N is 1.5: 1.06: 1 (Wang 2019). When treating corn deep processing wastewater (CDPW), the removal rate of PO_4^{3-}-P, NH_4^{+}-N, and Mg^{2+} can reach

Table 2. Phosphate removal efficiency under optimum n (Mg): n (P) ratios in different kinds of wastewater.

Sewage types	Magnesium source	Optimal n (Mg): n (P)	Phosphate removal rate (%)
Pig wastewater	$MgCl_2$	(0.8-1.0): 1	87-89
Semiconductor wastewater	$MgO/MgCl_2$	4.2: 1	97.5
Sludge fermentation liquid	$MgCl_2$	1.17-1: 1	90.7
Septic tank wastewater	$MgCl_2$	1.2: 1	73.4
Grain fermentation wastewater	$MgCl_2$	1.38: 1	92.25
Sludge digestion solution	$MgCl_2$	1.5: 1	95
Artificial water distribution	—	(1.3-1.5): 1	89~92
Synthetic aquaculture wastewater	$MgCl_2$	1.4: 1	97.0
Line panel wastewater	$MgCl_2$	1.1: 1	—

Table 3. Optimum pH of MAP in different sewage plants.

Sewage types	Optimal pH
Sewage treatment plant process water	8.5
Pig wastewater	9.0
Grain fermentation wastewater	9.0-9.6
Sludge fermentation liquid	7.5-9.0
Semiconductor wastewater	8-9
Fertilizer wastewater	10.5

84.43-91.32%, 8.96-11.22%, and 64.22-69.40%, respectively, and the purity of MAP is 91.5% (Xu 2018).

In addition to the conventional MAP method, new MAP composite methods have been continuously developed and utilized in recent years to improve processing efficiency and reduce costs. For example, pig biogas slurry has the characteristics of a high nitrogen-phosphorus ratio and neutral water environment, so it is difficult to achieve efficient removal of ammonia nitrogen. Tricalcium aluminate ($Ca_3Al_2O_6$, C_3A) can produce hydration and lattice replacement to form intermediate calcium and aluminum LDH, magnesium-rich C_3A (Mg-C_3A) can be synthesized by the solid phase reaction method, and guano and calcium and aluminum LDHs crystallization used for nitrogen and phosphorus removal of pig marsh slurry. The maximum amount of ammonia nitrogen and phosphate can reach 42.2 mg.g^{-1} and 20.2 mg.g^{-1}(Ouyang 2020), respectively. Natural zeolite and modified diatomite can be used for wastewater nitrogen and phosphorus removal (Li 2019). Bing Li (2020) removed phosphorus and nitrogen from phosphate leachate by magnesium phosphate precipitation method and recycled it. The removal rate of nitrogen and phosphorus exceeds 90%, and the P_2O_5 mass fraction of magnesium phosphate exceeds 25%, which can be used as sustained-release fertilizer of magnesium phosphate. Yin Zhichao (2020) studied the influencing factors of releasing calcium and magnesium ions under acidic conditions, and the order from large to small was reaction temperature, reaction time, the number of dolomite, and hydrochloric acid addition. The optimal reaction conditions are 100 mesh dolomite, 3.4 ml of hydrochloric acid injection, and 180 min at room temperature. The pH value is 9.0; the stirring speed is 200 r/min; Ca: P = 1.67: 1; and the phosphorus removal rate reaches 92%.

Wenqian Du (2019) believed that humic acid substances in landfill leachate affect the treatment effect of the MAP precipitation method, and the concentration of humic acid substances can be reduced by flocculation to improve the efficiency of nitrogen and phosphorus removal. Weiwei Liu et al. (2019) treated nitrogen and phosphorus wastewater with the "electrochemical sacrifice magnesium anode method+MAP method", and the purity of magnesium ammonium phosphate exceeded 80% when the voltage, initial pH, and electrolysis time were 7 V, 8, and 60 min, respectively. Peng Li et al. (2020) treated high concentration ammonium phosphate wastewater with sleeve mechanical stirring magnesium ammonium phosphate crystal-inclined plate. As the reactor speed is 250 r/min, the hydraulic retention time is 0.375 h, and the average GT value range is between 14000-20000, the average recovery of ammonia nitrogen is more than 76.00%, and the average recovery of phosphorus is more than 97.00%. Ci Fang (2018) uses hydrothermal treatment (HT), anaerobic fermentation technology (AD), and bird manure crystallization technology (MAP), the higher the HT temperature, the greater the decrease in fixation content, and the higher the proportion of phosphorus in solid phase (more than 75%). The "HT plus AD" process improves the crystallinity of HAP, and 77-86% of liquid soluble phosphate is successfully recovered in the form of bird manure. Zhi Cao et al. (2020) used the financial analysis method to study the economic feasibility of MAP technology and believed that the phosphorus by MAP technology is more suitable for the sewage treatment plant of 30,104 m^3/d.

3 CHEMICAL COAGULATION AND PRECIPITATION METHOD

Jianyong Che et al. (2021) used rare earth phosphorus-containing wastewater to generate $LaPO_4$ and $CePO_4$ precipitation within the range of 2.70 to 8.66 pH; $LaPO_4$ and $CePO_4$ can be converted into La $(OH)_3$ and Ce $(OH)_3$ within the range of 14.5 to 16.0 to realize the regeneration of precipitant and the recovery of phosphorus. When the phosphorus content is 30 mg/L, the phosphorus recovery can reach 99.72%. Lin Qiu et al. (2015) removed phosphorus in wastewater by iron-calcium complex precipitation, which was antagonistic when the molar ratio of iron and calcium was 1:10-1:1, but iron and calcium were synergistic when the iron molar ratio is 1:1-10:1, and the best when the Fe/Ca molar ratio is 7:3-4:1. When the phosphorus concentration was 100 mg/L. The optimization conditions for phosphorus removal were as follows. Iron:calcium = 2.4:1; calcium complex:phosphorus = 1.5:1; pH = 7; mixing speed (FMS) = 100 rpm. Compared with the traditional single iron salt precipitation technology, the phosphorus removal technology of iron and calcium complex has the advantage of low cost, and the phosphorus content in the precipitation (measured by P_2O_5) reaches 29.77%.

Phosphate wastewater is often treated by chemical precipitation, electrodialysis, micro electrolysis, and advanced oxidation. Electrophenton technology can be oxidized secondary phosphate $(H_2PO_2^-)$ into positive phosphate (PO_4^{3-}), and deposit to form iron phosphate products to achieve the purpose of oxidizing secondary phosphate in water and synchronous phosphorus removal (Han 2018). The Electro-Fenton (E-Fenton) technology, photoelectric-Fenton (UV/E-Fenton) technology, and UV/Fe^{2+} activated potassium sub-phosphate technology $(UV/Fe^{2+}-K_2S_2O_8)$ were introduced to treat the sub-phosphate wastewater, which provides the strong oxidation activity of OH radical and SO_4^- radical in the process of sub-phosphate oxidation, and provides Fe^{3+} for the formation of iron phosphate precipitation. The whole reaction system will not produce secondary pollution, which reduces the cost and operation complexity of agent injection and can realize the resource utilization of phosphorus.

4 ADSORPTION METHOD

4.1 *Active carbon adsorption*

Hongyan Nan (2017) reported the preparation of biochar from landscaping waste. The influence of pyrolysis temperature on physicochemical properties of biochar was studied. The adsorption mechanism of ammonia nitrogen and phosphorus in wastewater was explored, and the influence of different biochar on soil physicochemical properties and migration of nitrogen and phosphorus in soil was studied. The highest phosphorus adsorption amount of the prepared biochar was 3.21 mg/g. Qingchuan Jia (2020) prepared biochar composite material (MgCS-char) loaded with nano magnesium oxide from the raw material. Then, the biochar was used to remove the phosphate in the water. After that, the ammonia nitrogen in the water was removed from the adsorbed product MgCS-char-P. It was shown that the adsorption capacity of MgCS-char obtained at 5°C/min reached 399.6 mg/g. The coexisting anions HCO_3^-, NO_3^-, and Cl^- had no obvious effect on the phosphate-selective sorption and treatment capacity of MgCS-char. The adsorption of phosphate by MgCS-char is an endothermic reaction, and the maximum phosphorus adsorption capacity of the material reaches 399.6 mg/g. The MgCS-char and the phosphorus-removal product MgCS-char-P were used in domestic wastewater. Xi Gou (2020) prepared modified biochar (KMg-BC) with KOH and $MgCl_2$ modified orange peel. The best condition is: usage of KOH and $MgCl_2$ is 2wt%, the reaction time is 2 h, the reaction temperature is 600°C, the usage of KMg-BC is 4 g L^{-1} and the initial pH of the solution is 3-9. KMg-BC provides Mg^{2+} precipitated by phosphate via loads of MgO, $MgCl_2$, and Mg (OH) $_2$. It was concluded that KMg-BC/P provides soil improvement effects. Yifan Ding et al. (2021) prepared magnesium-modified biochar and magnesium silicate with corn cob and straw to study the effects of pH and alkalinity on phosphorus absorption by the material to recover phosphate fertilizer from animal waste. Studies have shown that abundant agricultural waste can be used to synthesize phosphorus-absorbing materials and can be used as fertilizers.

4.2 *Biomass adsorption method*

Banafsheh Faraji (2020) uses a modified walnut wood shell (MWWS) and an almond wood shell (MAWS) as a new anion exchanger to remove phosphorus from the water and then reuse it in the soil. New granular chitosan (gamma-AlOOH@CS) was prepared by Peigen Ma et al. (2020) to recover the phosphate. As the usage of the material is 45.82 mg/g, the semi-adsorption capacity is 0.5 hours. With sodium hydroxide solution as a solvent, phosphoric acid as precipitating agent, under the condition of hydrothermal reaction after the treatment of adsorbed material can be analyzed from chitosan, aluminum phosphate, sodium dihydrogen phosphate. The purity reaches the industrial standard and chitosan can be reused. Chen Kun (2020) chose wheat straw, corn straw, straw treatment aquaculture wastewater, and a tertiary biological matrix treatment system was established to remove phosphorus from pig wastewater, the total phosphorus (TP) and dissolution of state inorganic P phosphorus (DIP) removal rate can reach 33.03% and 28.62%. The chitosan iron (II) complex shows excellent adsorption capacity for phosphate. The adsorption capacity (70.5 mg/g) and pH value adaptability were excellent (Li et al. 2018). A beneficial element of iron, phosphorus, and nitrogen was successfully enriched in chitosan complexes.

4.3 *Metal oxide adsorption method*

Wei Zhu (2017) prepared the phosphorus removal sorbent (A-MDR) from the magnesium agent desulphurization waste residue (MDR) of the thermal power plant and treated the phosphorus-containing wastewater with a concentration of 50 mg/L under the condition of the initial solution pH of 9.0, the addition amount of 12 g/L and the reaction time of 120 min, and its phosphate removal rate could reach 98.4%. Meanwhile, A-MDR is reproducible to resist the effects of Cl^-, SO_4^{2-} and NO_3^{2-} isolations. The maximum adsorption amount of phosphorus by A-MDR is 27.548 mg/g, which is higher than the phosphorus removal adsorbent prepared by most modified waste residues. The mechanism of phosphorus removal adsorption is that the phosphate can form a chemical bond with the Mg^{2+} released by A-MDR and is firmly fixed to the A-MDR surface, or the phosphate has a ligand exchange with the hydroxyl group in the Mg-OH functional group and is adsorbed. Lingjun Kong et al. (2019) used the sludge containing inorganic minerals such as iron, aluminum, silicon, and calcium in the wastewater to achieve phosphorus removal. Tingjiao Liu (2019) prepared new mafic metal-based carbon nanomaterials from magnetic hydrotalcite and chlorella waste. MgFe-450 has the largest phosphorus adsorption capacity of 28.3 mg/g, and the order of coexisting ions on phosphorus adsorption capacity is $CO_3^{2-} > SO_4^{2-} > NO_3^- > Cl^-$. Binbin Zhou (2017) uses hydroxyl iron oxide to remove phosphorus in wastewater. The saturated adsorption amount of phosphorus can reach 19.56 mg/g, and the maximum phosphorus removal rate is 92.5%. As the concentration of NaOH was 1.5 mol/L, the phosphor-desorption rate was 98.57%. The Fe_3O_4, FeOOH adsorbent was prepared by the coprecipitation method with a desorption rate greater than 98%. The main component of the recovered products is HAP, the phosphorus content is 38.25% (calculated by P_2O_5), and the recovery rate is more than 90%. Yongqiu Li (2017) uses iron sulfate, ferrous iron sulfate, sodium hydroxide, and zirconium oxychloride to prepare four different iron magnetic nano zirconium/iron oxide, the adsorbent has the characteristics of easy separation and high adsorption capacity. After three times of desorption by 0.1MNaOH solution, the adsorption efficiency remained above 70%. Mengmeng Cui (2016) synthesized Ferror Mox (FM) to treat phosphorus-containing wastewater, and the removal rate of phosphate reached 99.14% under the adsorption time of 60 min, initial pH of 2, 7 g L^{-1} of FM concentration, and initial phosphorus concentration of 10 mg. L^{-1}.

LDHs are compounds formed by the accumulation of interlayer anions and positively charged laminate. The chemical composition is: $[M^{2+}{}_{1-x}M^{3+}x\,(OH)_2]\,x+ (A^{n-})_{x/n}.mH_2O$, with acid-base bifunctionality, interlayered ion exchangeability and memory effect. Ye Yuan (2018) prepared Mg-Al, Mg-Fe, Zn-Al, Zn-Fe, and Cu-Al to absorb phosphorus in the wastewater, and then recovered phosphorus resources by desorption. The Zn-Al hydrotalcite has the maximum phosphorus adsorption capacity (41mg.g^{-1}). The addition of coexisting anions will reduce their phosphorus adsorption capacity, according to $CO_3^{2-} > SO_3^{2-} > SO_4^{2-} > NO_3^- > Cl^-$, and the Zn-Al hydrotalcite phosphorus removal mechanism is the result of the joint action of surface adsorption and ion exchange.

4.4 *Adsorption method of inorganic materials such as clay minerals*

Clay mineral (clay minerals) refers to the hydrous aluminosilicate mineral with a layered structure, mainly concluding kaolinite, mormoronite, mica, hydrate, chlorite, etc. The chemical composition is SiO_2, Al_2O_3, and water, with a good adsorption effect. Diatomite mainly contains SiO_2 and a small amount of Fe_2O_3, CaO, MgO, Al_2O_3, and organic impurities. Ning Duan et al (2013) prepared composites with diatomite as the matrix material and zeolite as the compound. The phosphorus removal capacity reached 99.95%. The best composite conditions are diatomite: zeolite =6: 4, binder concentration is 15%, the roasting temperature is 800°C, and roasting time is 90 min. After 5 times of regeneration and reuse, the phosphorus removal capacity of the composite adsorbent can reach 99% of the fresh material, and the activity is almost completely recovered. The general chemical formula of zeolite is: $A_mB_pO_{2p}\,nH_2O$, and the structure formula is A (x/q) [(AlO_2) x (SiO_2) y] n (H_2O). Zhao Yan et al (2020) modified zeolite and steel slag and desulfurization gypsum in a certain proportion to remove phosphorus, as the raw material ratio of zeolite: steel slag: desulfurization gypsum is 4: 4: 1: 1 (mass ratio), pH=8, solid-liquid ratio of 1: 6, reaction time is 90 min, the removal effect of the mixture on ammonia nitrogen and total phosphorus in wastewater reached 29.41% and 78.44% respectively. Zhenyang Song (2018) reported the adsorbent of natural sheet zeolite, eight-sided zeolite, volcanite and melite was modified by four methods of high temperature roasting. Then, NaOH, $AlCl_3$ and $FeCl_3$ were dipped to remove phosphate. Marine foam ((Si_{12})(Mg_8) $O_{30}(OH)_4(OH_2)_4·8H_2O$) is a fibrous aqueous magnesium silicate with the largest specific surface area (up to 900 m^2/g) and a unique content pore structure with good adsorption, rheological and catalytic properties. The maximum adsorption amount of N and P (Dai 2014) was 1.165 mg/g and 1.121 mg/g, respectively, which can be regenerated with NaOH solution. Mingyang Xu (2020) prepared adsorbent material using red mud with weathered long-time and greatly reduced activity of the yard slope as raw material, and further improved by microwave and roasting method. Tobermolite (Tobermorite) is a hydration calcium silicate mineral with the chemical structure formula: $Ca_5Si_6O_{16}(OH)_24H_2O$. The tobermolite/$SiO_2$/$Fe_3O_4$ composites were synthesized by Shuhua Zhong (2014), which does not require complex pretreatment. These composites have good phosphorus removal efficiency and good magnetic properties, indicating an easy recovery.

5 CONCLUSION

The recovery and application method of phosphorus resources in wastewater vary according to the characteristics, technical level and reuse field of wastewater. Among them, high concentration phosphorus-containing wastewater is suitable for MAP crystallization method, and low concentration phosphorus-containing wastewater can be used by adsorption and chemical coagulation precipitation of phosphorus-containing wastewater. Recently, compound phosphorus removal methods have been widely studied and applied due to their high adaptability, high efficiency and low cost. The abolition of phosphorus is an important trend of phosphorus removal technology. At the same time, in order to promote the development of the phosphorus removal technology and the application of the phosphorus recovery resources, the relevant regulations and technical standards need to be improved.

ACKNOWLEDGMENTS

This work was financially supported by the Construction System Science and Technology Project of Jiangsu Province (China) and the Technological Innovation in Key Industries of Suzhou (NO. SYG201744).

REFERENCES

Alain Mollier, Kalimuthu Senthilkumar, Magalie Delmas, Sylvain Pellerin, Thomas Nesme. Phosphorus recovery and recycling from waste: An appraisal based on a French case study [J]. *Resources Conservation and Recycling*. 2014,87: 97–108.

Alexia Prskawetz, David Laner, Helmut Rechberger, Johanna Grames, Matthias Zessner, Miguel Sanchez-Romero, Ottavia Zoboli. Understanding feedbacks between economic decisions and the phosphorus resource cycle: A general equilibrium model including material flows [J]. *Resources Policy*, 2019, 61: 311–347.

Andrzej Matynia, Anna Stanclik, Krzysztof Piotrowski, Nina Hutnik. Comparison of the effect of copper (II) and zinc (II) ions in phosphorus mineral fertilizer industry wastewater on size -dependent linear growth rate of struvite crystals in phosphorus recycling process [J]. *Przemysl Chemiczny*. 2019,98 (11): 1751–6.

Andrzej Matynia, Anna Stanclik, Nina Hutnik. Size effect of struvite crystals on their linear growth rate in phosphate (V) and aluminum (III) or iron (III) ions-containing in phosphorus recycling aqueous solutions [J]. *Przemysl Chemiczny*. 2020,99 (2): 286–92.

Anming Yang, Bingyu Zheng, Shujun Zhang. Ammonium magnesium phosphate crystallization method is used for phosphorus recovery in wastewater treatment process [J]. *Environmental Engineering*, 2019,37 (06): 90–95.

Anyin Huang, Diansheng Wang, Feng Liu, Mengmeng Cui. Adsorption properties of synthetic water iron ore on phosphorus-containing wastewater [J]. *Environmental Science*, 2016,37 (09): 3498–3507.

Bailian Xiong, Jinzhong Zhang, Juan Dai, Yang Liu. Syntaneous treatment of nitrogen and phosphorus in wastewater with compound modified seolite [J]. *Journal of Environmental Engineering*, 2014,8 (05): 1732–1738.

Baiquan Zhu, Ming Zhu, Yan Zhao, Yanjuan Liu. Study on the synchronous removal of ammonia nitrogen and total phosphorus from wastewater by modified zeolite and desulfurization gypsum [J]. *New chemical materials*, 2020,48 (03): 265–268.

Banafsheh Faraji, Mahboubeh Zarabi, Zahra Kolahchi. Phosphorus removal from aqueous solution using modified walnut and almond wooden shell and recycling as soil amendment [J]. *Environmental Monitoring and Assessment*. 2020,192 (6): DOI: 10.1007/s10661-020-08326-x

Baozhong Ma, Jianyong Che, Wenjuan Zhang. Rare earth phosphorus removal and precipitant regeneration in phosphorus-containing wastewater [J]. *Chinese Journal of Nonferrous Metals*, 2021,31 (01): 114–124.

Binbin Zhou. Study on phosphorus removal by superconducting magnetic adsorption based on hydroxxy iron oxide [D]. Tutor: Li Yiran. East China Technology University, 2017.

Bing Li. Treatment of phosphogypsum leachate combined with by-production of magnesium ammonium phosphate sustained-release fertilizer [J]. *Non-metallic ore*, 2020,43 (05): 17–20.

Carsten Gellermann, Daniel Frank, lexander Maurer, Martin Bertau, Peter Fröhlich, Reinhard Lohmeier Dipl. -Ing. The vital Element Phosphorus and Phosphate Recycling [J]. *Chemie in Unserer Zeit.* 2018,52 (5): 350–8.

Changwang Jiang, Jing Li, Jon Zhang, Wansheng Shi, Wenqian Du, Wenquan Ruan. Study on removal characteristics of ammonia nitrogen in landfill leachate by flocculation reinforced magnesium phosphate precipitation method [J]. *Environmental Pollution and Prevention*, 2019,41 (11): 1304–1308.

Chen Baidi, Fu Shanming, Kong Lingjun, Liu Mingxiang, Zhang Fagen, Zhu Yuting. Guangzhou University,. Research on the key technology of regulating multistage porous biochar adsorption and induced crystallization recovery of urban wastewater phosphorus [Z]. Project approval No.: 201607010311. Appraisal unit: Guangzhou University. Identification Date: 2019-07-15.

Chun Zhao Junfeng Li, Weiwei Liu, Xinlin He, Zhaoyang Wang. Factors influencing the removal of phosphorus and the purity of recycling struvite in wastewater by the electrochemical sacrificial magnesium anode method [J]. *Science of Advanced Materials.* 2019,11 (1): 128–34.

Chunying Wang, Dazhao Zhang, Meng Wu, Min Chen, Shi Xu. Progress of wastewater phosphorus removal technology [J]. *Nonferrous Metals Science and Engineering*, 2019,10 (2): 97–103.

Ci Fang. *Research on phosphate transformation, crystallization and resource utilization of pig manure based on hydroheat treatment* [D]. Tutor: Jiang Rongfeng, Zhang Tao. China Agricultural University, 2018.

David A. Sabatini, Elizabeth C, Yifan Ding. Butler. Phosphorus recovery and recycling from model animal wastewaters using materials prepared from rice straw and corn cobs [J]. *Water Science and Technology.* 2021, 83 (4).

El Wali M, Golroudbary SRA, Kraslawski A. Environmental sustainability of phosphorus recycling from wastewater, manure and solid wastes [J]. *Science of the Total Environment.* 2019,672: 515–24.

EW Mohammad, GS Rahimpour, K Andrzej. Impact of Recycling Improvement on the Life Cycle of Phosphorus [J]. *Chinese Journal of Chemical Engineering.* 2019,27 (5): 1219–29.

Ghulam Abbas, Lin Qiu, Meng Zhang, Ping Zheng, Xiaoqing Yu. Phosphorus. removal using ferriccalcium complex as precipitant: Parameters optimization and phosphorus-recycling potential [J]. *Chemical Engineering Journal.* 2015,268: 230–5.

Gongwu Song, Huaili Sha, Shuhua Zhong, Yu He. Hydrothermal synthesis of easy-recycled tobermorite/SiO_2/Fe_3O_4 composites for efficient treatment of phosphorus in wastewater [J]. *Desalination and Water Treatment*, 2014,52 (22-24): 4305–4313.

Haitao Zhang, Jiongliang Yuan, Long Yi, Peigen Ma, Wenming Ding. Total recycle strategy of phosphorus recovery from wastewater using granule chitosan inlaid with gamma-AlOOH [J]. *Environmental Research*, 2020, 184.

Hongju Han. *Oxidation of hypophosphite by photo electro-fenton process for synchronous removalphosphatein-water* [D]. Tutor: Fu Min, Guan Wei. Chongqing Technology and Business University, 2018.

Hongyan Nan. The adsorption characteristics of garden waste biochar on nitrogen and phosphorus and its effect on soluaching of soil nitrogen and phosphorus [D]. Mentor: An Qiang, Gou Jizo. Chongqing University, 2017.

Hui Hua, Songyan Jiang, Xin Liu, Xuewei Liu, You Zhang, Zengwei Yuan. Historic Trends and Future Prospects of Waste Generation and Recycling in China's Phosphorus Cycle [J]. *Environmental Science & Technology.* 2020,54 (8): 5131–9.

Jian Wang. *Study on phosphorus phosphorus Process in Industrial Wastewater* [D]. Instructor: Wu Huadong. Wuhan University of Engineering, 2019.

Jiawen Li, Peng Li, Ruixia Hao, Tong Sun, Xuyuan Wu. Ammonium magnesium phosphate crystallization-precipitation reactor construction and flow simulation [J]. *Environmental Science in China*, 2020,40 (04): 1523–1530.

Keke Ma, Lv Zhou, Yilei Shi, Zhi Cao. Economic analysis of phosphorus recovery in biogas slurry by magnesium ammonium phosphate crystallization method [J]. *Industrial Water Treatment*, 2020,40 (10): 43–46.

Keming Wu, Ning Duan, Yaowen Gao, Yinfeng Zhang. Preparation of diatomite composite adsorbent and its properties of nitrogen removal in wastewater [J]. *Silicate Bulletin*, 2013,32 (08): 1528–1533.

Kun Chen. *Characteristics of phosphorus transformation of high-load pig wastewater treated with straw biological matrix material* [D]. Instructor: Lv Dianqing. Hunan Normal University, 2020.

Mingyang Xu. *Preparation, characterization and characterization of red mud-based granular phosphorus removal materials* [D]. Tutor: Wang Liping, Zhao Yaqin. China University of Mining and Technology, 2020.

Qingchuan Jia. *Removal of nitrogen and phosphorus in water by magnesium-modified halworms* [D]. Mentor: Sun Qina. Yanshan University, 2020.

Ruixia Li. *Study on nitrogen and phosphorus removal from wastewater by non-metallic ore-guite* [D]. Instructor: Zhu Jingping. Southwestern University of Science and Technology, 2019.

Sida Ouyang. *Preparation of magnesium-rich C₃A and the coremoval of ammonia nitrogen and phosphorus in pig biogas slurry* [D]. Tutor: Zhang Ping. Nanchang University, 2020.

Tingjiao Liu. *Remoof phosphorus and algae in water and its resource utilization* [D]. Instructor: Xu Yin. Xiangtan University, 2019.

Wei Zhu. *Study on phosphorus adsorption properties and mechanism of modified magnesium residue* [D]. Tutor: Zhao Dan. Suzhou University of Science and Technology, 2017.

Xi Gou. *Study on phosphorus removal properties and soil improvement of biochar* [D]. Tutor: Xie Yanhua, Zhu Ruigen. Chengdu University of Technology, 2020.

Xufeng Yang. Brief Analysis of wastewater phosphorus removal process [J]. *Energy saving and environmental protection*, 2020, (03): 48–49.

Yaru Xu. *Study on the resource treatment technology of corn deep processing wastewater based on nitrogen and phosphorus recovery* [D]. Instructor: Yu Xiaohua. Beijing Jiaotong University, 2018.

Ye Yuan. *Modification of hydrotalc preparation and deep phosphorus removal* [D]. Tutor: He Zhengguang. Zhengzhou University, 2018.

Yiran Li. *Removal of high concentration of inorganic nitrogen and enrichment of nitrogen and phosphorus in industrial wastewater* [D]. Instructor: Luo Xubiao. Nanchang Hangkong University, 2018.

Yongqiu Li. *Preparation of magnetic nanozirconium/iron oxide and its adsorption of phosphorus in wastewater* [D]. Tutor: Zhang Chang, Wang Qiao. Hunan University, 2017.

Zhenyang Song. *Study on the adsorption of phosphorus in wastewater by natural and modified adsorbent* [D]. Tutor: Feng Sumin, Liang Hao. Hebei University of Science and Technology, 2018.

Zhichao Yin. *Experimental study on phosphorus removal and recovery from aquaculture wastewater by dolomite* [D]. Tutor: Fu Ying, Chen Qingfeng. University of Jinan, 2020.

Advances in Energy, Environment and
Chemical Engineering – Abdullah & Osman (Eds)
© 2023 The Author(s), ISBN: 978-1-032-36083-6

Influence of internal resistance on underwater corrosion electrostatic field of submarine

Chun-Yang Liu, Wei-Ting Lin & Yu-Long Zhao*
College of Information and Communication, National University of Defense Technology, Wuhan, China

ABSTRACT: The internal resistance of the submarine is an important factor affecting the underwater corrosion electrostatic field. Based on the finite element method, the model of underwater corrosion electrostatic field is established, and the effects of internal resistance on corrosion current and underwater electric field at different resistivity of the coating were studied. The results show that the corrosion current and underwater corrosion electrostatic field decreases with the increase of internal resistance. In addition, the lower the resistivity of submarine coating, the greater the influence of internal resistance on corrosion current and underwater corrosion electrostatic field.

1 INTRODUCTION

Because seawater is a highly corrosive environment, galvanic corrosion occurs between different metal structures of the submarine. The current generated by the submarine's metal structure corrosion produces an underwater corrosion electric field, which is an important physical field of the submarine. It cannot only be used to track and locate the submarine, but also be used as the fuse of underwater weapons such as mines (Jiang 2017; Schaefer 2016). Studying the characteristics of a submarine's underwater corrosion electric field is conducive to controlling, weakening, and even eliminating the corrosion electric field in the process of submarine design and manufacturing. The study of the relationship between internal resistance and underwater corrosion electrostatic field is of great military significance. It is useful to improve the survival ability of submarines in modern war and give better play to its combat effectiveness.

The underwater corrosion electric field of the submarine is closely related to the corrosion current, and the internal resistance is an important factor affecting the submarine's surface corrosion current. On the other hand, with the increase in submarine service time, the coating resistivity of the submarine will also change, which will affect the corrosion current too. Numerical methods such as the finite element method have been widely used in the field of the submarine electric field with high accuracy (Gao 2015; Liu 2018; Xu 2019), so the commercial finite element software COMSOL is used to establish the submarine's underwater corrosion electrostatic field simulation model, and the influence of internal resistance on underwater corrosion electrostatic field in the whole life cycle of submarine be analyzed.

2 MODELING OF UNDERWATER CORROSION ELECTRIC FIELD

2.1 *Governing equations and boundary conditions*

In the seawater, the potential value ϕ_l meets the Laplace's equation, as shown in Equation (1).

$$\nabla^2 \phi_l = \frac{\partial^2 \phi_l}{\partial x^2} + \frac{\partial^2 \phi_l}{\partial y^2} + \frac{\partial^2 \phi_l}{\partial z^2} = 0 \tag{1}$$

---

*Corresponding Author: wuzhangyuanlt@163.com

The relationship between the current density vector $\boldsymbol{J}$ and the gradient of the potential $\nabla\phi_l$ is following:

$$\boldsymbol{J} = -\sigma_l \nabla\phi_l \tag{2}$$

where σ_l represents the electrical conductivity of seawater. The underwater corrosion electrostatic field of the submarine $\boldsymbol{E}$ satisfies the constitutive relation:

$$\boldsymbol{J} = -\sigma_l \boldsymbol{E} \tag{3}$$

On the surface of anode and cathode, the relationship between electrode potential and current density meets the polarization curve of anode and cathode materials respectively (Kim 2018):

$$J_a = \sigma_l \frac{\partial\phi}{\partial n} = f_a(\phi) \tag{4}$$

$$J_c = \sigma_l \frac{\partial\phi}{\partial n} = f_c(\phi) \tag{5}$$

where ϕ is the electrode potential; $\phi = \phi_s - \phi_l$. J_a and J_c are the anode current density and cathode current density, respectively. $f_a(\phi)$ and $f_b(\phi)$ represent the anode polarization equation and cathode polarization equation, respectively.

Because the air conductivity is 0, the reverse current density at the interface between air and seawater is 0, as shown in Equation (1).

$$J = \sigma_l \frac{\partial\phi_l}{\partial n} = 0 \tag{6}$$

The normal component of seawater-seabed interface current meets the natural boundary conditions:

$$\sigma_l \frac{\partial\phi_l}{\partial n} = \sigma_b \frac{\partial\phi_b}{\partial n} \tag{7}$$

where σ_b is seabed conductivity; ϕ_b is the potential value of seabed. And the current density of the seawater area far enough away from the submarine is 0.

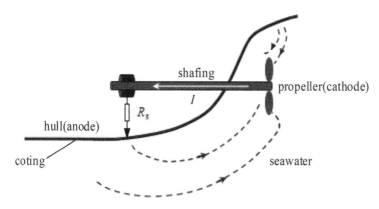

Figure 1. Schematic diagram of submarine corrosion current circuit.

As shown in Figure 1 (Zhao 2022), due to the existence of shafting between hull and propeller, there is internal resistance between hull and propeller, and the potential difference between hull and propeller is:

$$\phi_{sa} - \phi_{sc} = IR_g \tag{8}$$

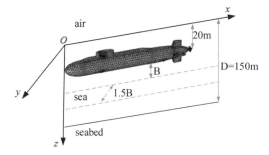

Figure 2.　Submarine model and coordinate.

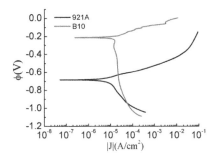

Figure 3.　Polarization curve of 921A and B10.

where ϕ_{sa} is the anode potential; ϕ_{sc} is the cathode potential; I is the submarine's surface corrosion current; R_g is the internal resistance between anode and cathode.

The potential difference between both sides of the coating is:

$$\Delta\phi_{s,c} = IR_c = J_n\rho_c \tag{9}$$

where J_n is the anode surface normal phase current density; R_c is the resistance of the coating; ρ_c is the resistivity of coating.

2.2　Model and test

The model of the submarine is established by COMSOL 5.4, as shown in Figure 2. The submarine model includes a hull, sail, stern rudder, shaft, and propeller. The impressed current cathodic protection system is closed. The sea level is the XOY plane, and the projection point of the bow on the XOY plane is the origin of the three-dimensional coordinate system. The positive direction of the x-axis is from the bow to the stern. The positive direction of the y-axis is from the starboard to the port, and the positive direction of the z-axis is vertically downward. The seawater depth is 150 m. The submergence depth of the submarine is 20 m. The conductivity of seawater and seabed are 4 S/m and 0.1 S/m, respectively.

The hull is made of alloy steel 921A, which is covered with an anti-corrosion coating. The propeller is exposed, and the material is nickel aluminum bronze B10. The polarization curves of the 921A and B10 are tested on CS310 electrochemical workstation, which is shown in Figure 3. It can be seen from Figure 3 the open circuit potentials of 921A and B10 are -0.68 V and -0.22 V, respectively. So, the hull is the anode and the propeller is the cathode, and the galvanic corrosion circuit "hull–seawater–propeller–shaft–hull" is formed in seawater. The COMSOL secondary current module is used to solve the corrosion current on the surface of the submarine and the current density of seawater, and then the electric field E in the seawater solution domain can be obtained: $J = -\sigma_l E$.

As we know, with the increase in submarine service time, the resistivity of hull coating will change greatly. The resistivity of the early, middle, and late service of the submarine are assumed 50 k$\Omega \cdot$ m^2, 5k$\Omega \cdot$ m^2, and 500$\Omega \cdot$ m^2, respectively. The internal resistance is set as 0.001–100Ω. The submarine surface corrosion current and underwater electrostatic field under different internal resistance are solved and compared.

3　RESULTS AND ANALYSIS

3.1　Effect on corrosion current

The corrosion current of the submarine I under different internal resistance R_g is shown in Table 1. It can be seen from Table 1 that the corrosion current decreases with the increase of internal resistance.

When the internal resistance increases from 0.001Ω to 0.1Ω, the current variation is small. But when the internal resistance increases from $1\ \Omega$ to 100Ω, the current variation is big. It shows that the current value is mainly affected by the absolute value of internal resistance.

The current change rate ΔI is defined as:

$$\Delta I = \frac{|I_0| - |I_{R_g}|}{|I_0|} \times 100\% \tag{10}$$

where $|I_0|$ is the corrosion current when the internal resistance is 0; $|I_{R_g}|$ is the corrosion current when the internal resistance is R_g. The resistivity of the coatings is 50 k$\Omega \cdot$ m^2, 5 k$\Omega \cdot$ m^2, and 500 $\Omega \cdot$ m^2, respectively. When the internal resistance is 0.001 Ω, the change rates of corrosion current ΔI are 0.03%, 0.05%, and 0.41% respectively. When the internal resistance is 1Ω, the Δ. It are 4.68%, 32.66% and 80.85% respectively. When the internal resistance is 100Ω, the ΔIt are 83.09%, 97.98%, and 99.76% respectively. It shows that with the decreases in coating resistivity, the effect of internal resistance on corrosion current increases.

Table 1. The corrosion current of submarine I under different internal resistance.

R_g/Ω	I/A			$\Delta I/\%$		
	$\rho_c = 50$k$\Omega\cdot$m^2	$\rho_c = 5$k$\Omega\cdot$m^2	$\rho_c = 500$k$\Omega\cdot$m^2	$\rho_c = 50$k$\Omega\cdot$m^2	$\rho_c = 5$k$\Omega\cdot$m^2	$\rho_c = 500$k$\Omega\cdot$m^2
0	0.023071	0.2278	1.9901	—	—	—
0.001	0.023065	0.2277	1.9819	0.03	0.05	0.41
0.01	0.023061	0.2267	1.9112	0.05	0.48	3.96
0.10	0.022959	0.2173	1.4050	0.49	4.63	29.40
1	0.021991	0.1534	0.3811	4.68	32.66	80.85
10	0.015471	0.0389	0.0459	32.94	82.91	97.69
100	0.003902	0.0046	0.0047	83.09	97.98	99.76

3.2 Impact on underwater electric field

The underwater electric field amplitude $|E|$ at the depth of 1.0 B (the boat width) under the submarine is shown in Figure 4. Due to the limitation of the length, only given the results that the internal resistances are 0.001Ω, 1Ω, and 100Ω. It can be seen from Figure 4 that the distribution of $|E|$ is the same under different internal resistances, but the intensity of $|E|$ differs by several orders of magnitude.

When the internal resistance is 0.001Ω, the maximum value of $|E|$ at different coating resistivity are about 2.4×10^{-6}V$\cdot$m^{-1}, 2.4×10^{-5}V$\cdot$m^{-1}, 2.0×10^{-4}V$\cdot$m^{-1}, respectively. When the internal resistance increases to 1Ω, the maximum values of $|E|$ at different coating resistivity are about 2.3×10^{-6}V$\cdot$m^{-1}, 1.6×10^{-5}V$\cdot$m^{-1}, 3.9×10^{-5}V$\cdot$m^{-1}, respectively. When the internal resistance increases to 100Ω, the maximum values of $|E|$ at different coating resistivity are about 4.0×10^{-7}V$\cdot$m^{-1}, 4.8×10^{-7}V$\cdot$m^{-1}, 4.8×10^{-7}V$\cdot$m^{-1}, respectively. It shows that the higher the coating resistivity, the smaller the electric field amplitude $|E|$. And the larger the internal resistance, the smaller the $|E|$.

The maximum of $|E|$ under different internal resistance R_g is shown in Table 2. The electric field change rate $\Delta|E_{\max}|$ is defined as:

$$\Delta|E_{\max}| = \frac{|E_{0\max}| - |E_{R\max}|}{|E_{0\max}|} \times 100\% \tag{11}$$

where $|E_{0\max}|$ refers to the maximum of $|E|$ when the internal resistance is 0; $|E_{R\max}|$ is the maximum of $|E|$ when the internal resistance is R_g. From Table 1 and Table 2, we can conclude that with the increase of coating resistance, the influence of internal resistance on the maximum decreases. And

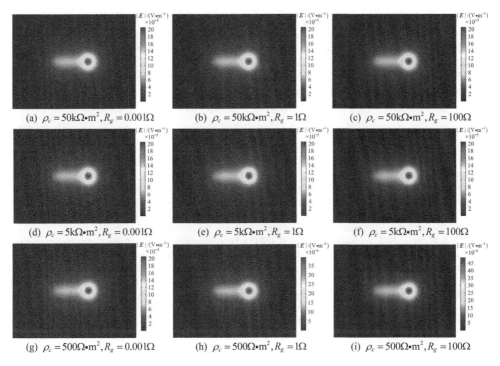

(a) $\rho_c = 50\mathrm{k}\Omega\cdot\mathrm{m}^2, R_g = 0.001\Omega$ (b) $\rho_c = 50\mathrm{k}\Omega\cdot\mathrm{m}^2, R_g = 1\Omega$ (c) $\rho_c = 50\mathrm{k}\Omega\cdot\mathrm{m}^2, R_g = 100\Omega$

(d) $\rho_c = 5\mathrm{k}\Omega\cdot\mathrm{m}^2, R_g = 0.001\Omega$ (e) $\rho_c = 5\mathrm{k}\Omega\cdot\mathrm{m}^2, R_g = 1\Omega$ (f) $\rho_c = 5\mathrm{k}\Omega\cdot\mathrm{m}^2, R_g = 100\Omega$

(g) $\rho_c = 500\Omega\cdot\mathrm{m}^2, R_g = 0.001\Omega$ (h) $\rho_c = 500\Omega\cdot\mathrm{m}^2, R_g = 1\Omega$ (i) $\rho_c = 500\Omega\cdot\mathrm{m}^2, R_g = 100\Omega$

Figure 4. Underwater electric field distribution at the depth of 1.0 B.

the variation law of $|\boldsymbol{E}|$ with R_g is consistent with that of I with R_g. This is because the $|\boldsymbol{E}|$ of the submarine can be equivalent to the superposition of the electric field of the point current source, and the electric field of the point current source is proportional to the current intensity.

Table 2. The maximum value of $|\boldsymbol{E}|$ at different internal resistance.

R_g/Ω	$E\|_{max}/(\mathrm{V/m})$			$\Delta|E|_{max}/(\%)$		
	$R_c = 50$ kΩ	$R_c = 5$ kΩ	$R_c = 500$ Ω	$R_c = 50$ kΩ	$R_c = 5$ kΩ	$R_c = 500\Omega$
0	2.3854×10^{-6}	2.3553×10^{-5}	2.0549×10^{-4}	—	—	—
0.001	2.3846×10^{-6}	2.3541×10^{-5}	2.0469×10^{-4}	0.03	0.05	0.39
0.01	2.3843×10^{-6}	2.3439×10^{-5}	1.9739×10^{-4}	0.05	0.48	3.94
0.10	2.3738×10^{-6}	2.2463×10^{-5}	1.4511×10^{-4}	0.49	4.63	29.38
1	2.2737×10^{-6}	1.5860×10^{-5}	3.9368×10^{-5}	4.68	32.66	80.84
10	1.5996×10^{-6}	4.0254×10^{-6}	4.7410×10^{-6}	32.94	82.91	97.69
100	4.0344×10^{-7}	4.7570×10^{-7}	4.8396×10^{-7}	83.09	97.98	99.76

4   CONCLUSIONS

The model of submarine underwater corrosion electrostatic field is established by the finite element method, and the influence of internal resistance on underwater corrosion electrostatic field is analyzed. The research results are as follows.

(1) The corrosion current decreases with the increase of internal resistance, and the values of the resistivity of coatings are $500\Omega \cdot \mathrm{m}^2$, $5\mathrm{k}\Omega \cdot \mathrm{m}^2$, and $50\mathrm{k}\Omega \cdot \mathrm{m}^2$, respectively. The maximum of ΔI can reach 99.76%, 97.98%, and 83.09%, respectively.

(2) The lower the resistivity of submarine coating, the greater the influence of internal resistance on underwater corrosion current. The internal resistance is 1Ω, the ΔI is 80.84% when the resistivity of coating is $500\Omega \cdot m^2$. But when the resistivity of coating is $50k\Omega \cdot m^2$, the ΔI is only 4.68%.

(3) The variation law of electric field with internal resistance is almost consistent with that of corrosion current with internal resistance.

REFERENCES

Cao, Y. & Ji, D. & Zhu, W. B. (2015). Finite element model simulation analysis of SE field of ship. *Ship Science and Technology*, 37 (7): 69–72.

Jiang, R. X. & Zhang, J. W. & Lin, C. H. (2017). Study of quick prediction method for ship corrosion related static electric field based on point charge source model. *Acta Armamentarii*, 38 (4): 735–743.

Kim, Y. S. & Lee, S. K. & Kim, J. G. (2018). Influence of anode location and quantity for the reduction of underwater electric fields under cathodic protection. *Ocean Engineering*, 163: 476–482.

Liu, C. Y & Wang, X. N. & Wang, X. J & Miao, H. (2018). Simulation analysis on influence of propeller coating on submarine's underwater electrostatic fields. *Chinese Journal of Ship Research*, 13 (S1): 182–188.

Schaefer, D. & Doose, J. & Pichlmaier, M. (2016). Conversion of UEP signatures between different environmental conditions using shaft currents. *IEEE Journal of Oceanic Engineering*, 41: 105–111.

Xu, Q. L. & Wang, X. J. & Zhang, J. C. (2019). Influence of temperature on the cathodic protection and corrosion electrostatic field of ships. *Journal of National University of Defense Technology*, 41 (4): 182–189.

Zhao, Y. L. &Liu, C. Y. & Zhou, D. (2022). Estimation of submarine underwater corrosion electric field based on equivalent circuit and point current source. *Journal of National University of Defense Technology*, 44 (3): 176–183.

*Advances in Energy, Environment and
Chemical Engineering – Abdullah & Osman (Eds)*
© 2023 The Author(s), ISBN: 978-1-032-36083-6

Determination of content of recycled polyethylene and polyethylene terephthalate blends

Chuan Luo
Ningbo Customs District Technology Center, Ningbo, China

Shujun Dai
Ningbo Cross-border E-commerce Promotion Center, Ningbo, China

Fengping Ni
Ningbo Institute of Inspection and Quarantine Science and Technology, Ningbo, China

Chen Ruan
Ningbo Cross-border E-commerce Promotion Center, Ningbo, China

Haisong Ying & Lifeng Yuan*
Ningbo Customs District Technology Center, Ningbo, China

ABSTRACT: A determination method of the composition of recycled polyethylene and polyethylene terephthalate blends was studied by differential scanning calorimetry and infrared spectroscopy. Rapid identification of polyethylene and polyethylene terephthalate blends by infrared spectroscopy, and determination of PE and PET content in polyethylene and polyethylene terephthalate blends by differential scanning calorimetry. Influencing factors such as heating rate, cooling rate, sample mass, and sample thickness were investigated. The standard enthalpy value method and the melting enthalpy area normalization method are compared. It is found that the area normalization method of melting enthalpy is more beneficial to the determination of the composition of polyethylene and polybutylene terephthalate blends. For the polyethylene and polyethylene terephthalate bands, the repeatability of polyethylene and polyethylene terephthalate was between 0.15% and 7.42%.

1 GENERAL INSTRUCTIONS

Recycled plastic is a green industry encouraged and advocated by Chinese national policies, and it is also an important green field where the entire society works to reduce and recycle a large amount of plastic solid waste. With the gradual closure of domestic supervision on imported waste plastics, major waste plastics exporting countries such as the United States and Europe have lost the world's largest solid waste place. Many countries have invested their production capability in Malaysia, Thailand, Vietnam, and other countries. After processing waste plastics into recycled plastic particles, they continue to be exported to the Chinese market. With the gradual implementation of the GB/T 40006 series of national standards for recycled plastics, further requirements are put forward for the quality of recycled plastics. The method of detecting polymer composition becomes the premise of judging whether the sample is suitable for the requirements of national standards, so it is necessary to establish a series of detection methods for the composition of different polymers.

According to the data of port inspections in recent years, it is found that some recycled polyethylene derived from the packaging bags of washing products will contain a small amount

---

*Corresponding Author: yuanlifeng57@126.com

DOI 10.1201/9781003330165-57

of polyethylene terephthalate, while some recycled polyester derived from composite films will contain a small amount of polyethylene terephthalate. polyethylene.

For the determination of the composition of polymers, Dai et al. have determined polyethylene and polypropylene blends by infrared spectrum, Yang and Zhang et al. used pyrolysis gas chromatography and DSC melting point test to determine the content of each component in the PA6 and PA66 blends, Yuan and Zhang et al. used pyrolysis gas chromatography-mass spectrometry with solid dispersion method to determine the composition of PC/ABS blends and PA6/PA66 blends, The composition detection of polyethylene and polyethylene terephthalate blends has not been reported. Considering that both are crystalline polymers, this study intends to use infrared spectroscopy to identify polyethylene and polyethylene terephthalate rapidly qualitatively. The specific content of each component in the glycol blend was quantitatively determined by differential scanning calorimetry.

2 EXPERIMENT

2.1 Instrument and reagents

Equipment and reagents are listed below.

Nexus Fourier Transform Infrared Spectrometer (with ATR accessory) (Thermo Fisher, U. S.); 35E injection molding machine (Boy, Germany); Differential Scanning Calorimeter (Switzerland METTLER TOLEDO, DSC 3+); 40 ul standard aluminum crucible (China); Sn standard material (METTLER TOLEDO, Switzerland); high purity nitrogen, LLDPE (Zhenhai Refinery 7042); PET (Yizheng Chemical Fiber, BG85).

2.2 FT-IR test

Experimental steps: Cut the sample, and check whether the cut surface was flat and whether there were defects. Put the sample under the infrared spectrum-ATR detector, and test after compaction. The wave number is from 400 cm^{-1} to 4000 cm^{-1}; scan times are larger than 32. Find the typical absorption peaks.

2.3 DSC test

Experimental steps: Flatten the sample to (0.7±0.1) mm, cut a sample of (5.0±1.0) mg, put it into a 40 ul standard aluminum crucible, punch the lid, press it, and perform the DSC test according to the above-mentioned procedure. Select the melting peak of the second heating section after eliminating the thermal history for analysis.

3 RESULTS AND DISCUSSION

3.1 Infrared spectrometer identification

Infrared spectrum, in addition to the typical absorption peaks of polyethylene (2,840 cm^{-1}, 1,468 m^{-1} and 720 cm^{-1}), there are also characteristic peaks of polyethylene terephthalate (1,720 cm^{-1}, 1,247 cm^{-1}, and 1,123 cm^{-1}) 1, and 725 cm^{-1}), or there are characteristic absorption peaks of polyethylene (2,840 cm^{-1} and 1,468 m^{-1}) in the characteristic spectrum of polyethylene terephthalate. The above characteristic absorption peaks can quickly identify polyethylene. Whether it contains polyethylene terephthalate, or whether polyethylene is present in polyethylene terephthalate.

3.2 Influence of temperature rising program on enthalpy determination in DSC test

According to the GB/T 19466.3-2004 standard, we established the temperature program, selected the same cooling for crystallization after the heating section, then selected different heating rates for heating, and studied the changes in enthalpy and peak type of polyethylene and polyethylene terephthalate. The heating program is as follows. Heat the samples to 50°C, then heat to 290°C at a rate of 10°C/min and stay constant for 2 min. Cool the samples to 50°C at a rate of 10°C/min, stay constant for 2 min, and then heat to 290°C at a rate of 1, 2, 5, 10, 20, and 30°C/min again. Stay constant for 2 min, and cool down to 50°C at a rate of 10°C/min.

Weigh about 5 mg of polyethylene and polyethylene terephthalate for analysis according to the above temperature program, and select LLDPE for polyethylene research. The normalized melting enthalpy is listed in Table 1.

Table 1. Effect of different heating rates on melting enthalpy.

| Sample | Raise temperature rate (°C/min) | | | | | |
	1	2	5	10	20	30
PE	73.02	64.73	90.54	91.15	97.81	90.42
PET	21.83	26.39	38.45	39.65	4.367	2.903

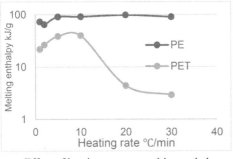 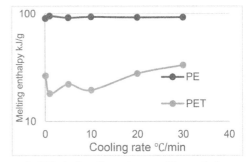

a. Effect of heating rates on melting enthalpy   b. Effect of cooling rate on melting enthalpy

Figure 1. Effects of heating and cooling rates on melting enthalpy.

It can be seen from the above Table 1 and Figure 1a that different heating rates have different effects on the melting enthalpy of different polymers. The melting enthalpy of polyolefin such as polyethylene decreases first and then increases with the increase in heating rate. The melting enthalpy of polyethylene terephthalate decreases and gradually fluctuates. And low heating rate will lead to detection efficiency. Based on the above analysis, the heating rate is selected as 10°C/min.

3.3 Effect of cooling rate on polymer crystallization in DSC test

Under different cooling programs, the regular arrangement of polymer molecular chains will change, resulting in different crystallinity of polymer, which will affect the subsequent melting determination.

Therefore, the study of the effect of cooling rate on melting enthalpy is tested according to the following procedures: the first and second raise temperature rate and the second cool temperature rate is as same as the temperature grogram in Section 2.1; the cool temperature rates are 0.1, 1, 5, 10, 20, and 30°C/min.

Weigh about 5 mg of various polyethylene and polyethylene terephthalate for analyzing them according to the above temperature programs. The normalized melting enthalpy is listed in Table 2.

Table 2.   Effect of different heating rates on melting enthalpy.

Sample	Cool temperature rate (°C/min)					
	0.1	1	5	10	20	30
PE	90.12	94.93	91.51	93.04	92.31	92.51
PET	26.50	18.06	22.02	19.45	27.83	33.32

From Table 2 and Figure 1b, we can conclude that a lower cooling rate is conducive to the regular arrangement of polymer molecular chains, but too low a cooling rate will affect the test efficiency. Therefore, the cooling procedure of 10°C/min should be selected.

3.4   *Effect of sample thickness in DSC test*

For the influence of the sample's thickness and mass, press samples of different thicknesses, and then cut another batch of samples of different mass on the samples for testing. The results are shown in Table 3.

Table 3.   Effect of different test plate thickness on melting enthalpy.

No.	PE		PET	
	Thickness (mm)	Melting enthalpy (kJ/g)	Thickness (mm)	Melting enthalpy (kJ/g)
1	0.829	89.10	0.857	13.28
2	0.706	88.70	0.721	16.79
3	0.574	88.08	0.599	13.54
4	0.475	88.79	0.485	11.85
5	0.401	89.17	0.392	9.97

The above data are plotted separately. It can be seen that the enthalpy of various polymers varies with the thickness of the test piece. Considering the above, when the thickness of the test piece is (0.7 ± 0.1) mm, most polymers have better enthalpy, so the sample thickness is set as (0.7 ± 0.1) mm.

3.5   *Effect of sample mass in DSC test*

Weigh four kinds of polymers with different mass, press them into (0.7 ± 0.1) mm sheets, test them according to the above method, and study the relationship between polymer mass and standard enthalpy. The results are shown in Table 4:

From the above results, the sample mass and melting enthalpy of polyethylene, and polyethylene terephthalate show a linear relationship, and the sample mass has little effect on the test results of melting enthalpy. When the sample mass is greater than 10 mg, the sample will occupy the whole crucible and cannot be placed. Therefore, the sample mass is 5 mg $\pm$ 1 mg.

3.6   *Accuracy testing of polyamide and polyethylene*

Since there is no suitable solvent to dissolve PE and PET at the same time, the standard sample is prepared by physical blending. Mix PE and PET with the appropriate weight and with a high-speed

Table 4. Effect of different sample amounts on melting enthalpy.

Sample	1	2	3	4	5	6
PE (mg)	1.14	2.01	4.99	8.04	9.47	11.54
Enthalpy	77.98	142.73	343.86	554.44	662.90	816.69
Standard enthalpy	68.40	71.01	68.91	68.96	70.00	70.77
PET (mg)	1.10	2.12	4.99	7.94	9.55	11.48
Enthalpy	16.38	36.36	96.56	211.44	256.13	316.73
Standard enthalpy	14.89	17.15	19.35	26.63	26.82	27.59

mixer and then use a small injection molding machine for injection molding. Take the standard sample after 10 molds for testing and analyzing according to the similar steps of PE/PET. The results are shown in Table 5 and Table 6.

Table 5. PE/PET standard mixture composition test results by standard enthalpy method.

Standard mixture	Actual ratio		Standard enthalpy method			
			PE (%)		PET (%)	
	PE	PET	Test result	Def.	Test result	Def.
PE/PET R1	11.4%	88.6%	4.45%	6.95%	139.15%	−50.55%
PE/PET R2	19.6%	80.4%	7.10%	12.50%	130.58%	−50.18%
PE/PET R3	39.6%	60.4%	14.77%	24.83%	98.28%	−37.88%
PE/PET R4	50.5%	49.5%	20.05%	30.45%	81.56%	−32.06%
PE/PET R5	59.7%	40.3%	23.58%	36.12%	65.43%	−25.13%
PE/PET R6	78.6%	21.4%	32.49%	46.11%	36.97%	−15.57%
PE/PET R7	88.4%	11.6%	37.51%	50.89%	15.76%	−4.16%

Table 6. PE/PET standard mixture composition test results by melting enthalpy normalization method.

Standard mixture	Actual ratio		Melting enthalpy normalization method			
			PE (%)		PET (%)	
	PE	PET	Test result	Def.	Test result	Def.
PE/PET R1	11.4%	88.6%	9.16%	−2.24%	90.84%	2.24%
PE/PET R2	19.6%	80.4%	14.65%	−4.95%	85.35%	4.95%
PE/PET R3	39.6%	60.4%	32.18%	−7.42%	67.82%	7.42%
PE/PET R4	50.5%	49.5%	43.70%	−6.80%	56.30%	6.80%
PE/PET R5	59.7%	40.3%	53.22%	−6.48%	46.78%	6.48%
PE/PET R6	78.6%	21.4%	73.51%	−5.09%	26.49%	5.09%
PE/PET R7	88.4%	11.6%	88.25%	−0.15%	11.75%	0.15%

It can be seen from Table 5 and Table 6 that using the standard enthalpy method, the absolute difference between the test value and the actual value of PE fluctuates between 6.95%-50.89% and the average absolute difference is 29.69%. Both are positive deviations without any rules. The absolute difference between the test value and the actual value of PET fluctuates between 4.16% and 50.55%, and the average absolute difference is 30.79%, both of which are positive deviations without any rules. Using the melting enthalpy normalization method, the absolute difference is between 0.15% and 7.42%, and the average absolute difference is 4.73%. When the PE/PET content is close, the deviation is large, and when it is low, the deviation is small. In conclusion, the accuracy of the melting enthalpy normalization method is better. For the PE/PET blend system, the repeatability of PE and PET should not exceed 8%.

4 CONCLUSION

A determination method of the composition of polyamide/polyethylene blends was established by differential scanning calorimetry. Through the study of the influence factors, the best thickness is (0.7 ± 0.1) mm, the best sample mass is (5 ± 1) mg, and 10°C/min is the optimum raise and cool temperature rate. The comparative study of actual blend samples found that the accuracy of the melting enthalpy normalization method is better. For the PE/PET blend system, the repeatability of PE and PET is between 0.15% and 7.42%.

ACKNOWLEDGEMENTS

This work was financially supported by the Science Research Project of the General Administration of Customs, P. R. China (Grant No. 2020HK245) and the Zhejiang Basis Public Science Research Project (Grant No. LGC20B040001).

REFERENCES

Dai S. J, Luo C, Zhang C., etc. (2021) Determination of Small-Amount Polypropylene in Imported Regenerated Polyethylene-Polypropylene Blends by Fourier Transform Infrared Spectroscopy. E3S Web of Conferences 261, 02067 (2021).

General Administration of Quality Supervision, Inspection and Quarantine of the People's Republic of China and Standardization Administration of the People's Republic of China and Standardization administration. (2004) Plastics – Differential scanning calorimetry (DSC) – Part 1: General principles: GB / T 19466.1-2004. Standards Press of China, Beijing.

General Administration of Quality Supervision, Inspection and Quarantine of the People's Republic of China and Standardization Administration of the People's Republic of China and Standardization administration. (2004) Plastics – Differential scanning calorimetry (DSC) – Part 3: Determination of temperature and enthalpy of melting and crystallization: GB / T 19466.3-2004. Standards Press of China, Beijing.

Liu YX, Yang Q, Zhao L, etc. (2006) Study on crystallization behaviour of LLDPE / BR, LLDPE / PP by SALS. *China Plastics Industry*. 34 (2): 48–51.

Yang L, Zhou ZC, Yao ZP, etc., (2019) Qualitative and quantitative analysis for PA6 / PA66 compounds and co-polymers. *China Synthetic Resin And Plastics*. 36 (2): 7–9,1.

Yuan LF, Luo C, Xu SH, etc., (2020) Quantitative analysis of blend ratio of polycarbonate to acrylonitrile-butadiene-styrene copolymer by pyrolysis gas chromatography – mass spectrometry using solid dispersant for sample preparation, *Journal of Instrumental Analysis* 39 (6): 69–773.

Zhang, Y., Luo, C., Yuan, L., Xu, SH, etc., (2021). *Pyrolysis-gas chromatography / mass spectrometry to identification of the solid waste characteristic of imported polyamide recycled plastics. In IOP Conference Series: Earth and Environmental Science* (Vol. 621, No. 1, p. 012038). IOP Publishing.

Zhou MY, Yan LK, Jiao GY. (2016) Combined test melted with IR and DSC for plastics. *Plastics SCI. & Technology*. 125 (3): 47–52.

Advances in Energy, Environment and
Chemical Engineering – Abdullah & Osman (Eds)
© 2023 The Author(s), ISBN: 978-1-032-36083-6

Preparation and properties of gelatin/hydrophobic sodium alginate carrier for microbial immobilization

Jianqiang Shi*
College of Merchant Marine, Shanghai Maritime University Shanghai, China

Shaojun Zhang & Mingyu Wang*
School of Navigation and Shipping, Shandong Jiaotong University Weihai, China

ABSTRACT: The removal of oil pollution by the biological method is one of the most critical worldwide concerns. To solve the problem of low biodegradation efficiency, a kind of
gelatin/hydrophobic sodium alginate carrier with high floatability, hydrophobicity, and biological
affinity was prepared in this article to enhance the degradation of oil pollutants. The experimental results show that the saturated oil adsorption and oil retention of heavy oil particles by
gelatin/hydrophobic sodium alginate carrier are as high as 9.07 g/g and 93.28%, respectively; the
contact angle is 160.434° , the average density is 0.563 g/cm^3, the average mechanical strength
is 3.04 mN, the penetration rate in 30 s is 92.15%, and the degradation rate of diesel oil in 7 d is
77.85%. *Bacillus sp.* has a better degradation effect on short-chain alkanes in diesel. In conclusion,
the carrier prepared is suitable for bacteria immobilization.

1 INTRODUCTION

The biological method has the advantage of clearing up oil film and is considered the fundamental
way to solve the pollution of petroleum hydrocarbons (Kadri et al. 2017). However, due to the
complex marine environment, diverse ecosystems, and other factors, the microorganisms used to
degrade oil have antagonistic effects with the microorganisms originally existing in the ocean, and
the degradation effect of the degrading bacteria is easily affected, which is not conducive to the
full degradation of petroleum-degrading bacteria (Wang et al. 2020). Immobilization technology
plays an important role in avoiding the loss of microorganisms in the marine environment (Sarkap
et al. 2017), improving the environmental tolerance and degradation efficiency of microorganisms
(Hou et al. 2019).

Non-toxic and non-antigenic gelatin is a natural polymer carrier material. Using gelatin as a
microbial carrier can improve the stability of microorganisms (Yuan et al. 2021). Moreover, as an
auxiliary carrier, sodium alginate (SA) has the characteristics of good biocompatibility, stability,
and biodegradability. It can enhance the interaction of the molecular surface through hydrophobic modification and increase the loading capacity of petroleum-degrading bacteria (Xiong et al.
2019). As early as 1997, researchers used calcium alginate/cellulose triacetate as carriers to embed
microorganisms in organic wastewater. Wang et al. (2016) synthesized 2–3 mm gel beads composed of SA/PVA and nitrifying bacteria to remove organic wastewater containing NH4-N. The
results showed that gel beads made of 10% PVA, 2% SA, and 2% CaCl$_2$ had the best removal effect
of NH4-N, reaching 73% COD and 85% ammonia nitrogen.

In this study, an oil-affinity carrier that used gelatin and SA as substrates was provided and
the microbial immobilization technology was applied to degrade marine diesel pollution. The

---

*Corresponding Authors: stoneseaman1228@163.com and 13792750903@163.com properties of the carrier were also investigated. At last, the degradation characteristics of diesel oil by immobilized microorganisms were studied.

2 EXPERIMENTAL METHODS

2.1 *Preparation of gelatin microspheres*

Gelatin solution (20%) was prepared as the water phase. A liquid paraffin solution containing 0.05 g/mL Span-80 was prepared as the oil phase. The water phase was added to the oil phase by stirring while dripping. After emulsification for 10 min, the three-necked flask was quickly transferred to an ice-water bath (0°C). After 20 min, 1.5 mL of glutaraldehyde (50%) was added dropwise. After crosslinking for 1 h, acetone/water solution (volume ratio of 5: 1) was added. The light-yellow microspheres were filtered out and then placed in an acetone solution for solidification at 5°C for 24 hrs. The microspheres were fully washed with acetone and distilled water, respectively, and dried at room temperature for standby.

2.2 *Preparation of gelatin/hydrophobic SA microspheres and microbial immobilization*

Add gelatin microspheres into isooctane solution containing Span-80 and Tween-80, stirring in the water bath (40°C) for 10 min, and then emulsifying at high speed for 10 min. After that, an equal volume of hydrophobic SA solution (the preparation method referred to (Zhao 2019)) and $CaCl_2$ solution (5 mg/mL) was added to it successively and kept stirring for 30 min. The gelatin/hydrophobic SA microspheres were prepared by fully washing with absolute ethanol and distilled water and drying at room temperature.

    Bacillus sp. with a good degrading effect on crude oil was obtained by incubation in Zobell 2216E medium at 24°C for 3 d at a constant temperature. The prepared carrier was weighed at 5.0 g and added to it. After shaking for 3 hrs, the *Bacillus sp.* was immobilized.

2.3 *Characterization test of samples*

Contact angle measuring instrument (Easy-Drop, RUSS company, Germany) measures surface contact angle. Ultraviolet spectrophotometer (UV-1800, Shimadzu, Japan) Measure residual oil concentration. Fourier transform infrared absorption spectrometer (FTIR-8400 S, Shimadzu, Japan) is used to examine the chemical functional group. GC-FID was used to determine the chromatogram of diesel oil: the carrier gas was high-purity N_2, the chromatographic column was Wonda Cap 5 (specification: 30 m × 0.25 mm × 0.25 μm), the initial column temperature was 50°C, and retained for 1 min. Then, the column temperature was increased to 280°C at the speed of 5°C per minute, the inlet temperature and the detector temperature were 280°C, and the collection time was 50 min. The oil absorption capacity and oil retention ability are measured by the mass difference ratio method. As seen in Formulas (1) and (2) (Zhang 2019), the mass of the sample before adsorbing oil was recorded as m_1. The mass of the sample after adsorbing oil was recorded as m_2. After 15 min, the sample was weighed again, and the mass of the sample was recorded as m_3.

$$G_a = \frac{m_2 - m_1}{m_1} \tag{1}$$

$$G_r = \frac{m_3 - m_1}{m_2 - m_1} \times 100\% \tag{2}$$

3 RESULTS AND ANALYSIS

3.1 *Adsorption performance and other basic physical properties*

According to Formulas (1) and (2), the final saturated oil absorption rate Ga (g/g) and the final sustained release oil retention rate Gr (%) are calculated (see Table 1). The prepared carrier has a strong adsorption capacity for heavy oil, crude oil, and diesel oil, and the oil retention rate is above 83%, which indicates that the prepared carrier is conducive to microbial adsorption and degradation of oil pollutants.

The average density of the carrier is 0.563 g/cm^3, meaning the carrier has good floating performance on the water surface, which facilitates the contact between the carrier and the oil film on the water surface. The average mechanical strength of the carrier is 3.04 mN, which provides an effective protective barrier for bacteria in the marine environment. The permeability of the carrier reached 92.15% in 30 s, which is beneficial to the entry and exit of oil pollutants.

Table 1. Adsorption test results of gelatin/hydrophobic SA microspheres for different oil products.

No.	Type	$m_1(g)$	$m_2(g)$	$m_3(g)$	$G_a(g/g)$	G_r (%)
1	Heavy oil	0.20 0	2.013 3	1.891 4	9.07	93.28
2	Lubricating oil	0.20 0	0.731 1	0.644 8	2.66	83.75
3	Diesel oil	0.20 0	1.796 3	1.595 6	7.98	87.43

3.2 *Static contact angle measurement*

The hydrophobicity of the gelatin/hydrophobic SA carrier is characterized by a static contact angle. The test results are shown in figure 1. The average contact angle is 160.434°C, which is over 150°C, showing super-hydrophobicity. Super-hydrophobic carrier is conducive to rapid contact and adsorption with oil pollutants, which provides a carbon source for the survival of bacteria inside the carrier, thus improving the biodegradation efficiency.

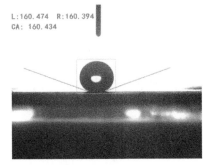

Figure 1. Static contact angle of gelatin/hydrophobic SA microspheres.

3.3 *FTIR spectrum of carrier*

The FTIR spectrum result is presented in Figure 2. There is a stretching vibration peak of -OH of gelatin at 3566 cm^{-1}, a stretching vibration peak of C-O at 1099 cm^{-1}, and an asymmetric stretching vibration peak of C-H at 2926 cm^{-1}. There is a symmetrical stretching vibration peak of –COOH of hydrophobically modified SA at 1652 cm^{-1}, and an asymmetric stretching vibration peak of –COOH at 1456 cm^{-1}. Gelatin/hydrophobic SA carrier has a wide absorption band in the range of 3200-3400 cm^{-1}, and it overlaps with C-H (3000-3100 cm^{-1}), which indicates that there is a hydrogen chain interaction between gelatin and hydrophobic SA.

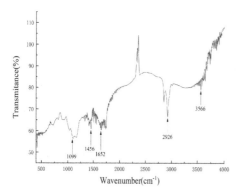

Figure 2.   FTIR Spectra of gelatin/hydrophobic SA microspheres.

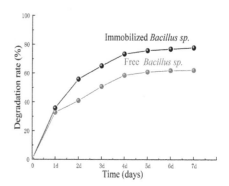

Figure 3.   The heavy oil degradation rate of immobilized and free bacteria.

3.4   *Comparisons of degradation rate between immobilized and free Bacillus sp.*

The degradation rates of immobilized and free bacteria were compared. The results were shown in Figure 3. The degradation rate of heavy oil by immobilized bacteria was 1.37 times that of free bacteria in 48 hrs. The degradation rate of immobilized and free bacteria reached 62.18% and 77.85% in 7 d. The results showed that the immobilized *Bacillus sp.* showed a higher degradation rate than the free *Bacillus sp.* which is consistent with the experimental results of Wu (2009), who compared the removal efficiency of COD between immobilized bacteria (SA embedding method) and free bacteria. The results showed that the immobilized bacteria had a higher removal efficiency, reaching more than 80%. Combining uncertainties such as wind, wave, current, and temperature, the preparation of immobilized natural carriers is of great practical value.

Diesel fuel is mainly composed of complex hydrocarbon mixtures with carbon atom numbers of $C_{10}-C_{22}$. Figure 4 shows that the effect of the immobilized *Bacillus sp.* on the degradation of short-chain alkanes in diesel components is higher than that of medium-chain alkanes and long-chain alkanes. The reason is that the emulsification and uptake ability of the immobilized *Bacillus sp.* on alkanes decreases with the increase of the carbon chain. The results showed that the immobilized *Bacillus sp.* was an effective way to degrade diesel pollutants. In addition, the strain *Bacillus sp.* presents a weak degradation of long-chain alkanes, which can play a role in promoting strengths and avoiding weaknesses through microbial consortium technology. It is also a major development direction of biodegradation in the future.

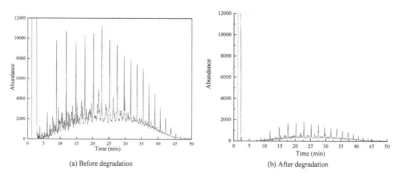

(a) Before degradation   (b) After degradation

Figure 4.   FC-FID spectrum of diesel oil.

4 CONCLUSION

A kind of gelatin/hydrophobic SA carrier was prepared. The carrier shows good hydrophobic and oleophilic properties, which can provide a safe and stable breeding environment for the internal embedded degrading bacteria, and then improve the degradation rate of oil pollutants by the degrading bacteria. Compared with free *Bacillus sp.*, the degradation rate of diesel oil by immobilized *Bacillus sp.* performs better. In addition, *Bacillus sp.* exhibits different degradation characteristics for alkanes with different carbon chain lengths, which provides a research basis for the future use of microbial consortium technology. The carrier material has the characteristics of low material price, environmental friendliness, and simple preparation process, and can be extended to the microbial immobilization application of other degrading bacteria.

ACKNOWLEDGMENTS

This work was financially supported by the Shandong Province Key Research and Development Program (Grant No. 2019GHY112018) and the Research Innovation Team Program of Shandong Jiaotong University (Grant No. SDJTUC1802).

REFERENCES

Hou, L. G., Li, J., Chen, G. G., et al. (2019) Application of Microorganism Immobilization Technology in Wastewater Treatment. *Tech. Water Tr.*, 45 (01): 1–5.

Kadri, T., Rouissi, T., Brar, S. K., et al. (2017) Biodegradation of polycyclic aromatic hydrocarbons (PAHs) by fungal enzymes: A review. *J. Environ. Sci.*, 29 (1): 52–74.

Sarkap P., Roy A., Pal S., et al. (2017) Enrichment and characterization of hydrocarbon-degrading bacteria from petroleum refinery waste as potent bioaugmentation agent for in situ bioremediation. Bioresour. *Technol.*, 242 (SI): 15–27.

Wang H., Zhao X., LiB. W., et al. (2020) Advances in the study of adsorptive immobilized microorganisms in wastewater treatment. *Applied Chemical Industry*, 49 (09): 2341–2345.

Wang W., Ding Y., Wang Y., et al (2016) Intensified nitrogen removal in immobilized nitrifier enhanced constructed wetlands with external carbon addition. *Bioresour. Technol.*, 218: 1261–1265.

Wu L., Ge G., Wan J. (2009) Biodegradation of oil wastewater by free and immobilized Yarrowia lipolytica W29. *J. Environ. Sci.*, 21 (2): 237–242.

Xiong, X. J., Xu, H., Zhang, B. P., et al. (2019) Floc structure and membrane fouling affected by sodium alginate interaction with Al species as model organic pollutants. *J. Environ. Sci.*, 82 (08): 1–13.

Yuan X. L., Li B. X., Huang Y. Y., et al. (2021) Progress in preparation and application of sodium alginate microcapsules. *Chemical Industry and Engineering Progress*, 1432: 1–12.

Zhang, S. J. (2019) *Study on degradation of marine oil spill by immobilized microflora on floatable carrier* [D]. Dalian Maritime University, 67–68.

Zhao, C.. (2019) *Study on the hydrophobic modification and application performance of sodium alginate* [D]. Qingdao University of science and technology, 15–16.

*Advances in Energy, Environment and
Chemical Engineering – Abdullah & Osman (Eds)
© 2023 The Author(s), ISBN: 978-1-032-36083-6*

Study on reasonable well patterns and fracturing parameter optimization in low permeability reservoir

Xiu-wei Wang*

*The Exploration and Development Research Institute of Petrochina Huabei Oilfield Company,
Renqiu, Hebei, China*

Yi Liu*

*The Production and Operation department of Petrochina Huabei Oilfield Company, Renqiu,
Hebei, China*

Shao-ning Wang*

*The Quality Safety and Environmental Supervision Center of Petrochina Huabei Oilfield Company,
Renqiu, Hebei, China*

Qi-hao Hu*

*The Exploration and Development Research Institute of Petrochina Huabei Oilfield Company,
Renqiu, Hebei, China*

He-chuan Zu*

*The Erlian Oil Production Company of Petrochina Huabei Oilfield Company, Xilinguole,
Inner Mongolia, China*

Peng-sheng Dang*

The No. 3 Oil Production Company of Petrochina Huabei Oilfield Company, Hejian, Hebei, China

Min Wan*

The No. 4 Oil Production Company of Petrochina Huabei Oilfield Company, Langfang, Hebei, China

Ying-shun Mu*

The No. 3 Oil Production Company of Petrochina Huabei Oilfield Company, Hejian, Hebei, China

Ya-li Jiang*

The No. 4 Oil Production Company of Petrochina Huabei Oilfield Company, Langfang, Hebei, China

ABSTRACT: Reasonable well patterns and fracturing parameters are the prerequisites for the economic and effective development of low permeability reservoirs. The well pattern models of rhombic anti-nine-spot, rectangular five-spot, square anti-nine-spot, and triangle anti-seven-spot under the same well pattern density are established for the target reservoir, and the triangle anti-seven-spot well pattern is optimized and deployed through numerical simulation results. The development effect is the best when the included angle of the triangle well pattern water injection well array is 45° with the fracture, that is, the included angle between the fracturing fracture orientation and the maximum principal stress direction is 45°. Through the optimization research on the matching relationship between reasonable well spacing and fracturing parameters, the reasonable well spacing of the target reservoir is 250 m, and the optimal fracture half-length is 125 m. Through the study of reasonable well patterns and fracturing parameter optimization, the development of a low permeability reservoir with fracture can be effectively guided.

*Corresponding Authors: yjy_wxw@petrochina.com.cn; scc_ly@petrochina.com.cn; jjj_wsn@petrochina.
com.cn; yjy_hqh@petrochina.com.cn; el_zuhechuan@petrochina.com.cn; cy3_dps@petrochina.com.cn;
cy4_wanm@petrochina.com.cn; cy3_mys@petrochina.com.cn and cy4_jiangyl@petrochina.com.cn

 DOI 10.1201/9781003330165-59

1 INTRODUCTION

In recent years, the proportion of low permeability oil fields in exploration and development is increasing with the rising difficulty of exploration (Chen 2012). How to improve the development effect of low permeability oil fields has become particularly important. The development effect of low permeability oil fields is closely related to fractures, whether natural fractures or artificial fractures caused by hydraulic fracturing (Gao 2011). Due to the existence of fractures in the formation, the permeability of the principal stress direction in a low permeability reservoir is dozens or even hundreds of times higher than that in other directions (Hou 2000). Therefore, the fracture direction should be fully considered in the well pattern of a low permeability reservoir (Liu 2003). The pressure conduction of a low permeability reservoir is slow, and the injection production well spacing should not be too large. The initial well pattern should be reasonably arranged to facilitate the later flexible adjustment (Li 2011). Due to the dual role of fractures in the low permeability reservoir, on the one hand, the injection water will cause water to break through along the fracture direction, which will cause the oil well prematurely water flooding; on the other hand, it can enhance the water absorption capacity of low permeability reservoir and improve the production capacity and oil production index of oil wells (Li 1998; Liu 2003; Ma 2009). Therefore, the key to efficiently developing a low permeability reservoir is the matching relationship between well pattern and fracture (Wu 2007).

2 OPTIMIZATION OF REASONABLE WELL PATTERN

In the process of oilfield development, the study of formation horizontal principal stress can provide a reference for reservoir fracturing and well pattern deployment. The fractures produced by hydraulic fracturing are usually generated and extended along the direction of the maximum horizontal principal stress in the formation. And natural fractures are easy to be generated along the direction of the maximum horizontal principal stress. Therefore, during water flooding development, the water injection wells should not be deployed in the direction of the maximum horizontal principal stress.

The formation dip logging data of 17 wells in the target reservoir show that the direction of maximum principal stress is almost the same, and the maximum horizontal principal stress orientation is 240° to 270°, which is near the east-west direction. The minimum horizontal principal stress ranges from 310° to 354°, which is the direction of northwest-southeast. In the process of well pattern deployment, the influence of natural fractures and hydraulic fractures on well pattern is fully considered. Referring to the domestic experience of well pattern deployment in low permeability and ultra-low permeability reservoirs, the well array direction of the water injection well is deployed parallel to the direction of maximum horizontal principal stress as far as possible to achieve better water flooding development effect.

Four types of well patterns are adopted for the scheme: rhombic anti-nine-spot well pattern, triangle anti-seven-spot well pattern, square anti-nine-spot well pattern, and rectangular five-spot well pattern. After numerical simulation prediction and comparison, the well patterns with relatively good oil production capacity are optimized.

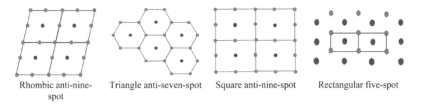

| Rhombic anti-nine-spot | Triangle anti-seven-spot | Square anti-nine-spot | Rectangular five-spot |

Figure 1.   Schematic diagram of different well patterns.

The well pattern models of rhombic anti-nine-spot, rectangular five-spot, square anti-nine-spot, and triangle anti-seven-spot under the same well pattern density are established. Each scheme ensures the same number of 31 oil wells are deployed, and 20 water wells are deployed by rhombic nine-spot well pattern; 20 water wells are deployed by triangle anti-seven-spot well pattern; 18 water wells are deployed by square anti-nine-spot well pattern; 20 water wells are deployed by rectangular five-spot well pattern.

To study the adaptability of different well patterns in low permeability reservoir water flooding development, the numerical simulation models of rhombic anti-nine-spot well pattern, rectangular five-spot well pattern, triangle anti-seven-spot well pattern, and square anti-nine-spot well pattern are established respectively for simulation calculation. The simulation figures and results are shown as follows.

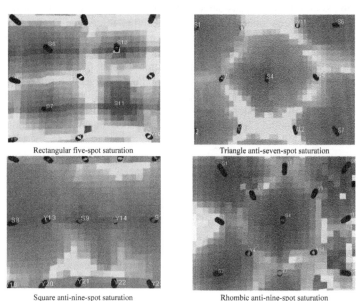

Rectangular five-spot saturation Triangle anti-seven-spot saturation

Square anti-nine-spot saturation Rhombic anti-nine-spot saturation

Figure 2. Oil saturation distribution of different well patterns.

According to the cumulative oil production comparison chart of different well patterns, the recovery degree of the triangle anti-seven-spot well pattern is higher than that of the other three well patterns. And the recovery degree of the square anti-nine-spot well pattern is the lowest. Therefore, the target reservoir is optimized with the triangle anti-seven-spot well pattern.

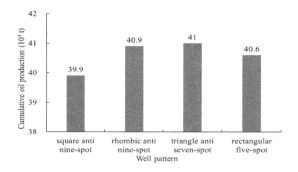

Figure 3. Cumulative oil production of different well patterns.

3 FRACTURE ORIENTATION OPTIMIZATION OF REASONABLE WELL PATTERN

Due to the different injection-production well positions, maximum principal stress direction, and fracture orientation, the influence of hydraulic fracturing on the production performance is also different. Therefore, the fracture orientation needs to be optimized to determine the most appropriate fracture orientation.

The simulation study is carried out by establishing one injection-production unit model when the included angle of water injection well array and fracture are 0°, 15°, 30°, 45°, 60°, 75°, and 90°. Combined with the actual fracture data provided on-site, the half-length of fracture is 125 m, the width of fracture is 0.005 m, and the fracture conductivity is 200 mD. At the same time, the maximum liquid production is limited according to the liquid production capacity of the oil production well. The software automatically divides the water injection volume according to the water absorption index of each layer. At the same time, the maximum water injection volume is limited according to the water absorption capacity of the water injection well. And the maximum bottom hole flowing pressure of the water injection well is limited to 35 MPa according to the formation breakdown pressure of 55 MPa. The fracture direction after fracturing that is processed by software is shown in Figure 4.

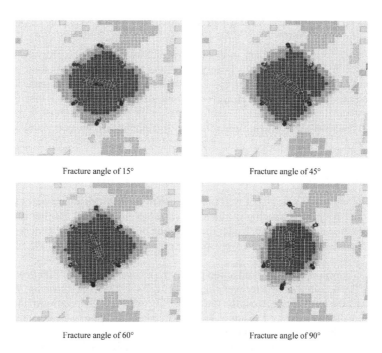

Fracture angle of 15°   Fracture angle of 45°

Fracture angle of 60°   Fracture angle of 90°

Figure 4. Water drive diagram of different fracture angle with water injection well.

The numerical simulation results show that the initial water content is the lowest when the fracture orientation is 0°, but the water content increases rapidly with the increase in production time. When the fracture orientation is 45°, the water content of the oil well is the lowest, while the cumulative oil production reaches the highest value and the recovery degree is the highest. When the fracture orientation is 90°, the analysis shows that some oil wells are water breakthrough prematurely, and the water drive sweep area is reduced, resulting in a low recovery degree. Therefore, the optimum water injection well array of triangle well pattern is 45° with the fracture, that is, the optimum included angle between the fracturing fracture orientation and the maximum principal stress direction is 45°.

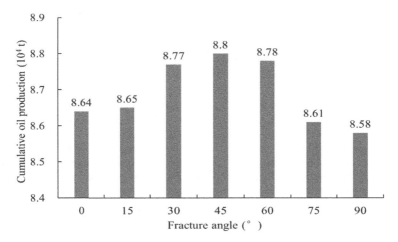

Figure 5.   Cumulative oil production of different fracture angles.

4   OPTIMIZATION MATCHING OF WELL SPACING AND FRACTURING PARAMETER

Due to the start-up pressure gradient in a low permeability oilfield, the pressure loss gradient between injection and production wells is very large, and the pressure loss is more serious under the condition of large well spacing. The most effective method to reduce pressure loss is to drill infill wells and narrow well spacing. However, a blind infill well pattern will inevitably increase investment cost and development risk. Therefore, reasonable injection-production well spacing is of great significance for the economic and effective development of low permeability oil fields.

4.1   *Study on reasonable well spacing in low permeability reservoirs*

Based on the triangle anti-seven-spot well pattern, different well spacing schemes are prepared. Considering the current well spacing of the research reservoir, the well spacing is selected as a 25 m interval to prepare the scheme. Under the condition of the triangle well pattern, 33 oil wells, and 16 water wells are arranged at the well spacing of 200 m, 225 m, 250 m, 275 m, and 300 m, respectively.

According to the cumulative oil production graph, the recovery degree of 300 m well spacing is the highest at 22.9%, followed by a recovery degree of 21.2% at 275 m well spacing. When the well spacing is 200 m, the recovery degree is the lowest at 15.3%. The recovery degree of the well pattern increases with the increase of well spacing. The water content of 300 m well spacing is the lowest at 78.2%, followed by the water content of 81.8% at 275 m well spacing. When the well spacing is 200 m, the water content is the highest at 88.3%. The water content of the well pattern decreases with the increase of well spacing. When the well spacing is less than 250 m, the water content rises faster, that is, the water content is higher under the same recovery degree. However, when the well spacing is greater than 250 m, the water content rising rate is relatively slow, indicating that the water content is low under the same recovery degree. As can be seen from the remaining oil distribution diagram, the remaining oil saturation gradually decreases with the increase of well spacing, and the remaining oil is mainly distributed in the narrow strip area between oil production wells.

The cumulative oil production of different well spacing shows that the net oil gain decreases obviously when the well spacing is greater than 250 m. And when the well spacing is greater than 250 m, the early water content increases slowly. Considering the recovery degree and water content rising rate, the best well spacing is 250 m.

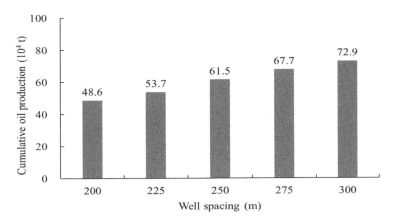

Figure 6.   Cumulative oil production of different well spacing.

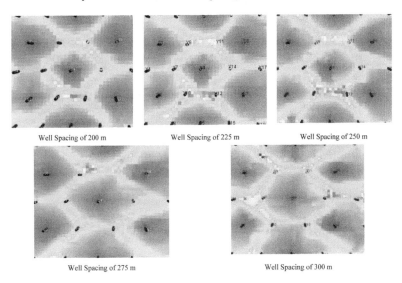

Well Spacing of 200 m   Well Spacing of 225 m   Well Spacing of 250 m

Well Spacing of 275 m   Well Spacing of 300 m

Figure 7.   The oil saturation of different well spacing.

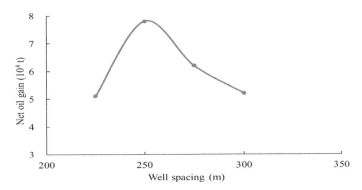

Figure 8.   Net oil gain of different well spacing.

In the development of low permeability reservoirs, most of them have been fracturing to increase production. Therefore, the fracture orientation, fracture half-length, and fracture conductivity need to be considered when studying the oil production well after fracturing. To give full play to the role of hydraulic fracture and avoid premature water flooding or water channeling caused by long fracture length, it is necessary to study the influence of hydraulic fracture length. The optimal fracture length of different well spacing is determined to guide the fracturing reconstruction of oil wells. Based on reasonable well spacing, the reasonable fracturing fracture half-length is studied for the target reservoir. And the simulation models are established by setting the fracture half-length as 75 m, 125 m, 175 m, and 225 m, respectively.

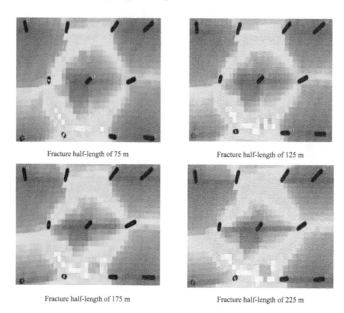

Fracture half-length of 75 m   Fracture half-length of 125 m

Fracture half-length of 175 m   Fracture half-length of 225 m

Figure 9. The oil saturation of different fracture half-lengths.

The numerical simulation results show that with the increase in fracture penetration ratio, the connectivity between the oil well and formation is better, and the cumulative oil production of the oil well is higher. However, when the fracture half-length exceeds 125 m, the net oil gain decreases gradually and tends to be gentle, while the water content increases quickly. To quantitatively evaluate the impact of fracture penetration ratio on oil well production, the cumulative oil production of different fracture half-lengths is made up. It can be seen that the net oil gain increase reaches the highest when the fracture half-length is 125 m, so the optimal fracture half-length is 125 m.

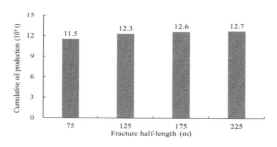

Figure 10. Cumulative oil production of different fracture half-lengths.

5 CONCLUSIONS

(1) The well pattern models of rhombic anti-nine-spot, rectangular five-spot, square anti-nine-spot, and triangle anti-seven-spot under the same well pattern density are established for the target reservoir. And the triangle anti-seven-spot well pattern is optimized and deployed through numerical simulation results.

(2) The fracture orientation of a reasonable well pattern is optimized. The development effect is the best when the included angle of the triangle well pattern water injection well array is 45° with the fracture, that is, the included angle between the fracturing fracture orientation and the maximum principal stress direction is 45°.

(3) Through the optimization research on the matching relationship between reasonable well spacing and fracturing parameters, the reasonable well spacing of the target reservoir is 250 m, and the optimal fracture half-length is 125 m.

REFERENCES

Chen Q J (2012). The adjustment of flood pattern in low-permeability reservoirs. *Oil-Gas field Surface Engineering* 31 (9), 71–76.

Gao M, Song K P, Zhang Y Z, et al. (2011). The study of well pattern adjustment in fractural low-permeability reservoirs. *Journal of South-west Petroleum University* 33 (2), 109–114.

Hou J F (2000). The study on reasonable development well pattern system of ultra-low permeability reservoirs in Ansai Oilfield. Petroleum Exploration and Development 27, 72–75.

Li N, Zhang L, Bian F X, et al (2011). The study on reasonable flood pattern of low-permeability reservoirs. *Liaoning Chemical Industry* 40 (9), 965–968.

Li S Q, Tang Z X (1998). The reasonable well pattern for the development of low-permeability reservoirs. *Acta Petrolei Sinica* 19 (3), 52–55.

Liu J J, Liu X G (2003). The influence of fracture distribution to the water flooding effect. *Xinjiang Petroleum Geology* 23 (2), 146–147.

Liu Z L, Wei Z S, Chen W L, et al (2003). The reasonable flood pattern of fractural low-permeability sandstone reservoirs. *Petroleum Exploration and Development* 30 (4), 85–88.

Ma J, Yang G H, Xu W T (2009). The study of reasonable well pattern in low-permeability reservoirs. *Inner Mongolia Petrochemical Industry* 20, 104–107.

Wu Y J, Li J, Zhao M, et al (2007). The study on residual oil caused by fracture water breakthrough. *Journal of Southwest Petroleum University* 29 (3), 36–38.

*Advances in Energy, Environment and
Chemical Engineering – Abdullah & Osman (Eds)*
© 2023 The Author(s), ISBN: 978-1-032-36083-6

Overview of advances in air filtration materials

Zhen Jin*, Zihan Liu, Fang Fang & Yujing Zou
Wuhan Second Ship Design and Research Institute, Wuhan, China

ABSTRACT: The development and exploration of fiber air filtration materials mainly include electret materials, electrospun nanofiber materials, gradient structure materials, and composite materials. Their common research and design goals are low resistance, high efficiency, long service life, and other excellent performance to meet the different needs of various industries for clean air. In this paper, the relevant theory, development, and market application of the above air filtration materials are described, and the latest progress, research hot spots, and application of gradient filtration materials are emphatically discussed. Better performance and economic benefits will be provided for the development of air filtration devices.

1   INTRODUCTION

Air filtration is a process in which fine suspended particles or harmful components in the air are purified and separated. In the field of air purification, the filtration device based on fiber material has low resistance and high efficiency, which is of great practical significance. The research on the filtration mechanism of this material and filter has a deep theoretical and experimental basis. And it has been widely used in engineering. The filtration mechanism of fiber filter media is the result of comprehensive and synergistic effects of various trapping effects, including interception effect, inertial effect, diffusion effect, gravity effect, and electrostatic effect (Xu 2016). At least these five kinds of filtration effects constitute the stable stage of fiber filtration particles. At this stage, the collection efficiency and resistance of the filter for particles do not change with time but are determined by the inherent structure of the filter, the properties of particles, and the characteristics of airflow. The efficiency of the whole filter is calculated by the logarithmic penetration law (Clyde 1977).

$$\eta = 1 - \exp{-4\alpha H \eta \Sigma \pi} 1 - \alpha df \eta = 1 - \exp\left[-\frac{4\alpha H \eta_\Sigma}{\pi\,(1-\alpha)\,d_f}\right]$$

Davis empirical formula (Davis 1979), which is widely used in practical applications, is used to calculate the resistance of air flowing through the fiber filter material

$$\Delta P = \frac{64 U_0 \mu H}{d_f^2} \alpha^{1.5}\left(1 + 56\alpha^3\right)$$

where U_0 is the filtration rate (M/S); μ is the gas viscosity (Pa·s); d_f is the fiber diameter; α is the fiber filling rate.

Based on the above theoretical basis, the filtration efficiency of the filter made of fiber material is related to the diameter of the fiber, the filling rate of filter material, the thickness of filter material, and the amount of electric charge. The filtration resistance is related to the diameter of the fiber, the filling rate of filter material, and the thickness of filter material. In recent years,

---

*Corresponding Author: 15623564370@163.com

                                    DOI 10.1201/9781003330165-60

to meet the demand for high efficiency and low resistance of air purification devices, the research and preparation of fiber filter materials usually focus on the above parameters. As a result, a series of new materials have been developed, such as electret material (an increase of charge), electrospun nanofiber material (decrease of fiber diameter and increase of fiber surface area), gradient material (gradient composite of coarse fiber and fine fiber, respectively, undertaking different functions) have been developed.

Fiber filter material is widely used in the air filtration industry because of its good filtration performance. Due to the adjustable diameter and pore size of fiber materials, it can be used in different levels of air filtration systems, which is the mainstream and research hotspot in the field of air filtration. At present, this paper will focus on the basic research of fiber filter materials.

2 GENERAL SITUATION OF CONVENTIONAL FILTER MEDIA AND ITS APPLICATION

The *Filter Materials for Air Filter* (JGT 404-2013) specifies the definition, classification, requirements, test methods, and inspection rules of filter materials for air filters (including devices, modules, and units), which is applicable to filter materials of glass fiber, synthetic fiber, natural fiber, composite material, or other materials (China 2013).

According to the standard, the air filtration materials are classified as follows: coarse, medium, medium, medium, sub-high, high, and ultra-high efficiency. The filtration performance is shown in Tables 1-3.

Table 1. Filtration performance of sub-high efficiency, medium efficiency, medium efficiency, and coarse efficiency filter media.

	Performance index		
Level	Rated filtration rate (M/S)	Efficiency (%)	Resistance (Pa)
Sub-high efficiency (YG)	0.053	Particle size $\geq 0.5\mu m$	$95 \leq E < 99.9$	≤ 120
High and medium efficiency (GZ)	0.100		$70 \leq E < 95$	≤ 100
Medium efficiency 1 (z1)	0.200		$60 \leq E < 70$	≤ 80
Medium efficiency 2 (z2)			$40 \leq E < 60$	
Medium efficiency 3 (Z3)			$20 \leq E < 40$	
Crude effect 1 (C1)	1.000	Particle size $\geq 2.0\mu m$	$50 \leq E$	≤ 50
Crude Effect 2 (C2)			$20 \leq E < 50$	
Crude effect 3 (C3)		Standard weighing	$50 \leq E$	
Crude effect 4 (C4)		efficiency for artificial dust	$10 \leq E < 50$	

Table 2. Filtration performance of high-efficiency filter media.

Level	Rated filtration rate (M/S)	Efficiency (%)	Resistance (Pa)
A	0.053	$99.9 \leq E < 99.99$	≤ 320
B	0.053	$99.99 \leq E < 99.999$	≤ 350
C	0.053	$99.999 \leq E$	≤ 380

Table 3. Filtration performance of ultra-high efficiency filter media.

Level	Rated filtration rate (M/S)	Efficiency (%)	Resistance (Pa)
D	0.025	$99.999 \leq E < 99.9999$	≤ 220
E	0.025	$99.9999 \leq E < 99.99999$	≤ 270
F	0.025	$99.99999 \leq E$	≤ 320

It is suitable for the above situations. For indoor air purification in buildings, the combination filter is often used, such as the combination of coarse filter, medium efficiency filter and sub-high efficiency or high-efficiency filter to realize the purification of indoor particles with different particle sizes and ensure certain economic performance.

At present, the research and application of conventional filter materials are mainly in the following aspects.

2.1 Nonwovens

The processing technology of fiber air filter material is mainly divided into woven filter material and non-woven filter material. The nonwoven-fiber material is a new kind of material developed from traditional textile fibers. It usually refers to the fiber web material directly obtained by needle punched, spun laced, spun-bonded, melt blowing, etc. Compared with the traditional fabric obtained by spinning and weaving, the nonwoven filter material has a complex fiber arrangement structure, thus forming a three-dimensional space channel, increasing the filtering path of dust-laden air flow, which is conducive to the improvement of filtration efficiency; At the same time, the disordered accumulation of fibers forms a large number of micropores, which provides a transport channel for the airflow and is conducive to the reduction of piezoresistance.

Because the fibers of nonwoven filter materials can be evenly separated and have no obvious directionality, the fibers are least covered by the upper fibers and have many fluffy pore structures. The filtered particles can contact the fibers extensively in the fluffy pores, which can give full play to various trapping mechanisms. It has the advantages of high filtration efficiency, small pressure loss, large dust holding capacity, and easy preparation of composite filter material, so it is widely used in the field of air filtration. Nonwoven materials account for about 70% of all fiber air filtration materials (Kim 2007).

The research on nonwovens filter materials mainly focuses on the combination of nonwovens and other technologies, such as electret technology, to improve its filtration performance.

2.2 Glass fiber materials

Glass fiber is the first used artificial inorganic fiber air filter material, its main component is silica. It is widely used in an air filtration system in the chemical industry, medical treatment, aviation, and metallurgy because of its excellent characteristics such as good temperature resistance, corrosion resistance, high strength, good dimensional stability, and dust stripping property.

Glass fiber material plays an important role in the production of high-efficiency and ultra-high-efficiency air filters. HEPA (high-efficiency particulate air) filter with glass fiber as the filter medium was 0.3μ. The results show that the filtration efficiency of M-sized particles is more than 99.9%, while the ultra-low penetration air (ULPA) prepared by glass fiber filter material has a good filtration efficiency of 0.1-0.2μ. The filtration efficiency of M particles, smoke, and microbial particles was more than 99.999%.

However, the glass fiber filter material has some defects, such as poor folding resistance and wear resistance, low peeling strength with the substrate, and low flexibility in subsequent processing, which makes it need frequent ash removal in the process of use, and is prone to wear and fracture, which greatly shortens its service life. Based on this, the researchers mixed glass fiber, synthetic fiber, and reinforcing agents in a certain proportion. After carding and chemical treatment, a new type of glass fiber composite filter material (Lin 2010) was prepared. The filtration performance, temperature resistance, and machinability of the composite glass fiber filter materials can meet the requirements of the processing process and practical application.

2.3 *Electret materials*

In the field of charged fiber, the electrostatic filter mainly consists of charged fiber and charged fiber. When the filtration efficiency is the same. The resistance of electret fiber filter material is several times lower than that of ordinary physical interception filter material. Therefore, the electret filter material can be made into a highly efficient and low-resistance air filter, which makes the operation cost of the filter lower and the noise lower. The long-term stability and environmental adaptability of electret are important research directions in the field of air filtration materials.

Due to its advantages of low resistance, high efficiency, and good economic performance, electret materials are widely used in the air filtration industry such as fresh air purification, ventilation, air conditioning, medical masks, and other civil air filtration industries. In practical application, the performance of the product is often affected by the poor stability and uniformity of its charge, so it is necessary to change the filter material regularly. The poor charge stability of electret is reflected in its electrostatic attenuation. At present, there are two kinds of electrostatic attenuation mechanisms: electrostatic neutralization and electrostatic shielding. The uniform surface charge distribution of electret filter material is beneficial to the stability of filtration efficiency the electrostatic potential difference on the surface of electret filter material should be small. Zhou Chen (2011) measured the surface electrostatic potential of polypropylene melt blown electret filter material and analyzed the distribution of positive and negative charges on the surface of the filter material. The results show that the uniformity of electrostatic potential on the surface of the filter material is poor. Improving this problem is the key to improving the filtration performance of electret filter material.

2.4 *Organic and inorganic nanofiber materials*

Nanofibers are also the research focus in the field of air filtration materials. The diameter of the fiber has a great influence on filtration performance. For micron-sized fiber, the other flow through the fiber can be regarded as a continuous fluid, but for the ultra-fine nanofiber, the difference between the fiber diameter and the average free path of air molecules is small, and the other cannot be regarded as a continuous fluid, resulting in sliding flow phenomenon. From the analysis of a single fiber, slipstream has several advantages: first, the number of air molecules colliding with fibers is less, the momentum exchange is less, the drag force of fluid on fiber is reduced, and the filtration resistance is lower. Second, because the air velocity on the fiber surface is not zero, the streamline is closer to the fiber surface, and the filtration effect of direct interception is better, so the filtration efficiency is provided.

The current methods include electrospinning, template separation, and nanofiber preparation. Due to its wide range of spinnable raw materials, strong structural adjustability, good combination of multiple technologies, and strong expansibility of the preparation process, electrospinning technology has gradually approached the stage of industrialization. It has become one of the main ways to effectively prepare nanofiber filtration materials. Due to the submicron and nano scale fiber diameter, electrospun nanofibers have unique pore channels and stacking structures, which make them have many excellent characteristics. The main characteristics of electrospun nanofibers are as follows:

a) The uniformity of the diameter distribution will directly affect the pore size distribution of the materials and the uniformity of the stacking structure, and then affect the filtration efficiency and air resistance;

b) It has an adjustable pore channel and stacking structure. Pore channel structure and fiber stacking structure are two very important indexes affecting the filtration performance of materials. Different pore channels and stacking structures have different filtration performances, which is conducive to the preparation of nanofiber air filtration membranes for different application fields;

c) It has a relatively uniform pore size structure and stacking structure, which can better meet the high-efficiency filtration requirements of air filtration materials.

At present, most nanofiber filter materials are only in the laboratory research, preparation, and testing stage, but their potential economic value is huge, with a wide range of market application prospects.

2.5 *Composite materials*

In recent years, composite filter media is a research hotspot of air filter materials. Generally, it refers to the combination of two or more kinds of fiber or other materials with different properties by chemical, thermal or mechanical methods. The composite filter material integrates the excellent properties of various materials, and through the complementary effect of the properties of various composite materials. At the same time, different composite filter materials can be developed according to different application places (such as high temperature, acid-base, and high humidity environment).

The development direction of the above air filter materials is usually a multi-dimensional combination. For example, the electrospun electret melt is blown composite filter material (Li 2011) prepared in the laboratory of Donghua University, and the electrospun microfiber electret filter material (Kang 2017) prepared by Cheng Boyan's team of Tianjin University of technology, combined with the non-woven fabric technology with the electret technology, the electret technology, and the nanofiber technology. A new type of air filtration material with excellent performance was then developed.

3   APPLICATION STATUS OF GRADIENT FILTER MEDIA

To meet the purification requirements of PM2.5 in the indoor environment of healthy buildings, control the purification energy efficiency and prolong the service life, a new gradient composite material is proposed in recent years, which is composed of one or more filter materials with different filtration grades. Gradient structure is aimed at 0.3-5 μm particles. Gradient filter media is an important development direction of high-performance air filter media, which can achieve both coarse and fine collection, layered filtration, and excellent performance of low resistance and high efficiency.

3.1 *Overview of existing gradient filter media*

The concept of "gradient filter material" is proposed for the first time in *Non-Woven Bonded Fiber Layer Filter Material for Gas Purification* (JB/T 10535-2006). According to the structure classification, the fiber layer filter material is divided into uniform density fiber layer filter material and non-uniform density fiber layer filter material. The non-uniform density fiber layer filter material is divided into graded density fiber layer filter material and step density fiber layer filter material (China 2006).

With the gradual research and application of gradient filter media in industrial gas purification, as well as the research, preparation, and performance test on indoor air purification in buildings, the concept and definition of "gradient filter material" (China 2017) are defined again in *Fiber Layer Filter Media for Gas Purification* (GB/T 35753-2017). The standard defines the gradient fiber layer filter material, which is composed of two or more fibers of different diameters, whose density varies in steps.

The existing gradient filter material usually refers to the filter material with a gradient structure, which means that the elements (such as composition and structure) of the material change in a certain direction. Each fiber layer has different fiber fineness, and the pore structure, pore size and distribution, and volume density are also different to achieve the goal of step filtration for different coarse and fine particles.

3.2 *Application of existing gradient filter media*

3.2.1 *Industrial dust removal field*

The existing gradient filter materials are mainly used in the field of industrial dust removal, and the gradient structure is composed of superfine fiber as a surface layer - mixed fiber/coarse fiber as skeleton layer 2, or superfine fiber as a surface layer - mixed fiber/coarse fiber as skeleton layer - fine fiber as base 3 gradient structure. The principle of gradient filtration is that the surface superfine fiber is used to filter dust particles, and the inner layer and inner layer are used to filter a small number of fine particles and provide mechanical strength. Due to the special structure of the filter material, the gradient structure filter material has high filtration efficiency and low-pressure drop.

Yan Changyong et al. (2007) designed and developed a gradient filter material for dust removal in power plants. It is composed of four layers, which are superfine fiber made by melt blowing and hot rolling technology as the surface layer, the high temperature resistant and corrosion resistant mixed fiber layer as the upper layer, the glass fiber with high-temperature resistance and high mechanical strength as the base layer, the fiber with certain temperature resistance and corrosion resistance, and the fiber with an appropriately increased density as the inner layer. The results show that compared with the conventional filter media, the gradient filter material has good physical and chemical properties (mechanical properties, temperature resistance, chemical resistance, etc.), high filtration efficiency (second only to the membrane filter material, much higher than the conventional filter material) and good air permeability (resistance far lower than the membrane covered filter material, equivalent to the conventional filter material). The material has been used in a power plant, and the system resistance is below 1,100 Pa. The emission concentration of the volume was kept below 10 mg/m^3, which was far lower than the design value. The dust removal efficiency was 99.98%.

Liu Wei et al. (2011) prepared fiber porous ceramics with gradient structure by using two different ceramic fibers. The results show that the gradient structure fiber porous ceramics can filter fine particles better under high-temperature dust removal conditions. The porous ceramics with gradient structures were prepared by the one-step forming method. The maximum porosity was 76%, and the air flow rate was 1 m/min at room temperature. The filtration resistance is 98 Pa and the flexural strength is 6.7 MPa.

The gradient structure filter material of multilayer carbon nanotubes/quartz wire was prepared by P Li et al. (2014), and its filtration performance and service life were studied. It was proved that the gradient structure filter material had high filtration efficiency.

3.2.2 *Air filtration field*

In the Research Report on the key technologies of indoor air purification product development and engineering application of the 12th Five-Year Plan, Li Xianting and others proposed two types of gradient filter media, namely, air filtration gradient composite filter material and high-temperature flue gas gradient composite filter material (Li 2019). The introduction is as follows.

The gradient composite filter media technology for air filtration, which takes electret filter media as the research object, is the composite technology of high efficiency and low resistance electret filter media with different filling rates and medium efficiency filter media. At the same time, a gradient filter material filter was developed based on the technology, which met the requirements of sub-high efficiency filter grade (98.1%), the filtration resistance was significantly reduced (18.1 Pa), and had high dust capacity.

The gradient composite filter material of high-temperature flue gas was prepared, which was composed of a fine denier fiber layer, fine fiber layer, base cloth layer, and coarse fiber layer. The fine denier fiber layer on the surface can effectively improve the filtration efficiency at the side of purifying flue gas and form the effective capture of particulate matter. The fiber is distributed on both sides of the fabric. Two layers of fiber with different filling rates can not only support the filter material but also effectively reduce the total fiber consumption and balance the economic performance of the filter material. The test results show that the filtration resistance after aging is 500pa, which is 64% of the conventional filter material. The cleaning cycle is 38 S, which is about 2.5 times the conventional filter material. The penetration rate of PM2.5 is 0.112%, which is 23.5% of that of conventional filter media.

According to the requirements of an indoor healthy environment for air purification PM10/PM2.5, Wu Yiren et al. (2012) selected and prepared several new gradient structure composite filter materials by using a filter material performance test bench to test and screen the external layer with large dust capacity and low resistance filter material with high efficiency and low resistance for the inner layer. The filtration efficiencies of PM2.5 and PM10 are 45%-90% and 60%-95%, respectively. It can meet the requirements of indoor particulate matter purification.

Li Jinglan et al. (2019) used PE (polyethylene)/PP (polypropylene) core fibers with different linear densities as raw materials to form a single-layer fiber mesh. Then, 2-3 layers of single-layer fibers were laminated to form composite fibers by using different gradient structures, and then hot-air bonding and corona electret treatment were used to obtain PE/PP core fiber air filter materials. The filtration performance, dust capacity, and electrostatic attenuation performance of the materials were tested. It is proved that the "fine coarse fine" gradient filter material has good filtration efficiency, static attenuation performance, and low filtration resistance. This kind of gradient structure filter material has good application potential in the field of air filtration.

With the increasing number of indoor pollution sources that affect environmental health, the requirements of indoor environmental health are constantly refined and deepened, the requirements for indoor air quality control products and devices are constantly improved, and the requirements for the core of indoor air filtration filter media are also constantly improved. The research and design of gradient filter media should consider the performance characteristics of high efficiency and low resistance, both coarse and fine, and filtration of various pollutants, which can help achieve the optimization of comprehensive filtration performance.

3.3 *Research hotspots of gradient filter media*

At present, electrospun nanofibers as the main filter layer and compounded with other gradient structure filter media and functional gradient composite filter media are the research hotspots of gradient filter media.

3.3.1 *Electrospun composite gradient filter material*

With the development and maturity of electrospinning nanofiber processing technology, the multi-level density gradient structure composite filter material is made with electrospun nanofiber mesh as the main filter layer and melt blown/electrostatic/spunbonded cotton as the auxiliary layer, which can effectively improve the comprehensive filtration performance of the product. Electrospun nanofiber composite gradient structure filter material has many advantages, such as high efficiency, low resistance, long life, easy to clean dust, large dust capacity, and so on. Strengthening the structure design, composite process and numerical simulation research of gradient filter material is the focus of product development in the future.

Wang Na et al. prepared electrospun/inorganic electret composite filter material for PM2.5 particles in the air. The filter material has strong charge storage stability, high efficiency, and low resistance, and the preparation process is simple, so it has a great application prospect in PM2.5 filtration; A kind of cobweb fiber material was prepared. The non-woven fabric was used as the base layer and the electrospun micro nanofiber material was used as the main filter layer (Wang 2017).

The porous polypropylene (PP) was prepared by electrospinning the PP2 with different proportions of pores. The filtration efficiency of particles with m and above reached 99.57%, and the filtration resistance was only 65 Pa, which reached the standard of first grade PM2.5 respirator (Chen 2019).

Ahn et al. studied and prepared a kind of electrospun composite material and its pressure drop is lower than that of HEPA (Ahn 2006).

Fan Jingjing et al. Used electrospinning technology to deposit cellulose acetate drug-loaded nanofibers on the surface of viscose spun laced nonwovens, and then covered the surface with polypropylene spunbonded nonwovens to make gradient structure composite filter materials, which were used as filter materials for protective masks (Fan 2015).

To prepare high-efficiency dust-proof mask filter material, Liu et al. created spun polyamide 6/chitosan (PA6/CS) blend nanofibers with a diameter of 88 nm by electrospinning technology and made composite filter material (Liu 2015) with polypropylene melt blown nonwovens as substrate.

Ni Bingxuan et al. (Ni 2018) prepared a three-stage density gradient structure composite material with spunbonded fabric as a support layer, melt blown cloth as an intermediate filter layer, and electrospun nanofiber as a surface filter layer. The filtration performance test was carried out. When the gas flow rate was 32 L/min, the median diameter of aerodynamic mass was 0.26 μ. The filtration efficiency of NaCl solid particles is 99.2%.

3.3.2 Functional gradient composite filter material

In addition to fine particles, indoor air pollution sources include formaldehyde, TVOC, and benzene series (Liu 2012). Classified indoor air pollution into physical, biological, and chemical pollution. Physical pollution mainly refers to unsuitable temperature and humidity, filtration rate, electromagnetic radiation, vibration, and noise. Biological pollution mainly includes bacteria, viruses, fungi, and other microbial pollution factors. Chemical pollution mainly refers to the pollution caused by a series of chemical substances, such as particulate matter, formaldehyde, benzene series, nitrogen oxide, ammonia, and so on. The purpose of functional gradient composite filter material is to realize the simultaneous purification of biological pollution, chemical pollution, or a variety of both pollution sources, such as filter particle layer, formaldehyde absorption layer, odor absorption layer, microbial purification, and so on.

At present, some air purification devices for simultaneous purification of indoor formaldehyde and particulate matter have come out. The basic principle is to combine the fiber filter material and the activated carbon filter material in structure and use the adsorption of formaldehyde by activated carbon and the filtration of particulate matter by fiber filter material to achieve the purpose of simultaneous purification. In addition, a functional gradient composite filter material was developed at the filter material level. It is also a research hotspot for multiple purifications of indoor air in buildings.

Xing Jincheng et al. (2018) prepared chitosan polypropylene fiber and chitosan Tencel composite filter media for *Staphylococcus Albus* and indoor particulate matter and carried out sterilization test and particulate matter filtration test in the air duct purification system. Through the test, when the mass fraction of chitosan was 55%, the filter material had the best sterilization efficiency, reaching 93.2%.

Ma Ruiyue studied the structured composite of polypropylene fiber and chitosan double-layer filter material. The amino group in chitosan can react with formaldehyde to purify formaldehyde to realize the dual purification of formaldehyde and PM2.5 in the building room. Gradient composite filter material can realize the simultaneous purification of different pollution sources and ensure the indoor air quality of buildings (Ma 2017).

4 CONCLUSION

The development of air filter materials is closely related to the development of various industries, especially the military and electronic industries. In recent years, with the increasingly severe air pollution situation and the aggravation of the complexity of pollutant components, people's demand for the indoor environmental quality of buildings is increasingly obvious. The research and development of control technology, devices, and products for PM2.5 particles, formaldehyde, and TVOC in the building room are widely concerned. As a kind of filter material, it has been widely used in the air filtration market.

The development and exploration of air filtration materials mainly have several directions: electret materials, nanofiber materials, gradient structure materials, composite materials, etc. The common research and design objectives are excellent performance such as low resistance, high efficiency, and long service life to meet the different requirements of clean air in various industries. Fiber filtration technology with higher performance is also in continuous exploration. And it will show broad application prospects in various industries in the field of air filtration.

REFERENCES

Ahn Y C, Parks K, Kim G T,et al. (2006). Development of high efficiency nanofilters made of nanofibers. *Current Applied Physics*, 6 (6): 1030–1035.

Chen Yajun, Wang Di, Li Dawei, etc. (2019). Preparation and filtration performance of diacetate fiber composite filter material with gradient pore structure. *Modern chemical industry*. 39 (2): 136–139.

China Academy of Building Sciences Filter media for air filtration. (2013) *JG / T 404-2013*. Beijing: China Standard Press.

China Machinery Industry Federation Non woven bonded fiber layer filter media for gas purification. (2006). *JB / T 10535-2006*. Beijing: China Standard Press.

China Textile Industry Federation Fiber layer filter media for gas purification. (2017). *GB / T 35754-2017*. Beijing: China Standard Press.

Clyde Orr. (1977). *Filtration Principles and Practices*. New York: Marcel Dekker, Inc.

Davis CN. (1979). *Air filtration*. Beijing: Atomic Energy Press.

Fan Jingjing, Zhou Li, Hu Jie, etc. (2015). Preparation and properties of composite structure respirator material. *Materials Guide*, 29 (4): 50–54.

Kang Weimin. (2017). A new tool to stop PM2.5-Electrospun nanofiber electret filter material. *Technical textiles*, 2017.

Kim S C, Harrington M S, Pui D Y. (2007). Experimental study of nanoparticles penetration through commercial filter media. *Journal of Nanoparticle Research*, 9 (1): 117–125.

Li Jinglan, Wu Haibo. (2019). Study on the performance of gradient structure PE / PP skin core fiber air filter media. *Textile Industry* (2019-14): 37.

Li Xianting, He Lumin, Li Jingguang, etc. (2019). Research Report on key technologies for development and engineering application of building indoor air purification products Beijing: Tsinghua University, 2019: 34–49.

Li Xiaoqi. (2015). *Application of electret polyetherimide silica nanofiber membrane in air filtration*. Shanghai: Donghua University.

Lin C, Song D, Li Y, etc. (2010). The preparation of high dimension stable glass fiber reinforced PPS composites. *China Plastics Industry*, 2010, 38 (10): 52–55.

Liu Jing, Liu Huiqing, Ren Xiaofen. (2012). *Indoor air pollution control*. China University of mining and Technology Press, Xuzhou, 2012: 26–27.

Liu Leigen, Shen Zhongan, Hong Jianhan. (2012). Preparation and properties of electrospun high efficiency dust-proof composite filter media. *Acta textile Sinica*, 2015, 36 (7): 12–16.

Liu Wei, Cui Yuanshan, Jin Jiang. (2011). Preparation and properties of fibrous porous ceramic materials for high temperature dust removal. *Journal of Nanjing University of Technology (NATURAL SCIENCE EDITION)*, 2011,33 (4): 107–110.

Ma Rui. (2017). Purification of formaldehyde and PM2.5 by chitosan polypropylene composite air filter media. Tianjin: Tianjin University, 2017.

Ni Bingxuan, Zhang Peng, Yang Xinhui, etc. (2018). Study on structure and filtration performance of density gradient composite filter media. *Synthetic fiber industry*, 2018,41 (1): 11–15.

P Li, C Y Wang, Z Li, et al. Hieraichical carbon-nanotube / quartz-fiber films with gradient nanostructures for high efficiency and long service life air filters. *RSC Advances*, 2014, 4 (96): 54115–54121.

Wang Na. (2017). Structure design and properties of electrospun micro / nano fiber materials for PM2.5 air filtration. Shanghai: Donghua University, 2017.

Wu Yiren, Shen Henggen, He Jin, Song Gang. (2012). Experimental study on composite filter media for purifying fine particles in building healthy environment. *Building science*, 2012,28 (2): 67–71.

Xing Jincheng, Ma Ruiyue, Li Yonggang, etc. (2018). Sterilization effect of chitosan based air filter media. *Journal of Tianjin University (Natural Science and Engineering Technology Edition)*, 2018,51 (6): 605–609.

Xu Zhonglin. (2016). *Principle of air cleaning technology*. Beijing: Science Press. 2016: 167,102–109.

Yan Changyong, Wang Chengbiao, Shen Henggen. (2007). Study on filtration performance of HBT 'gradient' composite filter media and its application. *Building thermal ventilation and air conditioning*, 2007,26 (4): 100–103.

Zhou Chen, Jin Xiangyu. (2011). Study on surface electrostatic potential of polypropylene melt blown electret filter material. *Filtration and separation*, 2011,21 (1): 16–19.

*Advances in Energy, Environment and
Chemical Engineering – Abdullah & Osman (Eds)
© 2023 The Author(s), ISBN: 978-1-032-36083-6*

The optimal scheduling method of hydrogen integrated energy system considering carbon emission cost

Nan Zheng*, Yinan Li, Wanqing Chen, Keren Chen, Han Chen & Jinchun Chen
Power Economic Research Institute of State Grid Fujian Electric Power Company, Fuzhou, China

ABSTRACT: As a kind of clean secondary energy, hydrogen can promote the consumption of renewable energy. Considering the access to hydrogen energy, the operation mode of the integrated energy system is more flexible, which can improve the utilization efficiency of clean energy, effectively reduce carbon emissions, and reduce the impact on the environment. However, in the existing research, the carbon emission cost of the operation of the integrated energy system has not been considered, and it is difficult to fully reflect the environmental protection benefits of the integrated energy. To this end, a steeped carbon emission cost mechanism is constructed, and then an optimal scheduling method for the integrated energy system containing hydrogen is proposed. Firstly, a stepped carbon emission cost mechanism is constructed considering the carbon emissions from the operation of the integrated energy system. Secondly, the optimal operation model aiming at the lowest carbon emission cost and energy purchase cost of the integrated energy system is established. Finally, the calculation example shows that considering the cost of carbon emission can effectively promote the local consumption of renewable energy, reduce the carbon emission of the operation of the integrated energy system, and take into account the economic and environmental benefits of the operation of the integrated energy system.

1 INTRODUCTION

With the development of the economy and society, energy problems such as oil shortages continue to intensify, and accelerating the utilization of renewable energy has become the main development trend of the future energy system (Jiang et al. 2020; Sun, et al. 2015). As the secondary clean energy, hydrogen has the characteristics of a wide source and long-term storage. After considering the access to hydrogen energy, there is a mutual conversion relationship among various energy networks such as electricity, heat, and hydrogen, and the operation mode of the integrated energy system becomes more flexible (Chen et al. 2014; Meng et al. 2019) which is more conducive to further promoting the consumption of renewable energy. Therefore, it is of great practical significance to consider the optimal scheduling method for a comprehensive energy system with multiple energy forms.

In the existing research, Pan G (2020) summarizes the development basis of hydrogen energy technologies and research, puts forward the structural framework of the electric hydrogen energy system, and analyzes the economy and future development path of the electric hydrogen energy system. Ren Z (2022) proposed the various utilization modes of hydrogen energy, and established the operation optimization model of a hydrogen-containing regional comprehensive energy system, considering the physical characteristics of hydrogen energy storage. Ge H (2022) proposed the operation strategy of the integrated energy system composed of wind, solar, and hydrogen, considering the characteristics of wind, solar, and load. And it shows that the integrated energy system containing hydrogen energy can effectively improve energy utilization efficiency. Jia C

---

*Corresponding Author: zhengnan@tju.edu.cn

DOI 10.1201/9781003330165-61

(2020) proposes the optimal operation mode of the integrated energy systems containing hydrogen energy, considering the cost benefits of comprehensive energy commercial operators and users, and puts forward the solution method based on NSGA-II and entropy weight method. The above literature mainly considers the economic cost of the integrated energy system and fails to consider the perspective of environmental protection. At present, China has paid more attention to carbon dioxide emission reduction, and the carbon emission trading market has been preliminarily established. In the future, the cost of carbon dioxide emission will gradually increase. Therefore, in the operation optimization of an integrated energy system, it is also necessary to consider the cost of carbon dioxide emission.

Aiming at the above problems, a stepped carbon emission cost calculation model is established, on this basis, aiming at the lowest carbon emission cost and energy purchase cost, and an operation optimization model of integrated energy system considering carbon emission cost is proposed. Through this model, the economics of the operation scheme can be taken into account while considering the environmental protection benefits of the comprehensive energy system.

2 STRUCTURE OF THE INTEGRATED ENERGY SYSTEMS CONTAINING HYDROGEN

The typical integrated energy-containing hydrogen system is mainly composed of a wind turbine, photovoltaic, electric boiler, gas boiler, electricity energy storage, and hydrogen production equipment. The structure of the IES is shown in Figure 1.

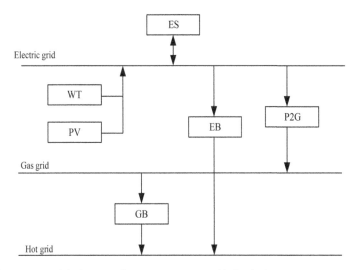

Figure 1. The structure of the integrated energy systems considering hydrogen.

On the one hand, the electric energy generated by a photovoltaic and wind turbine can meet the electrical demand of the users, on the other hand, it can meet the heat load demand through an electric boiler device. If there is still abundant electric energy, hydrogen can be produced through an electric hydrogen production device and stored through a hydrogen storage device. Otherwise, when there is an energy supply gap during the peak period of energy consumption, hydrogen can be converted into electric energy and heat energy through fuel cell devices, gas-fired boilers, and other equipment to supply energy to users in the integrated energy system. Through a variety of energy conversion equipment, we can make full use of renewable resources such as wind and solar, reduce the external energy purchase costs and improve the economic benefits of the integrated system. Meanwhile, it can also reduce the carbon dioxide emission during the operation of the integrated energy system, reduce the pollution to the environment, and reduce the carbon emission cost of the integrated energy systems.

3 THE CARBON EMISSION COST MODEL OF THE INTEGRATED ENERGY SYSTEM

3.1 *The carbon emission of the integrated energy system*

According to the existing research, the carbon dioxide produced by the operation of coal-fired units and gas boilers is a quadratic function of the output power (Zhang X et al. 2022). Based on that, the carbon dioxide emission model of coal-fired units and gas-fired boilers is constructed.

$$F_m = \sum_{t=1}^{8760} \left(a_m + b_m P_m + c_m P_m^2 \right) \tag{1}$$

$$F_g = \sum_{t=1}^{8760} \left(a_g + b_g H_g + c_g H_g^2 \right) \tag{2}$$

Where, a_m, b_m and c_m represents the carbon emission coefficient of the power planet, a_g, b_g and c_g represents the carbon emission coefficient of the gas boiler. P_m and H_g are the output of the power planet and the gas boiler.

A certain amount of carbon dioxide will be absorbed when the hydrogen is used to synthesize methane (Liu et al. 2021). According to that:

$$F_{p2g} = \eta \sum_{t=1}^{8760} H_t^{p2g} \tag{3}$$

Where η is the amount of carbon dioxide absorbed per unit of hydrogen into methane.

According to this, the carbon dioxide emission during the operation of the integrated energy system is:

$$F_{IES} = F_m + F_g - F_{p2g} \tag{4}$$

3.2 *The stepped carbon emission cost mechanism*

In order to promote the users to reduce carbon dioxide emissions independently, a stepped carbon emission cost trading mechanism is proposed, that is, when carbon dioxide emissions exceed a certain range, the unit carbon emission cost will gradually increase. It is shown as:

$$C^c = \begin{cases} \lambda_1 F_{IES} & F_{IES} \leq l_1 \\ \lambda_1 l_1 + \lambda_2 (F_{IES} - l_1) & l_1 < F_{IES} \leq l_2 \\ \lambda_1 l_1 + \lambda_2 (l_2 - l_1) + \lambda_3 (F_{IES} - l_1) & l_2 < F_{IES} \end{cases} \tag{5}$$

In which, λ_1, λ_2 and λ_3 are the unit carbon emission cost under different emission ranges. l_1, l_2 and l_3 are the carbon emission range.

4 THE OPERATION OPTIMIZATION MODEL OF INTEGRATED ENERGY SYSTEMS CONTAINING HYDROGEN

4.1 *The objective function*

Considering the economy and environmental protection of the integrated energy systems containing hydrogen, the objective function is constructed with the lowest carbon emission cost and energy purchase cost. It is shown in (8)

$$\min C = C^c + C^e + C^g \tag{6}$$

Where C^c represents the Carbon emission cost of the integrated energy system. C^e and C^g are the electricity and gas purchase cost of the integrated energy system:

(1) The electricity purchase cost

$$C^e = \sum_{t=1}^{8760} \rho_t P_t^{UP} \tag{7}$$

Where, ρ_t is the unit electricity cost of time t, and P_t^{UP} represents the electricity purchase from the power network of time t.

(2) The gas purchase cost

$$C^g = \sum_{t=1}^{8760} g_t G_t^{UP} \tag{8}$$

Where, g_t is the unit gas cost of time t, and G_t^{UP} represents the gas purchase from the gas network.

4.2 Constraints condition

(1) Electricity power balance constraint

$$\begin{cases} P_t^{PT} + P_t^{WT} - \lambda_{t,c}^{ES}\eta_c^{ES}P_t^{ES} + \left(1-\lambda_{t,c}^{ES}\right)\eta_d^{ES}P_t^{ES} + P_t^{up} - P_t^{P2G} - P_t^{ET} = P_t^L \\ Q_t^{PT} + Q_t^{WT} + Q_t^{up} = Q_t^L \end{cases} \tag{9}$$

Where, P_t^{PT} and P_t^{WT} are the active power of the wind turbine and the photovoltaic of time t, Q_t^{PT} and Q_t^{WT} are the reactive power of the wind turbine and the photovoltaic of time t. $\lambda_{t,c}^{ES}$ indicates the charging and discharging state of energy storage at time t. When the energy storage is in the charging state, the value is 1 and the value in the discharging state is 0. η_c^{ES} and η_d^{ES} are the charging and discharging efficiency of energy storage. P_t^{P2G} and P_t^{ET} are the active power of the hydrogen production unit electric boiler at time t. P_t^L and Q_t^L represent the active power and reactive power of load in time t.

(2) Hot power balance constraint

$$\sigma^{ET}P_t^{ET} + \sigma^{GT}G_t^{GT} = H_t^L \tag{10}$$

Where, σ^{ET} and σ^{GT} are the heat conversion efficiency of the electric boiler and gas boiler; G_t^{GT} represents the gas consumption of gas-fired boiler at time t; H_t^L is the heat load at time t.

(3) Gas power balance constraint

$$G_t^{up} + G_t^{p2g} = G_t^{GT} \tag{11}$$

Where G_t^{p2g} represents the amount of methane produced by P2G at time t.

(4) Operation constraints of the electric hydrogen production unit

$$\begin{cases} 0 \le P_t^{P2G} \le P^{P2G} \\ N_{h,t} = \eta_h P_t^{P2G}\Delta T/H_{LV} \end{cases} \tag{12}$$

Where P^{P2G} is the rated capacity of the electric hydrogen production unit; $N_{h,t}$ is the amount of hydrogen produced by the electrolytic cell at time t; H_{LV} represents the low calorific value of hydrogen; ΔT is the operation time of the electric hydrogen production device, which is 1h in the paper.

(5) Energy storage operation constraints

$$\begin{cases} P_t^{ES} \le P^{ES}, \\ 0 \le E_{t-1}^{ES} + \left[\lambda_{t,c}^{ES}\eta_c^{ES}P_t^{ES} - \left(1-\lambda_{t,c}^{ES}\right)\eta_d^{ES}P_t^{ES}\right] \le E^{ES} \end{cases} \tag{13}$$

Where P^{ES} represents the rated charging power of energy storage and E^{ES} is the rated capacity of energy storage.

5 CASE STUDY

5.1 *Example overview*

Taking a certain region as a typical example, the operation optimization scheme of an integrated energy system containing hydrogen considering the carbon emission cost proposed in this paper is analyzed. The output of the wind turbine, photovoltaic, and load of the typical days in this area are shown in Figure 2. The electricity price from 0 to 7 hours is 110 yuan/kWh, the electricity price from 18 to 21 hours is 800 yuan/kWh, the electricity price at other times is 600 yuan/kWh, and the gas purchase cost of natural gas is 2.5 yuan / m^3. The carbon emission ranges. l_1, l_2 and l_3 are 500kg, 1000kg, and 1500kg respectively. The unit cost λ_1, λ_2 and λ_3 are 0.35 yuan/kg, 0.5 yuan/kg, and 0.75 yuan/kg respectively. The parameters of the energy equipment are shown in Table 1.

Table 1. The parameters of the energy equipment.

Energy equipment	Conversion coefficient	Capacity/kW
P2G	0.6	100
Gas boiler	0.8	600
Electric boiler	0.8	600
Energy storage	0.9	200

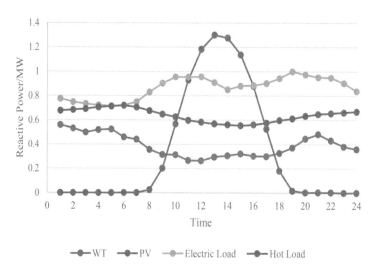

Figure 2. The output characteristic of the planning area.

5.2 *Result analysis*

In order to verify the effectiveness of the method proposed in this paper, the same algorithm is used to solve the following three operation schemes and compared. Scheme 1: optimize the scheduling without considering the cost of carbon emission. Scheme 2: carry out scheduling optimization aiming at the lowest carbon emission cost and operating cost.

The results of electric and thermal power on typical operation days of scheme 1 are shown in Figure 3.

428

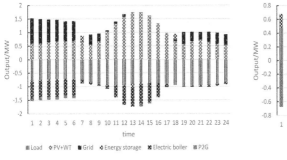

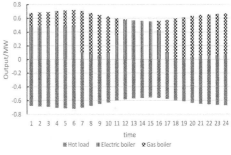

(a)The operation result of electric grid

(b)The operation result of hot grid

Figure 3. The operation result of scheme 1.

It can be seen from Figure 3 that when only the economic cost is considered, in the period of the low power purchase price, the energy system mainly purchases power from the outside, meets the heat load demand through the electric boiler, and works the energy storage device and p2g equipment at the same time. At the peak of the power purchase price, the wind turbine, photovoltaic, and energy storage discharge are mainly used for energy supply. The electric boiler device and p2g equipment stop working and mainly rely on the gas boiler for heat supply.

The results of electric and thermal power on typical operation days of scheme 2 are shown in Figure 4.

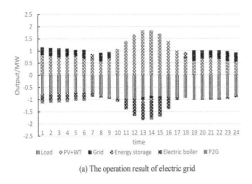

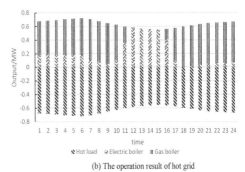

(a) The operation result of electric grid

(b) The operation result of hot grid

Figure 4. The operation result of scheme 2.

As can be seen from Figure 4, compared with scheme 1, after considering carbon dioxide emission, the power generation efficiency of renewable energy such as fans and photovoltaics is significantly improved. At the same time, due to the low unit carbon emission of the gas-fired boiler, the proportion of gas-fired boilers for heating is increased.

The operation cost of the integrated energy system is shown in Table 2.

Table 2. The parameters of the energy equipment (Million Yuan).

Scheme	Energy purchase cost	Carbon emission cost	Total cost
1	10.67	7.83	18.5
2	12.35	5.03	17.38

429

Through the comparison between scheme 1 and scheme 2, it can be seen that, after considering the carbon emission cost, the total cost of scheme 2 has decreased by 6.1%. Although the energy purchase cost has been increased by 1.68 million yuan, the carbon emission cost has been reduced by 2.81 million yuan, so the comprehensive operation cost of scheme 2 is lower than scheme 1. This is mainly because, without considering the cost of carbon emission, the system relies more on the power grid system with relatively cheap energy prices for energy supply, which reduces the use of renewable energy such as wind power and photovoltaic and relatively clean energy such as natural gas, resulting in greater environmental pollution. After considering the cost of carbon emission, it can effectively measure the environmental benefits of clean energy such as wind, light, hydrogen, and natural gas. Although it improves the operation cost of the system to a certain extent, it can better promote the use of low-carbon energy.

6 CONCLUSION

This paper considers the carbon emission of the operation of the integrated energy system, constructs a stepped carbon emission cost model, and puts forward the operation optimization model of a hydrogen-containing integrated energy system considering the carbon emission cost.

Through modeling and example analysis, the following conclusions can be drawn. Considering that the carbon trading cost of the operation of the integrated energy system can reduce the energy consumption with high carbon emissions such as coal, it is required to effectively reduce the carbon emission of the system, reduce the total operation cost of the integrated energy system, and improve the environmental protection benefits. In addition, considering the cost of carbon emission, the output of renewable energy has increased. It is considered to be conducive to promoting the local consumption of renewable energy, improving the utilization efficiency of clean energy, and promoting the energy conservation and emission reduction of a comprehensive energy system.

REFERENCES

Chen Z, Wang D, Jia H, et al. 2017. Research on optimal day-ahead economic dispatching strategy for microgrid considering P2G and multi-source energy storage system. *Proceedings of the CSEE*, 37(11): 3067–3077.
Ge H, Jia Y& Han X, 2022. Capacity Planning of Integrated Electricity-gas Energy Systems Considering Seasonal Complementarity. *Proceedings of the CSEE*, 2022: 1–13.
Jia C, Wang L, Meng E,et al. 2020. Optimal capacity configuration and day-ahead scheduling of wind-solar-hydrogen coupled power generation system. *Electric Power*, 53(10): 80–87.
Jiang H, Du E& Zhu G, et al. 2020. Review and prospect of seasonal energy storage for power systems with a high proportion of renewable energy. *Automation of Electric Power Systems*, 44(19): 194–207.
Liu H, Zhu H& Li F, Economic Operation Strategy of Electric-Gas-Heat-Hydrogen Integrated Energy System Considering Carbon Cost. 2021. *Electric Power Construction*, 42(12): 21–29.
Meng B, Guo F, Hu L, et al. 2019. Wind abandonment analysis of multi-energy systems considering gas-electricity coupling. *Electric Power Engineering Technology*, 38(6): 2–8.
Pan G, Gu W, Zhang H, et al. 2020. Electricity and Hydrogen Energy System Towards Accommodation of High Proportion of Renewable Energy. *Automation of Electric Power Systems*, 23(33): 1–10.
Ren Z, Luo X, Qin H, et al. 2022. Mid/long-term Optimal Operation Method of Reginal Integrated Energy Systems Considering Hydrogen Physical Characteristics. *Power System Technology*, 2022: 1–11.
Sun H, Guo Q& Pan Z. 2015. Energy Internet: concept, architecture, and frontier outlook. *Automation of Electric Power Systems*, 39(19): 1–8.
Zhang X, Xiong H, Wang C, et al. 2022. Economic and Flexible Dispatching of Perk-level Integrated Energy System Based on Optimal Output Interval and Carbon Trading. *Automation of Electric Power Systems*, 2022: 1–15.

Advances in Energy, Environment and Chemical Engineering – Abdullah & Osman (Eds)
© 2023 The Author(s), ISBN: 978-1-032-36083-6

Research on recovery performance strengthening of nitrogen and phosphorus in AnMBR based on crystallization reaction

Fengchao Wang & Shoubin Zhang*
School of Civil Engineering & Architecture, University of Jinan, Jinan, China

Yutian Liu
Jinan Municipal Engineering Design & Research Institute (Group) Co., Ltd., Jinan, China

Jingying Chen
ChinaShandong Jinnuo Construction Project Management Co., Ltd., Qingdao, China

Shikai Zhao
Shandong Industry Ceramics Research and Design Institute, Zibo, China

Wenhai Jiao
Jinan Municipal Engineering Design & Research Institute (Group) Co., Ltd., Jinan, China

ABSTRACT: Phosphorus is a non-renewable mineral resource. Recovering nitrogen and phosphorus from high concentration wastewater can not only alleviate the crisis of phosphorus resource shortage but also reduce the environmental pollution caused by excessive discharge of nitrogen and phosphorus. Anaerobic Membrane Bioreactor (AnMBR) has the ability to treat high concentration wastewater, and combining AnMBR with other processes can also realize the recycling of resources, but the serious membrane fouling still restricts its development. This paper reviewed the characteristics of the AnMBR and discussed the problem of membrane fouling. It was introduced that the carrier can be added to enhance the pollutant removal efficiency and slow down the membrane fouling. At the same time, the added carrier can be used as the crystal seed of magnesium ammonium phosphate (MAP) crystal to realize the recovery of nitrogen and phosphorus.

1 INTRODUCTION

AnMBR has the advantages of a high organic matter removal rate, low sludge output, interception of suspended solids, and low operation energy consumption. It has been widely used in urban sewage treatment and industrial wastewater treatment, but there are still some defects. Due to the low dissolved oxygen in its effluent, the removal rate of nitrogen, phosphorus, and other nutrient elements is low, and phosphorus is the nutrient for the growth of phytoplankton in most freshwater bodies. The discharge of excess nitrogen and phosphorus into natural water bodies can cause eutrophication. At the same time, phosphorus is a valuable non-renewable resource, and the technology of recovering phosphorus from wastewater is urgent. In addition, membrane bioreactor is often accompanied by serious membrane fouling, which makes the effluent flux of membrane module decline seriously.

Therefore, this paper introduces the process characteristics of AnMBR, and analyzes its membrane fouling mechanism and factors affecting membrane fouling. The application of carrier to

---

*Corresponding Author: zhangshoubin1980@vip.163.com

reduce membrane fouling and enhance nitrogen and phosphorus recovery was introduced, and the factors affecting nitrogen and phosphorus recovery were analyzed.

2 AnMBR

2.1 AnMBR process characteristics

AnMBR is a sewage treatment process combining anaerobic bioreactor and membrane separation technology. Therefore, it has the characteristics of high pollutant load, low energy consumption, and energy recovery. At the same time, the efficient separation of the membrane system realizes the separation of HRT and SRT, facilitates operation and management, and effectively solves the problem of sludge loss in an anaerobic activated sludge system (Ho & Sung 2010).

2.2 Influencing factors of membrane fouling

2.2.1 Membrane materials

AnMBR can be divided into the organic membrane and inorganic membrane according to different membrane materials. The membrane structure is closely related to surface properties and membrane fouling. The higher the membrane pore size or porosity, the faster the membrane flux decreases (Yue et al. 2015). The comparison between the organic membrane and inorganic membrane is shown in Table 1.

Table 1. Comparison between organic membrane and inorganic membrane.

	Common membrane materials	advantages	disadvantages
Organic membrane	Polyvinylidene fluoride membrane (PVDF), polyvinyl chloride membrane (PVC), polysulfone membrane (PSF)	Wide range of materials, low manufacturing cost, the high filling density of membrane modules	Easy to deform, unstable treatment effect
Inorganic membrane	Ceramic membrane, metal membrane, alloy membrane, glass membrane	High-temperature resistance, good chemical stability, high mechanical strength, good pollution resistance	High cost, easy to brittle, pore size can only achieve microfiltration and ultrafiltration

Gao et al. (2010) compared the membrane fouling characteristics of polyvinylidene fluoride membrane (cPVDF) and polyetherimide membrane (PEI) in AnMBR. The results showed that the adhesion characteristics of microorganisms of different species to different membranes were different. The fouling of the PEI membrane was faster than cPVDF membrane.

2.2.2 Characteristics of sludge mixture

The properties of the activated sludge mixture are complex. The pH, activated sludge concentration, sludge particle size, microbial flora structure, extracellular polymeric substances (EPS), and soluble microbial products (SMP) of the mixture will affect the membrane fouling to varying degrees. S. Rosenberger (Rosenberger et al. 2006) showed that the concentration of polysaccharides, protein, and organic colloid in sludge led to different degrees of membrane fouling. The higher the concentration of polysaccharides in sludge, the more serious the membrane fouling was. Wang et al (2008)'s research on the particle size distribution of membrane fouling showed that the fine particles in the mixed solution had a strong tendency to deposit on the membrane surface, and the membrane pollutants were much smaller than those in the mixed solution.

2.2.3 *Operating conditions*

The operation mode has the greatest impact on membrane fouling, and the appropriate operation mode can delay membrane fouling. Zhong (2019) concluded that the significance of average membrane flux was peristaltic pump speed (i.e. initial membrane flux) > intermittent time > filtration time > hydraulic recoil time, and under the optimal operating condition, the flat ceramic membrane could maintain an average membrane flux of more than $43.08 \text{ L·}(\text{m}^2\text{·h})^{-1}$ and operate stably for 16 days (384 h).

2.3 *Control of membrane fouling*

Membrane fouling can be divided into reversible fouling and irreversible fouling. Reversible fouling is mainly gel layer fouling caused by concentration polarization, and it can be removed by physical cleanings, such as hydraulic backwashing. Irreversible pollution is caused by irreversible adsorption and blockage, and it can only be removed by chemical cleaning. Common chemical cleaning agents include acid, alkali, oxidant, chelating agent, surfactant, etc.

The addition of carriers in AnMBR can not only strengthen the operational performance of the process, but also increase the particle size, reduce the probability of small particles blocking the membrane hole, and improve the characteristics of the sludge mixture, so as to reduce the filtration resistance, improve the membrane permeation rate, and help to slow down the membrane fouling. Carriers usually have a large specific surface area and rich pores, which can reduce the content of soluble substances to a certain extent, so as to slow down membrane fouling and further improve effluent quality (Xia et al. 2017). Muhammad Aslam (Aslam et al. 2019) used an anaerobic fluidized bed membrane bioreactor (AFBR) to treat municipal sewage. GAC particles were fluidized through the reflux of reactor effluent to control membrane fouling. The results showed that GAC as a fluidizing medium could effectively reduce the reversible pollution of the membrane.

3 STRENGTHENING NITROGEN AND PHOSPHORUS RECOVERY IN AnMBR

3.1 *Principle of MAP crystallization*

When there are Mg^{2+}, PO_4^{3-}, and NH_4^+ in the wastewater, and the ion concentration product is greater than the standard concentration product of MAP ($K_{SP}=2.5 \times 10^{-13}$), MAP precipitation reaction occurs, and nitrogen and phosphorus are recovered. The reaction formula can be expressed as follows:

$$NH_4^+ + H_2PO_4^- + Mg^{2+} + 6H_2O = MgNH_4PO_4 \cdot 6H_2O \downarrow + 2H^+ \tag{1}$$

$$NH_4^+ + HPO_4^{2-} + Mg^{2+} + 6H_2O = MgNH_4PO_4 \cdot 6H_2O \downarrow + H^+ \tag{2}$$

$$NH_4^+ + PO_4^{3-} + Mg^{2+} + 6H_2O = MgNH_4PO_4 \cdot 6H_2O \downarrow \tag{3}$$

MAP crystal has experienced three stages from occurrence to growth, namely nucleation stage, crystal growth stage, and crystal polymerization stage (Mehta & Batstone 2013).

3.2 *Adding carrier*

In the AnMBR, the carrier not only has the above functions of strengthening pollutant removal and reducing membrane fouling but also has a positive effect on the growth of MAP. The introduction of the carrier can reduce the saturation required for MAP crystallization, shorten the nucleation time of MAP crystallization, accelerate the nucleation speed, make MAP grow on the surface of crystal seed, facilitate the separation of crystal and water, reduce the loss due to fine grains, and improve the phosphorus removal efficiency and recovery rate. The formation process of granular sludge and MAIP is shown in Figure 1

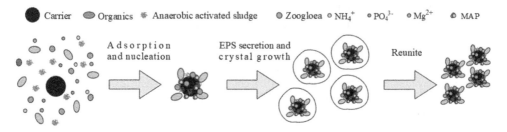

Figure 1.   The formation process of enhancement of anaerobic sludge granulation and MAP by adding carrier.

Zhao et al. (2022) prepared high-efficiency nitrogen and phosphorus recovery material (MgO-RM) by loading magnesium oxide on red mud to recover nitrogen and phosphorus from wastewater. It showed that under optimal conditions, the recovery of ammonia nitrogen and phosphate could reach more than 65% and 90% respectively. The main mechanism of nitrogen and phosphorus recovery was the formation of MAP precipitation. Lu et al.(2019) investigated the crystallization phosphorus removal effect of steel slag-based crystal seed filter (SCMF) on simulated septic tank wastewater. The results showed that the phosphorus removal rate of the system had been stable at about 95% under the optimal operating conditions and continuous operation for 30 days. According to the XRD spectrum, the phosphorus removal product obtained by the steel slag-based crystal seed filter was mainly MAP, with a good crystallization degree and high recovery value.

3.3   *Influencing factors of nitrogen and phosphorus recovery effect*

3.3.1   *Molar ratio of reactants*
From the crystallization reaction formula of MAP, it can be seen that the amount of magnesium salt directly affects the solubility product constant of MAP, and then affects the recovery rate of nitrogen and phosphorus in wastewater. Therefore, the ratio of magnesium salt to nitrogen and phosphorus in wastewater is the key factor to control the recovery rate of phosphorus. Theoretically, the required n (Mg): n (P): n (N) = 1:1:1 for MAP crystallization. However, in the process of treating actual agricultural and aquaculture wastewater, the molar recovery of phosphorus is not exactly the molar dosage of magnesium, but when the molar ratio is n (P): n (Mg) = 1:0.8 $\sim$ 1.6, the phosphorus recovery rate is higher (Fang et al. 2015).

3.3.2   *pH*
pH is an important parameter to control the formation of MAP. According to the MAP reaction formula, with the increase of basic ions, the reaction proceeds to the right to promote the crystallization of MAP, but too high pH will also affect the purity of MAP. Hao et al. (2008) showed that MAP with a larger crystal form could be obtained at lower pH. When pH = 7.0 $\sim$ 8.5, the crystal form was regular rod shape, with less adhesion and high purity. When pH > 9.0, impurities such as $Mg(OH)_2$ and $Mg_3(PO_4)_2$ appeared in the precipitate, with irregular crystal shape and impurity color.

3.3.3   *Hydraulic Retention Time (HRT)*
The fluidization of crystal in AnMBR is mainly driven by the rising flow of fluid, so the reactor is required to maintain an appropriate rising flow rate to ensure the normal growth of the crystal. If the flow rate is too high, the crystal can be fully fluidized, but the crystal itself is difficult to grow stably, and the new crystal still has the risk of loss with the outflow; if the flow rate is too small, it may not meet the fluidization requirements, and the generated crystals are deposited at the bottom and cannot be fully mixed with the water flow, resulting in low treatment efficiency of the reactor (Qiu et al. 2018). Li (2020) explored the effect of HRT on the phosphorus removal effect of MAP crystallization in a fluidized bed crystallization reactor. At 30–60 min, PO_4^{3-} in the water

fully reacted with Mg^{2+} with the extension of HRT, and the phosphorus removal rate showed an upward trend, and the peak value can reach about 75% at 30 min. When the HRT exceeded 90min, the reaction process was difficult to form a flow state due to the slow flow rate, which was not conducive to the growth of crystals, and the phosphorus removal rate began to decline.

4 CONCLUSION

AnMBR has the advantages of anaerobic biological treatment and membrane process and has great application potential. The addition of different types of carriers can not only strengthen the performance of AnMBR, but also effectively reduce the content of soluble substances and colloidal substances, so as to slow down membrane fouling and prolong the service life of the membrane, which is conducive to the sustainable, stable and efficient operation of the process. Combined with crystallization reaction, nitrogen and phosphorus can also be recycled, the reaction speed is fast, and the recovery rate is high. However, at present, the research is mainly concentrated in the laboratory stage, and the practical application needs further experiment and investigation.

ACKNOWLEDGMENTS

This work was financially supported by the Key R&D Program of Shandong Province (2017GSF17105).

REFERENCES

C. Fang, T. Zhang, R.F. Jiang, X.Y. Zhao. (2015) Research progress of magnesium ammonium phosphate crystallization for phosphorus recovery from livestock wastewater. *Technology of Water Treatment*, 41(09), 1–6+13.
C. Zhao, D.P. Peng, Q. Li, Q.L. Wan, T. Huang, R. Zhao. (2022) Simultaneous recovery of nitrogen and phosphorus in wastewater by MgO modified red mud composite material. *China Environmental Science*, 42(01), 135–145.
C.M. Mehta, D.J. Batstone. (2013) Nucleation and growth kinetics of struvite crystallization. *Water Research*, 47(8), 2890–2900.
D.W. Gao, Z. Tong, C. Tang, W.M. Wu, C.Y. Wong, Y.H. Lee, D.H. Yeh, C.S. Criddle. (2010) Membrane fouling in an anaerobic membrane bioreactor: Differences in the relative abundance of bacterial species in the membrane foulant layer and in suspension. *Journal of Membrane Science*, 364(1–2), 331–338.
J. Ho, S. Sung. (2010) Methanogenic activities in anaerobic membrane bioreactors (AnMBR) treating synthetic municipal wastewater. *Bioresource Technology*, 101(7), 2191–6.
J.X. Zhong. (2019) Experimental study on optimizing and efficient cleaning technology of plate ceramic membrane for municipal secondary effluent treatment. University of Jinan.
L.Q. Lu, L.P. Qiu, Q. Qiu, R.Z. Cheng, Z.Y. Tao, M.Z. Fan. (2019) Performance of Simulated Septic Tank Wastewater Treatment by Steel Slag Seed Crystal Media Filter. *China Water & Wastewater*, 35(21), 103–108.
M. Aslam, J. Kim. (2019) Investigating membrane fouling associated with GAC fluidization on the membrane with effluent from anaerobic fluidized bed bioreactor in domestic wastewater treatment. *Environmental Science and Pollution Research*, 26(2), 1170–1180.
Q. Qiu, L.Q. Lu, B.W. Zhao, R.Z. Cheng, G.C. Liu, L.P. Qiu. (2018) Recovery of Nitrogen and Phosphorus from Septic Tank Wastewater by Fluidized Bed Crystallization Reactor. *China Water & Wastewater*, 34(13), 14–19.
S. Rosenberger, C. Laabs, B. Lesjean, R. Gnirss, G. Amy, M. Jekel, J.C. Schrotter. (2006) Impact of colloidal and soluble organic material on membrane performance in membrane bioreactors for municipal wastewater treatment. *Water Research*, 40(4), 710–20.
T. Xia, X.Y. Gao, X.Y. Xu, L. Zhu. (2017) Review on the performance enhancement of an anaerobic membrane bioreactor by adding filters. *Chinese Journal of Applied and Environmental Biology*, 23(02), 392–399.

X. Yue, Y. Koh, H.Y. Ng. (2015) Treatment of domestic wastewater with an anaerobic ceramic membrane bioreactor (AnCMBR). *Water Science & Technology A Journal of the International Association on Water Pollution Research*, 72(12), 2301–7.

X.D. Hao, C.C. Wang, L. Lan, M.C.M. van Loosdrecht. (2008) Struvite formation, analytical methods, and effects of pH and Ca2+. *Water Science and Technology: a journal of the International Association on Water Pollution Research*, 58(8), 1687–92.

Y.Z. Li. (2020) *Study on side-flow technique system of A^2/O process to enhance phosphorus removal and reclaim phosphorus*. University of Jinan.

Z. Wang, Z. Wu, Y. Xing, L. Tian. (2008) Membrane fouling in a submerged membrane bioreactor (MBR) under sub-critical flux operation: Membrane foulant and gel layer characterization. *Journal of Membrane Science*, 325(1), 238–244.

*Advances in Energy, Environment and
Chemical Engineering – Abdullah & Osman (Eds)*
© 2023 The Author(s), ISBN: 978-1-032-36083-6

Silver chloride spherical electrode diffusion based on a two-dimensional continuous point source model

Dou Ji & Chen Li*
College of Electrical Engineering, Naval University of Engineering, India

ABSTRACT: A new method is provided for the measurement of underwater electric fields. A two-dimensional model of continuous point source was established by combining the continuous point source diffusion and the non-steady-state diffusion of spherical electrodes in static seawater. The changes in silver ion concentration and current density on the surface of electrodes over time were solved and simulated by combining Fick's second law, Laplace transformation, complementary error function, and so on. The simulation results show that the reaction ion concentration decreases with the increase of the electrode distance when the time is fixed. With the increase of time, the current density decreases rapidly in the start and then becomes stable. Within 0-5s, the current density range is 0.1–0.7mA/m^2.

1   INTRODUCTION

With the rapid development of today's social economy, the resources on land can no longer meet the needs of production, so people have moved their goals to the sea. The ocean is a huge resource pool, which promotes the development of the shipping industry in the process of exploring the ocean. The sea water is highly corrosive and the ship grts damaged through movement in seawater. At this time, an electric field will be generated. The sensor can detect the value of the electric field to ensure the safety of the ship. The sensor uses a silver-silver chloride spherical reference electrode.

In recent years, scholars at home and abroad have done extensive research on silver-silver chloride electrodes. The Ag/AgCl electrode developed by Wei Yunge et al. has stable electrode potential and large current exchange density, fast electrode reaction speed, and good reversibility, ensuring stability and sensitivity in the process of electric field detection (Wei 2012). Shen Zhen found through experiments that Ag/AgCl electrode has limitations, large contact impedance, and large noise threshold (Shen 2018). Xiong Lu et al. studied the self-noise of silver-silver chloride electrodes, the electrode range potential, and the response function to weak electric field signals. In the marine environment, the electrode self-noise performance is good, reaching the NV level (Xiong 2013). Wangyicheng developed high-performance silver chloride cathode material for seawater batteries through nanomaterial preparation technology (Wang 2018). Beck et al., Briggs et al. and Jin et al. Found that sintering the Ag/AgCl electrode after chlorination can reduce the body defects and crystal unevenness that induce electrode noise, so that electrons can be evenly distributed on the surface of the electrode without causing changes in the local anode and cathode reaction activities and increasing potential stability (Beck 1984; Briggs 1952; Jin 2003). The research on point source diffusion is relatively late. Sulfah et al. used the explicit finite difference method to simulate the point source diffusion model and came to the conclusion that the influence of atmospheric conditions alone on the pollutant concentration level is uncertain (Ulfah 2018). Wang Yang, Zhou Ying, et al. studied the influence of height ratio and wind direction of the street canyon

---

*Corresponding Author: 2314114541@qq.com

on pollutant point source diffusion concentration by using the computational fluid dynamics (CFD) model. The results show that vertical flow will lead to an increase in pollutant concentration. The increase of windward buildings is higher than that of leeward buildings. The point source diffusion of pollutants is closely related to the direction of the wind and the height ratio of the street canyon (Wang 2018). Dong Jinliang, Tan Zijing et al. conducted numerical research on the influence of seasonal variation on airflow and pollutant diffusion characteristics. The results show that the pollutant diffusion modes are different in the different seasons (Dong 2017).

Traditional underwater electric field measurements only use electrochemical principles to measure potential. Continuous point source diffusion models have been studied in life sciences such as brain science, and great achievements have been made. However, electric field measurement is negligible. This paper combines the continuous point source diffusion model with the non-steady-state diffusion model, establishes a two-dimensional continuous point source model with silver chloride spherical electrodes as an example, studies the important parameters of the electrodes, and provides a new method for underwater electric field measurement.

2 ESTABLISHMENT OF MATHEMATICAL MODEL

As the electrode reaction occurs at the solid-liquid interface, the so-called stationary electrolytes are mainly targeted at the surface of the electrodes, including all solutions in a stationary state. When there are a large number of reactive ions in a quiescent solution, the main mode of mass transfer from the reactive ions to the surface of the electrodes is diffusion. The model of a spherical electrode in stationary sea water is shown in Figure 1.

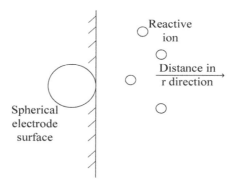

Figure 1.　Unsteady-state diffusion on the surface of a spherical electrode.

3 FORMULA DERIVATION

The distribution of reactive ion concentration in solution around a spherical electrode is spherical and symmetrical, that is, the concentration of reactive ion should be the same at all points on the spherical surface of the radius. The diffusion process of reactive ions on the surface of a spherical electrode satisfies Fick's second law:

$$\frac{1}{D}\frac{\partial c}{\partial t} = \frac{\partial^2 c}{\partial r^2} + \frac{2}{r}\frac{\partial c}{\partial r} \tag{1}$$

In the formula, c is the concentration of silver ions in solution, D is the diffusion coefficient of silver ions, and r is the distance in the polar coordinate system from the surface of the electrodes in the r direction.

3.1 Concentration boundary conditions

When the unsteady diffusion process is analyzed, the variation of ion concentration over time is obtained. According to Fick's second law and setting the boundary conditions, this is done. Its boundary and initial conditions are:

The ion diffusion coefficient D=constant, that is, the diffusion coefficient does not change with the concentration of the diffusion ion.

The ion concentration before the start of the electrode reaction is uniformly distributed in the solution, i.e., the initial condition is $c(r_0, 0) = cs$, cs which is the initial ion concentration in the solution.

When the reaction starts, the concentration of ions in the solution does not polarize, and the boundary condition can be set to $c(r_0, t) = cs$.

When the ions react completely on the surface of the electrodes, $c(\infty, t) = 0$

3.2 Continuous point source concentration equation

For ease of calculation, $r' = r - r_0$

In the formula, r_0 is the radius of the surface of the electrodes and r is the distance of the electrodes in the seawater space.

Then the equation is converted to

$$\frac{1}{D} \frac{\partial c(r', t)}{\partial t} = \frac{\partial^2 c(r', t)}{\partial (r' + r_0)^2} + \frac{2}{r_0 + r'} \frac{\partial c(r', t)}{\partial r'} \tag{2}$$

Set

$$\zeta(r', t) = (r' + r_0) \cdot c(r', t) \tag{3}$$

Then

$$\frac{\partial c(r', t)}{\partial t} = \frac{1}{r' + r_0} \frac{\partial \zeta}{\partial t} \tag{4}$$

The equation is

$$\frac{2}{r' + r_0} \frac{\partial c}{\partial r'} = \frac{2}{(r' + r_0)^2} \frac{\partial \zeta}{\partial r'} - \frac{2}{(r' + r_0)^3} \zeta \tag{5}$$

$$\begin{cases} \frac{\partial c}{\partial r'} = \frac{\partial \left(\frac{1}{r' + r_0} \right) \zeta}{\partial r'} = \frac{1}{r_0 + r'} \frac{\partial \zeta}{\partial r'} - \frac{1}{(r' + r_0)^2} \zeta \\ \frac{\partial^2 c}{\partial r'^2} = \frac{\partial}{\partial r'} \left[\frac{1}{r' + r_0} \frac{\partial \zeta}{\partial r'} - \frac{1}{(r' + r_0)^2} \zeta \right] \\ = \frac{\partial \zeta}{\partial r'} \left[-\frac{1}{(r' + r_0)^2} \right] + \frac{1}{r_0 + r'} \frac{\partial^2 \zeta}{\partial r'^2} + \frac{2}{(r' + r_0)^3} \zeta - \frac{1}{(r' + r_0)^2} \frac{\partial \zeta}{\partial r'} \end{cases} \tag{6}$$

By substituting and simplifying, the equation is

$$\frac{1}{D} \frac{\partial \zeta}{\partial t} = \frac{\partial^2 \zeta}{\partial r'^2} = \frac{d^2 \zeta}{dr'^2} \tag{7}$$

Boundary conditions

$$\begin{cases} (r' + r_0) \cdot c(r_0, t) = (r' + r_0) cs \\ \zeta(0, 0) = (r' + r_0) cs \end{cases} \tag{8}$$

The above formula can be obtained by taking Laplace to transform with respect to t

$$\overline{\zeta}(p, r') = \Lambda[\zeta(t, r)] = \int_0^\infty \zeta(t, r) \exp(-pt) \, dt \tag{9}$$

$$\frac{1}{D} p \overline{\zeta} = \frac{d \overline{\zeta}}{dr'^2} \tag{10}$$

$$\overline{\zeta} = a1 \exp\left(-\sqrt{\frac{p}{D}}r'\right) + a2 \exp\left(\sqrt{\frac{p}{D}}r'\right) \tag{11}$$

Where, a_1, a_2 is the undetermined constant.

If $\overline{\zeta}$ is bounded, then

$$a_2 = 0\overline{\zeta} = a1 \exp\left(-\sqrt{\frac{p}{D}}r'\right) \tag{12}$$

$$r' = 0, \quad \overline{\zeta}\big|_{r'=0} = \frac{r_0}{p}cs \tag{13}$$

At this time

$$a_1 = \frac{r_0}{p}cs \tag{14}$$

To solve the equation, the complementary error function is introduced and simplified as

$$c = \frac{r_0}{r}cs \cdot erfc\left(\frac{r - r_0}{2\sqrt{Dt}}\right) \tag{15}$$

Available after finishing

$$\frac{c}{cs} = \frac{r_0}{r} \cdot erfc\left(\frac{r - r_0}{2\sqrt{Dt}}\right) \tag{16}$$

3.3 Derivation of current density

To obtain the analytical solution of the current density, the derivative of the ion concentration distribution in the direction perpendicular to the electrode surface r should be obtained. Starting from Fick's second law, the concentration gradient of reactive ions can be applied.

$$\frac{\partial c}{\partial r}\bigg|_{r=r_0} = -cs\left(\frac{1}{r_0} + \frac{1}{\sqrt{\pi Dt}}\right) \tag{17}$$

According to Faraday's law in thermodynamics, the instantaneous diffusion current density caused by the reduction of reactive ions on the electrode surface under unsteady diffusion is

$$j = nFD\left(\frac{\partial c}{\partial r}\right)\bigg|_{r=r_0} = -nFDcs\left(\frac{1}{r_0} + \frac{1}{\sqrt{\pi Dt}}\right) \tag{18}$$

When there are enough reactive ions

$$t \to \infty, \frac{1}{\sqrt{\pi Dt}} \to 0 \tag{19}$$

At this time, the diffusion current density becomes a stable value independent of time t, i.e.,

$$j = -\frac{nFDcs}{r_0} \tag{20}$$

4   SIMULATION RESULTS AND ANALYSIS

The silver chloride spherical electrode was immersed in an electrolyte solution with a conductivity of 4 m/s and a 3.5% NaCl solution. In the process of unsteady diffusion, because the silver concentration required for reversible reaction on the electrode surface will change with time, the current density generated in the whole system is controlled by the reduction rate on the electrode surface.

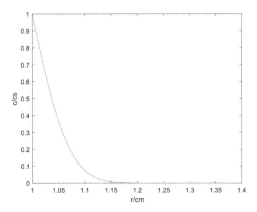

Figure 2. Reaction ion concentration distribution.

When the reaction ion reacts completely on the electrode surface, the diffusion coefficient $D = 1.648 \times 10^{-5} cm^2/s$ when $t = 100s$, then the development of reactive ion concentration polarization on the electrode surface is shown in Figure 2. It can be seen from Figure 2 that the ion is assumed to be on the electrode surface (i.e., $r = 1cm$ place). Before starting the reaction, $c = cs = 1$ With the increase of time t, the total thickness of ion distribution decreases gradually, and finally becomes 0. According to the properties of the complementary error function, at that time. $(r - r_0)/2\sqrt{Dt} \to 0, erfc(x) \approx 1$. Therefore, the diffusion layer thickness of concentration polarization is limited and will increase with the increase of reaction time. Before 1.1cm, the concentration of reactive ions decreased rapidly, and after 1.3 cm, the concentration of reactive ions tends to be stable. This is because in general, the mass transfer between the solid phase and liquid phase always exists with convection phenomenon, so the mass transfer process caused only by diffusion will not last too long. Once the instantaneous effective diffusion layer thickness is similar to or equal to the effective diffusion layer thickness generated by convection, the mass transfer process on the electrode surface gradually tends to be stable. It can also be seen from Figure 2 that the C value at any point decreases with the increase of time, resulting in the gradual increase of the thickness of the diffusion layer of concentration polarization. After a period of reaction time, the silver ion concentration on the electrode surface gradually tends to 0, and the silver ion concentration gradient in the solution near the electrode surface reaches the maximum value. At this time, the thickness of the diffusion layer does not change with the change in the silver concentration.

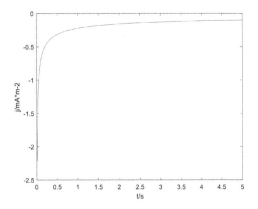

Figure 3. Change of current density with time.

It is observed from Figure 3 that the current density decreases rapidly with the increase of time before 1s, and gradually tends to be stable after 4s. The change with time is not great. This is mainly because the instantaneous effective diffusion thickness of ions gradually increases during the concentration polarization process, and it is more difficult for ions to reach the surface of the reaction electrode, which will inhibit the reaction and reduce the current density. In 0–5 s, the current density range is 0.1–0.7 ma/m^2, and the direction is opposite to the current generated by reactive ions.

5 CONCLUSION

In this paper, aiming at the problem of electric field measurement of ship hull in static seawater, combined with the theories of electrode unsteady diffusion and continuous point source diffusion, the expressions of concentration and current density of silver chloride spherical electrode in static seawater are derived. Based on the theory of dissolution equilibrium, the concentration boundary conditions and initial concentration are determined, a two-dimensional continuous point source diffusion model is established, and the equation is simplified by Laplace transform and complementary error function,.The variation of silver ion concentration and current density on the electrode surface with time was solved and simulated. The results show that:

1) When the time is fixed, the concentration of reactive ions decreases with the increase of electrode distance.
2) With the increase of time, the current density first decreases rapidly and then tends to be stable. Within 0–5 s, the current density range is found to be 0.1–0.7 ma/m^2.

ACKNOWLEDGMENT

The authors are grateful to support from the foundation strengthening plan technical field fund (2019-jcjq-jj-050).

REFERENCES

Beck T R, Rice D E. Conductivity of anodic silver chloride during formation [J]. J. Electrochem. Soc., 1984, 131: 89–93.
Briggs G, Thirsk H R. A study of the behavior of polarized electrodes. Part 3. The behavior of the silver/silver chloride system during electrolytic reduction [J]. Trans. Faraday Soc., 1952, 48:1171–1178.
Dong Jingliang, Tan Zijing, Xiao Mingyi, Seasonal Changing Effect on Airflow and Pollutant Dispersion Characteristics in Urban Street Canyons [J]. Atmosphere, 2017, 8(43): 1–8.
Jin X B, Lu J T, Liu P F, et al. The electrochemical formation and reduction of a thick AgCl deposition layer on a silver substrate [J]. J. Electroanal. Chem., 2003, 542: 85–96.
S Ulfah, S A Awalludin, Wahidin, et al.Advection-diffusion model for the simulation of air pollution distribution from a point source emission[C]. Conference Series 2018. Mathematics Education, Universitas Muhammadiyah Prof. DR. Hamka, Jakarta Indonesia, 2018: 1–9.
Shen Zhen. Study on detection mechanism and characteristics of carbon fiber underwater electric field electrode[D]. Wuhan: Naval University of engineering, 2018.
Wang Yang, Zhou Ying, Zuo Jian, et al. A Computational Fluid Dynamic (CFD) Simulation of PM10 Dispersion Caused by Rail Transit Construction Activity: A Real Urban Street Canyon Model [J]. Environmental Research and Public Health, 2018, 15(3):482.
Wangyicheng Synthesis of nano silver chloride and its performance as cathode material for seawater battery [D] Tianjin: Tianjin University, 2018.
Wei Yunge, Cao Quanxi, Huang Yunxia, et al. Study on the performance of ag/ag CL porous electrode based on underwater electric field measurement [J]. Rare metal materials and engineering, 2012, 41 (12): 2173–2177.
Xiong Lu, Gong Shenguang, Jia Yizhuo Ocean electric field signal detection performance of silver-silver chloride sensor [J] Journal of detection and control, 2013 (01): 76–79.

*Advances in Energy, Environment and
Chemical Engineering – Abdullah & Osman (Eds)
© 2023 The Author(s), ISBN: 978-1-032-36083-6*

Polystyrene nanoplastics transport in marine sediments

Rubi Zhao, Xia Liu* & Chu Wang
College of Environmental Science and Engineering, Ocean University of China, Qingdao, China

ABSTRACT: Marine microplastic pollution gradually becomes a global environmental problem. Nanoplastics are much smaller than microplastic and can migrate in the marine environment, bringing higher environmental risks. In this study, carboxyl-modified PS nanoplastics were selected to explore the transport mechanism of nanoplastics in marine sediments. Results showed that the recovery rate of PS nanoplastics in marine sediments ($C/C_0 = 0.43$) was significantly lower than that in quartz sand ($C/C_0 = 0.70$). Three reasons are given as follows: (1) the porosity (0.28) of the sediment was lower than that of saturated quartz sand (0.38), resulting in a stronger blocking effect in sediment; (2) the heteroaggregation of PS nanoplastics and marine sediments inhibited their transport; (3) the components of marine sediments (e.g., minerals) promoted the interaction between nanoplastics and sediments. This study gave insights into the transport behavior and mechanism of PS nanoplastics in sediments, which would help to further understand the transport and fate of nanoplastics in the marine environment.

1 INTRODUCTION

Marine plastic pollution has emerged as one of the most serious environmental problems (Gigault et al. 2016). Plastic wastes could break into microplastics or nanoplastics under the physical, chemical, and biological degradation in the marine environment (Gewert et al. 2015). Homoaggregation of nanoplastics and heteroaggregation of nanoplastics with mineral particles/natural organic matter led to the settlement of nanoplastics to sediments. Therefore, marine sediments became the sink of nanoplastics. There were some works of literature exploring the effects of particle properties, solution chemistry, flow rate, and other coexisting colloids on the transport behaviors of nanoplastics (Bradford et al. 2007; Dong et al. 2018; Sasidharan et al. 2014; Song et al. 2019), however, these studies used quartz sand or glass beads as porous media (Cornelis et al. 2013; Fisher-Power & Cheng 2018; Sun et al. 2015). The transport behavior and mechanism of nanoplastics in marine sediments are quite limited, and relevant research is urgently needed. Therefore, it is of significance to investigate the transport behavior and mechanism of nanoplastics in the marine environment for assessing their ecological risks and environmental fate.

In this study, marine sediments were selected to explore the transport process of nanoplastics in the marine environment. Carboxyl-modified polystyrene nanoplastics (PS-COOH) (200 nm) were selected to investigate the transport behavior of nanoplastics in marine sediments by marine sediment medium column experiments. Results would help better understand the transport and fate of nanoplastics in the marine environment, and assess the risks of nanoplastics in the marine environment.

---

*Corresponding Author: liuxia9396@ouc.edu.cn

2 MATERIALS AND METHODS

2.1 *Materials*

Green fluorescent PS-COOH nanoplastics were purchased from Shanghai Huge Biotech Co, China, with an average diameter of 200 nm. The seawater used in this research was artificially prepared, and the salinity of the seawater is 35.00‰ (Kester et al. 1967). The quartz sand in this study was purchased from Macklin (Shanghai, China), and the particle size ranged from 1–2 mm. The marine sediments used in this experiment were sampled from the Bohai Sea. It was mainly composed of quartz (53.6%), albite (21.8%), muscovite (13.2%), sanidine (6.8%), calcite (4.29%), kaolinite (0.40%), and a variety of other silicates. Before using in experiments, the sediments were dried at 40°C for 48 h, then sieved with stainless steel sieve, and stored in a sealed bag.

2.2 *Characterization*

PS-COOH nanoplastics were diluted with ultrapure water and artificial seawater respectively to obtain the stock solution with a concentration of 100 mg/L. PS-COOH nanoplastic suspensions (5 μL) were dropped onto a carbon-coated grid sample holder to observe using Transmission Electron Microscope (TEM) JEM-2100PLUS (Japanese Electron Optics Laboratory, JEOL, Japan). The hydrodynamic diameters and zeta potential of PS-COOH nanoplastics were characterized using a Zetasizer Nano ZS90 (Malvern Instruments, UK).

2.3 *Transport experiment of nanoplastics in quartz sand*

The columns were made of Plexiglas (10 cm in length and 2.6 cm inner diameter). Stainless steel screens of 100-mesh were used at both ends to avoid sand loss and ensure uniform flow. During the experiment, the packed column was configured vertically to reduce the influence of gravity. The column was packed with quartz sand and tapped with a rubber mallet to induce uniform packing. The packed column was saturated with ultrapure water using a peristaltic pump (BT100F-1, Lead Fluid, China) in an upward direction at a flow rate of 4 mL/min, and the porosity was gravimetrically determined.

Prior to PS-COOH nanoplastic injection, background solution (artificial seawater) was injected into the column for approximately 4 h to equilibrate it. PS-COOH nanoplastic suspension (10 mg/L) was then pumped into the column for 5 pore volumes (PVs), followed by flushing with 5 PVs of background solution. The column effluent samples were collected every 10 min using an automatic partial collector (EBS-20, Shanghai Huxi Analytical Instrument, China) and the concentrations of nanoplastics were analyzed by a fluorescence spectrophotometer (F4600, Hitachi, Japan).

2.4 *Transport experiment of nanoplastics in natural marine sediments*

To determine the transport of PS nanoplastic in natural marine sediments, column experiments were performed using marine sediment as transport media. To maintain sufficient hydraulic conductivity and ensure the particles at a low level so as to not clog column stoppers, the column was packed with sediment and quartz sand in a mass ratio of 1:4. The steps of the marine sediment column experiment were the same as that of the quartz sand column experiment.

3 RESULTS AND DISCUSSION

3.1 *Characterization*

TEM image showed that PS-COOH nanoplastics were regular spheres (Figure 1A). Moreover, PS-COOH nanoplastics were dispersed in ultrapure water well (Figure 1A), but aggregated in seawater (Figure 1B). The particle size of the nanoplastic was measured by dynamic light scattering

(Figure 1C), the average particle size in ultrapure water and seawater were 211 ± 3 nm and 2638 ± 66 nm, respectively. In addition, the zeta potentials of PS-COOH nanoplastics in ultrapure water and seawater were measured, and the results are shown in Figure 1D. The zeta potential of PS-COOH nanoplastics possessed more negative in ultrapure water (–35.17 mV) than that in seawater (–8.27 mV).

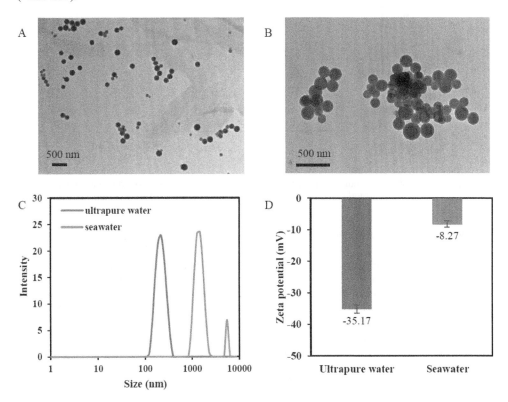

Figure 1. Characterization of PS-COOH nanoplastics. (A) TEM images of PS-COOH nanoparticles in ultrapure water; (B) TEM images of PS-COOH nanoparticles in seawater; (C) Hydrodynamic diameters of PS-COOH nanoplastics in ultrapure water and seawater; (D) Zeta potentials of PS-COOH nanoplastics in ultrapure water and seawater.

3.2 Transport experiment of nanoplastics

To understand the transport of nanoplastics, the transport experiments of 200 nm PS-COOH nanoplastics were conducted both in quartz sand and sediment, and the results were shown in Figure 2. In quartz sand, the breakthrough curves of PS-COOH nanoplastics (the highest C/C_0 was 0.70) were higher than that in the sediment (the highest C/C_0 was 0.43). This result indicated that the presence of sediment decreased the transport of PS-COOH nanoplastics.

3.3 Mechanisms of nanoplastic transport in sediment

The porosity of the quartz sand column was 0.38, while the porosity of the marine sediment column was only 0.28, which results in the nanoplastics being more easily trapped in the porous medium.

The zeta potentials of quartz sand and sediment in seawater were determined. There was no significant difference in the zeta potential of sediment (–11.94 mV) and quartz sand (–10.36 mV). The observation suggested that the surface charge of the transport medium was not the main factor for the low transport of the nanoplastics in sediments.

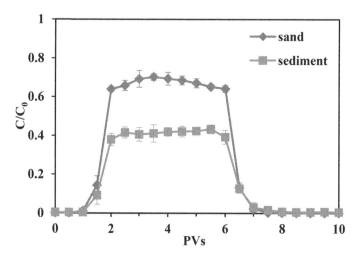

Figure 2. The breakthrough curves of PS-COOH nanoplastics in saturated quartz sand and in marine sediments. The background solution was artificial seawater, pH = 8.1 ± 0.1, PS = 10 mg/L.

Additionally, the homoaggregation of PS-COOH nanoplastics was enhanced under high ionic strength in seawater (Figure 1B) due to the compressed double electric layer and reduced negative charge on the PS-COOH surface (Figure 1D), which facilitated the heteroaggregation between PS-COOH nanoplastics and sediments (Liu et al. 2022) and inhibited the transport.

Previous studies demonstrated that iron oxide, a common ubiquitous component of natural sediment, could enhance the deposition of nanoplastics (200 and 2000 nm), and inhibit their transport (Li et al. 2019). In this study, the components of marine sediment are different from quartz sand, including organic matters and minerals. These components may play a crucial role in the transport of nanoplastics, and we will further study them in the future.

4   CONCLUSIONS

This study focused on the transport of nanoplastics in marine sediments. The transport of PS-COOH nanoplastics in marine sediments was lower than that in quartz sand, which was due to the low porosity of sediments, stronger aggregation of nanoplastics, and the complicated components of marine sediments.

ACKNOWLEDGMENTS

This work was financially supported by China National Postdoctoral Program for Innovative Talents (BX20190306), China Postdoctoral Science Foundation (2019M660168), and Qingdao Postdoctoral Applied Research Project.

REFERENCES

Bradford, S. A., Torkzaban, S., Walker, S. L. (2007) Coupling of physical and chemical mechanisms of colloid straining in saturated porous media. *Water Res.*, 41(13): 3012–3024.
Cornelis, G., Pang, L., Doolette, C., Kirby, J. K., & McLaughlin, M. J. (2013) Transport of silver nanoparticles in saturated columns of natural soils. *Sci. Total Environ.*, 463: 120–130.

Dong, Z., Qiu, Y., Zhang, W., Yang, Z., & Wei, L. (2018) Size-dependent transport and retention of micron-sized plastic spheres in natural sand saturated with seawater. *Water Res.*, 143: 518–526.

Fisher-Power, L. M., Cheng, T. (2018) Nanoscale titanium dioxide (nTiO$_2$) transport in natural sediments: the importance of soil organic matter and Fe/Al oxyhydroxides. *Environ. Sci. Technol.*, 52(5): 2668–2676.

Gewert, B., Plassmann, M. M., MacLeod, M. (2015) Pathways for degradation of plastic polymers floating in the marine environment. *Environ. Sci.: Processes Impacts.*, 17(9): 1513–1521.

Gigault, J., Pedrono, B., Maxit, B., & Ter Halle, A. (2016) Marine plastic litter: the unanalyzed nano-fraction. *Environ. Sci.: Nano.*, 3: 346–50.

Kester, D. R., Duedall, I. W., Connors, D. N., Pytkowicz, R. M. (1967) Reparation of artificial seawater. *Limnol. Oceanogr.*, 1(12): 176–179.

Li, M., He, L., Zhang, M., Liu, X., Tong, M., & Kim, H. (2019) Cotransport and deposition of iron oxides with different-sized plastic particles in saturated quartz sand. Environ. *Sci. Technol.*, 53(7): 3547–3557.

Liu, X., Song, P., Lan, R., Zhao, R., Xue, R., Zhao, J., & Xing, B. (2022) Heteroaggregation between graphene oxide and titanium dioxide particles of different shapes in the aqueous phase. *J. Hazard. Mater.*, 128146.

Sasidharan, S., Torkzaban, S., Bradford, S. A., Dillon, P. J., & Cook, P. G. (2014) Coupled effects of hydro-dynamic and solution chemistry on long-term nanoparticle transport and deposition in saturated porous media. *Colloids Surf. A Physicochem. Eng. Asp.*, 457: 169–179.

Song, Z., Yang, X., Chen, F., Zhao, F., Zhao, Y., Ruan, L., Wang, Y., & Yang, Y. (2019) Fate and transport of nanoplastics in complex natural aquifer media: Effect of particle size and surface functionalization. *Sci. Total Environ.*, 669: 120–128.

Sun, P., Shijirbaatar, A., Fang, J., Owens, G., Lin, D., & Zhang, K. (2015) Distinguishable transport behavior of zinc oxide nanoparticles in silica sand and soil columns. *Sci. Total Environ.*, 505: 189–198.

*Advances in Energy, Environment and
Chemical Engineering – Abdullah & Osman (Eds)
© 2023 The Author(s), ISBN: 978-1-032-36083-6*

Study on the influencing factors of fluoride removal from groundwater by aluminum sulfate coagulation

Kun You*, Kai Cao* & Ning Kang*
Shenyang Jianzhu University, Shenyang, Liaoning Province, China

Weiwei Zhou*
Shandong Urban Construction Vocational College, Jinan, Shandong Province, China

Xin Wen*, Fahui Qin* & Jinxiang Fu*
Shenyang Jianzhu University, Shenyang, Liaoning Province, China

ABSTRACT: In view of the excessive fluoride content in groundwater, the effect of coagulation on fluoride removal was analyzed, and the mechanism of fluoride removal was explored. It provides a reference for practical application. Methods: Static experiment was carried out to investigate the effects of aluminum sulfate dosage, pH value, coagulation time, stirring speed, and other factors on the fluoride removal effect of aluminum sulfate. The results show that when the concentration of fluoride ion is 2 mg/L, the dosage of aluminum sulfate is 20mg, the pH of the water sample is 5.5, the rotational speed is 70–80 r/min, and the coagulation time is 15 min, the fluorine removal efficiency is the best, and the removal rate reaches 61.3%. When fluoride ion concentration is 5 mg/L, aluminum sulfate dosage is 35 mg, water pH is 5.5, rotating speed is 70–80 r/min, coagulation time is 30 min, fluoride removal efficiency is the best, and the removal rate reaches 43.8%.

1   INTRODUCTION

According to the National Sanitary Standards for Drinking Water (GB5749-2006), the mass concentration of fluoride in groundwater should be less than 1 mg/L. However, fluorine-containing drinking water is widely distributed in China, and the range of fluoride ions in drinking water exceeds the standard is roughly 2-5mg /L, and a few areas are as high as 10 mg/L. Long-term drinking water with excessive fluoride content may lead to dental fluorosis, skeletal fluorosis, osteosclerosis, and other diseases (Liao 2018).

The research content of fluorine removal technology at home and abroad mainly includes adsorption, ion exchange, membrane separation, and coagulation precipitation, and its advantages and disadvantages and application scope are not the same. At present, the research focus of the adsorption method is modification and regeneration, mainly aiming at the problems of small adsorption capacity and greatly reduced adsorption performance after adsorption regeneration, which is suitable for small-scale water plants to treat fluorine-containing water. Although the ion exchange technology and membrane separation technology are very effective in removing fluoride, they only stay in the theoretical and experimental stage. For the construction and use of water plants, they face the problems of large investment, high operation costs, and high requirements for water quality, so it is difficult to realize.

Based on the actual situation of underground treatment, coagulation and precipitation fluoride removal not only fits the existing process but also meets the needs of fluorine removal in most sewage treatment plants with the advantages of economic efficiency and simple operation. In this study, based on the traditional coagulation process of the water plant, aluminum sulfate was used

---

*Corresponding Authors: 466755@qq.com; 201732432@qq.com; 784286836@qq.com;
201732432@qq.com; 945579586@qq.com; fahuiqin@163.com and 201732432@qq.com

DOI 10.1201/9781003330165-65

as a coagulant to investigate the influence of various factors on the effect of fluoride removal by coagulation, and explore the mechanism of fluoride removal by coagulation, so as to provide a reference for fluoride removal by groundwater coagulation.

2 MATERIALS AND METHODS

2.1 *Experimental device*

Zr4-6 type coagulation stirrer, Electronic scale, PH-3C digital pH meter, Ultraviolet spectrophotometer, Magnetic stirrer, etc.

2.2 *Experimental reagents*

The reagents used in this test include: aluminum sulfate, sodium fluoride, fluorine reagent, sodium hydroxide, hydrochloric acid, etc., and the reagents are analytical pure reagents.

2.3 *Experimental methods*

2.3.1 *Simulated water sample configuration*
Distilled water and sodium fluoride were used to prepare fluoride water samples with mass concentrations of 2 mg/L and 5 mg/L. Sodium hydroxide solution or hydrochloric acid solution was used to adjust the pH of the water samples to the desired value.

2.3.2 *Coagulation scheme*
The single-factor experiment was used to set different dosages, pH, stirring time, stirring speed, and other influencing factors. At the initial stage of mixing, coagulant was added to the water sample. After the coagulation and mixing, the supernatant was taken after standing for 30 min. At the same time, three parallel experiments were carried out under each test condition, and its average value was analyzed and determined.

3 RESULTS AND DISCUSSION

3.1 *Influence of aluminum sulfate coagulant dosage on fluorine removal effect*

The mass concentrations of fluoride ions in the experimental raw water were 2 mg/L and 5 mg/L. Under the condition of 25°C and pH 5.5, the mixture was placed in a coagulant and stirred at a speed of 70–80 rpm for 30 min. 5, 10, 15, 20, 25, 30, 35, and 40 mg/L aluminum sulfate were added to explore the influence of aluminum sulfate dosage on the fluorine removal effect, as shown in Figure 1.

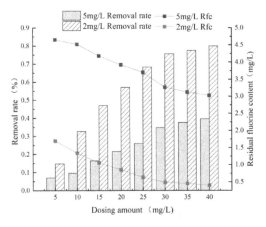

Figure 1.   Influence of aluminum sulfate dosage on fluorine removal effect.

As can be seen from Figure 1, with the increase of aluminum sulfate dosage, the fluorine ion removal rate keeps improving, while fluorine residue decreases. For the water sample with the initial fluoride ion concentration of 2 mg/L, when the aluminum sulfate dosage is 20 mg/L, the removal rate is 57%. In the later stage, with the increase of aluminum sulfate dosage, the removal rate is still rising significantly, but at this time, the fluoride ion concentration in the effluent is 0.86 mg/L, lower than 1 m/L, which meets the standard. For water samples with an initial fluoride concentration of 5 mg/L, when aluminum sulfate dosage is 35 mg/L, the removal rate is 37.6%. However, as the dosage increases to 40 mg/L, the removal rate increases only 2%, indicating that the removal effect tends to be stable. This is mainly because when aluminum sulfate dosing quantity is higher, because of acidic aluminum sulfate solution, and aluminum sulphate hydrolysis to produce hydrogen ions, all these make water pH value drops rapidly, aluminum sulfate dosing quantity, the more, the smaller the coagulation reaction system of pH, it affected the aluminum with fluoride or form of the combination of hydrolysate and fluorine and stability (Yuan 2013).

For the effect of fluoride removal and economic consideration, 20 mg/L aluminum sulfate was selected for groundwater with 2 mg/L fluoride ion mass concentration, and 30 mg/L aluminum sulfate was selected for groundwater with 5 mg/L fluoride ion initial mass concentration.

3.2 *Influence of pH value on fluoride removal*

The mass concentration of fluoride ions in the experimental raw water was 2 mg/L and 5 mg/L. Under the condition of 25°C, the pH value was adjusted to 3.5, 4.5, 5.5, 6.5, 7.5, 8.5, and 9.5. Aluminum sulfate was added to the solution, and the solution was placed in a coagulant and stirred for 30 min at a speed of 70–80rpm to explore the influence of pH value on fluoride removal, as shown in Figure 2.

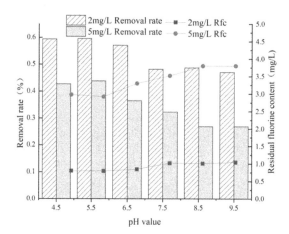

Figure 2.   Influence of pH value on fluoride removal.

As can be seen from Figure 2, weak acidity facilitates fluoride removal. Under the condition of weak acidity, the monomer aluminum ions hydrolyze into a large number of mesomer aluminum forms, especially the Al13 polymer can become the dominant aluminum form, which is considered the most effective flocculating component of aluminum salts. For raw water with an initial fluoride ion concentration of 5 mg/L and pH of 3.5–5.5, there is little difference in fluoride removal effect, and the fluoride ion removal rate shows an increasing trend between 41.4% and 43.8%. When pH increased from 5.5 to 9.5, the removal rate of fluoride decreased from 43.8% to 26.9%.

This is because when pH is high, Al(OH)3 flocs in the solution begin to dissolve gradually, and the morphology of flocs is gradually transformed towards the formation of complex anion [Al(OH)4]-, and the fluoride ions are adsorbed, and complexed by flocs are desorbed, resulting in the reduction of fluoride ion removal rate. At the same time, when pH is too high, OH- in water will compete with fluoride ion adsorption. Even replacing the granular fluoride that has been adsorbed on the surface of aluminum hydroxide will lead to a decrease in fluorine removal efficiency (Gong 2012; Liu 1993; Pablo 2006).

3.3 *Influence of coagulation stirring time on fluorine removal effect*

The mass concentrations of fluoride ions in the raw water were 2 mg/L and 5 mg/L. Under the condition of 25°C and pH 5.5, aluminum sulfate was added to the solution, and the solution was placed in a coagulant and stirred at a speed of 70-80rpm for 5, 10, 15, 30, 45, 60, and 90 min to explore the influence of coagulation stirring time on the fluorine removal effect, as shown in Figure 3.

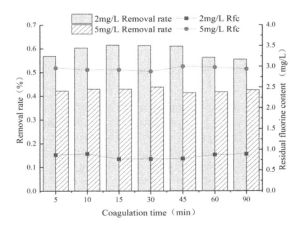

Figure 3.   Influence of coagulation time on fluorine removal effect.

As can be seen from Figure 3, when the coagulation time is 30 min, the flocs are in good shape and react fully with fluoride ions in water, and the fluoride removal effect is the best. At this time, the fluoride ion removal rate of raw water with an initial fluoride ion concentration of 2 mg/L and 5 mg/L is 61.3% and 43.8%, respectively. However, with the increase in coagulation stirring time, the fluoride ion removal rate is relatively stable and has a downward trend. At this time, the flocs begin to age and dehydrate, and excessive stirring will lead to the fragmentation of the increasing flocs due to shear force, which is not conducive to flocculation, resulting in the reduction of fluoride ion adsorption capacity in water, thus reducing the effect of fluoride removal by coagulation (Tie 2013). When the coagulation time is 90 min, the fluorine ion removal rates are 55.5% and 42.6%, respectively. Therefore, in order to ensure the full progress of the reaction, 30 min coagulation time was selected as the optimal reaction condition in this study.

3.4 *Influence of stirring intensity on fluorine removal effect*

Stirring intensity is closely related to ion diffusion, formation, and destruction of flocs in coagulation reactions. Too much or too little stirring intensity is not conducive to coagulation precipitation reaction. For the raw water with fluoride ion mass concentration of 2 mg/L and 5 mg/L, it is needed to adjust the pH value of the solution to 5.5 at 25°C, add aluminum sulfate, put it in the coagulant, and stir at 70–80 rpm, 100–110 rpm, 200–210 rpm for 30 min. The influence of stirring intensity on fluorine removal was explored. The experimental results are shown in Figure 4.

As can be seen from Figure 4, with the increase of stirring speed, the mass concentration of fluoride ion in effluent decreases gradually, fluoride ion constantly collides with small flocs, and the particle size of flocs increases gradually, indicating that low-speed stirring is conducive to ion

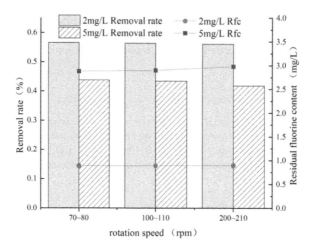

Figure 4.　Influence of stirring intensity on fluorine removal effect.

diffusion and flocs generation. When the stirring speed is 70–80 rpm, the initial concentration of fluoride ion is 2 mg/L of raw water, the mass concentration of fluoride ion effluent is 0.86 mg/L, and the removal rate is 56.6%. The fluoride ion removal rate of raw water with an initial fluoride ion concentration of 5 mg/L is 43.8%. In this case, regardless of the initial fluoride ion concentration of 2 mg/L or 5 mg/L, the fluoride ion concentration of effluent water reaches the lowest and the fluoride ion removal rate is the highest. When the stirring speed continues to increase, the effluent fluoride ion concentration gradually increases. When the stirring speed increases to 200-210rpm, the initial fluoride ion concentration of the raw water is 2 mg/L, the effluent fluoride ion mass concentration is 0.9 mg/L, and the removal rate is 56.1%. The removal rate of fluoride ions in raw water with an initial fluoride ion concentration of 5 mg/L is 41.8%.

This is because the speed, the greater the hydraulic gradient is, the greater the hydraulic gradient and determine the size of the floc size, smaller flocs and more open structure and better ability of adsorption, flocculation, sediment particle boundary layer thickness decrease at the same time, the external diffusion resistance is reduced, solubility increases, so the water quality of fluorine ion concentration increased gradually (Li 2006). It can be seen that maintaining a low stirring speed is beneficial to the formation of flocs. Therefore, the stirring speed should be 70–80 rpm.

4　CONCLUSIONS

1. The proportion relation between the influencing factors of aluminum sulfate coagulation and fluorine removal effect is dosage > pH value > stirring intensity > stirring time.
2. The optimal conditions for fluoride removal by aluminum sulfate coagulation are as follows: aluminum sulfate dosage 20mg, pH=5.5, rotating speed 70–80 rpm, coagulation time 15 min, the removal rate of fluoride concentration 2 mg/L water sample reaches 61.3%; Aluminum sulfate dosage 35 mg, pH value 5.5, rotating speed 70–80 RPM, coagulation time 30 min, fluoride concentration 5 mg/L water sample removal rate reached 43.8%.

REFERENCES

Gong Wenxin. (2012) Study on the mechanism of aluminum-fluorine morphology interaction and coagulation/adsorption fluoride removal. Beijing: *Doctoral dissertation of Graduate School of Chinese Academy of Sciences.*

Liao Y Q. (2018) Study on adsorption and regeneration of fluoride ions in water by activated alumina [D]. *Beijing University of Chemical Technology*.

Liu Shirong, Yang Aiyun. (1993) Effect of ph of water on fluorine removal by aluminum sulfate coagulation and its mechanism [J]. *Journal of Natural Science of Hunan Normal University*, (02):146–151.

Pablo Cañizares, Fabiola Martínez, Carlos Jiménez, Carlos Jiménez, Justo Lobato et al. *Comparison of the Aluminum Speciation in Chemical and Electrochemical Dosing Processes.*

Tao Li, Zhe Zhu, Dongsheng Wang, Chonghua Yao, Hongxiao Tang. (2006) Characterization of floc size, strength, and structure under various coagulation mechanisms[J]. *Powder Technology*, 168(2).

Tie Zhong. (2013) *Research on groundwater hardness removal Technology* [D]. Shandong Jianzhu University.

Yuan XIxin. (2013) *Study on enhanced coagulation for fluoride removal from drinking water* [D]. Jinan University.

*Advances in Energy, Environment and
Chemical Engineering – Abdullah & Osman (Eds)*
© 2023 The Author(s), ISBN: 978-1-032-36083-6

Environmental functional materials in soils

XinXin Yu
College of Environmental Science and Engineering, Guilin University of Technology, Guilin, China

ShaoHong You
Guilin University of Technology, Guilin, China
Technical Innovation Center of Mine Geological Environment Restoration Engineering in Southern
Stony Hill Area, Nanning, China

Yu-Cui Shi, You-Ran Lv, Bing-Fang Pang, Yuan Long & Ping-Ping Jiang*
Guilin University of Technology, Guilin, China

ABSTRACT: Along with the development and progress of society, soil environmental problems have become prominent and have posed a threat to the survival of human beings. The application of environmental materials has drawn attention to the fact that they have attracted widespread attention in the soil environmental community, and more and more research has been conducted on the use of environmental functional materials to improve soil or remediate soil pollution. This paper mainly reviews the research on the environmental effects of different environmental functional materials in soil fertility maintenance and pollution management.

1   INTRODUCTION

China has a serious problem of drought and water shortage, with more than 50% of the country's land area in arid and semi-arid zones, and agricultural development has been greatly constrained (Ren 2011). Moreover, large areas of soil are heavily contaminated by organic pollutants and heavy metals. Therefore, remediation of soil pollution is a matter of urgency.

Environmentally functional materials are materials that have both special functions to improve the environment and environmentally friendly features. In recent years, due to the advantages of enhancing soil water retention, improving soil structure, reducing soil nutrient loss, easy recycling or regeneration of the material itself, and no pollution, environmental functional materials have been more commonly used in agricultural production to improve crop yield and water and fertilizer use efficiency, becoming a new technology and a new way to increase agricultural chemical water conservation and yield (Li et al.2018). In this paper, we review the environmental effects of different EFMs on soil fertility maintenance and pollution control.

2   ENVIRONMENTALLY FUNCTIONAL MATERIALS AND THEIR TYPES

Environmental remediation technology is the technology applied to contaminated sites, and functional materials for environmental remediation are the materials applied to the remediation technology. Environmentally functional materials, also known as ecological materials or environmental materials, are a general term for a class of materials that are functional, environmentally compatible, and economical (Huang & Sun 2013). Environmentally functional materials mainly include natural environmental mineral materials, artificial materials, oxidation-reduction materials, and catalyst-based remediation materials.

---

*Corresponding Author: 1228472817@qq.com

      DOI 10.1201/9781003330165-66

2.1 Natural environmental mineral materials

Natural environmental mineral materials can be subdivided into silicate mineral materials, apatite mineral materials, metal oxide materials, etc.

Silicate minerals have a unique structure and excellent properties and have been used in many studies for the remediation of heavy metal contaminated soils. Silicate minerals commonly used for remediation of heavy metal contaminated soils include bentonite, albite clay, seafoam, kaolin, and zeolite. Gianotti et al. investigated the adsorption and desorption of 2,4,6-trichlorobenzene and 4-chlorophenol on montmorillonite and kaolinite with good adsorption results.

Apatite minerals can be used to remediate heavy metal contaminated soils such as Pb and Cd and are also used in permeable reaction walls. Sneddon et al. showed that the application of apatite artificially synthesized from the fish bone in Pb, Zn and Cd contaminated soils had a strong fixation capacity for Pb and Cd.

The interaction of metal oxides, especially iron and manganese oxides, with heavy metal ions is one of the main focuses of research in soil chemistry and environmental chemistry and has received extensive attention from experts at home and abroad. Metal oxides adsorb organic matter through ligand exchange, electrostatic forces, cation bridging, water repellency, entropic interactions, and hydrogen bonding (Guo et al. 2006).

2.2 Artificial materials

Artificial materials can include synthetic, artificially modified, or even some agricultural or industrial wastes. Synthetic zeolites are well-researched and applied materials that can form (hydro)oxide precipitates with heavy metals, effectively reducing the mobility and bio effectiveness of heavy metals in soil (Terzano et al. 2007). Czurda et al. studied the use of fly ash synthetic zeolites in permeable reaction walls, comparing them with several other clay minerals, and found fly ash to be the superior material for permeable reaction walls.

2.3 Oxidation-reduction materials

This type of material is itself involved in redox reactions, where the material itself is the redox agent. For organically contaminated sites, oxidants are often added to oxidize the organic material to end products such as CO_2 and H_2O. Commonly used chemical oxidizing agents include ozone, hydrogen peroxide, Fenton's reagent, potassium permanganate, chlorine dioxide, and so on (Liu & Shi 2005).

2.4 Catalyst-based restoration materials

Catalyst-based remediation materials are mainly used on sites contaminated by organic pollutants, such as organic pesticides, chemical industrial materials, and other POPs. They can be divided into catalytic oxidation and catalytic reduction materials according to the type of reaction they catalyze. Unlike redox materials, however, catalyst materials do not undergo permanent changes in their own structure. Photodegradation reactions that occur at the soil surface are an important degradation pathway for various organic pollutants in soil. The photodegradation of organic matter in soil is influenced by the combination of light intensity, the composition and physicochemical properties of the soil, the nature of the organic pollutants, and various environmental factors.

3 ENVIRONMENTAL FUNCTIONAL MATERIALS IN SOILS

3.1 Environmental effects of biomass char on soil

Biochar, also known as biochar, is a solid, stable, highly aromatized material containing char obtained by thermal cracking ($< 700°C$) of biomass under fully or partially anoxic conditions (Gaunt J L & Lehmamnn J 2008). Carbon materials are generally electrically and thermally conductive, light in mass, low in density, resistant to acids and alkalis, and have an excellent pore structure that allows them to be fully exposed to electrolytes, making them widely used in electrochemistry (Beesley 2011). Biomass charcoal has attracted a lot of attention in the soil and environmental science community as a new class of environmental functional materials (Tang et al. 2013).

3.1.1 *Study on the improvement of soil fertility by biochar*

Biochar can significantly increase soil pH, alter soil texture and increase salt base exchange, thereby causing an increase in soil CEC. Lehmann et al. suggest that the CEC level of the soil is significantly increased after biochar application due to its rich aromatic ring structure and hydroxy carbon groups and higher surface exchange activity, which affects the uptake of nutrients by plants. At the same time, the water-holding capacity and water supply of the soil are improved by the application of biochar, which avoids the leaching of nutrients by reducing the dissolution and migration of water-soluble nutrient ions, and is released continuously and slowly into the soil, acting as a slow-release carrier for nutrients, thus achieving fertility maintenance (Woods 2008). Biochar is used in combination with other organic or inorganic fertilizers to increase crop yields. Chan et al. compared the interaction between biochar and N fertilizer on radish and found that the crop yield increased by 120% when biochar was added under N fertilizer conditions.

3.1.2 *Biochar remediation of contaminated soil*

Due to its rich pore structure and oxygenated active groups on its surface, biomass char can be added to the soil to adsorb pollutants and make them immobilized within the biomass char, reducing the toxicity and chemical activity of pollutants in the soil. As a new remediation material, biomass char is highly effective in the remediation of contaminated soils. Compared to traditional remediation methods, biomass charcoal is not only low cost and easy to use, but also maintains soil nutrients and reduces nitrogen leaching, making it a new and highly sought-after technology for soil remediation. Beesley et al. found that the application of biomass char to soil increased the activity of soil microorganisms and promoted the degradation of organic pollutants by microorganisms.

3.1.3 *The effect of biochar on soil microorganisms*

The complex pore structure and huge surface area of biomass char provide excellent sites for the survival and reproduction of microorganisms (Luo et al. 2019). The small amount of soluble substances such as organic carbon and minerals in biomass charcoal can provide certain nutrients for soil microorganisms and help them to grow. At the same time, due to its unique physicochemical properties, biomass char has many fine pores, which can also regulate the environmental structure of the soil, change the physicochemical properties of the soil, and even create a soil microbiological environment suitable for microbial growth, thus allowing microorganisms to grow better (Zhu & Sheng 2018). In addition, changes in the plant growth environment after the application of biomass charcoal to the soil can lead to changes in root growth and root secretions; changes in soil animal behavior can also lead to changes in the soil environment, all of which can influence the interaction between biomass charcoal and soil microorganisms. Biomass charcoal improves crop yields mainly through interactions between biomass charcoal and soil organisms, for example, by promoting the growth of slave-like mycorrhizae and influencing the water holding capacity of the soil. The nature of the biomass charcoal and the amount applied to the soil affect the response of the soil microbial community. The application of biomass charcoal gives some microbial groups in the soil more space and nutrients to grow, promoting their growth and development; at the same time, some microorganisms are inhibited by it, thus changing the structure of the soil microbial community.

4 CONCLUSION

The development and application of environmental functional materials have received increasingly widespread attention as a new way to save water and increase production chemically. Exploring the combination and proportioning of multiple environmental materials and developing composite materials to more fully exploit the advantages of individual environmental materials and the effects of composite materials is a new growth area for environmental materials research. Therefore, we need to carry out research on pilot and demonstration projects for engineering applications from the perspective of practical applications, as well as tracking and evaluating their remediation effects and their ecological and environmental effects, and establishing technical specifications regarding the implementation of contaminated soil remediation projects.

ACKNOWLEDGMENTS

This research is supported by the Natural Science Foundation of China (51868010, 52170154); Guangxi Natural Science Foundation Program (2021GXNSFBA196023); Guilin Science and Technology Development Program (20190219-3); The project of improving the basic scientific research ability of young and middle-aged college teachers in Guangxi(2021KY0275).

REFERENCES

Beesley l. & Moreno J E. & Gomez E J l. (2010) Effects of biochar and greenwaste compost amendments on mobility, bioavailability, and toxicity of inorganic and organic contaminants in a multielement polluted soil [J]. *Environmental Pollution*, 158 (6):2282–2287.

Beesley Luke. & Eduardo Moreno-Jimenez. & Jose L Gomez-Eyles, et al. (2011) A review of biochars' potential role in the remediation, revegetation, and restoration of contaminated soils[J]. *Environmental Pollution*, 159(12):3269–3282.

Chan K Y. & Van Zwieten L. & Meszaros I, et al. (2007) Agronomic values of greenwaste biochars a soil amendment[J]. *Aust J Soil Res*, 45(8):629–634.

Czurdaka. & Hausr. (2002) Reactive barriers with fly ash zeolites for in situ groundwater remediation[J]. *Applied Clay Science*, 21(1):13–20.

Gaunt J L. & Lehmamnn J. (2008) Energy balance and emissions associated with biochar sequestration and pyrolysis bioenergy production[J]. *Environ Sci Technol*, 42(11):4152–4158.

Gianottiv. & Benzim. & Croceg, et al. (2008) The use of clays to sequestrate organic pollutants. Leaching experiments[J]. *Chemosphere*,73(11):1731–1736.

Guo Jin. & Ma Jun. & Liu Song, et al. (2006) Study on the adsorption mechanism of natural organic substances on alumina surface[J]. *Journal of Environmental Science*, 26 (1):111–117.

Huang Zhanbin. & Sun Zaijin. (2013) The application of environmental materials in agricultural production and its environmental management[J]. *Chinese Journal of Ecological Agriculture*, 21(01):88–95.

Lehmann J. & Kern D C. & Glaser B. (2003) *Amazonian dark earth: origin properties management*[M]. Dordrecht: Kluwer Academic Publishers, 125–139.

Li Jiazhu. & Huang Zhanbin. & Bao Fang. (2018) Comprehensive evaluation of water and fertilizer conservation performance of environmental functional materials[J]. *China Soil and Water Conservation Science*, 16(03):125–133.

Liu Zq. & Shi Lil. (2005) Risk management and remediation techniques for contaminated land in the UK[J]. *Environmental Protection*, 33 (10): 69–73.

Luo Fei. & He Qingwen. & Tian Dan, et al. (2019) Research progress and applications of environmental functional materials[J]. *Information Record Materials*, 20(10):24–25.

Ren Chong. & Dong Yuanhua. & Liu Yun. (2011) Functional materials for environmental remediation of contaminated sites[J]. *Environmental Monitoring Management and Technology*, 23(03):63–70

Sneddon IR. & Orueetxbarriam. & Hodsonm E, et al. (2006) Use of bone meal amendments to immobilize Pb, Zn and Cd in soil: A leaching column study[J]. *Environ Mental Pollution*, 144 (3):816–825.

Tang J C. &Zhu W Y. &Kookana R, et al. (2013) Characteristics of biochar and its application in remediation of contaminated soil[J]. *Journal of Bioscience and Bioengineering*, 116(6): 653–659.

Terzano R. & Spagnuolo. & Medicil, et al. (2007) Microscopic single-particle characterization of zeolites synthesized in a soil polluted by copper or cadmium and treated with coal fly ash[J]. *Applied Clay Science*, 35(1): 128–138.

Woods W I. (2008) *Amazonian dark earth: Wim Sombroek's vision* [M]. Berlin: Springer, 8–14, 213–228, 325–328.

Zhu Lanbao. & Sheng Ti. (2018) Biomass carbon preparation technology and its environmental effects on soil[J]. *Journal of Bengbu College*, 7 (02):23–27.

Advances in Energy, Environment and
Chemical Engineering – Abdullah & Osman (Eds)
© 2023 The Author(s), ISBN: 978-1-032-36083-6

Nutrients releasing rule and denitrification performance of seven kinds of solid carbon sources in continuous flow bioreactors

Xiaohong Hong
Guanlan River Watershed Management Center of Shenzhen River, Longgang District, Shenzhen, Guangdong, China

Bohan Chen
Civil and Environmental Engineering, Harbin Institute of Technology, Shenzhen, Shenzhen, Guangdong, China

Haixia Feng* & Xianqiong Hu
Shenzhen Municipal Engineering Consulting Center CO., LTD, Shenzhen, Guangdong, China

ABSTRACT: This study investigated the nutrient releasing rules of three kinds of natural cellulose (NC), including wheat straw (WS), rice straw (RS), corncob (CC), and the static releasing nutrients experiments indicated that CC had a slower rate on releasing organic matter and released less nitrogen and phosphorus compared with WS and RS. Meanwhile, the denitrification performance of CC and four kinds of synthetic biodegradable polymers (SBP), including polycaprolactone (PCL), polybutylene succinate (PBS), polylactic acid (PLA), and polyhydroxyalkanoate (PHA), were also evaluated. In continuous-flow bioreactors (CFBs), only CC and PCL reached an 80% nitrogen removal rate (NRR) when the hydraulic retention time (HRT) was 4h. In addition, when the HRT was reduced to 2 h, the NRR of CC decreased to about 36%. For PCL, the NRR could be maintained at above 98% when HRT was 1h and decreased to about 60% when the HRT was 0.5 h.

1 INTRODUCTION

With social awareness of environmental protection improved, the standard of effluent in wastewater treatment plants (WWTPs) has become stricter in China. Research has indicated that inorganic nitrogen pollution in aquatic ecosystems can be prevented when the total nitrogen concentrate is below 0.5–1.0 mg/L (Camargo & Alonso 2006). Bioheterotrophic denitrification is the most common and effective technology to remove nitrate. However, the deep removal of total nitrogen in the effluent, mainly including nitrate, is still the most difficult challenge for WWTPs in China (Jin et al. 2014). The low organic carbon in wastewater limits the nitrogen removal capacity of bioheterotrophic denitrification, which means some soluble organic matter such as methanol, sodium acetate, and glucose are usually used as extra carbon sources to support the denitrification process (Wu et al. 2012). However, due to the fluctuation of wastewater quality, the dosage of carbon sources is hard to control to ensure stable COD in the effluent (Qi et al. 2020).

Recently, solid carbon sources (SCS) used in solid-phase denitrification have been considered a substitute for the soluble carbon sources to complete bio-heterotrophic denitrification (Wang & Chu 2016). Compared with soluble carbon sources, the characteristics of SCS such as appropriate

---

*Corresponding Author: 59253553@qq.com

 DOI 10.1201/9781003330165-67

carbon release rate, long carbon release time, ease of control, and good bioloading capacity make sure the denitrification effect and effluent quality (Zhang et al. 2021). The commonly used SCS could be classified into two main categories: natural cellulose (NC) and synthetic biodegradable polymers (SBP). NC such as cotton, straw, and corncob is mainly composed of cellulose, lignin, and hemicellulose, with advantages of abundant stock, low price, and rich variety (Della Rocca et al. 2007; Soares & Abeliovich 1998). Nonetheless, the complex spatial structure and chemical properties of NC affect the carbon utilization rate in denitrification (Behera et al. 2014). In addition, using NC may bring high dissolved organic carbon and color to effluent (Chu & Wang 2016). On the contrary, SBP, represented by poly(3-hydroxybutyrate-co-3-hydroxyvalerate) (PHBV), has the advantages of easy biodegradation and a more stable carbon release rate without secondary pollution. However, high prices limit their practical use (Xu & Chai 2017).

Therefore, this study aims to select the SCS with the best performance from three kinds of NC: rice straw (RS), wheat straw (WS), corncob (CC), and four kinds of SBP: polycaprolactone (PCL), polybutylene succinate (PBS), polylactic acid (PLA), polyhydroxyalkanoate (PHA). The static nutrients release rules of them, as well as the organic composition of three NC extracts, were analyzed and the best one of NC was selected for the denitrification experiment. The denitrification performance of four SBPs and the selected NC were evaluated in continuous flow conditions. Meanwhile, the influence of hydraulic retention time (HRT) on denitrification performance was also studied.

2 MATERIALS AND METHODS

2.1 *Materials*

The particle diameter of RS, WS, and CC (Gongyi Hengrun Water Treatment Material, China) were 5–7 mm and PCL (Dongguan Zhanyang Polymer Material, China), PBS (Dongguan Zhanyang Polymer Material, China), PLA (Dongguan Baoqian Plastic Technology, China) and PHA (Dow Group, America) was 3 mm. The seed sludge gained from the secondary treatment aerobic tank in a sewage treatment plant in Shenzhen was preserved at four degrees centigrade. The SCS dipped in the seed sludge for 24 hours.

Synthetic wastewater had the following characteristics: NO_3^--N 45–55 mg / L, TP 13 mg /, Lalkalinity 600 mg / L, Mg^{2+} 10 mg / L, Ca^{2+} 5 mg / L, trace element solution 1 mL / L. Trace microelement solution included: boric acid 0.015 g / L, manganese chloride tetrahydrate 0.99 g / L, nickel chloride hexahydrate 0.19 g / L, EDTA-Na 25 g / L, cobalt chloride hexahydrate 0.24 g / L, ferrous sulfate 9 g / L, copper sulfate pentahydrate 0.25 g / L, sodium molybdate 0.22 g / L, zinc sulfate 0.44 g / L.

2.2 *Static experiments*

The SBP would not release nutrients without microbial (Li et al. 2014). Therefore, static experiments were only performed in NC. The static release of nutrients experiments was operated in beakers. 10 g SCS were put into conical flasks with 200 mL distilled water (DW). 10 mL water samples were taken at 1, 3, 6, 12, 24, 36, 48, 60, 72, 96, and 120 h from conical flasks, and 10 mL DW were added after each sampling. The COD, TN, NO_3^--N, NO_2^--N, NH_4^+-N, and TP of every sample were detected.

2.3 *Denitrification experiments*

The continuous flow bioreactor (CFR) was made of cylindrical polyvinyl chloride with a diameter of 25.6 mm, a height of 280 mm, and an effective volume of 100 liters. The water inlet and outlet were set at the bottom and the top of the reactor, respectively. Four kinds of SBP and CC loaded with sludge-filled the reactor until the height of 260 mm respectively. The wastewater was pumped by a peristaltic pump continuously and the water sample was obtained from a sampling port in the middle of the reactor. The initial HRT was 4 hours, and the membrane could be considered successful when the denitrification rate of the reactor reached over 80%. Then the HRT was gradually reduced to 2 hours and 1 hour.

When the gas generated by denitrification and the biofilm were observed to accumulate in the filler void and the effluent NO_3^- -N concentrate exceeded 5 mg / L, the gas-liquid backwashing started to run. The backwashing operates as follows: (1) aeration in the reactor for 5 min without water; (2) stop aeration and pump clear water into the reactor for 5–10 min; (3) end the backwashing when there is no sludge in the effluent.

2.4 Analytical methods

COD, TN, NO_3^- -N, NO_2^- -N, NH_4^+ -N, and TP were analyzed according to the Standard Methods. The releasing carbon component of three cellulose carbon sources was analyzed by EEMs technology. RF-5301PC fluorescence spectrophotometer (Shimazu, Japan) was used for the fluorescence spectrum, and a xenon lamp was used as the light source. The scanning range of excitation E_x and emission E_m is 225–450 nm and 225–800 nm, respectively, with a scanning interval of 5 nm. All data were analyzed by Origin Pro 2016 software. The average nutrient release rate, inorganic nitrogen concentrate, and denitrification rate of the reactors were calculated based on the following equations:

$$R_n = 0.95C/(T_i - T_e) \tag{1}$$

$$C_{in} = C_{NO_3^- -N} + C_{NO_2^- -N} + C_{NH_4^+ -N} \tag{2}$$

$$E_e = (C_{in}C_{ef})/C_{in} \times 100\% \tag{3}$$

where R_n, C_{in} and E_e were average nutrients (COD, TN, TP) release rate, the inorganic nitrogen concentrate, and the nutrients (COD, TN) removal rate, respectively. 0.95 was the diluted multiples in static experiments. $C_{NO_3^- -N}$, $C_{NO_2^- -N}$, $C_{NH_4^+ -N}$ were the concentrate of NO_3^- -N, NO_2^- -N, NH_4^+ -N in the effluent. C_{in} and C_{ef} were the concentrate of nutrient removal percentage in the influent and effluent, respectively.

3 RESULTS AND DISCUSSION

3.1 Static release of C/N/P from various SCSs

The COD concentrate of each NC in static conditions was shown in Figure 1(a). In the first hour, the COD concentrations of WS, RS, and CC were 206.4 mg / L, 261.6 mg / L, and 156 mg / L respectively, which indicated that the WS and RS released many organic matters in the initial. The WS leaching solution reached a maximum COD of 244.8 mg / g at 3 h. While RS and CC took a longer time to reach their maximum COD of 316.8 mg / g at 36 h and 297.6 mg / g at 24 h, respectively. After 12 h, the COD concentrate of three kinds of NC showed a stable or little decreasing trend due to the dilution effect. In addition, the average carbon release rates of CC were all higher than WS and RS in 3 h, 6 h, and 12 h. Hence, the static experiment suggested that the CC had a better carbon release performance than WS and RS in static conditions.

The release rule of nitrogen and phosphorus was similar to COD in static experiments. The TN and TP concentrate of WS and RS were both higher than 10 mg / g at 24 h and CC was lower than 5 mg / g (Figure 1(b)). Similarly, the maximum TP concentrate of WS was 4.28 mg / g, which was much larger than 1.35 mg / g of RS and 1.17 mg / g of CC (Figure 1(c)). After 24 h, the release rate of nitrogen and phosphorus showed that no more nitrogen and phosphorus was released for all NC. Therefore, due to the stable carbon release performance and less nitrogen and phosphorus release of CC, it was selected as the NC filler for subsequent denitrification experiments.

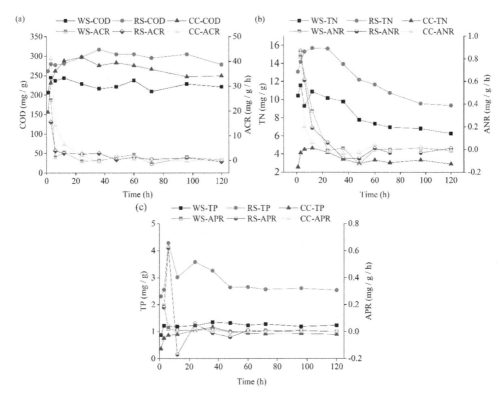

Figure 1. Static nutrients release of NC: (a) COD; (b) TN; (c) TP.

3.2 *Organic composition analysis of leach solution*

The distribution of various substances in leach solution in the static experiment was determined by 3D fluorescence spectroscopy, as shown in Figure 2. It can be seen from Figure 2 that the characteristic fluorescence peaks of humic acid (Ex/Em = 300-370/400-500 nm) and tyrosine (Ex/Em = 275/310 nm) appeared in all three cellulosic carbon sources. It was speculated that the humic acids in the leached solution are lignin with complex structures and difficult to biodegrade. Compared with RS and CC, the humic acid light peak of WS was the strongest, which indicated that the releasing organic matters of WS were more difficult to be biodegradable. In addition, it was found that RS contained the least tyrosine, and the available amino acids were less, which was not conducive to the denitrification process. Therefore, it was believed that CC was the more suitable carbon source for three kinds of NC. Three-dimensional (3D) fluorescence spectroscopy was employed to reveal the composition of the dissolved organic carbon and the results were shown in Figure 3. The characteristic fluorescence peaks of humic acid (Ex/Em = 300-370/400-500 nm) and tyrosine (Ex/Em = 275/310 nm) appeared in all three cellulosic carbon sources. It was speculated that the humic acids in the leached solution are lignin with complex structures and difficult to be biodegraded. Compared with rice straw and corncob, wheat straw showed the strongest humic acid light peak, which indicates the releasing organic matters of wheat straw were more difficult to be biodegradable. In addition, it was found that rice straw contained the least tyrosine which is easily utilized by microorganisms. It suggests that rice straw was not conducive to the denitrification process. Therefore, corncob was considered more suitable as a carbon source for heterotrophic denitrification among all.

461

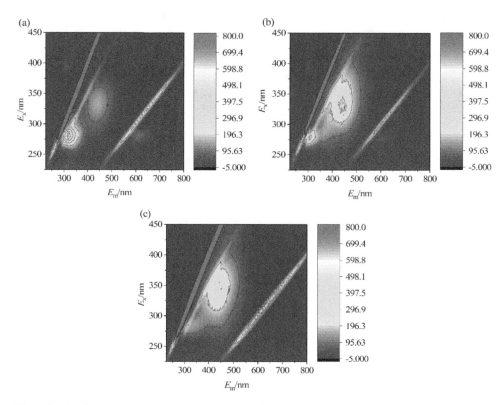

Figure 2.   3D fluorescence spectrogram of cellulose solid carbon source leaching solution: (a) CC; (b) WS; (c) RS.

3.3   Comparison of denitrification efficiency of different SCSs

CC was considered to be the best carbon source among the three cellulose carbon sources after analyzing the composition of the leaching solution. CC and four SBP were used as fillers respectively to construct a denitrification reactor for a solid carbon source. At the start-up stage, the influent nitrate concentration was 51.97 mg / L, and the initial HRT was 4 h. After the stable operation, the HRT was gradually reduced to compare the denitrification effect. The experimental results are shown in Figure 3.

After 8 days, the NO_3^- -N removal rate was stable above 95%, indicating that the membrane was successfully attached to the fillers. The HRT was reduced to 2 h after stable operation for a week, which led to the effluent NO_3^- -N increasing rapidly. After continuous operation for 45 d, the NO_3^- -N removal rate of CC was maintained at about 36%, indicating that the nitrogen removal effect was relatively stable when using CC as SCS (Figure 3(a)).

The denitrification effect of PLA, PHA, and PBS was studied under the condition of HRT = 4 h with degradable polymers as fillers. The results showed that the average NO_3^- -N removal rates of the three bacteria were 23.68%, 7.78%, and 5.10%, respectively. It was speculated that the denitrifying bacteria could not adapt to the PLA, PHA, and PBS carbon sources, leading to the unsatisfactory denitrification effect (Figure 3(b)).

When PCL was used as a carbon source and HRT is 4h, it is found that the NO_3^- -N removal rate can reach 100% on the 4th day, indicating that PCL is a suitable carbon source for denitrification. Then, HRT was adjusted to 2 h and 1 h on the 20th day and the 33rd day respectively, and it was found that the removal rate of NO_3^- -N remained above 98%. However, after the 39th day, the denitrification effect gradually deteriorated. Considering that it might be due to the hardening problem of reactor packing, backwashing was carried out on the 51st day. Four days later, the NO_3^--N removal rate recovered to 95%. On the 60th day, HRT was adjusted to 0.5 h, and the largest denitrification rate of 1.44 kg N / (m³·d) was obtained after proper backwashing (Figure 3(c)).

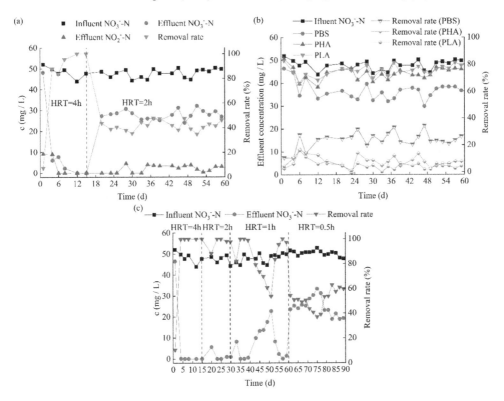

Figure 3.   Denitrification performances under each carbon source: (a)CC; (b)PLA, PHA, PBS; (c)PCL.

4   CONCLUSIONS

The static experiments and organic analysis indicated that CC was a more suitable filler for bio-denitrification in three kinds of NC and the denitrification performance of PCL was better than PBS, PLA, and PHA. Low HRT had a significant influence on the denitrification performance of CC and PCL. The nitrogen removal rate of CC and PCL both dropped from 100% to 40% and 60% when the HRT was reduced to 2 h and 0.5 h, respectively.

ACKNOWLEDGMENTS

This work was supported by Shenzhen Science and Technology Innovation Commission (Grant No. JCYJ20180306172051662, KQJSCX20180328165658476, JCYJ202001109113006046, KCX FZ20201221173602008, and KCXFZ202002011006362).

REFERENCES

Behera, S., Arora, R., Nandhagopal, N. and Kumar, S. (2014). Importance of chemical pretreatment for bioconversion of lignocellulosic biomass. *Renewable and Sustainable Energy Reviews*. 36, 91–106.

Camargo, J.A. and Alonso, A. (2006). Ecological and toxicological effects of inorganic nitrogen pollution in aquatic ecosystems: A global assessment. *Environ Int.* 32, 831–849.

Chu, L. and Wang, J. (2016). Denitrification of groundwater using PHBV blends in packed bed reactors and the microbial diversity. *Chemosphere.* 155, 463–470.

Della Rocca, C., Belgiorno, V. and Meric, S. (2007). Overview of in-situ applicable nitrate removal processes. Desalination. 204, 46–62.

Jin, L., Zhang, G. and Tian, H. (2014). The current state of sewage treatment in China. *Water Res.* 66, 85–98.

Li, P., Tang, L., Zuo, J.-e., Yuan, L. and Li, Z.-x. (2014). Tertiary nitrogen removal of the municipal secondary effluent using PHAs as solid carbon sources. *China Environmental Science.* 34, 331–6.

Qi, W., Taherzadeh, M.J., Ruan, Y., Deng, Y., Chen, J.S., Lu, H.F. and Xu, X.Y. (2020). Denitrification performance and microbial communities of solid-phase denitrifying reactors using poly (butylene succinate)/bamboo powder composite. *Bioresour Technol.* 305, 123033.

Soares, M.I.M. and Abeliovich, A. (1998). Wheat straw as a substrate for water denitrification. *Water Research.* 32, 3790–3794.

Wang, J. and Chu, L. (2016). Biological nitrate removal from water and wastewater by solid-phase denitrification process. *Biotechnol Adv.* 34, 1103–1112.

Wu, W., Yang, F. and Yang, L. (2012). Biological denitrification with a novel biodegradable polymer as carbon source and biofilm carrier. *Bioresource Technol.* 118, 136–140.

Xu, Z. and Chai, X. (2017). Effect of weight ratios of PHBV/PLA polymer blends on nitrate removal efficiency and microbial community during solid-phase denitrification. *Int Biodeter Biodegr.* 116, 175–183.

Zhang, F., Ma, C., Huang, X., Liu, J., Lu, L., Peng, K., and Li, S. (2021). Research progress in solid carbon source–based denitrification technologies for different target water bodies. *Science of The Total Environment.* 782, 146669.

Advances in Energy, Environment and Chemical Engineering – Abdullah & Osman (Eds)
© 2023 The Author(s), ISBN: 978-1-032-36083-6

Application of coiled tubing foam sand flushing technology in tight gas reservoir horizontal wells

Fang Zhang
Exploration & Development Research Institute, Petro China Changqing Oilfield Company, Xi'an, Shaanxi, China
National Engineering Laboratory for Low~permeability Petroleum Exploration and Development, Xi'an, Shaanxi, China

Xiaoling Meng*
Exploration & Development Research Institute, Petro China Changqing Oilfield Company, Xi'an, Shaanxi, China
National Engineering Laboratory for Low~permeability Petroleum Exploration and Development, Xi'an, Shaanxi, China
Evaluation & Potential Excavation Project Team of Thousand-gas Well, Changqing Oil Field Branch Company Ltd., PetroChina, Xi'an, Shaanxi, China

Min Xu
Evaluation & Potential Excavation Project Team of Thousand-gas Well, Changqing Oil Field Branch Company Ltd., PetroChina, Xi'an, Shaanxi, China

Hongbo Zhang
CNPC Chuanqing Drilling Engineering Company Limited, Chengdu, Sichuan, China

ABSTRACT: Horizontal well technology has become an important way for the effective development of Upper Paleozoic tight sandstone reservoirs in Ordos Basin. Affected by reservoir characteristics, measures, and processes, the formation is prone to sand production, resulting in burying the horizontal section with sand and gas well shutdowns. Compared with the traditional technology, using coiled tubing flushing sand with foam has the advantages of "without shut-in and moving string, low reservoir damage and efficient sand flushing". By optimizing downhole tools, sand flushing fluids, and coiled tubing parameters, such as length and diameter, 39 wells were operated in the eastern part of the basin with good effect.

1 INTRODUCTION

The Upper Paleozoic Reservoir in the east of Ordos Basin is a typical tight sandstone reservoir, being an effective technology where a horizontal well is widely used to increase single well production. Because of the special structure of the horizontal well, sand production is easy to form sand bed at the bottom of the Wellbore, which seriously affects the normal production of the gas well. In order to solve the problem of poor effect and high cost of traditional sand-flushing operation, a new method—the coiled tubing process is proposed. In recent years, the application field of coiled tubing technology has been expanding and widely used in gas field workover, drilling, completion, logging and other operations. This technology can realize continuous sand washing under pressure, with a simple process, short construction time, efficient and safe sand washing operation, and achieved good economic benefits (Liang 2017; Li et al. 2001; Yu & Jiang 2011; Zhang et al 2020; 2018; 2007).

---

*Corresponding Author: Mengxl1_cq@petrochina.com.cn

DOI 10.1201/9781003330165-68

2   CAUSE ANALYSIS OF SAND PRODUCTION IN GAS WELLS

Sand production in gas wells will greatly affect normal production. There are two main reasons for sand production: (1) During the process of gas well production, on account of loosen reservoir cementation and fluid erosion, the sandstone layer structure near the perforation channel or near the well is damaged, so that the sand particles migrate from the oil and gas reservoir with the fluid; (2) At present, large-scale hydraulic jet fracturing technology is widely used in tight reservoirs. This technology uses high-speed and high-pressure fluid to carry sand particles for perforation. However, when the fracture opens, a large number of fractured sand particles are retained in the formation and will enter the wellbore in later production. Sand production of gas wells leads to a sharp decline in production, a surge in operating costs and serious economic losses.

3   RESEARCH OF PARAMETER OPTIMIZATION

3.1   *Principle of coiled tubing sand washing*

Given the sand blockage in the gas well bore, coiled tubing sand flushing technology is adopted. A sand flushing fluid circulation channel is established between the coiled tubing and tubing without shut-in and moving string. After the sand flushing fluid is injected into the coiled tubing, the fractured sand and foreign matter are carried out from the coiled tubing and the tubing annulus, so as to return the productivity of the gas well.

The coiled tubing truck runs a whole thread-free tubing into the casing. With the running process of the tubing, the sand-washing fluid will be pumped to the bottom of the well with the power provided by the supercharger and the cement truck. The high-speed sand flushing fluid will wash away the sand plug. At the same time, the sediment at the bottom of the well will be brought to the ground with the help of the upper return fluid—until the wellbore is clean.

The outer diameter of the coiled tubing string can be divided into three sizes: $\Phi 31.75$ mm, $\phi 38.1$ mm, and $\Phi 50.8$ mm, the minimum tensile strength is 205.1 KN and the minimum internal pressure resistance is 72.88 MPa, the length can be adjusted according to different well conditions. Nitrogen-making vehicles use the difference between the permeability of various components in the air to separate nitrogen through the separation membrane tube. Nitrogen is added to the foam generator distribution pipe through the supercharger and mixed with the liquid. After passing through the fixed impeller, the initially mixed foam liquid changed its direction several times and was mixed with water and crushed to form microbubbles in water. Figure 2 shows the construction site of coiled tubing sand flushing.

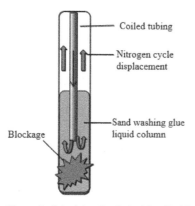

Figure 1. Principle of coiled tubing flushing and sand washing process.

Figure 2. Coiled tubing construction site.

3.2 Optimize sand flushing process parameters

3.2.1 Coiled tubing length and pipe diameter

A coiled tubing drum truck can carry coiled tubing with different diameters and lengths for construction. The well depth of horizontal wells in the east of the basin is 5300 ~ 5800m. Considering the increases in friction and coiled tubing truck quality with the increases in coiled tubing length, the road and site requirements must be improved, 6000m coiled tubing is selected for construction. Coiled tubing has ϕ 31.75 mm ϕ 38.1mm and ϕ 50.8mm specifications. In order to reduce friction, larger displacement shall be selected as much as possible during sand flushing. Sand flushing operation with 50.8mm coiled tubing is shown in Table 1

Table 1. Roll to traffic to carry different coiled tubing.

Rolling depth	Total mass /t	Climbing angle /(°)	Turning radius /m
4500 m of coiled tubing	48	30	16
6000 m of coiled tubing	55	30	16

3.2.2 Downhole treatment tools

According to the process flow of coiled tubing measures: firstly, kill the hydraulic well, dismantle the wellhead gas flow tree, install the blowout preventer and test the pressure, then pull out the tubing string of the original well, and then explore the sand surface and wash the sand. In view of the complex downhole conditions, some wells need to be grind-drill due to serious scaling during flushing. It is preferred to use pneumatic and hydraulic drilling tools + five-blade flat bottom grinding shoes for grinding.

3.2.3 Sand flushing fluid

Due to the poor productivity of horizontal wells in some tight reservoirs in this area, in the continuous flushing and sand flushing operation, the simple use of clean water and guanidine gum sand flushing fluid will lead to large flushing fluid leakage, difficult circulation establishment and low sand flushing efficiency, affecting the productivity recovery of gas wells. After multi-well construction effect analysis and field verification, a nitrogen foam solution was selected for sand washing. Meanwhile, the Na_2CO_3 solution of nitrogen foam solution +3% was used for sand flushing in a high H_2S gas well.

Corresponding to the minimum sand carrying capacity of continuous coiled tubing (Figure 3), the type of nitrogen foam is selected. As the mass fraction of the foaming agent raise, the foaming volume, and half-life of the foam liquid increase with the increase in production. However, after the mass fraction exceeds 0.5%, the foaming volume and half-life almost no longer change, indicating that the critical micelle mass fraction has reached the critical mass fraction. Finally, through experiments and on-site verification, the YFP-10 foaming agent with a mass fraction of 0.5% was prepared to prepare foam liquid with active water (Figure 4).

3.2.4 Sand flushing parameters

In the sand flushing process, the wellbore condition and casing flowback rate are fully considered, and the tripping speed is adjusted to no more than 3m / min to ensure that the nitrogen volume meets the standard. The pump truck displacement is determined to be 210 ~ 285L / min, which provides a more reasonable casing flowback carrying efficiency while ensuring the flushing efficiency. At the same time, the coiled tubing negative pressure sand flushing technology without killing the well and moving the pipe string is adopted to establish a circulation channel of sand flushing fluid between the coiled tubing and the production tubing in the well, so that the pressure generated by the sand flushing fluid on the bottom of the well is less than the formation pressure. The sand flushing fluid flushes the bottom sand out of the ground without entering the formation, so as to avoid reservoir damage.

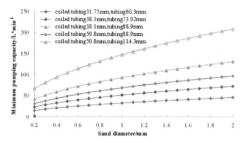

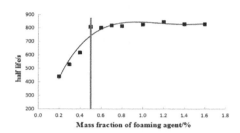

Figure 3. Calculation of minimum sand carrying capacity of coiled tubing sand flushing.

Figure 4. Half-life of foam under different foaming agent mass fraction.

4 FIELD APPLICATION EFFECT EVALUATION

The coiled tubing sand flushing process has been implemented in 39 horizontal wells with an average daily production raises of 15000 m³ and a cumulative increase of 152.1 million m³. The effect of production increases after the measures is good. The open flow potential (Q_{AOF}) of horizontal well SD6 in this area is 71.2310×10^4 m³/d, put into operation in 2020, and the gas well was shut in due to wellbore sand production on March 23, 2021. The normal production was resumed after the coiled tubing sand flushing measures, and the average output per day was 1.0216×10^4 m³/d (Figure 5).

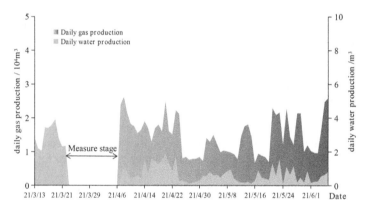

Figure 5. Comparison of effects before and after measures.

5 CONCLUSION

(1) In view of the sand production problem of horizontal wells in tight sandstone reservoir in Ordos Basin, the coiled tubing sand flushing process can selectively treat the horizontal well section, which can not only effectively remove the formation pollution, but also save the amount of treatment fluid.

(2) The coiled tubing negative pressure sand flushing technology is adopted to make the pressure generated by the sand flushing liquid on the bottom of the well less than the formation pressure. The sand flushing liquid flushes the bottom of the well sand out of the ground without entering the formation, which has the advantages of small reservoir damage and high construction efficiency.

(3) Coiled tubing sand flushing can be used for highly deviated wells and horizontal wells, especially for sand flushing in wells with complex wellbore trajectories. Coiled tubing sand flushing

achieves a good application effect in horizontal wells in tight reservoirs in the east of the basin, and can be popularized in analogous gas fields.

FUND PROJECT

National Science and Technology Major Special Project "Key Technology of complex Natural Gas Reservoir Development" (No.: 2016ZX05015).

REFERENCES

Liang Huzhan. Application of Coiled Tubing Sand Washing and Flushing Technology in Horizontal Well[J].*Guangdong chemical industry*, 2017; 17(44):236–239

Li Aifen, Wang Shihu et al Study on the Settlement Law of Formation Sand in Liquid[J]. *Petroleum Geology and Recovery Efficiency*, 2001; 8(1):70–73

Yu Hongwei, Jiang Yichao. Application of Coiled Tubing Sand Washing and Flushing Technology in Horizontal Well[J]. *Science Technology and Engineering*, 2011; 11(32):8018–8021

Zhang Qian Wang Fangxiang Hao Huasong, et al. The Optimal Sand Flushing Displacement with Coiled Tubing[J]. *Well Testing*2020; 29(4): 63–67

Zhang Shuo, Wang Fangxiang, Liu Dezheng, Ni Qinghuai. Application of Continuous Tubing Sand Washing Technology in Horizontal Wells[J]. *Petrochemical Industry Application*, 2018; 37(10):34–36

Zhang Shuping, Wang Xiaorong, Fan Lianlian, et al.Research and Application of Sand Carrying Theory in Gas Wells[J]. *Fault-Block Oil&Gasfield*, 2007; 14(1):50–52

Advances in Energy, Environment and
Chemical Engineering – Abdullah & Osman (Eds)
© 2023 The Author(s), ISBN: 978-1-032-36083-6

Study of renewable energy consumption under energy internet

Yu Fu*
Liaoning Zhengxin Green Energy Industry Management Co. Ltd., Liaoning, China

Chao Yang, Tong Li & Heyang Sun
State Grid Liaoning Electric Power Supply Co., Ltd., Liaoning, China

ABSTRACT: Energy Internet is a heterogeneous energy interconnection and sharing network formed by using electricity as the core, using renewable energy power generation technology and information technology, integrating multiple energy networks such as power network, natural gas network, heating/cooling network, and electric transportation network. It is an important way to absorb renewable energy, improve energy efficiency, and achieve China's "dual carbon" goal. At present, new energy generation is extremely unbalanced in terms of time and power demand, requiring the energy Internet to use a mature power transmission and distribution network as the carrier of new energy transmission to realize the aggregation, transmission, and distribution of new energy. And wind charge photovoltaic power generation has gradually become the main way of new energy consumption. Based on the relationship between the various dimensions of the energy Internet's vertical "source, network, load, and storage" and renewable energy consumption, this paper deeply perceives the development trend of the energy internet for renewable energy consumption and points out that for the energy supply-side and demand-side, the energy internet and renewable energy consumption form the development trend of mutual promotion and coexistence. Based on the examples of renewable energy wind power and photovoltaics, the structure and output model of the renewable energy unit is constructed, and the maximum output of the model unit is calculated.

1 INTRODUCTION

Energy is the basic resource for social and national development. As long as the energy resources are mastered, the progress of human society can be grasped. Energy is very important to the development of various countries in the world. Many major countries attach great importance to the reform and development of the energy system. At the same time, all countries are developing their economies, the use of fossil energy has reached its peak, and fossil energy has been decreasing, which not only increases the pressure of energy shortage but also brings about various problems. The depletion of traditional fossil energy will cause huge resistance to human society and national economic development, and will also make the country's environmental problems more and more prominent. Therefore, the development of renewable energy has become a necessary task at present. Renewable energy, which is inexhaustible and inexhaustible, can also find a solution to the current social energy shortage, and the pollution to the ecological environment is as small as possible. It will become the main source of energy system construction in various countries in the future. However, since most energy systems are thermal power systems, the use of traditional energy cannot be completely avoided. At present, the use of renewable energy can only be increased while reducing the use of fossil energy, thus giving birth to the emergence of the energy Internet system.

At present, there are still many defects in the power generation system of renewable energy, resulting in a very high construction cost of renewable energy, and the power generation technology

---

*Corresponding Author: newlakeliver@163.com

DOI 10.1201/9781003330165-69

of renewable energy is not enough to support the overall power consumption of the energy system, and the stability of power generation is at great risk. The energy shock may occur at any time in the energy system, leading to the paralysis of the entire energy generation system, and the power quality of renewable energy power generation is not high, accounting for only 3% of my country's total power generation. Therefore, in the reform of the energy Internet system in the future, new energy technology will become the main focus of research in various countries around the world, and the reform and innovation of new energy technology will become huge progress for human society.

Therefore, the development of renewable energy is an important measure to achieve my country's energy revolution strategy and the medium and long-term "carbon peaking and carbon neutrality" goals. On the basis of strengthening the clean, efficient, and flexible utilization of coal power, it is important to promote clean energy to become the main body of energy increment, realize large-scale renewable energy grid-connected consumption, build a new power system with new energy as the main body, and open a new era of low-carbon supply. It is the only way to promote the energy revolution and achieve carbon emission reduction goals.

By analyzing the architecture of the Energy Internet, this paper deeply understands the relationship and development trend between the different dimensions of the Energy Internet and the consumption of renewable energy; and takes the wind turbine model and photovoltaic unit model as examples to study the basic structure and economic model of renewable energy output.

2 THE RELATIONSHIP BETWEEN ENERGY INTERNET AND RENEWABLE ENERGY CONSUMPTION

There is no widely accepted definition of the Internet of Energy, and different organizations have proposed different concepts and names, each with its own focus, and the existing concepts can be broadly classified into 3 categories according to their characteristics.

(1) Focusing on global power interconnection, represented by the State Grid Global Energy Internet, the main feature is the spatial expansion of the power network, interconnection of different regional grids, and realization of cross-regional consumption of different types of new energy in different regions.
(2) Focusing on multiple energy integration, represented by an integrated energy system, the main feature is the interconnection between different energy systems such as electricity, heat, cooling, gas, transportation, etc. On the one hand, energy efficiency is improved through integrated energy development and utilization, and on the other hand, renewable energy consumption is realized by converting electricity into heat, cooling, natural gas, and electric vehicle energy storage.
(3) Focusing on energy information integration, represented by the U.S. FREEDM, Germany E-Energy, Japan Digital Grid, and Rifkin Energy Internet, the main feature is the use of power electronics, information and communication, and Internet technologies for energy control and real-time information sharing to achieve energy sharing and supply and demand matching, so as to consume renewable energy.

Energy Internet is a new open system of information-energy integration built based on Internet concept and technology, which will transform or even subvert the existing energy industry, break the industry monopoly, realize decentralization, and make the huge traditional industry of energy a fertile ground for innovation and entrepreneurship, which can significantly improve the efficiency of energy use and promote the large-scale development of renewable energy.

2.1 *Energy internet architecture*

Energy Internet is a next-generation energy system with a power system as the center, a smart grid as the backbone, Internet, big data, cloud computing, and other cutting-edge information and communication technologies as the link, and the integrated use of advanced power electronics and intelligent management technologies, which can realize horizontal multi-energy complementary, vertical source-grid-load-storage coordination. The next-generation energy system with high integration of energy and information. At the same time, it is flat, socially-oriented, commercial and customer service oriented. Specifically, on the basis of the backbone grid, a new power network composed of a large number of distributed energy collection and storage devices is used as a linking hub to interconnect energy nodes such as electricity, oil, natural gas, and transportation networks, forming a multi-layer coupled network architecture, realizing personalized and customized energy production and application through virtual power plants and grids, and using energy routers to transfer the disorderly and low-entropy energy flow Energy routers are used to flow disorderly, low-entropy energy flows to controllable, high-entropy loads according to the best path, realizing comprehensive regulation, optimization, interaction and sharing of energy flows. The proposed Energy Internet breaks the boundaries of supply and demand between traditional energy industries and maximizes the interconnection, interoperability, and complementarity of primary and secondary energy types such as coal, oil, natural gas, heat, and electricity; supports the large-scale access of various new energy sources and distributed energy sources on the user side to realize plug-and-play use of power equipment; realizes optimal regulation and efficient energy flow through local autonomous consumption and wide-area peer-to-peer interconnection; builds open and flexible industries and applications. Through local autonomous consumption and wide-area peer-to-peer interconnection, we can realize optimal regulation and efficient utilization of energy flow, and build an open and flexible industry and business form. Energy Internet is a product of the deep integration of energy and the Internet and has received wide attention from academia and industry.

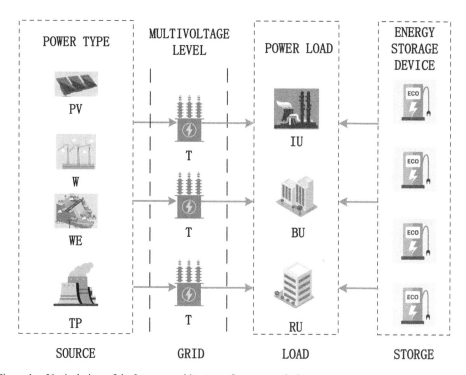

Figure 1. Vertical view of the Internet architecture of energy evolution.

Figure 1 gives a simple architecture of the Energy Internet from a vertical perspective. However, the overall concept of the Energy Internet involves the integration of technical elements in the physical foundation layer, information fusion layer, application practice layer, and institutional mechanism layer, and specifically, it includes multi-energy complementarity, integrated energy system, flexible power transmission, virtual power plant, distributed energy, energy+big data, energy+blockchain, energy router, energy+ Distributed trading, energy storage, electric vehicles, and other concepts. Examining and constructing all the links of the energy internet can provide a background basis for renewable energy consumption.

2.2 *The development of energy internet for renewable energy consumption*

New energy generation converts the film of energy covering the earth into electricity. As the proportion of new energy continues to rise, there is an urgent need for power grids to cover a larger and larger spatial area in order to achieve the convergence of new energy generation. Due to the extreme imbalance of new energy in space, there is a need for the power grid to deliver the electricity from the new energy surplus areas to the geographical areas with high power demand and lack of power generation resources to realize the distribution of new energy. Due to the extreme imbalance between new energy generation and power demand in time, the grid needs to have strong energy storage capacity on different time scales. One of the important ways to achieve the goal of "double carbon" is to build a new power system with new energy as the main body. The future energy supply is based on new energy generation, and the power grid is the platform for convergence, transmission, and distribution of electric energy. It will become an important link of the energy internet. The energy internet will use the mature transmission and distribution network as the carrier of new energy transmission to realize the aggregation, transmission, and distribution of new energy. When there is a surplus of new energy, "green power to produce hydrogen" will be carried out at the location of the former thermal power plant, and when there is a shortage of new energy generation, "hydrogen will be re-generated", thus solving the problem of power balance on a long time scale, and at the same time, the green hydrogen obtained can also meet the demand for deep carbon reduction in steel and chemical industries. At the same time, the green hydrogen obtained can also meet the demand for deep carbon reduction in steel and chemical industries.

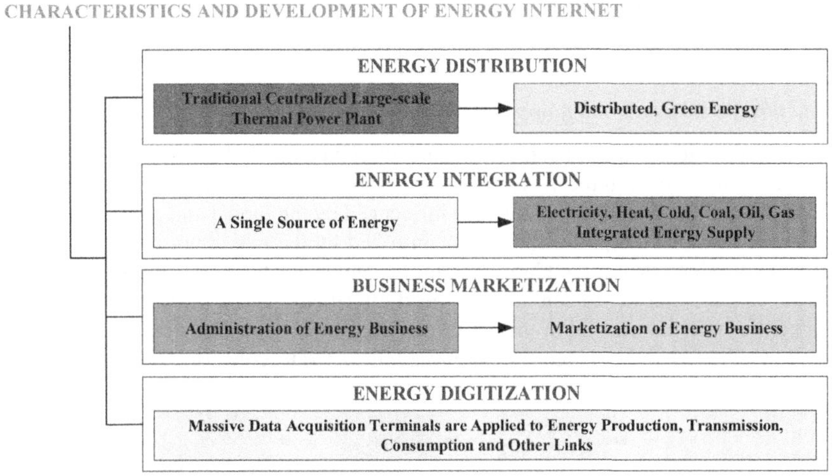

Figure 2. Energy internet characteristics and development trends.

1) Energy Internet development from the energy supply side

In February 2021, the State Council issued the "guidance on accelerating the establishment of a sound green low-carbon cycle development of the economic system", and put forward the development path "to promote the green low-carbon transformation of the energy system, to enhance the proportion of renewable energy use, vigorously promote the development of wind power, photovoltaic power generation, the development of water energy, geothermal energy, ocean energy, hydrogen energy, biomass, solar thermal power generation according to local conditions". The core of decarbonization and low-carbonization of electricity is to build a "multi-energy integration" power system with renewable energy as the main body. The randomness and volatility of high proportion of renewable energy brings great challenges to the balance and safe operation of the power grid, and it is urgent to promote the traditional "source follows the load" to "source and load interaction" transformation, improve the resilience of the power system, and through the big data-based power supply side and The power system needs to be transformed from traditional "source follows load" to "source and load interaction", to improve the resilience of the power system, and to realize the safe and stable operation of the power grid through big data-based power supply-side and demand-side forecasting and management, as well as Internet-based power trading and service platform, and finally to build a "multi-energy fusion" power system with renewable energy as the main body and energy storage and CCUS thermal power as the guarantee, so that the conventional thermal power generation can be transformed from the current base-load power to peak-load power and realize the decarbonization and decarbonization of power. and low carbonization.

2) Energy Internet development from the energy demand side

Re-electrification of energy use refers to a high level of electrification based on zero-carbon electricity, building on traditional electrification. The future energy of a carbon-neutral society must revolve around zero-carbon electricity, and the global electrification level is expected to be higher than 50% in 2050. Due to the inherent randomness, intermittency, and instability of new energy sources such as photovoltaic and wind power, the safe and stable operation of the power grid brings many adverse effects, which require the power system to be controllable and able to "unconditionally" accept new energy sources. To achieve this goal, it is necessary to build a new power system with information technology and digitalization, and build a highly intelligent and powerful software platform with cloud resource storage, big data processing, and analysis, so that the power grid can be visible, knowable and controllable, and realize the "source network, load, and storage" intelligent development.

3 RENEWABLE ENERGY OUTPUT MODEL UNDER ENERGY INTERNET

3.1 *Wind turbine structure and power output model*

Wind power has the advantages of being clean and efficient and wind power generation technology has been greatly developed in recent years. The output power of a wind turbine is not only related to the wind direction and wind speed but also influenced by the model and parameters of the wind turbine device, in which the power system is controlled by the maximum power tracking control (MPPT) system, which drives the blade rotation by capturing the wind fluid kinetic energy, and then drives the generator set to run at high speed after being accelerated by the variable speed system. Finally, under the control of the active/reactive power control system, the power output of the generator set is fed to the power system through the AC/DC converter.

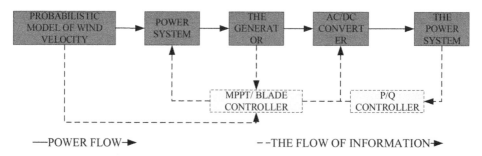

Figure 3. Wind turbine structure.

The power generation cost of a wind turbine includes basic operation and maintenance costs and equivalent investment costs, and the cost of the wind turbine group is smaller compared with thermal power and energy storage units. The wind turbine output is considered as a segmental function of wind speed, and the specific formula is as follows.

$$P_w = \begin{cases} 0, & v_w < v_{ci} \, or \, v_w \geq v_{ci} \\ \frac{v_w^3 - v_{ci}^3}{v_r^3 - v_{ci}^3} P_{w.max}, & v_{ci} \leq v_w < v_r \\ P_{w.max}, & v_r \leq v_w < v_{co} \end{cases} \tag{1}$$

where, v_{ci}, v_{co} and are the cut-in wind speed, cut-out wind speed, and rated wind speed, respectively. $P_{w.max}$ is the maximum output of a wind turbine, i.e., rated output. When the wind speed is less than the cut-in wind speed or greater than the cut-out wind speed, the wind turbine is not put into operation and the output is 0; when the wind speed is greater than the cut-in wind speed and less than the rated wind speed, the output of the wind turbine increases by three times the wind speed until it reaches the rated output; when the wind speed is greater than the rated wind speed and less than the cut-out wind speed, the output of the wind turbine is maintained at the rated output.

3.2 Photovoltaic generator set structure and power output model

The principle of photovoltaic power generation is to convert the received light into DC electricity through the photovoltaic effect, and deliver the electricity to the load according to the system load demand under the system controller and AC/DC converter.

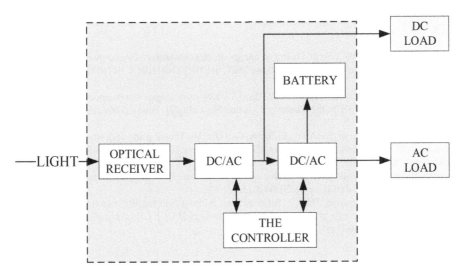

Figure 4.   Structure of photovoltaic generator set.

The PV unit output is approximated as a segmented function of solar radiation illumination by the following equation.

$$P_{pv} = \begin{cases} P_{pv.max} \frac{I^2}{I_{std} I_r}, & 0 \leq I \leq I_r \\ P_{pv.max} \frac{I}{I_{std}}, & I_r < I \leq I_{std} \\ P_{pv.max}, & I > I_{std} \end{cases} \tag{2}$$

Where I_r and I_{std} are the solar radiation illuminance at specific and standard radiation illuminance points, respectively. $P_{pv.max}$ is the maximum output of the PV unit, i.e., the rated output.

4 CONCLUSION

Through the analysis of the energy Internet architecture and the development trend of the energy Internet for renewable energy consumption, this paper studies the relationship between the energy Internet and renewable energy consumption. Based on the basic structure and economic model, the main conclusions can be summarized as follows:

(1) Energy Internet and renewable energy consumption form a development trend that promotes each other and you have me. On the energy supply side, realizing multi-source complementarity can promote the green and low-carbon transformation of the energy system; on the energy demand side, energy electrification will become an unstoppable mainstream energy use model.

(2) By constructing the structure and output model of the renewable energy unit, the maximum output of the model wind turbine and the maximum output of the model wind power photovoltaic generator are calculated.

At present, the consumption of renewable energy is an issue worthy of consideration on both the power generation side and the user side. At the same time, the installed capacity of large wind power, large photovoltaics, and hydropower generation at the source has gradually and will soon exceed the distance of thermal power. The construction of microgrids such as small wind power and small photovoltaics at the load end will be the main way to consume new energy. Therefore, an in-depth analysis of the relationship between the various dimensions of the energy Internet and renewable energy consumption, and the study of the typical unit structure and output patterns of renewable energy sources are necessary stages in the development of new power systems in the future.

REFERENCES

Cao JW, Zheng Y, Zhang WQ, Yu B. Hydrogen energy development driven by energy internet[J]. *Journal of Tsinghua University* (Natural Science Edition), 2021, 61(04):302–311.DOI:10.16511/j.cnki.qhdxxb.2021. 25.007.

Shen H, Zhou QY, Liu Y, Sun W, He Q, Ren DW, Zhang YT. Key Technologies and Prospects for the Construction of Global Energy Internet in the Context of Carbon Neutrality[J]. *Power Generation Technology*, 2021, 42(01):8–19.

Sun Hongbin, Pan Zhaoguang, Sun Yong, Li Baoju, Guo Qinglai. Thinking and understanding of cross-border thinking in the application of energy Internet[J]. *Power System Automation*, 2021, 45(16):63–72.

Tian Kunpeng, Sun Weiqing, Han Dong, Yang Ce, Zhang Wei. Coordinated source-grid-storage planning to meet the target of non-water renewable energy generation[J]. *Power Automation Equipment*, 2021, 41(01):98–108. DOI:10.16081/j.epae.202012015.

Xu L, Wang QG, Yang Moucun, Zhu YZ. An evaluation index system for all-renewable multi-energy complementary systems taking into account resource endowments[J/OL]. *Grid Technology*:1–8 [2022-04-08]. DOI:10.13335/j.1000-3673.pst.2021.2087.

Advances in Energy, Environment and Chemical Engineering – Abdullah & Osman (Eds)
© 2023 The Author(s), ISBN: 978-1-032-36083-6

Preparation and synthesis of hyperbranched small molecule pentaerythramine

Han Ren*, Yuanzhi Qu, Shifeng Gao, Ren Wang, Zhilei Zhang, Rongchao Cheng, Zhiyuan Yan & Yuehui Yuan
CNPC Engineering Technology R&D Company Limited, Beijing, China

ABSTRACT: Nowadays, with the rapid development of science and technology and the continuous enhancement of innovation, the research direction of the oilfield chemical industry has also shifted from the previous focus on the research and development of chemicals to the research and preparation of such chemicals with a wide range of uses and diversified application directions. Among them, amine inhibitor products represented by pentatetramine with the dendritic structure are in increasing demand in the market, so they have great development potential. And there is no synthetic preparation route suitable for large-scale industrial production of pentatetramine in the current international and domestic markets. In this context, a novel synthetic route of pentatetramine with milder reaction conditions, higher yield, and more suitable for large-scale production was explored in this paper. The main research ideas are as follows: Firstly, the PCC reagent is synthesized from chromium trioxide (CrO_3), pyridine, and hydrochloric acid (HCl) as raw materials. Then, by means of the Oppenauer oxidation reaction, pentaerythritol and PCC reagent are used as raw materials to synthesize pentaerythraldehyde. The Leuckart reaction was carried out with pentaerythraldehyde, ammonium formate ($HCOONH_4$), formic acid (HCOOH), etc. as raw materials, and finally, pentaerythramine was synthesized.

1 INTRODUCTION

1.1 *Research background*

In petrochemical production operations, on-site formation oil recovery plays a vital role. In the process of on-site construction, in order to stabilize the clay minerals in the formation, the use of high-efficiency treatment agents is a necessary condition to ensure smooth and effective oil extraction. At present, in the field operations of oilfields, inorganic salt shale inhibitors, bitumen, and polymer alcohol shale inhibitors belong to a class of inhibitors with mature production routes and stable field applications. However, these inhibitors have various shortcomings and complicated constraints. For example, inorganic salt inhibitors are not satisfactory in performance, asphalt inhibitors do not meet the premise of environmental protection because of their fluorescence, and polyethylene glycol inhibitors are currently used in the treatment of activated shale, showing poor performance in drilling problems.

Therefore, with the current understanding and implementation of the environmentally sustainable development concept that "golden mountains and silver mountains are not as good as lucid waters and lush mountains" and the increase in the mining volume of deep and ultra-deep wells, amine-based treatment agents are becoming more and more popular. At present, amine inhibitors have become the most commonly used inhibitors in the petrochemical field. Amine inhibitors have many advantages over other types of inhibitors. For example, its toxicity is smaller than that of polymer inhibitors and asphalt inhibitors, and it has less environmental pollution to oil field construction and less harm to the human body. Meanwhile, inorganic cationic and low molecular weight cationic

---

*Corresponding Author: renhandr@cnpc.com.cn

DOI 10.1201/9781003330165-70

polymer-based inhibitors were compared. Amine inhibitors have strong adaptability and high compatibility, which are beneficial to exploiting deep and ultra-deep wells in production and reducing economic and labor costs in petrochemical field applications. Among them, aliphatic diamines and polyamines belong to amine inhibitors, which can be divided into linear and branched types according to their molecular structures. Among them, linear aliphatic diamines and polyamines have certain effects, but it is difficult to balance the inhibition and rheological properties of water-based drilling fluids. Pentatetramine with a branched structure can increase the water solubility of the inhibitor due to its special structural characteristics and large density of amine groups, and at the same time adhere to the adjacent clay crystal layer to play a role in pulling. The adsorption of the inhibitor is stronger and the clay crystal layer is stronger.

Figure 1. Pentaerythritamine formula.

1.2 Significance

So far, the domestic and foreign attention to pentatetramine is still very low, and the research on its synthetic route is still in the development stage. At present, no matter domestic or foreign, indoor research or outdoor production, the synthetic route of pentaerythramine still has many steps and long processes, and most of them are complicated and tedious in terms of the synthesis process. And most of the existing pentatetraamine synthesis routes will produce toxic, harmful, and irritating gases. With the current focus on environmental protection and sustainable development, the current pentatetramine synthesis route has been eliminated and has become a kind of Inevitably, the research and exploration of new synthetic routes is an imminent and inevitable trend.

2 EXPERIMENT

2.1 Target product

Pentaerythritol, whose systematic name is 2,2-diaminomethyl-1,3-diaminopropane, was first studied and synthesized by Litherand and Mann in 1928. They first used pentaerythritol to react with PBr_3 to generate tetramine. Bromopentane then performs a further condensation reaction with p-toluenesulfonamide sodium, and finally uses acid hydrolyzate to obtain pentatetramine.

2.2 Research ideas

The main research ideas are as follows: Firstly, the PCC reagent is synthesized from chromium trioxide (CrO_3), pyridine, and hydrochloric acid (HCl) as raw materials. Then, by means of the Oppenauer oxidation reaction, pentaerythritol and PCC reagent are used as raw materials to synthesize pentaerythraldehyde. The Leuckart reaction was carried out with pentaerythraldehyde, ammonium formate ($HCOONH_4$), formic acid (HCOOH), etc. as raw materials, and finally, pentaerythramine was synthesized. The structure of the synthesized hyperbranched small molecule pentatetramine was characterized by infrared spectroscopy (FTIR) and nuclear magnetic resonance (^1H NMR/^{13}C NMR).

2.3 Synthesis of pentaerythraldehyde

The dichloromethane solution is first dried. A small amount and an appropriate amount of anhydrous sodium sulfate were added to the dichloromethane solution, and after stirring and soaking for 15 minutes, the supernatant was filtered out using a vacuum suction filtration funnel. Under the condition that the reaction temperature is 20°C, PCC (0.06 mol, 12.9 g) and 12.9 g of silica gel powder are thoroughly mixed and then dispersed in the dichloromethane solution, and then transferred into a three-necked flask, and the mechanical stirring is turned on. After the solute was dispersed and stirred evenly, pentaerythritol (0.01 mol, 1.36 g) was added in batches. After 1.5 h of reaction, additional PCC (0.02 mol, 4.3 g) was added. The reaction continued for about 3 hours. During this period, the TLC central control point plate was performed until the raw material point disappeared and the reaction was completed (the reaction conditions are to be optimized later). The reaction mixture in the three-necked flask was subjected to vacuum filtration, the three-necked flask was washed twice with a dichloromethane solution, and the washings were also poured into a vacuum filtration device. The filter cake in the funnel is soaked in dichloromethane solution, and then filtered to dry under normal pressure, and the soaking and filtering steps are repeated 2 to 3 times. The obtained mother liquor was repeatedly washed with saturated sodium bicarbonate solution (4–5 times) and separated with a separating funnel until the water layer was nearly light-colored and neutral. (CH_2Cl_2 is heavier than water, so our desired organic layer is below). Finally, anhydrous sodium sulfate was added to the organic layer for soaking and drying for about 30 minutes, suction filtration was performed, and the filtrate was subjected to reduced pressure rotary evaporation.

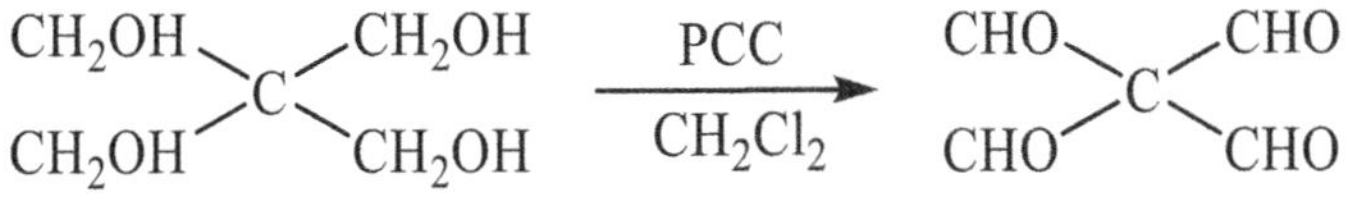

Figure 2.   Synthesis route of pentaerythritaldehyde.

2.4   *Synthesis of pentatetramine*

A reaction device was built, and 0.004 mol of pentaerythraldehyde and 0.032 mol of ammonium formate were dispersed in 40 mL of toluene and transferred into a three-necked flask. Adjust the oil bath to gradually heat up to 130–150°C in stages. At this time, there is obvious condensation and reflux phenomenon. A water separator is used to separate and remove water, and the reaction ends after 24 hours. The product was filtered under reduced pressure, the filter cake was soaked in toluene under normal pressure for 20 min, vacuum filtered again, and the mother liquor was combined. It is required to wash the mother liquor 2–3 times with saturated sodium carbonate solution, and use a separatory funnel for liquid separation until the water layer is nearly neutral. The obtained organic layers were combined, dried over anhydrous sodium sulfate, and rotary-evaporated under reduced pressure to obtain a dark yellow-brown pentaerythramine liquid. The final synthesized product is a yellow-brown oily liquid with no obvious irritating odor, and a black-gray powdery substance after being refrigerated in the refrigerator.

Its synthetic route is shown in Figure 3:

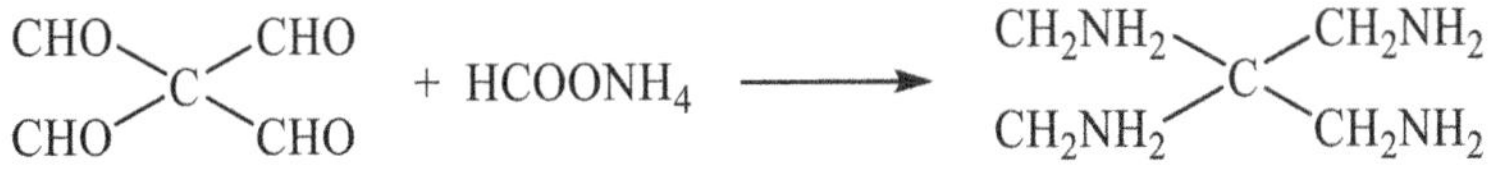

Figure 3.   Synthesis route of pentaerythramine.

3   CHARACTERIZATION OF PENTAERYTHRALDEHYDE

3.1   *Infrared spectroscopy (FTIR) characterization*

The intermediate pentaerythraldehyde was scanned by infrared spectrum by reflection method, and the product was characterized by infrared spectrum using Nicolet 200SXV Fourier transform infrared spectrometer of Nicolet Company of Japan.

FTIR test was performed on the synthesized product to obtain the curve of transmittance versus wavenumber. It can be seen that the absorption peak of the sample at the corresponding position is obvious.

It can be seen from this infrared spectrum that the absorption peak at 3225 cm^{-1}–2848 cm^{-1} is the stretching vibration peak of aldehyde group C-H; 1636 cm^{-1} is the absorption peak of the aldehyde group C=O; 1538 cm^{-1}–985 cm^{-1} C-C skeleton stretching vibration; 904 cm^{-1}–500 cm^{-1} is the vibration absorption peak of C-H bond.

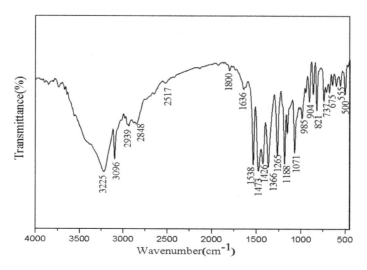

Figure 4.   IR spectra of the product pentaerythritaldehyde.

3.2   Nuclear magnetic resonance (1H NMR) characterization

The intermediate synthesized and purified product, pentaerythraldehyde, uses deuterated chloroform (CDCl$_3$) and deuterated methanol (MeOD) as solvents and adopts AV II-400 (400MHz) high-resolution nuclear magnetic resonance instrument from Bruker Company, Germany to conduct nuclear magnetic resonance. Hydrogen spectrum characterization is used.

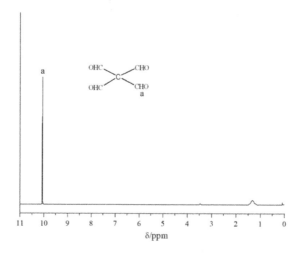

Figure 5.   1H NMR of pentaerythritaldehyde.

The figure is the ^{1}H NMR spectrum of pentaerythraldehyde. There is one kind of H in the figure, the corresponding peak at the chemical shift of 10.11 ppm is H in -CHO linked to the quaternary carbon, and the chemical shift of 1.41 ppm is the solvent peak. To sum up, the ^{1}H NMR spectrum characterization indicates that the functional group characteristics contained in the product are consistent with the characteristics of the target product.

4   CHARACTERIZATION OF PENTATETRAMINE

4.1   *Infrared spectroscopy (FTIR) characterization*

The final synthesized and purified product, pentatetramine, was prepared by tableting with potassium bromide (KBr). The products were characterized by infrared spectroscopy using a Nicolet 200SXV Fourier transform infrared spectrometer from Nicolet, Japan.

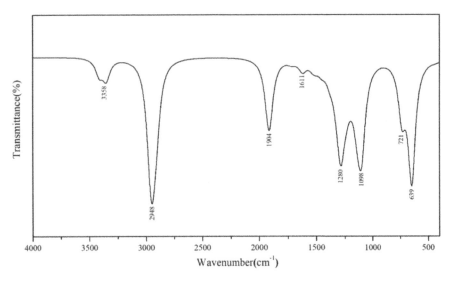

Figure 6.   IR spectra of the product pentaerythritamine.

FTIR test was performed on the synthesized product to obtain the curve of transmittance versus wavenumber. It can be seen that the absorption peak of the sample at the corresponding position is obvious.

3358 cm^{-1} is the stretching vibration peak of the pentatetramine-NH$_2$ bond, 1904 cm^{-1} is the C-C frequency doubling stretching vibration peak of quaternary carbon in pentatetramine, and this peak is a weak double peak. 1611 cm^{-1}, 721 cm^{-1}, 639 cm^{-1} are the -N-H deformation vibration peaks in pentatetramine, 1280 cm^{-1} is the -CH$_2$ stretching vibration peak in pentatetramine, 1098 cm^{-1} is -NH$_2$ in pentatetramine stretching vibration peak. The absorption peaks of other reactants do not appear in the figure, so there are no unreacted monomers and other side reaction components in the test sample.

4.2   *Nuclear magnetic resonance (1HNMR/13C NMR) characterization*

The final synthesized and purified product, pentatetramine, was carried out using deuterated chloroform (CDCl$_3$) and deuterated methanol (MeOD) as solvents, and the AV II-400 (400MHz) high-resolution NMR instrument of Bruker, Germany was used for nuclear magnetic resonance. Nuclear Magnetic Resonance (^{1}HNMR/^{13}C NMR) Characterization.

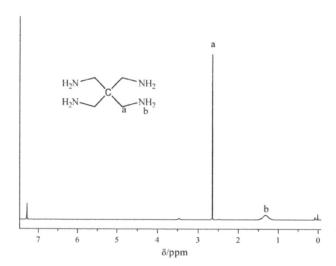

Figure 7. ^{1}H NMR of pentaerythramine.

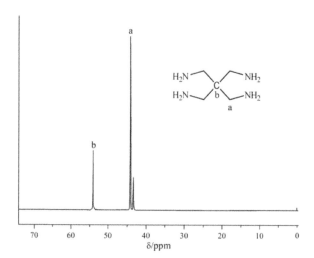

Figure 8. ^{13}C NMR of pentaerythramine.

Figure 7 is the ^{1}H NMR of pentatetramine, using deuterated methanol solvent, it can be seen that the chemical shift at δ=2.65 is the characteristic peak of H on the methylene group of pentatetramine, and Figure 8 is pentatetramine. From the ^{13}C NMR of tetramine, it can be seen from the figure that the chemical shift of C on the upper methylene group of pentatetramine is $\delta = 47.5$, and the chemical shift of quaternary carbon is $\delta = 44.8$. From the ^{1}H NMR and ^{13}C NMR of pentatetramine, it can be seen that the pentaerythramine has high purity and the structure is confirmed.

5 CONCLUSION

Because of its unique branched structure, pentatetramine can be used in the field of medicine. Moreover, as an important raw material for the synthesis of new high-energy density materials, a monomer for molecular architecture, and dendrimer synthesis, it is also used as an inhibitor

482

of petrochemical production, etc. But there are all kinds of problems with the synthetic route of pentaerythramine both at home and abroad. For example, the synthetic method not only produces many raw materials, requiring a complex production process, but also requires high temperature and high pressure in the production process, which is unsafe. Therefore, this paper tries to propose a new synthetic route of hyperbranched small molecule pentaerythramine through investigation and continuous exploration and has passed the verification. The synthetic route requires less raw materials and a reasonable price, the preparation process is simple and easy to operate, the whole experimental operation process is clean and environmentally friendly, and the preparation is safe, non-toxic, and harmless.

ACKNOWLEDGMENTS

Funding: This research was funded by the CNPC Scientific research and technology development projects (Project No.:2020A-3913,2021DJ4407,2021DJ4404)

REFERENCES

Anderson W S, Hyer H J, Sundberg J E,et al. Pentaery-thrityltetramine[J].*Industrial Engineering Chemistry*. 2000(39):4011–4013.

Bastos, M.B.R., Moreira.J.C., et al. Adsorptive stripping voltammetric behavior of UO_2 (II) complexed with the Schiff base N, Nft-ethylenebis (salicylidenimine) in aqueous 4-(2-hydroxyethyl)-1-piperazine ethanesulfonic acid medium [J]. *Anal Chim. Acta*, 2000, 408:83–88.

Bo Meng, Qiaoxia Guo, Xiaoping Men, et al. Modified bentonite by polyhedral oligomeric silsesquioxane and quaternary ammonium salt and adsorption characteristics for dye[J]. *Journal of Saudi Chemical Society*, 2020, 24(3).

Chi-Hsien Liu, Guan-Wei Lee, Wei-Chi Wu, et al. Encapsulating curcumin in ethylene diamine-β-cyclodextrin nanoparticle improves topical cornea delivery[J]. *Colloids and Surfaces B: Biointerfaces*, 2020, 186.

Dorofeeva Olga V., Filimonova Marina A., Cyclic aliphatic amines: A critical analysis of the experimental enthalpies of formation by comparison with theoretical calculations[J]. The *Journal of Chemical Thermodynamics*, 2020(prepublish).

Edwin R, Buchman A C. *pentaerythrityl sulfonates*[P]. USP 2703808, 1951.

Egor V. Verbitskiy, Yuriy A. Kvashnin, Anna A. Baranova, Konstantin O. Khokhlov, et al. Synthesis and characterization of linear 1,4-diazine-triphenylamine–based selective chemosensors for recognition of nitroaromatic compounds and aliphatic amines[J]. *Dyes and Pigments*, 2020, 178.

Fleischer G.Conversion of aliphatic and alicyclic polyalcohols to the corresponding primary polyamines [J]. *J.Org.Chem.*, 1971, 36(20): 3042–3044.

HongLin Hu, Lu Zhang, RuiLian Yu, et al. Microencapsulation of ethylenediamine and its application in binary self-healing system using dual-microcapsule[J]. *Materials & Design*, 2020, 189.

Lihua Zhou, Yang He, Shaohua Gou, et al. Efficient inhibition of montmorillonite swelling through controlling flexibly structure of piperazine-based polyether Gemini quaternary ammonium salts[J]. *Chemical Engineering Journal*, 2020, 383.

Ruoheng Li, Zhixia Wang, Qiwei Xu, et al. Synthesis, characterization and physicochemical properties of new chiral quinuclidinol quaternary ammonium salts[J]. *Journal of Molecular Structure*, 2020, 1209.

*Advances in Energy, Environment and
Chemical Engineering – Abdullah & Osman (Eds)
© 2023 The Author(s), ISBN: 978-1-032-36083-6*

Study on rheological properties of montmorillonite drilling fluid under different influencing factors

Zhilei Zhang, Hong Yang, Rongchao Cheng, Yan Zhang, Zheng Yang, Han Ren &
Zhiyuan Yan*
CNPC Engineering Technology R&D Company Limited, Beijing, China

ABSTRACT: The exploitation and utilization of geothermal resources is a major research work in earth science and the energy industry. It is of great significance to solve the increasingly severe energy shortage and global environmental problems. In view of the complex drilling conditions of geothermal drilling, it is of great scientific significance and engineering practical value to understand the rheological properties of montmorillonite as drilling fluid base slurry under different conditions. In this paper, the effect of montmorillonite as a base slurry of drilling fluid on the rheological property at different temperatures, pH, and salt concentrations was studied through a lot of experiments. The experimental results show that the influence of high temperature on the rheological properties of base slurry is complex, mainly affecting the clay structure, resulting in high-temperature thickening, thinning, and curing of montmorillonite. Proper pH value can improve the rheological property of drilling fluid. Keeping the pH in the 9-10 range reduces plastic viscosity and helps maintain drilling operations. Adding electrolyte to montmorillonite suspension, Na^+ and K^+ in the electrolyte will press the counter-ions in the diffusion layer into the adsorption layer, so that the charge of montmorillonite is reduced and particles are easy to gather.

1 INTRODUCTION

With the rapid development of China's economy, the demand and exploitation of geothermal resources are growing. In order to ensure the full exploitation and utilization of resources, the exploitation of geothermal resources has gradually increased. China is rich in geothermal resources, some of which have high temperatures or are buried in deep and ultra-deep layers. Therefore, the exploration and development of geothermal resources are gradually transferred to more complex formations, which leads to the emergence of more and more ultra-high temperature deep Wells. Montmorillonite, as an important water-based drilling fluid slurry material, has been widely used in oil drilling because of its excellent properties such as expansion, dispersion suspension, pulping, cation exchange and adsorption (Peretyazhko et al. 2018). At present, due to the complex drilling conditions, most scholars study and use the slurry mineral (mainly montmorillonite), but its research is mostly conducted at room temperature and low temperature.

In the process of high temperature and deep well drilling, drilling fluid must effectively meet the requirements of drilling engineering and geological conditions at high temperature and high density (Zhang et al. 2020). Therefore, it is necessary to study the effects of temperature, pH and salt concentration on the rheological properties of montmorillonite as a drilling fluid base slurry. In this paper, the rheological properties of montmorillonite as drilling fluid base slurry were studied from three performance parameters of apparent viscosity (AV), plastic viscosity (PV) and dynamic shear force (YP) at different temperatures, pH and salt concentrations (Fahd et al. 2021).

---

*Corresponding Author: yanzhydr@cnpc.com.cn

DOI 10.1201/9781003330165-71

2 MATERIALS AND METHODS

2.1 *Materials*

Montmorillonite was obtained from Inner Mongolia, China. Its cation exchange capacity (CEC) was 85 ± 3 mmol/100 g, with a layer charge of 0.32 eq/mol per $(Si, Al)_4O_{10}$ and an external surface area of 23 m^2/g, respectively. NaOH, HCl, NaCl and KCl were obtained from Shanghai Aladdin Chemicals Co. Ltd. All reagents in this study were Analytic-grade and used without further purification.

2.2 *Formulation of montmorillonite dispersions*

The base montmorillonite dispersion was prepared by mixing 6 wt% montmorillonite in deionized water by using an ultra-high-speed disperser for 120 min and the montmorillonite dispersion was standing for 24 h.

2.3 *Measurement of rheological properties*

The montmorillonite dispersion was tested after aging for 16 h under different conditions (temperature from 160 to 240°C, pH from 7 to 10, salinity of NaCl/KCl from 7% to 30%). Rheological experiments were performed using Discovery Hybrid Rheometer following the reference (Ettehadi et al. 2020).

3 RESULTS AND DISCUSSION

3.1 *Effect of high temperature on rheological properties*

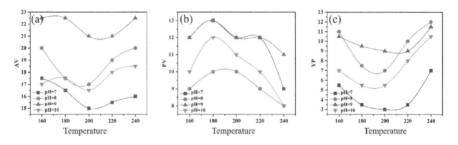

Figure 1. Effect of temperature on apparent viscosity (AV) (a), plastic viscosity (PV) (b) and yield point (YP) (c) of montmorillonite slurry at different pH values.

In Figure 1, the value of apparent viscosity (AV) decreases with the growth of temperature from 160°C to 200°C and increases slightly from 200°C to 240°C. The plastic viscosity (PV) increases with the temperature from 160°C to 180°C, and then decreases with the temperature from 180°C to 240°C, reaching the maximum value at 180°C. The yield point (YP) decreases first and then increases with the increase of temperature, showing an opposite trend.

From 160°C to 180°C, high-temperature passivation is dominant, and high-temperature dispersion also exists. The further increase of temperature not only intensifies the thermal motion of various particles in drilling fluid but also intensifies the thermal motion of clay minerals, which enhances the ability of the layer structure of clay minerals to separate from each other and reduces the granularity of particles in the drilling fluid. High temperature reduces the surface activity of clay particles, thereby reducing the ability to form stable structures by clay particles, causing thinning behavior after high temperature. However, as the temperature increases from 180 to 240°C, the effect of high-temperature aggregation on clay particles is dominant. The results of high-temperature dehydration decrease the stability of clay particles, which increases the particle size of clay in drilling fluid and decreases the specific surface area (Liu et al. 2022).

In general, drilling fluids show different phenomena after high-temperature action, such as thickening, thickening, and curing. These phenomena not only occur in different drilling fluid systems but also occur in the same drilling fluid system under different conditions, revealing that the influence of high temperature on the thermal stability of drilling fluid rheology is complicated.

3.2 Effect of pH on rheological properties

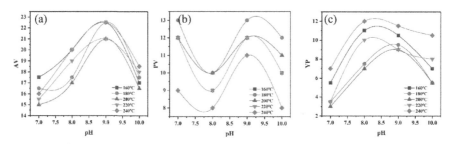

Figure 2.   The effect of pH on the apparent viscosity (AV) (a), plastic viscosity (PV) (b) and yield point (YP) (c) of montmorillonite slurry at different temperatures.

In Figure 2a, the results of AV are increased firstly in pH 7∼ 9 and decreased dramatically in pH 9∼10. As shown in Figure 2c, the YP also shows an increase at first and then decreases with the increase of pH, indicating that the spatial network structure of montmorillonite slurry weakens with the increase of pH, resulting in the decrease of YP. However, under the influence of pH, the results of PV display a trend of ups and downs, and finally decreased as pH 9 ∼ 10. The dispersion degree of montmorillonite is the main factor affecting the PV, and with the increase in pH, the dispersion degree becomes stronger.

When pH<pHIep (isoelectric point of pH), all edges are positively charged, leading to the dominance of edge-layer interactions. The mechanism for the formation of three-dimensional networks is the connection between the edge and the layer, hence YP increases dramatically. When pH>pHIep, all edges are negatively charged, which leads to the destruction of the three-dimensional networks, and Van der Waals is dominant, making the results of YP reduced (Huang et al. 2016).

After high temperature and alkalinity, the surface of the clay is passivated, which reduces the surface activity of clay and intensifies the thermal motion of water molecules, weakening the surface hydration ability and the outer hydration film. Therefore, keeping the pH in the range of 9∼ 10 increase permeability and reduce pump pressure, which helps maintain mud circulation in drilling operations (Hany et al. 2019).

3.3 Effects of KCl/NaCl concentrations on rheological properties

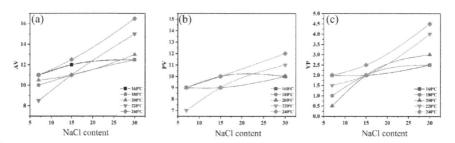

Figure 3.   Effects of NaCl content on the apparent viscosity (AV) (a), plastic viscosity (PV) (b) and yield point (YP) (c) of montmorillonite slurry at different temperatures.

The test results of rheological properties of drilling fluid under different NaCl concentrations were shown in Fig. 3. At the same temperature, the AV, PV, and YP results of drilling fluid increased with the increase of NaCl concentration. With the increase of Na+ concentration in drilling fluid, the number of cations in the diffusion layer of the double electric layer increases, and the diffusion layer of the double electric layer is compressed, resulting in the thickness of the diffusion layer decreasing. At this time, the electrostatic repulsion between clay decreases, the hydration film becomes thinner, the dispersion degree of particles becomes weaker, and the end-to-end and face-end connection between particles is also strengthened, resulting in the increase in viscosity of drilling fluid (Laura et al. 2019).

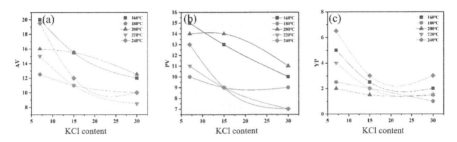

Figure 4. Effects of KCl content on the apparent viscosity (AV) (a), plastic viscosity (PV) (b) and yield point (YP) (c) of montmorillonite slurry at different temperatures.

Figure 4 shows the results of rheological properties at different KCl salt concentrations. The experimental results show that AV, PV, and YP of drilling fluid show a decreasing trend with the increase of KCl salt concentration under the same temperature. Since the hydration energy of K^+ is lower than that of Na^+ and clay is selective for cation adsorption, K^+ is preferentially adsorbed by clay than Na^+. After KCl is adsorbed by clay, on account of low hydration energy, it will promote dehydration between crystal layers and compress the crystal layers, forming a compact structure, reducing particle size, and affecting the rheological property of drilling fluid (Ali & Ahmad 2020).

4 CONCLUSION

In this paper, the rheological properties of montmorillonite as a drilling fluid base slurry under different conditions were studied. The main conclusions can be summarized as follows:

(1) The influence of high temperature on water-based drilling fluid is complex, mainly affecting clay structure, resulting in high-temperature thickening, thinning, and curing of montmorillonite.
(2) Proper pH value can improve the rheological properties of drilling fluid. Keeping the pH in the 9-10 range reduces plastic viscosity and helps maintain drilling operations.
(3) Electrolyte is added to montmorillonite suspension. Na^+ and K^+ in the electrolyte press the counter-ions in the diffusion layer into the adsorption layer, so that the charge of montmorillonite is reduced and particles are easy to gather.

In future work, more research should be carried out on the performance of drilling fluid base slurry to strengthen the summary and progress of its performance changes.

ACKNOWLEDGMENTS

Funding: This research was funded by the CNPC Scientific research and technology development projects (2020A-3913, 2020D-5008-04, 2020F-45, 2021DJ2003).

REFERENCES

Ali, K., Ahmad, R. (2020) Mesoscopic rheological modeling of drilling fluids: Effects of the electrolyte. *Journal of Petroleum Science and Engineering*, 195:107880.

Ettehadi, A., Tezcan, M., Altun, G. (2020) Rheological behavior of water-clay suspensions under large amplitude oscillatory shear. *Rheol Acta*, 59: 665–683.

Fahd, S. A., Mysara, E. M., Mohammed, A. A., Ali, S. M., Anas, H. (2021) Apparent and plastic viscosities prediction of water-based drilling fluid using response surface methodology. *Colloids and Surfaces A: Physicochemical and Engineering Aspects*, 616:126278.

Hany, G., Salaheldin, E., Salem, B., Abdulaziz, A. (2019) Effect of pH on Rheological and Filtration Properties of Water-Based Drilling Fluid Based on Bentonite. *Sustainability*,11(23):6714.

Huang, W., Yee-Kwong, L., Chen, T., Pek-Ing, A., Liu, X., Qiu, Z. (2016) Surface chemistry and rheological properties of API bentonite drilling fluid: pH effect, yield stress, zeta potential, and aging behavior. *Journal of Petroleum Science and Engineering*, 146:561–569.

Liu, J., Cheng, Y., Zhou, F., Amutenya Evelina, L.M., Long, W., Chen, S., He, L., Yi, X., Yang, X. (2022) Evaluation method of thermal stability of bentonite for water-based drilling fluids. *Journal of Petroleum Science and Engineering*, 208:109239.

Laura, V., Rodrigo, F., Beatriz, R., Cláudia, M., Luis, A. (2019) Study on the dissolution kinetics of NaCl in environmentally friendly drilling fluids containing glycerin. *Journal of Petroleum Science and Engineering*, 182:106165.

Peretyazhko, T.S., Niles, P.B., Sutter, B., Morris, R.V., Agresti, D.G., Ming, D.W. (2018) Smectite formation in the presence of sulfuric acid: Implications for acidic smectite formation on early Mars. *Geochimica et Cosmochimica Acta*, 220:248–260.

Zhang, J., Xu, M., Christidis, G., Zhou, C. (2020). Clay minerals in drilling fluids: Functions and challenges. *Clay Minerals*, 55: 1–11.

*Advances in Energy, Environment and
Chemical Engineering – Abdullah & Osman (Eds)
© 2023 The Author(s), ISBN: 978-1-032-36083-6*

Analysis of characteristics of mineral spontaneous combustion indicators gas in open pit mines

Yangyang Zhang*
CCTEG Ecological Environment Technology Co., Ltd., Tangshan, China

ABSTRACT: Taking the minerals of Fushun West Open Pit Mine as the research object, in order to study the index gas release law of the minerals in the process of spontaneous combustion, the high-temperature program temperature system was used to conduct heating experiments on samples with the different mixing modes, and the gas chromatograph was used to measure the consumption during the heating process. Oxygen characteristics and the laws of CO, CO_2, CH_4, C_2H_4, C_2H_6, and other gases in the low-temperature oxidation stage, rapid heating stage, and high-temperature combustion stage were analyzed, and the critical parameters and the high-temperature combustion of the tested mineral samples from the low-temperature oxidation stage to the rapid heating stage were determined. The peak parameter index of the stage was of great significance to the monitoring and prevention of the spontaneous combustion of minerals in open pit mines.

1 INTRODUCTION

Mineral spontaneous combustion is a very complex physical and chemical change process, and it is a changeable self-accelerating exothermic process. In this process, carbon-containing minerals are mainly compounded with oxygen, and a large amount of carbon-containing compounds will be produced during the entire change process, such as CO, CO_2, CH_4, C_2H_6, and C_2H_4, there are certain regularities in the concentration of indicator gases in different temperature stages of coal samples during the oxidation process (An et al. 2018; Liu et al. 2013; Li et al. 2016; Wu et al. 2012; Yang et al. 2018; Zhao et al. 2019), and the variation laws of indicator gases in the spontaneous combustion process of different coal types can be used for the early prediction of the coal spontaneous combustion (Wang et al. 2017, 2020, 2021; Xiao et al. 2008; Zhu et al. 2020).

In this paper, through the high temperature programmed heating experiments on minerals, the critical values and peak values of the indicator gas were found at different temperature stages by studying the oxygen consumption characteristics of the minerals during the heating process through comparative analysis, and analyzing the changing laws of various gases in different oxidation stages comprehensively to predict the spontaneous combustion and ignition law of minerals in open pit mines accurately.

2 ANALYSIS OF MINERAL SAMPLES

Mineral samples were collected from the open pit in West Open Pit, and the coal, oil shale, and oil shale after dry distillation were crushed to 80~120 mesh respectively, and the coal, oil shale, and coal gangue were mixed in a ratio of 1:1:1 to form mixtures. Samples were mixed with the coal, oil shale, coal gangue, and oil shale after dry distillation in a ratio of 1:1:1:1 to form the mixtures. Five samples were weighed 0.9-1.1g, the industrial analysis experimental instrument was 5E-MAG6700 Kaiyuan Automatic Industrial Analyzer, and the experiment was executed at room temperature and normal pressure. An experimental instrument for the carbon and sulfur analysis was CS-8820

---

*Corresponding Author: zhangyy8510@126.com

High-frequency Infrared Carbon and Sulfur analyzer. Under the high-temperature combustion of 1650°, CO, CO_2, and SO_2 were generated. After the processing and calculation, the carbon and sulfur contents in the sample were obtained. Industrial analysis and the carbon and sulfur analysis results of minerals are shown in Table 1.

Table 1. Industrial and elemental analysis results of minerals.

Samples	Industrial analysis (%)				Total heat release (J/g) Q	Elemental analysis (%)	
	Mad	Aad	Vad	Fcad		C	S
Coal	2.94	8.12	42.01	46.93	4714.403	52.24	0.33
Oil shale after dry Distillation	1.63	83.99	8.55	5.83	2195.078	15.25	0.68
Oil shale	2.04	77.34	16.56	4.05	2573.635	10.59	0.62
Three mixtures	2.17	55.28	22.53	20.01	2530.522	27.78	0.43
Four mixtures	2.09	63.6	19.15	15.15	1795.625	25.66	0.49

3 EXPERIMENTAL PROCESS

Tested mineral samples were the coal, oil shale, oil shale after dry distillation, three mixtures, and four mixtures. Each sample was crushed in the air and sieved out 5 different particle size samples of 0–0.9mm, 0.9–3mm, 3–5mm, 5–7mm, and 7–10mm, and 1kg of each sample was taken to form the mixed particle size sample. Temperature-programmed experiment was executed in the XKGW-1 Briquette Self-ignition high temperature programmed experimental system. Experimental conditions are shown in Table 2.

Table 2. Coal spontaneous combustion temperatures program experimental conditions.

Particle-size (mm)	Coal weight (kg)	Air flow (ml/min)	Temperatures range (°C)	Heating-up speed (°C/min)
Mixed size	5	120	30–600	1

4 ANALYSIS OF EXPERIMENTAL RESULTS

4.1 *Variation laws of oxygen concentration*

The consumption law of the oxygen is one of the important macroscopic parameters of the reaction spontaneous combustion intensity. Change trends of the oxygen consumption with temperatures are shown in Figure 1.

Oxygen concentrations decrease very slowly at the initial stage of combustion. Between 150°C and 200°C, the oxygen concentrations increase sharply. Except for the oil shale, the other four samples all reach the lowest point of the oxygen concentrations at about 200°C. At 350°C–400°C, the oxygen concentrations rebound to varying degrees and continue to heat to about 450°C, and the oxygen concentrations gradually begin to decrease.

4.2 *Variation laws of oxygen consumption rates*

Trends of the oxygen consumption rates with the temperatures are shown in Figure 2. In the low-temperature stage of the reaction (before 200°C), the oxygen consumption rates are low relatively, and the oxygen consumption rates of the coal reach the peak at 330°C, and the oxygen consumption rates of the oil shale after dry distillation reach the peak at 300°C. Oxygen consumption rates of the oil shale start to increase at 200°C, and the highest point is 405°C, the time point when the

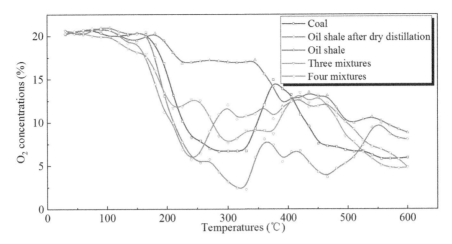

Figure 1. Variation trends of O_2 concentrations with temperatures.

oxygen consumption rates of the three mixtures begin to increase was earlier, but the upward trend is relatively slow, and the oxygen consumption rates at the highest point were lower, while the oxygen consumption rates of four mixtures rise rapidly between 165−240°C.

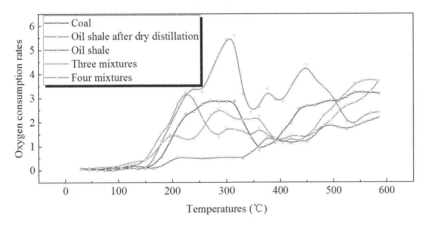

Figure 2. Variation trends of oxygen consumption rate with temperatures.

4.3 Analysis of the indicators gas release laws of mine minerals

The release of gaseous products is the main macroscopic feature of the high-temperature oxidation of the samples. Gas generated by the spontaneous combustion of the samples is a spontaneous reaction. Gaseous products of each sample mainly include CO, CO_2, CH_4, C_2H_4, and C_2H_6.

4.3.1 Analysis of the change laws of CO concentrations

CO gas was measured at the beginning of the experiment, and the CO concentrations of the coal showed an upward trend generally, reaching the peak values at 555°C. Variation trends of CO concentrations are in the form of an obvious parabola in the oil shale after dry distillation. Growth rates increase rapidly after 165°C and reach the peak value at 330°C. Oil shale begins to show a rapid increase after 210°C, and reaches the peak value at 300°C. Variation trends of CO concentrations in the two mixed samples are basically between the coal and rock samples, and both start to increase rapidly around 180°C and reach the peak values at 390°C. Variation trends of CO concentrations with temperatures are shown in Figure 3.

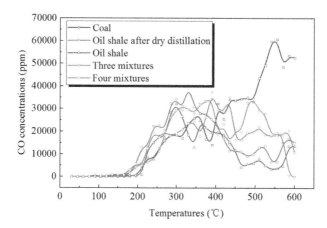

Figure 3.   Variation trends of CO concentrations with temperatures.

By analyzing the change of CO concentrations of each sample by the growth rate analysis method, the characteristic temperature points of each sample could be obtained, as shown in Table 3. The period from the start of the experiment to the activation temperature was divided into the low-temperature oxidation stage, the period from the activation temperature to the ignition temperature was divided into the rapid heating stage, and the period after the ignition point was called the high-temperature combustion stage.

Table 3.   Mineral characteristic temperatures of the open pit mine.

	Characteristic temperature				
Samples	Critical temperatures (°C)	Cracking temperatures (°C)	Active temperatures (°C)	Acceleration temperatures (°C)	Ignition temperatures (°C)
Coal	75	120	195	240	315
Oil shale after Dry distillation	135	165	270	315	360
Oil shale	75	150	210	270	345
Three mixtures	90	120	180	255	345
Four mixtures	90	120	165	210	360

4.3.2   Analysis of the change law of CO_2 concentrations

Variations of CO_2 concentrations with temperatures are shown in Figure 4. CO_2 concentrations produced by the spontaneous combustion of the coal are small relatively in the low-temperature oxidation stage, rise rapidly in the rapid heating stage, and reach the peak value at 465°C.

CO_2 concentrations of the oil shale and oil after dry distillation begin to increase gradually in the low-temperature oxidation stage, and the growth rate continues to accelerate after entering the rapid heating stage, reaching the peak at 510°C. CO_2 concentrations of three mixtures increase rapidly during the rapid heating stage and the peak value at 360°C. CO_2 concentrations of four mixtures reach the peak at the beginning of the high-temperature combustion stage.

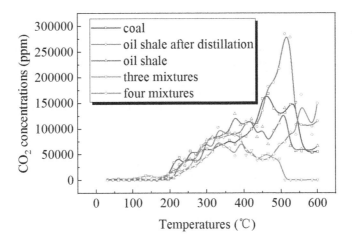

Figure 4.   Variation trends of CO2 concentrations with temperatures.

4.3.3   *Analysis of the change laws of CH4 concentrations*

CH_4 gas is the most easily detected gas species in the hydrocarbon and alkenes gas, and it is the main gaseous product in the oxidative pyrolysis process of the sample, it is formed throughout the oxidative pyrolysis process of the sample. Variations of the concentrations with temperatures are shown in Figure 5.

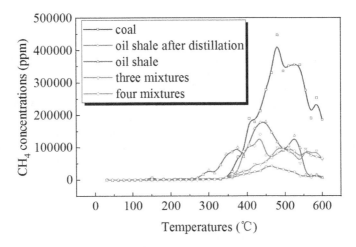

Figure 5.   Variation trends of CH4 concentrations with temperatures.

CH_4 gas could be detected from the beginning of the experiment. During the high-temperature combustion stage of the coal, the concentrations of CH_4 gas rise sharply until they reach the peak value of around 480°C. Amounts of CH_4 precipitation are less in the oil shale after dry distillation, and reach the peak at 465°C; the amounts of CH_4 precipitation are larger in the oil shale and reach the peak at 450°C, and the CH_4 concentrations of all samples reach the peak at the high-temperature combustion stage.

4.3.4   *Analysis of the change laws of C_2H_6 concentrations*

Samples begin to precipitate C_2H_6 in the low-temperature oxidation stage, but the contents are low extremely. With the continuous increase of temperature, the curves appear in the form of a parabola.

Before the high-temperature combustion stage, the contents of C_2H_6 gas are small relatively. After entering the high-temperature combustion stage, the generations of C_2H_6 gas grow rapidly, and the amounts of the coal and the four mixtures reach the peak at 435°C. The oil shale after dry distillation and the oil shale both reach the peaks at 420°C, and the three mixtures reach the peak at 480°C. Changes in C_2H_6 concentrations with temperatures are shown in Figure 6. Amounts of C_2H_6 in five samples in the low-temperature oxidation stage and the rapid heating stage are less, and large amounts of C_2H_6 begin to produce after entering the high-temperature combustion stage, and the coal has the largest precipitation.

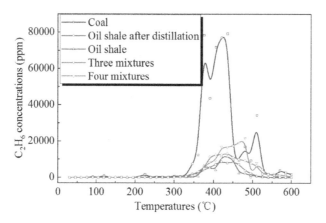

Figure 6.   Variation trends of C_2H_6 concentrations with temperature.

4.3.5 *Analysis of the change laws of C_2H_4 concentrations*

Variation trends of C_2H_4 of the samples are in the form of a parabola, as shown in Figure 7. Precipitations of C_2H_4 gas are low extremely in the low-temperature oxidation stage and the rapid heating stage, but C_2H_4 gas is detected between 100−150°C, indicating that C_2H_4 gas is the high-temperature reaction product rather than occurring in the samples. After entering the high-temperature combustion stage, C_2H_4 begins to come out in large quantities, and its peak concentration time is the same as that of C_2H_6, but its peak concentration is much lower than that of C_2H_6, and the peak concentration of C_2H_4 gas is the highest in the coal.

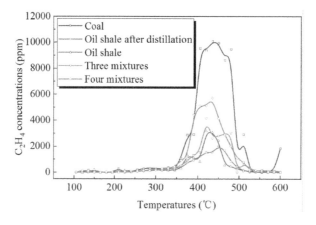

Figure 7.   Variation trends of C2H4 concentrations with temperature.

In the low-temperature oxidation stage, the indexes of the five different samples all increase slowly, reach the rapid heating stage, the indexes increase rapidly, and reach the peak quickly, entering the high-temperature combustion stage. According to the analysis, it is determined that the rapid heating stage is the dangerous stage of the mineral spontaneous combustion process. Entering this stage, spontaneous combustion will be difficult to control, and eventually, the combustion will be caused by the activation of the active structure. For this reason, spontaneous combustion prevention should be carried out in the low-temperature oxidation stage, and the critical parameter values and peak values system should be established, as shown in Table 4.

Table 4. Critical values and peak values system.

Samples	CO/ppm		CO_2/ppm		CH_4/ppm		C_2H_4/ppm		C_2H_6/ppm	
	Critical values	Peak values	Critical values	Peak values	Critical values	Peak values	Critical values	Peak values	Critical values	Peak values
Coal	540	60000	1930	168000	3580	440000	196	10000	570	79000
Oil shale after dry distillation	760	40000	3170	285000	5800	42000	300	2000	570	8800
Oil shale	540	32000	3700	140000	480	178000	190	3100	1100	11000
Three mixtures	1540	34000	3100	89000	1720	97000	330	4100	573	21000
Four mixtures	760	37000	3170	170000	1700	140000	300	5700	570	13000

5 CONCLUSIONS

(1) Consumption law of oxygen is one of the important macro parameters for replying to the severity of spontaneous combustion. Oxygen concentrations of the five samples all changed slowly before the critical temperature. After the critical temperature, except for the oil shale, the oxygen concentrations of the other four samples showed a significant downward trend, and the oxygen consumption rate increased sharply. Between 300–500°C, the oxygen concentrations of the five samples all showed different degrees of recovery, and the oxygen consumption rate decreased.

(2) Combined with the change laws of the index gas, the spontaneous combustion process was divided into three stages, the first stage was the low-temperature oxidation stage before the activation temperature, the second stage was the rapid heating stage before the activation temperature to the ignition temperature, and the third stage was the high-temperature combustion stage after the ignition point. The rapid heating stage was determined as the dangerous stage of the mineral spontaneous combustion process. Spontaneous combustion prevention should be carried out in the low-temperature oxidation stage, and the critical parameter index and peak index system had been established.

ACKNOWLEDGMENTS

This research was supported by China Coal Technology & Engineering Group Co., Ltd. Science and Innovation and Entrepreneurship Fund Special Key Project (2019-ZD004).

REFERENCES

An B, Wang J F, Wang Y Y, Song S. Experimental Study on Formation Laws of Composite Index Gases for Coal Spontaneous Combustion[J]. *Safety in Coal Mines* 2018, 49(2):23–26.

Li L, Chen J C, Jiang D Y, Fan J Y, Gao Y L. Experimental study on temporal variation of the high-temperature region and index gas of coal spontaneous combustion [J]. *Journal of China Coal Society*, 2016, 41(2): 444–450.

Liu Q, Wang D M, Zhong X X, Jiao X M, Zhang H J. Testing on indicator gases of coal spontaneous combustion based on temperature program [J]. *Journal of Liaoning Technical University* (Natural Science), 2013, 32(3): 362–366.

Wang H T, Liu Y L, Shen Bi, Fan C H, Zhang L J. Analysis on Index Gas Characteristics of Long-flame Coal Spontaneous Combustion [J]. *Coal Technology*, 2021, 40(11):167–170.

Wang Y H, Zhou N, Zhang J L, Zhu Z. Study on gas optimization of spontaneous combustion index for different coal types [J]. *Journal of Henan Polytechnic University* (Natural Science), 2020, 39(6):10–15.

Wang Ya, Wang J F, Wu J M, Tang Y B. Experiment study on gas indexes and characteristic of oxidation temperature for low metamorphism coal [J]. *Coal Technology*, 2017, 36(3): 207–209.

Wu J M, Peng J, Wu Y G. Experiment study on index gas selection of coal spontaneous combustion in Pingshuo Mining Area [J]. *Coal Science and Technology*, 2012, 40(2):67–69.

Xiao Y, Wang Z P, Ma L, Zhai X W. Research on correspondence relation ship between coal spontaneous combustion index gas and feature temperature[J]. *Coal Science and Technology*, 2008, 36(6):47–51.

Yang S, Dai G L, Tang M Y. Experimental study on marker gases of spontaneous combustion of coal based on programmed warming[J]. *Coal Mine Safety*, 2018, 49(7): 24–27.

Zhao J Y, Zhang Y X, Song J J, Zhang Y N, Wang K. Gas test of coal spontaneous combustion index at different temper-ature stages under high temperature and poor oxygen[J]. *Journal of Xi'an University of Science and Technology*, 2019, 39(2):189–193.

Zhu J G, Dai G L, Tang M Y, Ye Q S, Li P. Experimental study on spontaneous combustion prediction index gas of water immersed long flame coal [J]. *Coal Science and Technology*, 2020, 48(5): 89–94.

*Advances in Energy, Environment and
Chemical Engineering – Abdullah & Osman (Eds)*
© 2023 The Author(s), ISBN: 978-1-032-36083-6

Environmental, physical, and structural characterization of fly ash from municipal solid waste incineration in South China

Xiuya Ye*

School of Chemical Engineering and Technology, Guangdong Industry Polytechnic, Guangzhou, China

ABSTRACT: The fundamental properties of fly ashes are analyzed using the following characterization techniques: low-temperature nitrogen adsorption-desorption, infrared spectroscopy, X-ray diffraction, and scanning electron microscopy. Experiments indicate that fly ash has a complex structure and changeable properties. The study results show that up to 75% of fly ash particles are concentrated in the range of 38.5–75 μm. The main chemical compositions are O, Ca, Cl, Si, S, K, Al, Na, Fe, Mg, and P, the main heavy metals are Zn, Cr, Cu, Pb, and Cd. The Cl content ranges from 6.2% to 29.56%, whereas that of SiO_2 ranges from 5.76% to 29.91%. BET results show that the fly ash particles' specific surface area and pore volume increase as their diameter increases. The main structures of fly ash are irregular amorphous morphology and polycrystalline aggregates.

1 INTRODUCTION

Among the solid residues from MSW incineration, fly ash is considered a highly toxic substance due to its high concentration of leachable heavy metals and, in some cases, chlorinated organic compounds (Nie et al. 2011). Although incineration can effectively reduce the mass and volume of waste, at the same time heavy metals may escape into the flue gas in the form of soot or vapor. Potential heavy metal emissions from these municipal solid waste incinerators (MSWI) have raised some public concerns (Zheng et al. 2020). The emission characteristics of heavy metals in MSWI are affected by various operating parameters, including the concentration of metals in the waste, combustion temperature, and performance of pollutant control devices (Chen et al. 2020; Xiao et al. 2020).

2 MATERIALS AND METHODS

2.1 Samples

The fly ash samples used in this paper were collected from cloth bags from a large-scale incinerator in southern China. The incinerator has a capacity of 1040 tons/day and is equipped with an air pollution control system. The fly ash extracted from the furnace was homogenized, placed in a muffle furnace, and dried at 105°C $\pm$ 0.5°C for 24 hours. After drying and cooling, the samples were ground and homogenized, and finally screened into 6 different size fractions: < 37.5μm (FA1), 37.5–75μm (FA2), 75–125μm (FA3), 125–187.5μm (FA4), 187.5–375μm (FA5) and > 375μm. The characteristics of each component of fly ash except particle size > 375μm were analyzed.).

2.2 Chemical analysis

Chemical analysis was performed using the Philips-type PW 2400 fluorescence X-ray spectrometer. The crystalline phases of the fly ash were investigated by X-ray diffraction (XRD, Bruker D8-Advance diffractometer).

---

*Corresponding Author: 274621048@qq.com

2.3 *Physical characteristics*

FT-IR was performed using a Nicolet Nexus670 spectrometer within the range of 4000–400 cm^{-1}, using 1.5 mg of the sample mixed with 200 mg KBr and pressed into thin transparent flakes.

To grasp the surface area, specific volume, and average pore width of the samples, the samples were first degassed at 350°C and then subjected to N_2 adsorption. The study was conducted using a Quantachrome Autosorb 6 apparatus.

Microscopic observations were performed using scanning electron microscopy (SEM, Hitachi S-3000N) equipped with Quantax SDD energy-dispersive X-ray spectroscopy (EDS) (Bruker, resolution 123 eV, Germany) at a 15-kV accelerating voltage. First, the samples were suspended and covered with a thin film smear of gold in a Balzer-type SCD 050 evaporator under an argon atmosphere.

3 RESULTS AND DISCUSSION

3.1 *Particle size distribution*

Figure 1 illustrates the size distributions of a particle in the fly ash. A wide range of particle size distribution of the fly ash exists. The fly ash particles are mainly distributed between 38.5 μm and 75 μm. More than half of fly ash particles have a particle size <75μm. 75.5% of fly ash particle size is less than 75 microns, 87.9% of fly ash particle size is less than 125 microns, and 92.8% of fly ash particle size is less than 200 microns.

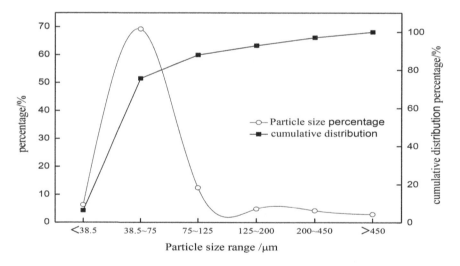

Figure 1.   Weight fraction distribution and cumulative mass fraction of fly ash with different particle sizes.

3.2 *XRF analysis*

Table 1 shows the experimental results. It could be seen that the major elements (>10,000 mg/kg, listed in decreasing order of abundance) in fly ash are O, Ca, Cl, Si, S, K, Al, Na, Fe, Mg, and P, which account for 87.6%–97.2% of the total weight of fly ash.

Cl content ranges from 6.2% to 29.56%, whereas the SiO_2 content ranges from 5.76% to 29.91%. The minor elements (1,000–10,000 mg/kg) include Zn, Cr, Cu, Pb, and Cd. These results are similar to those reported in other studies (Yao et al. 2020; Zhao et al. 2020). The high content of calcium oxide comes from the excess lime solution injected into the scrubber, which is used to remove acid gases from the incineration flue gas. It is beneficial to the stabilization and fixation of some dangerous heavy metals, such as cadmium and copper (Dongwoo Kang et al. 2020; Mitali Nag et al. 2020). The Cl concentration in the fly ash is also high (7.18%–29.56%), and generally increases as the particle size decreases. Cl in the fly ash mainly came from the garbage, including PVC and kitchen waste. The waste composition in the presence of chlorine-containing substances in waste incineration is a significant source of dioxins (Chen et al. 2020; Zheng et al. 2020). The smaller the particle size of the fly ash, the easier the chlorine substances to be absorbed (Qadeer Alam et al. 2020). The SO_3, K_2O, and Na_2O content reflect the higher flue gas purification system in good working conditions.

Table 1. Chemical composition of the fly ash in wt.%.

	SiO_2	CaO	Cl	SO_3	K_2O	Na_2O	Al_2O_3	MgO	Fe_2O_3	P_2O_5	ZnO	Cr_2O_3	CuO	PbO	CdO
FA	8.82	37.55	24.31	6.98	6.19	5.47	3.48	1.84	1.68	0.83	0.69	0.42	0.08	0.04	0.01
FA1	5.76	33.99	29.56	6.61	13.56	2.04	2.14	1.46	1.13	0.56	1.04	0.41	0.102	0.303	0.025
FA2	7.70	39.51	26.76	6.99	6.55	3.36	2.82	1.47	1.34	0.63	0.90	0.39	0.089	0.148	0.015
FA3	9.38	35.40	20.93	8.08	6.76	8.06	3.96	1.89	1.79	0.98	0.64	0.44	0.065	0.054	0.007
FA4	23.36	26.78	6.20	8.14	2.69	3.74	8.90	2.67	3.32	1.80	0.55	0.45	0.037	0.036	0.004
FA5	29.91	27.14	7.18	6.8	3.40	4.17	10.0	2.58	3.56	1.91	0.62	0.49	0.05	0.04	N.D.[a]

[a]N.D: non-detected

3.3 *XRD Mineralogy*

Figure 2 shows the main component of the fly ash KCl, NaCl, $CaCO_3$, $CaSO_4$ and SiO_2, which were consistent with the XRF results. Fly ash samples with sizes of <125 μm contained more NaCl and KCl (Tong et al. 2020).

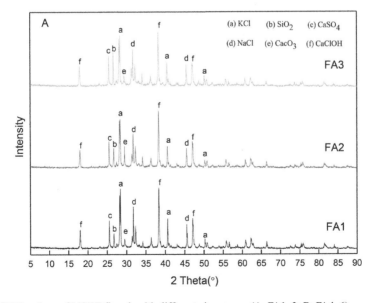

Figure 2. XRD pattern of MSWI fly ash with different size ranges (A: FA1–3; B: FA4–6).

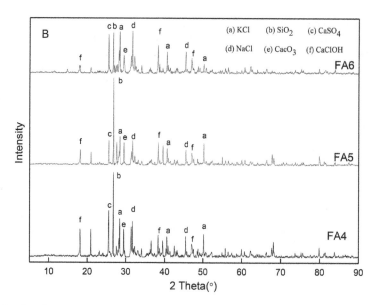

Figure 2.   Continued.

3.4   *FT-IR*

The FT-IR spectrum of fly ash (Figure 3) shows broadband around 3,435 cm^{-1}, which is attributed to the surface -OH groups and adsorbed water molecules on the surface. The broadness of the band at 3,643 cm^{-1} is due to the strong hydroxyl bonding. The peaks at 1,650 cm^{-1} in the spectra of both samples are attributed to the bending mode of the water molecule (Fernández-Jiménez and Palomo 2005). The adsorption bands at 595.5–1,637.8 cm^{-1} were assigned to SO$_3$ and–N=N–groups on the samples (Dizge et al. 2008); 2,340 cm^{-1} is caused by CO$_2$ because the testing process cannot be completely isolated with air mixed with CO$_2$.

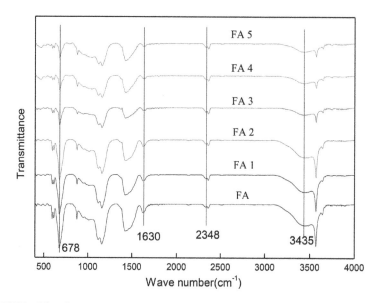

Figure 3.   FT-IR of fly ash.

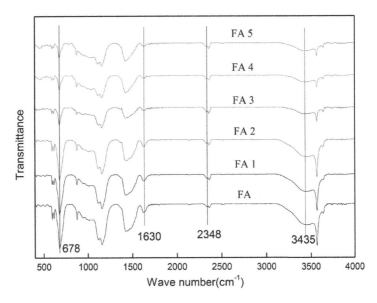

Figure 3.　Continued.

3.5　*Surface characteristics*

Figure 4 shows the nitrogen adsorption and desorption isotherms of the fly ash samples. According to the IUPAC classification, the adsorption curve follows a type III behavior. The N2 adsorption-desorption curves showed a hysteretic trend, probably due to the extremely irregular solid pores of the fly ash. The authors observed that the BET surface of fly ash was positively low (Table 2), and the surface area of BET decreased with increasing diameter. Similar values were reported (Astryd Viandila Dahlan et al. 2020), suggesting that the particles have negligible porosity.

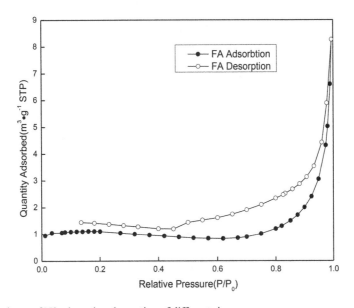

Figure 4.　Isotherm of N2 adsorption-desorption of different size ranges.

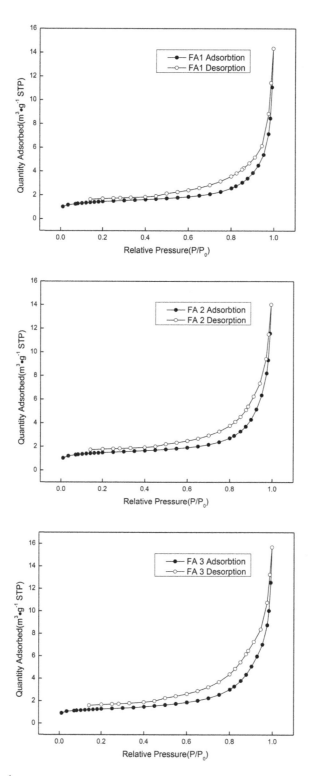

Figure 4.   Continued.

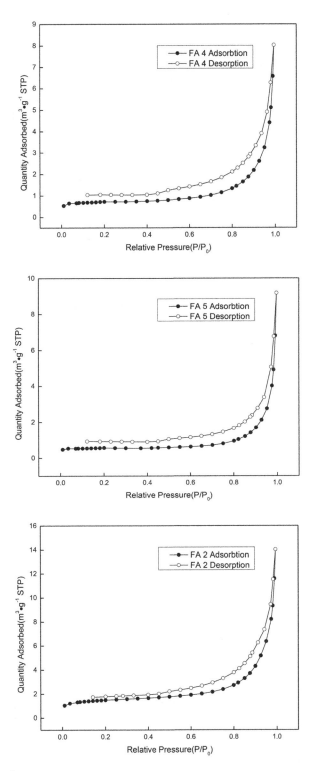

Figure 4.   Continued.

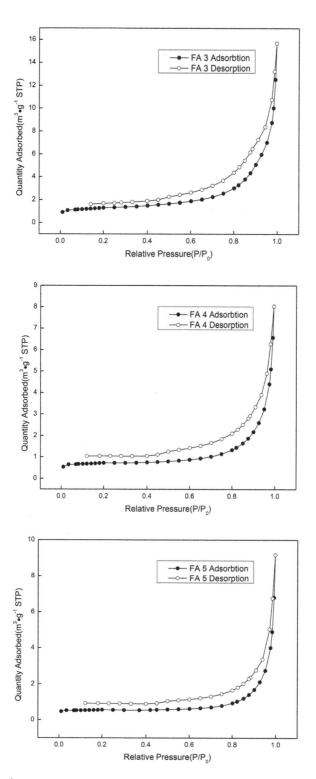

Figure 4.   Continued.

Table 2. Pore structure parameters of fly ash particles of different size ranges.

	BET surface area (S_{bet}) ($m^2 \cdot g^{-1}$)	t-Plot micropore area (S_{mic}) ($m^2 \cdot g^{-1}$)	S_{mic}/S_{bet} (%)	BJH adsorption cumulative volume of pores (V_t) ($cm^3 \cdot g^{-1}$)	t-Plot micropore volume (V_{mic}) ($cm^3 \cdot g^{-1}$)	V_{mic}/V_t (%)	Average pore width (nm)
FA1	5.1176	2.1586	42.18	0.021275	0.000964	4.53	8.65115
FA2	5.1657	2.4985	48.37	0.020805	0.001135	5.45	9.84613
FA3	4.4635	2.0102	45.04	0.024073	0.000911	3.78	12.13617
FA4	2.4433	1.6820	68.84	0.012038	0.000767	6.37	11.15494
FA5	1.8974	1.4691	77.43	0.013907	0.000670	4.82	13.07920

3.6 *SEM-EDS*

Figure 5 shows the SEM images. The fly ash sample consists of polycrystalline flakes <0.5 μm and aggregated into spheroids between 20 and 100 μm. The crystallinity of these materials may be due to the high temperature in the flue gas stream. These images confirm the low porosity suggested by the aforementioned surface area values, and possibly resynthesis on the platelet surface. No significant differences were observed with different size ranges (Geng et al. 2020). Figure 6 shows the picture of the analysis position in the fly ash. As shown by the EDS results, the main elements in FA were Ca, Cl, Si, Zn, O, Al, and Mg (Zhang et al. 2020).

Figure 5. SEM images of fly ash samples.

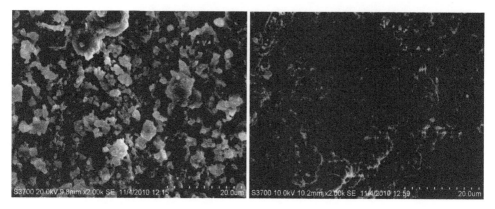

Figure 5.   Continued.

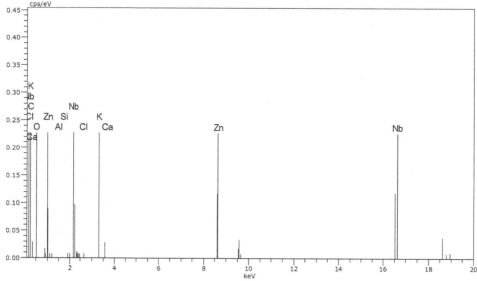

Figure 6.   SEM image of fly ash particles and EDX image of the analysis position.

4 CONCLUSION

Better forecast trends and characterization of fly ash can help understand metal behavior, and analyze waste, coupled with more data.

The fly ash consists of a wide range of particle size distributions. The highest peak in the size range is 75 and 38.5 μm. The main components of the fly ash are CaO, Cl, SiO_2, SO_3, K_2O, Na_2O, Al_2O_3, MgO, Fe_2O_3, P_2O_5, and ZnO. The XRD and XRF results agree excellently. The FT-IR spectrum of the fly ash (Figure 3) shows broadband between 3400 and 3000 cm^{-1}, which is attributed to the surface -OH groups and adsorbed water molecules on the surface. The nitrogen adsorption and desorption isotherms of fly ash samples revealed that the fly ash is a type III isotherm.

REFERENCES

Astryd Viandila Dahlan, Hiroki Kitamura, Yu Tian,Hirofumi Sakanakura, Takayuki Shimaoka, Takashi Yamamoto, Fumitake Takahashi. Heterogeneities of fly ash particles generated from a fluidized bed combustor of municipal solid waste incineration [J]. *Journal of Material Cycles and Waste Management*, 2020 (prepublish).

Chao Geng, Chao Chen, Xianfeng Shi, Shichao Wu, Yufeng Jia, Bing Du, Jianguo Liu. Recovery of metals from municipal solid waste incineration flies ash and red mud via a co-reduction process [J]. *Resources, Conservation & Recycling*, 2020, 154.

Dizge, N., Aydiner, C., Demirbas, E., Kobya, M., Kara, S., 2008. Adsorption of reactive dyes from aqueous solutions by fly ash: Kinetic and equilibrium studies [J]. *Journal of Hazardous Materials* 150, 737–746.

Dongwoo Kang, Juhee Son, Yunsung Yoo, Sangwon Park, Il-Sang Huh, Jinwon Park. Heavy-metal reduction and solidification in municipal solid waste incineration (MSWI) fly ash using water, NaOH, KOH, and NH_4OH in combination with the CO_2 uptake procedure[J]. *Chemical Engineering Journal*, 2020, 380.

Fernández-Jiménez, A., Palomo, A., 2005. Mid-infrared spectroscopic studies of alkali-activated fly ash structure [J]. *Microporous and Mesoporous Materials* 86, 207–214.

Lizhi Tong, Jinyong He, Feng Wang, Yan Wang, Lei Wang, Daniel C.W. Tsang, Qing Hu,Bin Hu, Yi Tang. Evaluation of the BCR sequential extraction scheme for trace metal fractionation of alkaline municipal solid waste incineration fly ash [J]. *Chemosphere*, 2020, 249.

Mitali Nag, Amirhomayoun Saffarzadeh, Takeshi Nomichi, Takayuki Shimaoka, Hirofumi Nakayama. Enhanced Pb and Zn stabilization in municipal solid waste incineration fly ash using waste fishbone hydroxyapatite [J]. *Waste Management*, 2020, 118.

Nie, Z., Zheng, M., Liu, W., Zhang, B., Liu, G., Su, G., Lv, P., Xiao, K., 2011. Estimation and characterization of PCDD/Fs, dl-PCBs, PCNs, HxCBz and PeCBz emissions from magnesium metallurgy facilities in China. *Chemosphere* 85, 1707–1712.

Peng Zhao, Minghai Jing, Lei Feng, Bai Min. The heavy metal leaching property and cementitious material preparation by treating municipal solid waste incineration fly ash through the molten salt process [J]. *Waste Management & Research*, 2020, 38(1).

Qadeer Alam, Alberto Lazaro, Katrin Schollbach et al. Chemical speciation, distribution and leaching behavior of chlorides from municipal solid waste incineration bottom ash [J] *Chemosphere*, 2020, 241

Xiao Haiping,Cheng Qiyong,Liu Meijia,Li Li,Ru Yu,Yan Dahai. Industrial disposal processes for treatment of polychlorinated dibenzo-p-dioxins and dibenzofurans in municipal solid waste incineration fly ash. [J]. *Chemosphere*, 2020, 243.

Zhang Yuying, Wang Lei, Chen Liang, Ma Bin, Zhang Yike, Ni Wen, Tsang Daniel C W. Treatment of municipal solid waste incineration fly ash: State-of-the-art technologies and future perspectives. [J]. *Journal of hazardous materials*, 2021, 411.

Zheng Peng, Roland Weber, Yong Ren, Jianwei Wang, Yangzhao Sun, Lifang Wang. Characterization of PCDD/Fs and heavy metal distribution from municipal solid waste incinerator fly ash sintering process [J]. *Waste Management*, 2020, 103.

Zhiliang Chen, Sheng Zhang, Xiaoqing Lin, Xiaodong Li. Decomposition and reformation pathways of PCDD/Fs during thermal treatment of municipal solid waste incineration fly ash[J]. *Journal of Hazardous Materials*, 2020, 394.

*Advances in Energy, Environment and
Chemical Engineering – Abdullah & Osman (Eds)
© 2023 The Author(s), ISBN: 978-1-032-36083-6*

Research on the development path of methanol fuel powered ships in China

Lipeng Wang
China Waterborne Transport Research Institute, Beijing, China

Jiang Lei*
China Waterborne Transport Research Institute, Beijing, China
Wuhan University of Technology, Wuhan, Hubei, China

He Xing
Jiu Jiang Port & shipping Admin, Jiujiang, China

Yongbo Ji
China Waterborne Transport Research Institute, Beijing, China

Chen Pan*
China Ship Development and Design Center, National Key Laboratory on Ship Vibration & Noise, Wuhan, Hubei, China

ABSTRACT: Based on the analysis of the basic characteristics, raw material composition, production and supply characteristics of methanol fuel, as well as the development status of methanol fuel powered ships, engine technology, and fuel filling, this paper studies the safety, environmental protection, technical applicability and economic feasibility of methanol as the marine fuel, and summarizes the problems in development. The development path and relevant suggestions for the application and promotion of methanol fuel on ships in China are put forward.

1 INTRODUCTION

The world is facing two major problems, the oil energy crisis and environmental pollution. Energy and environmental issues have become important factors affecting China's economic and social development. In the context of ensuring national energy security, easing the pressure on ecological and environmental protection, and realizing green and sustainable development, the development of green, efficient, and clean alternative fuels for petroleum has become a top priority. As a major energy consumer, shipping has attracted wide attention from all walks of life in recent years for its clean alternative energy applications. The promotion of methanol fuel application is conducive to the protection of national energy security. China is rich in coal and short of oil and gas. The energy structure is dominated by coal. Crude oil and natural gas are highly dependent on foreign sources. In 2019, the external dependence on crude oil and natural gas resources reached 70.8% and 43% respectively. China is the country with the largest methanol production in the world and is one of the main countries that master coal-to-methanol technology. Domestic coal-based-methanol accounts for about 75%. From the perspective of China's energy structure and resource endowment, promoting the development of methanol fuel-powered ships in the transportation industry is not only conducive to giving full play to the advantages of China's coal resource and promoting the

*Corresponding Authors: jianglei@wti.ac.cn and panda3267@126.com

DOI 10.1201/9781003330165-75

transformation and upgrading of traditional industries but also conducive to promoting green circular development, realizing energy diversification and ensuring national energy security. Chinese shipping companies want to use methanol fuel, but the progress is relatively slow. In this context, this paper systematically analyzes the feasibility of methanol as ship fuel, summarizes the problems faced by development, and puts forward development paths and suggestions, in order to share with the industry and jointly promote the popularization and application of methanol in Chinese ships.

2 SAFETY CHARACTERISTICS OF METHANOL AS A FUEL

Methanol is a colorless, transparent, toxic, and volatile flammable liquid. Methanol, gasoline, diesel, and ethanol belong to dangerous goods of category III (GB 13690-1992, 2005). The flash point of methanol is higher than gasoline but lower than diesel. The spontaneous ignition point of methanol is close to gasoline but higher than diesel. The physical and chemical properties of methanol are close to that of gasoline. However, methanol does not contain sulfur, and the SO_x emission of methanol fuel can be reduced by almost 100% compared with gasoline and diesel (or fuel oil).

From the safety perspective, compared with traditional energy (heavy oil, diesel), the main risks of methanol include low flash point, strong corrosion, wide explosion range, and so on. In view of the risk of methanol fuel used on ships.

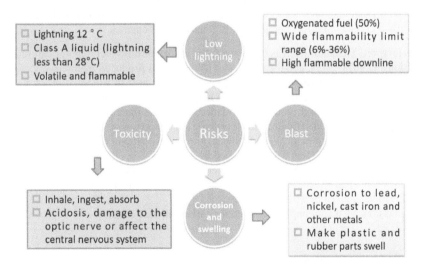

Figure 1.   Main risks of methanol fuel.

The IMO Interim Guidelines for the Safety of Ships Using Methyl/Ethyl as Fuel and CCS Guidelines for Ships Using Alternative Fuels (2017) have proposed specific risk control measures, including the safety requirements and measures for corrosion resistance of materials used, ship layout, ventilation, fire protection, monitoring (CCS, 2017), etc. From the supervision perspective, relevant departments have issued a number of regulations and guidelines to manage and prevent fire hazards and ensure the safety of methanol bulk transportation by land and sea. The International Code for the Construction and Equipment of Ships carrying Dangerous Chemicals in Bulk (IBC Code) provides detailed provisions for the safe transportation of methanol. IMO and classification societies believe that various standards and safety technical measures can ensure the safety of using methanol on ships. At present, there are a total of 17 methanol fuelled ships in operation worldwide, the earliest methanol-powered ship "Stena Germanica" in the world was put into operation between Gothenburg, Sweden, and Kiel, Germany in March 2015, and has been operating safely for 6 years.

The first 50000-ton methanol-powered chemical tank "LINDANGER" in the world was delivered in April 2016, and has been operating safely for more than 5 years.

Figure 2.   Methanol-powered retrofit ferry and chemical carrier.

3   ENVIRONMENTAL PROTECTION PERFORMANCE OF METHANOL FUEL

In the case of both NO_x and SO_x, the emissions reductions are due to the fact that methanol results in lower emissions during the combustion phase (IMO, 2016). According to the test results of MAN, an engine manufacturer, the methanol dual fuel engines (95% methanol and 5% diesel) have a good emission reduction effect compared with Tier II diesel engines.

(1) SO_x emission. Methanol itself is sulfur-free, the methanol dual fuel engines can reduce SO_x emissions by 90% - 97%, and the rest 3% - 10% of SO_x is produced by micro pilot diesel.
(2) NOx emission. Compared with diesel engines, the latent heat of vaporization of methanol is higher than that of gasoline and diesel, which helps to reduce the intake temperature, to improve the charge coefficient and power performance, to reduce the combustion temperature, and improve the combustion efficiency to a certain extent. High oxygen content, low combustion temperature, reduced air demand, and reduced nitrogen oxidation rate also helps to reduce NO_x emissions, the NO_x emission can be reduced by 30–50%.
(3) PM emission. Methanol has high oxygen content (50%), which is conducive to complete combustion and can effectively reduce the emissions of PM. The PM emission of the methanol engine can be reduced by 90%.
(4) CO_2 emission. Methanol engines can reduce carbon dioxide emissions by 15%.

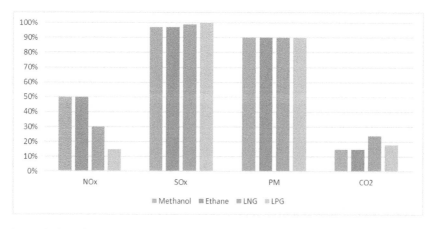

Figure 3.   Emission reduction effects of different fuels.

4 RELIABILITY OF METHANOL SUPPLY SYSTEM

(1) Sufficient methanol supply capacity. Since 2000, China's methanol industry has been on a trend of continuous growth. In 2019, China's methanol production capacity was nearly 90 million tons, higher than the national consumption. With the continuous growth of China's methanol production capacity, methanol production is also increasing year by year. In 2019, China's methanol production reached 69.97 million tons, accounting for 87.5% of the national consumption. Coal is the main raw material for methanol production in China, and the plants put into operation in recent years are also mainly coal-to-methanol. By 2019, the capacity of coal-to-methanol has reached 67.791 million tons, accounting for 75.4% of the total capacity. China is rich in coal resources and can provide sufficient raw materials.

(2) Abundant water transportation capacity for methanol. China's methanol transportation modes are diversified, including tank truck transportation, railway transportation, water transportation, etc. At present, China's land tank truck transportation and water transportation of methanol have been running for many years, and the relevant standards and regulations system is basically completed.

(3) Diversified methanol bunkering modes. Tank car bunkering is the most convenient way in the early stage. The cost of a tank truck is low, and the bunkering process is flexible, but the bunkering amount is limited. Due to the influence of domestic transportation specifications, vehicle specifications, and road facilities, the capacity of tank trucks is limited. At present, the capacity of a domestic large tank is about 60 m^3. In the future, the bunkering modes of methanol will also be diversified to ensure the fuel supply for ships. In China, the main ways of oil bunkering and LNG bunkering include tank truck bunkering, shore base station bunkering, pontoon bunkering, bunkering ship, and so on. From the perspective of fuel characteristics, the above ways can also be used for methanol fuel bunkering.

Figure 4.　Tank truck bunkering.

5 TECHNICAL FEASIBILITY OF METHANOL AS A MARINE FUEL

At present, several types of methanol fuel engines have been developed at home and abroad. There are mainly low-speed two-stroke engines of MAN 50ME-LGIM onboard ocean-going chemical carriers and medium-speed four-stroke engines of Wartsila ZA40S onboard offshore ro-ro passenger carriers in foreign countries, which all adopt high-pressure direct injection dual-fuel (DF) technology route. The methanol substitution rate is 95%. The power performance is equivalent to diesel engines, and the emission meets IMO Tier II.

The domestic research and development of methanol engines are mostly small-bore high-speed engines with power less than 600kW, including pure methanol engines and diesel-methanol dual-fuel engines. The ignition route adopted by the pure methanol engine is similar to the technical route of the methanol engine used on heavy trucks. Diesel-methanol dual-fuel engines (methanol substitution rate can reach 40%) need to meet the requirements of phase II of GB 15097, and the substitution rate of medium and high load will be reduced. The medium- and low-speed methanol marine engines with medium and large bores are currently being studied by Shanghai Marine Diesel Engine Research Institute.

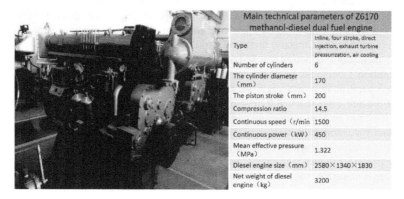

Main technical parameters of Z6170 methanol-diesel dual fuel engine	
Type	Inline, four stroke, direct injection, exhaust turbine pressurization, air cooling
Number of cylinders	6
The cylinder diameter （mm）	170
The piston stroke （mm）	200
Compression ratio	14.5
Continuous speed （r/min）	1500
Continuous power （kW）	450
Mean effective pressure （MPa）	1.322
Diesel engine size （mm）	2580×1340×1830
Net weight of diesel engine （kg）	3200

Figure 5.　Zichai Z6170 methanol-diesel dual-fuel engine.

6　ECONOMIC FEASIBILITY OF METHANOL AS A MARINE FUEL

According to FCBI Energy's statistics on the construction cost of existing methanol-fueled ships in Europe, the cost of engine fuel system increases by 10 ~20%, and the cost of fuel supply pipeline increases by 30 ~ 50% (for engine power of 10MW and 24MW). The overall retrofit cost is lower than that of LNG-powered ships. FCBI Energy report has made an economic comparison of HFO + SCR, MGO, methanol, LNG, and other fuels used on European coastal ships. The results show that for ships sailing only in the ECA, methanol is superior to LNG, which is equivalent to heavy oil with after-treatment (IMO, 2016). Compared with high sulfur heavy fuel oil, methanol has no price advantage. Using methanol fuel only in SO_x ECAs (low sulfur oil is required) can save fuel cost, and the longer the sailing time in ECAs, the greater the possibility of fuel cost savings.

Methanol fuel-powered ships mainly reduce LNG tanks compared with LNG-powered ships, and other increased equipment and costs are close to LNG-powered ships. In comparison, the conversion cost of LNG-powered ships with the same tonnage and type increases by 26% ~ 50% compared with that of methanol-powered ships (Gan et al. 2016). From the perspective of the ship life cycle, the fuel cost, reconstruction cost, operation, and maintenance cost of the whole operation cycle are considered. Compared with diesel fuel, methanol fuel has obvious economic advantages, and the economy of methanol fuel and LNG fuel is close. The price fluctuation of methanol fuel and LNG fuel will have a great impact on the economy, so we should pay attention to the price trend of methanol fuel and LNG fuel in the future.

7　PROBLEMS WITH METHANOL APPLICATION ON SHIPS

According to the analysis of the development of methanol-powered ships at home and abroad, combined with the extensive investigation of the domestic methanol-powered ship industry chain, considering the key technology development level, standards and regulations, regulatory policies, fuel production and transportation, storage and supply and other infrastructure conditions of domestic methanol powered ships, methanol fuel application on ships in China is in the initial stage and there are still some problems as follows:

(1) Matching of high power engine and parts is not mature
 • Domestic medium-and-low-speed high-power engine technology still needs to be improved;
 • There are no alcohol-related special parts for high-power engines;

- Key injection system components are monopolized by foreign technology;
- Engine dynamic response is relatively poor, affecting the maneuverability

(2) Standards and specifications are incomplete and design and operation experience are lacked
  - Domestic statutory regulations have not been issued
  - Lack of standard for methanol as a marine fuel
  - Lack of standards for marine products and special lubricants, etc.

(3) Lack of methanol bunkering stations and bunkering standards
  - The rule for the construction of marine methanol fuel bunkering station
  - Regulatory standards for methanol fuel transportation and supply
  - Operation guidelines for methanol fuel bunkering

8 DEVELOPMENT PATH OF METHANOL FUEL-POWERED SHIPS IN CHINA

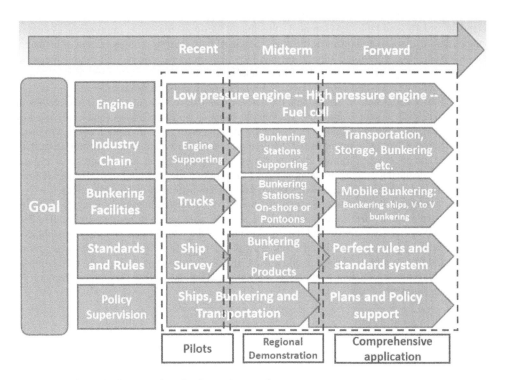

Figure 6. Schematic diagram of application and promotion route.

The application of methanol fuel on ships in China is at an early stage, and a lot of work still needs to be carried out in the application and promotion stage, including technology research and development, pilot projects application, standard formulation, and policy support, etc.

In the near future, the application is mainly focused on carrying out pilot projects for specific ship types to accumulate experience and to develop relevant standards and specifications, which is help to make a good foundation for the promotion of methanol-powered ships.

On the basis of pilot projects, the representative areas with good shipping conditions shall be chosen to carry out the regional demonstration of the integration of methanol fuel powered ships, methanol fuel transportation, and bunkering, so as to form good demonstration benefits and drive the enthusiasm of market promotion. Through regional demonstration, standards and regulations will be further improved, and the introduction of relevant planning and national support policies and measures will be promoted.

On the basis of pilot demonstration, further expand the application of ship types to realize the comprehensive promotion of main ship types; promote the diversification of bunkering models, establish and form a bunkering supply system network; promote the research on the application and promotion of methanol-reforming fuel cell ships to realize the diversified application of methanol fuel on ships and to broaden the application scope.

9 CONCLUSION AND SUGGESTIONS

The application of methanol on Chinese ships has technical and economic feasibility, good emission reduction benefits, and broad development prospects in the future. However, it also faces some bottlenecks. The application and promotion of methanol on Chinese ships is a systematic project, which requires joint efforts of all parties:

First, it is suggested that the government should strengthen the top-level design and guide the scientific, orderly, and rapid development of methanol fuel-powered ships; Improve the system of policies and regulations and promote the development of high-quality standardization of methanol fuel-powered ships; Formulate supporting facilities planning and promote the establishment of supporting service guarantee system for filling supply; Optimize the market environment, improve the level of safety supervision and service quality, and create a good market order.

Second, it is suggested to strengthen the research and development of key technologies at the industrial level, combine industry, University, research and application, and carry out key research on marine engine technology; Master the manufacturing of core equipment and improve the localization of key equipment in the whole industrial chain of methanol transportation, filling, storage and operation; Promote the pilot demonstration application, and comprehensively and orderly promote the integrated pilot demonstration of methanol fuel powered ships, filling facilities construction and operation supervision, so as to form a demonstration driving the effect.

Third, it is suggested that enterprises explore innovative development models, integrate industry resources, explore and form innovative business models with win-win results, and realize cost reduction and efficiency increase and the development of the whole industrial chain; Establish a long-term service system, establish and improve the ship construction, filling and after-sales service network, and form a long-term sustainable service and profitability; Strengthen the cultivation of professionals, build an efficient talent team with strong technical ability, master core technology and enhance competitiveness.

REFERENCES

CCS. *Guidelines for Ships Using Alternative Fuels*[z]. 2017.
FCBI energy. *Methanol as a marine fuel report* 2015[J]. 2015.
IMO. *Methanol as a marine fuel: Environmental benefits, technology readiness, and economic feasibility*[J]. 2016.
Shaowei Gan, Lei Wei, Zhou Guoqiang. Methanol: future marine fuel. *China ship inspection*, 2016(08).
The national standards of GB 13690-1992 General rule for classification and hazard communication of chemicals and GB 6944-2005[Z]. 2005.

*Advances in Energy, Environment and
Chemical Engineering – Abdullah & Osman (Eds)
© 2023 The Author(s), ISBN: 978-1-032-36083-6*

Application of unstable well testing in the late stage of oilfield development

Ma Ning*
No. 2 Oil Production Plant Fifth Operation Area Geotechnical Team, China

ABSTRACT: Unstable well testing is an important method to study reservoir performance in oilfield development. In the later stage of development, unstable well testing is easy to carry out, and the data obtained are more reliable. Using various formation parameters obtained from well testing, it is possible to understand the dynamic changes of the oil reservoir during the development process, determine a reasonable working system for a single well, select a measure well, and provide a basis for quantitative management of oil wells.

1 INTRODUCTION

Unstable well testing is an important method to dynamically study reservoir properties. The formation seepage parameters obtained from the pressure recovery curve can predict the productivity of a single well, determine a reasonable working system, select a measure well, improve the efficiency of the measure, and provide a basis for quantitative management of oil well production.

2 DETERMINING FORMATION PARAMETERS IN OILFIELD DEVELOPMENT

When the fluid flow in the reservoir is in a balanced state, changing the working system of one of the wells, that is, changing the flow rate, will cause a pressure disturbance at the bottom of the well, and this disturbance will expand radially to the strata around the wellbore over time, and finally reach a new equilibrium state. This unstable process of pressure perturbation is related to the properties of the reservoir, the well, and the fluid. Therefore, in this well and other wells, the variation law of the bottom hole pressure with time is measured with instruments, and through analysis, the basic characteristic parameters of the well and the reservoir can be judged and determined, and the nature, distance, and type of the boundary and the type of the reservoir can be obtained. area and reserves.

The pressure recovery curve can reflect the formation seepage parameters, and obtain parameters such as formation flow coefficient, permeability, skin coefficient, average pressure, and boundary distance.

According to the permeability and flow coefficient obtained from the test, the current formation permeability of block XX is: the average flow coefficient of the Gaotaizi oil layer is $62.9 \times 10^{-3} \mu m^2 \cdot m/mPa.s$, and the variation range of permeability is $264.9 \times 10^{-3} \mu m^2 \cdot m/mPa \cdot s$, the average flow coefficient in SD area is $41.8 \times 10^{-3} \mu m^2 \cdot m/m \cdot Pa.s$, the permeability is $316.7 \times 10^{-3} \mu m^2$, and the flow coefficient of N5 block is $102.4 \times 10^{-3} \mu m^2 \cdot m/mPa \cdot s$, the permeability is $461.5 \times 10^{-3} \mu m^2$.

From the above data, it can be seen that the N5 block reservoir has a large flow coefficient and high permeability. The SD area is a set of gray, gray-green sand-mud interbeds, mainly composed of complex cyclic and thin-layered combination of anti-cyclic sediments, belonging to inland lakes Basin fluvial-delta deposition, there are some low permeability reservoirs; C oil layer Anpeng

---

*Corresponding Author: 44499821@qq.com

DOI 10.1201/9781003330165-76

middle-deep reservoirs are deeply buried, the oil layer is tight, and the permeability is low. The test results are consistent with the geological conditions.

3 PREDICT THE PRODUCTIVITY OF A SINGLE WELL AND DETERMINE THE WORKING SYSTEM

From the flow pressure and average formation pressure obtained by the test, as well as the current production, the liquid production index can be obtained, and the production under different pressure differences can be calculated from the liquid production index, so as to determine a reasonable working system. It is also possible to calculate the specific increase in oil volume after liquid extraction, so as to manage the production of oil wells quantitatively. The parameters obtained through testing can be used to obtain the specific production of oil wells by formulas.

The formula for the liquid production index in oilfield development and production is:

$$J = Q_L / (P_R - P_{wf}) \tag{1}$$

Among them:

J: oil well production index $(m^3 / MPa \cdot d)$

Q_L: oil well daily liquid production (m^3 / d)

P_R: formation pressure (MPa)

P_{wf}: flow pressure (MPa)

For example, the well test data of Well NX1 on June 17, 2007 showed that when the formation pressure was 10.85MPa and the production pressure difference was 1.59MPa, the daily fluid production was 35t, and the fluid production index was $22m^3 / MPa \cdot d$. The flow pressure of this well is as high as 9.27MPa, which has the potential to further amplify the differential pressure extraction. When the production pressure differential increases to 3.5MPa, the daily liquid production will increase to 77t. Based on the well-water cut of 93% at that time, the predicted daily production The oil is 5.4t, and the daily increase is 2.9t. However, in the later stage of development, due to the high water content of oil wells, it is easy to cause water cut to rise after liquid extraction. Therefore, it is more cautious when selecting wells. Liquid extraction wells must not only have sufficient energy, but also must have at least two effective directions of water injection, and the water injection on the plane should be balanced. Well NX1 is connected to three water injection wells NXX1, NXX2, and NXX3 on the plane, and all of them can complete the injection well and basically meet the liquid extraction conditions. On August 3, 2007, the well was replaced with a larger pump, and the pump diameter was increased from Φ44mm to Φ70mm, the stroke is increased from 3m to 4.2m. After the pump was replaced, the daily fluid production of the well increased from 30t to 71t, the daily oil production increased from 2.1t to 7.9t, and the water cut decreased from 93.0% to 88.9%.

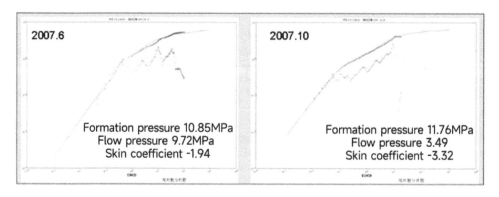

Figure 1. Pressure measurement data before and after pump replacement.

4 SELECTION OF MEASURE WELLS ACCORDING TO FORMATION DYNAMIC SEEPAGE PARAMETERS

The size of the skin factor reflects the degree of contamination of the oil layer near the wellbore. The difference between the bottom-hole flow pressure of the ideal well and the actual well represents the additional pressure loss due to formation damage, stimulation measures in the vicinity of the well, and other flow resistance at the well inlet. This additional pressure loss is usually called the skin pressure drop, while the skin pressure coefficient is defined to be proportional to this pressure drop.

$$S = \frac{Kh}{1.842 \times 10^3 q\mu B} \left(p'_{wf} - p_{wf} \right) \tag{2}$$

Since 2000, 6,448 oil production wells have been tested, of which 597 have skin coefficients greater than zero, accounting for 9.26% of the number of wells tested. 2,607 intervals have been tested in water injection wells, and 387 layers have skin coefficients greater than zero, accounting for 9.26% of the tested intervals. 14.84% are polluted wells, and acidizing and fracturing stimulation measures should be carried out in combination with the characteristics of the reservoir. In actual work, based on the well test results, a reliable basis for measures to tap the potential is provided. For example, on May 13, 2006, the fracturing of the GXX-X well received good results. The daily increase of fluid was 57t, the daily increase of oil was 6.4t, and the water cut decreased by 7.6%. Fractures are used.

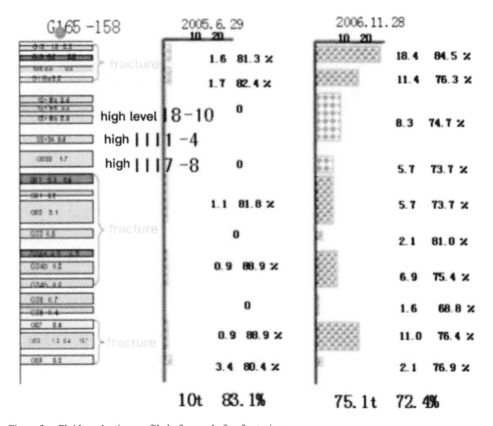

Figure 2. Fluid production profile before and after fracturing.

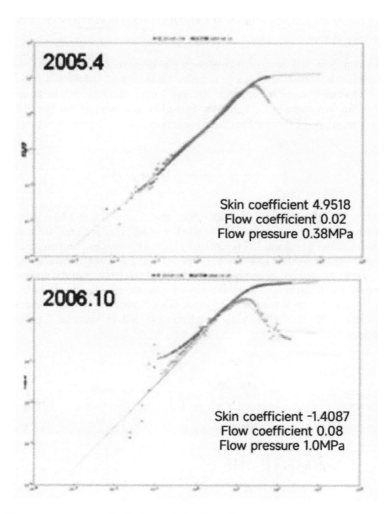

Figure 3.   Pressure measurement data before and after fracturing.

Since 2000, water injection wells have tested 2,607 intervals, and 387 intervals have a skin coefficient greater than zero, accounting for 14.84% of the tested intervals. They are polluted wells and require acidizing and fracturing stimulation measures based on reservoir characteristics.

For example, in September 2005 of Well NX2, the injection pressure was 11.70MPa, the oil allowable pressure difference was 0.43MPa, the daily injection was 65m³, and the daily injection was 32m³. The stratified static pressure data in May 2005 shows that the formation pressure of the A31-B210 section is 13.58MPa, the flow coefficient is $0 \times 10{-}3\mu m^2$.m/mPa.s, and the skin coefficient is 3.98. This section is a polluted section. In October 2005, the A31-A34 and A38-A310 sections of the well were fracturing, with a sandstone thickness of 12.5m and an effective thickness of 3.1m. After fracturing, the injection pressure of the well was 9.78MPa, and the daily injection was 61m³. Compared before and after fracturing, the injection pressure decreased by 1.92MPa, and the daily injection increased by 29m³. Among them, the daily injection in the A31-B210 interval increased by 27m³. The data show that the formation pressure of the A31-B210 interval is 18.31MPa, the flow coefficient is $53 \times 10^{-3} \mu m^2$.m/mPa.s, and the skin coefficient is −3.79.

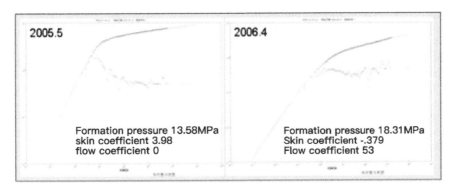

Figure 4.   Pressure measurement data before and after fracturing.

Table 1.   Layered pressure measurement data before and after fracturing in Well NX2.

Test Time	interval	Formation pressure (MPa)	Flow Coefficient $(10^{-3}\mu m^2 . m/mPa.s)$	Skin Coefficient
2005/5/1	A11-A14+5	17.18	51	−2.83
	A24-A213-14	16.71	57	-0.13
	A31-B210	13.58		3.98
2006/4/1	A11-A14+5	18.14	42	-3.34
	A24-A213-14	17.91	52	-3.40
	A31-B210	18.31	53	-3.79

The oil-water well test data can quantitatively reflect the pollution degree of the oil layer near the wellbore, provide a reliable basis for the oil-water well measures to increase the injection and production, and ensure that the oil-water well is effective. Through the implementation of oil and water well measures to increase injection and production, the pollution degree of oil layers in the near-wellbore area has shown a downward trend year by year. Since 2009, 57 production wells have been tested with a skin coefficient greater than zero, accounting for 8.58% of the tested wells. In the layered test of water injection wells, there are 16 intervals with a skin coefficient greater than zero, accounting for 12.62% of the tested intervals. are below the average since 2000.

5   A METHOD OF JUDGING FAULTS USING WELL-TEST DATA

When a fault exists, the pressure double logarithmic curve obtained by the well testing method passes through a straight section with a value of 0.5, and both the pressure curve and the pressure derivative curve appear upturned. According to the actual boundary conditions, determine the stable value of the curve after warping. However, due to the influence of various factors, the characteristics of each stage of the curve are not obvious in the actual situation. According to the derivation of the above formula, when there is a fault, due to the situation of the production well being mapped by the fault, the later curve of the semi-logarithmic curve, that is, after being affected by the fault, the slope of the curve and the slope of the middle straight line of the previous pressure curve become a certain value The specific relationship between the two is determined by the actual fault situation.

For example: Well NX3 is an electric pump well in block N6, about 20 meters away from the No. 187 fault at 827 meters. The stable fluid production before well testing was 160t/d, and the well was shut in for 72 hours for well testing. The obtained pressure data is used to draw double logarithmic and semi-logarithmic curves, the standard chart is used to fit, and the obtained double logarithmic and semi-logarithmic curves are shown in the following two figures. On the double logarithmic curve of the well, it can be seen that the pressure derivative curve does not have a clear peak (the peak of the pressure derivative is mainly affected by the skin coefficient), but after a slow rise for a certain distance, the pressure suddenly drops, and it at a certain pressure boundary may be encountered in the process of pressure transmission. After the curve returns to about 0.5, the curve turns upward, which indicates that the well is also affected by a low permeability layer or a closed fault. From the double logarithmic curve, the starting point of the mid-term curve of the semilogarithmic curve is determined. According to the shape of the curve, the slopes of the middle and late curves are determined, $i_1=1.1879647$, and $i_2=2.2464036$, respectively, which can indicate that there is a closed fault in the displacement area of this well.

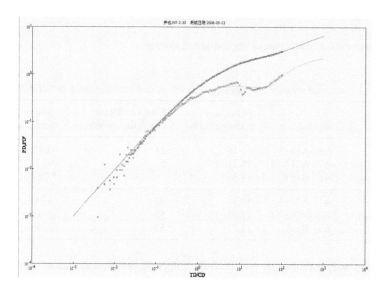

Figure 5.　South 7-2-30 double logarithmic curve.

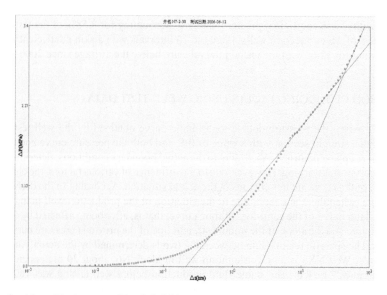

Figure 6.　Semi-logarithmic curve of South 7-2-30.

6 CONCLUSIONS AND UNDERSTANDING

1. In the later stage of oilfield development, the distribution of remaining oil is scattered, and it is more and more difficult to understand and tap potentials. Well testing provides a lot of valuable information for understanding the existing oil reservoirs.
2. Formation parameters are used to predict single well productivity, determine a reasonable working system, and provide a basis for quantitative management of oil well production.
3. The obtained formation parameters and select measure wells are tested to improve the effectiveness of the measures.
4. When there is a closed fault near the oil well, in the double logarithmic curve, the derivative curve of the pressure starts to warp after the end of the 0.5 horizontal line and starts to stabilize when it reaches 1. When there are two or even more faults around the oil well, the ratio of the stable value reached by the curve upturn and the two slopes of the semi-logarithmic curve will also become uncertain. At this point, more qualitative judgments can be made.

REFERENCES

Hu Shuyong, Chen Jun, Ren Dexiong, et al. Well test diagnosis and application of deep reservoir pollution during water injection[J]. *Journal of Southwest Petroleum Institute*, 2002, 24(1):67

Tong Xianzhang. *Application of pressure recovery curve in oil and gas field development* [M]. Beijing: Petroleum Industry Press, 1983

Well Testing Manual Compilation Group. *Well Testing Manual* [M]. Beijing: Petroleum Industry Press. 1992

Xu Xianzhong. *Basis of Petroleum Seepage Mechanics* [M]. Wuhan: China University of Geosciences Press, 1992: 1572163

Advances in Energy, Environment and
Chemical Engineering – Abdullah & Osman (Eds)
© 2023 The Author(s), ISBN: 978-1-032-36083-6

Medium and long-term prediction on power generation cost of renewable energy in China within the context of carbon peaking and carbon neutrality

Qian Zhao*, Qixing Sun & Chao Zhang
State Grid Energy Research Institute Co., Ltd, Beijing, China

ABSTRACT: A revolutionary leap in the development of renewable energy has become a general trend and an important support for realizing the "dual carbon" (carbon peaking and carbon neutrality) goal. The medium- and long-term prediction of renewable energy power generation cost is not only the basis and key to price prediction but also a major challenge. Based on the technical characteristics and cost behaviors of wind and solar power generation, this paper considers the changes in incremental and stock installed structure and subsequently constructs a medium- and long-term prediction model of wind and solar power generation cost within the context of carbon peaking and carbon neutrality. It further forecasts the power generation costs of wind and solar power generation from 2025 to 2060. The results can provide support for revealing the medium- and long-term trend of electricity price, guiding investment in renewable energy power generation, and promoting sound supporting policies.

1 INTRODUCTION

China has been striving to achieve carbon peaking by 2030 and carbon neutrality by 2060, referred to collectively as China's "dual carbon" goal. Building a new power system with renewable energy as a central element will be an important means for China to achieve the "dual carbon" goal, which will lead to considerable developments in China's renewable energy power generation, especially the substantial increase in the installed capacity of wind power and photovoltaic power generation.

The cost prediction of wind and solar power generation is the basis and a key link in price prediction. As such, it is essential to predict the cost of wind and solar power generation to assess the medium- and long-term trends associated with the price of electricity.

Many studies have shown that the learning curve model is suitable for predicting renewable energy generation costs by considering different factors, such as Iniyan (2006) (Iniyan et al. 2006), Collantes (2007) (Collantes 2007), and Kahouli (2011) (Kahouli 2011). Based on reasonable assumptions about the future installed capacity, Yin and Chen (2012) (Yin & Chen 2012) analyzed the changes in future power generation costs based on a learning curve model. Zhang et al. (2013) (Zhang et al. 2013) established a learning curve model for predicting the future trend of photovoltaic power generation cost. Based on the analysis of historical changes in photovoltaic installed capacity and power generation cost, they formulated three scenarios for learning rates and predicted the corresponding photovoltaic power generation cost. Song (2017) (Song & He 2017) developed a two-factor learning curve model and provided feasible suggestions to reduce the wind power generation cost by studying the effects of cumulative installed capacity and investment cost on the cost of wind power generation. What's more, Li et al. (2020) (Li et al. 2019) constructed both single and two-factor learning curve models for the wind power industry to make cost-level predictions

---

*Corresponding Author: zhaoqian@sgeri.sgcc.com.cn

DOI 10.1201/9781003330165-77

for the next five years and concluded that the wind power generation cost gradually decreases with the increase in cumulative installed capacity. However, the learning curve model does not adequately take the change in installed structure into account, which is the main factor that can affect renewable energy generation costs.

Moreover, many other studies have analyzed the cost components of wind and solar power generation. For example, Sun (2010) (Sun 2011) conducted an analysis of the cost of wind power using the hierarchical analysis method from the perspective of total social cost, combined with examples to conduct a sensitivity analysis of relevant factors affecting the cost of wind power. He and Zhao (2019) (He & Zhao 2019) argued that wind power benefits fundamentally depend on the amount of feed-in power and the feed-in tariff. It was concluded that the greatest impact on cost-effectiveness is the feed-in tariff by sensitivity analysis. Chen et al. (2022) (Chen & Xi 2022) analyzed the cost components of offshore wind power and the pricing mechanism of wind power feed-in tariffs in various countries. Sun et al. (2017) (Sun et al. 2017) developed a model for calculating the cost of electricity (LCOE) applicable to photothermal power projects in China and used the model to calculate the level of LCOE for photothermal power projects in China and predict its trend in the next 4 years. Jin (2019) (Jin 2019) pursued the idea of the LCOE to build a wind power cost-calculating model and comprehensively calculate the cost of wind power generation based on the results of wind power load prediction. Zhou et al. (2020) (Zhou & Yuan 2020) conducted a comparative analysis of the cost of coal power generation and photovoltaic power generation and concluded that coal power still has a cost advantage in most regions of China.

To sum up, a review of the existing research reveals that there has yet to be a medium- and long-term prediction model of renewable energy power generation cost constructed within the context of the "dual carbon" goal. Furthermore, the cost learning curve model does not fully consider the main cost influencing factor involved in the change of installed structure, thus making it difficult to adapt to the medium- and long-term power generation cost prediction as it relates to the dual-carbon strategy.

Therefore, this paper focuses on the problem of wind and solar power generation costs medium- and long-term prediction helping to keep track of price movements, guide investment, and promote sound supporting policies on renewable energy power generation.

The rest is organized as follows. Section 2 introduces the prediction model of renewable energy power generation costs. Section 3 gives the prediction results and analysis. Conclusions are given in the end.

2 PREDICTION MODEL OF RENEWABLE ENERGY POWER GENERATION COSTS

2.1 Cost structure and main influencing factors

The wind power discussed in this paper includes onshore and offshore wind power; solar power generation includes photovoltaic and photothermal power generation.

(1) Cost structure

The costs involved in renewable energy power generation projects mainly include construction investment and operation costs. Specifically, the construction investment primarily encompasses engineering costs (unit or component purchase costs, construction, and installation costs), site costs, and other engineering construction costs as well as the budget reserve. The operation costs include material expenses, repair charges, labor wages, welfare expenses, and other expenses.

(2) Cost-influencing factors and trend analysis

The overall cost of power generation for renewable energy projects is mainly affected by factors such as cost reduction, utilization hours, and operation and maintenance costs stemming from technological progress. The average power generation cost is also affected by the power supply structure.

The substantial increase in the scale of wind and solar power generation will promote technological progress, resulting in a decline in investment costs to varying degrees. In 2020, the installed capacity of offshore wind power accounted for about 3% of the total installed capacity of wind power. Estimates suggest that it will reach 20% in 2025 after a 5-year period of rapid development; the installed capacity of photothermal power generation is expected to rise from about 1 million kW in 2020 to about 10 million kW by 2030 and 350 million kW by 2060.

The following is a technical and economic analysis of power generation from various renewable energy sources.

Onshore wind power: At present, the average total investment cost of onshore wind power in China is 7,000–8,000 RMB/kW, of which the cost of the generator set accounts for about 65%, and the cost of installation and site accounts for 35%. During the "14th Five-Year Plan" period (2021–2025), the single unit capacity of wind power will increase from the current 3–4 MW to 6 MW to reduce the cost of installation and supporting equipment, resulting in lower kWh power cost. Over the long run, the conversion efficiency of the turbine impeller can gradually increase from about 0.43 to 0.5, in step with the progress of technology, thus improving the utilization hours of wind power and reducing the kWh cost. However, because the costs of towers and other equipment are connected to commodity prices, the long-term cost decline will be limited.

Offshore wind power: The total investment in offshore wind power in China is 14,000–16,000 RMB/kW, of which the cost of wind turbine units accounts for about 40%, and the cost of installation and supporting equipment accounts for about 60%. At present, the main installed unit capacity is 5–6 MW. During the 14th Five-Year Plan period, offshore wind power will be installed on a large scale, accounting for about 20% of the total installed capacity of wind power. Wind turbines of 10 MW and above are expected to be installed on a large scale. The increase of single unit capacity will reduce the cost of installation and supporting equipment, which will, in turn, reduce the kWh cost of wind power. In the long run, the nearby offshore wind power resources will be limited. Offshore wind power will gradually develop in deep sea areas that are located farther from land, and the technology of "floating" distant offshore wind power stations will gradually achieve a breakthrough, which will prove conducive to increasing utilization hours. However, this development will slow the downward trend of offshore wind power costs to a certain extent.

Photovoltaic power generation: The average total cost of photovoltaic power generation in China is 3,500–4,000 RMB/kW, of which the cost of silicon modules accounts for 45%, and the cost of non-silicon materials, installation, operation, and maintenance accounts for 55%. In the future, the kWh cost will decline mainly due to the reduction in silicon module costs and the improvement of utilization efficiency. It is understood that there is no room for decrease in photovoltaic costs this year due to the rise of commodity prices in 2021. It is estimated that during the 14th Five-Year Plan period, due to the gradual reduction of the cost of crystalline silicon materials, the price of photovoltaic modules will also have a downward trend, and the costs associated with these modules are also expected to decline to some extent. In the long term, through the continuous development of three photovoltaic power generation technology routes, namely crystalline silicon cells, copper-zinc-tin-sulfur thin-film cells, and perovskite cells, energy conversion efficiency for photovoltaic power generation is expected to rise from the current maximum of 24% to more than 30%, which will improve the utilization efficiency for this form of power generation and reduce the kWh cost to a certain extent.

Photothermal power generation: Since 2018, China's large-scale photothermal power stations have been steadily put into operation, and their kWh cost is about 1–1.2 RMB/kWh. At present, photothermal power generation mainly utilizes technical methods, such as linear Fresnel type, trough type, and tower type. Of these methods, linear Fresnel technology is relatively new. It features a simple structure, low construction and operation cost, and high theoretical utilization efficiency. At the end of the 14th Five-Year Plan period, the linear Fresnel type will become the main power supply used for newly added photothermal power generators, bringing the kWh cost down. At present, the equipment localization rate of China's photothermal power generation projects that have been put into operation exceeds 90%, which can enable large-scale application and development. In later stages, the kWh cost will continue to decline due to large-scale development and the increase in utilization hours.

2.2 *Method and model*

Given the impact of the project cost reduction of new units caused by the progress of power generation technology, utilization hours, and power supply structure on cost, one method that could be adopted to calculate the kWh cost of new units is the operation period method. On this basis, the comprehensive kWh cost of new energy can be calculated by comprehensively accounting for the stock and incremental power generation installation structure.

The comprehensive kWh cost of new energy power generation is determined by the weighted average of the new energy power generation power supply stock and new power generation in the forecast year. The newly added generating capacity in the forecast year includes the newly added installed power generation and the electricity reduced due to the decommissioning of existing installed capacity. The kWh cost of the newly added units is calculated by the operation period method; the existing power generation is the remaining power after the decommissioning of the existing installed capacity, and the kWh cost is the on-grid price of new energy power generation after considering subsidies.

Comprehensive kWh cost of new energy power generation in the nth year (considering the electricity price structure) = (the generating capacity of the stock * the kWh cost of the stock + the newly added generating capacity * the newly added kWh cost) in the nth year/(the generating capacity of the stock + the newly added generating capacity) in the nth year (1).

The operation period method is meant to make the net cash flow of self-owned funds of the project in the economic life cycle meet a certain financial Internal Rate of Return (IRR) by investigating the yearly cash flow in the economic life cycle of the project. When calculating, the kWh cost is to be continuously adjusted until the IRR has reached the agreed level.

Cash inflows include sales revenue, fixed asset recovery, working capital recovery, and other cash inflows, whereas cash outflows include capital investment, operating costs, loan repayment amount, taxes, and other expenses.

$$0 = \sum_{n=1}^{N} \frac{I_n(C, Q, S, \cdots) - O_n(V_n, \rho_n, T_n, \cdots)}{(1 + r_e)^n} \tag{1}$$

In the above formula, N is the length of the project operation period. $I_n(C, Q, S, \cdots)$ refers to the cash inflow obtained by the project in the nth year, and the decisive factors include kWh cost C, project annual power consumption Q, residual value S, and other factors. $O_n(V_n, \rho_n, T_n, \cdots)$ refers to the cash outflow of the project in the nth year, and the decisive factors include capital investment and operating cost V_n, loan repayment amount ρ_n, tax T_n, and other factors. r_e is the expected internal rate of return on capital.

3 PREDICTION RESULTS AND ANALYSIS

3.1 *Wind power kWh cost*

The comprehensive kWh cost of wind power is calculated according to the stock of various power sources as well as the structure of incremental generating units and the generation price[1]. From the calculated results, the comprehensive kWh cost of wind power shows a trend increase and then decrease. The comprehensive kWh cost of wind power will rise to 0.416 RMB/kWh in 2025 and 0.347 RMB/kWh in 2060. The wind power cost mainly changed due to technological progress that has enabled lower construction costs, increased utilization hours, and also somewhat lowered the cost of new onshore and offshore wind power. Due to the significant increase in the installed proportion of offshore wind power with high power generation cost, the proportion of offshore wind

---

[1] The subsidy is considered for the stock cost, but the subsidy is not considered for the new cost, resulting in the cost of newly installed kWh in previous years being higher than the stock cost.

power, however, will increase from 3% in 2020 to 34% in 2060, which will drive the comprehensive kWh cost.

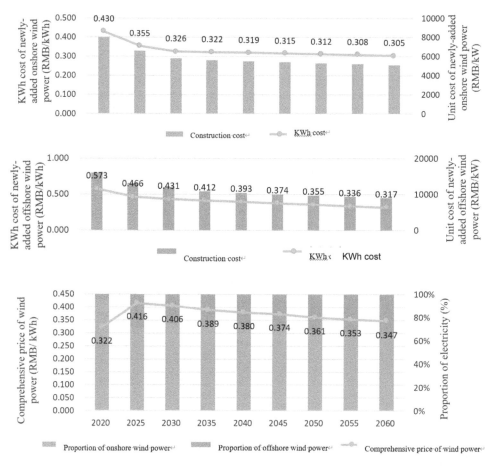

Figure 1.   Prediction results of wind power construction cost and kWh cost.
Note: The construction cost of wind power is determined by the change in trends seen in technological development and installed capacity as well as the comprehensive industry research and prediction of authoritative institutions.

3.2   kWh cost of solar power generation

The comprehensive kWh cost of solar power is calculated according to the stock of various power sources as well as the structure of incremental generating units and the generation price. From the calculation results, the comprehensive cost of solar power generation also shows a trend of an initial rise followed by a fall. The comprehensive kWh cost of solar power generation will rise to 0.349 RMB/kWh by 2030 and 0.263 RMB/kWh by 2060. The cost of solar power has changed mainly because technological progress has allowed for lower construction costs, increased utilization hours, and also slightly lowered the kWh cost of new solar power generation. However, due to the significant increase in the installed proportion of photothermal power generation with high power generation cost, the proportion of photothermal power generation will increase from 1% in 2020 to 34% in 2060, which will drive the comprehensive kWh cost.

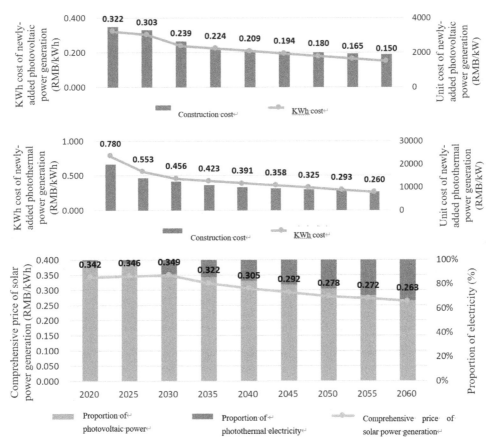

Figure 2. Prediction results of solar power generation cost and kWh cost.
Note: The construction cost of solar power generation is determined by the change in trends seen in techno-logical development and installed capacity as well as the comprehensive industry research and prediction of authoritative institutions.

4 CONCLUSION

Based on the technical characteristics and cost behaviors of wind and solar power generation, this paper analyzed the changes in incremental and stock installed structure and proposed a method and model for calculating the cost of wind and solar power generation within the context of carbon peaking and carbon neutrality. Additionally, it conducted an empirical calculation of various power generation costs of wind and solar power generation.

The results of this research indicate that, on the one hand, due to the reduction of the cost of new units and the increase of utilization hours caused by technological progress, the kWh cost of renewable energy power generation has decreased to various degrees; on the other hand, the comprehensive cost of renewable energy power generation has risen due to the increase in the installed proportion of high-cost renewable energy power generation and the decline of subsidies. Generally, the comprehensive kWh cost of wind and solar power generation in China generally will likely see an initial rise and then fall. Specifically, the costs of wind and solar power will experience an inflection point in 2025 and 2030, respectively.

According to the report on *Renewable Power Generation Costs in 2020* released by IRANA, onshore wind power and solar photovoltaic power generation are still weakening the market share of even the cheapest new coal power generation without any financial support. In such a market environment, the results of this paper can serve as a reference for studying and determining the trend of electricity prices, guiding investment in renewable energy power generation, and promoting sound price policies.

ACKNOWLEDGMENTS

This research is supported by the Science and Technology Project of State Grid Corporation of China "Research of key models for energy Internet enterprise decision support" (Grant No. SGSDJY00GPJS1900057).

REFERENCES

Chen H, Xi S. (2022). Offshore wind power cost components and the price mechanism. *Wind Energy*, 1: 12–15.
Collantes G O. (2007). Incorporation stakeholders' perspectives into models of new technology diffusion: The case of fuel-cell vehicles. *Technological Forecasting and Social Change* 74(3): 267–280.
He T, Zhao X. (2019). Cost-benefit Analysis of Wind Power Generation—Taking Zhangjiakou as an Example. *Sino-Global Energy*, 24(05): 21–25.
Iniyan S, Suganthi L, Samuel A. (2006). Energy models for commercial energy prediction and substitution of renewable energy sources. *Energy Policy*, 34(17): 2640–2653.
Jin C. *The research on wind power cost forecast and pricing game model*. North China Electric Power University, 2019.
Kahouli S. (2011). Effects of technological learning and uranium price on nuclear cost: Preliminary insights from multiple factors learning curve and uranium market modeling. *Energy Economics*, 33: 840–852.
Li Z, Qin Z, Wang H, et al. (2019). Research on the cost level and influencing factors of wind power in China. Price: *Theory & Practice*, 10: 24–29+166.
Song D, He Y. (2017). Study on the cost of wind generation based on the double-factors learning curve. *Northeast Electric Power Technology*, 38(09): 1–3.
Sun K, Quan P, Tang Y, et al. (2017). LCOE modeling and key impact factor analysis. Solar Energy, 10: 21–23+27.
Sun L. *Evaluation model of the economic costs of wind power business and its application*. North China Electric Power University (Beijing), 2011.
Yin X, Chen W. (2012). Cost of carbon capture and storage and renewable energy generation based on the learning curve method. *Journal of Tsinghua University* (Science and Technology) 52(02): 243–248.
Zhang W, Liu R, Liu J, Pan W, Chen T. (2013). Probe into Cost and Trend Forecast Model of Photovoltaic Power Generation Based on Multiple Factors Analysis, *Smart Power*, 41(11): 17–20.
Zhou Y, Yuan H. (2020). Economic Evaluation of Coal Power Generation and Photovoltaic Power Generation Technologies in China. *Journal of Technical Economics & Management*, 12: 97–102.

Advances in Energy, Environment and
Chemical Engineering – Abdullah & Osman (Eds)
© 2023 The Author(s), ISBN: 978-1-032-36083-6

Solid-state preparation and electrical properties of the marcasite type compound $FeTe_{2-x}Bi_x$

Zhiling Bai, Yonghua Ji, Sha Wang & Bingke Qin*
Liupanshui Normal University, Liupanshui, Guizhou, China

Yong Li
School of Data Science, Tongren University, Tongren, China

ABSTRACT: Thermoelectric material is a new type of energy conversion material that can use temperature differences to generate electricity, and the research on thermoelectric material has important significance in the energy crisis today. In this paper, the compounds $FeTe_{2-x}Bi_x$ (x=0.05, 0.1, 0.3, 0.5, 0.7) were prepared by the solid-state reaction method, the preparation conditions with the temperature interval of 653~863 K, and the holding time within 30 min. The physical structure and microstructure of $FeTe_{2-x}Bi_x$ were determined by XRD and SEM. The results show that the main substances synthesized by solid-state reaction for $FeTe_{2-x}Bi_x$ are Bi_2Te_3 and $FeTe_2$, and the sample crystals were found to have a lamellar structure by SEM testing. With the increase of Bi doping coefficient, the power factor shows an increasing trend. The power factor of the components with large doping coefficients decreased with the increase of the synthesis temperature, while the power factor of the components with low doping coefficients did not change significantly. When the preparation temperature was 653 K, the sample $FeTe_{1.3}Bi_{0.7}$ obtained the highest absolute Seebeck coefficient $\sim$145 μV/K at room temperature, when the preparation temperature was 833 K, the sample $FeTe_{1.5}Bi_{0.5}$ obtained the lowest resistivity of about 1.4 mΩ·cm at room temperature, and when the preparation temperature was 693 K the sample $FeTe_{1.5}Bi_{0.5}$ obtained a maximum power factor of about 417.88 μW·m^{-1}K^{-2}.

1 INTRODUCTION

The ordinary marcasite is a mineral widely found in nature and is homogeneous with pyrite, which can be transformed into pyrite at high temperatures. $FeTe_2$ belongs to the albite structure, an orthorhombic crystal system with high symmetry, the lattice constants a=5.26Å, b=6.27Å, c=3.87Å. Due to the different atomic radii of the metal elements, the crystal structure differs when forming compounds with the sulfur group element Te (Huang et al. 2017; Kishimoto et al. 2006; Lutz & Miiller 1991; Sales et al. 1996; Takashi 1998). The crystal structure of the marcasite type compound $FeTe_2$ has a narrow-forbidden bandwidth, thus leading to the special properties of the material of this system in electrical, optical, magnetic and superconductivity, etc (Cao et al. 2017; Kishimoto et al. 2006; Zhang et al. 2001). The corresponding research is being carried out by the scholars of materials research related to these fields in recent years.

The marcasite type compound $FeTe_2$ with its excellent physicochemical properties has often been studied in recent years for its potential use in hydrolytic electrocatalysts, lithium-ion batteries, electromagnetism, superconductivity, electrode materials, biology, solar cells, and supercapacitors. It has been shown that $FeTe_2$ has a narrow band gap and exhibits three-dimensional magnetic ordering and semiconductivity, and $FeTe_2$ prepared at high pressure has a low resistivity (Di et al. 2017; Jia et al. 2020; Parthasarathy et al. 2013; Messaoud et al. 2013; Takashi 1998).

---

*Corresponding Author: qinbingke@126.com

With the development of today's world, energy and environmental issues have become one of the biggest challenges facing mankind. Then nowadays a new energy material, thermoelectric material, which can convert heat and electricity to each other, has become one of the hot spots of current research by material experts (Guo et al. 2021; Hosayn et al. 2022; Qin et al. 2019; Takako et al. 2014; Yang et al. 2021). The high-temperature solid-state reaction method is a fast and efficient way to prepare materials, and the experimental process is relatively simple. Firstly, the reaction materials are mixed thoroughly, the samples are pressed and shaped, and then sintered at high temperature to make them fully react and form compounds. In this experiment, Bi-doped $FeTe_2$ compounds are prepared by the solid-state reaction method, which has a low resistivity and a low thermal conductivity among many thermoelectric materials. In this paper, Bi was selected for the doping of $FeTe_2$ to investigate the effect of preparation conditions and doping coefficient on the thermoelectric properties of the compound $FeTe_2$.

2 EXPERIMENT

The experimental raw materials are Fe, Bi, and Te metal powders with 99.99% purity and 300 mesh size. The raw materials were mixed by ball milling for 3h under nitrogen atmosphere protection. After homogeneous mixing, the raw materials were cold pressed into cylinders to be sintered, which were then sealed in sintering molds and placed in a vacuum furnace for solid-state reactions under different preparation conditions. Then sealed in a sintering mold and placed in a vacuum furnace for solid-state sintering. The compounds $FeTe_{2-x}Bi_x$ (x = 0.05, 0.1, 0.3, 0.5, 0.7) were prepared by the solid-state reaction method at the temperature range of 653~863 K, and the holding time within 30 min. The Seebeck coefficients and resistivity of t $FeTe_{2-x}Bi_x$ were tested and analyzed near room temperature. The vacuum furnace selected model is HMZ-1700-20. Phases of the synthesized samples were detected by X-Ray diffraction (TD, 2500). The microscopic morphology of the sample was observed with a scanning electron microscope (SEM, ZEISS Gemini 300). The resistivity of the samples at room temperature was tested using an RTS-9 dual electrodynamic four-probe tester. the Seebeck coefficient was tested using a homemade calibrated Seebeck tester, and the temperature difference electric potential was measured by maintaining a temperature difference of 8-10 K at both ends of the sample to calculate the Seebeck coefficient of the sample with a measurement error of ±5%. The power factor of the sample was obtained by the formula $PF = S^2/\rho$.

3 RESULTS AND DISCUSSIONS

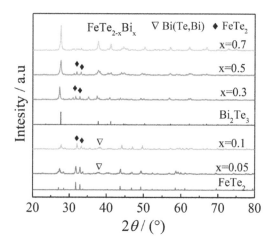

Figure 1.   XRD patterns of $FeTe_{2-x}Bi_x$.

Figure 1 shows the XRD spectra of $FeTe_{2-x}Bi_x$. It can be observed that for Bi doping amount less than 0.3, the synthesized phase is mainly $FeTe_2$, and the XRD spectra of the prepared $FeTe_2$ samples show peaks of Bi_2Te_3 compared to the standard spectrum of $FeTe_2$. The XRD spectra of the samples were compared with the standard spectra of Bi_2Te_3, and $FeTe_2$ became the impurity peak. The samples synthesized with Bi doping amounts between 0.05 and 0.7 had both Bi_2Te_3 and $FeTe_2$ phases, so the synthesized samples were multi-phase composite thermoelectric materials. It can be seen that the peak shape was sharper when the doping amount was 0.7, indicating that the samples were better crystalline.

Figure 2 shows the SEM of the internal section of the sample $FeTe_{1.5}Bi_{0.5}$. From the SEM photographs, it can be seen that there are many holes inside the synthesized sample, and the reason for the holes is due to the solid-state reaction. The sample is mainly synthesized from two different elements, which have a large difference in their melting points, making the partial volatilization of the tellurium element of the sample during high-temperature sintering cause more holes. The grain boundaries of the sample are not obvious and the crystallinity of the grains is poor, and these characteristics are consistent with the test results of XRD.

Figure 2.   SEM images of the internal section of sample $FeTe_{1.5}Bi_{0.5}$.

Figure 3(a) shows the resistivity curve of the sample $FeTe_{2-x}Bi_x$ with the preparation temperature. It can be seen from the figure that the resistivity of the sample decreases with increasing Bi doping and then increases. As the preparation temperature increases, the resistivity of the sample with the same amount of Bi doping tends to decrease. The minimum resistivity was obtained for a doping amount of 0.5 in all samples, and the lowest resistivity of 1.4 $m\Omega\cdot cm$ was obtained for sample $FeTe_{1.5}Bi_{0.5}$ when the preparation temperature was 833 K.

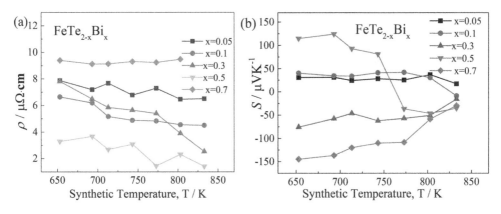

Figure 3.   Temperature dependence of resistivity (a), Seebeck coefficient (b) for $FeTe_{2-x}Bi_x$.

The Seebeck coefficients of the prepared samples $FeTe_{2-x}Bi_x$ were tested at room temperature, and the measured data were plotted to obtain a graph of the relationship between the Seebeck coefficients and the preparation temperature for different Bi doping amounts as shown in Figure 3(b). From Figure 3(b), it can be seen that when the Bi doping amount <0.5, the Seebeck coefficient of the prepared sample is positive, indicating that the semiconductor type of the sample is p-type. At this time, the Seebeck coefficient of the sample does not change significantly with the preparation temperature. When the Bi doping amount is <0.5, the Seebeck coefficient of the prepared sample is negative, indicating that the semiconductor type of the sample is n-type. The absolute value of the Seebeck coefficient of the sample decreases with the increase of the preparation temperature; when the Bi doping amount~0.5, the Seebeck coefficient of the sample gradually changes from a positive value at the beginning to a negative value with the increase of the preparation temperature.

This phenomenon is mainly due to the narrow band gap of the $FeTe_2$ energy band, and the variation of the Bi doping amount as well as the synthesis temperature can easily affect the type of its carriers. The largest absolute value of the Seebeck coefficient 145 $\mu V/K$ was obtained for the sample $FeTe_{1.3}Bi_{0.7}$ when the preparation temperature was 653 K. This value represents an improvement of about 64% compared to the maximum value of the Seebeck coefficient in the literature [14].

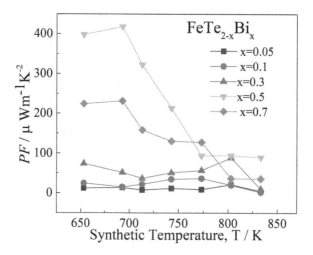

Figure 4.   Temperature dependence of the power factor for $FeTe_{2-x}Bi_x$.

The power factor of the samples is shown in Figure 4. The power factor is obtained from the Seebeck coefficient and resistivity calculation. As can be seen from the figure that at the same preparation temperature, the power factor tends to increase with the increase of Bi doping. The power factor of the samples with high Bi doping tends to decrease with the increase of the synthesis temperature. This situation is mainly due to the fact that the material phase in the sample is dominated by Bi_2Te_3 at a higher preparation temperature, which increases the internal carrier concentration of the sample and leads to a decrease in the absolute value of the Seebeck coefficient and resistivity. The highest power factor of 417.88 $\mu W \cdot m^{-1} K^{-2}$ was obtained at a Bi doping level of 0.5.

4   CONCLUSION

Based on the results and discussions presented above, the conclusions are obtained as: Multi-phase composite thermoelectric compound $FeTe_{2-x}Bi_x$ was prepared by solid-state reaction method, and the composition of the synthesized phase is related to the Bi doping amount, the main phase is

FeTe$_2$ when the Bi doping amount is small (≤ 0.5) and Bi$_2$Te$_3$ when the Bi doping amount is large (>0.5). The minimum resistivity of 1.4 mΩ·cm was obtained for Bi doping of 0.5, and the absolute maximum Seebeck coefficient of 145 μV/K was obtained for Bi doping of 0.7. The maximum power of 417.88 μW·m^{-1}K^{-2} was obtained for FeTe$_{1.5}$Bi$_{0.5}$ at a preparation temperature of 693 K.

ACKNOWLEDGMENTS

This work was financially supported by Liupanshui Normal UniversityNational Natural Science Foundation (LPSSYKYJJ201902), the Science and Technology Innovation Group of Liupanshui Normal University (LPSSYKJTD201905), the Liupanshui Key Laboratory of thermoelectric and electrode materials (52020-2020-0903) and the National Natural Science Foundation of China (52062031, 12064038).

REFERENCES

H. D. Lutz, B. Miiller. Lattice vibration spectra. LXVIII. Single-crystal Raman spectra of marcasite-type iron chalcogenides and pnictides, FeX$_2$ (X=S, Se, Te; P, As, Sb) [J]. *Physics and Chemistry of Minerals*, 1991, 18: 265–268.

Hosayn Chibani, Hakima Yahi, ChaoukiOuettar. Tuning the magnetic properties of FeTe$_2$ monolayer doped by (TM: V, Mn, and Co) [J]. *Journal of Magnetism and Magnetic Materials*, 2022, 522(15): 169204.

Jiaxin Di, Hongtao Li, Guiying Xu. Thermoelectric Properties of Fe$_{1+y}$ Te Prepared by a High-Pressure Sintering[J]. *Method Journal of Electronic Materials*, 2017, 46: 2969–2975.

K Kishimoto, K Kondo, T Koyanagi. Preparation and thermoelectric properties of sintered Fe$_{1-x}$Co$_x$Te$_2$ (0$\leq$x$\leq$0.4) [J]. *Journal of Applied Physics*, 2006, 100 (9):57–87.

K. Ben Messaoud, J. Ouerfelli, K. Boubaker, et al. Structural properties of FeTe$_2$ thin films synthesized by cellularization of amorphous iron oxide thin films [J]. *Materials Science in Semiconductor Processing*, 2013, 16(6): 1912–1917.

Parthasarathy, G; Sharma, D.K. Sharma, et al. High-pressure Electrical Resistivity Studies on FeSe$_2$ and FeTe$_2$. *Solid State Physics*[J], 2013, (1512): 40–41.

Qing-Chao Jia, Hu-Jun Zhang, Ling-Bin Kong. Nanostructure-modified in-situ synthesis of nitrogen-doped porous carbon microspheres (NPCM) loaded with FeTe$_2$ nanocrystals and NPCM as superior anodes to construct high-performance lithium-ion capacitors[J]. *Electrochimica Acta*, 2020, 337(20): 135749.

Qin Bingke, Ji Yonghua, Su Taichao, et al. Rapid Preparation and Electrical Transport Properties of Compound FeTe$_2$[J]. *Rare Metal Materials and Engineering*, 2019, 48(03): 931–935.

Sales B. C., Mandrus D., and Williams R. K. Filled Skutterudite Antimonides: A new class of thermoelectric materials [J]. *Science*, 1996, 272(5) : 1325–1328.

Takashi Harada. Transport Properties of Iron Dichalcogenides FeX_2 (X=S, Se and Te) [J]. *Journal of Physical Society of Japan*, 1998, 67(4): 1352–1358.

Takako Kikegawa, Kazuki Sato, Keisuke Ishikawa. Thermal Conductivity and Electrical Resistivity of FeTe$_{1-x}$S$_x$ Sintered Samples [J]. *Physics Procedia*, 2014, (58): 86–89.

Weixin Zhang, Youwei Cheng, Jinhua Zhan, et al. Synthesis of nanocrystalline marcasite iron ditelluride FeTe$_2$ in aqueous solution [J]. *Materials Science and Engineering*, 2001, 79(3): 244–246.

X. L. Cao, W. Cai, H. D. Deng, et al. Preparation and Thermoelectric Properties of Pb$_{1-x}$ Fe$_x$Te Alloys Doped with Iodine[J]. *Journal of Electronic Materials*, 2017,(46)5: 2645–2651.

Yang Jina, Tao Hong, Dongyang Wang, et al. Band structure and microstructure modulations enable high-quality factors to elevate thermoelectric performance in Ge$_{0.9}$Sb$_{0.1}$Te-x%FeTe$_2$[J]. *Materials Today Physics*, 2021, 20:100444.

Yali Guo, Yonghua Cheng, Qingqing Li, et al. FeTe$_2$ as an earth-abundant metal telluride catalyst for electrocatalytic nitrogen fixation[J]. *Journal of Energy Chemistry*, 2021, 56:259–263.

Zhiwei Huang, Samuel A. Miller, Binghui Ge, et al.High thermoelectric performance of new rhombohedral phase of GeSe stabilized via alloying with AgSbSe$_2$[J]. *Energy.Environ* 2017, 56(45): 14113–14118.

*Advances in Energy, Environment and
Chemical Engineering – Abdullah & Osman (Eds)*
© 2023 The Author(s), ISBN: 978-1-032-36083-6

Analysis of tight oil accumulation characteristics and resource potential of Fuyu reservoir in the Shuangcheng area

Zheng Jiang*

Daqing Oilfield Co. Ltd., Daqing, Heilongjiang, China

ABSTRACT: The tight oil in the Fuyu reservoir in northern Songliao Basin is a realistic replacement field for the increase of reservoir and production in the Daqing oilfield, which has important strategic significance. In this paper, the accumulation conditions of the Fuyu reservoir in the Shuangcheng area are systematically studied. It is concluded that the dark mudstone in the Qingyi formation is the material basis for the formation of tight oil because of its wide distribution, large thickness, and good hydrocarbon generation index. The reservoir is mainly composed of siltstone formed by a fluvial and shallow delta sedimentary system, with an average porosity of 9% and permeability of 0.9mD, which is a typical tight reservoir. The top faults of the Fuyu reservoir are well developed, and the oil and gas are distributed under the overpressure of the Qingyi formation. The distribution of oil and gas is controlled by the distribution system matched by fault and sand. The Fuyu reservoir in the Shuangcheng area has great potential for tight oil resources, and the volume method is used to calculate the resource amount of 0.25 million tons. Through comprehensive analysis of reservoir formation, favorable areas such as the northwest Schuangcheng depression are further selected, which provides resources and target support for tight oil exploration and deployment.

1 INTRODUCTION

After more than 60 years of exploration and development, Daqing Oilfield is faced with the problems of rising exploration costs and insufficient replacement of effective resources. Tight oil resource is a realistic replacement field for the increase of reservoir and production of Daqing oilfield, which has important strategic significance. The tight oil resources in the Fuyu oil layer in the Shuangcheng area have good exploration prospects. The study area has shown good potential in the past exploration practice, but with the improvement of exploration degree, the exploration difficulty also increases, the control effect of structure on oil and gas accumulation needs to be recognized, the control factors of oil and water plane are not clear, and the law of oil and gas accumulation is not clear (Chen et al. 2008; Fu et al 2016; Yang et al. 2007). In this paper, the favorable zones for further exploration are predicted based on the accumulation model and reservoir type in the study area.

2 GEOCHEMICAL CHARACTERISTICS OF SOURCE ROCKS

The interbedding of mudstone and oil shale in the Qingyi Formation in the Shuangcheng depression is the main source rock of the Fuyu reservoir in the study area, which is widely distributed, thick, with a good hydrocarbon generation index and high maturity. The thickness is between 90m and 110m.

---

*Corresponding Author: 598639824@qq.com

                                     DOI 10.1201/9781003330165-79

2.1 Organic matter abundance

The geochemical characteristics of 47 sample points from 35 wells drilled in the Qingshankou formation in the Shuangcheng depression were analyzed. The total organic carbon content (TOC) of source rocks in the Qingyi formation ranges from 2% to 6.22%, with an average of 3.87%. According to the analysis of hydrocarbon generation potential ($S_1 + S_2$), the value of it in source rocks in the Qingyi formation is $(9.24–45.35) \times 10^{-3}$, with an average value of 23.07×10^{-3}. Analysis of chloroform bitumen "A" shows that the content of chloroform bitumen "A" in the source rocks of Qingyi formation ranges from 0.01 to 1.7, with an average value of 0.61. According to the evaluation criteria of continental facies source rocks proposed by Huang Di-fan et al., the organic matter abundance of source rocks in the Qingyi formation is high, reaching the standard of good quality source rocks (Chen et al 2013; Fu & Pang 2008; Liu 2009; Lu et al. 2010).

2.2 Types of organic matter

According to the organic elemental analysis for statistical data of 35 wells in Shuangcheng depression we made a relation graph of the atomic ratio between O/C and H/C in the organic matter of source rocks in Qingyi formation, and confirm that the type of organic matter in source rocks are mainly II-1 type. In terms of the accumulation of oil and gas, the abundance of organic matter in source rocks in the Qingyi formation is high and reaches the standard of high-quality hydrocarbon source rocks.

2.3 Maturity of organic matter

The vitrinite reflectance (Ro) is the main reference index of organic matter maturity. This paper analyzes the Ro values of 35 wells in Shuangcheng depression, and the Ro values of Qingyi formation range from 0.41% to 0.87% with an average value of 0.61%, which is in the immature to low maturity evolution stage. The relation of the atomic ratio between H/C and O/C also indicates that the source rocks of the Qingyi Formation in the Shuangcheng depression are in a lower stage of thermal evolution.

Table 1. Geochemical characteristics of source rocks in Shuangcheng depression.

Structure zone	Toc(%)	S1(mg/g)	S2(mg/g)	Chloroform bitumen"A"(%)	R_o(%)
Shuangcheng depression	2–6.22 3.87	0.18–6.8 2.56	9.06–38.55 20.51	0.01–1.7 0.38	0.41–0.87 0.61

3 RESERVOIR CHARACTERISTICS

The reservoirs in the study area are mainly controlled by the south-north provenance system and are mainly sandbodies deposited in shallow water delta on water and underwater distributary channels. The lithology is mainly siltstone and fine sandstone, with fine grain size and good sorting. The reservoir sandbodies change quickly in both transverse and vertical directions, and the reservoir connectivity is poor. There are 6–10 layers in the Fuyu reservoir, and the thickness of a single sand layer is between 0.8 and 3.4m.

3.1 Reservoir lithology

The sand body type of the Fuyu reservoir in Shuangcheng depression is mainly distributary channel sand, and the rock type is mainly feldspar lithic sandstone and lithic feldspar sandstone, which are close to the provenance. According to the thin section identification data and scanning electron microscope analysis data, there are five types of pore roar in the study area, and the diagenesis is strong.

3.2 *Reservoir physical properties*

According to the reservoir data statistics of 35 wells in Shuangcheng depression, the main distribution range of porosity is 7%–19%, the average porosity is 9%, the main distribution range of permeability is 0.1–3mD, and the average permeability is 0.9MD. Fuyu oil reservoir in the study area belongs to medium and low porosity and low permeability reservoir. Due to the large burial depth, the overall physical properties of the reservoir area in the depression and slope area become worse, and the tight reservoir is relatively developed. Under the influence of burial depth, the physical properties of reservoirs on the north slope of Shuangcheng depression are better, while those in the depression and the south slope become worse.

Most of the porosity in the study area is less than 12.5%, and the average permeability is less than 1mD. The reservoirs with porosity greater than 8% are mainly located on the north slope of the depression, and the porosity of less than 8% is mainly in the center and surrounding of the depression.

4 STRUCTURAL CHARACTERISTICS

4.1 *Fault distribution*

Shuangcheng depression is adjacent to Changchunling anticline in the northwest, and adjacent to Qingshankou anticline in the southeast, forming a ring syncline with a long axis of NNE, steep northwest wing, and slow southeast wing. The study area is high in the southeast and low in the northwest, with a two-uplift and one-depression tectonic pattern as a whole. The structure amplitude varies greatly, with the elevation of the high part -500m and the elevation of the low part -1800m, with a structure elevation difference of 1300m.

The top surface of the Fuyu reservoir in the study area is full of faults, all of which are normal faults. The distribution direction of faults can be roughly divided into north-east and south-north. The faults in the high part of the structure are densely distributed, while the faults in the low part of the structure are sparse, but the scale of faults is large. Among them, the nearly north-south faults are the most developed and have the most number, and have a skew relationship with the structure strike. In the longitudinal direction, the faults are mainly distributed in the T2 layer, a few of them break through the T1 layer, and some of them break through the T3, T4, and T5 layers. In the study area, the NNW and NWW trending faults are the main faults, and the NE and Near East trending minor faults are derived. The maximum length of fracture extension is 17km, generally 3–5km; The maximum vertical fault distance is 120m, generally 20~60m.

4.2 *Fault period*

Early developed faults, cut through the T2 layer, upward disappeared inside the Qingshankou formation, faults of this period are characterized by fewer cut formations and small slips, but the number of faults is large. This period is the main developmental period of faults. The faults are mainly controlled by basement faults and distributed in dense strips. These faults were mainly formed during the structure movement at the end of the Qingyi formation. Under the action of the regional extensional stress field, in order to adjust the extension of the basement fault, the fault occurred in the late Qingyi formation, the oil and gas is not mature, and the main hydrocarbon expulsion period occurred in the late period of the Nenjiang formation, MingShui formation end and the end in Paleogene, and a large number of T2 faults in oil and gas migration period resurrect and open. It is the most favorable channel for downward migration of oil and gas in the Qingyi

Formation, and the good disposition of distributary channel sand bodies of the Fuyu reservoir between faults is conducive to the formation of various traps related to faults. The widely developed T2 layer faults create necessary conditions for the formation of the Fuyu reservoir in the study area (Wei et al 2006; Wang et al 2009).

5 ACCUMULATION CHARACTERISTICS

5.1 *Reservoir type*

Oil and gas show a lot in the Shuangcheng depression, but its distribution is complex, reservoir types are mainly structure-lithologic reservoirs and lithologic reservoirs. Structure-lithologic reservoirs are mainly distributed on the slope around the Shuangcheng depression and in the densely fractured area. Each oil layer has a relatively large distribution area and is mainly controlled by faults, lithology, and updip pinching out of sandstone. Lithologic reservoirs are mainly distributed in the syncline area.

5.2 *Accumulation mode*

The analysis of accumulation conditions shows that the oil and gas of the Fuyu reservoir in the study area mainly come from the source rock of the overlying Qingyi Formation. The caprock is also the mudstone of the Qingyi Formation. The accumulation mode belongs to a typical apical reservoir from the point of view of the combination types of source, reservoir, and caprock.

The Qingyi formation source rocks overlying on Fuyu reservoir have good hydrocarbon generation and expulsion capacity, reaching the threshold of hydrocarbon expulsion at the end of the Mingshui Formation, and the overpressure of the Qingyi formation provides the driving force for hydrocarbon accumulation in the Fuyu reservoir. And the period of the fault activity and hydrocarbon expulsion phase matching the north-south fault communicate the Qingshankou formation hydrocarbon source rocks and Fuyu reservoir sandstone reservoir, which provides conditions for the migration and accumulation of mature oil and gas from the depression center to the surrounding areas, distributary channel sand bodies provide good reservoir space for oil and gas gathering, a large number of faults formed before hydrocarbon expulsion period are favorable for the accumulation and preservation of oil and gas, different combinations of faults and distributary channel sand bodies formed various types of traps and provide good storage space for oil and gas. Most faults in the Shuangcheng depression do not break through the Qingshankou formation upwards, that is, faults do not destroy the sealing effect of the regional cap layer, which enables oil and gas to accumulate and preserve in the Fuyu reservoir of Quantou formation.

6 RESOURCE POTENTIAL

6.1 *Calculation method*

The resource quantity is calculated by the volume method, and the calculation formula is as follows:

$$N = 100A_o h \phi \rho_o S_{oi} / B_{oi}$$

N–Resource($\times 10^4$t);
A_o–Oil-bearing area(km^2);
H–Effective thickness(m);
Φ–Effective porosity(%);
ρ_o–Oil density(t/m^3);
S_{oi}–Oil saturation;
B_{oi}–Volume coefficient.

6.2 Selection of calculation parameters and results

Through the reexamination of 35 wells in the Fuyu reservoir in the study area, the accumulative oil layer thickness is relatively thick, generally 2.2–19.8m, with an average of 10.7m. Core and logging data show that the effective porosity of the Fuyu reservoir is generally 7%-19% and the average porosity is 9%. Due to the lack of saturation data, the oil saturation of the reservoir mainly refers to the value of the reserve area of the adjacent oil field. The crude oil density is $0.850t/m^3$, the oil saturation is 54%, and the volume coefficient is 1.093. The volume method was used to calculate the resource potential of $0.25 \times 10^8 t$.

6.3 Favorable area optimization

The horizontal distribution of oil and gas is mainly controlled by the combination of oil source, sand body, and structure. Current oil and gas exploration practice shows that the oil and gas in the northwest slope is more concentrated, and the exploration degree is higher, while the oil and gas in the center of depression and other areas are more scattered, with low yield, and irregular and sporadic distribution of oil and dry layers. The reservoir thickness on the north slope of the depression is large and the physical property is good, followed by the south slope, and the area and thickness of the east and west sides of the depression are small and thin. The favorable reservoirs in the Shuangcheng depression are mainly developed on the north slope of the depression, followed by the south slope, and the development area on the east and west sides is relatively small. The tight reservoirs are mainly developed in the center and adjacent areas of the depression. It is predicted that the favorable hydrocarbon accumulation areas are mainly distributed on the north slope of the Shuangcheng depression and the eastern area adjacent to the depression center.

7 CONCLUSION

In this paper, the geochemical characteristics, reservoir characteristics, structural characteristics, reservoir types, and accumulation modes of tight oil in the Shuangcheng depression are systematically analyzed, and favorable areas are selected. The results show that the source rocks of Qingyi formation have high maturity and high abundance and good type, which provide a material foundation for the target layer. The porosity of the reservoir is 9% on average, and the permeability is 0.9mD on average. Affected by the buried depth, the physical properties around the depression are good, but the physical properties in the center of the depression are poor. The top surface of the Fuyu reservoir is densely fractured, and most faults are north-south trending, which is the channel of oil and gas migration. The reservoir types are mostly structure-lithologic reservoirs and lithologic reservoirs. The predicted resource potential is $0.25 \times 10^8 t$, and the north slope of the Shuangcheng depression is the favorable area.

REFERENCES

Chen Fang-wen, Lu Shuang-fang, Li Ji-jun, et al. Oil-source identification and resource evaluation of the oil-source area of Changchunling anticline in Northern Songliao Basin[J]. *Journal of the China University of Petroleum*, 2013, 36(3):26–31.

Chen Zhao-nian, Wang Xiao-min, Chen Shan, et al. Structural evolution of Fuyu reservoir in the Chaochang Area of Songliao Basin[J]. *Geoscience*, 2008, 22(4):512–519.

Fu li, Liang Jiang-ping, Bai Xue-feng, et al. Resources evaluations of the tight oil in Fuyu reservoirs of north Songliao basin[J]. *Petroleum Geology and Oilfield Development in Daqing*, 2016, 35(3):168–174.

Fu Xiu-li, Pang Xiong-qi. Hydrocarbon expulsion characters of Qnl sourcerocks in the Northern Songliao Basin [J]. *Journal of Oil and Gas Technology*, 2008, 30(1):166–169.

Liu Hui-min. Hydrocarbon migration and accumulation direction and distribution of Lin nan sag in Jiyang depres-sion [J]. *Geoscience*, 2009, 23(5):894–901.

Lu Shuang-fang, Hu Hui-ring, Liu Hai-ying, et a1. Gassource conditions and exploration potential of deep Layer in Yingtai fault depression[J]. *Journal of Jilin University* (Earth Science Edition), 2010, 40(4): 912–920.

Wang Jian-wei, Song Guo-qi, Song Shu-jun,eta1. Controlling factors for petroleum dominant lateral irrigation along Eogene carrier beds in the southern slope of Dongying sag[J]. *Journal of China University of Petroleum* (Edition of Natural Science), 2009, 33(5):3640.

Wei Zhao-sheng, Miao Hong-bo, Wang Yan-qing,eta1. Reservoir formation process in the Changchunling anticline of Songliao Basin[J]. *Petroleum Exploration and Development*, 2006, 33(3):351–355.

Yang Qing-jie, Lin Jing-ye, Wang Ya-feng. On reservoir forming process of Changchunling anticline belt in Songliao Basin[J]. *Journal of Oil and Gas Technology*, 2007, 29(3): 11–14.

*Advances in Energy, Environment and
Chemical Engineering – Abdullah & Osman (Eds)
© 2023 The Author(s), ISBN: 978-1-032-36083-6*

Prevention method of hydrate formation in the wellbore during deep-water gas field development

Yanfang Zhang, Panfeng Zhang*, Shubing Zeng, Wenfeng Chen & Zhenyou Zhang
Offshore Oil Engineering Co., Ltd, Tianjin, China

ABSTRACT: With the development of oil and gas resources gradually moving towards deep water areas, hydrate management at high pressure and the low-temperature condition is becoming the key to ensuring flow safety in deep water gas field development. In this study, the hydrate risk of the wellbore was first judged during the gas well emergency shut-in process and production restart process based on the hydrate equilibrium line and the P-T (Pressure and Temperature) line of the wellbore, and then the subcooling degree and the location of hydrate risk in the wellbore were calculated. Finally, the different measures were formulated corresponding to different hydrate risk conditions. After long-term shut-in, hydrate risk exists in the wellbore due to the low temperature, the maximum subcooling degree of the wellbore is 21°C at the subsea wellhead, and the risk section is located 472m below the wellhead, injection with 100wt% methanol solution into the wellbore can prevent hydrate formation in the inner wall of the wellbore. While the hydrate risk exits 78min during the restart process at a given production style. Methanol is injected with a rate of 0.66m³/h for 120 minutes from the location near the downhole safety valve, which can prevent hydrate formation in the wellbore.

1 INTRODUCTION

There are abundant offshore hydrocarbon resources in the South China Sea, and they are mainly located in Qiongdongnan, Yinggehai, and Pearl River Mouth basins (Jiang et al. 2021; Li et al. 2020). While about 70% of the resources are located in areas with water depth exceeding 300m. So far, the deep-water gas fields that have been built include the Liwan 3-1 gas field group and the Lingshui 17-2 gas field (Xie & Zeng 2021). Liwan 3-1 gas field is the first deepwater gas field in China, and it adopts the development method with an underwater production system and floating platform (Hu et al. 2014). In the LS17-2 gas field, the depth of seawater is 1200–1550 m, the pressure of the reservoir is nearly 40 MPa and the temperature of the reservoir is over 90°C, and the temperature of the subsea mudline is about 3.5°C. The production fluid is gathered by the underwater manifold, then it is transported to a semi-submersible production platform by the subsea pipeline, after treatment and drying in the platform, the dry gas is supplied to the user terminal (Jia et al. 2018; You et al. 2020; Yu et al. 2020; Zhu et al. 2018).

In the process of gas development in deep water, the wellbore and subsea pipelines all have serious hydrate plug risks, so it is necessary to carry out hydrate risk identification and prevention. Injection of thermodynamic inhibitors (THI), including methanol (MEG) and ethylene glycol (TEG), can completely prevent hydrate formation and is the most reliable hydrate management method in deepwater gas field development. In this study, the hydrate risk of the gas well during the shut-in and restart process was judged, and then the different measures were formulated to prevent hydrate plug and ensure production safety, which provided a method for hydrate management in deep water gas fields.

---

*Corresponding Author: panfengzhang@126.com

DOI 10.1201/9781003330165-80

2 HYDRATE RISK IN THE WELLBORE

2.1 *The judgment method of hydrate risk*

The hydrate equilibrium curve of the natural gas-water (black line) and the P-T (Pressure and Temperature) condition along with the depth of the wellbore A are shown in Figure 1. In this study, the P-T of the wellbore is calculated by the model in the present study (Dong et al. 2022). With a higher concentration of THI, the temperature of the hydrate formation decreases. When the P-T of the wellbore exists at the left side of the equilibrium curve, there is a risk of hydrate formation. A certain concentration of THI must be added to prevent hydrate formation. The green line and the blue line are the P-T of the wellbore with long-term shut-in and stable production, respectively. If the P-T condition of the wellhead is in the hydrate formation zone, it has a risk of hydrate formation and the maximum subcooling degree. As shown in Figure 1, there is no hydrate risk in the stable production process, while hydrate prevention measures must be taken during the shut-in process and restart process.

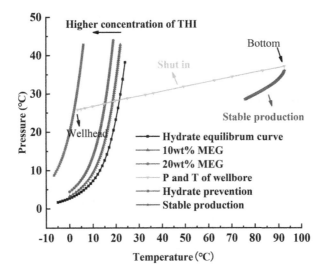

Figure 1. The risk analysis of the hydrate formation at different production styles for wellbore A.

2.2 *Hydrate risk during the shut-in process*

When Well A is stable production with the gas production of 100×10^4 Sm³/d and water production of 5 m³/d, and after emergency shut-in by the safety valve at the subsea wellhead, the temperature and subcooling degree at the subsea wellhead with shut-in time are shown in Figure 2.

With the shut-in time increasing, the temperature of the subsea wellhead gradually decreases. There is no hydrate risk at the early period of shut-in at the subsea wellhead, and the hydrate subcooling degree at the wellhead is always 0°C. After a long-term shut-in, the P-T of the subsea wellhead reaches the hydrate formation conditions, and the risk of hydrate formation begins to appear. With enough shut-in time (greater than 3600 min), the temperature of the subsea wellhead approaches the mudline temperature (3.5°C), and the subcooling degree at the subsea wellhead is 20.97°C, which is the maximum subcooling degree of hydrate risk after well shut-in. The hydrate risk section of the wellbore is below 472m from the subsea wellhead.

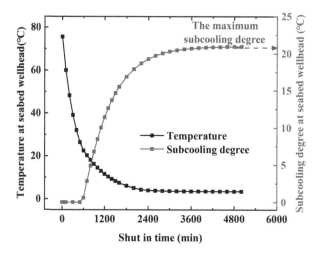

Figure 2. The temperature and subcooling degree at the subsea wellhead with the shut-in time at the production of 100×10^4 Sm^3/d.

2.3 *Hydrate risk during the restart*

After the long-term shut-in of the well, the temperature of the inner wall of the subsea wellhead is near the mudline temperature (3.5°C), therefore, it exists the hydrate risk at the early period of the restart process. After the restart, the temperature of the wellbore gradually increases for the heating to the high temperature of the fluid from the reservoir, so the length of the hydrate risk well section gradually decreases. When the temperature of the subsea wellhead exceeds the temperature of hydrate phase equilibrium, all the walls of the wellbore would have not the conditions for hydrate formation, and the risk of hydrate formation in the wellbore disappears.

The existing time of hydrate risk after the restart is related to the production method of the restart. The temperature of the subsea wellhead with restart time at different production methods is shown in Figure 3. The gas well is stable production with 100×10^4 Sm^3/d at the end, and the water-gas ratio (WGR) is always at 0.005%. When the gas well is restarted with $\Delta(5 \times 10^4$ Sm^3/d)/60min,

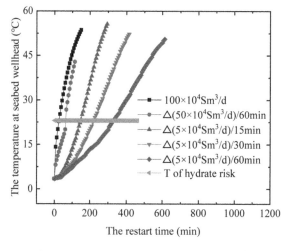

Figure 3. The temperature of the subsea wellhead with the restart time at different restart methods.

it means that the gas production increases by 5×10^4 Sm³/d every 60 min. For example, the gas production is 5×10^4 Sm³/d in the first 60min after the restart, then gas production is 10×10^4 Sm³/d from 60min to 120min, until 19 hours later, the gas well is in stable production at 100×10^4 Sm³/d. To production method of the restart with 100×10^4 Sm³/d, it means the gas well restart at the production of 100×10^4 Sm³/d.

The different production method of restart has different average gas production. The higher the average production, the faster the heating rate of the wellbore and the shorter the risk of hydrate formation. When the wells restarted with different production rates, the risk of hydrate formation was 329 min, 219 min, 148 min, 62 min, and 25 min, respectively. If the well is directly restarted with gas production of 100×10^4 Sm³/d, the inner wall of the wellbore is rapidly heated, and the risk of hydrate formation is only 25 minutes. However, it takes a certain time for the gas production rapidly to increase from 0 to 100×10^4 Sm³/d. Therefore, the actual time of hydrate risk is higher than 25 min. Besides, the fluid will damage an underwater production facility at high sudden gas production. Therefore, for choosing the optimal production method for the restart, the time of hydrate risk and fluid damage to the facility must be comprehensively considered.

3 THI PERFORMANCE AND INJECTION METHOD

3.1 Selection of thermodynamic hydrate inhibitors

Methanol is an important hydrate inhibitor in deepwater gas development. The required mass concentration of methanol for hydrate prevention can be calculated as follow:

$$\Delta T = -72 \times \ln\left[\frac{32(1 - W)}{18W + 32(1 - W)}\right] \tag{1}$$

where W is the mass fraction of the required methanol solution; ΔT is the subcooling degree required for prevention and control, °C

3.2 Parameter design of inhibitor injection

3.2.1 Inhibitor injection location

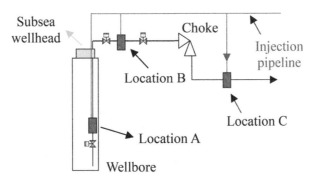

Figure 4. The location of the hydrate inhibitors injection port.

During the process of deepwater gas field development, the hydrate inhibitor is mainly transported from the offshore floating platform to the seabed wellhead through the umbilical cable (Xie et al. 2021). The injection ports of hydrate inhibitor are mainly located at position A) near the downhole safety valve, B) before the choke at the subsea wellhead, and C) behind the choke at the subsea wellhead, as shown in Figure 4. The downhole injection port A is located 700 meters below the subsea wellhead.

3.2.2 *Hydrate inhibitor injection rate*

The injection rate of hydrate inhibitor is related to the water production, inhibitor loss in the gas phase, and the required concentration, the methanol is used as a 100% methanol solution.

The loss of methanol in the gas phase is calculated as follows:

$$\alpha = 0.197 \times P^{-0.7} \cdot W \cdot \exp\left(6.054 \times 10^{-2} T - 11.28\right) \tag{2}$$

where α is the methanol mass lost per 10^4 Sm^3 of natural gas, kg; P is the pressure at injection position, MPa; T is the fluid temperature at injection position, K; W is the required concentration of methanol.

The injection rate of methanol can be calculated as follow:

$$Q_{MeOH} = \frac{1}{24\rho_{MeOH}} \times \left(Q_g \cdot f_w \rho_{H_2O} \cdot \frac{W}{1-W} + \alpha \cdot Q_g\right) \tag{3}$$

where Q_{MeOH} is the injection rate of the methanol solution, m^3/h; Q_g is the gas production, 10^4 Sm^3/d; f_w is the water-gas ratio, m^3/(10^4 Sm^3/d); W is the required concentration of methanol; ρ_{MeOH} is the density of methanol, 791.8 kg/m^3; ρ_{H_2O} is the density of water, 1000 kg /m^3.

4   HYDRATE INHIBITION METHOD

4.1   *Hydrate control methods during shut-in*

4.1.1   *Shut in by the downhole safety valve*

After shut in by the downhole safety valve, it is a low temperature in the upper part of the wellbore, so the upper part of the wellbore must be filled with methanol to prevent damage to the fluid underwater facilities during the restart process. This part of methanol will be mixed with the produced water and can prevent hydrate formation during the restart process. The total injection amount of methanol solution can be calculated as follow:

$$V_{MeOH} = \frac{\pi D^2}{4} \cdot H \tag{4}$$

where V_{MeOH} is the injection amount of methanol solution, m^3; D is the inner diameter of the tubing, 0.0964 m; H is the height of the methanol injection section (not more than 700 m), m. The height of the methanol injection section is comprehensively determined by the water production and the shut-in pressure of the wellbore.

4.1.2   *Shut-in by the safety valve at subsea wellhead*

When the gas well is shut down by the safety valve at the subsea wellhead, the water at the bottom of the wellbore can diffuse to the upper part of the wellbore due to evaporation, and then water vapor can condense at the inner wall in the upper part of the wellbore to form hydrate with natural gas and deposit on the inner wall of the tubing. Although the growth rate of the thickness of the hydrate sedimentary layer caused by the evaporation and diffusion of water vapor is about 0.452 mm/year, these hydrates can rapidly accelerate hydrate formation during the restart process. Methanol is volatile, and it can volatilize into the gas phase of the wellbore to prevent the risk of hydrate in the upper part of the wellbore. After shutting in, the methanol solution is injected into the wellbore from location A (in Figure 4) near the downhole safety valve. The maximum injection amount of methanol can be calculated as follow:

$$V_{MeOH} = \frac{\pi D^2}{4} \cdot H_L \cdot \frac{\rho_{H_2O}}{\rho_{MeOH}} \cdot \frac{W}{1-W} \tag{5}$$

where H_L is the height of the water column at the bottom of the wellbore hole after shut-in, m.

In order to simplify the calculation process, the fluid accumulation in the wellbore after shut-in is calculated according to the fluid amount in the wellbore during stable production; in addition, the influence of the wellbore fluid accumulation on the wellhead pressure is not considered. The methanol evaporation is calculated according to the P-T condition at the bottom of the wellbore. After shut-in at different gas production (WGR=0.01%), the injection parameters of methanol solution are shown in Table 1.

Table 1. The injection parameters of methanol for hydrate inhibition after shut-in at different gas production (WGR= 0.01%).

Gas production/ $(10^4 \ Sm^3/d)$	Water column height /m	Wellhead subcooling /°C	Required concentration of methanol /wt%	Methanol injection volume /m^3
30	86.4	21	37.6	0.480
50	84.3	21	37.6	0.468
70	81.8	21	37.6	0.454
100	78.3	21	37.6	0.435

4.2 *Hydrate management during the restart process*

Although methanol is toxic compared to ethylene glycol, methanol has a lower freezing point and a better hydrate prevention ability (Wu 2016; Yi et al. 2012). In the gas field, the methanol solution as hydrate inhibitor is injected first for hydrate prevention at the restart process, and then the ethylene glycol solution is injected later.

During the restart process of well A, the gas production increased from 0 to $40 \times 10^4 \ Sm^3/d$ within 30 min, and the production was stable for 90 min, and then the production was increased by $15 \times 10^4 \ Sm^3/d$ within 30 min, and the production was stable for 90 minutes until the target production ($105 \times 10^4 \ Sm^3/d$) was finally reached. The maximum subcooling degree is 21 °C and the time of hydrate risk is 75min at the seabed wellhead, while they are 33.47°C and 108min near location C (in Figure 4). The detailed calculation process of hydrate risk at choke will be shown in the future study. For preventing hydrate formation at the wellbore and the choke, the required concentration of methanol is 55wt%, and the injection time of methanol is set as 120 min. The methanol injection rate is 0.66 m^3/h from location A.

5 CONCLUSION

There is a serious hydrate risk during the shut-in and restart process. However, it can be prevented by injecting methanol or ethylene glycol solution.

During the shut-in process, a certain amount of methanol solution should be injected into the wellbore, which can not only prevent the formation of hydrate but also reduce damage impact on underwater facilities.

During the restart process, the methanol solution should be injected at the rate of 0.66 m^3/h for 120min, as a result, hydrate formation at the wellbore and the choke can be prevented.

ACKNOWLEDGMENTS

Financial support from the Dongfang 1-1 gas field underwater production system engineering project (2019GXB01-08) is greatly acknowledged.

REFERENCES

Dong Z., Diao Y., Li Z., Jiang D., Zhang P. (2022)Kinetic Modeling of gas hydrate deposition and blockage at annular-mist flow state in production wells of deep water gas fields. *Journal of Southwest Petroleum University*(Science &Technology Edition), 44(01): 132–142.

Hu X., Zhou M., Zhang F., Cao W., Fu J. (2014) The General Design technological research on subsea Manifold in LW3-1 Gas Field. *China offshore Platform*, 29(06): 10–14.

Jia X., Da L., Yi C., Zhu H., Bai X., Cao P., Bian X. (2018) *Technology Development and Application of Deep-Draft Semi-Submersible Production Platform With Condensate Storage*. The 28th International Ocean and Polar Engineering Conference. Sapporo, Japan. pp. 1087–1093

Jiang P., Lei X., Wang W., Zhang J., Wang Q., Lu R., Huang H. (2021) Progress and future development direction of oil and gas development technologies for western South China Sea oilfields. *China offshore oil and gas*, 33(01): 85–92.

Li H., Zhang M., Lau H. C., Fu S. (2020) China's deepwater development: subsurface challenges and opportunities. *J. Pet. Sci. Eng.*, 195: 107761.

Wu Q. (2016) Natural gas hydrate prohibition method for wellheads in Panyu 35-2 natural gas field during the commission. *Contemporary Chemical Industry*, 45(04): 746–748.

Xie B., Zeng H. (2021) Research advancement in offshore deepwater oil and gas development engineering technologies in China. *China offshore oil and gas*, 33(01): 166–176.

Xie J., Huang D., Li H. (2021) Study on subsea production system layout of LS17-2 gas field. *Guangdong Chemical Industry*, 48(02): 138–143.

Yi H., Zhou X., Zhu H., Hao Y., Chen H., Huang Z. (2012) Research on hydrate inhibition in deep water gas field development by subsea wellheads. *China offshore oil and gas*, 24(05): 54–57.

You X., Chen B., Liu K., Liu X., Wu Y., Zeng D. (2020) Study on the closure scheme of semi-submersible production and storage platform in LS17-2 gas field. *China offshore oil and gas*, 32(06): 141–149.

Yu C. A., Song H, Cheng Y, et al.(2020) Riser Integrity Management Plan for Lingshui 17-2 Project[C]. The 30th International Ocean and Polar Engineering Conference.

Zhu H., Li D., Wei C., Li Q. (2018) Research on LS17-2 deepwater gas field development engineering scenario in the South China Sea. *China offshore oil and gas*, 30(04): 170–177+214.

*Advances in Energy, Environment and
Chemical Engineering – Abdullah & Osman (Eds)*
© 2023 The Author(s), ISBN: 978-1-032-36083-6

Design of distributed energy supply system based on green fuel

Ruobing Duan, Xinyu Qiu, Jinjie Cheng, Ruilin Qiu, Hang Zeng & Zhiming Xu*
Guangzhou Xinhua University, China

ABSTRACT: With industrial upgrading and changes in energy supply methods, distributed energy is widely used in power generation production. In building a distributed energy power generation system, the exploration of new energy supply methods has been gradually carried out. In this paper, the mainframe of the distributed new energy system is designed first, and the structure of each part of the power generation system is specifically designed. This paper firstly designs the circulation unit of the distributed energy system and designs the coupling characteristics in the circulation process and the application characteristics of the energy gradient for the distributed small and medium-sized circulation unit of the combined cycle. On this basis, this paper redesigns the large-scale distributed energy system and constructs the system environment, calculates the degree of energy dissipation between each part of the system, and analyzes the working characteristics of each part.

1   INTRODUCTION

1.1   *Introduction of large-scale distributed energy supply system at the core of gas turbine*

The distributed energy supply system of gas turbines has inherent advantages for the supply of industry mainly relies on combined cooling, heating, and power to supply power to high energy-consuming enterprises. In addition, this distributed energy system is also applied in parks and commercial electricity. With the development of new energy power grid construction, the scale of the power grid is expanding day by day, and the failure of the power grid and the accidents often caused by the collapse of the whole network are increasing. Therefore, it is necessary to evaluate the power generation capacity of the distributed power supply system of the power grid and customize quantitative indicators based on the evaluation content (Chen & Gao 2019). The distributed power generation system needs to be reconstructed to ensure that it meets the new industrial user power consumption characteristics and is green and environmentally friendly.

1.2   *Introduction of small distributed energy supply system at internal combustion engine core*

The operation flow chart of the small distributed energy supply system at the core of the internal combustion engine is shown in Figure 1 below. The main structural feature of this system is that it can provide users with three energy sources of cold, heat, and electricity simultaneously, and the consumption fuel is green natural gas. The characteristic of this internal combustion engine is to provide electricity to users through the power grid. At the same time, during the power generation process, the heat in the flue gas and steam of the internal combustion engine is recovered to ensure that its heat can also be used for power generation and heating. The system's refrigeration is mainly carried out by using compression refrigeration machines, and the power source is electricity consumption in the grid (Xiao et al. 2022).

---

*Corresponding Author: xujay@xhsysu.edu.cn

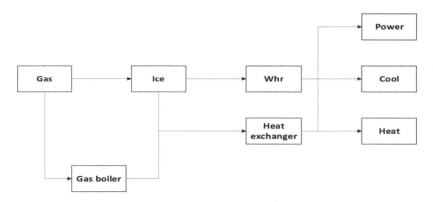

Figure 1. The operation flow chart of the small distributed energy supply system at the core of the internal combustion engine.

With the extensive advancement of the informatization process, distributed energy supply and power generation technology are gradually applied to all aspects of the power system. With the development of the new energy grid construction process, the scale of the power grid is expanding day by day, followed by the operation of the power system. More and more loads are carried, which leads to grid failures and accidents caused by the collapse of the entire grid. Most of these accidents are caused by the grid's voltage drop caused by the grid's immature distributed generation technology (Luo & Qing 2018). Therefore, it is necessary to evaluate the grid's distributed generation power and voltage support capability based on the evaluation.

1.3 *Development trend of distributed energy supply system*

The development trends of distributed energy supply systems mainly include the following:

(1) Renewable Energy Coupling Power Generation System
This distributed energy supply system mainly relies on solar coupled power supply panels to supply distributed energy to the system. This power supply device generally adopts the mode of combined cooling, heating, and power to supply power to users. At the same time, when designing power supply parameters, the design parameters are usually optimized and constructed according to the particle swarm algorithm. In the current smart grid context, the characteristics of power grid data are very large in magnitude, and most of them are automatically generated low-value data. Therefore, it is necessary to perform a certain type of feature screening on these massive low-value data through particle swarm algorithm and, on this basis, perform specific visualization operations on the screened data. The extraction technology of power grid data based on the above ideas and particle swarm algorithm is based on the massive automatically generated data in the smart grid and is driven by data algorithms and technical software to realize automatic data collection and information processing to generate value for the power grid. The coupled renewable energy distributed power supply system usually uses the waste heat collection of the prime mover and solar energy to power the system (Yang et al. 2019). In the power supply process, the heat storage device is usually used to distribute, dispatch and store the heat energy to meet the user's heating and cooling needs.

(2) Green Functional System of Biomass Gasification Distributed
This distributed energy supply system mainly uses biomass for power generation operations, and the main form is the centralized processing of crop wastes, including crop wastes and industrial leftovers. The production process relies mainly on converting energy, including the combustion of biomass fuels in power plants to meet their electricity and heat supply needs.

But this method will produce a certain amount of exhaust pollution. Therefore, the mainstream method adopted in the distributed energy power generation system is to gasify biomass energy. After the treatment, the relatively inferior impurity components in the biomass can be removed, and the greening fuel can be further purified.

2 WORKING MECHANISM OF CYCLE UNIT ON DISTRIBUTED ENERGY SUPPLY SYSTEM

2.1 *Components of a distributed energy supply system*

The constituent unit of the distributed energy supply system mainly includes the prime mover and the waste heat circulation system. These two parts can be subdivided again. The main function of the prime mover is to provide the system with stable load power output needs, thereby driving the generator of the distributed energy system to generate electricity for the user. The prime mover included in the distributed energy system is mainly composed of a gas turbine and an internal combustion engine. The distinction is mainly based on the load required by the user.

2.2 *Cycle unit of distributed energy supply system*

The circulation unit of the distributed energy supply system includes the following processes. The core method of the isolated forest system of the distributed energy circulation system is to separate the data samples of the distributed energy circulation system through the distributed energy circulation system and the random plane. And until each distributed energy circulation system subsample space has only one type of distributed data points of energy circulation system. Normal data needs to be separated by cutting green energy multiple times. In contrast, the outlier data used in the distributed energy circulation system can be achieved with only a small cutting. Because of the unique characteristics of green energy, such as infinity and disorder in the modern power system, it is difficult to directly use this green energy in data processing related to green energy. In the green energy processing system, each PMU receives a large amount of green energy power information samples, analyzes the green energy in these power data in real-time, and makes quick decisions. Compared with the sliding window, data processing technology designed for green energy came into being.

The simple cycle four working processes of the cycle unit of the distributed energy supply system can be expressed as:

Isentropic compression $1 - 2s$;
Isobaric heating $2s - 3$;
Isentropic expansion $3 - 4s$:
Isobaric heat release $4s - 1$;

2.3 *The integration mechanism of distributed energy supply system*

The integration process of distributed energy supply systems is mainly used to mine the huge value of massive energy supply systems, so their quality becomes the key to its mining technology. The excavation process of the energy supply system is extremely susceptible to external factors. Integrating the distributed energy supply system with the energy supply system is necessary to eliminate these problems. The integration process of the energy supply system's distributed system mainly includes the multi-dimensional deconstruction and verification of the energy supply system. The main points are as follows: (1) Standardization of energy supply system: this system is a standard-setting measure proposed by the International Electrotechnical Commission and dedicated to developing information interface standards for distributed power grids and smart grids, which helps to achieve power information integration. And the quality of the energy supply system is greatly improved. (2) Rapid identification of abnormal energy supply systems: For abnormal energy supply systems, it is necessary to use the engineering mathematical powder method to identify them from a mathematical point of view and to identify and process the energy supply system with missing values and out-of-limit situations; from a time point of view, starting from the analysis and regression of the energy supply system for the time series, the abnormal energy supply system with greater volatility for detection is found out; from the perspective of spatial law, it is needed to jointly judge those multi-source energy supply systems, and find out the mapping conditions in the energy supply system Insufficient relevant content is filtered. (3) The energy supply system type of the abnormal energy supply system is reconstructed: For the abnormal energy supply system, if the single-point energy supply system is abnormal, the constant mean value is used for reconstruction.

The integrated cycle process of the distributed energy supply system is shown in Figure 2 below. First, the condensed steam is generated in the generator by the water. Then the absorption capacity is lowered in the cooling device and returned to the storage device to complete a cycle (Bohlayer & Zoettl 2018).

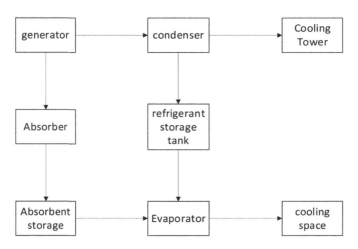

Figure 2.   The flow chart of the integration cycle of the distributed energy supply system.

3   DISTRIBUTED ENERGY SUPPLY SYSTEM BASED ON GREEN FUEL

3.1   *Preparation and characterization of green fuel*

The preparation process of green fuel is shown in Figure 3 below. First, biomass energy needs to be input, and then through the energy conversion of syngas, it is ensured that biomass energy becomes storable energy in distributed energy systems. Then, the green fuel is stored by the liquid gas storage device and converted into the storage device for preparation. Finally, the DME device is used to convert and output green energy.

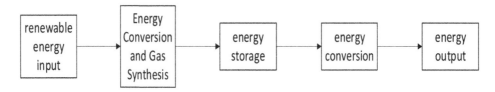

Figure 3.   Preparation process of green fuel.

Since the system collected by the green energy fuel pretreatment strategy adopted by the current distributed energy supply system is not perfect, the green energy fuel must be pretreated first. In this chapter, the selected system to collect distributed energy supply systems has various problems such as missing attribute values, missing values, and abnormal green energy fuels when preprocessing green energy fuels. Various methods are often used in intelligent distributed energy supply systems, and the preprocessing process of abnormal green energy fuel is performed on this green energy fuel (Nidhis et al. 2018).

3.2 *Parameter design of distributed energy supply system*

The parameter design method of the distributed energy supply system is as follows.

(1) Judging the parameter indicators of the distributed energy supply system refers to normalizing the parameter indicators of the distributed energy supply system by evaluating the type of target reactive power support capability as the parameter benefit index of the distributed energy supply system and forming the parameter evaluation matrix of the energy supply system with original distribution.
(2) The distributed energy supply system parameter reference matrix is selected according to the optimal value of the distributed energy supply system parameters to form the absolute difference distributed energy supply system parameter matrix.
(3) The parameter resolution coefficients of distributed energy supply system are calculated according to the dynamic resolution coefficient method of distributed energy supply system parameters.
(4) The distributed energy supply system parameter entropy value method is used to determine the objective weight value of the distributed energy supply system parameters.
(5) The gray correlation degree of the comprehensive distributed energy supply system parameters are calculated according to the resolution coefficient obtained from the above content and the determined objective weight value of the distributed energy supply system parameters.

4 CONCLUSION

This paper firstly summarizes the existing distributed energy supply system and introduces the core technology. The relevant research background points out the shortcomings of the distributed energy supply technology in the current industrial environment and aims at the shortcomings of this part. Innovations in efficiency improvements and green fuel applications are made. Then this paper proposes a set of process methods for the conversion of green fuels and the parameter design of the distributed energy supply system, which has certain guiding significance for the direction of research.

ACKNOWLEDGMENTS

This work was financially supported by the 2021 Guangdong University Student Innovation and Entrepreneurship Training Program (202113902108).

REFERENCES

Bohlayer M, Zoettl G. Low-grade waste heat integration in distributed energy generation systems - An economic optimization approach[J]. *Energy*, 2018, 159(sep.15):327–343.
Chen Y, Gao S. Design of Distributed Energy Supply System for Small Gas Turbine and Discussion on Economic Benefit[J]. *Modern Information Technology*, 2019.

Luo N, Qing H. Application of Distributed Energy Supply System in Modern Agricultural Greenhouse[J]. *Electric Power Science and Engineering*, 2018.

Nidhis A D, Pardhu C, Reddy K C, et al. Smart Crop Health Diagnosis and Treatment Unit Powered by Green Fuel[J]. *Journal of Green Engineering*, 2018, 8(3):389–410.

Xiao J, Qiu Z, Jiang C, et al. Optimal Design of Energy Supply System Based on River Water Source Heat Pump in Load Intensive Area[J]. 2022.

Yang C, Wang P, Liu H, et al. Thermo-economic Analysis on Gas-steam Combined Cycle-based Distributed Energy Supply System[J]. *Proceedings of the CSEE*, 2019.

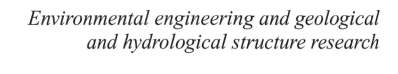

Environmental engineering and geological and hydrological structure research

Advances in Energy, Environment and
Chemical Engineering – Abdullah & Osman (Eds)
© 2023 The Author(s), ISBN: 978-1-032-36083-6

Changes in the relationship between Yangtze River and Poyang Lake and its impact on the water level of Poyang Lake in non-flood seasons

Bo Yan
Changjiang River Scientific Research Institute, Wuhan, China
Hubei Key Laboratory of Water Resources & Eco-Environmental Sciences, Wuhan, China

Zhixiang Min
College of Hydrology and Water Resources, Hohai University, Nanjing, China

Bin Xu & Jijun Xu*
Changjiang River Scientific Research Institute, Wuhan, China
Hubei Key Laboratory of Water Resources & Eco-Environmental Sciences, Wuhan, China

Jiqiong Li
Changjiang River Scientific Research Institute, Wuhan, China
Wuhan Changjiang Kechuang Technology Development Co. Ltd., Wuhan, China

Zhigao Li
Haikou Water Supply and Drainage Water Quality Monitoring Station, Haikou, China

ABSTRACT: Changes in the relationship between rivers and lakes are an important part of the study on changes in the hydrological regime in the Yangtze River Basin, which can provide technical support for the protection of the Poyang Lake Basin. This paper selects the water level of the mainstream from the Jiujiang to Hukou section and the Xingzi to Hukou section of Poyang Lake to perform the analysis of the intra-annual water level drop change, revealing the relationship between the Yangtze River and Poyang Lake and its change characteristics. The analysis results show that the smaller the proportion of the inflow of Poyang Lake in the mainstream of the Yangtze River, the stronger the supporting effect of the mainstream on the lake; the larger the proportion of Poyang Lake's inflow, the stronger the supporting effect of the lake on the mainstream of the Yangtze River. After the impoundment of the Three Gorges Project, under the conditions of similar confluence ratio, the mainstream water level difference is not significantly reduced, which is related to the erosion of the mainstream of the Yangtze River and the water channel into the river. After the normal storage of the Three Gorges Reservoir, the average water level of Xingzi Station during the non-flood season dropped significantly, and the largest drop occurred in September, which was 2.09m.

1 INTRODUCTION

The Poyang Lake is located in the north of Jiangxi Province and on the south bank of the middle reach of the Yangtze River, which is the largest freshwater lake in China, an internationally important wetland, and an important flood regulation site and water source in the middle and lower Yangtze River (Feng et al. 2015; Xu & Chen 2013). The Poyang Lake water system is radial and accepts water from the Ganjiang River, Fuhe River, Raohe River, Xinjiang River, Xiuhe River ('Five Rivers' for short), Boyang River, Zhangtian River, Tongjin River and other small tributaries. After being stored

---

*Corresponding Author: xujj@mail.crsri.cn

DOI 10.1201/9781003330165-82

in Poyang Lake, water is injected into the Yangtze River from the mouth of the lake. In brief, Poyang Lake is a water-passing, throughput and seasonal lake. The Poyang Lake basin flows through an area of 162,200 square kilometers, which is 156,000 square kilometers in Jiangxi Province, accounting for 96.6% of the total area. The Poyang Lake basin has flat terrain, a humid climate, numerous rivers and lakes, and good natural conditions. As an important national granary, the Poyang Lake basin has abundant water resources and concentrated cultivated land. And its economic growth is full of life, with a large population and industrial parks. As we know, the Poyang River Basin is rich in ecological resources and important in its ecological location. It undertakes the 'Five Rivers' and 'Three Defenses' in the province's ecological security pattern.

As the largest lake in the Yangtze River's middle and lower reaches, Poyang Lake plays an important part in maintaining the water balance of the Yangtze River and the ecological balance of regional waters. On the other hand, under the combined influence of the amount of water entering the lake and the Yangtze River mainstream, the water regime characteristics of Poyang Lake are very complex. Compared with other freshwater lakes, the water regime changes have significantly different characteristics. Its water level variation, water-sand process and ecological environment evolution are closely related to the hydrological regime of the Yangtze River mainstream (Wang et al. 2017). After the Three Gorges Project is constructed, water storage at the end of the flood season will change the water amount coming from the Yangtze River, which in turn will change the supporting impact on the Yangtze River during the end of the flood season, and affect the hydrological condition of Poyang Lake (Lai et al. 2012). Studies have shown that the Three Gorges Project operation will have certain effects on hydrology, sediment deposition, water environment, wetland ecology, and so on. The relationship between rivers and lakes is affected by natural factors such as climate change, tectonic subsidence, soil erosion, sediment deposition, and changes in the Yangtze River regime. However, human activities since the 20th century have accelerated the adjustment of the relationship between rivers and lakes (Poyang Lake Water Conservancy Project Construction Office, 2016). Analyzing the law of river and lake water regime change, the identification of the law of river and lake water regime variation can provide scientific support for revealing the driving mechanism of the interaction of rivers and lakes.

The paper statistically analyzes the changes in the mutual support relation between the Yangtze River and Poyang Lake and, combined with the background of the relationship changes, analyzes the characteristics of the water level changes in Poyang Lake in the non-flood season.

2 STUDY AREA

The Poyang Lake Basin, located on the right bank of the Yangtze River's middle and lower reaches, is bounded by Huaiyu Mountain, Wuyi Mountains, Fushui River, and Minjiang River in the east, and Dayu Ridge, Jiulian Mountain, Dongjiang and Beijiang River in the south. In the west, it is bordered by Luoxiao Mountains and Dongting Lake, and Mufu Mountain and the Yangtze River mainstream in the north. Its location is between $113°30'\sim118°31'$E and $24°2'\sim30°02'$N (Jiangxi Hydrological Bureau, 2008).

3 DATA AND METHODOLOGY

3.1 *Data*

According to the Hydrological Yearbook of the Yangtze River Basin, this paper collects the multi-year daily water level data of Jiujiang Station (on the Yangtze River mainstream), Xingzi Station (the typical water level station on Poyang Lake), and Hukou Station (lake outlet control station).

3.2 *Methodology*

In the confluence section of rivers and lakes, when one side forms a strong supporting effect on the other side, the water level drop will decrease locally in the area that is supported. This study selected the water levels of the mainstream from Jiujiang to Hukou and from Xingzi to Hukou in the Poyang Lake basin and performed an analysis of water level variation within the year, revealing the relationship between the Yangtze River and the Poyang Lake and its variation characteristics.

4 RESULTS AND DISCUSSION

4.1 *Changes in the support relation between the rivers and lakes*

(1) The supporting effect of the Yangtze River on Poyang Lake

The Poyang Lake basin enters the flood season in March and April, from the perspective of monthly variation of the water level difference between Jiujiang Station to Hukou Station and Xingzi Station to Hukou Station in different periods (Figure 1). The flood season in Hukou comes earlier than the Yangtze River mainstream. Correspondingly, the supporting effect of the Yangtze River mainstream on Poyang Lake is not obvious from January to April, and the water level difference between Xingzi Station and Hukou Station is the largest in this period. Since then, as the flow of the Yangtze River mainstream increases, the supporting effect of the mainstream on the lake becomes stronger and stronger. The Yangtze River shows a strong supporting impact on Poyang Lake from July to October. Then the mainstream enters the dry season in November. During the dry season, the supporting impact on Poyang Lake gradually weakens. And the water surface slope from Xingzi to the Hukou enters a period of recovery until it reaches the maximum in March and April of the following year.

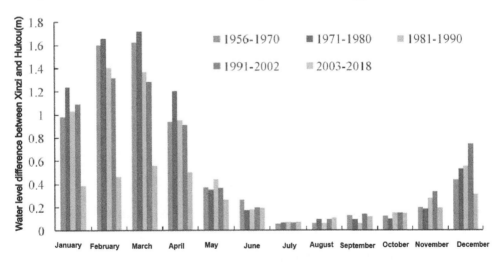

Figure 1. Monthly change of the water level difference from Xingzi to Hukou Station in Poyang Lake in different periods.

The supporting effect relationship of the confluence of rivers and lakes generally depends on the size of the incoming flow from the two sides. To study the influence of incoming flow conditions on the supporting impact, Figure 2 establishes respectively the confluence ratio of Poyang Lake from November to April and May to October (the confluence ratio uses the ratio of the flow from Hukou Station to the flow from Datong Station, due to the lack of flow observation data before 1988 at Jiujiang Station) and correlation of water level difference between Xingzi and Hukou. From the figure, the water level difference of the lake entrance channel is positively correlated with the confluence ratio during the flood season and non-flood season. That is to say, the smaller the proportion of the incoming flow of Poyang Lake to the Yangtze River mainstream, the stronger the supporting impact of the mainstream on the lake, and the smaller the slope of the waterway entering Yangtze River. Furthermore, in the flood season, the correlation has no obvious trend change before and after the impoundment of the Three Gorges Reservoir. However, in the non-flood season, the water level is greatly affected by the boundary conditions of the river channel. Under the condition of a similar confluence ratio, the water level difference of the lake mouth is obviously smaller than before the impoundment, which indirectly reflects the obvious adjustment of the riverbed morphology of the watercourse into Yangtze River.

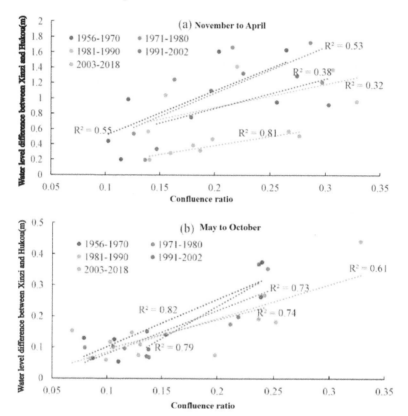

Figure 2.   Correlation changes between the water level difference between Xingzi and Hukou Station in Poyang Lake area and the inflow and confluence ratio of Poyang Lake.

(2) The supporting effect of Poyang Lake on the Yangtze River
The supporting impact of Poyang Lake on the Yangtze River mainstream is generally weaker than that of the mainstream on the lake, and the strong supporting period is mainly from April to June of the year. From January to June during the year, Poyang Lake's supporting impact on the Yangtze River mainstream gradually increases. And from July to December, the supporting impact gradually weakens (Figure 3). There exists a negative correlation between the water level difference of the mainstream of the confluence of rivers and lakes and the proportion of the inflow of Poyang Lake (Figure 4). That is, the larger the proportion of the inflow of Poyang Lake, the stronger the supporting impact of the lake on the Yangtze River mainstream, the smaller the water level difference between Jiujiang Station and Hukou Station, and the smaller the slope of water surface. Similarly, the relation between water level

difference and the confluence ratio in the mainstream confluence of rivers and lakes during the flood season has no obvious trend adjustment before and after the impoundment of the Three Gorges Reservoir. The changes in non-flood seasons are mainly manifested in that after the

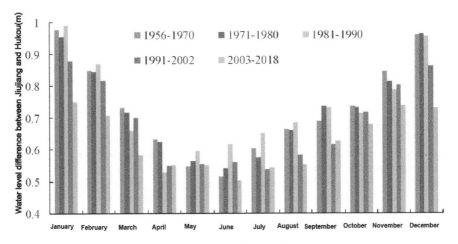

Figure 3. Monthly variation of the water level difference between Jiujiang and Hukou Station at the junction of the Yangtze River and Poyang Lake in different periods.

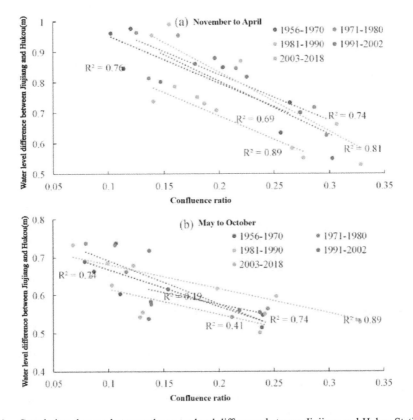

Figure 4. Correlation changes between the water level difference between Jiujiang and Hukou Station and the inflow and confluence ratio of Poyang Lake.

Three Gorges Reservoir is impounded, the mainstream water level difference will decrease to a certain extent under the conditions of a similar confluence ratio. This is related to the erosion and cutting of the Yangtze River mainstream and the waterways into the river. But the absolute change ranges are not large.

4.2 *Variation of water level in non-flood season in Poyang Lake*

The most significant time period for water level variation in the Poyang Lake area is from September to November, according to the results of the analysis of variations in hydrological regimes in the Poyang Lake area. This period also happened to be a critical period for the development of the Poyang Lake wetland ecosystem and agricultural irrigation. Considering that the Three Gorges Reservoir has been tentatively impounding water since 2008, this study used data from 1959-2002 and 2008-2019. We conducted a comparative analysis of the influencing factors of the hydrological regime in Poyang Lake during the dry season from September to February of the following year. The monthly average water level variation of Xingzi Station from September to February of the following year during the two statistical periods is shown in Table 1.

Under the influence of variations in the relationship between rivers and lakes, compared with before the construction of the Three Gorges Reservoir, the average water level declined by 2.09 m, 2.66 m, and 1.46 m in September, October and November of 2008-2019, respectively. From December to February of the next year, the number is 0.36 m. As we can see, the water level of Poyang Lake drops significantly during the non-flood season.

Table 1. Comparison of monthly average water levels at Xingzi Station from 2008 to 2019 and from 1959 to 2002.

Series	Monthly average water levels (m, Freeze Wusong Coordinate System)			
	September	October	November	December-February of the next year
①1959~2002	16.14	14.67	12.28	9.55
②2008~2019	14.05	12.01	10.82	9.19
②-①	-2.09	-2.66	-1.46	-0.36

5 CONCLUSIONS

(1) Under the influence of human activities, the relationship between the Yangtze River and Poyang Lake has undergone a series of complex changes. From a statistical analysis point of view, the smaller the proportion of Poyang Lake's incoming flow in the Yangtze River mainstream, the stronger the supporting effect of the mainstream on the lake, and the smaller the slope of the waterway into the river. In the non-flood season, the water level is greatly affected by the boundary conditions of the river. At a similar confluence ratio, the water level difference of the lake mouth channel is obviously smaller than before the impoundment of the Three Gorges Reservoir, which indirectly reflects the obvious adjustment of the river bed morphology of the water channel into the river.

(2) The larger the proportion of Poyang Lake's inflow is, the stronger the supporting effect of Poyang Lake on the Yangtze River mainstream is. The change in non-flood season is mainly manifested in the decrease of the water level difference of the mainstream at a similar confluence ratio after the impoundment of the Three Gorges Reservoir. This phenomenon is related to the scour and cutting of the mainstream and the channel into the Yangtze River, but the absolute change ranges are not large.

(3) Many factors affect the change of Poyang Lake's dry water hydrological regime, such as the change of natural runoff, water intake pumping station, water diversion, scheduling of small and medium-sized reservoirs and other projects, and the increase in water consumption. The influence of these factors on runoff is difficult to be reduced one by one. This study only shows the statistical changes in Poyang Lake water level during the non-flood season after the experimental impoundage of the Three Gorges Reservoir. The specific influencing factors will be further discussed in the future.

ACKNOWLEDGMENTS

This work is funded by the National Key R&D Program of China (2017YFC1502404), the National Natural Science Foundation of China (U2040206, 41890824), National Public Research Institutes for Basic R & D Operating Expenses Special Project (CKSF2017061/SZ).

REFERENCES

Feng, W., Ligang, X. U., Fan, H., Xinyan, L. I., & Dong, L. (2015). Analysis of the relationship of water level and water exchange between Meixi lake and Poyang lake. *Journal of Shaanxi Normal University*(Natural Science Edition), 43(4):83–88.
Jiangxi Hydrological Bureau. (2008). *Poyang Lake Water Conservancy Project's impact on flood control, sedimentation, water quality, and dry season water replenishment in the lake area and its countermeasures.* Jiangxi Hydrological Bureau, Nanchang.
Lai, X., Jiang, J., & Huang, Q. (2012). Water storage effects of three gorges project on the water regime of Poyang lake. *Journal of Hydroelectric Engineering*, 31(6), 132–136+148.
Poyang Lake Water Conservancy Project Construction Office, Jiangxi Province. (2016). For "a lake with clear water"-Poyang Lake Water Conservancy Project Introduction. Poyang Lake Water Conservancy Project Construction Office, Jiangxi Province, Nanchang.
Wang, J., Guo, S. L., Tan, G. L. (2017). *Research and Application of Hydrology and Water Resources in Poyang Lake Area under Changing Environment*, China Water Conservancy and Hydropower Press, Beijing.
Xu, J. J., & Chen, J. . (2013). Study on the impact of three gorges reservoirs on Poyang lake and some proposals. *Journal of Hydraulic Engineering*, 44(7), 757–763.

Advances in Energy, Environment and
Chemical Engineering – Abdullah & Osman (Eds)
© 2023 The Author(s), ISBN: 978-1-032-36083-6

Influence of cementation on fracture propagation in glutenite reservoirs

Jiantong Liu*
The Unconventional Oil and Gas Institute, China University of Petroleum (Beijing), Beijing, China

Jianbo Wang
Department of Petroleum Engineering, China University of Petroleum (Beijing) at Karamay, Karamay, Xinjiang, China

Hongkui Ge
The Unconventional Oil and Gas Institute, China University of Petroleum (Beijing), Beijing, China

ABSTRACT: Glutenite formations are essential for oil and gas reservation. Glutenite comprises gravel, matrix, and cement surfaces, and each phase influences fracture propagation. To study the influence of cementing characteristics on fracture propagation, we performed three-point bending tests on semi-circular glutenite samples with different cementation. It was found that cementation significantly affects fracture propagation. The fracture propagation process is similar to sandstone in the firmly cemented glutenite. Not until the load was close to the peak load would acoustic emission (AE) and fracture process zone (FPZ) appear. In this kind of glutenite, the FPZ is small in size. The fracture penetrated gravel and formed a single simple fracture when it hit the gravel. However, the glutenite with weak cementation is different. In the weakly cemented glutenite, cracks appeared at a lower load ratio, and the FPZ was large. Instead of penetrating the gravel, the fractures extended around the gravel to form a tortuous and complex fracture. Therefore, the firmly cemented glutenite can be treated as a homogeneous rock in hydraulic fracturing. In contrast, the weakly cemented glutenite cannot be treated as a homogeneous rock, as the tortuous fracture increases flow resistance and hinder proppant migration.

1 INTRODUCTION

Unconventional resources have received more and more concerns with the increased demand for oil and gas resources. Glutenite reservoirs were also concerned because of their potential reserves. Oil reservoirs of 17.6 billion tons in glutenite formations have been discovered in the Mahu oilfield in Junggar Basin (Zhi et al. 2021). However, the glutenite reservoirs are usually of low porosity and permeability. Hydraulic fracturing has become necessary to develop glutenite reservoirs effectively (Chen et al. 2020). Multistaged hydraulic fracturing in horizontal wells is a common technology in glutenite, just like the sandstone and shale reservoirs. This method has been proved effective in glutenite reservoir propagation. However, some problems exist, such as interference between wells and drastic production attenuation. There is still a lack of clear understanding of these problems due to the insufficient understanding of the fracture propagation in glutenite reservoirs.

Much research has been performed on fracture propagation in glutenite reservoirs. These studies are mainly based on compression tests and numerical simulations (Ma et al. 2017; Rui et al. 2018; Zhao et al. 2015). These studies found that the fracture propagation in glutenite reservoirs differs from that in fine-grained rocks like sandstone and shale (Hou et al. 2017). Gravels in glutenite

---

*Corresponding Author: ardow100@163.com

 DOI 10.1201/9781003330165-83

significantly influence fracture propagation (Ma et al. 2017; Rui et al. 2018). When a fracture hits gravel, behaviors such as penetration, deflection, and bifurcation may form tortuous fractures (Shi et al. 2021). In addition, gravel size and content, in-situ stress, and fracturing fluid influence the fracture pattern in glutenite (Hou et al. 2017; Li et al. 2013). These studies mainly focus on the effect of gravel on mechanical properties and fracture propagation in glutenite. However, the effect of cementation was seldom concerned. However, we have to face the fact that the cementation of glutenite reservoirs is different. For instance, the cementation in the Mahu oilfield is siliceous, calcareous, and argillaceous (Jia et al. 2017; Kang et al. 2019; Li et al. 2020). Therefore, it is of great significance to study the influence of cementation on fracture propagation in glutenite.

In order to study the fracture propagation in glutenite reservoirs with different cementation, we performed a study on the fracture propagation on glutenite samples from the Junggar Basin. The fracture propagation was studied through three-point bending tests on semi-circular samples (Kuruppu et al. 2014; Su et al. 2019; Wei et al. 2016). The cementation of the glutenite was analyzed by X-ray fluorescence (XRF), electron microscopy, and micron indentation. To monitor the fracture propagation, we applied acoustic emission (AE) and digital image correlation (DIC) (Doll et al. 2017; Sharafisafa et al. 2020) to monitor the cracking process. The CT scanning was also applied to detect the fractures within the samples. Finally, the influence of cementation on fracture propagation was analyzed based on the experimental results.

2 MATERIALS AND METHODS

2.1 *Sample preparation*

This study conducted the three-point bending test on the semi-circular disks as they are easy to machine from the core plugs. The core plugs were collected from the Triassic Baikouquan (T_1b) and Permian Wuerhe (P_3w) formations in four wells in Junggar Basin. For comparison, the outcrops of glutenite and sandstone were also studied. The core plugs were cut into disks of approximately 100 mm diameter and 40~ 60 mm thickness. Each disk was cut into two semi-circular disks, and a crack perpendicular to the diameter plane was cut. In order to avoid the influence of water and heat on the sample's mechanical properties, the sample was cut with 0.3~0.5 diamond wirelines under air cooling. The sample sizes refer to the shape recommended by Kuruppu (Kuruppu et al. 2014). In order to increase the fracture propagation path, the crack length was shorter than recommended. These samples (numbered G1-G6) used in this study are of different textures and appearances, as shown in figure 1.

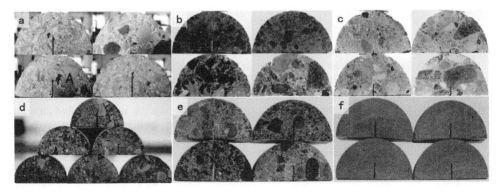

Figure 1. Samples tested in this study (numbered G1~ G6). The samples in (a) and (d) are glutenite with loose cementation. The samples in (c) and (e) are glutenite with dense cementation. (c) shows the glutenite with moderate dense cementation. (f) shows the samples made from sandstone outcrops.

2.2 Sample characteristics

Gravel occupies the most rock volume in glutenite. However, most gravels do not break during fracture propagation. Compared with the mineral constituent of the whole rock, we believe the cement constituent better reflects the cementation characteristics. The sample offcuts were first immersed in distilled water for 120 h to reduce their strength and were then ground with a bowl to eliminate gravel. The matrix was finally dried, ground, and screened with a 200-mesh sieve to collect the minerals in the cement. The minerals were analyzed by X-ray diffraction to determine the mineral type. The cementation of each group is shown in Table 1. The cementation in G1 is of much plagioclase and relatively loose. In contrast, the cementation in G3 is very dense and much quartz. In G4, the cementation contains many clay minerals and is soft. Formation G5 contains much calcite and is densely cemented.

Table 1. Cementation and hardness of the glutenite samples.

	Clay mineral %	Quartz %	Plagioclase Feldspar %	Calcite %	Hematite %	Potassium feldspar %	Matrix hardness GPa	Gravel hardness GPa
G1	28.0	22.4	42.2			7.4	0.85	3.95
G2	33.8	31.5	11.7	12.7		10.3	0.97	4.25
G3	10.9	52.9	20.6	9.2		6.4	1.29	5.08
G4	50.2	31.3	6.4		9.6	2.5	0.42	4.89
G5	18.8	21.1	19.6	22.4		18.1	1.06	2.97

The cementing properties are not only related to the cementation minerals but also the dense degree. It is inaccurate to evaluate the cementing properties only from the mineral constituent. An ideal method is to test the cementing strength, if possible. However, this approach is inefficient and complex, as the rock was naturally formed, and the cementing surface is irregular. One possible approach is to compare the mechanical properties of the matrix, which reflect the cementing strength to a certain extent. Therefore, we also evaluate the hardness of gravel and matrix through nanoindentation. Indentation samples were cut and polished under air-cooled conditions to avoid the influence of water sensitivity. Detailed methods were given in the literature (Liu et al. 2022). The hardness of the matrix and gravel is shown in Table 1.

2.3 Experimental procedure

An RTR1500 from GCTS was applied for loading, as shown in Figure 2(a). This loading system provides a maximum load of 1000 kN and a frame stiffness of 1750 kN/s. A special fixture that comprises two bottom supporters and a top indenter was designed for sample fixing, as shown in Figure 2(b). The two bottom rollers of the fixture are steel rods with a diameter of 5 mm, while the top indenter is a half-cylinder with a diameter of 5mm. A load sensor of 3 tons and an LVDT with a measuring range of 0~25 mm were used for load and displacement gathering. In addition, a clip gauge was used to gather the open distance of the crack mouth. Loading was controlled through displacement control at a 0.02 mm/min loading rate. The loading was terminated after the peak load when the load dropped to half of the peak load to avoid rock fragmentation.

DIC and AE were used to monitor the formation of the cracks. The sample surface was polished and painted to generate a random speckle pattern. The pictures were shot every five seconds with a camera and macro-lens, as shown in Figure 2(c). Acoustic emissions were also acquired to monitor fracture propagation. An acoustic emission monitoring system from ASC and an RS-54B sensor from Softland Times were used to acquire the acoustic emission. The nominal center frequency of the transducers covers 0.1~0.9 MHz. The sensor was stuck on the semi-circular face near the fracture tip. During the experiment, the acquisition settings remained unchanged.

Figure 2. Experimental facilities. (a) RTR 1500 loading system. (b) Installation of sensors. (c) Diagram of image acquisition for DIC.

3 RESULTS AND DISCUSSION

3.1 *Cracking process*

In order to better reflect the process of fracture propagation, we analyzed the change of crack mouth opening distance (CMOD) with the load. In this process, acoustic emission and deformation on the sample surface were also analyzed. The speckle images collected during the tests were processed with Vic-2D to determine the sample surface deformation. The strain field perpendicular to the prefabricated crack better reflected the deformation of the sample as the type I fracture mainly extends along the original crack. Acoustic emissions were also used to analyze the propagation of cracks. The amplitude of the AE signal and the accumulative amplitude of all AE signals before a specific time were used to characterize the cracking process. To analyze the fracture propagation process in different glutenites, we selected representatives in G4, G4, G5, and G6 to characterize the fracture propagation.

The glutenite in G3 is siliceous cemented with gravel content of about 40.0% and an average gravel size of about 5.1mm. The experimental results in Figure 3 show that the peak load is 4715.5 N. The AE results show that severe stress concentration and continuous acoustic emission were detected when the load reached 82% of the peak load. For ease of expression, we defined the ratio of the load at a given moment to the peak load as the load ratio. DIC analysis results show that microcracks have formed around gravel before fracture propagation. In G3, cracks appeared around the gravel when the load ratio reached 90%.

The glutenite in G4 is argillaceous cemented with gravel content of about 36.7% and an average gravel size of about 6.7mm. The experimental results in Figure 4 show that the peak load is 1785.4 N, much lower than the glutenite in G3. Continuous AE signals were detected when the load ratio reached 60%. DIC analysis results show that microcracks have formed around gravel before fracture propagation. Apparent fracture zones had developed around the gravel when the loading ratio reached 72%. Compared with the siliceous cemented rocks in G3, argillaceous cemented glutenites were cracked at a lower load ratio.

The glutenite in G5 is calcareous cemented with gravel content of about 19.5 % and an average gravel size of about 4.6mm. Figure 5 shows that the peak load is 3740.5 N, at the same level as G3. AE signals indicated that the initial crack began at the load ratio of 80%. DIC results showed no apparent microcracks around gravel before fracture propagation.

The sandstone in G6 is homogeneous with a uniform texture and a maximum grain diameter of less than 0.3mm. The experimental results in Figure 6 show that the peak load is 4808.7 N. The acoustic emission of such samples appeared very late, and sustained acoustic emission occurred only when the load ratio exceeded 97%. DIC results show no apparent microcracks near the fracture before fracture propagation.

Cracking processes in groups of G3-G6 are pretty different. The cracking process of glutenite is different from that of homogeneous sandstone. The microcracks in glutenite are generated at a lower load ratio, indicating that the presence of gravel promotes microcracking. In our opinion, the

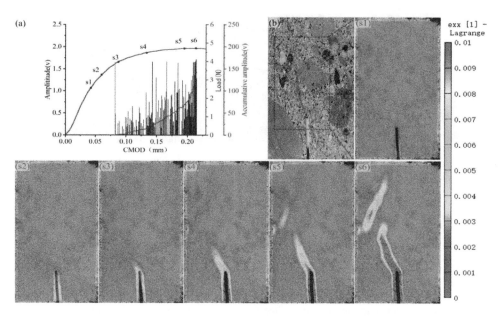

Figure 3. Fracture propagation in a sample of G3. (a) Load-CMOD-acoustic emission curve. (b) Fracture morphology and DIC analysis area. S1~ S8 shows the horizontal strain at different loading stages.

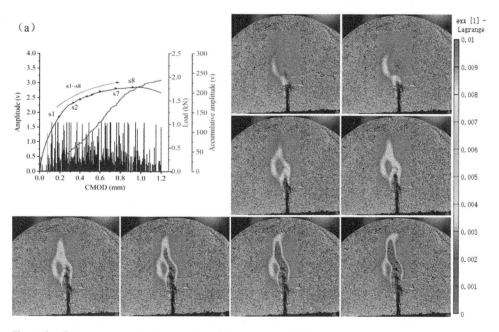

Figure 4. Fracture propagation in a sample of G4. (a) Load-CMOD-acoustic emission curve. (b) Fracture morphology and DIC analysis area. S1~ S8 shows the horizontal strain at different loading stages.

additional stress between gravels leads to early cracking. According to Orowan, additional stress exists between heterogeneous bodies in heterogeneous materials. This stress is determined by the distance between gravels, the matrix's shear modulus, and the gravel shape. Due to the high gravel content and large gravel size, the additional stress between gravel is more prominent in glutenite.

In the weakly cemented glutenite, the matrix and cementing surface are easier to crack due to weak cementing, and the microcracking is initiated at a low load ratio, as shown in Figure 4. After the microcracking, a long loading stage is still needed before the unstable cracking. This process is apparent in the glutenite with argillaceous cement. According to the fracture mechanics in concrete, this fracture characteristic is caused by the cohesive stress between gravels. This fracturing process shows strong plasticity and is not conducive to forming the complex fracture network in hydraulic fracturing.

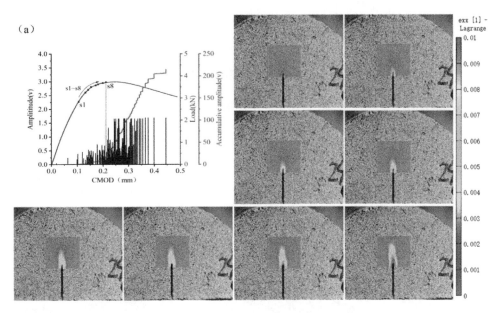

Figure 5.   Fracture propagation in a sample of G5. (a) Load-CMOD-acoustic emission curve. (b) Fracture morphology and DIC analysis area. S1~ S8 show the horizontal strain at different loading stages.

On the other hand, masses of cracks will be induced around gravel during fracture propagation. These cracks lead to more branching near the fracture surface and form the fracture process zone. This fracture process zone has important implications for hydraulic fracturing. On the one hand, this fracture pattern requires more energy to form cracks, resulting in more energy consumption and reducing the fracture length in hydraulic fracturing. On the other hand, branching fractures in the fracture process zone can improve rock permeability. The premise, of course, is that these microcracks are not easy to close.

3.2   Fracture pattern

3.2.1   Fractures within the samples

The cracks within the samples were detected through the CT scan with a resolution of $8.45{\sim}24.19\mu$m, as shown in Figure 7. It was found that the fracture pattern in glutenite fell into two types: 1) simple fractures with little branching, as shown in Figures 7(a), (d), (g), and (f), and 2) complex fractures with masses of branching, as shown in Figures 7(b), (b) and (e). The first type of fracture pattern has an apparent primary fracture, which is tortuous in the fracture surface. This fracture pattern is mainly developed in strongly cemented glutenite. Siliceous cemented glutenite in G3, and calcareous cemented glutenite in G2 and G5 belong to this type. The second fracture pattern occurs in weakly cemented glutenite. The argillaceous cemented glutenite in G4 and the glutenite cemented by feldspar in G1 belong to the second type. In other words, a wide range of fracture process zone and complex fractures are produced in the weakly cemented glutenite. In

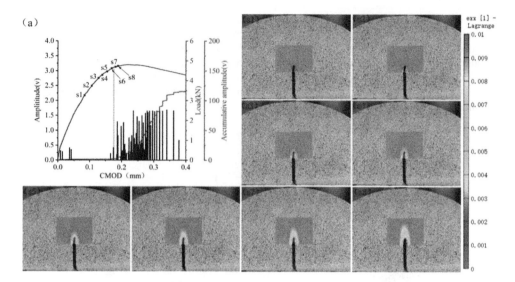

Figure 6. Fracture propagation in a sample of G6. (a) Load-CMOD-acoustic emission curve. (b) Fracture morphology and DIC analysis area. S1~ S8 show the horizontal strain at different loading stages.

contrast, the fracture process area is small in the strongly cemented glutenite like the siliceous and calcareous cemented, and the fracture morphology is simple.

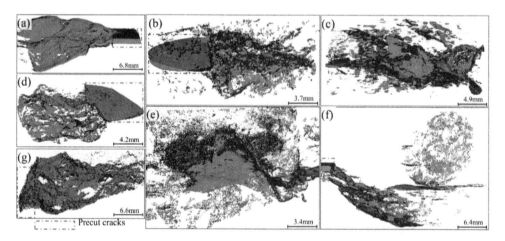

Figure 7. Fractures within the samples. (a) and (d) show the fractures in the siliceous cemented glutenite in G3. (b) and (c) show the fractures in the glutenite cemented by feldspar in G1. (e) shows the fractures in the argillaceous cemented glutenite in G4. (f) and (g) show the fractures in the calcareous cemented glutenite in G2 and G5.

3.2.2 Failure patterns on the fracture surface

As mentioned above, two fracture patterns exist in glutenite. The fracture morphology is related to the behavior when the fracture hits the gravel. Behaviors such as penetration and deflection were observed on the fracture surfaces, as shown in Figure 8. In Figures 8(a) and (b), the fracture propagated around the gravel, and no gravel was fragmented. In Figure 8(c), a small amount of gravel was penetrated. In contrast, most gravel was penetrated in Figures 8(d) and (e). The fracture behavior is related to the cementation of the glutenite. The fractures are more likely to extend around gravel in glutenite with cementation of clay and feldspar. In contrast, cracks are more likely to penetrate gravel in glutenite with siliceous and calcareous cementation. The fracture behavior is also affected by the mechanical properties between the gravel and matrix. When the gravel is much harder than the matrix, the high gravel strength prevents the gravel from being penetrated. From the view of energy consumption, fracture penetrating the gravel requires more energy consumption than extending around gravel, although the expansion path is increased. When the gravel is not tough enough, penetration requires less energy consumption. In G5, the hardness of gravel is lower than in other groups, resulting in minor mechanical properties difference between the matrix and gravel and the penetration of gravel.

The tortuosity of the fracture surface is also related to fracture behavior. The deflection by the gravel can significantly increase the roughness of the fracture surface, as shown in Figures 8(a) and (b). In particular, fracture propagation around large gravel significantly increased fracture surface roughness in argillaceous cemented glutenite, as shown in Figure 8(b). The roughness of the fracture surface will increase the flow resistance of fracturing fluid during hydraulic fracturing and is not conducive to the migration of proppant. In contrast, penetrating the gravel does not increase the roughness and tortuosity of the fracture surface. As shown in Figures 8(d) and(e), penetrated gravel did not increase the complexity of producing the fracture surfaces as in sandstone in Figure 8(f). That is because the fracture direction is not greatly changed, resulting in a lower roughness of the fracture surface.

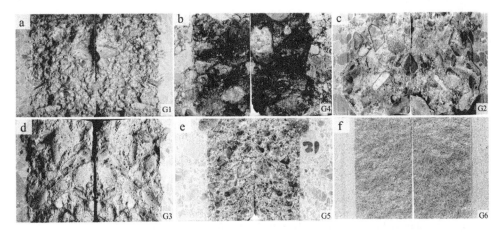

Figure 8.   Fractures on glutenite surface. (a) and (b) show the fractures on glutenite samples with cementation of clay and feldspar. Fractures extending around gravel are visible. (c) and (e) show the fractures on glutenite samples with calcareous cementation. (d) shows the fractures on glutenite samples with siliceous cementation. (f) shows the fractures on sandstone samples.

4   CONCLUSIONS

To study the influence of cementation on fracture propagation in glutenite, we performed three-point bending tests on glutenite samples with different cementation. The results show that cementation significantly affects fracture propagation in glutenite. Glutenites with different cementation are different in the fracture propagation process and fracture morphology. The fracture propagation process is more similar to sandstone in the glutenite with strong cementation, such as siliceous cement and calcareous cement. Not until the load is close to the peak load will acoustic emission and fracture process zone appear. In this kind of glutenite, the fracture process zone is small in size. The fracture will penetrate gravel and forms a single simple fracture when it hits the gravel. However, the fracture propagation process and fracture morphology in glutenite with soft cementation differ from the homogeneous sandstone. According to acoustic emissions and fracture process zone, cracks appeared at a lower load ratio. The fracture process zone in the weakly cemented glutenite is much larger than the strongly cemented glutenite. The gravel was not penetrated, fractures extended around gravel, and the fractures formed are more tortuous and complex. Therefore, from the perspective of hydraulic fracturing, the glutenite with strong cementation can be treated as a homogeneous rock. In contrast, the glutenite with weak cementation cannot be treated as a homogeneous rock in hydraulic fracturing. The fractures are tortuous and complex, increasing flow resistance and hindering proppant migration.

REFERENCES

Chen B, Xu B, Li B, Kong M, Wang W, Chen H (2020) Understanding the performance of hydraulically fractured wells in the laumontite-rich tight glutenite formation. *J Petrol Sci Eng* 185: 106600

Doll B, Ozer H, Rivera-Perez J, Al-Qadi IL, Lambros J (2017) Damage zone development in heterogeneous asphalt concrete. *Eng Fract Mech* 182: 356–371

Hou B, Zeng C, Chen D, Fan M, Chen M (2017) Prediction of Wellbore Stability in Conglomerate Formation Using Discrete Element Method. *Arab J Sci Eng* 42: 1609–1619

Jia H, Ji H, Wang L, Gao Y, Li X, Zhou H (2017) Reservoir quality variations within a conglomeratic fan-delta system in the Mahu sag, northwestern Junggar Basin: Characteristics and controlling factors. *J Petrol Sci Eng* 152: 165–181

Kang X, Hu W, Cao J, Wu H, Xiang B, Wang J (2019) Controls on reservoir quality in fan-deltaic conglomerates: Insight from the Lower Triassic Baikouquan Formation, Junggar Basin, China. *Mar Petrol Geol* 103: 55–75

Kuruppu MD, Obara Y, Ayatollahi MR, Chong KP, Funatsu T (2014) ISRM-Suggested Method for Determining the Mode I Static Fracture Toughness Using Semi-Circular Bend Specimen. *Rock Mech Rock Eng* 47: 267–274

Li L, Meng Q, Wang S, Li G, Tang C (2013) A numerical investigation of the hydraulic fracturing behavior of conglomerate in Glutenite formation. *Acta Geotech* 8: 597–618

Li S, He H, Hao R, Chen H, Bie H, Liu P (2020) Depositional regimes and reservoir architecture characterization of alluvial fans of Karamay oilfield in Junggar basin, Western China. *J Petrol Sci Eng* 186: 106730

Liu J, Ge H, Zhang Z, Wang X (2022) Influence of mechanical contrast between the matrix and gravel on fracture propagation of glutenite. *J Petrol Sci Eng* 208

Ma X, Zou Y, Li N, Chen M, Zhang Y, Liu Z (2017) Experimental study on the mechanism of hydraulic fracture growth in a glutenite reservoir. *J Struct Geol* 97: 37–47

Rui Z, Guo T, Feng Q, Qu Z, Qi N, Gong F (2018) Influence of gravel on the propagation pattern of hydraulic fracture in the glutenite reservoir. *J Petrol Sci Eng* 165: 627–639

Sharafisafa M, Aliabadian Z, Shen L (2020) Crack initiation and failure development in bimrocks using digital image correlation under dynamic load. *Theor Appl Fract Mec* 109: 102688

Shi X, Qin Y, Xu H, Feng Q, Wang S, Xu P, et al. (2021) Numerical simulation of hydraulic fracture propagation in conglomerate reservoirs. *Eng Fract Mech* 248: 107738

Su C, Wu Q, Weng L, Chang X (2019) Experimental investigation of mode I fracture features of steel fiber-reinforced reactive powder concrete using semi-circular bend test. *Eng Fract Mech* 209: 187–199

Wei MD, Dai F, Xu NW, Zhao T, Xia KW (2016) Experimental and numerical study on the fracture process zone and fracture toughness determination for ISRM-suggested semi-circular bend rock specimen. *Eng Fract Mech* 154: 43–56

Zhao Z, Guo J, Ma S (2015) The Experimental Investigation of Hydraulic Fracture Propagation Characteristics in Glutenite Formation. *Adv Mater Sci Eng* 2015: 1–5

Zhi D, Tang Y, Wenjun HE, Guo X, Huang L (2021) Orderly coexistence and accumulation models of conventional and unconventional hydrocarbons in Lower Permian Fengcheng Formation, Mahu sag, Junggar Basin. *Petrol Explor Dev+* 48: 43–59

Advances in Energy, Environment and
Chemical Engineering – Abdullah & Osman (Eds)
© 2023 The Author(s), ISBN: 978-1-032-36083-6

Analysis of water quality improvement of Qiantang River Xianghu water transfer project based on two-dimensional flow dynamics and numerical simulations

Chengjie Tu
Zhejiang Province Qiantang River Basin Center, China

Jie Zhang*
Puyangjiang River Basin Management Office, Xiaoshan District, Hangzhou, China

Dongfeng Li, Han Ye, Zihao Li, Xianzhe Sheng & Xiongwei Zhang
Key Laboratory for Technology in Rural Water Management of Zhejiang Province Zhejiang University of Water Resources and Electric Power, Hangzhou, China

ABSTRACT: Qiantang River water diversion and enhancing the fluidity of the river network and Xianghu Lake water body are important countermeasures to improve the water quality of rivers and lakes. In this paper, the flow and water environment of Xianghu Lake and the river network after diversion are calculated and simulated using the plane two-dimensional hydrodynamic water environment mathematical model. The three indicators BOD, DO, and vertical oxygen curves of environmental water quality from the diversion outlet to Zheshanwan gate are selected. By analyzing the temporal and spatial changes of these three indicators, the impact of the water transfer project on the water environment can be predicted, which is of great significance to the guidance of the diversion project and the improvement of the water environment.

1 INTRODUCTION

The Nansha plain area south of the Qiantang River in Hangzhou, including Binjiang District, Xiaoshan District and Qiantang District, is an important part of Hangzhou's future urban development. But the water quality of the river network and Xianghu Lake is poor, and the water environment still can not meet the needs of social development. There is an urgent need to improve the water environment. The two-dimensional hydrodynamic and water environment mathematical model is an important tool and means (Gao et al. 2015; Yang et al. 2018). By the established two-dimensional numerical model (Liu et al. 2014; Tong et al. 2016; Wu & He 2019), the hydrodynamic and water environment indexes BOD and DO are calculated, and the temporal and spatial changes of the indexes are analyzed the improvement effect of the river network and lake water environment is predicted (Chen & Ying 2016).

2 ESTABLISHMENT AND VERIFICATION OF PLANE TWO-DIMENSIONAL WATER ENVIRONMENT MATHEMATICAL MODEL

2.1 *Basic equation of plane two-dimensional water environment mathematical model*

Water flow continuity equation:

$$\frac{\partial z}{\partial t} + \frac{\partial (hu)}{\partial x} + \frac{\partial (hv)}{\partial y} = 0 \tag{1}$$

*Corresponding Author: zhangjie11012022@126.com

DOI 10.1201/9781003330165-84

X- and Y-direction flow equation:

$$\frac{\partial u}{\partial t} + u\frac{\partial u}{\partial x} + \upsilon\frac{\partial u}{\partial y} - f\upsilon + g\frac{\partial z}{\partial x} + g\frac{u\sqrt{u^2 + \upsilon^2}}{c^2 h} = \lambda\Delta u \tag{2}$$

$$\frac{\partial u}{\partial t} + u\frac{\partial \upsilon}{\partial x} + \upsilon\frac{\partial \beta\upsilon}{\partial y} + fu + g\frac{\partial z}{\partial y} + g\frac{\upsilon\sqrt{u^2 + \upsilon^2}}{c^2 h} = \lambda\Delta\upsilon \tag{3}$$

Where: x and y are spatial coordinates; z is the water level; t is the time; u,v is the velocity component in the X direction and Y direction, respectively; h is the water depth; f is Coriolis force, $f = 2W\sin\Phi$; W is the rotation velocity of the earth; Φ is the latitude of the earth; c is Xie Cai coefficient; g is the gravitational acceleration.

Water quality equation:

$$\frac{\partial hC}{\partial t} + \frac{\partial uhC}{\partial x} + \frac{\partial vhC}{\partial y} = \frac{\partial}{\partial y}\left(E_x h\frac{\partial C}{\partial x}\right) + \frac{\partial}{\partial y}\left(E_x h\frac{\partial C}{\partial}\right) + S + F(C) \tag{4}$$

Where C is the concentration (mg /L); E_x and E_y are diffusion coefficients in X and Y directions, respectively (m/s^2); S is the source and sink item ($g/m^2/ s$); F (C) is biochemical reaction term.

2.2 Establishment of model and verification

The recent relatively disaster "20131006" Typhoon Fitow, with typical rainfalls and complete measured data, is used for verification calculation. The maximum water level and water level process of each representative station are relatively consistent with the actual situation, so it is considered that the calculation model is reliable and can be used to calculate various design conditions.

3 MODEL SCOPE AND BOUNDARY CONDITIONS

The calculation scope of the two-dimensional hydrodynamic and water environment mathematical model of the river network and Xianghu Lake are shown in Figure 1, the river water flow inlet named from No.21 to No.25, the river water flow outlet named from No.301 to No.316. Three typical diversion lines are selected. The inlet boundary condition is that No.25 is given discharge

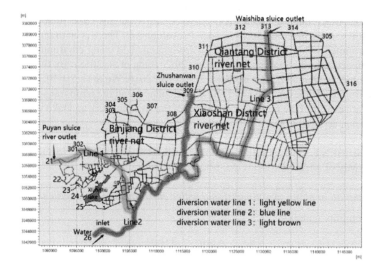

Figure 1. The river network and Xianghu Lake model and inlets and outlets.

of 50 m³/s, the others are zero; the outlet boundary condition is that the from No.301 to No.316 is given discharge -5m³/s; initial condition: in all waters, u=0, v=0, the water level is 3.7m, BOD and DO is respectively 10 mg/L and 2 mg/L.

4 CALCULATION DISCUSSION AND ANALYSIS

4.1 *Water flow hydrodynamic strengthening of Qiantang River diversion*

4.1.1 *Velocity vector and size distribution*

As can be seen from Figure 2, before the diversion of the Qiantang River, the flow waters of Xianghu lake and the river network were relatively weak, and the flow velocity was almost 0. After the diversion, the flow velocity increased, and velocity increased from 0.025 m/s to 0.25 m/s.

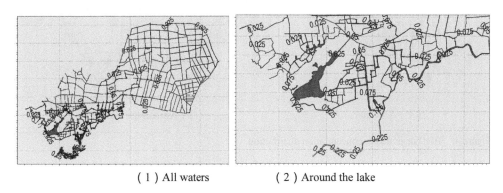

(1) All waters　　　　(2) Around the lake

Figure 2.　Velocity vector and size distribution.

4.1.2 *Velocity vector and water level distribution*

As can be seen from Figure 3, before the diversion of the Qiantang River, the initial water level condition for the calculation of the water area is given as 3.7 m. After two days of water diversion, the water level at the diversion inlet increases to 3.93 m, and the water level at Xianghu Lake increases to 3.753 m.

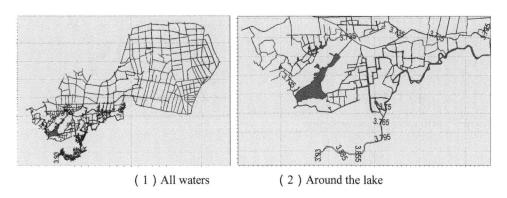

(1) All waters　　　　(2) Around the lake

Figure 3.　Velocity vector and water level distribution.

4.2 *Water flow water quality improvement of Qiantang River diversion*

4.2.1 *BOD improvement analysis*

Before the water diversion of the Qiantang River, it is given that the initial water quality of the calculated waters is fives classification, and the water quality index BOD is 10 g / L.

4.2.1.1 BOD degradation analysis

As seen from Figure 4, after two days of water diversion, BOD in the whole waters attenuates, among which the attenuation in Xianghu Lake waters is the largest, with the maximum value of 1.26 g/L, and that in other waters is 0.66-1.14 g/L.

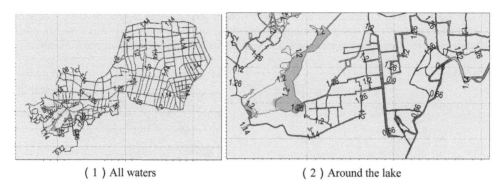

(1) All waters (2) Around the lake

Figure 4. BOD degradation concentration distribution.

4.2.1.2 BOD improvement analysis

As can be seen from Figure 5, due to the accelerated flow speed of the water body and the increase of fluid turbulence, the BOD in the whole water area is attenuated, and the attenuation near the water intake is the largest. The BOD concentration is reduced to 3.2 g/L-7.6 g/L, and the BOD concentration in Xianghu Lake and other water areas is reduced to 7.6 g/L.

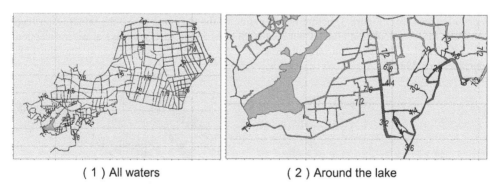

(1) All waters (2) Around the lake

Figure 5. BOD concentration distribution.

4.2.2 *DO improvement analysis*

4.2.2.1 Reaeration analysis

As can be seen from Figure 6, with the increase in water flow speed and fluid turbulence, the air enters the water body, and the water flow is aerated. The reaeration concentration in the whole water area increases from 1.2 g/L to 3.4 g/L. The reaeration concentration in the water area near the water intake increases the most, with an increase of 3.4 g/L. The reaeration concentration in the water area near Xianghu Lake increases from 1.4 g/L to 2.2 g/L.

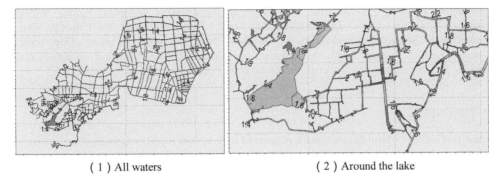

| (1) All waters | (2) Around the lake |

Figure 6.   Reaeration concentration distribution.

4.2.2.2 DO improvement distribution
As can be seen from Figure 7, before the diversion of the Qiantang River, the initial dissolved oxygen concentration in the calculated water area is given as 2. Due to the aeration of the water flow, the reoxygenation concentration in the whole water area increases, and the dissolved oxygen DO concentration increases from 2.8 g/L to 7.2 g/L. The dissolved oxygen concentration in the water area near the diversion outlet is the largest, with an increased value of 7.2 g/L, and the dissolved oxygen concentration in the water area near Xianghu Lake increases from 3.2 g/L to 4.4 g/L.

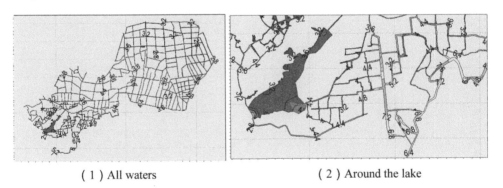

| (1) All waters | (2) Around the lake |

Figure 7.   DO concentration distribution.

5   CONCLUSION

In this paper, the flow and water environment of Xianghu Lake and the river network after diversion are calculated and simulated using the plane two-dimensional hydrodynamic water environment mathematical model. The analysis shows that after the diversion, the fluidity of rivers and lakes is greatly improved, both the BOD index and DO index are improved to varying degrees, and the water quality after the diversion is improved from class V water body to class IV water body. In conclusion, the improvement effect of diversion water quality is obvious, and the diversion scheme is feasible.

ACKNOWLEDGMENTS

This research was supported by the Funds Key Laboratory for Technology in Rural Water Management of Zhejiang Province, the Joint Funds of the Zhejiang Provincial Natural Science Foundation of China (No. LZJWZ22C030001, No. LZJWZ22E090004), the Funds of Water Resources of Science and Technology of Zhejiang Provincial Water Resources Department, China (No.RB2115, No.RC2040), the National Key Research and Development Program of China (No.2016YFC0402502), the National Natural Science Foundation of China (51979249).

REFERENCES

Chen, F., Ying, X.L. (2016) Research and application of river network model in Shaoxing Plain. *Journal of Zhejiang University of water resources and hydropower.*,28(05): 43–47.
Gao, Q., Tang, Q.H., Meng, Q.Q. (2015) Evaluation on improvement effect of connecting water environment of tidal river and lake water system. *People's Yangtze River.*,46(15): 38–40 +50.
Liu, B.J., Deng, Q.L., Zou, C.W. (2014) Study on the necessity of river lake water system connection project People's Yangtze River.,45(16): 5–6 + 11.
Tong, Y.Y., Li, D.F., Nie,H. (2016) Study on hydrodynamic and plane two-dimensional mathematical model of river network in Datian Plain. *Journal of Zhejiang University of Water Resources and Hydropower.*,28(01): 14–17.
Wu, Y, He, L.Q. (2019) Discussion on constructing hydrodynamic model of Cao'e River flow area in Shaoxing City. *Zhejiang Water Conservancy Science and Technology.*, 47(03): 21–23 + 31.
Yang, W., Zhang, L.P., Li Z.L., Zhang, Y.J., Xiao, Y., Xia, J. (2018) Study on river lake connection scheme of urban lake group based on water environment improvement. *Journal of Geography.*,73(01):115-128.

*Advances in Energy, Environment and
Chemical Engineering – Abdullah & Osman (Eds)
© 2023 The Author(s), ISBN: 978-1-032-36083-6*

Study on the construction of defense system of Qiantang River estuary seawall project in the new era

Yanfen Yu
Zhejiang Province Qiantang River Basin Center, China

Pingping Mao*
Zhejiang Qiantang River Seawall Property Management Co., Ltd, China

Zhihao Fang
*Key Laboratory for Technology in Rural Water Management of Zhejiang Province, Zhejiang University of
Water Resources and Electric Power, Hangzhou, China*
School of Naval Architecture and Maritime, Zhejiang Ocean University, Zhoushan, China

Yishan Chen, Mei Chen, Jianqing Zhao & Linjie Wang
*Key Laboratory for Technology in Rural Water Management of Zhejiang Province, Zhejiang University of
Water Resources and Electric Power, Hangzhou, China*

ABSTRACT: After more than 20 years of operation and repeated typhoon attacks, the original standard seawall of Qiantang River has gradually exposed some problems, such as the obvious settlement of some seawalls, inability to meet the original design standards, thin or unclosed structure of some seawalls, mismatch between the seawall construction standards, and the economic and social development of the reserve. In order to explore the seawall process construction management system of the Qiantang River estuary under the new situation, we inspected the first-line seawalls and reserve ponds in Hangzhou, Jiaxing, and Shaoxing, comprehensively surveyed the current situation of seawall defense, organized and held a symposium for seawall management units in all districts along the river in Hangzhou, and studied and analyzed the aspects of making up the shortcomings of seawall engineering, solving the weak problems of seawall management and further improving the flood control and tide resistance ability of seawall engineering. The research results of the defense system of the Qiantang River estuary seawall project in the new period are obtained, which improves the decision-making basis for the construction and management of the Qiantang River.

1   INTRODUCTION

The Qiantang River estuary seawall project is the lifeline of the two banks to prevent water disasters. Its defense ability is becoming more and more important to ensure the economic development along the river, social harmony and stability, and the safety of people's lives and property. The Zhejiang provincial party committee and government attach great importance to the construction and management of seawalls. After Typhoon named "9711", they successively implemented the construction of standard seawalls and strong seawalls, forming the current seawall anti typhoon and tide engineering system, and successfully withstood the attacks of previous typhoons with remarkable results (Li & Shen 2003). It has been more than 20 years since the Qiantang River standard seawall was built. After years of operation and repeated typhoon attacks, the seawall that originally

---

*Corresponding Author: maopingping@hzsteel.com

met the standard has gradually exposed problems. Some seawalls have an obvious settlement of the pond body, which can not meet the original design standard, some seawall structures are thin or not closed, and the standard does not match the economic and social development of the reserve (Guo & Hou 2022).

2 OVERVIEW OF ESTUARY AND SEAWALL

2.1 Overall situation of seawall

2.1.1 First line seawall.
There are 91 sections of seawall along the Qiantang River estuary, with a total length of 373.8km, including 181.5km on the north bank and 192.3km on the south bank. Among the first-line seawalls, the provincial seawall is 96.2km long (51.8km in Hangzhou and 44.4km in Jiaxing). At present, the standard 100-year return period and above of the first-line seawall reaches 254.9km, accounting for 68.2%. See Figure 1.

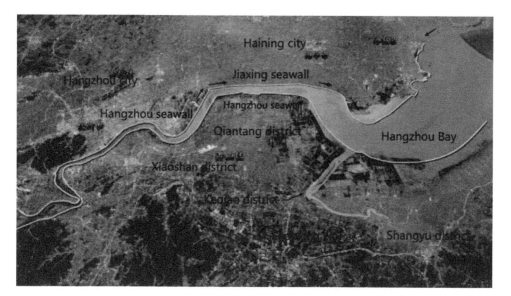

Figure 1. Distribution map of seawalls in the first line of Qiantang River estuary by region.

2.1.2 Pond preparation
Based on the investigation report of Qiantang River estuary pond preparation in 2015 and through a comprehensive review, the total length of Qiantang River estuary pond preparation is 714.21km, including 232.95km of provincial pond preparation (108.64km in Hangzhou, 69.98km in Jiaxing and 54.33km in Shaoxing) and 481.26km of local pond preparation (158.72km in Hangzhou, 52.76km in Jiaxing, 180.12km in Shaoxing and 89.66km in Ningbo).

2.2 Operation management

The management responsibility subjects of Qiantang River estuary seawall are different, including provincial direct management, local water conservancy department management, other local government departments management, enterprise management, and other management subjects, involving four prefectures and cities, 15 counties (cities, districts) and 34 management units in Hangzhou, Ningbo, Shaoxing, and Jiaxing. Among them, 15 sections in the enterprise pond are managed by many enterprises such as port administration and nuclear power. The water administration departments at all levels shall be responsible for the safety supervision and management and technical guidance of seawalls according to their management authority.

2.3 *Current defense system of seawall*

The seawall defense system on the north and south banks of the Qiantang River estuary consists of two types: the basic defense line within the design standard and the super standard defense system.

2.3.1 *Defense baseline within*

The basic defense line in the design standard refers to the basic defense assurance line formed by the main body of the first-line seawall according to the design standard. The basic defense line of the seawall is mainly the first-line seawall near the river. In some sections, the first-line seawall near the river and the second-line seawall jointly meet the design defense standard and jointly form a defense line (in this case, there are only Changshan Qingshan standard seawall and Qinshan Lantianmiao seawall in Haiyan County, Jiaxing City, with a total length of 4.505km).

2.3.2 *Super standard defense system*

It is composed of the basic defense line within the design standard and the rear reserve pond, which can basically form a closed super standard defense system to jointly resist extraordinary conditions such as super standard flood and tide disasters and reduce the disaster situation.

3 WEAK LINKS OF SEAWALL DEFENSE SYSTEM

3.1 *First line seawall defense shortboard*

3.1.1 *The defense standards of some seawalls do not match the development needs*
Currently, the fortification standards of the first-line seawall in reserve are uneven, and the fortification standards of 118.9km local or enterprise ponds are once in 50 years or less. The existing seawall flood control standards can no longer meet the development requirements of economic factors in the reserve. According to the analysis of the length of the first-line seawall, the mismatch rate between the current defense standard of the seawall and the current development demand is 31.8%.

3.1.2 *There are hidden dangers in some seawall structures*
After the completion and operation of the first line seawall project of Qiantang River for many years, the original seawall meeting the standard has gradually exposed potential safety hazards. Some seawalls have settled to varying degrees, and there are hidden dangers such as unclosed gaps, scouring of embankment feet, and unsatisfied overall stability. The actual moisture-proof capacity of the seawall has not reached the original design standard.

3.1.3 *Different seawall management capabilities*
According to the main body of construction and management and protection, the seawall is divided into the provincial directly managed seawall, local water conservancy department managed seawall, other local government departments managed seawall, and enterprise managed seawall. The management model of different ownership and management subjects of Qiantang River seawall and decentralized management, protection, and flood control responsibility subjects restricts the effectiveness of seawall management. Most of the seawall property rights are not clear, the connection between construction and operation is insufficient, and the participation of social capital is insufficient.

3.2 The defense function of pond preparation is gradually weakened

3.2.1 The defense function of pond preparation gradually disappears
In recent years, due to the development of new urban areas along the river and the construction of road traffic and other infrastructure, some prepared ponds have been disposed of or destroyed section by section, forming a gap, making the pond line discontinuous and losing its complete defense capacity.

3.2.2 The management and protection mechanism of most reserve ponds is not smooth
The reserve pond line retired from the front line has a wide range, scattered distribution, and old pond body, its functions are not clarified in time, and the pattern of the second line of defense has not been effectively confirmed. There is no basis for relevant laws, regulations, and planning, and the total amount of daily maintenance funds is small.

3.2.3 The contradiction between pond preparation protection and development and utilization is prominent
The reserve pond is separated between the cities and connected development zones, and there are many contradictions in the layout of local economic and social development along the river. The local authorities believe that the functions of the second and third-tier seawalls have been degraded, the preparation of seawalls has affected the overall layout of cities and towns, restricted the land use efficiency, and the demand for the preparation of seawalls is very urgent.

4 DEFENSE DEMAND ANALYSIS

4.1 Requirements for improving the safety performance of seawall

Under the condition of meeting the flood control and tide control standard, the seawall at the Qiantang River estuary still has a low safety margin, such as small pond top width.

4.2 Improvement of flood control and tide control standards

First, seawalls on both sides are still not met compared with the existing planning standards. Second, urban function and urban energy level need to be further strengthened. Third, there are power plants and other key energy bases near the seawall in the protected area along the seawall. The defense standards of important bases with low defense standards need to be gradually improved.

4.3 Management requirements

Due to the relative dispersion of management subjects and uneven management levels, there is still a large gap in seawall management in terms of specialization, intensification, and informatization, which still needs further improvement.

5 COUNTERMEASURES FOR IMPROVING THE DEFENSE SYSTEM CONSTRUCTION

5.1 Accelerate the construction of first-line high-standard seawall projects

In order to ensure the safety and reliability of the seawall defense line, based on improving the tide control standard, the safety performance of the seawall is improved by comprehensively adopting various methods such as wave elimination in front of the pond, widening of the pond body, anti wave scouring, smooth passage, intelligent management, and protection, etc. (Chen & Zhou 2003)

5.2 Rationalize the seawall management mechanism

In order to improve the intensive and efficient seawall management, we must implement the seawall flood control responsibility system, promote the construction of the seawall "10 billion investment" project, and promote the ownership replacement of provincial and local seawalls.

In combination with the construction of the seawall "10 billion investment" project, the provincial managed seawall and enterprise seawall are integrated into the regional development and demand positioning, the bid raising construction is incorporated into the local overall planning and construction, the establishment of financing platforms is encouraged, and social capital is attracted to actively participate in the development and construction (Liang 2011).

5.3 Strengthen pond preparation management and comprehensive utilization

Reserve pond is an important part of the estuary flood control and tide control system, and it is an effective defense line to limit the expansion of the disaster scope of super standard storm surge. The first-line seawall and the second-line seawall jointly form a safety barrier for flood control and typhoon prevention, strengthen the functional evaluation of the second and third-line seawall in the defense system, and clarify the role orientation of pond preparation and its "storage or waste" conditions (Huang 2007). Other reserve ponds shall be properly disposed of according to the standard raising and reinforcement of the first-line seawall to comply with and support the development of local economic construction. We must improve the concept, give full play to the comprehensive functions of municipal roads, highways, and other infrastructure in the hinterland along the river, deeply study the main response measures when the Qiantang River seawall encounters storm surges of all levels, and effectively deal with the prevention of super standard storm surges (Chen 2018; Gong & Cui 2016).

5.4 Promote the "three modernizations" management of seawall

Focusing on property rights and digital management of seawall management. The property owner of the seawall project shall fully perform the main responsibility of construction and management, comprehensively delimit the scope of seawall project management and protection according to law and regulations, take the preparation of land and space planning as an opportunity to verify and adjust the permanent basic farmland and sea area within the scope of seawall management, and promote the confirmation of rights within the scope of seawall management.

6 CONCLUSION

Through a comprehensive analysis of the current situation of seawall defense, this paper puts forward the countermeasures for the construction of the defense system of the seawall project. (1) Adopt various methods such as wave elimination in front of the pond, widening of the pond body, anti-wave scouring, smooth passage, and intelligent management and protection, and accelerate the construction of first-line high-standard seawall projects. (2) Incorporate provincial seawall and enterprise seawall into local overall planning and construction and attract social capital to participate in development and construction. (3) Strengthen pond preparation management and comprehensive utilization. (4) Promote property rights and digital management of seawall, and build a digital seawall intelligent application system integrating the automatic control system of seawall sluice pump, intelligent dispatching system, seawall safety monitoring system, seawall road automatic management system, seawall patrol system, seawall digital identification system and seawall public interaction system. It improves the decision-making basis for the construction and management of the Qiantang River.

ACKNOWLEDGMENTS

This research was supported by the Funds of Key Laboratory for Technology in Rural Water Management of Zhejiang Province, the Joint Funds of the Zhejiang Provincial Natural Science Foundation of China (No. LZJWZ22C030001, No. LZJWZ22E090004), the Funds of Water Resources of Science and Technology of Zhejiang Provincial Water Resources Department, China (No.RB2115, No.RC2040), the National Key Research and Development Program of China (No.2016YFC0402502), and the National Natural Science Foundation of China (51979249).

REFERENCES

Chen, B.B. (2018) Technical management of seawall engineering maintenance. *Shaanxi water resources.*, S1: 216–217.
Chen, M.L., Zhou, X.J. (2003) Practice and exploration of property management in Qiantang River seawall. *Zhejiang water conservancy science and technology.*, 01: 25–26.
Gong, R.P, Cui, C. (2016) Thoughts on strengthening market-oriented operation and management of seawall facilities. *China water resources.*, 08: 37–38 + 50.
Guo, X.J., Hou, X.M. (2022) Practice of the construction idea of Haitang Anlan 100 billion projects in the project. *Zhejiang water conservancy science and technology.*,50 (01): 5–9.
Huang, H.Z.(2007) Practice and Exploration on separation of management and maintenance of Qiantang River seawall project. *Water conservancy construction and management.*,27 (04): 58–59.
Li, S.X, Shen, S.T. (2003) Current situation and countermeasures of seawall management in Zhejiang Province. *Zhejiang water conservancy science and technology.*, S1: 43–45.
Liang, J.C. (2011) Discussion on the contract mode of water conservancy project maintenance property. *Hydropower and new energy.*, 01: 48–49.

Advances in Energy, Environment and
Chemical Engineering – Abdullah & Osman (Eds)
© 2023 The Author(s), ISBN: 978-1-032-36083-6

Energy evolution characteristic of rock mass under shallowly buried explosion loading

Qindong Lin
Xi'an Modern Chemistry Research Institute, Xi'an, Shaanxi, China

Chun Feng
Key Laboratory for Mechanics in Fluid Solid Coupling Systems, Institute of Mechanics, Chinese Academy of Sciences, Beijing, China

Yundan Gan, Jianfei Yuan*, Wenjun Jiao, Yulei Zhang & Haiyan Jiang
Xi'an Modern Chemistry Research Institute, Xi'an, Shaanxi, China

ABSTRACT: During the shallowly buried explosion process, a complicated energy evolves between the explosive and the rock mass. Based on the continuum-discontinuum element method and energy statistics algorithm, the energy evolution characteristic of rock mass under shallowly buried explosion loading is investigated. First, the energy statistics code is parallelized using the OpenMP technology to enhance computational efficiency, and the accuracy is verified. Then, a full-time numerical simulation of rock mass under shallowly buried explosion loading is conducted, and the change trends of various constituent energy are investigated. The numerical results show that the various constituent energy have different change trends, and the change trends are closely related to the detonation gas pressure, rock movement characteristic, and crack characteristic.

1   INTRODUCTION

The shallowly buried explosion technology is widely used in the civilian field and military fields. In the civilian field, it is used in the excavation of rivers and mining minerals. In the military field, it is used to destroy shallowly buried fortifications. The current research methods include theoretical analysis, experimental observation, and numerical simulation. Due to the complexity of the shallowly buried explosion process, it is difficult to conduct a detailed theoretical analysis and to obtain experimental observation.

With rapid developments in the numerical method, scholars have begun studying the dynamic mechanical response of geotechnical bodies under shallowly buried explosion loading based on numerical simulation. To overcome the element distortion under explosion loading, the arbitrary Lagrange-Euler (ALE) algorithm is widely used (Liang & Cai 2022; Liu et al. 2021; Wang et al. 2021). In addition, with rapid developments in the meshless algorithm, smoothed particle hydrodynamics (SPH) is being widely used because of its unique advantage in solving explosion problems (Chen et al. 2021; Mao et al 2017).

Ambrosini (Ambrosini & Luccioni 2006) presented a numerical study related to craters produced by explosive charges located on the soil surface and validated the material constitutive model. Grujicic (Grujicic et al. 2006) developed a new material model and studied various phenomena associated with the explosion of shallow-buried and ground-laid mines. Liu (Liu et al. 2020) conducted the close-in blasting simulation and obtained the time-history curves of overpressure

---

*Corresponding Author: yjfdldx@126.com

and impulse at various points on the surface. Qian (Qian et al. 2021) performed a comprehensive numerical investigation on the blast performance of utility tunnels subjected to ground surface explosion with small-scaled distance. Wei (Wei et al. 2009) evaluated the effect of loading density, rock mass rating, and weight of charge on the rock mass damage induced by an underground explosion. Zhu (Zhu et al. 2018) investigated the blast-induced shock wave propagation in the jointed rock mass and soil cover, and studied the dynamic response of surrounding structures and ground motions.

Currently, scholars are conducting several numerical studies on the dynamic mechanical response of geotechnical bodies under shallowly buried explosion loading. However, most studies focus on the displacement and fracture evolution characteristics. During the explosion process, there is a complicated energy evolution between explosive and rock mass, while the change law is rarely investigated. To investigate the energy evolution characteristic of rock mass under shallowly buried explosion loading accurately, a full-time numerical simulation is conducted based on the continuum-discontinuum element method (CDEM) and energy statistics algorithm, and the change laws of various constituent energy are analyzed.

2 BASIC CONCEPT

As a dynamic explicit numerical method, the CDEM is established under the Lagrange system, and it can simulate the whole process of material from continuous deformation to crack and movement. The basic model in CDEM is composed of block and interface, which are used to characterize the continuous and discontinuous features of the material, respectively. The interface is composed of a real interface and a virtual interface, and the real discontinuous feature of the material is characterized by the real interface. The virtual interface not only connects two blocks and transfers mechanical information; but also provides potential space for crack expansion (Wang et al. 2019; Zhang et al. 2019).

To investigate the energy evolution characteristic of the material, Lin (Lin et al. 2021) established the energy statistics algorithm, which achieved the accurate calculation of all constituent energy, and the correctness was verified. As the energy statistics code is a serial code, it is difficult to be applied to large-scale engineering calculations. To enhance computational efficiency, the serial code is parallelized using OpenMP technology. To validate the accuracy of the parallel energy statistics code, a numerical simulation of the uniaxial compression test is conducted.

The change curves of external work W_O, cumulative energy W_T, and various constituent energy are plotted in Figure 1. From Figure 1(a), it is observed that the change trends of W_O and W_T are the same. When strain $\varepsilon = 3‰$, external work $W_O = 28.66$ J, cumulative energy $W_T = 28.92$ J, and the error is 0.9%. The change trends of various constituent energy in Figure 1(b) are closely related to the damage state of the specimen. When the compressive stress does not reach the peak value, W_{EE} and W_{PE} gradually increase with the increasing strain. When the compressive stress reaches the peak value, W_{EE} decreases rapidly, while W_R and W_D increase sharply. The change trends of various constituent energy and the agreement between W_O and W_T verify the correctness of the parallel energy statistics code.

3 NUMERICAL SIMULATION

3.1 *Numerical model*

The numerical model of explosive and rock mass is plotted in Figure 2. The vertical height of rock mass is 8 m, and the horizontal length of rock mass is 20 m. The model is meshed by triangular mesh, and there are 26,810 elements in the numerical model. The JWL equation of state is adopted to describe the explosion process, and the mechanical parameters of the explosive are listed in Table 1. The mechanical parameters of rock mass are listed in Table 2. The left, right, and

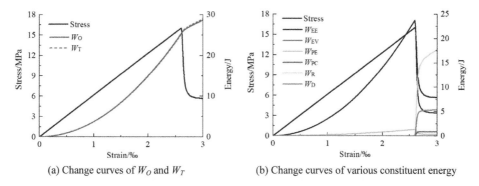

(a) Change curves of W_O and W_T (b) Change curves of various constituent energy

Figure 1. Change curves of energy and stress.

bottom boundaries of the numerical model are artificial truncation boundaries, which are set as the reflection-free boundary to avoid stress wave reflection. The top boundary is ground and set as the free boundary.

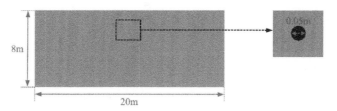

Figure 2. Numerical model of explosive and rock mass.

Table 1. Mechanical parameters of JWL equation of state.

Material	Charge density (kg/m^3)	Initial internal energy (J/m^3)	CJ pressure (GPa)	Detonation velocity (m/s)
TNT	1630	7e9	20	6930

Table 2. Mechanical parameters of rock mass.

Density (kg/m^3)	Elastic modulus (GPa)	Cohesive strength (MPa)	Tensile strength (MPa)	Friction angle (°)
2500	50	6	2	40

3.2 Numerical results

The time-history curves of various constituent energy are obtained based on the energy statistics algorithm, which are plotted in Figure 3.

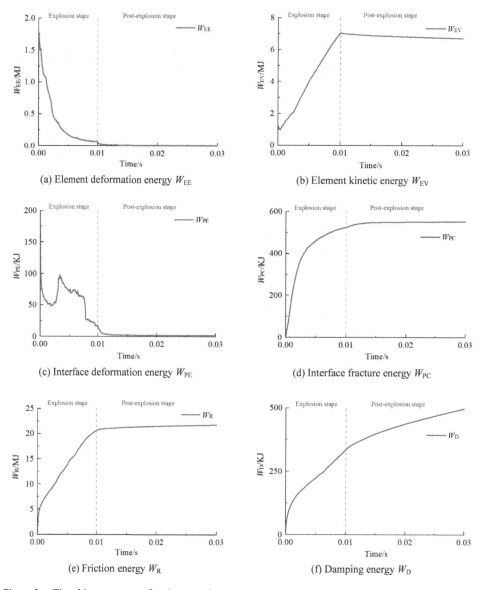

Figure 3.　Time-history curves of various consistent energy.

The element deformation energy W_{EE} increases sharply to the maximum at the initial moment. As the pressure of explosion gas gradually decreases, the fracture of interface and the slippage between elements, W_{EE}, decreases sharply, and the decreasing rate decays. When the model enters the post-explosion stage, W_{EE} decreases sharply again due to the disappearance of explosion gas, and then the decreasing rate slows down.

For the element kinetic energy W_{EV}, it can be seen that it has an obvious stage characteristic; W_{EV} gradually increases in the explosion stage, and then gradually decreases in the post-explosion stage. In the explosion stage, W_{EV} increases sharply due to the high pressure of explosion gas at the initial moment. Although the element acceleration gradually decreases with the decrease in explosion gas pressure, W_{EV} still increases slowly. In the post-explosion stage, due to the disappearance of explosion gas and the influence of gravity, W_{EV} decreases slowly.

For the interface deformation energy W_{PE}, it is observed that the changing trend is complicated. In the explosion stage, due to the effect of explosion gas, the rock around the explosive is squeezed and dislocated at the initial moment, resulting in W_{PE} increasing sharply to the maximum and then decreasing sharply. However, since the rock near the top boundary has not been significantly displaced, when the rock moves upward, the extrusion between elements occurs again, resulting in increased W_{PE}. With the movement of elements and the fracture of the interface, the extrusion between elements is weakened, and W_{PE} gradually decreases. When the model enters the post-explosion stage, the disappearance of explosion gas causes W_{PE} to decrease gradually.

For the interface fracture energy W_{PC}, it can be seen that with the increase in time, W_{PC} first increases sharply, then increases slowly, and finally remains unchanged. In the explosion stage, the cracked area increases sharply due to the high pressure of explosion gas, resulting in sharp increase in W_{PC}. Then, with the decrease in explosion gas pressure, the growth trend of W_{PC} slows down. In the post-explosion stage, due to the disappearance of explosion gas, the growth rate of the cracked area slows down again, and finally, no new cracked plane appears, which causes the growth trend of W_{PC} to slow down until it drops to zero.

For the friction energy W_R, it is observed that W_R gradually increases, and the changing trend has an obvious stage characteristic. At the initial moment of the explosion stage, the explosion gas results in large shear force and frictional displacement at the interface, so W_R increases sharply. Subsequently, due to the decrease in explosion gas pressure, the shear force gradually decreases, resulting in a slow increase of W_R. In the post-explosion stage, due to the disappearance of explosion gas, the value of shear force and frictional displacement decreases to a small value, so the growth rate of W_R decreases sharply.

For the damping energy W_D, it can be seen that W_D gradually increases, and there are differences in the growth trend of W_D at different stages. At the initial moment of the explosion stage, the explosion gas pressure is high, resulting in the nodal damping force being large. Although the nodal displacement increment is small, W_D still increases sharply. With the decrease of explosion gas pressure, the nodal damping force decreases, and the growth rate of W_D slows down. In the post-explosion stage, although the explosive gas disappears, there are still nodal damping forces and displacement increments, and W_R gradually increases.

4 CONCLUSION

Based on the numerical results presented above, the following conclusions can be drawn:

1) According to the uniaxial compression results, the correctness of the parallel energy statistics code is verified. The external work W_O is mainly transformed into the element deformation energy W_{EE} in the pre-peak stage. Once the compressive stress reaches the peak stress, W_{EE} decreases rapidly, while W_{PC}, W_R, and W_D increase rapidly.
2) The various constituent energy of rock mass have different change trends, and the change trends are closely related to the explosion gas pressure, rock movement characteristic, and rock crack characteristic. With the increase in time, W_{EE} gradually decreases; W_{EV} first increases and then decreases. The change trends of W_{PC}, W_R, and W_D are similar, which increase gradually, and the growth rate during the explosion stage is larger than that during the post-explosion stage.

During the explosion process, it is better to convert the chemical energy of the explosive into the W_{PC} and W_{EV} of rock as much as possible. The energy statistics algorithm can obtain the utilization ratio of explosive energy to evaluate the pros and cons of different explosion schemes. The research results indicate that the energy evolution characteristic of rock mass is closely related to the fracture evolution characteristic. Therefore, the corresponding relationship between the energy and fracture characteristics needs to be established in future works.

ACKNOWLEDGMENTS

The authors would like to acknowledge the financial support received from the National Key Research and Development Project of China (Project No. 2018YFC1505504).

REFERENCES

Ambrosini R D, Luccioni B M. (2006) Craters Produced by Explosions on the Soil Surface. *J Appl Mech*, 73: 890-900.

Chen J Y, Feng D L, Lien F S, et al. (2021) Numerical modeling of interaction between aluminum structure and explosion in soil. *Appl. Math. Model.*, 99: 760–784.

Grujicic M, Pandurangan B, Cheeseman B A. (2006) The Effect of Degree of Saturation of Sand on Detonation Phenomena Associated with Shallow-Buried and Ground-Laid Mines. *Shock. Vib.*, 13: 41–61.

Liang L Y, Cai X Y. (2022) Time-sequencing European options and pricing with deep learning–Analyzing based on interpretable ALE method. *Expert Syst. Appl.*, 187: 115951.

Lin Q D, Feng C, Zhu X G, et al. (2021) Evolution characteristics of crack and energy of low-grade highway under impact load. *Int. J. Pavement Eng.*, 1–16.

Liu G K, Wang W, Liu R C, et al. (2020) Deriving formulas of loading distribution on underground arch structure surface under close-in explosion. *Eng Fail Anal*, 115: 104608.

Liu S, Tang X W, Li J. (2021) Extension of ALE method in large deformation analysis of saturated soil under earthquake loading. *Comput Geotech*, 133: 104056.

Mao Z R, Liu G R, Dong X W. (2017) A comprehensive study on the parameters setting in smoothed particle hydrodynamics (SPH) method applied to hydrodynamics problems. *Comput Geotech*, 92: 77–95.

Qian H M, Zong Z H, Wu C Q, et al. (2021) Numerical study on the behavior of utility tunnel subjected to ground surface explosion. *THIN WALL STRUCT*, 161: 107422.

Wang H Z, Bai C H, Feng C, et al. (2019) An efficient CDEM-based method to calculate full-scale fragment field of warhead. *Int J Impact Eng*, 133: 103331.

Wang S, Islam H, Soares C G. (2021) Uncertainty due to discretization on the ALE algorithm for predicting water slamming loads. *Mar. Struct.*, 80: 103086.

Wei X Y, Zhao Z Y, Gu J. (2009) Numerical simulations of rock mass damage induced by underground explosion. *Int. J. Rock Mech. Min. Sci.*, 46: 1206.

Zhang Q L, Yue J C, Liu C, et al. (2019) Study of automated top-coal caving in extra-thick coal seams using the continuum-discontinuum element method. *Int. J. Rock Mech. Min. Sci.*, 122: 104033.

Zhu J B, Li Y S, Wu S Y, et al. (2018) Decoupled explosion in an underground opening and dynamic responses of surrounding rock masses and structures and induced ground motions: A FEM-DEM numerical study. *Tunn. Undergr. Space Technol.*, 82: 442–454

Advances in Energy, Environment and
Chemical Engineering – Abdullah & Osman (Eds)
© 2023 The Author(s), ISBN: 978-1-032-36083-6

Structural analysis of pressure tunnel of hydropower station based on extended finite element method

Yunfeng Peng* & Hailin Wu
College of Hydraulic & Environmental Engineering, China Three Gorges University, Hubei, Yichang, China

ABSTRACT: The lining of the tunnel generally cracks under the action of internal water pressure, which will threaten structural safety. It is difficult to simulate the discontinuity problem like cracking by the traditional finite element method. In order to study the influence of different tunnel lining structure types on structural safety and to find the better structure type, two types of tunnel lining models (circular and horseshoe) are simulated in this paper. The extended finite element method is used to calculate the stress distribution, crack distribution law, and expansion process of tunnel lining under different conditions. The results show that the lining cracks penetrate the horseshoe section, but not the round section under the same load and lining thickness. By improving the lining of the horseshoe section and analyzing it again, the results showed that the crack length was shortened, and the structural safety was improved. The study can provide a reference for the structural type selection of pressurized tunnel lining.

1 GENERAL INSTRUCTIONS

With the construction of conventional hydropower stations and pumped storage power stations, the construction of high water pressure tunnels will usher in a peak period (Li 2018). Figure 1 illustrates the relationship between the hydropower station and the pressure tunnel (Zou 2020).

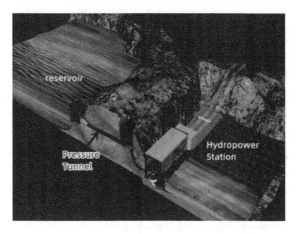

Figure 1. The relationship between the hydropower station and the pressure tunnel.

Due to the limited construction site, the application of steel plate lining in the tunnel is limited. Although the construction site does not limit the reinforced concrete lining type, it also faces the

---

*Corresponding Author: whpyf@qq.com

problem of concrete cracking under the action of water pressure. Choosing an appropriate tunnel cross-section shape can make the stress state of the tunnel lining more reasonable, thereby avoiding and reducing structural damage, which is the primary problem in the structural design of hydraulic tunnels (Huang 2015). The finite element method has been widely used in the analysis of tunnel lining structures. The traditional finite element method is based on the principle of continuity, which cannot analyze discontinuities such as structural cracking (Daux 2000). The extended finite element method (XFEM) can make the discontinuity description mesh independent, increasing the flexibility of describing discontinuity problems. It improves the shortcomings of the conventional finite element method in dealing with discontinuous problems (Duflot 2008). Xia (2019) conducted a numerical study on hydraulic fracturing of concrete gravity dams using XFEM. Yang (2019) studied the cracking characteristics of sluices under cold waves based on XFEM and the thermal-mechanical coupling method. The calculation results are consistent with the field conditions. In this paper, XFEM is used to calculate and analyze the tunnel lining with different section forms, and the generation and expansion process of lining cracks are simulated. The law of crack propagation is obtained, which shows the feasibility and unique advantages of XFEM in studying crack problems, and provides a reference for the selection of tunnel section form in practical projects.

2 IMPLEMENTATION AND VERIFICATION OF XFEM IN CONCRETE STRUCTURE

The finite element method usually uses the separation crack model and the dispersion crack model to simulate the crack generation and propagation process. In practice, it is found that these two methods have limitations. The separation crack model regards the crack as the boundary of the element. Once the crack occurs, it is necessary to adjust the node displacement or add new nodes and re-divide the mesh, which greatly increases the modeling workload and lowers the calculation efficiency. Although the dispersed crack model does not need to be re-meshed, it needs to closely mesh near the crack, which is very difficult to model. XFEM inherits the characteristics of the conventional finite element method, and its basic idea is based on element decomposition, which ensures its convergence (Xia 2019). On this basis, it uses an enrichment technique to improve the displacement function of the conventional finite element method. Two functions are added to the displacement approximation function. One is an asymptotic function that can reflect the singularity of the crack tip, and the other is a step function that can reflect the discontinuity of the crack surface, which makes the discontinuity characteristics described. At the same time, XFEM uses the level set method to locate the cracks and track the growth process of the cracks. XFEM is a kind of fixed mesh method, which avoids the trouble of mesh reconstruction in calculating cracks, improves the efficiency of calculation, and can be calculated in combination with general finite element software.

The extended XFEM displacement function consists of three parts: the general finite element, the crack crossing, and the crack tip. The node distribution after the extension of degrees of freedom is shown in Figure 2.

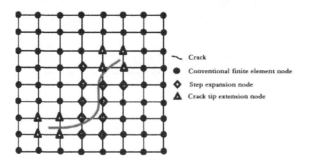

Figure 2. The XFEM element node.

At present, there is no precedent for the application of XFEM in the light structure of reinforced concrete lining, so it is necessary to explore the feasibility of the application of XFEM in reinforced concrete structures. As shown in Figure 3, the crack propagation path and distribution of the concrete beam calculated by the test and XFEM are the same. It starts from the crack tip of the initial crack and then extends towards the loading point until the destruction of the whole beam, which conforms to the characteristics of type I fracture. Therefore, it is feasible to use XFEM to simulate the crack development of the concrete structure.

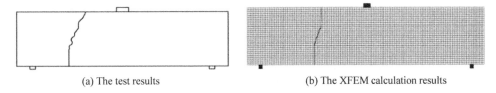

(a) The test results   (b) The XFEM calculation results

Figure 3. Comparison of crack distribution between concrete beam test and XFEM calculation.

3 ANALYSIS OF TUNNEL LINING STRUCTURE WITH DIFFERENT SECTIONS

3.1 Calculation model and parameters

A tunnel is a pressurized water diversion and power generation tunnel with a total length of 15km. The surrounding rock is mainly class II rock and class III rock. The designer proposes two different excavation section schemes. Scheme I is a circular section with an internal radius of 2.35m after lining. Scheme II is a horseshoe-shaped section with a top arch radius of 2.35 m and a bottom arch and side arch radius of 4.7m.

In this model, the calculation range of the surrounding rock boundary of the tunnel is five times the tunnel diameter. It is assumed that the concrete lining is added after a long period for the initial support, the influence of in-situ stress is not considered, and only the structural stress after the lining is considered. The model is a plane strain problem. Figure 4 shows the finite element model of tunnel lining with different sections.

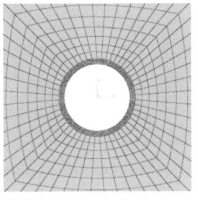
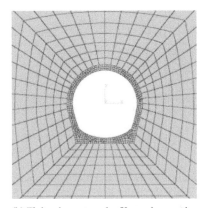

(a) Finite element mesh of circular section   (b) Finite element mesh of horseshoe section

Figure 4. The finite element model of tunnel lining with different sections.

The surrounding rock of the hydraulic tunnel adopts the mechanical parameters of class II surrounding rock, and the lining adopts C20 concrete. Table 1 shows the mechanical parameters of the model.

Table 1. Mechanical parameters of tunnel surrounding rock and lining

Type	Density/ KN/m^3	Modulus of Elasticity/ GPa	Poissons ratio	Tensile Strength/ MPa
Surrounding rock	26.0	18.0	0.20	
Concrete (C20)	24.0	28.0	0.167	1.3

The bottom edge of the tunnel surrounding rock is constrained in the Y direction, and the left and right sides are constrained in the X direction. In order to compare and analyze the stress characteristics of tunnel lining with different sections, the two schemes adopt the same working condition, that is, the load combination of water pressure (0.914MPa), lining self-weight, and surrounding rock pressure during the operation period.

3.2 *Analysis of different cross-sections*

Figure 5 shows the stress distribution and crack propagation of 40cm thick horseshoe-shaped and circular section lining under the same working conditions. The maximum value of the first principal stress of the lining of the horseshoe-shaped section is larger than that of the circular section. The stress distribution of the circular section is relatively regular. The stress of the inner arch at the left and right sides is larger, and the stress of the horseshoe-shaped section at the bottom corner transition is larger. The lining of the horseshoe-shaped section has a crack at the transition of the left and right bottom corners, and the crack at the transition of the right bottom corner runs through, while the circular section lining has many small cracks at the left and right inner arches, but it does not run through.

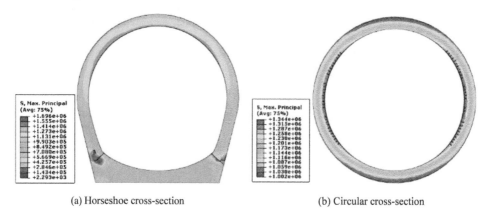

(a) Horseshoe cross-section   (b) Circular cross-section

Figure 5. Calculation results of the first principal stress of lining with different sections (unit: Pa).

Figure 6 shows the crack development process of the horseshoe section lining the underwater load. When the water load increases to 0.525 MPa, the maximum stress appears in the transition section at the bottom corner of the section, and the crack may occur in the lower right corner of the lining. When the water load is increased to 0.547 MPa, the initial crack is initiated, and a small crack is generated at the turning point of the right bottom corner. When the water load increased to 0.749 MPa, the crack continued to expand, and the crack at the bottom corner released part of the stress. The crack penetrates when the water load increases to the operating load of 0.914 MPa.

Figure 7 shows the crack development process of the circular section lining underwater load. As the circular section bears relatively uniform stress and the crack propagation at the left and right

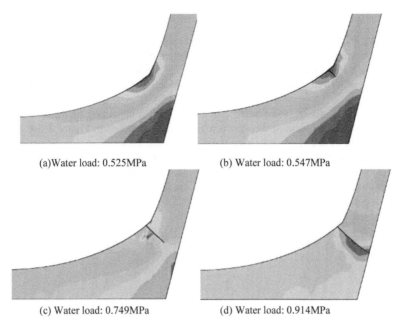

(a)Water load: 0.525MPa   (b) Water load: 0.547MPa

(c) Water load: 0.749MPa   (d) Water load: 0.914MPa

Figure 6.   The expansion process of lining cracks with water load in horseshoe cross-section.

inner arches are similar, a part of the right side of the circular section is taken to observe the crack propagation process. When the water load increases to 0.525 MPa, the stress at the right inner arch is large, and cracks may occur here. When the water load increases to 0.547 MPa and 0.749 MPa, the stress of the right inner arch continues to increase, and the possibility of cracks increases. When the water load increases to 0.914 MPa, a number of small cracks appear at the right inner arch.

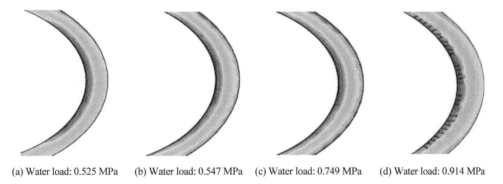

(a) Water load: 0.525 MPa   (b) Water load: 0.547 MPa   (c) Water load: 0.749 MPa   (d) Water load: 0.914 MPa

Figure 7.   The expansion process of lining cracks with water load in a circular cross-section.

According to the analysis of the above calculation results, the lining stress of the tunnel with a circular section is evenly distributed, and the initial crack load is large. However, due to the geometric corner of the tunnel lining with a horseshoe section, the stress concentration is easy to occur at the transition of the footing, and the initial crack load is small. Moreover, the lining of the horseshoe-shaped section is prone to produce penetrating cracks at the transition of the footing, which will greatly affect the safety of the structure. According to the above analysis, the circular section lining is more suitable for the lining of the pressure diversion tunnel.

3.3 *Optimization of tunnel lining with horseshoe section*

From the above analysis results, it can be seen that the tunnel lining with the horseshoe section can easily produce stress concentration and form through cracks at the transition turning point of the bottom corner. However, the construction process of tunnel lining with a horseshoe section is simpler than that of a circular section, and it can also be selected as a suitable section form in some cases.

We can consider improving the turning part of the transition section of the horseshoe-shaped section to make the stress state of the structure uniform. As shown in Figure 8, we can improve the angle transition of the horseshoe section to smooth the curve transition. Then the improved section lining is calculated, and the results show that the maximum value of the first principal stress of the lining after optimization is slightly higher than that before optimization due to the stress concentration at the crack tip; however, the cracks in the transition section of the bottom corner are not penetrated.

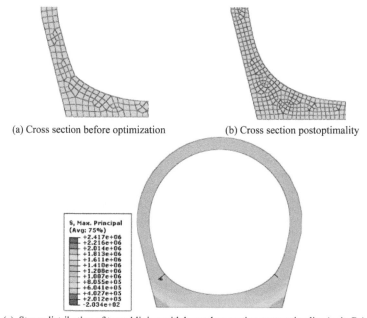

(a) Cross section before optimization   (b) Cross section postoptimality

(c) Stress distribution of tunnel lining with horseshoe section post optimality (unit: Pa)

Figure 8. Optimization of lining structure of horseshoe-shaped tunnel and its calculation results.

From the analysis of the above calculation results, it can be seen that under the same water load, the cracks in the optimized horseshoe section do not penetrate, the initial crack load is increased, the length of the longest crack is shortened, and the structural safety is enhanced, achieving the optimization effect.

4 CONCLUSIONS

This paper establishes the finite element model of pressure tunnel lining with different section forms. XFEM is used to study the generation and propagation process of tunnel lining cracks. The stress distribution and crack propagation law of tunnel lining with different sections are obtained.

XFEM is used to calculate and analyze the crack propagation law of pressure tunnel lining with different section forms, which shows that XFEM is feasible and applicable to deal with discontinuous problems such as cracking. Under the design load, the lining of the tunnel with horseshoe-shaped section cracks and cracks penetrate, while the lining of the tunnel with circular section cracks but cracks do not penetrate. In the stress state, the circular section is also better than the horseshoe-shaped section. Therefore, the circular section lining is more suitable as the lining form of the pressure diversion tunnel.

ACKNOWLEDGMENTS

This work was supported by a grant from the National Natural Science Foundation of China (No.51879146)

REFERENCES

Daux, C., Moës, N., Dolbow, J., Sukumar, N., Belytschko, T., (2000). Arbitrary branched and intersecting cracks with the extended finite element method[J]. *Int. J. Numer. Meth. Eng.* 48 (12), 1741–1760.

Duflot, M., (2008). The extended finite element method in thermoelastic fracture mechanics. *Int. J. Numer. Meth. Eng.* 74 (5), 827–847.

Huang, Y., Fu, Z., Chen, J., Zhou, Z., Wang, J., (2015). The external water pressure on a deeply buried tunnel in fractured rock. *Tunn. Undergr. Space Technol.* 48, 58–66.

Li, P., Liu, H., Zhao, Y., Li, Z., (2018). A bottom-to-up drainage and water pressure reduction system for railway tunnels. *Tunn. Undergr. Space Technol.* 81, 296–305.

Xia, Y., Jin, Y., Huang, Z., et al., (2019). *An Extended Finite Element Method for Hydro- Mechanically Coupled Analysis of Mud Loss in Naturally Fractured Formations*. 53rd US Rock Mechanics/Geomechanics Symposium. OnePetro.

Xia, Y., Jin, Y., Oswald, J., Chen, M., Chen, K., (2018). Extended finite element modeling of production from a reservoir embedded with an arbitrary fracture network. *Int. J. Numer. Meth. Fluids* 86 (5), 329–345.

Yang, D., Zhou, Y., Xia, X., Gu, S., Xiong, Q., Chen, W., (2019). Extended finite element modeling nonlinear hydro-mechanical process in saturated porous media containing crossing fractures. *Comput. Geotech.* 111, 209–221.

Zou, J.F., Wei, A., Liang, L., (2020). Analytical solution for steady seepage and groundwater inflow into an underwater tunnel. *Geomechanics and Engineering* 20, 267–273.

*Advances in Energy, Environment and
Chemical Engineering – Abdullah & Osman (Eds)
© 2023 The Author(s), ISBN: 978-1-032-36083-6*

Soil and water conservation benefits of two economic forests under different types of rainfall

Gang Tian, Gang Li* & Tao Qiu
Zhejiang Guangchuan Engineering Consulting Co., Ltd., Hangzhou, China
Zhejiang Key Laboratory of Water Disaster Prevention and Reduction, Hangzhou, China

Xiang-chao Liu & Qing-yuan Li
Zhejiang Guangchuan Engineering Consulting Co., Ltd., Hangzhou, China

ABSTRACT: The rapid development of the economy in low mountain regions has resulted in a shortage of land resources and serious reclamation of steep slopes of land, as well as regional soil and water loss. To explore the soil and water loss control mode of economic forest in low mountainous and hilly regions, and provide a reference for the high-quality development of economic forest ecosystem, this study selected two kinds of economic forests to build runoff plots in typical low mountainous and hilly regions of eastern Zhejiang, carried out runoff and soil loss monitoring, and compared and analyzed the differences of soil and water conservation functions under different rainfall conditions by using the benefit analysis method of soil and water conservation. The results showed that the average water conservation benefit of natural *Phyllostachys* forest was 54% and the average soil conservation benefit was 58%. The average water conservation benefit of natural *Myrica rubra* forest was 27% and the average soil conservation benefit was 53%. It can be seen that disturbance activities are an important inducement leading to soil and water loss of economic forests in low mountain regions, and reducing unnecessary production disturbance can enhance the function of soil and water conservation of economic forests.

1 INTRODUCTION

Human development activities in low mountain regions have a great impact on water and soil loss of economic forests. Soil and water conservation functions degenerate. A large number of previous studies showed that the growth of trees, shrubs and herbaceous vegetation is of great significance to increase runoff infiltration and reduce soil and water loss (Degens 1997). The interception and buffering effect of vegetation on rainfall reduce the kinetic energy of raindrops (Mosley 2010). Different vegetation structures have different abilities to weaken rainfall kinetic energy. The results show that canopy interception can reduce rainfall kinetic energy, and surface coverage can reduce rainfall kinetic energy and runoff energy (Liu et al. 1994). Other studies show that the canopy density of trees and shrubs is related to weakening rainfall energy, while herbs mainly weaken runoff energy (Liu et al. 2001; Li et al. 2005). The effect of the litter layer on reducing the total energy of rainfall is the largest, followed by the shrub and grass layer, and the canopy layer is the smallest (Liu et al. 1994). It can be seen that low vegetation and litter on the surface play a very important role in soil and water conservation.

As economic forest species, *Phyllostachys* and *Myrica rubra* are more frequently developed and planted in hilly and mountainous areas in the eastern part of Zhejiang Province. In the early stage, due to the lack of scientific planning, extensive management, and predatory management,

---

*Corresponding Author: happylglove@126.com

DOI 10.1201/9781003330165-88

the function of the economic forest ecosystem was fragile and mountain resources were destroyed. This study analyzed and studied the differences in soil and water conservation functions between the above two natural economic forests and exploitative economic forests, which can provide a reference for the restoration of the economic forest ecosystem.

2 MATERIALS AND METHODS

Two kinds of economic forests, including *Phyllostachys* forest and *Myrica rubra* forest, were selected in the study area. Natural runoff plots (growth of shrubs and herbs under the forest, with high surface coverage) and exploitative runoff plots (serious surface damage by digging bamboo shoots and weeding, which lacked surface coverage) were built in the two kinds of economic forests. Among them, in *Phyllostachys* forest there are two natural runoff plots and two exploitative runoff plots. And one for each of the two types of plots in *Myrica rubra* forest.

A rainfall process data is collected every 5 minutes by using a telemetry rain gauge. The runoff bucket is used at the outlet of the runoff plots to collect the runoff and sediment generated by each rainfall. Sediment content in runoff is determined by the drying method. The study refers to the classification standard of rainfall intensity issued by the China Meteorological Department (Table 1).

Table 1. Classification standard of precipitation rainfall.

Rainfall intensity grade		Rainfall within 12 hours (mm)	Rainfall within 24 hours (mm)
SP	Sprinkle	0.1–0.5	0.1–10.0
MR	Moderate rain	5.0–15.0	10.0–25.0
HR	Heavy rain	15.0–30.0	25.0–50.0
RS	Rainstorm	30.0–70.0	50.0–100.0
LR	Large rainstorm	70.0–140.0	100.0–200.0
TR	Torrential rain	$\geq$140.0	$\geq$200.0

The runoff and sediment data collected and measured manually were sorted and calculated, and matched with the rainfall data, and the soil and water loss database of two economic forests under different types of rainfall was established. According to the above data, the difference in soil and water loss between the two economic forests under different types of rainfall is analyzed, and the soil and water conservation effects of natural and exploitative economic forests are compared.

3 RESULTS AND ANALYSIS

3.1 *Water conservation benefits*

3.1.1 *Phyllostachys forests*
The regulating effect of vegetation on rainfall is affected by vegetation structure, coverage and rainfall. Table 2 arranges the rainfall-runoff of *Phyllostachys* forests under different rainfall intensity levels. It can be seen from Table 2 that among the observed rainfall events, there are four heavy rain events, three large rainstorm events, two moderate rain events, two rainstorm events and two torrential rain events, respectively. However, in terms of rainfall, the total rainfall of torrential rain events is the most, followed by large rainstorm events.

Under different rainfall intensity levels, the runoff of natural *Phyllostachys* forest is less than that of exploitative *Phyllostachys* forest. According to the variation relationship between rainfall and runoff (see Figure 1), the runoff of two types of *Phyllostachys* forests increases with the increase

Table 2.   Rainfall and runoff statistics of two *Phyllostachys* forests.

Rainfall intensity grade	Rainfall events	Total rainfall (mm)	Total runoff (mm)			
			P1	P2	P3	P4
Moderate rain	2	39.5	0.74	0.73	1.92	2.45
Heavy rain	4	144.0	7.09	7.48	9.39	8.44
Rainstorm	2	164.5	8.06	9.19	14.06	8.06
Large rainstorm	3	486.0	16.99	17.27	21.39	27.09
Torrential rain	2	565.5	13.56	13.66	29.13	30.46

Note: P1 and P2 are natural runoff plots, while P3 and P4 are exploitative runoff plots.

of rainfall. The runoff rate (slope) from moderate rain to rainstorm is greater than that of a large rainstorm to torrential rain.

According to statistical analysis, the runoff rate of natural *Phyllostachys* forest under five rainfall intensity levels is between 0.02 to 0.05, and that of exploitative *Phyllostachys* forest is between 0.05 to 0.07. The water conservation benefit of natural *Phyllostachys* forest is 18% to 54%. At the same time, with the increase in rainfall, the benefit increases as a whole.

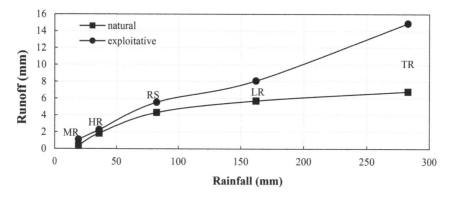

Figure 1.   Variation characteristics of runoff with rainfall of two *Phyllostachys* forests.

3.1.2 *Myrica rubra forests*
Table 3 summarizes the rainfall-runoff of two types of *Myrica rubra* forests under different rainfall intensity levels.

Table 3.   Rainfall and runoff statistics of two *Myrica rubra* forests.

Rainfall intensity grade	Rainfall events	Total rainfall (mm)	Total runoff (mm)	
			M1	M2
Moderate rain	2	30	0.78	1.36
Heavy rain	3	109.5	4.23	5.82
Rainstorm	4	314.8	15.44	21.24
Large rainstorm	2	263.2	12.78	15.38
Torrential rain	1	216.5	20.51	25.86

Note: M1 is a natural runoff plot, while M2 is an exploitative runoff plot.

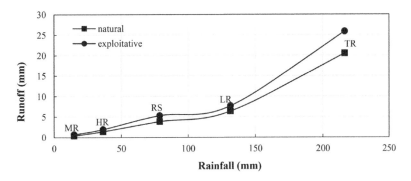

Figure 2.   Variation characteristics of runoff with rainfall of two *Myrica rubra* forests.

It can be seen from Table 3 that among the observed rainfall events, there are 12 times water and soil loss monitoring data. There was one moderate rain event, five heavy rain events, four large rainstorm events, and one torrential rain. The runoff of exploitative *Myrica rubra* forest is higher than that of natural *Myrica rubra* forest.

According to the changing relationship between rainfall and soil loss (see Figure 2), the runoff of the two kinds of *Myrica rubra* forest increased with the increase in rainfall. The runoff change rate of exploitative *Myrica rubra* forest from moderate rain to heavy rain is significantly lower than that from large rainstorms to torrential rain.

Through the comparative analysis of runoff data, the mean runoff coefficients of exploitative *Myrica rubra* forest and natural *Myrica rubra* forest are 0.07 and 0.05, respectively. Under different rainfall intensity levels, the runoff coefficient of natural *Myrica rubra* forest is between 0.03 to 0.09, and that of exploitative *Myrica rubra* forest is between 0.04 to 0.12. The water conservation benefit of natural *Myrica rubra* forest is between 19% to 34%, and the mean value is 27%.

From the above comparative analysis, it can be concluded that the natural economic forest absorbs rainwater due to the shrub and grass vegetation under the forest, and the developed root system of vegetation in the soil increased the infiltration channel. Especially under the condition of moderate rain and heavy rain, the water conservation benefit is very obvious. Therefore, surface vegetation restoration in the exploitative economic forest is conducive to rainfall interception and runoff infiltration.

3.2   *Soil conservation benefits*

3.2.1   *Phyllostachys forests*
Table 4 summarizes the soil loss of two types of *Phyllostachys* forest under different rainfall intensity levels. Under different rainfall intensity levels, the soil loss of natural *Phyllostachys* forest is less than that of exploitative *Phyllostachys* forest.

Table 4.   Rainfall and soil loss statistics of two *Phyllostachys* forests.

Rainfall intensity grade	Rainfall events	Total rainfall (mm)	Total soil loss (t·km^{-2})			
			P1	P2	P3	P4
Moderate rain	2	39.5	0.204	0.199	0.59	0.634
Heavy rain	4	144.0	1.607	1.418	3.107	3.246
Rainstorm	2	164.5	1.483	1.904	3.642	3.922
Large rainstorm	3	486.0	9.211	9.807	32.33	33.809
Torrential rain	2	565.5	26.843	26.182	61.292	63.566

Note: P1 and P2 are natural runoff plots, while P3 and P4 are exploitative runoff plots.

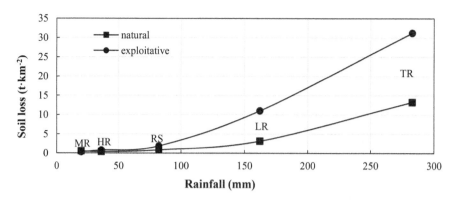

Figure 3.   Variation characteristics of soil loss with rainfall of two *Phyllostachys* forests.

Figure 3 reflects the changes in soil loss under different rainfall intensity levels, from which it can be concluded that the soil loss of the two *Phyllostachys* forests increases with the increase of rainfall. The change rate of soil loss from moderate rain to rainstorm is significantly higher than that from rainstorm to a torrential rainstorm. The change rate of soil loss in exploitative *Phyllostachys* forest is higher than that in natural *Phyllostachys* forest.

Through calculation, it is found that the average soil conservation benefit of natural *Phyllostachys* forest is 61%. Under the condition of moderate rain, compared with exploitative *Phyllostachys* forest, the soil loss of natural *Phyllostachys* forest was reduced by 67%. Under the condition of heavy rain, rainstorm, large rainstorm, and torrential rain, the benefit is 52%, 55%, 71%, and 58%.

3.2.2 *Myrica rubra forests*

The soil loss data of two types of *Myrica rubra* forest with different rainfall intensity levels are shown in Table 5. It can be seen from the data analysis in the table that the soil loss of natural *Myrica rubra* forest is significantly less than that of exploitative *Myrica rubra* forest. Figure 4 reflects the relationship between rainfall and soil loss. With the increase in rainfall, the soil loss of two types of *Myrica rubra* forest increases. The soil loss rate of *Myrica rubra* forest from moderate rain to heavy rain was significantly lower than that from heavy rain to torrential rain, while the change of soil loss rate of natural *Myrica rubra* forest was relatively small. Under the condition of a rainstorm and above, the soil loss of natural *Myrica rubra* forest was significantly less than that of exploitative *Myrica rubra* forest.

Table 5.   Rainfall and soil loss statistics of two *Myrica rubra* forests.

Rainfall intensity grade	Rainfall events	Total rainfall (mm)	Total soil loss (t·km^{-2})	
			M1	M2
Moderate rain	2	30.0	0.151	0.18
Heavy rain	3	109.5	0.342	0.701
Rainstorm	4	314.8	1.098	2.263
Large rainstorm	2	263.2	2.643	4.343
Torrential rain	1	216.5	6.979	10.602

Note: M1 refers to natural runoff plots, while M2 refers to exploitative runoff plots.

Through benefit calculation and analysis, the average soil conservation benefit of natural bayberry forest is 44%. Under the five rainfall types, the soil conservation benefits of natural bayberry forest were 36%, 37%, 55%, 51%, and 53% respectively. It can be seen that under the condition

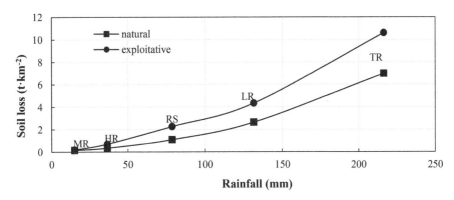

Figure 4.   Variation characteristics of soil loss with rainfall of two *Myrica rubra* forests.

of a rainstorm and above, the soil conservation benefit of the natural *Myrica rubra* forest is more obvious.

4   CONCLUSIONS

Through the collection and collation of rainfall, runoff and soil loss data of two different economic forest plots, the soil and water conservation benefits and their change regular pattern of the two economic forests under different rainfall intensity levels were analyzed and studied, and the change differences of regular patterns between natural and exploitative economic forest were compared. At the same time, according to the calculation method of soil and water conservation benefits, the soil and water conservation benefits of natural economic forests were analyzed compared with exploitative economic forests. The main conclusions are as follows.

Using the precipitation intensity classification standard issued by China Meteorological Administration Department, the rainfall is classified, the differences of soil and water loss between the two economic forests under different rainfall levels are studied and compared, and the soil and water conservation benefits of natural economic forests are calculated. Among them, (1) the average water conservation benefit of natural *Phyllostachys* forest is 54% and the average soil conservation benefit is 58%. (2) The average water conservation benefit of natural *Myrica rubra* forest is 27% and the average soil conservation benefit is 53%. (3) However, under the conditions of different rainfall levels, the soil and water conservation benefits of natural economic forests are different. In other words, the soil and water conservation benefits are not fixed, and the soil and water conservation benefits change under the influence of measure types and rainfall types.

ACKNOWLEDGMENTS

This work was supported by the Science and Technology Planning Project of Zhejiang Province, China (Grant No. 2018F10030) and the Hydraulic Technological Program of Zhejiang Province (Grant Nos. RB1609 and RC1723).

REFERENCES

Degens B. P. (1997) Macro-aggregation of soils by biological bonding and binding mechanisms and the factors affecting these: a review [J]. *Soil Research*, 35 (3): 431–459.
Mosley M. P. (2010) The effect of a New Zealand beech forest canopy on the kinetic energy of water drops and on surface erosion [J]. *Earth Surface Processes & Landforms*, 7 (2): 103–107.

Liu X. D., Wu Q. X., Wang Y. K., Zhao H. Y., Han B. (1994) Study on interception and evaporation of litter in Chinese Pine Forest [J]. *Research of Soil and Water Conservation*, (3): 19–23. (in Chinese).

Liu S. R., Sun P. S., Wang J. X., Chen L. W. (2001) Hydrological functions of forest vegetation in upper reaches of the Yangtze River [J]. *Journal of Natural Resources*, 16 (5): 451–456. (in Chinese).

Li M., Yao W. Y., Ding W. F., Yang J. F., Chen J. N. (2005) Effect of grass coverage on sediment yield in the hillslope-gully side erosion system [J]. *Journal of Geographical Sciences*, 60 (5): 725–732.

Liu X. D., Wu Q. Z., Zhao H. Y. (1994) The vertical interception function of forest vegetation and soil and water conservation [J]. *Research of Soil and Water Conservation*, 1 (3): 8–13. (in Chinese).

*Advances in Energy, Environment and
Chemical Engineering – Abdullah & Osman (Eds)
© 2023 The Author(s), ISBN: 978-1-032-36083-6*

Influence of rainfall on soil and water loss of typical economic forest in the hilly area of eastern Zhejiang Province

Xi-jiong Zhou*
Yuyao Water Resources Bureau, Yuyao, Ningbo, China

Gang Li* & Gang Tian
Zhejiang Guangchuan Engineering Consulting Co., Ltd., Hangzhou, China

ABSTRACT: In order to better guide the rational utilization of regional soil resources, it is very important to study the regular pattern of soil and water loss of economic forests and the impact of rainfall on soil and water loss. Taking the typical economic forest as the experimental object, this paper collects the data on rainfall, runoff, and soil loss, and makes a statistical analysis of the correlation and quantitative relationship between rainfall and runoff, and rainfall and soil loss. The results show that the economic forest with single tree species has serious soil and water loss due to disturbance and lack of surface cover, and the impact of rainfall on soil and water loss is also very obvious. The water and soil loss of economic forests with a high surface cover or high multi-layer cover is relatively slight, and the impact of rainfall on water and soil loss is relatively weak. Therefore, in the process of economic forest development, we should pay attention to the allocation of tree species with different heights to form a complex overburden, to achieve a good effect on soil and water conservation.

1   INTRODUCTION

Soil erosion is an environmental problem. Researchers have been studying soil erosion for many years (Lu & Yang 2003). There are many factors causing water and soil loss, such as soil type, rainfall-runoff, geological and topographic conditions, vegetation coverage, and rainfall. Among them, rainfall is one of the important causes of soil and water loss, especially in the red soil hilly areas in southern China, rainfall is the main natural influencing factor of soil and water loss (Liang et al. 2008). Therefore, it is necessary to study the internal regular pattern and relationship between soil erosion and rainfall, help policy-maker analyze the actual situation of rainfall and soil erosion, and then formulate scientific soil erosion control measures to protect the ecological environment.

Under the influence of climate change and human activities, regional rainfall, runoff, and erosion sediment have changed significantly (Peng et al. 2008; Liu et al. 2016). This has a serious impact on the land resources, the allocation of water resources, and the health of the forest ecosystem (Shang et al. 2009; Wang et al. 2002; Zuo et al. 2010). Therefore, the study on the impact of rainfall change on runoff, erosion, and sediment yield can better guide the rational utilization and protection of regional soil resources. At the same time, it can also provide an important scientific basis for watershed land use planning and soil and water conservation. Taking the typical economic forest as the test object, this study collected the rainfall, runoff, and sediment data from 2016 to 2017, and statistically analyzed the quantitative relationship between rainfall and soil loss of several economic forest lands.

---

*Corresponding Authors: 185854445@qq.com and happylglove@126.com

DOI 10.1201/9781003330165-89

2 MATERIALS AND METHODS

The runoff plot method is used to collect soil and water loss data. With slope length and soil texture as relatively fixed factors and different economic forest vegetation as the main research factors, the runoff and sediment yield of runoff plots under the condition of natural rainfall were observed. SPSS24.0 and ORIGIN8.5 software were used for mathematical statistics analysis and Curve fitting analysis separately. The quantitative relationship between rainfall and soil erosion was studied by correlation analysis.

3 RESULTS AND ANALYSIS

3.1 *Correlation analysis*

Table 1 reflects the correlation between rainfall and soil loss. It can be obtained from the table1 that the average Pearson correlation coefficient of rainfall and runoff in different economic forest plots is 0.809, the maximum is 0.957, the minimum is 0.565, and the coefficient of variation is 0.145. Through the Two-tailed test, there is a significant linear relationship between rainfall and runoff in 21 plots at the level of 0.01 or 0.05, and the significant proportion of plots is 91%.

Table 1. The Pearson correlation coefficient between rainfall and soil loss.

Runoff plots	Runoff	Soil loss	Samples number
Camellia sinensis-1	0.631	0.836**	13
Camellia sinensis-2	0.816**	0.697**	13
Camellia sinensis-3 *(seedling)*	0.957**	0.953**	20
Phyllostachys-1	0.637*	0.890**	13
Phyllostachys-2	0.770*	0.861**	13
Phyllostachys-3	0.728**	0.923**	13
Phyllostachys-4	0.737**	0.929**	13
Osmanthus-1	0.923**	0.929**	20
Osmanthus-2	0.924**	0.981**	20
Osmanthus-3	0.948**	0.956**	20
bare	0.951**	0.956**	20
Cerasus-1	0.780**	0.946**	12
Cerasus-2	0.767**	0.937**	12
Myrica rubra-1	0.759**	0.902**	11
Myrica rubra-2	0.720**	0.863**	11
Myrica rubra-3	0.799**	0.961**	11
Myrica rubra-4	0.767**	0.968**	11
Myrica rubra-5	0.719*	0.939**	11
Myrica rubra-6	0.565	0.918**	11
Prunus yedoensis-1	0.942**	0.879**	19
Prunus yedoensis-2	0.928**	0.914**	19
Atropurpureum-1	0.933**	0.858**	19
Atropurpureum-2	0.908**	0.905**	19

Note: *Phyllostachys*-1 and 2 were Steep slopes, undisturbed.
   Phyllostachys-3 and 4 were Gentle slopes, disturbed.
   Myrica rubra-1, 2, 3 and 4 were disturbed.
   Myrica rubra-5 and 6 were undisturbed.
   * and ** are significant at the level of 0.05 and 0.01, respectively.

The average Pearson correlation coefficient of rainfall and soil loss in different economic forest plots is 0.909, the maximum is 0.981, the minimum is 0.697, and the coefficient of variation is

0.067. Through the Two-tailed test, there is a significant linear relationship between rainfall and soil loss in all runoff plots at the level of 0.01 or 0.05.

3.2 *Comparative analysis of fitting curves*

To screen relatively better equations to express the relationship between rainfall and soil loss of different economic forests, and compare the rainfall erosion rates of different economic forests with the same type of equations, Figure 1 counts the determination coefficients of 11 curve equations in different runoff plots.

Figure 1 reveals that in the curve estimation of rainfall and runoff in different economic forest plots, the average determination coefficient is more than 0.7 for quadratic, cubic, and power curves, among which the average determination coefficient of quadratic and cubic curves is more than 0.85. The determination coefficient of these three curves is relatively better than that of other types of curves. From the variation of the determination coefficients of these three curves, the variation coefficients of the determination coefficients of the quadratic curve, cubic curve, and power function are 0.136, 0.079, and 0.139 respectively. It can be seen that the relationship between rainfall and runoff in different economic forest plots is the most stable in a cubic curve, followed by the quadratic function and power function.

Figure 1 shows that in the fitting curve between rainfall and soil loss, the determination coefficients of the cubic curve, quadratic curve, and linear equation are significantly higher than other types of curves, all above 0.80. The average values of the determination coefficients of quadratic and cubic curves are above 0.9 respectively. In terms of determination coefficient variability, the variation coefficients of the cubic curve, quadratic curve, and linear equation in different economic forest plots are 0.037, 0.040, and 0.125 respectively.

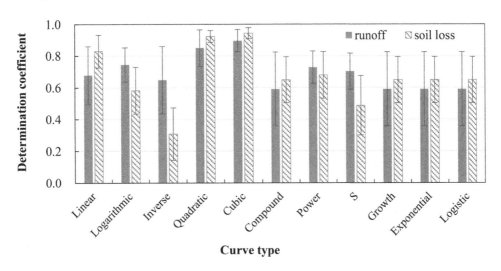

Figure 1.   Statistical curves between rainfall and soil loss.

Through the comparison of the determination coefficients of the fitting curve between rainfall and soil loss in different economic forest plots, it can be known that the quadratic curve and the cubic curve can reflect the quantitative relationship between runoff and soil loss to a certain extent, and the cubic curve is the most stable relationship between them. However, considering the influence mechanism of rainfall on soil loss and convenient for practical analysis and comparison, using a quadratic curve or linear equation to express the relationship between them is more conform to the development regular pattern of soil loss.

3.3 Influence of rainfall on soil loss

In order to further analyze and compare the differences in soil loss of different types of economic forests, the quantitative relationship between rainfall and soil loss of different types of economic forests was obtained by integrating the data of the same type of plots (as shown in Table 2).

It can be seen from Table 2 that the determination coefficients of linear equations between rainfall and runoff of different types of economic forests are relatively high, and the equation test is significant. The slope of the linear equation reflects the impact of rainfall on runoff and the rate of runoff produced by rainfall. The runoff rate of different types of economic forests is as follows: $K_B > K_{CS-S} > K_O > K_{CS} > K_{MR-UD} > K_{P-GD} > K_{P-SUD} > K_{MR-D} > K_C > K_{PY} > K_{AP}$.

From the above relationship, it can be seen that the runoff rates of different economic forests can be divided into three categories. The first category includes bare, *Camellia sinensis*-S, and *Osmanthus*. The runoff rate is between 0.4 to 0.6, which belongs to a high runoff rate. The second category is *Camellia sinensis*, *Phyllostachys*-SUD, *Phyllostachys*-GD, *Cerasus*, *Myrica rubra*-D, and *Myrica rubra*-UD, with a runoff rate of 0.01 to 0.03. The third category includes *Prunus yedoensis* and *Atropurpureum*. The runoff rate is between 0.005 to 0.01, which belongs to a small runoff rate.

According to the field investigation, the economic forest with a large runoff rate is due to the lack of surface coverage, the artificial removal of surface vegetation, very sparse, soil surface crusting, and hardening. Except for canopy closure, most rainfall forms surface runoff, resulting in a large runoff rate; at the same time, because of not covered, the runoff rate of the bare plot is the largest.

For the economic forests with a medium runoff rate, some of these plots belong to the type of low shrub coverage (*Camellia sinensis*), some belong to the type of high canopy density (*Phyllostachys*), and some belong to the type of partial surface coverage and interference (*Myrica rubra* and *Cerasus*). These differences lead to little difference in runoff during the observation period. However, their runoff rate is smaller than that of the type without surface cover.

The third category of economic forest with a small runoff rate (*Prunus yedoensis* and *Atropurpureum*) had been planted for many years, with very high coverage of litter and weeds under the forest, loose soil, and large rainfall infiltration. These comprehensive factors lead to the low runoff rate of the two types of economic forests.

Table 2. The fitting equation between rainfall and soil loss.

Code	Economic forest types	Runoff-Rainfall		Soil loss-Rainfall	
		Fitting equation	R^2	Fitting equation	R^2
CS	Camellia sinensis	Y=0.0287X+3.3836	0.7947	Y=0.0162X-0.1009	0.9599
CS-S	Camellia sinensis-S	Y=0.4778X-21.005	0.9153	Y=1.1195X-48.7150	0.9121
P-SUD	Phyllostachys-SUD	Y=0.0159X+2.8538	0.7769	Y=0.0264X-0.5185	0.9521
P-GD	Phyllostachys-GD	Y=0.0175X+3.3553	0.887	Y=0.0770X-2.1676	0.9731
O	Osmanthus	Y=0.3974X-16.4227	0.8926	Y=0.4784X-21.7730	0.9461
B	bare	Y=0.5133X-22.2942	0.9047	Y=1.7251X-88.3100	0.8898
C	Cerasus	Y=0.0144X+6.5049	0.7228	Y=0.0130X-0.1744	0.9121
MR-D	Myrica rubra-D	Y=0.0148X+4.5418	0.8326	Y=0.0114X+0.2145	0.8976
MR-UD	Myrica rubra-UD	Y=0.0179X+3.8314	0.8333	Y=0.0077X+1.9419	0.8099
PY	Prunus yedoensis	Y=0.0084X+0.1786	0.8018	Y=0.0050X+0.1469	0.8209
AP	Atropurpureum	Y=0.0052X+0.2551	0.7614	Y=0.0028X-0.0244	0.9383

note: -S, -D, -UD, -SUD, and -GD represent seedling, disturbed, undisturbed, Steep slope and undisturbed, Gentle slope and disturbed, respectively.

The relationship between soil loss rate (slope of linear equation) of different economic forests is as follows: $K_B > K_{CS-S} > K_O > K_{P-GD} > K_{P-SUD} > K_{CS} > K_C > K_{MR-D} > K_{MR-UD} > K_{PY} > K_{AP}$. From the slope relationship, it can be seen that the soil loss rates of different economic forests can be divided into four categories. The first category includes bare and *Camellia sinensis*-S. The soil loss rate is greater than 1.0, which belongs to the category with a particularly large soil loss rate; The second category is *Osmanthus* economic forest, and the soil loss rate is about 0.5; The third category includes *Phyllostachys*-SUD, *Phyllostachys*-GD, *Camellia sinensis*, *Myrica rubra*-D, and *Cerasus* economic forests, and the soil loss rate is between 0.01 to 0.09; the fourth category includes *Prunus yedoensis*, *Atropurpureum*, and *Myrica rubra*-UD economic forest, and the soil loss rate is between 0.002 to 0.008.

Through the above classification and comparison of soil loss rates, it can be found that the soil loss rates and runoff rates of different types of economic forests have been basically the same, indicating that the vegetation structure has a very obvious impact on soil and water loss. The rate of soil and water loss in the economic forest with a two-layer covering structure (forest cover and surface cover) is small, and the rate of soil and water loss in the economic forest with surface layer cover or high canopy cover is in the middle. When there is no surface cover or the canopy density is low, the water and soil loss rate is high.

4   CONCLUSIONS

Through the collation of rainfall, runoff and soil loss data of different economic forests, the relationship and change regular pattern between rainfall and runoff, rainfall and soil loss of different economic forests were analyzed, and their differences were compared and analyzed. The main conclusions are as follows:

(1) The Pearson correlation coefficients between rainfall and runoff, rainfall, and soil loss of different economic forests were analyzed using SPSS24.0 software. The results showed that there were significant correlations between rainfall and runoff, rainfall, and soil loss. Through curve fitting, comprehensive consideration is given to the occurrence mechanism of water and soil loss, using a quadratic curve or power function to express the relationships between rainfall and runoff, and rainfall and soil loss are more in accord with the regular pattern of water and soil loss, and the determination coefficient of the fitting curve is also high, which can explain the contribution of rainfall to water and soil loss to a certain extent.
(2) Based on the integration of rainfall, runoff, and soil loss data of economic forests, the differences in soil and water loss rates of different types of economic forests are analyzed and compared, and it is concluded that the water and soil loss rates of different economic forests show differences of different orders of magnitude. The runoff rate of economic forests with completely bare surface or low upper layer coverage is between 0.4 to 0.6, and the soil loss rate is greater than 1.0. The runoff rate of the economic forest of partially bare surface or high upper layer coverage or dense planting shrub is between 0.01 to 0.03, and the soil loss rate is between 0.01 to 0.5. The runoff rate of economic forests with good surface coverage or high multilayer coverage is between 0.005 to 0.01, and the soil loss rate is between 0.002 to 0.003.

ACKNOWLEDGMENTS

This work is supported by the Science and Technology Planning Project of Zhejiang Province, China (Under Grant No. 2018F10030) and also supported by the Hydraulic Technological Program of Zhejiang Province (Under Grant No. RB1609).

REFERENCES

Liang Y., Zhang B., Pan X.Z., Shi D.M. (2008) Current status and comprehensive control strategies of soil erosion for the hilly region in Southern China. *Science of Soil and Water Conservation*, 6(1):22–27.

Liu J.Y., Zhang Q., Chen X., Gu X.H. (2016) Quantitative evaluations of human- and climate-induced impacts on hydrological processes of China, *Acta Geographica Sinica*, 71(11):1875–1885.

Lu J.J., Yang H. (2003) Progress in research on effects of soil and water loss on the water environment, Soils, 35(3): 198~203.

Peng T, Tian H, Qin Z.X., Wang G.X (2008) Impacts of climate change and human activities on flow discharge and sediment load in the Yangtze River, *Journal of Sediment Research*, 43(6):54–60.

Shang Z.H., Lin P.S., Peng X.L. (2009) Ecosystem system health evaluation on soil erosion area in south China — A case study of Wuhua county. *Journal of Geological Hazards and Environment Preservation*, 20(2):116–120.

Wang N., Xie J.C., Ma B. (2002) Effect of Soil and Water Loss on Water Resources Utilization. *Research of Soil and Water Conservation*, 9(4):33–35.

Zuo T.A., Su W.C., Ma J.N., Xiao Y. (2010) Ecological Security of Land Evaluation in the Three Gorges Reservoir Area of Chongqing for Water and Soil Erosion. *Journal of Soil and Water Conservation*, 24(2): 74–78.

*Advances in Energy, Environment and
Chemical Engineering – Abdullah & Osman (Eds)*
© 2023 The Author(s), ISBN: 978-1-032-36083-6

The calculation method and application of water demand of urban water system ecosystem

Cen Wang[#]
*Wuhan Centre of China Geological Survey (Central South China Innovation Center for Geosciences),
Wuhan, China*
China University of Geosciences (Wuhan), Wuhan, China

Hongmei Bu[#]
*Key Laboratory of Water Cycle and Related Land Surface Processes, Institute of Geographic Sciences and
Natural Resources Research, Chinese Academy of Sciences, Beijing, China*

Jietao Wang & Xuehao Liu
*Wuhan Centre of China Geological Survey (Central South China Innovation Center for Geosciences),
Wuhan, China*

Shihua Qi
China University of Geosciences, Wuhan, China

Yulei Li*
Hubei University of Arts and Science, Xiangyang, Xiangyang, China

ABSTRACT: Water resources are the material basis for the survival and development of human society, and the lack of water resources has become an important factor limiting the healthy and sustainable development of urban economy and society. In recent years, China's urbanization process is accelerating, the degree of industrialization is gradually increasing, and the waste and destruction of the urban water system are also becoming more and more serious, due to the population expansion and the increasingly prominent contradiction between man and land caused by water pollution problem is increasingly serious, and China's uneven distribution of water resources, low utilization rate, coupled with the rapid development of urban economy, and the increasing demand for the water system lead to serious river pollution problems. Therefore, the calculation of urban ecological water demand is the premise and basis for the optimal allocation of urban water resources, and it is of great significance to realize the sustainable use, optimal allocation, and efficient management of water resources. In this paper, the main urban ecological water demand calculation scope should include the ecological water demand of urban rivers, lakes and green areas, etc. The respective calculation methods are discussed from the perspective of water quantity and quality, and this paper also summarizes and analyzes the related research.

1   INTRODUCTION

Excessive waste of water resources is a serious problem facing mankind, and China's economy has developed rapidly since the reform and opening up in 1978. However, with the accelerated social and industrial growth and the dramatic increase in population, the problem of water pollution has

---

[#]On behalf of the co-first author
*Corresponding Author: liyulei@126.com

become increasingly serious. Urban water system as one of the important regional water sources has also received more and more attention and attention, and many scholars at home and abroad on the water quality of lakes and other water bodies have made great achievements, but have not yet formed a systematic and perfect theoretical system. The purpose of this paper is to calculate the ecological water demand of the urban water system by investigating and analyzing the soil hydrological characteristics, land use and ecological environment around the urban water system and combining the practical experience of water resources utilization in China, which can provide a basis for the ecological design of urban water system in the future, and also provide experience reference for the technical work of sponge town construction planning and implementation, water pollution control, etc.

The ecological function of an urban water system refers to a natural landscape formed by the interaction between water, air, and soil in the water resources system. In this process, water bodies themselves have the role of regulating water quality, buffering load, and purifying pollutants. At present, domestic and foreign scholars on rainwater ecosystem research mainly focus on rainwater management and utilization, and foreign rainfall-runoff characteristics analysis focuses on the spatial and temporal distribution of precipitation and its influence factors change the law. The analysis of domestic urban rainfall-runoff characteristics focuses on the comprehensive water quantity and socio-economic aspects of regional watersheds, mainly from several perspectives such as the spatial and temporal distribution characteristics of precipitation, water resource utilization, and regional carrying capacity of rainfall runoff. This paper starts from the composition of ecological water demand of the urban water system and analyzes the changes in ecological water demand by combining the hydrological characteristics of the urban greenland system and different areas, different seasons, and different time periods in the region, and also analyzes and partitions the urban greenland system by combining relevant ecological water demand information, which can provide a basis for the application of sponge in the urban water system.

2 THE COMPOSITION OF THE URBAN ECOLOGICAL ENVIRONMENT WATER DEMAND

The urban water system is an important material basis for people to survive, it also provides a basic guarantee for the quality of the regional water environment. At present, China's urban river ecological water demand research started late. In the calculation process, we use traditional methods for design, but the heightening method and full-flow prediction and gray correlation analysis have certain shortcomings. The multi-stage multi-layer recursive relationship is more reflective of the dynamic trends of water ecosystems and water quality characteristics of the law, and it can better reveal the water system water environment quality conditions and its influencing factors, so as to provide a theoretical basis for the study of regional river ecological water demand.

2.1 *The connotation of water demand in urban ecological environment*

The ecological environmental water demand of an urban water system refers to the degree of water resources available in a certain time and space range. Since human activities have a great impact on the water environment, it is necessary to determine the ecological water demand according to different regions, time periods and spatial and temporal characteristics. Urban water ecosystems have strong flood storage and purification function, but there are some areas where water quality deteriorates due to human damage or natural factors. In addition, there are also some areas where the amount of water available in a certain time and space is small and relatively concentrated due to the special characteristics of the geographical location, resulting in serious water pollution (Lu 2013).

2.2 *Components of urban ecological environment water demand*

At present, the urban ecological environment water demand research has been carried out mainly refers to the urban natural environment system water demand, that is, the urban ecological environment system in its development process of water demand, as well as the matching water supply and water quality standards, which is a comprehensive embodiment of urban ecosystem service value, social and economic benefits and environmental quality and other factors. The water demand of the urban ecological environment should include the sum of urban living water demand, industrial water demand and natural environment water demand. The urban living water demand includes urban sewage, industrial wastewater and domestic waste discharge. Natural environment water demand refers to the total water demand of the urban ecological system in its development process.

2.3 *Ecological environment water demand and ecological water use*

If the city is regarded as an independent ecosystem, the composition of the urban ecological environment water demand has 3 parts: input, output and internal circulation; the urban water system environmental water demand mainly includes water, electricity and related energy consumption and domestic water, etc., which provides a large amount of electricity for external power equipment when the city is short of water resources, and will further increase the demand for electricity when people have planned or built production activities. The demand for electricity will be further increased when people plan or build their production activities. In order to meet the residents' demand for electricity, the proportion of electricity demand in the total water consumption of the city will gradually increase, therefore, reasonable planning and design and the development and utilization of renewable energy are important measures to ensure adequate water supply (Xu 2013).

3 CALCULATION METHOD OF WATER DEMAND OF URBAN ECOLOGICAL ENVIRONMENT

The urban water system refers to the urban construction area with a certain ecological service function of the interconnected surface water system, it is an important basis for human survival. The ecological water demand of the water system occupies a large proportion of the total water resources, but with the accelerated urbanization process, the urban scale is expanding, a large number of people gathered and the rapid growth of industrial land and other factors under the influence of its variability is also increasing. In order to make the urban water system give full play to its ecological service function, it is necessary to meet both water health and water quality health requirements. Water health refers to the change of water quantity, and mainly refers to the impact on ecological water demand due to the large difference in density and concentration between surface water and groundwater or its proportion in different water bodies within a certain time and space range. The water health test method and process are shown in Figure 1. Water quality health refers to the function and service quality of the urban water system ecosystem, mainly referring to the water body in a certain time and space range, due to the influence of various factors in its water quality conditions and trends. Water quality health testing methods and processes are shown in Figure 2 (Zhao 2020). According to the water demand in the process of the urban hydrological cycle, integration is divided into two parts: consumptive water demand and non-consumptive water demand, consumptive water demand refers to the total water content and water quality status of the water body in the urban water system, under different water bodies, due to the influence of various factors, its total water demand and change trends vary. Non-consumable water demand refers to the large difference between the water quality status and changing trend of water in the urban water system and the pollution load, which is mainly caused by the different water demands of water bodies in different environments.

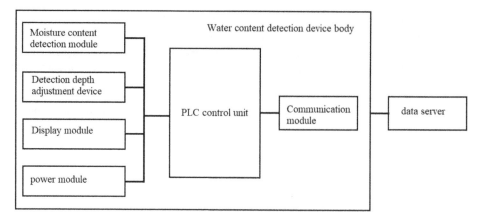

Figure 1. Water health testing method and process.

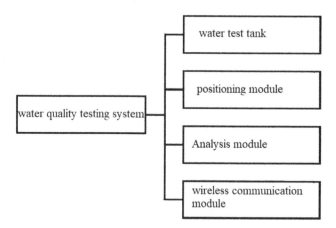

Figure 2. Water quality health testing methods and processes.

3.1 *Urban water demand*

Urban domestic water demand includes two parts: domestic water demand and municipal public water demand. Domestic water demand refers to residents' daily life, domestic water, production and other non-agricultural wastewater, etc. Municipal public water demand refers to the water needed for the construction of urban drainage facilities. Municipal public water demand refers to the construction needs of urban drainage system water, it refers to the municipal water supply sources and sewage network and other areas, as well as residents in their daily lives, production and domestic water contained in the water. At present, the most important influencing factors are the population and the comprehensive water quota per capita. The legal index of comprehensive water quota per capita in cities is given in the Planning Standard for Urban Water Supply Projects (GB50282-98) issued by the Ministry of Construction of China in 1998. The calculation method of urban domestic water demand in the i-th year is shown in Formula (1).

$$W_{li} = P_0(1 + R_1)^n K_i \qquad (1)$$

3.2 Urban industrial water demand

The urban industrial layout is an important part of urban planning, its distribution and planning directly affect the water environment of the region around the water system. Generally, large-scale factories or industrial zones are built in areas with flat terrain, convenient transportation and densely populated areas, while for large cities with complex terrain and high population density, the number of factories needs to be minimized and the scale of production needs to be increased to reduce transportation costs to meet demand. The method of calculating industrial water demand is the water demand growth trend analysis method, which is combined with the current situation of water resources and planning in Yunnan to calculate the water demand of each region, and then is combined with the water quality requirements of each design phase and relevant standards to determine the value. This method is based on the growth rate of industrial water use in the past years, and the industrial water demand in the i-th year is calculated as shown in Formula (2).

$$W_i = W_0(1 + R_2)^n \tag{2}$$

3.3 Urban environmental water demand

Urban environmental water demand refers to a certain area (or time period) in a certain time frame, a certain area as the object of study and other surrounding unrestricted change.

As an important part and place of human activities, material circulation, and information exchange center, the urban water system has a great influence on water resources. The number of pollutants in the water body and complex and diverse, different forms of water interaction formed a variety of types of the water quality evaluation index system. As there is a certain link between various indicators, it is necessary to study the interrelationship between various water quality standards and control factors and their regularity to determine their degree of merit. The water demand of the urban environment includes water demand of green areas, water replenishment of rivers and lakes, water for leisure and tourism, etc (Penru 2016).

(1) Green space water demand. Green space water demand includes green space vegetation evapotranspiration water demand, which is an important indicator to reflect the distribution of urban green space, vegetation growth water demand is the minimum soil water content to maintain vegetation growth, it is an important indicator to reflect the service function of the urban ecosystem, it is also the most direct and effective means to measure the distribution of green space and plant growth potential. The ecological water demand of the urban water system is an important indicator to characterize the distribution of green areas, and it is also the most intuitive and direct reflection of the overall evaluation of human activities on the quality of regional surface water environment and ecosystem service functions. The calculation of the evapotranspiration water demand of green space vegetation is shown in Formula (3).

$$W_E = AE_p \tag{3}$$

The evapotranspiration of different vegetation is different, so the water demand is also different, and the calculation of the water demand for vegetation growth is shown in Formula (4).

$$W_p = W_E/99 \tag{4}$$

The water absorbed by plants in the process of growth is the most direct source of nutrients required for their own growth and development. Vegetation absorbs water and nutrients during its growth and also secretes some biological factors, which have an important impact on the ecological carrying capacity of urban water systems. The minimum soil water content to maintain the growth of vegetation is calculated as shown in Formula (5).

$$W_S = Ah\xi \tag{5}$$

(2) Ecological environmental water demand of rivers and lakes. The ecological environment water demand of the urban water system refers to the distribution process of water resources in the water body, such as residential water, industrial production, urban living drainage, etc. which need to consume a certain amount of water. The river ecosystem formed under human activities interacts with the natural system, among which the artificial wetland has an important role in the water body, while the natural river has an important protective function for aquatic organisms. The rivers and lakes referred to here are rivers and lakes in the city, and their water demand is the amount of water needed to maintain the base flow of rivers and lakes in the city with a certain water surface area, meet the landscape conditions, water navigation, and protect biodiversity. The water demand of river and lake ecological environment includes evaporation water, seepage water, baseflow water, pollutant dilution, and purification water. Among them, the water surface evaporation water demand is calculated as shown in Formula (6).

$$W_e = A_1 E_e \tag{6}$$

River and lake seepage water demand means that when the river and the lake water level is higher than the groundwater level, surface runoff, groundwater percolation, and purification treatment can be used to achieve the water demand of the river and lake environmental water function area. The amount of water that will be replenished to groundwater through seepage at the bottom of the river and lake and lateral seepage at the bank is calculated as shown in Formula (7).

$$W_1 = KITW \tag{7}$$

River baseflow water demand is the amount of water required to maintain a certain flow rate and flow rate of the river, which is one of the important indicators of water environment quality and a key factor affecting the healthy operation and economic development of the river. The calculation of river baseflow water demand is shown in Formula (8).

$$W_1 = K_1 A_1 \tag{8}$$

The amount of water required to maintain the lake surface is the amount of water storage to maintain the normal existence and function of the lake, and the water system is an important part of the water cycle system, water resources as a non-renewable resource, which itself has a natural and fixed nature. The calculation of water storage is shown in Formula (9).

$$W_R = A_2 v \tag{9}$$

Pollutant dilution and purification water demand is an important part of the ecological water demand of urban water system, but due to the influence of human factors and natural laws, the concentration of pollutants in water bodies is generally high, which leads to serious eutrophication of water bodies, thus posing a great threat to the sustainable development of the ecological environment of urban water systems. The calculation of water demand for pollutant dilution and purification is shown in formula (10).

$$Q = (C_i / C_{oi}) Q_i \tag{10}$$

4 CONCLUSION

To sum up, the collection and analysis of water demand data of the aquatic ecosystem is the most basic and critical part of the current situation of water resource utilization. An ecosystem is a relatively stable natural body in a certain time and space range. It is a complex system composed of multiple factors that exist in some form and interact with each other in a certain time and space range. The urban ecosystem is an artificial ecosystem, and the water demand of the urban ecological environment is rising with the increase in population, industrial growth and green area, etc. The construction of an urban water system is a complex system, which involves many subsystems such as water, heat and air. The calculation of urban ecological environment water demand should be based on the historical and current water demand and water use, but also consider the future water demand development trend, so as to provide a scientific basis for the construction of urban water system, in addition, we also need to consider the rapid improvement of living standards, municipal facilities more perfect, and continuous innovation of production processes, the reuse rate of water resources will increase significantly, and the degree of damage to water resources will further intensify. This makes the urban water system ecosystem face serious challenges.

FUND SUPPORT:

National Key R&D Program of China (No.2020YFC1512402) & (No.2018YFC1800800; The project of the China Geological Survey (No.DD20221729): Monitoring and evaluation of resource and environmental carrying capacity in the Guangdong-Hong Kong-Macao Greater Bay Area.

REFERENCES

Chaojun Wang, Hongrui Zhao.The Assessment of Urban Ecological Environment in Watershed Scale[J]. *Procedia Environmental Sciences*, 2016:169–169.

Hao, Tian, Du, Pengfei, Gao, Yun. Water environment security indicator system for urban water management[J]. *Frontiers Of Environmental Science & Engineering*,2012(8):678–691.

Liu, Huazhi, Wang, Lin. Construction of Urban Water Ecological Civilization System[J]. *IOP Conference Series Earth and Environmental Science*, 2018(5):14–17.

Lu, Meng; Nie, Yan Qing. Beidaihe Area Water Ecological Environment Protection[J].*Advanced Materials Research*, 2013:55–59.

Penru, Y.; Antoniucci, D.; Barrero, M. J. Amores; Chevauché, C. .Water footprint calculation: application to urban water cycle[J].*International Journal on Interactive Design and Manufacturing* (IJIDeM), 2016:99–103.

Xu, Wen Jie, Chen, Wei Guo, Zhang, Xiao Ping, Gong, Hui Ling.Study on Urban Water Ecological Security Assessment[J]. *Applied Mechanics and Materials*,2013:829-832.

Xu, Wen Jie; Chen, Wei Guo; Zhang, Xiao Ping; Gong, Hui Ling. (2013). Study on Urban Water Ecological Security Assessment. *Applied Mechanics and Materials*doi:10.4028/www.scientific.net/AMM.295-298.

Zhao, Min, Li, Dian Yang, Cong, Chun Chun, Cai, Yuan Cheng.Research on Water Ecological Environment Evaluation System of Urban Rivers and Lakes[J].*Advanced Materials Research*, 2012:1444–1450.

Zhao, Li, Yang, Yusi, Wang, Tong, Zhou, Liang, Li, Yong, Zhang, Miao. A Simulation Calculation Method of a Water Hammer with Multpoint Collapsing[J].*Energies*, 2020:11–15.

Advances in Energy, Environment and
Chemical Engineering – Abdullah & Osman (Eds)
© 2023 The Author(s), ISBN: 978-1-032-36083-6

Stability analysis of unsaturated soil slope under fluid-structure coupling

Jiachen Xu & Qingwen Zhang*

School of Civil Engineering, Southwest Forestry University, Kunming, China

ABSTRACT: In order to study the slope stability under rainfall conditions, this paper simulates and analyzes the stability of front, center, back, and average conditions based on geo studio finite element analysis software and obtains the law of slope instability by analyzing the effective stress, pore water pressure and displacement changes at different positions of the slope. Through the slope module and Monte Carlo method, the slope instability probability under four rainfall conditions is front > back > average > center. The research results can provide suggestions for slope treatment and protection and provide a reference for slope instability probability analysis.

1 INTRODUCTION

The stability of unsaturated soil slope has always been the focus of research. Moreover, most of the excavated slopes and natural slopes are unsaturated. When the slope is unsaturated, the stability is better, and the shear strength is higher than that in the saturated state due to matrix suction. When rainwater infiltrates, the pore pressure of the slope increases and the effective stress decreases, and the slope may appear in a temporary saturation state, resulting in the reduction of shear strength and the increase of slope instability probability.

The calculation of slope instability probability can more accurately predict the slope instability probability and provide a reference for practical engineering, which is of great significance. Scholars at home and abroad have carried out a lot of research. LV Yuhua et al. 2021 used geo studio to analyze the numerical model of seepage stress coupling of unsaturated soil slope and studied the instability principle and evolution law of unsaturated soil slope. Lin Ya et al. 2019 obtained the pore water pressure, the change of safety factor, and the probability of instability through the analysis of the monitoring section of the slope with a weak interlayer. Zhang Xingui et al. 2016 comprehensively considered the influence of excavation effect and stress seepage coupling on slope stability and deduced the finite element solution equation of unsaturated soil under seepage stress coupling according to the seepage shear characteristics of unsaturated soil, and introduced the finite element strength reduction method for the stability of unsaturated soil excavation slope. Li Kan et al. 2014 used Matlab and geo studio to sort out and analyze the slope data and obtained the reliability and instability probability of the slope. Tao Mei et al. 2018 combined the orthogonal and Monte Carlo methods to determine the main factors affecting slope stability and the corresponding influence degree. In this paper, the slope under fluid-structure coupling is combined with Monte Carlo probability analysis by geo studio finite element analysis software, and the probability analysis is carried out to determine the slope instability probability.

---

*Corresponding Author: 904829146@qq.com

DOI 10.1201/9781003330165-91

2 THEORETICAL ANALYSIS

2.1 *Unsaturated shear strength theory*

This paper adopts the widely accepted Fredlund double stress theory formula. It is also a shear strength formula that combines the traditional Mohr-Coulomb failure criterion with the effective stress formula. The first two terms represent the traditional Coulomb criterion for the shear strength of saturated soil, and the third term represents that the shear strength of unsaturated soil increases with the increase of matrix suction.

$$\tau_f = c' + (\sigma - u_a)_f \tan\emptyset' + (u_a - u_w)_f \tan\emptyset^b \tag{1}$$

In the above formula, τ_f is the shear strength; c' is the cohesion when the matrix suction and net normal stress are 0; S_e is effective saturation; $(\sigma - u_a)_f$ is the net normal stress on the failure surface when the soil is damaged; $\emptyset'$ is the internal friction angle corresponding to the net normal stress; $(\sigma - u_w)_f$ is the matrix suction when the soil is damaged; $\emptyset^b$ is the internal friction angle corresponding to matrix suction.

2.2 *Monte Carlo theory*

The Monte Carlo method is a statistical method. In this paper, the slope module is used for Monte Carlo analysis. Firstly, n factors $(x_1, x_2 \ldots x_n)$ that will affect the slope stability are selected, such as cohesion and internal friction angle. Then we establish a normal distribution for the selected factors, and random sampling generates sample values. The safety factor F is obtained by substituting it into the equation obtained from the limit equilibrium state of the slope.

$$F = f(x_1, x_2, \ldots, x_n) \tag{2}$$

Then we repeat the above process m times to obtain m safety factors, of which m is less than 1. When m is large enough, we can obtain the slope instability probability P.

$$P = p(f(x)1) = {}^m/_M \tag{3}$$

3 WORK CONDITION DESIGN

3.1 *Model parameter selection and working condition design*

Kunming has a low latitude subtropical plateau mountain monsoon climate in the north latitude. It is dry and wet. The rainfall is concentrated from May to October, accounting for 85% of the whole year. It will peak in July and August, and most of it is short-term rainfall. This paper selects the daily rainfall data of Kunming from 2015 to 2020 and finds that the rainfall on July 20, 2019, reached 126.8mm/h, the largest in six years. It is dispersed within six hours as the rainfall boundary condition of the simulated slope.

Table 1. Design table of calculation conditions.

Working condition design	Rainfall type	Duration	Soil type
1	Backcourt type	6 h	Peat soil
2	Forward type	6 h	Peat soil
3	Center type	6 h	Peat soil
4	Average type	6 h	Peat soil

4 RESULT ANALYSIS

4.1 *Seepage stress coupling analysis*

According to the working condition design, the maximum shear stress, pore water pressure, and XY displacement of slope soil at the slope toe, slope center, and slope shoulder under the working condition I design condition are selected, as shown in the figure.

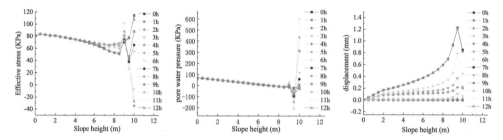

Figure 1.   Variation in the slope toe under the working condition I.

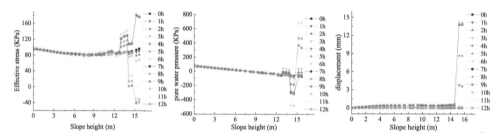

Figure 2.   Variation in the first slope under the working condition I.

As shown in Figure 2, in the first two hours of frontal rainfall, the average effective stress at the cross-section of the slope toe first decreases and then increases with the increase of the height of the slope model. From the third hour, the effective stress at the slope toe has decreased significantly compared with the first two hours, and the average effective stress gradually steepens with the increase of rainfall intensity from the height of 8.5m of the slope model. Then it drops rapidly at 9m. The pore water pressure also fluctuates from the height of 8.5m of the slope model, rises slightly with the increase of rainfall intensity, decreases rapidly at 9m, and then rises rapidly. When rainfall reaches the maximum in the sixth hour, the fluctuation is also the most intense, falling to - 207.4 kPa, then rising rapidly to 600 kPa. The change of pore water pressure caused by rainfall infiltration leads to the change of seepage field, which destroys the original stress balance of the slope and causes the change of effective stress.

When the rainfall stops, the pore water pressure decreases rapidly and gradually tends to be stable, while the average effective stress at the surface layer of the slope toe also rises rapidly and gradually tends to be stable. However, due to rainfall infiltration, the effective stress at the deeper slope toe will decrease and have no recovery trend. The change in seepage field and stress field also caused the soil displacement of the slope. The change of displacement at 9m was the most obvious, reaching 0.0079 m at the maximum rainfall in the 6th hour and continued to move in the subsequent 6-hour interval, up to 0.011m. With the increase of rainfall intensity, the pore water pressure increases within 1m of the surface depth of the slope, and the average effective stress decreases. At this time, it is also the position where the soil displacement is the largest. As the temporary saturation of the surface soil will compress the soil near the surface, the pore water pressure will decrease here so that the average effective stress will increase here. Overall, the coupling process between the average effective stress and pore water pressure at the slope toe has a certain lag. When the pore water pressure changes, the effective stress at the corresponding position does not change immediately, and vice versa, but the overall trend of the two is negatively correlated, which is consistent with the theory. Moreover, the effective stress at the toe of the slope will recover much slower than that at the center and shoulder of the slope, which also proves that the toe of the slope is indeed a dangerous area of the slope.

The front type is different from the rear type. The fluctuation of the average effective stress of the front type rainfall at the foot of the slope is mainly concentrated in the surface soil of the slope, and there is little change inside the slope. Moreover, because the front type rainfall is easier to form slope runoff on the slope surface, the maximum fluctuation value of pore water pressure of the slope under the front type condition is nearly half lower than that of the rear front type. The corresponding displacement at the toe of the slope only reaches 0.0032m when the change is the largest. The changes at the center and shoulder of the front slope are similar to those at the toe of the slope.

4.2 *Monte Carlo analysis*

Monte Carlo slope probabilistic stability analysis is carried out through a single slope/W module. In this paper, the in-situ seepage field and stress field of the slope are analyzed through the seep/W module and sigma/W module, and then the coupling analysis of the two is carried out under the condition of rainfall. The coupled stress value and pore water pressure value are brought into the slope / W module as the original data for instability probability analysis.

Through 2000 sampling tests on three probability parameters through the slope/W module, the slope instability probability under four working conditions (rear front type, front type, center front type, and average type) can be obtained. Monte Carlo sampling test under condition 1 shows that the fluctuation value at the slope toe, fracture center, and slope shoulder in the sixth hour is the largest and most representative. In this paper, the safety factor of the critical slip surface in the sixth hour is selected, the maximum can reach 1.206, and the minimum is only 0.681. In this case, the instability probability of the slope is as high as 81.6%, which belongs to a highly dangerous slope, and the stability grade is grade 2. According to figure 4 to figure 6, the fluctuation values of three sections under condition 2 are the most intense and close in the third and fourth hours. Therefore, this paper calculates the critical slope interface in the third and fourth hours. The instability probability in the third hour is 3.5%, the fourth hour is 5.05%, the minimum safety coefficient is 0.698, and the maximum is 2.03. Therefore, the fourth hour under condition 2 is the most representative. It belongs to low risk, and the stability level is level 4. Under condition 3, the fluctuation values of pore water pressure and effective stress are the largest in the fourth and fifth hours. Therefore, the instability probability of the two is measured respectively. The instability probability in the fourth hour is 0 and the instability probability in the fifth hour is 1.1%. It is divided into the stable slope, and the stability grade is grade 5. Under condition 4, the critical slope in the sixth hour is directly selected for prediction. The instability probability is 2.95%, which is also a stable slope. The stability grade is grade 5, the minimum safety factor is 0.746, and the maximum safety factor is 2.08. Therefore, under different rainfall conditions, the slope instability probability from large to small is front type, back type, average type, and center type.

5   CONCLUSION

Through the slope/W module in geo studio, the Monte Carlo instability probability stability analysis of the slope after seepage stress coupling is carried out, the instability probability of the slope under four working conditions is calculated, and the slope grade is divided: the instability probability of working condition 1 is 81.6%, which belongs to high dangerous slope, and the stability grade is grade 2. The instability probability of condition 2 is 5.05%, which belongs to low risk and the stability level is level 4. The instability probability of condition 3 is 1.1%, which belongs to a stable slope and the stability grade is grade 5. The instability probability of condition 4 is 2.95%, which is also a stable slope, and the stability grade is grade 5.

REFERENCES

Li Kan, Great power climbs slope reliability evaluation based on Monte Carlo method [J] *Chinese Journal of Geological Hazards and Prevention*, 2014, 25 01: 23–27.

Lin, Study on the rainfall failure probability of slope with weak interlayer based on limit equilibrium and Monte Carlo method [J] *Hydropower and Energy Science*, 2019, 37 12: 103–107.

LV Yuhua, Liang Dexian, Wang Ying, Huang Xiang Seepage stress coupling analysis of unsaturated soil slope under rainfall [J] *Journal of Guilin University of Technology*, 2021, 41 02: 318–324.

Tao Mei, Ren Qingwen, Study on slope failure probability based on orthogonal test Monte Carlo method [J] *Hydropower and Energy Science*, 2018, 36 03: 132–135.

Zhang Xingui, Xu Shengcai, Yi Nianping, Stability analysis of unsaturated soil excavation slope based on fluid-structure coupling theory [J] *Journal of Water Conservancy and Transportation Engineering*, 2016 03: 10–19.

Advances in Energy, Environment and
Chemical Engineering – Abdullah & Osman (Eds)
© 2023 The Author(s), ISBN: 978-1-032-36083-6

Theoretical implementation of extreme precipitation in Zhengzhou based on the Extreme Value Theory

Yujing Fan*

School of Atmospheric Sciences, Nanjing University of Information Science & Technology, Nanjing, China

ABSTRACT: In July 2021, extremely heavy rainfall in Zhengzhou, a city in China, caused serious loss of life and property. The max amount of daily precipitation in this catastrophe exceeded Zhengzhou's total amount of precipitation in 2019. In this research, the daily precipitation data are used to study the extremity of the event. It can help predict the extreme event's frequency occurrence and minimize the loss in the future. Additionally, we find out whether the intensity of the Western Pacific Subtropical High will affect the precipitation in Zhengzhou. The chosen methods in this research are mainly the technical approaches from Extreme Value Theory, such as the Block Maxima approach and Peak Over Threshold approach. The first part of this paper is a theoretical elaboration on the Extreme Value Theory. Next, an implementation based on this theory to assess heavy rainfall in Zhengzhou in 2021 is given. It is concluded that the extreme precipitation event in Zhengzhou is a once-in-a-millennium event, and there is a correlation between Western Pacific Subtropical High and the precipitation in Zhengzhou.

1 INTRODUCTION

From July 16, 2021, to July 21, 2021, heavy rain occurred in the central and northern part of Henan province, China, and heavy rain ($250 \sim 350$ mm) dropped in some areas of Henan, like Zhengzhou, Xinxiang, Kaifeng, Zhoukou and Jiaozuo. The cumulative average precipitation of Zhengzhou even reached 600 mm. As an inland city, Zhengzhou suffered severe urban waterlogging due to such excessive precipitation. The urban power grid and water supply system was unstable and collapsed completely, which significantly disturbed people's lives. According to the 10th press conference on Flood Control and Disaster Relief in Henan Province (Henan Government Net, 2021), up to 12 o'clock on August 2, 150 counties (cities, districts), 1,663 towns and townships in the province, 14.5316 million people had been affected. Farmlands were hit seriously, and direct economic losses totaled 114.269 billion yuan. Henan province locates in the boundary of warm temperate and subtropical zone, and the climate has obvious transitional characteristics (Yu, Liu, Chang, et al., 2008). Zhengzhou is a typical continental monsoon humid climate, near the Yellow River in the north, Songshan in the south, located north-northwest of central Henan Province. The research focuses on the changes in extreme climate prediction and finding some correlated factors in coping with climate change.

The Extreme Value Theory has been developed in parallel with the central limit theory (De Haan, Ferreira, 2006), and the two theories bear some resemblance. Let $X_1, X_2, X_3 \dots$ be independent and identically distributed random variables. The central limit theory is concerned with the limit behavior of the partial sums $X_1 + X_2 + \dots X_n$ as $n \to \infty$, whereas the theory of sample extremes is concerned with the limit behavior of the sample extremes $\max (X_1, X_2, \dots X_n)$ or $\min (X_1, X_2, \dots X_n)$ as $n \to \infty$. Unlike classical probability theory, like the central limit theorem, which describes the average behavior of a sequence of random variables, Extreme Value Theory (EVT) focuses on extreme and rare events. Such a method is more suitable for extreme precipitation events.

---

*Corresponding Author: 201883301121@nuist.edu.cn

Extreme Value Theory is a better chance to foresee extreme events that will happen for years, decades, or even centuries in the future.

In 2005 researchers (Liu, Wang, et al., 2005) found that the annual precipitation of Zhengzhou has the characteristics of periodic change at multiple time scales by Morlet wavelet transform. Some other researchers (Liu, Dai, 2007) found that heavy rains are mainly concentrated in summer, and typhoons and subtropical anticyclones may be factors that affect the precipitation of Zhengzhou during summertime, especially from July to August. However, few probabilistic or statistical methods are being applied in the field of extreme precipitation prediction. The use of the Extreme Value Theory is innovative. Additionally, although some factors like a typhoon are proven to have a relationship with precipitation in Zhengzhou, little research indicates that Western Pacific Subtropical High has that relationship too. Our research on the correlation between Western Pacific Subtropical High and precipitation in Zhengzhou will also be innovative.

The origin of the Extreme Value Theory (Matthews, 2005) can be traced back to the 18th century by mathematician Nicolas Bernoulli. However, it was not until the 1920s that the idea of predicting new records came to people's minds. Early mathematicians used the Gaussian distribution to calculate the probability of extreme events, which is a common distribution that many events follow. For example, people's heights follow the Gaussian distribution. Nevertheless, some heights far from the average look like narrow "tails" in the distribution. In the 1920s, mathematicians felt that the "tails" were predictors for low-probability events. In 1928, Cambridge mathematician R. A. Fisher and his colleague L.H.C. Piper published their famous paper on Extreme Value Theory, showing that low-probability events obeyed an alternative probability distribution. Now, EVT has become more and more widely used and trusted. EVT can be applied in many fields like finance, hydrology, and meteorology. For instance, EVT can be used in finance to estimate the value-at-risk (Ramazan Gençay, et al., 2003), by which the maximal daily loss will be calculated. When severe hydrological events like floods happen in hydrology, EVT may be one appropriate method to protect against floods (Canfield, Olsen, Chen, 1981). Most important, in meteorology, such theory can be applied to many extreme climate events, like hurricanes, droughts, heatwaves, or risk assessments.

2 MAIN WORKS

2.1 Basic theory

Definition 2.1.1
In the Block Maxima approach, Mn stands for a collection of the maximums of independent and identically distributed random variables. Mn follows a Generalized Extreme Value distribution (GEV) as $n \to \infty$.

In the distribution, for extreme value z, it is defined as

$$G(z) = \exp\left[-\left\{ 1 + \xi \left(\frac{z - \mu}{\sigma} \right) \right\}^{-\frac{1}{\xi}} \right] \tag{1}$$

Where ξ means shape, which is a parameter that describes the relative distribution of the probabilities, is the location that stands for the position of the mean value and is the scale of the function. In the paper, we assume that and are the functions of the index of Western Pacific Subtropical High.

Properties 2.1.2
The data will be separated into different blocks, and after the approach, the maximums of every block will be collected and calculated afterward.

Definition 2.1.3
In the Peak Over Threshold (Pandey, van Gelder, Vrijling, 2001), let $X_1, X_2, \ldots, X_n$ represent independent and identically distributed random variables. Considering k exceeds X over a threshold u and let $Y_1, Y_2, \ldots, Y_k$ be the peaks; for example, $Y_i = X_i - u$. The probability of exceeding the

threshold in the Peak Over Threshold approach is called Generalized Pareto distribution (GP). The distribution of $Y_i = [(X_i - u)|X_i > u]$ converges to (as $n \to \infty$):

$$H(y) = 1 - \left(1 + \xi \frac{y}{\gamma}\right)^{-\frac{1}{\xi}}, \tag{2}$$

where ξ have the same meanings as that in Generalized Extreme Value distribution, and $\hat{\sigma}$ is the function of u. The parameter u means threshold, which is one reference value.

Remark 2.1.4
The Peak Over Threshold Approach is similar to the Block Maxima approach. However, the Block Maxima approach only collects the largest numbers of every block. The Peak Over threshold does not have the parameter location, but the parameter threshold collects numbers bigger than the threshold of every block. Block Maxima approach uses max numbers alone, which is wasteful of data, and most of the information in the sample is ignored. With Peak Over Threshold approach, the data quantity will be larger, and the large database will help to reduce errors. As in the small number of data, some extreme values or some wrong data exist. When calculating, the small quantity may not symbolize the overall characteristic, let alone being influenced by wrong or extreme data.

Definition 2.1.5
A random process $X_1, X_2, \ldots, X_n$ is said to be stationary if, given any set of integers $\{i_1 \ldots, i_k\}$ and any integer m, the joint distributions of $(X_{i1}, \ldots, X_{ik})$ and $(X_{i1+m}, \ldots, X_{ik+m})$ are identical.

Remark 2.1.6
Stationarity implies that, given any subset of variables, the joint distribution of the same subset viewed m time points later remains unchanged. For example, if $X_i, X_2, \ldots, X_n$ is a stationary series, then X_1 must have the same distribution as X_{101}. And the joint distribution of (X_1, X_2) must be the same as that of (X_{101}, X_{102}), though X_1 needs not to be independent of X_2 or X_{102}.

In order to characterize stationary dependence, the concept of the auto-tail dependence function is introduced as follows.

Definition 2.1.7
The auto-tail dependence function is defined as

$$\lim_{u \to 1} \rho(u, h) \tag{3}$$

where u stands for threshold and h means interval.

If the value of ρ is very small, it means the data meet the stationarity requirement and can be directly used. If not, some decluttering processes should be done.

Remark 2.1.8
When doing decluttering, firstly, we identify clusters that can be the year, month, etc. Secondly, we fit the point process like the GP model to cluster maxima. For example, as the GEV is automatically fitted to block maxima, parameters are automatically adjusted for any temporal clustering.

Definition 2.1.9
The mean residual life plot is defined as

$$\left\{ \left(u, \frac{1}{n_u} \sum_{i=1}^{n_u} \left(x_{(i)} - u \right) \right) : u < x_{\max} \right\} \tag{4}$$

Above a threshold u_0 at which the distribution provides a valid approximation to the excess distribution, the mean residual life plot should be approximately linear in u (Coles, 2001).

Definition 2.1.10
According to the Empirical Cumulative Distribution Function (ECDF), for a given ordered sample of independent observation the empirical distribution function is defined as

$$\widetilde{F}(x) = \frac{i}{n+1}, \quad \text{for} \quad x_i \leq x < x_{i+1} \tag{5}$$

Definition 2.1.11
Given an ordered sample of independent observations $X_1 \leq X_2 \leq \ldots \leq x_n$ from a sample with an estimated distribution function $\widehat{F}$ (Waller, Turnbull, 2012). $\widehat{F}$ is the sample cumulative distribution function which is defined as:

$$\widetilde{F}(x) = \frac{i}{n}, \quad \text{for} \quad x_i \leq x < x_{i+1} \tag{6}$$

A probability plot consists of the points

$$\left\{ \left(\widehat{F}(x_i), \left(\frac{i}{n+1} \right), x(i) \right) : i = 1, \ldots, n \right\} \tag{7}$$

A quantile plot consists of the points

$$\left\{ \left(\widehat{F}^{-1} \left(\frac{i}{n+1} \right), x(i) \right) : i = 1, \ldots, n \right\} \tag{8}$$

Remark 2.1.12
Because the last point can be off the scale when $i = n$ for both a probability spot and a quantile plot, we use $\widehat{F}(x) = \frac{i}{n+1}$ instead of $\widehat{F}(x) = \frac{i}{n}$ for exact data. Both the probability plot and quantile plot are used to do model diagnostics. After determining one model with some collected parameters, we can use the two plots to check the bias and variance and modify our model to achieve the state-of-the-art level and help us do further analysis.

Definition 2.1.13
The Return Level function is defined as

$$P(M > z_m) = \frac{1}{m}, \tag{9}$$

where z_m is a given return level, and M is an independent variable.
  The Return Level function is a characterization of the possibility of every observation exceeding the level z_m.

Definition 2.1.14
The function of the Maximum Likelihood Estimate is

$$L(\mu, \sigma, \xi) = \prod_{i=1}^{n} f(y_i \mid \mu, \sigma, \xi), \tag{10}$$

where the three parameters ξ, μ and σ are the same as those defined in the function of the Block Maxima approach.

Remark 2.1.15
In the function of the Block Maxima approach, ξ, μ and σ can be many values. The Maximum Likelihood Estimate is used to estimate these three parameters and find the most optimal ones to make the likelihood function achieve the max value.

2.2 *Application*

2.2.1 *Data collection*

We use daily precipitation data from the China ground international exchange climatic data set (V3.0) in China Meteorological Data Service Center from Jan 1951 to Dec 2020. The China ground international exchange climatic daily data set (V3.0) contains the daily data of atmospheric pressure, air temperature, precipitation, evaporation, relative humidity, wind direction and speed, sunshine duration, and 0 cm ground temperature at 194 stations in China since January 1951. In this paper, we focus on the precipitation data of Station Zhengzhou, whose station number is 57083. In addition, we use NCEP/NCAR daily reanalysis data of geopotential high on pressure level 500 MB from Jan 1951 to Dec 2020 with a horizontal resolution of 2.5 degrees × 2.5 degrees to calculate the Western Pacific Subtropical High intense index.

2.2.2 *Research parameter*

Precipitation is one of the parameters in our research. Precipitation is when water vapor in the atmosphere condenses and falls to the ground as liquid water or solid water. It is a general term for rain, snow, dew, frost, and other natural phenomena. The data we use in the paper ranges from 0.0 to 2000.0 (0.1 mm) and includes some special forms. For example, "32700", as a usual special form, means the precipitation amount is too little to be calculated, which may be less than 0.1 mm. "32XXX" means there is pure fog, dew, and frost. "31XXX" means there is a mixture of rainfall and snow. "30XXX" means it snows on that day. Some data processes will be made on data in these special forms. When doing the data process, we turned all "32700" s into "0" s, which means that day had approximately no precipitation. For data in the form of "30XXX", "31XXX", and "32XXX", we used the tail number "XXX" s as the precipitation amount of that day.

Western Pacific Subtropical High Intense Index, another parameter in this research, represents the strength of Western Pacific Subtropical High. Western Pacific Subtropical High (10−80°N, 110−180°E) intense index is defined as (Liu, Liang, Sun, 2019)

$$\sum_{j=1}^{29} \sum_{i=1}^{29} \left(n_{i,j} \times \left(H_{i,j} - 587.0 \right) \times \cos \phi_j \right) \cdot \mathrm{d}x \cdot \mathrm{d}y,$$

$$n_{i,j} = \begin{cases} 1, H_{i,j} \geq 588 \\ 0, H_{i,j} \geq 588 \end{cases}, \tag{11}$$

where d_x means the value of latitudinal grid interval, d_y means the value of longitudinal grid interval, i means the serial number of latitudinal grids, j means the serial number of the longitudinal grid, $H_{i,j}$ stands for the geopotential high of one grid on 500 MB and ϕ_j is the latitude of the grid. The index ranges from 0 to 2500.

2.2.3 *Statistical analysis and model selection*

We first focus on the correlation between precipitation in Zhengzhou, China, and Western Pacific Subtropical High (WPSH) intense index. The first step is calculating the Western Pacific Subtropical High intense index. Secondly, we use the auto-tail dependence function for both the Block Maxima approach and Peak Threshold approach to check whether the data are stationary. The third step is to choose an appropriate approach and use a maximum likelihood estimate to find out whether there is any correlation between precipitation in Zhengzhou and the Western Pacific Subtropical High (WPSH) intense index based on the estimated parameters and their standard errors.

Next, we concentrate on the extremity of the precipitation in Zhengzhou, China, in 2021. The first step is to collect the parameter threshold to use the Peak Over Threshold approach through a mean residual life plot. Secondly, a high threshold will reduce the number of exceedances and thus lead to high variance, whereas a low threshold will induce a bias because the GPD will fit the exceedances poorly. Thus, we use probability plot and quantile plot to do model diagnostics. Thirdly, we use Peak Over Threshold to collect the data we need. Lastly, we use the return level to estimate the extremity of the precipitation in Zhengzhou, China, in 2021. The software we use in this research is RStudio (x64 4.1.0).

2.2.4 *Implementation*

2.2.4.1 *Calculate western pacific subtropical high intense index*

Using the geopotential high data from NCEP/NCAR, we put the data into the formula (11) that calculates the intense index and gets the index's distribution over time (Figure 1). We can conclude that the index ranges from 0.0 to 2500.0 and have an increasing trend over time. Since 1951, the intensity of Western Pacific Subtropical High has been strengthened and tends to continue. However, in the paper, we do not focus on the trends of the index, and we only care about whether the index has some effect on the precipitation. The calculation of the Western Pacific Subtropical High intense index is one preparatory work.

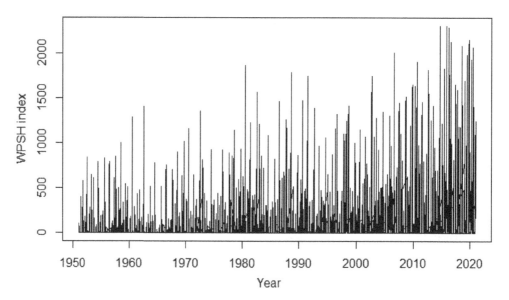

Figure 1. The distribution of the Western Pacific Subtropical High Intense Index.

2.2.4.2 *Check the independence among the data*

After applying formula (3), from the auto-tail dependence figure (Figure 2), we can vividly see that the ρ values are near zero, which means the precipitation data of Zhengzhou meet the basic assumption and can be directly used without other extra processing.

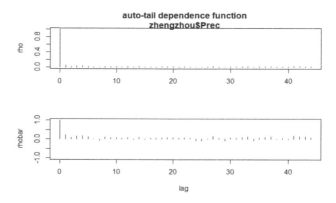

Figure 2. Auto-tail dependence of the top 1% precipitation data of Zhengzhou, China.

2.2.4.3 *The correlation between precipitation in Zhengzhou and Western Pacific Subtropical High (WPSH) intense index*

In this case, we assume that the location parameter μ and scale parameter σ are functions of the WPSH intense index. The distribution function of the Peak Over Threshold approach does not have the parameter of location. In addition, we only focus on the correlation but not the extreme events, so we do not choose the maximums of every block and whether choosing a threshold is unnecessary. For these reasons, we use the Block Maxima approach to find whether WPSH affects the precipitation. Using the maximum likelihood estimation by formula (10), we can get the parameters' estimation and standard errors (Table 1). Within two times standard errors, the parameters are all shown significantly. The results show that Western Pacific Subtropical High does have some effect on the precipitation.

Table 1. The values of parameters estimation and their standard errors.

	Estimated parameters	Standard Error Estimates
mu0	9.370593e-07	9.999850e-09
mu1	6.464666e-09	9.999791e-09
sigma0	2.288051e-06	9.999793e-09
sigma1	1.585251e-08	9.999762e-09
shape	2.317291e+00	1.216390e-03

2.2.4.4 *Collect an appropriate threshold by mean residual life plot*

From the mean residual life plot, it is easy to see the plot is linear in u, where the threshold lies. From Figure 3, it is clear that the plot begins to be linear when the threshold is approximately 250.

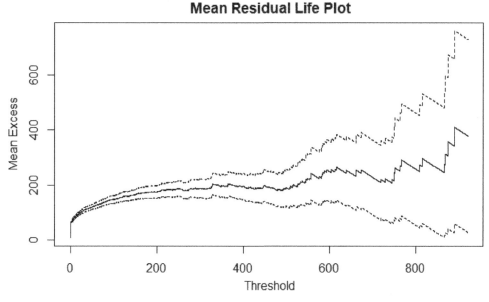

Figure 3.   Mean Residual Life Plot of the threshold. The solid black line stands for the mean residual life plot. The black dotted line stands for 95% confidence bands.

2.2.4.5 *Model diagnostics*

From the probability plot (Figure 4) and quantile plot (Figure 5), it is concluded after choosing the threshold numbered 250. The probability plot has good fitting with the 1-1 line. It means the estimated values fit well with empirical values, and it is reasonable to employ the model we choose. The quantile plot shows the bias and variances of the data. The figure shows that the regression line is close to the 1-1 line, which shows a small bias. The black dots are distributed near the line, and it shows a small variance. All the results show that the threshold we choose makes the model suitable for us to use.

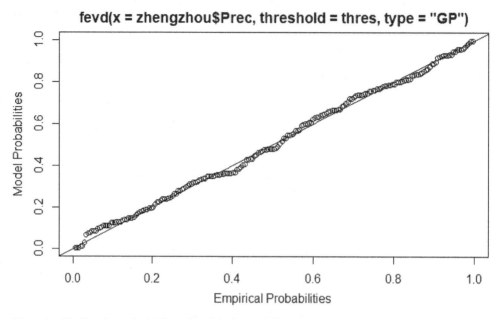

Figure 4. The line shows the 1-1 line of empirical probabilities and estimates probabilities. The hollow dots show the distribution of the data collected above the threshold.

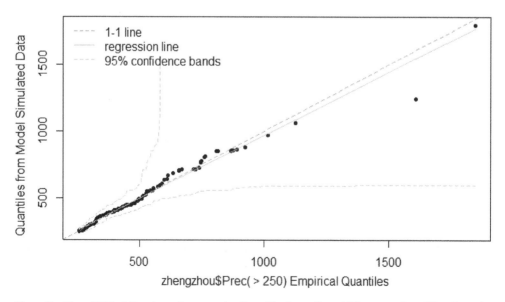

Figure 5. The solid black line shows the regression line of the data collected. The orange dotted line shows the 1-1 line of empirical quantiles and estimates quantiles. The black dots are data collected above the threshold.

2.2.4.6 *Collect extreme data*

After determining the threshold, we collect data whose precipitation amount is larger than the threshold as the basic data in the following steps. After choosing the threshold, we collect bigger data than the threshold of every block (Figure 6). According to the Peak Over Threshold approach, the block stands for the year and 70 blocks.

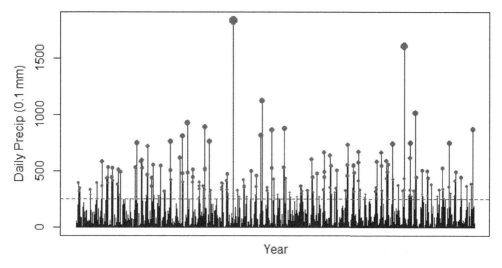

Figure 6.   The black vertical lines show the daily precipitation amount for 70 years. The blue dotted line shows the threshold. The blue dots are the data over the threshold that we choose.

2.2.4.7 *Calculate return levels*

We choose 5, 10, 100, 500 and 1000 as observed numbers. From the table, we can conclude the return levels of every probability (Table 2). For example, the case that the amount of precipitation is over 76.24579 mm will approximately occur once every five years. From the rough data we collected about the precipitation in Zhengzhou, China, in July 2021, the maximum daily precipitation exceeds 400 mm, significantly over the return level of the 1000-year level. From these results, we can roughly estimate that the precipitation in Zhengzhou, China, in July 2021 is a once-in-a-millennium event.

Table 2.   The years and their return levels.

m (year)	Z_m(0.1 mm)
5-year level	762.4579
10-year level	913.8312
100-year level	1482.8321
500-year level	1949.0366
1000-year level	2169.4721

3   CONCLUSION

By applying the Extreme Value Theory, especially Peak Over Threshold approach and Block Maxima approach on the precipitation data and Western Pacific Subtropical High intense index, we can conclude that the intense of Western Pacific Subtropical High does have some influence on the precipitation in Zhengzhou. In addition, the extreme precipitation event that happened in July 2021 is a once-in-a-millennium event.

However, there is some future work that needs to be done in the future.

(1) The knowledge of the correlation between the intensification of Western Pacific Subtropic High and precipitation in Zhengzhou is limited. More statistical methods will be applied in future work, for example, invoking the correlation coefficient between the precipitation and the index.

(2) In the future, some other elements that possibly influence the extremely heavy precipitation in Zhengzhou in July 2021, like the frequency occurrence of typhoons in the western Pacific, will be considered.

(3) We assume that the precipitation data are stationary. These data are non-stationary since there is some annual cycle in the precipitation. In the future, more work will be reported based on precipitation data only in the summertime.

(4) A further investigation will be conducted to check whether any other method could predict the occurrence of the extreme event.

REFERENCES

A. Ramazan Gençay et al. (2003) High volatility, thick tails and extreme value theory in value-at-risk estimation. Insurance: Mathematics & Economics, **33**: 337–356.

Henan Government Net. (2021) The 10th press conference on Flood control and Disaster Relief in Henan Province. http://www.henan.gov.cn/2021/08-03/2194608.html.

H. Liu, P. Dai. (2007) Synoptic and Climatic Analysis of Torrential Rain in Zhengzhou. Meteorological and Environmental Sciences, **S1**: 45–47.

Lance A. Waller, Bruce W. Turnbull. (2012) Probability Plotting with Censored Data. The American Statistician, **46**: 5–12

L. De Haan, A. Ferreira. (2006) Extreme Value Theory: An Introduction. Springer.

M. D. Pandey, P. H. A. J. M. van Gelder, J. K. Vrijling. (2001) The Use of L-Moments in the Peak Over Threshold Approach for Estimating Extreme Quantiles of Wind Velocity. International conference on structural safety and reliability. ICOSSAR'01. 0.

R. Matthews. (2005) 25 Big Ideas. Oneworld Publications, London.

R. V. Canfield, D. R. Olsen, T. L. Chen. (1981) Extreme Value Theory with Application to Hydrology, Statistical Distributions in Scientific Work. Springer, Netherlands. pp 337-350.

S. Coles. (2001) An Introduction to Statistical Modeling of Extreme Values. Springer- Verlag, London, UK.

W. Yu, J. Liu, J. Chang, et al. (2008) Changes in Extreme Temperature and Precipitation in Henan Province During 1957-2005, Advances in Climate Change Research, **02**: 78-83.

Y. Liu, P. Liang, and Y. Sun. (2019) The Asian summer monsoon: characteristics, variability, teleconnections and projection. Elsevier, Amsterdam.

Z. Liu, Y. Wang, et al. (2005) Multiple Time-Scale Analysis of Precipitation Variation in Zhengzhou during Last 54 Years. Scientia Meteorological Sinica, **S1**: 123–126.

Advances in Energy, Environment and
Chemical Engineering – Abdullah & Osman (Eds)
© 2023 The Author(s), ISBN: 978-1-032-36083-6

Influence of terrains and weather on wildfires in Australia

Yingkun Shi[†]
Environmental Science, Hebei University of Science and Technology, Shijiazhuang, China

Ruili Xu[*,†]
Geospatial Information Engineering, Shenzhen University, Shenzhen, China

ABSTRACT: Wildfires are uncontrollable and catastrophic. Therefore, monitoring wildfires and analyzing the impact factors of wildfire are of great significance for reducing the damage. Based on Landsat8 remote sensing imagery and atmosphere data, this study mainly uses delta normalized burn ratio (dNBR) and data correlation analysis to discuss the factors influencing wildfires in Australia. It is mainly studied from both weather and topography. In terms of topography, the dNBR index was utilized to study the evolution of the Australian wildfires from 2019 to 2020 and analyze the relationship between wildfire intensity and environmental factors such as topography, climate, and vegetation conditions. The wildfire burns more severely in areas with large terrain differences and lighter in areas with flat terrain. In addition, altitude is the main influence factor of the fire. Fires at high altitudes will be more serious. It means that places with high altitudes and steep terrain need attention to strengthen wildfire prevention. In terms of weather, the correlation analysis method mainly analyzes the correlation between wind speed, temperature and humidity and the occurrence and development of fire. Wind intensity is the main influence factor, positively correlated with forest fires. The correlation between temperature and humidity is not as significant as wind speed, but it can also be seen that most areas are positively correlated.

1 INTRODUCTION

Wildfire is a kind of natural disaster with strong suddenness, great destructiveness, and difficulty in handling and rescue. Once the forest suffers a fire, the most intuitive hazard is to burn the trees. On the one hand, forest accumulation has declined, and on the other hand, forest growth has been severely affected. It also will burn down forest plant resources, endanger wild animals, cause soil erosion, reduce the water quality of downstream rivers, cause air pollution, and threaten people's lives and property. Many countries have suffered serious losses, including the 2016 wildfire in Inner Mongolia, China, the 2017 Canadian Senli fire, and the 2018 Sichuan wildfire in China. However, the most representative and serious one in Australia. In 2019, a huge disaster happened in Australia. Since Australia entered the wildfire season in July 2019, almost everything has gone from the southeast coastal areas where the economy is the most developed, most densely populated, New South Wales and Victoria are located, to Tasmania, Western Australia, and the Northern Territory. Wildfires are burning in all states (Xinhuanet 2020). In November 2019, wildfires in eastern Australia raged, and the disaster intensified. On December 31, 2019, severe wildfires broke out in New South Wales, Victoria, South Australia, and other places in southeastern Australia, covering more than 6 million hectares. Thunderstorms occurred in Melbourne on January 15 and 16, 2020, which eased local air pollution compared to previous days. The local air quality has also changed from "harmful" to "medium." As of July 28, 2020, the Australian bushfires may have killed 3 billion animals and caused the local government huge losses of life and property.

---

*Corresponding Author: xuruili2018@email.szu.edu.cn
[†]These authors contributed equally.

However, research on the impacts of wildfires in Australia is rare. Many studies have used remote sensing data to map or classify the wildfire severity by the influence of vegetation (Hammill & Bradstock 2006) and normalized burned index (Gibson et al. 2020). Also, using remote sensing to monitor wildfire regimes and suggest wildfire managements are common in studies; however, only a few studies based on remote sensing data study the factors associated with wildfires in Australia. Meanwhile, the factors influencing the spread of wildfires are mostly studied from the perspectives of atmosphere, rainfall, and moisture degrees. Since there is less research on the impact factors of wildfire spread, this study will focus on Australia's southeast coast and research the impacts of atmospheric, terrain, and vegetation factors on the spread of wildfires, and the damage caused by wildfires also will be counted.

Therefore, this study takes the Australian wildfire from August 2019 to January 2020 as an example, discusses its relationship with the fire from the aspects of the terrain, wind direction, wind speed, temperature, and humidity, and analyses the main controlling factors.

2 STUDY AREA

The study area is on Australia's southeast coast, the worst-hit area in Australia's wildfires. The Landsat-8 image includes the Pine Mountain National Park, and Woomargama National Park was chosen as a typical area for observation. This area is the most cloudless area on Australia's southeast coast, which can provide more accurate data.

In order to observe how wind and climate affect wildfires, which is on a larger scale, the studying area of the atmosphere will be larger than the studying area of wildfire monitoring. A smaller area with obvious terrain changes in the studying area will be selected as the studying area to observe the terrains' changes effects. The overview of the study area is shown in Figure 1.

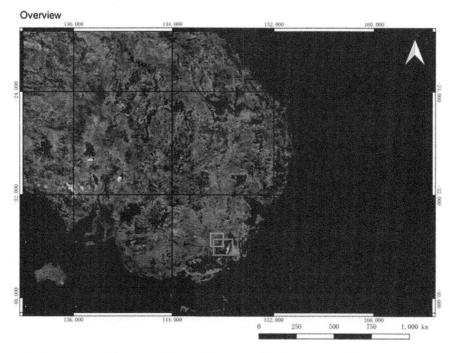

Figure 1.   Various research data processing constituencies in Southeastern Australia.
orange: terrain, blue: vegetation, panoramic view: atmospheric.

3 DATA AND PREPROCESSING

Landsat-8 imagery was used for this study. It is equipped with sensors of one panchromatic (PAN) band with a resolution of 15 m, eight visible, near-infrared, and short-wave infrared bands with 30 m.

In order to obtain accurate wildfire information using Landsat-8 imagery, an atmospheric correction was conducted to remove the atmospheric effect. The FLAASH (Fast Line-of-sight Atmospheric Analysis of Spectral Hyper-cubes), a widely used atmospheric correction module of the ENVI 5.3 software package, was used in this study. Meanwhile, Landsat 8/STRM C1V1 (a period of data:2019.8-2020.1) data from USGS are used in terrain research.

For the weather research, specific humidity(kg*kg-1), temperature (K), fire weather index, and wind(m/s) data are obtained from European Commission. Specific humidity refers to the ratio of the mass of water vapor to the total mass of air (the mass of water vapor plus the mass of dry air) in a mass of humid air. The Fire weather index combines the Initial spread index and Build-up index, and is a numerical rating of the potential frontal fire intensity. In effect, it indicates fire intensity by combining the rate of fire spread with the amount of fuel consumed. Fire weather index values are not upper-bounded; however, a value of 50 is considered extreme in many places. The Fire weather index is used for general public information about fire danger conditions.

4 DATA AND METHODS

Fire trail extraction and damage assessment of the burned area require the accurate detection of fire scars and detection of vegetation changes after the fire. Ratio-based indices can minimize topographic-induced variance. Thus the normalized burn ratio (NBR) and the normalized difference vegetation index (NDVI) has considered for detecting the fire scars and the damage to the vegetation of Australia's southeast coast, respectively:

$$NDVI = \frac{(NIR-RED)}{(NIR+RED)} \tag{1}$$

Where NIR and RED are the reflectivity of near-infrared bands and visible red bands, respectively, which correspond to the band5 and band4 of the Landsat-8 imagery.

$$NBR = \frac{(NIR-SWIR)}{(NIR+SWIR)} \tag{2}$$

Where SWIR is the reflectivity of short-wave infrared bands, respectively, which correspond to band7 of the Landsat-8 imagery. NBR is formulated as NDVI, and its value is between -1 to 1. Burned areas are generally negative, and unburned areas are generally positive. This study used the delta NBR (Dnbr) since it can better capture the spatial complexity of severity within fire perimeter than other indexes (Miller & Thode 2007).

$$dNBR = NBR_{pre-fire} - NBR_{post-fire}$$

Where $NBR_{pre-fire}$ is the NBR index before fire and $NBR_{post-fire}$ is the NBR index after the fire.

Figure 2 shows the minimum Dnbr value in the studying area, the maximum Dnbr value in the studying area, the average Dnbr values, and the standard deviation for Dnbr values. Statistic result refers to the Dnbr fixed threshold method rules (García & Caselles 1991) and the threshold classification result of some forest areas (Liu et al. 2018), the value of Dnbr in the studying area and the mainland cover of the studying area conform to the existing threshold method rules. Therefore, the classification rules of victimization of fires in the studying area are obtained based on the existing classification rules, as shown in Table 1. The areas moderately burned and severely burned will be extracted as the results of burned areas. Then, NDVI values were calculated and reclassified into three groups: no plant cover, weak plants, and healthy plants, which were designed in order of vegetation vigor. Based on the existing vegetation classification rule (El-Gammal et al. 2014) and the visual interpretation method, the vegetation cover classification rules are obtained in Table 2.

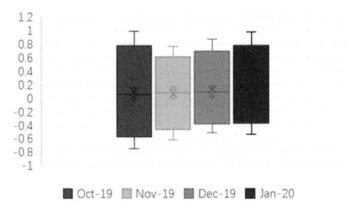

Figure 2.　Statistics of Dnbr in a different month.
Different color of boxes in the box diagram illustrates the data for different months.

Table 1.　Fire severity classification rule.

Fire severity classification rule	
Not burned	≤ 0.1
Slightly burned	0.1~0.38
Moderately burned	0.38~0.72
Severely burned	≥ 0.72

Table 2.　Vegetation classification rule.

Vegetation classification rule	
No plant cover	-0.38~0
Weak plants	0.0.1~0.3
Healthy plants	0.31~0.61

For the study of impact factors, which influence wildfire spread and relate to terrain and weather, QGIS and Python are mainly used for data processing and visualization. Based on Landsat 8 B1-B7 layers, this study applies Semi-Automatic Classification Plugin from QGIS to produce the true color images. Based on the selected area in layers, this study applies QuickMapServices from QGIS to produce an Overview Image of Southeastern Australia. Based on STRM C1L1 layers, this study applies Terrain Profile Plugin and Hillshade Properties from QGIS to produce the Hillshade Image of Southeastern Australia. Python language, including numpy, pandas, and matplotlib packages, is used for data processing and plotting.

5　RESULT

5.1　*Fire trails and vegetation changes*

Figure 3 shows the changes in vegetation cover area and burned area over the four months. In the first two months, the wildfires didn't spread significantly. In November, a new fire point was added in the northwest direction while the wildfires in the northeast terrains spread slightly. However, from November to December, the wildfires spread gradually. The two fire points spread to the middle area simultaneously, showing a southeast movement trend, and the vegetation cover of the central area and southeast corner has been destroyed. The completely spread of wildfires happened between December 2019 to January 2020. Until January, most of the studying areas are under fire. Meanwhile, more than half of the vegetation was destroyed by wildfires, with the most severe forest losses being in this area's central and eastern parts.

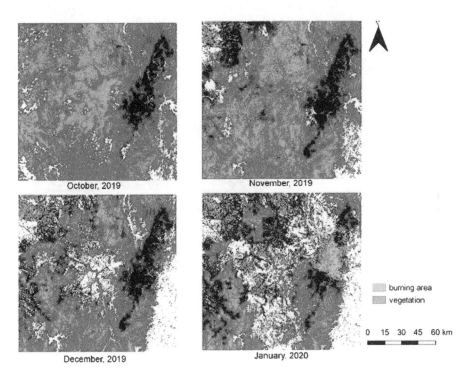

Figure 3.   Changes in wildfires and vegetation during October. 2019 to January. 2020.

5.2 *Terrain*

Figure 4 shows the altitude of southeastern Australia. It uses a solar altitude angle of 65 and an azimuth of 305. The graph below shows the orange line above. Light-colored places have low elevations, and dark places have high elevations. The highest place in this zone is about 850 meters, and the lowest is about 250 meters. The topography and changes are obvious.

5.3 *Weather*

Based on wind data, wind field maps. Figure 5 shows that the wind field distribution in August and September 2019 is similar. There is a large wind area in the southeast of the land in the figure, and the wind speed changed from west to southwest wind from south to north. In the northwest, the wind was southeast. In October and November, the wind field in the southeast of the land gradually weakened to the southwest. By November, the two windy areas were connected, and only the east of the land had a windless area. In December and January, the windless zone moved southward and shrank to a linear shape in January.

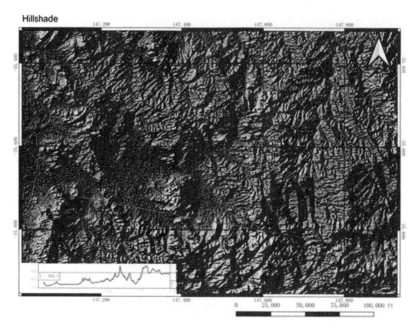

Figure 4.   Hillshade of southeastern Australia

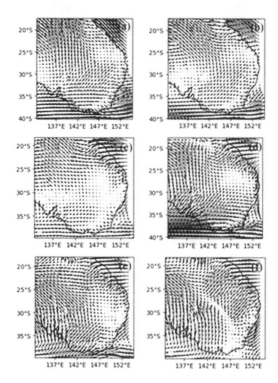

Figure 5.   Wind field in southeastern Australia. (a)–(f) are the monthly average wind fields from August 2019 to January 2020.

In terms of intensity, the humidity in southeastern Australia has increased month by month from August 2019 to January 2020. In Figure 6, the darker the red area, the smaller the humidity, and the darker the green area, the greater the humidity. In August 2019, the minimum value in the northwestern part of the land reached 0.004; in January 2020, the minimum value was only 0.007. From the perspective of the low-value center, the northwestern part of the land in the first figure is relatively low in humidity. As time went by, the lowest-value area gradually moved to the southeast.

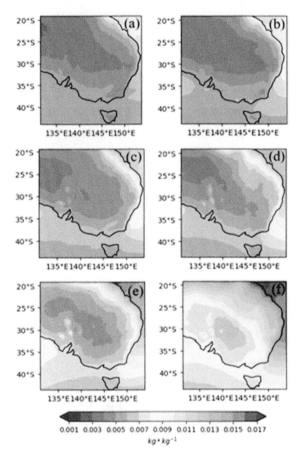

Figure 6.　Distribution of surface specific humidity in Southeastern Australia. (a)–(f) are the monthly averages from August 2019 to January 2020.

The temperature distribution increased month by month, gradually decreasing from northwest to southeast, and the increase scale was 10K. In August 2019, the highest temperature was 295K in the northern part of the land, the lowest temperature was about 280K in the north, and the monthly increase was about 2.5K. As of January 2020, the highest temperature was about 307K in the northwest of the land, and the lowest temperature was about 290K in the southeast.

In Figure 8, the darker the area, the more likely it is to fire. The fire started in August and appeared in the western part of the land, more severe on the southern shore, and gradually spread north to the east in September. In October, the fire reached its extreme value in the northwest, and areas with high indices were contiguous and large in scope. The fire was mainly distributed on the north side by November and developed to the south in December and January.

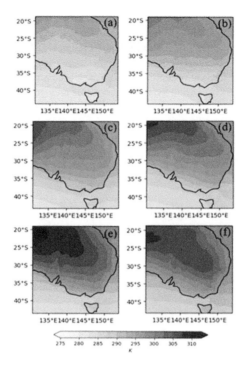

Figure 7.　Surface temperature distribution in southeastern Australia. (a)–(f) are the monthly averages from August 2019 to January 2020.

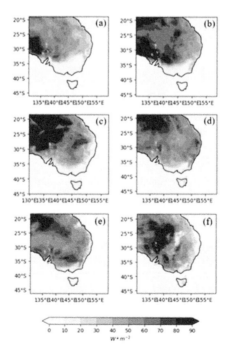

Figure 8.　Distribution of fire weather index in Southeastern Australia. (a)–(f) are the monthly averages from August 2019 to January 2020.

6 DISCUSSION

6.1 *Overall changes in the study area*

According to statistics, the total area of wildfire is mainly composed of moderately and severely burned areas. The first three months of the wildfire spreading was limited to a lesser scope, mainly concentrated in the north, northeast, and east. However, the spread speed of wildfires significantly changed between December 2019 and January 2020, except for the southwest and south, which spread from 0.14km^2 to 20.52 km^2 and 1.18 km^2 to 124.74 km^2, respectively, while the fire scars of this area have increased rapidly, in other directions the fire scars have spread at least 200km^2, with the wildfires in north direction spread the most seriously, which from 1.99km^2 to 1238.61km^2. Finally, the area of fire scars reached 4688.79 km^2, a nearly 7.8 times increase from 599.01 km^2 in October 2019 when the wildfires began.

The vegetation cover is composed of medium coverage and high coverage area. In Figure 9, the vegetation cover area showed a downward trend during the four months, which dropped from 12569 km^2 to 4757 km^2; about 62.1% of the vegetation was burned in the wildfires.

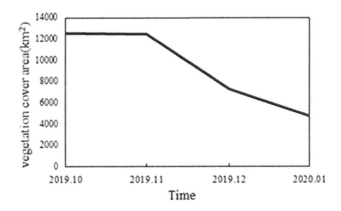

Figure 9. Changes in vegetation cover area from October 2019 to January 2020.

6.2 *The influence of terrains on wildfires*

Places with different terrain are steep slopes or cliffs, and with flat terrain are plains; lakes are less affected by fires.

Altitude and slope have an obvious influence on the burning situation of fire. Altitude plays a major role in the fire. The wildfire situation becomes heavier as the altitude increases. There are differences in temperature, humidity, soil, and vegetation due to different exposure to sunlight. Generally, it is easier to burn in the southeast of Australia than on the north slope and spread faster than on the south slope. Different slope positions will lead to the different moisture content in the forest land. This, in turn, can affect the spread of forest fires.

As shown in Figure 10, there is a fluctuant trend of terrain along the direction of the selected path, yet the overall trend shows an upward trend. While the wildfire trails' changes in October 2019 and January 2020 on the same path maintained a similar curve trend, between -0.5 and 0.7. During these three months of wildfires, the wildfire trails have increased apparently. However, the whole wildfires in this study area are moderate. The most serious burned part along the study path has not reached severely burned severity (only about 0.65).

According to the three curves in Figure 10, the effect of terrain altitude on wildfires trails is not obvious, and there is no significant difference between the value of Dnbr at high altitude and low altitude. A certain correlation has been shown between wildfire trails and slope changes. Before peaks of the terrains, the value of Dnbr has an apparent increase where the slope is also climbing sharply, and the steeper slope shows that the value of Dnbr between valley and peak has a significant change (in the third peak of the altitude at Figure 10 shows a sharp change).

In conclusion, the change of terrain altitude will not significantly impact the wildfires, and there is a positive correlation between the change of slope and the wildfires. The slope change is more obvious, and the severity of the wildfire trails is more serious; meanwhile, the slope changes faster, and the spread of wildfires also becomes faster.

As shown in Figure 9, the influence of altitude on wildfire burning is obvious in areas above 500 meters. The influence of altitude on fire burning is more obvious in areas above 500 meters. Altitude plays an important role in the spatial pattern of forest fire intensity (Carlson et al. 2011; Wimberly & Reilly 2007). As the altitude increases, the proportion of severely burnt areas in the landscape area increases, and the shape of the area tends to be simple. This may be because there is more vegetation in high-altitude areas, strong solar radiation, larger slopes, and a faster spread of forest fires, so the possibility of severe burns is greater. In addition, the fires in this study mostly occurred in the spring in southeastern Australia, and there may be more shrubs or herbs in high-altitude areas (Di et al. 2015; Li et al. 2017). The higher the altitude, the lower the air pressure, and due to the accumulation of combustibles, the fire is easy to spread, which may be the main reason for the large area of severe burning. The influence of slope and aspect is weaker than that of height. However, the slope and aspect will indirectly affect the combustion intensity when the fire spreads by affecting the water content of the combustibles (Fu et al. 2020). However, the influence of altitude below 500 on fire burning is not obvious. The dNBR value is closely related to the fire's burning degree and affects the total carbon. Related stUDIEs have shown that the dNBR value is significantly related to the duration of forest fires and the biomass loss of forest ecosystems (WoosterMJ & Drake NA et al. 2005). The classification and evaluation of forest fire intensity can estimate the biomass loss caused by forest fire disturbance and provide a reference for global fire prevention. Combined with Dnbr data, the classification, evaluation, and management of large-scale forest fire intensity can be widely studied in the future (Tan & Yi 2004).

Similar inland fires have occurred in the Greater Xing'an Mountains in my country, and the situation is relatively similar. The spread of forest fires is greatly affected by terrain and vegetation. The terrain of the Daxinganling forest area is mostly gentle slopes and hilly areas. The forest area of Daxinganling has an altitude of 137–1740m, with an average altitude of 678.6m (standard deviation 246.78). The slope is 0–38 degrees, the average slope is 4.4 degrees (standard deviation 3.64), and 75% of the area is below 13.3 degrees. This topography has little effect on the prevailing wind direction, although the prevailing wind direction will vary with the local topography. Under strong wind conditions, forest fires will spread rapidly with the prevailing wind direction (Wang et al. 2013).

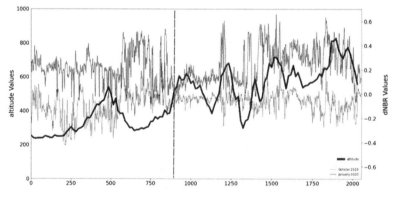

Figure 10.   The comparison between altitude and Dnbr value. All data in the figure are based on the same line in, and Figure 9 shows the common trend of altitude and wildfire severity simultaneously. The green line shows the Dnbr value in October 2019. The red line showed the Dnbr value in January 2020. The black line shows the altitude value. The blue line divides the chart into two sections. The left side represents terrains below 500 meters, and the right represents terrains above 500 meters.

6.3 The influence of weather on wildfires

Since the wildfire has just started in August and September, the distribution area is small, so October to January of the following year is selected for correlation analysis. From the correlation index between wind speed and fire, it can be found that there are mainly three areas with high correlation, which are in the southeast, southwest, and north of the land area in Figure 11. The wind speed in the southeast is lower in October, and wildfire is not easy to happen on edge. In November, the wind speed in the southeast increases, and the possibility of fires also increases. In December and January, the wind speed decreases slightly, and the possibility of fires also decreases slightly. At the same time, there is a similar positive correlation between the wind speed and the probability of wildfire in the southwest and north. From the correlation index between humidity and fire, it can be found that a large northern area has a higher positive correlation. From the evolution of humidity, it can be found that the humidity increases month by month from October to January of the following year. Due to multiple factors, the area spreading every month has irregularities, from October to November and mid-December. The wildfire weakened and spread to the coast, so a large part of the regional correlation performance was not good. It can be seen from the correlation diagram between temperature and wildfire that the southern wildfire has a strong correlation with temperature. Still, the correlation in the other northern regions is not large. The reason is that the wildfire conditions are met after reaching a certain temperature. The temperature has little effect on the fire.

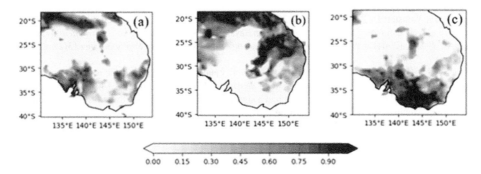

Figure 11. Correlation coefficient graph (a) is the correlation coefficient between wind speed and fire weather index, (b) is the correlation coefficient between humidity and fire weather index and (c) is the correlation coefficient between temperature and fire weather index.

7 CONCLUSION

This study analyzes the Australian wildfire from August 2019 to January 2020 and discusses its relationship with the terrain, wind direction, wind speed, temperature, and humidity.

The slope has obvious impacts on wildfire distribution. The wildfire is bigger in the area of the terrain (steep slopes or cliffs) and smaller in the area of the flat land (plains). High-altitude areas have a more serious impact on fires. There is more vegetation in high-altitude areas. The wind may increase the fire, and the influence of altitude on fire is also related to air pressure.

Of the three climatic factors considered in the study, the intensity of the wind has the greatest impact on the occurrence and development of the fire. The greater the wind speed, the greater the fire. Although temperature and humidity also positively impact the occurrence of fires, since wildfires are affected by many factors, their correlations are different in different regions.

It is necessary to pay attention to fire prevention in some places with significant terrain differences. Inland fire prevention needs to consider its terrain, climate, and vegetation types to take reasonable measures to adapt to local conditions more effectively and comprehensively.

REFERENCES

Carlson D J, Reich P B, Frelich L E. (2011) Fine-scale heterogeneity in overstory composition contributes to heterogeneity of wildfire severity in southern boreal forest. *Journal of Forest Research*, 16(3): 203–214.

Di X Y, Chu X, Yang G, Wu H.(2015) Distribution of forest fires in summer from 2000 to 2012 in China. *World Forestry Research*, 28 (4) : 72–75.

El-Gammal, M. I., Ali, R. R., Samra, R. A. (2014). NDVI threshold classification for detecting vegetation cover in Damietta governorate, Egypt. *Journal of American Science*, 10(8), 108–113.

Fu J J, Wu Z W, Yan S J, Zhang Y J, Gu X L, Du L H. (2020) Effects of climate vegetation and topography on spatial patterns of burn severity in the Great Xing'an Mountains. *Acta Ecologica Sinica* 40(5) : 1672–1682.

García, M. L., Caselles, V. (1991). Mapping burns and natural reforestation using Thematic Mapper data. *Geocarto International*, 6(1), 31–37.

Gibson, R., Danaher, T., Hehir, W., Collins, L. (2020). A remote sensing approach to mapping fire severity in southeastern Australia using sentinel 2 and random forest. *Remote Sensing of Environment*, 240, 111702.

Hammill, K. A., Bradstock, R. A. (2006). Remote sensing of fire severity in the Blue Mountains: influence of vegetation type and inferring fire intensity. *International Journal of Wildland Fire*, 15(2), 213–226.

Li M Z, Kang X R, Fan W Y. (2017) Remote sensing extraction and spatial analysis of forest fire intensity in Huzhong. *Forest Science and Technology*, 53 (3)

Miller, J. D., Thode, A. E. (2007). Quantifying burn severity in a heterogeneous landscape with a relative version of the delta Normalized Burn Ratio (dNBR). *Remote Sensing of Environment*, 109(1), 66–80.

Shuchao, L. I. U., Xiaozhong, C. H. E. N., Xianlin, Q. I. N., Guifen, S. U. N., Xiaotong, L. I. (2018). Remote Sensing Assessment of Forest Fire Damage Degree in Bilahe Forest Farm, Inner Mongolia. *FOREST RESOURCES MANAGEMENT*, (1), 90.

Smith AMS, WoosterMJ, Drake NA, et al.(2005) Testing the potential of multispectral remote sensing for retrospectively estimating fire severity in African Savannahs, *Remote Sensing of Environment*,97:92-115

Tan X-L, Yi H-R (2004) A method to identify forest on MODIS data. *Fire Safety Science*13(2), 83–91 (in Chinese)

Wimberly M C, Reilly M J, (2007) Assessment of fire severity and species diversity in the southern Appalachians using Landsat TM and ETM + imagery, *Remote Sensing of Environment*, 108(2): 189–197.

Xinhuanet. (2020) The recent rainfall has not had a significant impact on the Australian forest fire situation

Xiaoli Wang, Wenjuan Wang, Yu Chang, Yuting Feng, Yuanman Hu, Jianguo Chi(2013) Analysis of forest fire intensity in the Great Xing'an Mountains based on NBR, *Chinese Journal of Applied Ecology*, April 24 (4): 967–974

Advances in Energy, Environment and
Chemical Engineering – Abdullah & Osman (Eds)
© 2023 The Author(s), ISBN: 978-1-032-36083-6

Analysis and research on the influence of the research input level of the grid engineering cost standard project

Yan Zhang*
State Grid Jibei Electric Power Company Limited, Beijing, China

Yan Lu & Xu Cheng
State Grid Jibei Electric Power Company Limited Economic Research Institute, Beijing, China

Chao Wei
State Grid Jibei Electric Power Company Limited, Beijing, China

Pengyun Geng
State Grid Jibei Electric Power Company Limited Economic Research Institute, Beijing, China

ABSTRACT: With the increase of scientific research investment in China year by year, the concept of project budget management has undergone major changes. This paper summarizes and analyzes the many influencing factors of the research input level of power grid engineering cost standard projects and uses the fuzzy threshold method to identify the key influencing factors.

1 INTRODUCTION

In recent years, China's scientific and technological investment has shown an upward trend year by year. In 2020, the national research and experimental development (R&D) expenditure was 2,439.31 billion yuan (Statistics Bureau of the People's Republic of China, China Statistical Yearbook 2021), an increase of 55.60% compared with 2016. With the substantial increase in scientific and technological investment, the management of scientific research funds has been paid more and more attention, and it is no longer advisable to set budget control limits in advance. The management method of putting forward funding requirements according to the actual situation of the project is a major change in the concept of project budget management, and it is also a concrete manifestation of respecting the laws and characteristics of scientific research (Huang 2018; Fu & Sun 2017).

Power grid engineering construction is an important support for building a "new power system" and serving "dual carbon." The innovative application of new technologies, new processes, new equipment, and new materials has put forward updated requirements for the current cost pricing standards. At the same time, to promote the high-quality development of power grid engineering construction and ensure the accurate, efficient, and dynamic improvement of the price system for power grid engineering construction, a lot of research on cost standard calculation has been carried out. However, the current industry has not yet unified the standards for the calculation of related topics. Therefore, this paper summarizes and analyzes the many influencing factors of the research input level of power grid engineering cost standard projects and uses the fuzzy threshold method to identify the key influencing factors, so as to provide directions for trying to form a unified project budget and charging criteria, and scientifically and rationally allocate scientific research funds.

---

*Corresponding Author: yzhang202207@163.com

DOI 10.1201/9781003330165-94

2 ANALYSIS OF INFLUENCING FACTORS OF PROJECT RESEARCH INVESTMENT

There are many factors that affect the research input of power grid engineering cost estimation, such as the type of research object, sample support, the relationship between cost and current standards, the form of cost results, the number and application of research results, whether there is support from relevant research results, different topics. The situation of the implementation unit, the terrain situation of different regions, etc.

(1) Scope of cost measurement objects

The research on cost-related topics has the characteristics of many points and wide areas, involving a wide range of fields and many complex contractors. However, cluster analysis is carried out from the perspective of the research object, and such topics are summarized into five categories based on the closeness of the object and the power grid project, respectively. It is the main construction-related category, the auxiliary construction-related category, the site management-related category, the construction management-related category, and other categories, as shown in Figure 1. With the increase in the degree of divergence of the subject, the research workload and measurement factors also increase, thus affecting the research input of the subject.

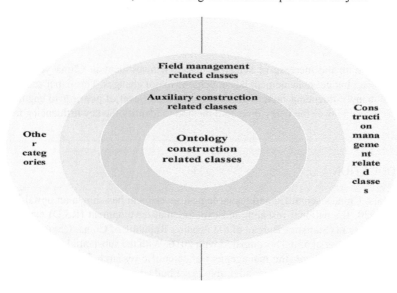

Figure 1.   Analysis of research object types.

(2) The complexity of the cost measurement object

The main consideration is the complexity of the cost measurement object and the geographical characteristics, which leads to the increase in the difficulty of cost calculation, which increases the investment of resources, such as special projects involving high altitude and UHV. The subject research needs to choose the appropriate measurement method according to the characteristics of the natural conditions of the location. The difficulty of the work also varies with the harshness of the natural environment, ultimately having an indirect impact on research investment.

(3) Sample support for cost research

According to the attributes of the research samples of the project cost of the grid project, it can be divided into the class with sample support and the class without sample support. Among them, there is no historical engineering data and samples for reference during the cost standard calculation period for the non-sample-supported projects. Project researchers are required to use actual measurements. The calculation and formulation of cost standards by technical means such as theoretical analysis are more difficult than that of sample support, so it affects the research input of the project.

(4) Relationship between cost research and current standards

According to the relationship between the cost research of power grid engineering cost and the current standard, it can be divided into the proposed new cost and the adjustment cost. Among them, the adjustment expenses refer to the pre-existing expense items under the current pricing standards and quota system, but due to various reasons such as technological innovation, progress in management methods, and changes in national pricing-related laws and regulations, the current standards cannot expense types adjusted to meet the actual project needs. The proposed new expenses refer to the temporary missing items in the pre-planned and the cost management documents under the current pricing standards and quota system that have not been covered by the current pricing standards. It is used to supplement the current pricing standards according to the needs of the project or refine the item's expense type. Considering that there is no result support in the early stage of the calculation of the new cost standard, to ensure the project's smooth implementation, the relevant resource input needs to be increased, thus affecting the research investment of the project.

(5) Form of expression of cost research results

According to the expression form of the cost research results of power grid engineering cost subjects, it can be divided into relative value attribute saliency class and absolute value attribute saliency class. Among them, the relative value attribute salient category means that the calculation results rely on the existing cost standards. It is determined through comparative analysis, analogy analysis, and other methods, according to a certain proportional relationship or set rate, and there is no exact value; the absolute value attributes salient category. It is a type of cost that obtains the measurement results through independent accounting and on-site technical measurement. The formulation of such cost standards relies on the calculation of consumption, so the calculation requirements for the consumption level must be accurate, scientific, and reasonable.

(6) Promotion and application of research results

According to the popularization and application of the research results of power grid engineering expenses, it can be divided into five categories: national promotion and application, industry promotion and application, state grid company promotion and application, provincial company promotion and application, and basic research not promoted. The promotion and application of research results reflect the technical complexity of the subject and the urgency of demand; that is, subjects with a wide range of promotion and application have relatively high research investment. Among them, national promotion and application> industry promotion and application> state grid corporation promotion and application > promotion and application by provincial companies > basic research not promoted.

3 IDENTIFICATION OF KEY INFLUENCING FACTORS BASED ON THE FUZZY THRESHOLD METHOD

3.1 *Fuzzy threshold method theory*

Set the domain of factors that have a key impact on the research input level of power grid engineering cost standards as:

$$U = \{u_1, u_2, ..., u_n\} \tag{1}$$

The evaluation level domain is:

$$V = \{v_1, v_2, ..., v_m\} \tag{2}$$

The membership degree r_{ij} of the $n - th$ factor corresponding to the m evaluation levels can be determined by the expert scoring method, and the evaluation matrix R can be obtained as:

$$R = (r_{i1}, r_{i2}, ..., r_{ip}), 0 \leq r_{ij} \leq 1, \sum_{j=1}^{p} r_{ij} = 1 \tag{3}$$

Determine the weight vector according to the importance of each level during evaluation:

$$W = \{w_1, w_2, ..., w_m\}^T \tag{4}$$

Where: $w_i \geq 0, \sum w_i = 1$. Among them, the different weight settings have a greater impact on the identification results of key influencing factors (Yang et al. 2019; Zhang 2021). Thus, the fuzzy comprehensive evaluation result vector $B = W \bullet R$ is obtained. Finally, based on experts and practical experience, the threshold is reasonably set, and the key factors affecting the cost are screened and determined.

3.2 *Determination of evaluation grade weight based on analytic hierarchy process*

The main principles of the Analytic Hierarchy Process (AHP) are as follows: https://baike.so.com/doc/6012611-6225598.html

First, construct pairwise comparison judgment matrix is constructed on importance scaling theory A:

$$A = \left(a_{ij} \right)_{n \times n} (i, j = 1, 2, \cdots, n) \tag{5}$$

The judgment matrix A is then normalized, and the calculation formula is:

$$\bar{a}_{ij} = a_{ij} / \sum_{k=1}^{n} a_{kj} \ (i, j = 1, 2, \cdots, n) \tag{6}$$

The calculation formula of the weights is:

$$w_i = \bar{w}_i / \sum_{i=1}^{n} \bar{w}_i \ (i = 1, 2, \cdots, n) \tag{7}$$

Finally, judge the consistency; if the results are valid according to the consistency test, do not adjust the results.

3.3 *Identification of key influencing factors and analysis of results*

Taking the above-mentioned factors as the domain of input level influencing factors, the five evaluation levels of the key influencing factors of cost are determined based on expert opinions as $V = \{$unrelated, unimportant, average, important, very important$\}$ (see the calculation results below Table 1) and the blur threshold is set to 0.16.

Table 1. Calculation results of evaluation grade weights based on AHP.

Serial number	Evaluation level	Weight calculation result
1	unrelated	0.05
2	unimportant	0.1
3	average	0.15
4	important	0.3
5	very important	0.4

Since there are many factors affecting the input level of research on grid engineering cost standards, only the one-time fuzzy threshold method can only initially identify the key factors affecting cost, which lacks accuracy. Therefore, this paper uses the fuzzy threshold method for secondary screening to narrow the range of key factors affecting costs and help achieve precise management and control of resource investment. According to the fuzzy threshold method theory, the secondary fuzzy comprehensive recognition results are obtained as shown in the following Table 2.

Table 2. The results of secondary fuzzy comprehensive recognition.

Serial number	Analysis dimension	Fuzzy evaluation value
1	Measured object range	0.2740
2	Object complexity	0.3438
3	Sample support	0.3625
4	Relationship with current standards	0.4000
5	The form of achievement	0.3250
6	Promotion and application	0.2940
7	Implementation unit	0.2688

It can be seen from the above table that among the many influencing factors, the relationship between cost research and current standards, the support of cost research samples, and the complexity of research objects have a greater impact on the level of research investment.

4 CONCLUSIONS AND RECOMMENDATIONS

In this paper, task decomposition, resource input analysis, and influencing factor analysis are carried out on the standard subjects of power grid engineering costs; the fuzzy threshold method is used to effectively identify the key influencing factors on the research input level of the subject research based on the expert scoring opinions.

Considering the typical complexity, uncertainty, and difference of research on cost-related projects, in order to better form a unified project budget and charging criteria, the following strategies are proposed:

(1) Further deepen the analysis of water impact factors of research investment in combination with subject classification

As different types of projects have different resource input at each stage and are affected by various factors, the statistical analysis of the reasonable input level of different types of projects should be further strengthened in the future, so as to better dynamically improve the resource input level according to the project needs.

(2) Improve the matching relationship between the actually incurred accounts and the listed expense accounts

Regarding the cost composition table, although the listed subjects have been considered as comprehensively as possible, including most of the content, it still does not exclude the situation of special projects. There may be situations where the actual expenditure of the cost and the budget are difficult to match. Therefore, it is necessary to strengthen and supplement the items that cannot be listed. For the cases that do not occur frequently or have occurred but the matching relationship is unreasonable, the scope of interpretation and adjustment should be continued to be strengthened to improve the rationality of the cost table.

(3) Establish and improve scientific research project funding management regulations and supervision mechanisms

A scientific research project funding management system with a scientific structure, complete system, clear powers and responsibilities, and efficient operation is an important carrier to realize the modernization of the scientific research project funding governance system and governance capacity. On the basis of the post-event supervision of the audit department, continue to strengthen the responsibilities of the project undertaking units and project cooperation units and strengthen their supervision functions in scientific research activities.

REFERENCES

Fu Ye, Sun Qiaoping. Analysis of key influencing factors of scientific research funding use behavior [J]. *Scientific Research*, 2017, 35(05): 729–736.

Huang Lingyun. Influencing factors and countermeasures of scientific research project funding management risks [J]. *Business Accounting*, 2018(14):81–84.

Statistics Bureau of the People's Republic of China. *China Statistical Yearbook* [M]. Beijing: China Statistics Press, 2021.

Yang Zhen, Zhao Weibo, Wu Luofei, Xu Xiaomin. Research on Influencing Factors of Basic Test Costs of Transmission Lines [J]. *Modern Economic Information*, 2019(20): 298–299.

Zhang Jie. *Evaluation and Analysis of Power Grid Investment Capability and Investment Strategy* [D]. Shanxi University, 2021.

Advances in Energy, Environment and
Chemical Engineering – Abdullah & Osman (Eds)
© 2023 The Author(s), ISBN: 978-1-032-36083-6

Experimental study on hydraulic gradient on groundwater cross strata pollution in coastal aquifers

Mogeng Cheng*, Guoqing Lin*, Yaru Xu, Yuwei Ping & Di Zhao
College of Environmental Science and Engineering, Ocean University of China, Qingdao, China
Shandong Provincial Key Laboratory of Marine Environment and Geological Engineering, Ocean University of China, Qingdao, China

ABSTRACT: Cross strata pollution can easily lead to salinization pollution of freshwater aquifers in coastal areas. Existing studies mainly focus on the analysis of field monitoring of cross-strata pollution. The mechanism of the impact of hydraulic gradient on cross strata pollution in freshwater aquifers is still unclear. This paper used a seepage tank test to explore the process of cross strata pollution under different hydraulic gradient conditions. The colorimetric tracer technology was utilized to analyze the impact on the dynamics change of saltwater plumes, and the spatial distribution of concentration was also explored. The results showed that the longitudinal diffusion velocity of the saltwater plume was positively correlated with the hydraulic gradient; the longitudinal dispersion velocity under the head difference of 4 cm is 2.4 times that under the head difference of 2 cm. And the lateral dispersion velocity slowed down with the increase of hydraulic gradient. The concentration of saltwater plumes showed a decreasing trend in the direction of longitudinal dispersion and an increasing trend in the direction of lateral dispersion.

1 INTRODUCTION

The Littoral aquifer is the link between marine and terrestrial hydro-ecosystems. Affected by various factors such as geological tectonic movement, climate change, and sea level fluctuations, complex and changeable strata with alternating deposition of continental, marine, and lake facies are formed in coastal areas. Usually, the upper part is saltwater or brackish water, and the lower part is freshwater. Groundwater is in its natural state when it is not exploited and utilized. Due to the action of aquitards and baffles, there is almost no hydraulic connection between shallow underground saltwater and deep freshwater. The shallow saltwater rarely causes the pollution of the deep fresh water. In order to obtain a larger amount of water inflow, water will be drawn through multiple aquifers during the completion of most wells. The aquifers and water barriers that originally played the role of water isolation are artificially penetrated, and the wells become the connection of each aquifer. The wellbore forms an artificial skylight that connects the aquifers that were originally isolated from each other. Wells may become channels for vertical flow and rapid transport of pollutants between aquifers (Achang et al. 2020; Gailey 2017; Casasso et al. 2020; Jiménez-Martínez et al. 2011). This paper aims to study the effect of changes in the hydraulic gradient on cross-strata pollution. Groundwater cross strata pollution was simulated by a seepage tank test. The dynamic changes and solute migration process of cross strata pollution under different hydraulic gradients were visually observed by dye tracers, and the quantitative relationship between the solute concentration and the RGB values of the pictures was established by using the colorimetric tracer technology. The spatial distribution of pollutant concentration was analyzed, and the dynamic change of cross strata pollution plume was systematically studied.

---

*Corresponding Authors: 1214105641@qq.com and lingq@ouc.edu.cn

DOI 10.1201/9781003330165-95

2 MATERIALS AND METHOD

2.1 *Experimental materials*

The experimental device is shown in Figure 1. The experimental device was made of transparent acrylic plate, including three parts: sand tank, fresh water tank, and saltwater inlet tank. The sand tank is 106 cm long, 55 cm high, and 10 cm wide. The length of the freshwater tanks on both sides is 7 cm. Saltwater flows into freshwater aquifers through cross strata pipes to simulate an underground aquifer scenario with upper saltwater and lower freshwater. The sand tank was filled with 36 cm thick homogeneous white quartz sand to simulate the lower freshwater confined aquifer. A transparent acrylic plate with a length of 105 cm and a width of 9 cm was placed on top of the quartz sand. Before the start of the experiment, glass glue was utilized to ensure that the device would not leak when the freshwater level on the left and right sides was higher than the filling height of the quartz sand.

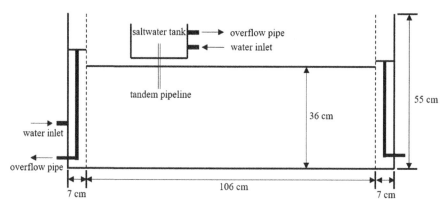

Figure 1.   Experimental setup.

Dissolve 33.4 g of sodium chloride (NaCl) and 1.6 g of edible dye (Strawberry Red) in 1 L of distilled water to prepare salt water with a density of 1.025×10^3 kg/m^3 and a concentration of 35 g/L. Edible dyes can be used as tracers to visually observe the dynamic characteristics of saltwater plumes (Yu et al. 2019), which will not be adsorbed by quartz sand (Jr et al. 2009; Kuan et al. 2012; Kalejaiye & Cardoso 2005).

2.2 *Experimental method*

Turn on the peristaltic pump and inject freshwater into the left and right freshwater tanks. The freshwater head on the right side was 38 cm. By changing the height of the overflow port in the left freshwater tank, the head of the aquifer was adjusted. The sand tank was saturated with saltwater. After the hydrodynamic field in the sand tank became stable, the saltwater peristaltic pump was turned on, and the configured saltwater was injected into the saltwater inlet tank. The timing starts after the saltwater enters the freshwater aquifer through cross strata pipe. During the experiment, video equipment was used for real-time photography to record the dynamic migration process of the saltwater plume.

While using chromogenic tracer technology to analyze the concentration of saltwater pollution plume, it is necessary to configure a series of salt water with a concentration gradient, inject salt water of known concentration into the freshwater aquifer in turn, and record the pictures of each concentration, ImageJ software was performed to convert the known concentrations of saltwater pollution plumes in the images into R+G+B values, R values, G values, and B values. A concentration-RGB standard curve was established. The concentration of saltwater was obtained by using image analysis software to convert the known RGB values.

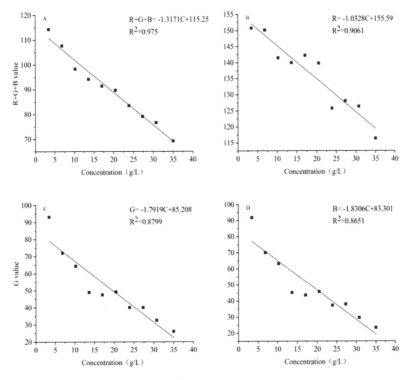

Figure 2. Standard curve of R+G+B value, R value, G value, B value, and saltwater concentration.

As shown in Figure 2, Figures A, B, C, and D represent the standard curve of R+G+B value, R value, G value, B value, and saltwater concentration, respectively. The figure shows that the R+G+B value has the best linear relationship with the salt water concentration, and the correlation coefficient R^2 is 0.975, while the R value, G value, and B value have a poor linear relationship with the salt water concentration. In the concentration analysis of the plume, the R+G+B value is selected to reflect the concentration change of the salt water plume.

2.3 Experimental parameters

To study the effects of different hydraulic gradients on the dynamic transport characteristics and concentrations of saltwater, cross strata pollution experiments were carried out under three different freshwater head differences (2 cm, 3 cm, and 4 cm).

Table 1. Experimental group.

Group	h_{f1} (cm)	h_{f2} (cm)	Δh (cm)	D (mm)	ρ_f (g/cm^3)	ρ_s (g/cm^3)
A1	40.0	38.0	2.0	3.0	1	1.025
A2	41.0	38.0	3.0	3.0	1	1.025
A3	42.0	38.0	4.0	3.0	1	1.025

Where h_{f1} is the freshwater level on the left, h_{f2} is the freshwater level on the right, Δh is the head difference of the freshwater aquifer, D is the diameter of the cross strata pipeline, ρ_f is the freshwater density, and ρ_s is the saltwater density.

3 RESULTS AND DISCUSSIONS

3.1 *Dynamic characteristics of the saltwater plume*

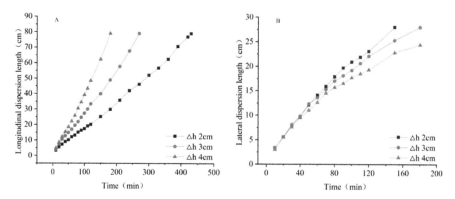

Figure 3. Dynamic change diagram of saltwater plume under different head differences (A is longitudinal dispersion length, B is lateral dispersion length).

Observing the dynamic changes of saltwater plumes, it was found that changes in aquifer hydraulic gradients have a more pronounced effect on longitudinal dispersion than lateral dispersion. As shown in Figure 3, when the head difference is 2 cm, 3 cm, and 4 cm, the average longitudinal dispersion velocity of the plume is 0.18 cm/min, 0.29 cm/min, and 0.44 cm/min, respectively. The longitudinal dispersion velocity under the head difference of 4 cm is 2.4 times that under the head difference of 2 cm. The results showed that the larger the hydraulic gradient of the aquifer, the faster the longitudinal dispersion velocity of the saltwater plume in cross strata pollution, and the larger the range of freshwater polluted by saltwater at the same time.

Contrary to the longitudinal dispersion of the saltwater plume, the lateral dispersion speed of the saltwater plume is slower with the increase of the hydraulic gradient of the aquifer. When the head difference of the aquifer is 2 cm, 3 cm, and 4 cm, the average velocities of the lateral dispersion of the saltwater plume are 0.19 cm/min, 0.16 cm/min, and 0.14 cm/min, respectively. The lateral dispersion velocity under the head difference of 2 cm is 1.4 times that under the head difference of 4 cm.

In the process of saltwater dispersion in freshwater aquifers, the longitudinal dispersion velocity of saltwater plumes in cross strata pollution becomes faster with the increase of the aquifer hydraulic gradient, and the lateral dispersion velocity is negatively correlated with the hydraulic gradient.

3.2 *Characteristics of the saltwater plume*

As shown in Figure 4, the spatial distribution of the concentration of the saltwater plume after three h of migration under different head differences was obtained by using the colorimetric tracing technique. Figure 4 shows that the saltwater plume has the characteristics of longitudinal and lateral concentration differences, and the concentration of the saltwater plume shows a decreasing trend in the longitudinal dispersion direction and shows a continuous increase in the lateral dispersion direction. In the process of solute dispersion, the saltwater is continuously diluted, which leads to the concentration of the saltwater plume in the longitudinal dispersion direction being continuously reduced. In the process of lateral dispersion, due to the action of density flow, saltwater concentrates and migrates downward, so the concentration of salt water in the lower part of the saltwater pollution plume is higher than that in the upper part. At the same time, we found that the concentration of the saltwater plume will increase with the increase of the hydraulic gradient of the aquifer, and the hydraulic gradient will affect the saltwater flux flowing into the freshwater aquifer. The head difference is 2 cm, 3 cm, and 4 cm. Under the conditions, the saltwater fluxes flowing into the freshwater aquifer per unit time were 8 ml/h, 18 ml/h, and 33 ml/h, respectively. When the head difference is large, the flux of saltwater flowing into the freshwater aquifer per unit time is large, and saltwater suffers limited dilution, so the concentration of saltwater plume will increase as a whole with the increase of hydraulic gradient.

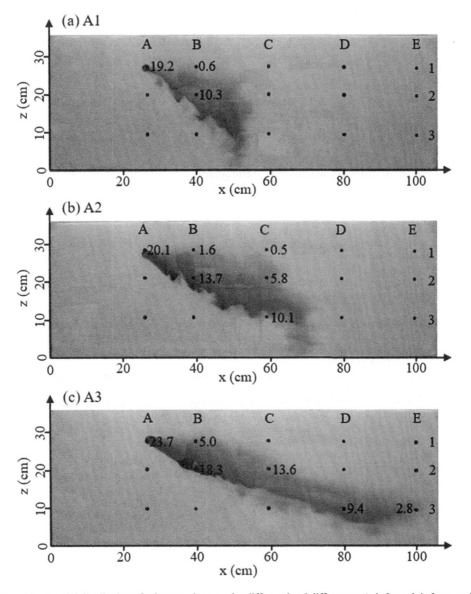

Figure 4.   Spatial distribution of saltwater plume under different head differences. (a is 2 cm, b is 3 cm, and c is 4 cm).

4   CONCLUSION

The effects of hydraulic gradient on cross strata pollution through indoor physical sand tanks were studied. By changing the head difference of the freshwater aquifer, the dynamic changes of the saltwater plume and the spatial distribution of the saltwater during the process of contamination were explored. The conclusions were as follows:

(1) The longitudinal dispersion velocity of the saltwater plume increases with the increase of the hydraulic gradient of the aquifer, while the lateral dispersion velocity is negatively correlated with the head difference. The larger the head difference, the slower the lateral dispersion velocity.

(2) With the progress of cross strata pollution, the concentration of the saltwater plume shows a decreasing trend in the direction of longitudinal dispersion. Due to the higher density of saltwater, under the action of the density flow, the concentration of saltwater plume shows an increasing trend in the direction of lateral dispersion.

ACKNOWLEDGMENTS

This work was supported by the Joint Funds of the National Natural Science Foundation of China-Shandong Province, China (grant number U1806210).

REFERENCES

Achang, M., Li, Y.Y., Radonjic, M. (2020) A Review of Past, Present, and Future Technologies for Permanent Plugging and Abandonment of Wellbores and Restoration of Subsurface Geologic Barriers. *Environmental Engineering Science.*, 37(6): 395–408.

Casasso, A., Ferrantello, N., Pescarmona, S. (2020) Can Borehole Heat Exchangers Trigger Cross-Contamination between Aquifers? *Water.*, 12(4): 1174.

Gailey, R.M. (2017) Inactive supply wells as conduits for flow and contaminant migration: conditions of occurrence and suggestions for management. *Hydrogeology Journal.*, 25(7): 2163–2183.

Jiménez-Martínez, J., Aravena, R., Candela, L. (2011) The Role of Leaky Boreholes in the Contamination of a Regional Confined Aquifer. A Case Study: The Campo de Cartagena Region, Spain. *Water Air & Soil Pollution.*, 215(1–4): 311–327.

Jr, R.L., Momii, K., Nakagawa, K. (2009) Laboratory-scale saltwater behavior due to subsurface cutoff wall. *Journal of Hydrology.*, 377(3): 227–236.

Kalejaiye, B.O., Cardoso, S.S.S. (2005) Specification of the dispersion coefficient in the modeling of gravity-driven flow in porous media. *Water Resources Research.*, 41(10): W10407.

Kuan, W.K., Jin, G., Xin, P., et al. (2012) Tidal influence on seawater intrusion in unconfined coastal aquifers. *Water Resources Research.*, 48(2): W02502.

Yu, X., Xin, P., Lu, C. (2019) Seawater intrusion and retreat in tidally-affected unconfined aquifers: Laboratory experiments and numerical simulations. *Advances in Water Resources.*, 132: 103393.

*Advances in Energy, Environment and
Chemical Engineering – Abdullah & Osman (Eds)
© 2023 The Author(s), ISBN: 978-1-032-36083-6*

Surface subsidence monitoring and prediction in mining area based on SBAS-InSAR and ARIMA-SVR combination model

Shuaishuai Yue*
School of Highway, Chang'an University, Xi'an, Shaanxi, China

Xiang Yin & Tao Wang
Shaanxi Nuclear Industry Engineering Survey Institute Co., Ltd, Xi'an, Shaanxi, China

Shaoliang Hou
China Power Construction Road and Bridge Group Co., Ltd, Beijing, China

Yang Liu
School of Highway, Chang'an University, Xi'an, Shaanxi, China

ABSTRACT: Based on 12 ALOS Image data, this paper uses a small baseline set of synthetic aperture radar interferometry (SBAS-InSAR) technology to extract the surface deformation information of the Meimeiba mining area in Ningxiang city. Particle swarm optimization (PSO) is used to optimize the core parameters of the support vector machine regression (SVR) model, and the difference autoregressive moving average model (ARIMA) is used to correct the residual sequence values. The accumulated surface deformation sequence values obtained in the mining area are reconstructed in phase space, and the ARIMA-SVR combined model is used to predict them. The research shows that the ARIMA-SVR combined forecasting model is superior to the single SVR model in accuracy and improves the accuracy of forecasting.

1 INTRODUCTION

At present, the traditional methods for surface deformation monitoring in mining areas are mainly leveling, electronic ranging, and GPS (Wei et al. 2018). Although the accuracy is high, due to the scattered distribution monitoring, the monitoring range is limited, and it is difficult to meet the needs of time and space scales. It is difficult to achieve high-precision ground disaster identification, tracking, and monitoring in large areas, and it is not ideal for predicting the development trend of surface deformation. Synthetic Aperture Radar Interferometry (InSAR) technology, as a new space-to-ground observation method, can theoretically reach millimeter-level deformation accuracy and effectively compensate for the shortcomings of traditional monitoring methods. In order to solve the spatial-temporal decoherence problem of traditional D-InSAR technology, scholars in various countries are committed to the research of temporal InSAR technology and have successfully applied it to the early identification and monitoring of geological disasters (Xu et al. 2019), such as landslide (Lu et al. 2019) and land subsidence (Li et al. 2021). In recent years, the intelligent algorithm has developed rapidly (Fan et al. 2019) and is widely used in time series prediction of surface deformation, such as the application of the grey Verhulst model to monitoring and forecasting ground settlement value in the mining area (Yang et al. 2015), combined with PS-InSAR technology and neural network to predict the deformation of the mining area (Li et al. 2020) combined with SBAS-InSAR technology and PSO-BP model to monitor and effectively predict the long-term surface subsidence of the mining area (Zhou et al. 2021).

---

*Corresponding Author: 157027629@qq.com

Based on ALOS satellite image data, this paper uses SBAS-InSAR technology to extract the surface deformation information of Meitanba goaf in Ningxiang City, Hunan Province, from 2007 to 2011. According to the surface deformation characteristics of the mining area, the ARIMA-SVR combined prediction model is established to predict the time series of nonlinear settlement results of large deformation in the mining area.

2 RESEARCH AREA

Meitanba area is located in the west of Ningxiang city and about 15 km south of Ningxiang county town, including Meitanba town, Dachengqiao town, Yujiaao town, and Huilongpu town. The east-west length of the mining area is about 28.73 Km, and the north-south width is about 18.77 Km, with a total area of 24.60 km^2 and a horseshoe-shaped range. There are five pairs of large-scale mines in the mining area, from north to south: Yuejin Mine, Xifenglun Mine, Zhushantang Mine, Wumuchong Mine, and Hejiawan Mine (Figure 1).

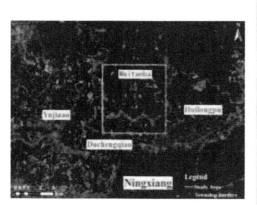

Figure 1. Location and distribution of the Meitanba mining area.

3 METHODOLOGY AND DATA PROCESSING

3.1 *ARIMA-SVR model*

The modeling process of the SVR-ARMA combined prediction model based on SBAS-InSAR results is shown in Figure 2.

(1) The non-equidistant surface temporal settlement obtained by SBAS-InSAR technology is interpolated by cubic spline, and the results are normalized. The G-P algorithm is used to analyze the correlation dimension of the time series settlement data. After the embedding dimension m of the prediction model is obtained, the original time series data can be reconstructed into matrix form in phase space and divided into the training set and verification set of the model.

(2) The SVR model is established, and the PSO algorithm is used to optimize the core parameters (γ, c, ε) of the model, so as to obtain the best core parameters of the SVR model.

(3) The residual sequence of the SVR model is obtained from the fitting value sequence of the original sequence and SVR model. The ARIMA model is established by Python software to

test the stability of the residual sequence and its difference sequence data, and the d value of ARIMA (p, d, q) is determined. Then the p and q values are determined by ACF and PACF.

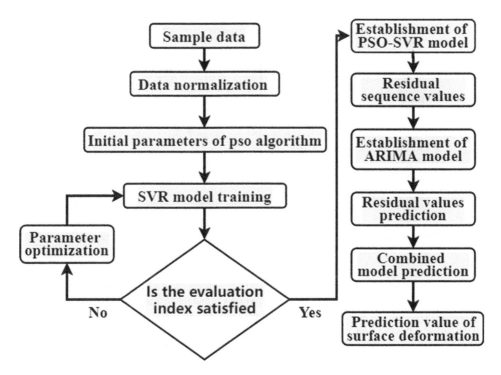

Figure 2.   The flow chart of the ARIMA-SVR model was applied to the prediction of surface settlement.

3.2   *Data processing*

In this paper, 12 ALOS PALSAR images were obtained, and the imaging angle is 34.3°, as shown in Table 1. Reference DEM selected SRTM 1 data with a resolution of 30 m, as shown in Figure 1.

Table 1.   Basic parameters of acquired PALSAR image data.

ID	Imaging date	Polarization	Mode	Level	Track number
1	2007-01-08	HH	FBS	1.1	05099
2	2007-07-11	HH+HV	FBD	1.1	07783
3	2007-10-11	HH+HV	FBD	1.0	09125
4	2008-01-11	HH	FBS	1.0	10467
5	2008-02-26	HH	FBS	1.0	11138
6	2008-05-28	HH+HV	FBD	1.1	12480
7	2008-07-13	HH+HV	FBD	1.1	13151
8	2009-10-16	HH+HV	FBD	1.1	19861
9	2010-01-16	HH	FBS	1.1	21203
10	2010-07-19	HH+HV	FBD	1.1	23887
11	2010-10-19	HH+HV	FBD	1.1	25229
12	2011-01-19	HH	FBS	1.1	26571

4 RESULT ANALYSIS

4.1 *Analysis of land subsidence rate in the mining area*

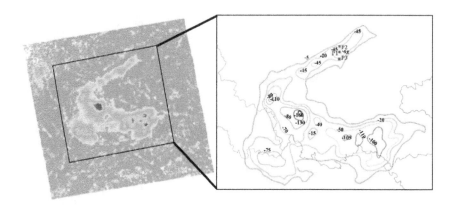

Figure 3.   The annual average deformation rate and contour map of surface in the Meitanba mining area.

Figure 3 shows that during 2007–2011, the surface deformation of each mining area of Meitanba increased year by year, forming many obvious subsidence funnels. The land subsidence area is concentrated on the working surface of each mining area, and the subsidence is not equal, and the intersection of mining areas is the most serious. The most serious subsidence area is located in the Wumuchong and Oujiaping areas at the intersection of the Wumuchong, Hejiawan, and Zhushantang mining areas. The annual average deformation rate at the center of the subsidence funnel reaches 150 mm/a, and the range is the largest. Located near Changchongli in the southwestern Wumuchong mining area, the annual average rate is also 150 mm/a, with a small range. The annual average deformation rate of the center of the subsidence funnel near Xintangpo in the southeast Xifenglun mining area is small, about 100 mm/a, and the range is the smallest. Due to the comprehensive shutdown of the Yuejin mining area in 2006, only small subsidence fields were found during the monitoring period, and the annual average deformation rate was up to 65 mm/a with the smallest range, but the surface still had subsidence.

4.2 *Prediction of surface subsidence in the mining area*

Since the surface subsidence data of the mining area obtained by SBAS-InSAR technology are relatively large, this paper selects the center point P1 of the settlement funnel in the Yuejin mining area (the cumulative deformations are all less than 300 mm), the northern point P2 of the dip line, and the southern point P3, and takes the time series cumulative subsidence of the three points as the data sample of the surface subsidence prediction model of the mining area.

The cubic spline interpolation was carried out on the 12-stage non-equidistant surface subsidence time series data to obtain the 34-stage isodistant surface subsidence time series data. The first 27 periods (January 8, 2007–March 3, 2010) were used as the training samples of the model, and the cumulative settlement of the last seven periods (April 18, 2007–January 19, 2011) was used as the validation set sample of the prediction model to test the accuracy of the ARIMA-SVR combination model. The results are shown in Figure 4.

By comparing the prediction curves of the SVR model and the ARIMA-SVR combination model at P1, P2, and P3 settlement points, the advantages of the ARIMA-SVR combination prediction

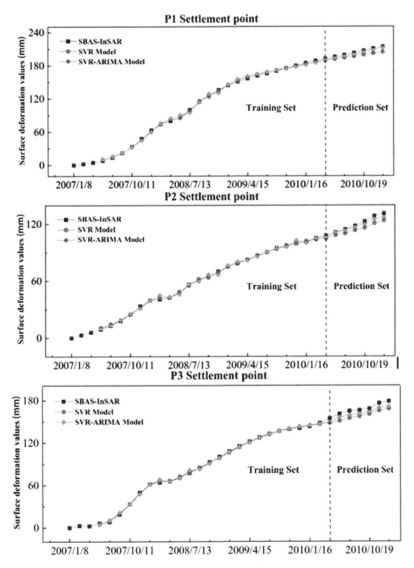

Figure 4.   Prediction curve of SVR model and ARIMA-SVR model at P1, P2, and P3 points.

model can be intuitively seen. Although the prediction curves of the two models are consistent with the surface temporal settlement curves extracted by SBAS-InSAR technology in the trend, with the increase of time, the prediction effect of the SVR model gradually becomes worse, and the residual value gradually increases. After residual correction using the ARIMA model, the time series curve shows a better prediction effect.

4.3   *Accuracy test of the prediction model*

In order to test the accuracy of the ARIMA-SVR combined prediction model, the statistical evaluation indexes of relative error δ, root mean square error (RMSE), and mean absolute error (MAPE) are used for verification as shown in Tables 2, 3, and 4.

Table 2. Comparison of prediction results of various models at P_1.

Date	SBAS-InSAR subsidence value(mm)	SVR Model		ARIMA-SVR Model	
		Predicted value (mm)	δ (%)	Predicted value (mm)	δ(%)
2010-04-18	192.1	188.99	1.62	189.59	1.31
2010-06-03	195.37	191.56	1.95	192.91	1.26
2010-07-19	198.76	193.95	2.42	196.09	1.35
2010-09-03	202.51	196.40	3.02	199.61	1.43
2010-10-19	206.46	199.05	3.59	203.66	1.35
2010-12-04	210.32	201.70	4.10	207.77	1.21
2011-01-19	213.83	204.21	4.50	211.61	1.04

Table 3. Comparison of prediction results of various models at P_2.

Date	SBAS-InSAR subsidence value(mm)	SVR Model		ARIMA-SVR Model	
		Predicted value (mm)	δ(%)	Predicted value (mm)	δ(%)
2010-04-18	107.38	104.55	2.63%	105.86	1.42%
2010-06-03	111.17	108.04	2.82%	110.96	0.19%
2010-07-19	114.13	109.88	3.72%	113.01	0.98%
2010-09-03	117.71	113.12	3.90%	116.54	1.00%
2010-10-19	122.39	115.72	5.45%	119.58	2.30%
2010-12-04	128.39	120.49	6.15%	124.90	2.72%
2011-01-19	130.93	123.42	5.74%	126.73	3.21%

Table 4. Comparison of prediction results of various models at P_3.

Date	SBAS-InSAR subsidence value(mm)	SVR Model		ARIMA-SVR Model	
		Predicted value (mm)	δ(%)	Predicted value (mm)	δ(%)
2010-04-18	154.41	147.94	4.19%	149.14	3.41%
2010-06-03	160.74	150.72	6.23%	156.12	2.87%
2010-07-19	164.99	154.79	6.18%	158.65	3.84%
2010-09-03	165.87	157.30	5.16%	160.88	3.01%
2010-10-19	168.11	160.49	4.54%	163.44	2.78%
2010-12-04	175.98	165.47	5.97%	169.20	3.85%
2011-01-19	178.72	168.18	5.90%	170.81	4.42%

Table 5. Accuracy comparison of two models.

Evaluation Index	Point Positions	SVR Model	ARIMA-SVR Model
RMSE	P_1	6.6161	2.5956
	P_2	5.6091	2.4696
	P_3	9.2533	5.9107
MAPE	P_1	0.0303	0.0128
	P_2	0.0434	0.0169
	P_3	0.0545	0.0346

Table 5 shows that the prediction results of the ARIMA-SVR model are close to the expected values, and the RMSE and MAPE values of the ARIMA-SVR model are lower than those of SVR. It shows that the ARIMA-SVR combined prediction model has good applicability in the prediction of nonlinear settlement of large surface deformation in mining areas.

5 CONCLUSION

Based on the surface deformation information of the Meitanba mining area obtained by SBAS-InSAR technology, this paper proposes an ARIMA-SVR combined prediction model to realize the monitoring and prediction of surface subsidence in the mining area.

(1) InSAR monitoring results show that: from 2007 to 2011, the Meitanba mining area is mainly characterized by land subsidence, and there are many obvious subsidence funnels on the working surface of each mining area.
(2) Based on the PSO algorithm, the stability test of the ARIMA model for the SVR model residual sequence and difference results are realized.
(3) Based on SBAS-InSAR technology, the study shows that the ARIMA-SVR combination model is superior to the single SVR model in prediction accuracy and improves the accuracy of surface subsidence prediction in the mining area.

REFERENCES

Fan Z L & Zhang Y H (2019). Research progress on intelligent algorithms based on ground subsidence prediction. *J. Geomatics & Spatial Information Technology*, 42 (05), 183–188.
Li B F & Li G E (2022). Identification of surface deformation caused by geothermal resource extraction using SBAS time-series technology. *J. Bulletin of Surveying and Mapping*, (2), 43–49.
Li Y F & Zuo X Q (2020). Surface subsidence monitoring and prediction based on PS-InSAR technology and Genetic Neural Network Algorithm. *J. Progress in Geophysics*, 35 (03), 845–851.
Lu H Y & Li W L (2019). Early Detection of Landslides in the Upstream and Downstream Areas of the Baige Landslide, the Jinsha River Based on Optical Remote Sensing and InSAR Technologies. *J. Geomatics and Information Science of Wuhan University*, 44 (9), 1342–1354.
Wei J & Hu X D (2018). Monitoring Data Analysis and Subsidence Features Recognition Based on GPS Static Positioning. *J. Coal Geology of China*, 30 (03), 53–58+81.
Xu Q & Dong X J (2019). Integrated Space-Air-Ground Early Detection, Monitoring and Warning System for Potential Catastrophic Geohazards. *J. Geomatics and Information Science of Wuhan University*, 44 (07), 957–966.
Yang J K & Fan H D (2015). Monitoring and prediction of mining subsidence based on D-InSAR and Gray Verhulst Model. *J. Metal Mine*, (03), 143–147.
Zhang M M & Fan X T (2021). Monitoring and interpretation of land subsidence in mining areas in Xuzhou City during 2016—2018. *J. Remote Sensing for Natural Resources*, 33 (4), 43–54.
Zhou D Y & Zuo X Q (2021). Surface subsidence monitoring and prediction in mining area based on SBAS-InSAR and PSO-BP neural network algorithm. *J. Journal of Yunnan University* (Natural Sciences Edition), 43 (5), 895–905.

*Advances in Energy, Environment and
Chemical Engineering – Abdullah & Osman (Eds)
© 2023 The Author(s), ISBN: 978-1-032-36083-6*

Research on the comparison and calibration of pump-gateway engineering based on precise dispatch requirements

Wenjia Sha*
Department of Water Engineering, Hohai University, Nanjing, China

Jin Xu*
Scientific Institute of Pearl River Water Resouces Protection, Zhujiang, China

Junwei Zhu* & Yiping Zhu*
Zhejiang Hohai Control Information Technology Co., Ltd, Jiaxing, China

Cheng Gao*
College of Hydrology and Water Resouces, Hohai University, Nanjing, China

ABSTRACT: In order to accurately obtain the over-water flow of the sluice gate and pumping station and to discuss its significance to flood control and drainage, such as flood control scheduling, water resource management, and river regulation, the study took the project called "the flood control of the central urban area of Suzhou" as the experimental area. The flow rate ratio is determined by establishing the correlation between the measured gate flow and relevant hydraulic factors. Through GPRS information collection, real-time transmission of pump and sluice station information is realized, and more economical and efficient flood control scheduling, water resources management, and river regulation approaches are found. This provides a scientific basis and lays the foundation for the improvement of urban flood control and disaster reduction system.

1 INTRODUCTION

A flood disaster is a natural disaster that has a great impact on urban construction. It destroys the living environment of urban residents, endangers people's personal and property safety, and hinders social development. Currently, in the process of flood control project management, due to the lack of information construction, the water output cannot be counted in a timely and accurate manner, and the ability to transmit water in real time has not yet been realized. Therefore, this study selected the project called 'the flood control of the central urban area of Suzhou' as the experimental area, 11 key water control projects, and 16 small pumps and small gates along the urban area as the main research objects. By establishing the correlation between the flow rate and the relevant hydraulic factors, the flow rate of the overwater can be accurately obtained. While the operating efficiency of each pumping station is calibrated and analyzed, the operating conditions and time of the pumping station are collected through the GPRS information. The online real-time transmission of the pumping station discharge is realized. This way improves refined and scientific scheduling of water conservancy projects in the process of flood control and drainage, which can reduce the frequency of urban waterlogging to a certain extent, protect urban water resources and ecosystems, and improve urban living happiness. This laid the foundation for the construction of a flood control and disaster reduction system for water conservancy projects with complete facilities, advanced equipment, and complete functions.

*Corresponding Authors: 2433056069@qq.com; 94151658@qq.com; hhcc888@163.com; jx-zhtech@ 163.com and cgao@hhu.edu.cn

DOI 10.1201/9781003330165-97

Figure 1.   Pump, gate flow test, and hub location.

Table 1.   Statistical table of parameters of various hub gates and pumping stations in the urban area.

		Gate station		Pumping station			
S/N	Name	Gate (hole)	Overall width (m)	Pump set (set)	Design flow (m³/s)	Installed power (kw)	Remark
1	Loujiang	1	14	3	15	750	
2	Waitanghe	2	28	3	15	750	
3	Yuanhetang	2	16	6	30	1890	1 hole ship lock
4	Dongfengxin	1	8	4	20	1000	
5	Nanzhuang	1	8	2	10	500	
6	Qinglong	1	8	4	20	1160	
7	Xujiang	3	30	4	20	1000	
8	Xianrendagang	1	8	3	15	750	
9	Peijiaxu	3	30	5	40	1250	
10	Dalonggang	1	12	4	20	1000	
11	Tantai Lake	3	30	3	60	2400	
	Total	19	192	41	265	12450	

2 PROJECT OVERVIEW

2.1 *The core question of the research*

Discussing the efficiency and role of sluices and pumping stations in flood control and drought relief, study how to grasp and understand the regional water inflow and outflow promptly, and how to improve the means of collecting the operating conditions of the sluices and pumping stations in the hub, so as to improve the operation monitoring system and management system, and promote the development of informatization and modernization of water conservancy project management.

2.2 *Basic information about the study area*

The central urban area of Suzhou has a developed economy and a large population. The annual flood control and drought prevention work are under great pressure, and the sluices and pumping stations of various hubs operate frequently.

The project called 'the flood control of the central urban area of Suzhou' consists of 11 hubs, such as Yuanhetang and Waitang River, and 16 small pumps and small gates along the urban area, with a total area of about 84km² and a river length of more than 200 km. To a certain extent, the functions of running water in Suzhou urban area, flood control, drainage during flood season, and navigation of ships have been realized. At the same time, the water conservancy projects and pump gates included in the project have their characteristics and are typical and representative. Therefore, the study area has valuable research value in urban flood control and drainage.

3 PROJECT IMPLEMENTATION

3.1 *Specific research process*

3.1.1 *On-the-spot investigation*
At the end of March 2017, a comprehensive survey and understanding of the equipment conditions of 11 pumping stations and 3 gate stations; water level and flow test conditions were completed.

3.1.2 *Run scheduling and test section location*
In order to fully understand and master the relationship between the single pump flow and the multi-pump superimposed flow during the operation of each pump station, single pump, double pump, or three pumps were used to run simultaneously without affecting the guaranteed water level of the regional river.

The location of the flow test section selected for the hub was different, and the flow measurement instruments and equipment were not the same: the flow test of the pumping station of the Loujiang hub, the Qinglongqiao hub, and the Xujiang hub adopted the navigation ADCP full-section flow monitoring, the average round-trip flow was measured as the average flow of a section. The other eight hub pumping station cross-section flow tests used the bridge hole or approach bridge section, adopted the suspension cable suspension method, laid out the speed measurement vertical line and the depth of the measurement point to monitor the flow rate, the flow measurement equipment adopted the LSH10-1Y hand-held ultrasonic flowmeter. The measuring range of the device is 0.02 ~ 7.00 m/s, and the measurement accuracy is 1.0%±1 cm/s. In order to find the factors that affect the efficiency of the operation, each pumping station was individually studied.

3.1.3 *Water flow test*
A pumping station flow test was conducted after a stable period after the pumping station was turned on. During the operation period, the flow measurement times of each working condition should be no less than five times. The flow test of each pumping station is as follows:

(1) Loujiang Pumping Station

After 20 minutes, we started measuring the flow. During this measurement period, the right bank side of the cross-section obviously produced a backflow. When one pump was running, the maximum return flow was 0.36 m³/s, and the average return flow was 0.18 m³/s. Affected by the backflow, the measured cross-section flow was somewhat larger. The water level process of the inner and outer rivers basically reflected the water potential of the pumping station.

(2) Waitang River Pumping Station

The riverbed of the flow measuring section was an irregular slope, and the bank coefficient was taken as 0.7. The pumping station started to measure the flow after 30 minutes of operation, and the water level process of the outer river was distorted.

(3)Yuanhetang Pumping Station

The bridge hole in the current measuring section was a stone bank protection, and the bank coefficient was taken as 0.8. After 20 minutes of operation, measuring flow started. The water level process of the inland river could reflect the current situation of the pumping station, and the water level process of the outer river was distorted.

(4) Dongfeng New Pumping Station

The flow measuring section was the bridge hole of the pump station, the concrete steep bank, and the bank coefficient was taken as 0.9. The pumping station started to measure the flow after 20 minutes of operation. The flow measurement section was too close, and turbulence was obviously generated. The water level process of the inner and outer rivers basically reflected the current status of the pumping station.

(5) Nanzhuang Pumping Station

The bridge hole of the current measuring section was a steep-slope block stone bank revetment, and the bank coefficient was 0.8. The pumping station was running for 10 minutes and began to measure the flow. Since the power of the pump had not yet reached the normal efficiency, the flow rate was obviously low. After 20 minutes of operation, the cross-section flow tended to be stable.

(6) Qinglongqiao Pumping Station

A pump started to measure the flow after 25 minutes of operation, and the water level of the inner and outer rivers could reflect the operation change process of the pumping station.

(7) Xujiang Pumping Station

One pump started to measure flow after 30 minutes of operation, and there was no measured water level data in Xujiang Junction.

(8) Xianren Dagang Pumping Station

The two sides of the approach bridge of the pump station at the flow measuring section were concrete steep slopes, and the bank coefficient was taken as 0.9. The pumping station started to measure the flow after 20 minutes of operation. Due to the abnormal operation of the No. 1 pump and the shutdown at 14:00, the measured flow was obviously small. At 14:00, when the No. 3 pump was turned on, the flow measurement section was too close, resulting in a turbulent flow. The water level of the inner and outer rivers could reflect the operation change process of the pumping station, and the water level of the outer river was greatly affected by waves and ships.

(9) Peijiawei Pumping Station

The two sides of the bridge hole in the current measuring section were steep rock slopes, and the bank coefficient was taken as 0.8. The flow measurement was started after the pumping station had been running for 20 minutes.

(10) Dalonggang Pumping Station

The two sides of the approach bridge of the pump station at the flow measuring section were concrete steep slopes, and the bank coefficient was taken as 0.9. The pumping station started to measure the flow after 20 minutes of operation. The water level of the inner and outer rivers could basically reflect the operation status of the pumping station, and the water level of the outer river fluctuated greatly.

(11) Tantai Lake Pumping Station

The test results were roughly consistent with the Dalonggang Pumping Station.

The relevant analysis of the flow monitoring of the sluice station mainly included the monitoring and correlation analysis of the flow of the three sluice stations at the Peijiawei Junction, the Waitang River Junction, and the Yuanhetang Junction. The purpose of flow monitoring and related analysis at the gate station was to establish a correlation between the measured flow rate and the water level factor inside and outside the gate and to calculate the water flow through the gate through the correlation formula from the water level.

3.1.4 *Ratio determination*

In this project, a cross-sectional flow test was conducted under various operating conditions, which could more objectively reflect the diversion and drainage water volume of each hub pumping station in actual operation. At the same time, the design flow rate of the pumping station was also checked and calibrated. In order to ensure the accuracy of the research results, the flow test was started when the pump was running at rated power, and the cross-sectional flow was basically stable. The flow at this time basically represented the design flow of the pumping station. The relative error statistics of the design flow and the measured section flow of each hub pumping station are shown in Table 2.

Table 2. Relative error between design flow and measured section flow of pumping station.

		Running one pump			Running two pumps			Running three pumps			
S/N	Hub pumping station	Design flow m³/s	Measured flow m³/s	Relatively poor %	Design flow m³/s	Measured flow m³/s	Relatively poor %	Design flow m³/s	Measured flow m³/s	Relatively poor %	Remark
1	Loujiang	5.00	7.01	28.7	10.0	11.9	16.3				Turbulent, reflux
2	Waitanghe	5.00	5.65	11.5	10.0	11.0	9.3	15.0	16.7	10.2	
3	Yuanhetang	5.00	6.03	17.1	10.0	11.7	14.8				
4	Dongfengxin	5.00	5.48	8.8	10.0	11.6	14.0	15.0	22.4	33.1	turbulent, reflux
5	Nanzhuang	5.00	5.14	2.8	10.0	10.6	5.6				
6	Qinglong	5.00	4.82	−3.8	10.0	9.72	−2.9				
7	Xujiang	5.00	4.86	−2.8	10.0	11.8	15.2				
8	Xianrendagang	5.00	6.70	25.4							turbulent, reflux tidal
9	Peijiaxu	8.00	9.82	18.5	16.0	20.4	21.4				influence turbulent, reflux
10	Dalonggang	5.00	5.76	13.1	10.0	9.61	−4.1	15.0	16.8	10.5	
11	Tantai Lake	20.0	19.9	−0.5	40.0	42.8	6.5				
	Average			10.8			7.9			10.5	

There are many reasons for the large or small flow of the measured section. Generally speaking, the main reason is related to the section control conditions. The large flow rate of the measured section is mainly because the test section was too close to the pumping station, which was prone to oblique flow and backflow such as horizontal drop, vortex, etc., which led to the general large flow velocity on the water surface. The small flow is mainly because the cross-section was located in the outer estuary, the waters of the outer estuary are open, and ships came and went frequently.

The reason why the flow rate of the measured section at Peijiaxu is too large is special and needs to be analyzed separately. The water system of the Peijiaxu hub is developed, and the water channels are crisscrossed. The situation is complicated. Due to the influence of the downstream tide, the measured flow is too large.

Figure 2. Flood control enclosure of Peijiaxu hub.

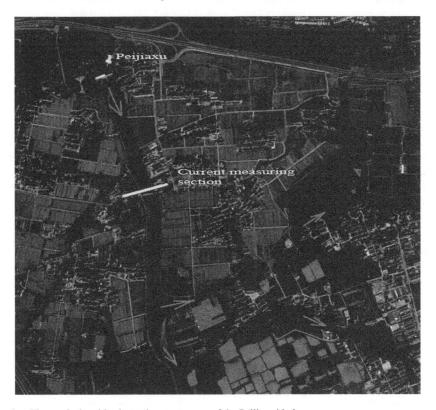

Figure 3. The vertical and horizontal water system of the Peijiawei hub.

3.1.5 Correlation analysis of pumping station flow and hydraulic factors

The traditional pumping station flow calculation usually adopts the exponential function method, but this method requires a lot of measured data for point drawing analysis, and the workload is large. Therefore, in this project, a representative flow test under various working conditions was selected, and the method of multiple regression analysis was adopted with the water level factor, which is simple and more accurate and can meet certain accuracy requirements. The relevant analysis results are shown in Table 3.

$$Q = f(\Delta Z)$$

In the formula:
Q—Section flow(m^3/s)
ΔZ—water level difference(m)

Table 3. List of water level differences to flow rate correlation in the pumping stations.

S/N	Pumping station	Correlation	Correlation coefficient	Remark
1	Loujiang	$Q = 153.34\Delta Z + 0.7108$	0.817	
2	Waitanghe	$Q = 35.477\Delta Z + 5.2357$	0.943	
3	Yuanhetang	$Q = -2309.6\Delta Z^2 + 413.15\Delta Z - 6.0575$	0.949	
4	Dongfengxin	$Q = -302.67\Delta Z^2 + 192.07\Delta Z + 1.8488$	0.976	
5	Nanzhuang	$Q = 34.262\Delta Z + 5.2296$	0.974	
6	Qinglong	$Q = 35.477\Delta Z + 5.2357$	0.926	
7	Xujiang	$Q = 95.269\Delta Z + 2.8094$	0.977	
8	Xianrendagang	$Q = -218.33\Delta Z^2 + 31.383\Delta Z + 5.841$	0.989	
9	Peijiaxu	$Q = 435.41\Delta Z + 1.2084$	0.917	
10	Dalonggang	$Q = 86.914\Delta Z + 2.8858$	0.983	
11	Tantai Lake	$Q = -2793.6\Delta Z^2 + 31.383\Delta Z + 5.841$	0.966	

3.1.6 Correlation analysis of flow and hydraulic factors in sluice station

Considering the traditional hydraulic method to determine the flow rate of the weir and gate, it is generally difficult to obtain relatively stable coefficients due to various influencing factors. Therefore, this project used the measured data, combined with the multiple regression method of hydrological statistics, and used the water level drop to conduct a correlation analysis on the flow through the gate. The correlation analysis results are shown in Table 4, which greatly improves the accuracy of the results.

$$Q = f(\Delta Z)$$

In the formula:
Q—flow (m^3/s)
ΔZ—water level difference (m)

Table 4. List of correlations between water level difference and flow at the gate station.

S/N	Gate station	Correlation	Correlation coefficient	Remark
1	Yuanhe tang	$Q = 11.530\Delta Z - 0.5731$	0.850	
2	Waitanghe	$Q = 93.022\Delta Z + 0.0288$	0.859	

3.2 Research results

3.2.1 Determination of flow stabilization time for pump running
Through the test situation, it can be seen that the cross-section flow basically tends to be stable after the pumping station is turned on for an average of about 25 minutes.

3.2.2 Pump efficiency
Through the relative error analysis, it can be seen that under the rated power, the measured section flow can reach the design value of flow; that is, the pump efficiency meets the design requirements.

3.2.3 The operation of each working condition satisfies the superposition of flow
In the flow comparison stage, the pump station operation adopted the single, two, and three pump operation modes to conduct flow tests. From the analysis of the measured flow data, the flow rate of the section has a linear relationship with the design flow of the pump involved in the operation, which is a multiple and superposition relationship. Take the Waitang River as an example:

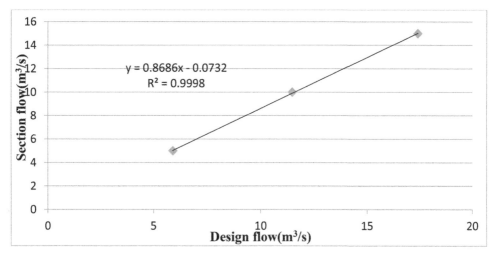

Figure 4.   The relationship between the cross-section flow and the design flow of the Waitang River.

4   CONCLUSION

This paper mainly introduces the specific process of the research on the flow rate ratio measurement of the large-scale flood control pump and gate station in Suzhou City. The conclusions are drawn through the correlation analysis and a water diversion scheduling method that can realize the information management of the pivot pump and the sluice station is proposed. The innovation lies in realizing the real-time transmission, calculation, and statistical analysis of the drainage volume of the pumping station according to the design flow of the pumping station and the running time of the working conditions; according to the relevant factors such as water level and drop, the discharge volume of the gate station can be calculated and obtained, and through the collection of operating conditions and GPRS water level information, the real-time transmission, calculation and statistical analysis of the discharge volume of the sluice station are realized. It improves the accuracy of statistical water output, realizes real-time water transmission, optimizes the means of collecting the operating conditions of the sluices and pumping stations of the hub, improves the operation monitoring system and management system, and promotes the process of informatization and modernization of water conservancy project management.

REFERENCES

Code for Flow Measurement of Hydraulic Structures and Weirs(SL537-2011) (Chinese)

Deng Jinyun, Liu Congcong, Gao Haoran, et al. Impact of drainage system construction on urban flood disasters[J]. *Journal of Yangtze River Academy of Sciences*, 2020, 37(3): 51–56; 69. (Chinese)

Guo Yuanyu. Farmland Hydraulic Science [M]. Beijing: *China Water Resources and Hydropower* Press, 1992. (Chinese)

Hydrological Survey Specifications (SL195-97) (Chinese)

River Flow Test Specification (GB50179-93) (Chinese)

Specification for Acoustic Doppler Flow Test (SL337-2006) (Chinese)

Zhou Fang. Application of information collection management system in flood control pumping station [J]. *Information Technology and Informatization*, 2015, (06): 54–55 (Chinese)

Zhu Qifu, He Liangyi, Jin Liping, The impact of urbanization on flood control and drainage[J]. *Henan Water Conservancy and South-to-North Water Transfer*, 2019, 48(3): 16–17 (Chinese)

*Advances in Energy, Environment and
Chemical Engineering – Abdullah & Osman (Eds)
© 2023 The Author(s), ISBN: 978-1-032-36083-6*

Application research and analysis of common problems of desulfurization wastewater zero discharge system in thermal power plants

Jingjuan Guo

Northwest Electric Power Research Institute of China Datang Group Science and Technology Research Institute Co., Ltd. Xian, Shanxi, China

ABSTRACT: At present, the quality of desulfurization wastewater is poor, the source is unstable, and there are many kinds of treatment technologies. There are many problems in the process of zero-discharge transformation of desulfurization wastewater. Through the investigation and analysis of the actual operation of the current desulfurization wastewater treatment system, the precautions in the design stage and the dangerous points in the operation process are pointed out, and suggestions are put forward for the next step in the study of zero discharge treatment of desulfurization wastewater.

1   INTRODUCTION

Desulfurization wastewater is one of the most difficult wastewaters to treat in thermal power plants, and it is also the key to realizing zero discharge of wastewater from coal-fired power plants. High, easy to scale, there are heavy metals. The following table shows the test data of the desulfurization wastewater quality indicators of a power plant.

Table 1.   Desulfurization wastewater quality of a power plant.

Item	Concentration (mg/L)	Item	Concentration (mg/L)	Item	Concentration (mg/L)
pH	6~9	Ca^{2+}	800~2500	COD_{Cr}	80~150
all solids	26000~50200	Mg^{2+}	3000~8000	Cl^-	<20000
alkalinity	240~800	Na^+	160~400	SO_4^{2-}	1000~7000
full hardness	8000~15000	K^+	50~180	NO_3^-	110~300
SiO_2(mg/L)	80~105	Fe^{3+}	10~100	total N	60~100
total Cd	0.05~0.06	Total Hg	0.08~0.15	total Zn	0.01~0.03
total Pb	0~0.03	Total As	0.02~0.04	total Ni	0.3~0.59

In response to the problem of water pollution, the Feasible Technical Guidelines for Pollution Prevention and Control of Thermal Power Plants (HJ 2301–2017) pointed out that power plants should proceed from the overall situation and strengthen the water management of the whole plant. Comprehensively plan and manage the water source, water use, and drainage of the power plant, select the optimal water distribution plan for the whole plant, treat various wastewater economically and reasonably, and maximize the recovery rate of wastewater. The release of "Design Specifications for Wastewater Treatment in Power Plants" (DL/T 5046-2018) regulates the discharge requirements of wastewater, strengthens the treatment technology, and fully guarantees the realization of the goal of zero discharge (He et al. 2019; Ma et al. 2017; Yuan, 2018).

2 SELECTION OF TECHNICAL ROUTE FOR DESULFURIZATION WASTEWATER TREATMENT

The technical route of zero discharge of desulfurization wastewater is a process of cross-use of technologies. Generally, it is divided into different processes according to the treatment process. In each unit, professional technologies are used to treat wastewater in specific aspects, so that the water quality of wastewater can be divided into steps and grades. The treatment process gradually approached the target water quality.

In the selection of desulfurization wastewater treatment technology for coal-fired power plants, it is necessary to consider both treatment efficiency and technology maturity, fully evaluate one-time investment and long-term operating costs, the impact on other systems, and reasonably handle capital, resources, and other inputs and energy-saving emission reduction output.

3 TECHNOLOGICAL CHARACTERISTICS OF ZERO DISCHARGE TREATMENT OF DESULFURIZATION WASTEWATER

At present, the zero discharge treatment system of desulfurization wastewater can be divided into pretreatment unit, concentration reduction unit, and solidification unit according to its process characteristics (Ye et al. 2019). The selection of each unit process should be based on the water volume and water quality conditions of the system, and the appropriate treatment technology and the process should be selected.

The pretreatment unit removes suspended solids, alkalinity, hardness, organic matter, etc., in the water source produced by the power plant through processes such as coagulation (softening), clarification, and filtration. The desulfurization wastewater pretreatment technology is mature and can generally be realized using domestic technology (Zhang 2019).

The concentration and weight reduction unit is the core unit of the whole desulfurization wastewater zero discharge treatment system, which can be divided into two types: membrane concentration and thermal concentration. Membrane chemical concentration technologies mainly include reverse osmosis (RO), forward osmosis (FO), electrodialysis (ED), and electrosorption (EST). The thermal concentration process can be divided into two categories: steam heat source concentration and flue gas waste heat concentration. The steam heat source concentration technology mainly includes mechanical vapor recompression (MVR), multi-effect evaporation (MED), and so on; the flue gas waste heat concentration technology mainly includes low-temperature flue gas concentration, flue gas waste heat flash evaporation, and so on. At present, the most widely used technology in the market is thermal concentration technology. Although the membrane concentration technology has the advantages of low system energy consumption, strong concentration capacity, and selective ion removal, it requires a lot of capital for investment, operation, and maintenance. In addition, there are problems such as easy scaling and blockage and high pretreatment requirements. The development direction of the membrane concentration process is to develop domestically produced new membrane materials, reduce operation and maintenance costs and diversify configuration.

After the concentration and weight reduction unit, the salt content of the wastewater concentrate is concentrated to more than 10%, which can be solidified by techniques such as evaporative crystallization, sidestream flue gas, and flue spray.

The three units are arranged and combined in sequence, and one-step, two-step, and three-step processes can be selected to achieve zero wastewater discharge. Currently, they are used in zero-discharge desulfurization wastewater projects at home and abroad.

4 PRECAUTIONS IN THE DESIGN STAGE OF ZERO DISCHARGE OF DESULFURIZATION WASTEWATER

(1) The newly added process sites are generally limited, and solutions with a small footprint and high process efficiency should be adopted.

(2) In the design stage, the technology of each processing unit should be mature, and advanced, reliable, and feasible process equipment should be selected as far as possible. The requirements for the influent water quality indicators are broad, the process flow is simple, the system buffer capacity is strong, and the maintenance workload is small.

(3) If the three-step process route is selected, to ensure the safety and stability of the curing unit operation, the water volume of the curing unit should be reduced as much as possible. In the concentration reduction unit, ensure that the concentration ratio of desulfurization wastewater is not less than two times, and consider setting up a bypass in the concentration unit to improve the operating stability of the system.

(4) The two-step or one-step process route should be chosen, that is, omit the triple box pretreatment unit or omit the concentration reduction unit, and the triple box system can be considered canceled depending on the configuration of the terminal treatment system. When choosing the two-step process route without the concentration and reduction unit, the extraction of high-temperature flue gas has a certain impact on the coal consumption of the unit, but it has the advantages of a simple process route and low maintenance workload.

(5) The steam heat source crystallization process system is adopted, and the designed evaporative wastewater volume per 100 MW capacity should not exceed 1.2 t/h, and the need for salt separation should be comprehensively considered according to the resource utilization of crystalline salt and the impact on the environment and economic benefits.

(6) The designed evaporation wastewater volume per 100 MW capacity of the high-temperature bypass flue evaporating process should not exceed 1t/h; the total amount of high-temperature flue gas extracted should not be greater than 3% of the flue gas volume in the boiler BMCR condition, and the extraction flue gas temperature should be $\geq 280°C$, the evaporative flue gas temperature (the flue gas temperature at the outlet of the high-temperature bypass flue) should be higher than the acid dew point by more than 10°C. The crystalline salt mixed into the ash should not affect the comprehensive utilization of fly ash.

5 ANALYSIS OF RISK POINTS IN THE OPERATION PROCESS

5.1 *Risk points in the operation of the membrane concentration process*

The outstanding problems in the operation of the membrane concentration process have high requirements on the quality of the incoming water, which is easy to cause pipeline blockage, which is easy to cause membrane blockage, and the membrane has a short service life; for the electrodialysis process, the power consumption is high, and the system operation is not stable.

5.2 *Risk points in operation in thermal concentration process*

The main risk points of thermal concentration are high cost, high energy consumption, evaporator fouling, and equipment corrosion. The desulfurization wastewater using the thermal concentration process is softened, clarified, and filtered, and most of the calcium and magnesium ions in the water are removed, but it still contains a small amount of hardness. The wastewater contains high salt content and chloride ions. Reasonable, causing corrosion, perforation, and scaling of the evaporator heat exchange tube, affecting the heat exchange efficiency.

5.3 *Flue drying and curing*

(1) The resistance of the bypass flue is high, the flue gas volume is too small, and the processing capacity cannot reach the design value

The bypass flue is generally connected to the flue gas from the front of the air preheater, and the flue gas passing through the evaporator returns to the front of the dust collector. When the resistance of the bypass flue is greater than the resistance of the air preheater, the amount of flue gas entering the bypass flue is reduced, which is lower than the designed flue gas volume, the

Figure 1. Corrosion and blockage of multi-effect evaporative heat exchanger in a power plant.

bypass flue gas flow rate is low, and the flow velocity is slow, which affects the system output on the one hand. On the other hand, the fly ash settles, accumulates, and hardens, which makes the wastewater evaporation system unable to operate normally.

(2) The diameter of the bypass flue evaporator is unreasonable, and the inner wall of the evaporator is scaled

When the diameter of the evaporator is designed to be too large or too small, due to key parameters such as wastewater flow, droplet size, flue gas temperature at the outlet of the air preheater, flue gas distribution, and gas-liquid ratio, the material and liquid cannot reach the evaporator wall before contacting the wall. Completely evaporated to dryness, the phenomenon of sticking to the wall appears (Lian & Wang 2018).

Figure 2. Scaling on the inner wall of the evaporator.

(3) The ash conveying system cannot be put into operation normally, or the design is unreasonable

The ash conveying pipeline of the wastewater evaporation system is merged into the ash conveying pipeline in a unit system. The ash conveying interval is about 30 minutes. The interval is too long, which makes the ash conveying untimely. The phenomenon of overload shutdown often occurs, resulting in the failure of the ash conveying system to be put into operation normally, causing the bottom plate and scaling problems of the evaporation tower.

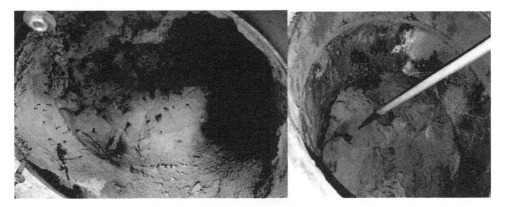

Figure 3. Fly ash hardening and scaling at the bottom of the evaporator.

(4) The atomization effect of the nozzle is poor, which affects the physicochemical efficiency

The desulfurization wastewater is broken up into fine droplets by the atomizer, and the atomized droplets are completely evaporated after mass and heat transfer with the high-temperature flue gas. In the field application of rotary atomization technology, the atomization shaft is easy to break, and the mechanical wear of the atomization disc is serious. The rotary atomization nozzle body is made of carbon steel, which is easy to corrode when the humidity of the flue gas is high. The water leakage phenomenon of the atomizing disc and the splashing phenomenon during the atomization process will cause scaling and scorching on the surface of the atomizer and the surface of the gearbox.

The atomization effect of the two-phase flow depends on the nozzle diameter, compressed air pressure, etc.; the nozzle is easy to wear, but the maintenance and replacement are simple; in the actual use process, the atomization effect is not easy to monitor.

(5) The problem of uneven mixing of flue gas flow field

The bypass flue is introduced into the flue from the front of the air preheater. During the design and installation stage, due to the small cross-sectional area of the flue, most of the projects did not consider the flue gas flow field, and no deflectors were installed; If the distance between the dust collector flue is short, the conventional air preheater flue has two outlet flues, and the dust collector inlet flue has three or four. The flue gas enters the dust collector from the air preheater, and there is a flow field and temperature field. The bypass flue and the inlet flue of the dust collector are only flues, and the hot flue gas has a high and small flow rate. After entering the horizontal flue from top to bottom, it cannot mix well with a large amount of low-temperature main flue gas. As a result, the temperature of part of the flue gas after entering the dust collector is too high, which affects the dust removal efficiency. For the unit using the bag filter, the filter bag will shrink, deform, harden, become brittle, and carbonized in severe cases, which will accelerate the strength loss and shorten the life of the dust filter bag.

5.4 *Other questions*

(1) Influence on fly ash

After the flue drying and solidification technology are adopted, the residual impurities in the desulfurization wastewater will enter the ash together with the ash in the flue gas, so that all the impurities from coal combustion will be collected into the ash, and the quality of some of the ash will be degraded and must be downgraded use.

(2) Solid salt treatment after evaporation and crystallization

Due to the complex salt content of desulfurization wastewater, the final solid miscellaneous salt obtained after the evaporation crystallization process is difficult to identify whether the solid miscellaneous salt belongs to general solid waste or hazardous waste and power plants often dump

675

it as solid hazardous waste for disposal. However, the current domestic cost of solid hazardous waste treatment is basically more than 3000 RMB per ton, and the cost is even higher than the sum of the costs of the previous sections of membrane concentration, evaporation concentration, and evaporation and crystallization. Due to the instability of the incoming water quality, whether it is thermal salt separation or membrane salt separation, the process flow is complex, the equipment is various, and there are many uncertainties in reliability involving investment and operation. The high maintenance costs and the need to consume high-quality steam or electricity have put a lot of pressure on coal-fired power plants, generating fewer and fewer hours in recent years.

Figure 4.   Solid salt content after evaporation and crystallization.

6   CONCLUSIONS

Zero discharge of desulfurization wastewater is imperative, and the core of achieving zero discharge of desulfurization wastewater lies in zero discharge of terminal process wastewater (Zhang et al. 2020). Combining the characteristics of desulfurization wastewater to select an appropriate treatment and disposal process will become the key to the safety, reliability, and economic feasibility of the power plant. To achieve low-cost and stable zero-discharge treatment of desulfurization wastewater, key research work can be carried out in the following aspects, establishing a chlorine balance network in the whole plant water system, incorporating chlorine in flue gas into the balance system, and reducing the use of chemicals, Sulfuric acid replaces hydrochloric acid, dechlorination before desulfurization tower, etc. to achieve source reduction; wet desulfurization system operates with high chlorine to reduce the discharge of desulfurization wastewater; develop a new type of electricity salt separation and integration system to achieve low-cost desulfurization wastewater return to tower reuse and resources Recycling; develop comprehensive utilization technology of zero-emission high-salt solid by-products, improve the economic value of by-products, and reduce system operating costs.

REFERENCES

Chunsong Ye, Rong Cao, Wei Liang, et al. Decomposition of chloride and ammonium salts during high-temperature flue gas evaporation of desulfurization wastewater [J]. *Power Generation Technology*, 2019, 40(4): 362–366.

Jianhua Zhang, Yufei Chi, Yijin Zou, et al. Status and the prospect of desulfurization wastewater treatment technology engineering in coal-fired power plants [J]. *Industrial Water Treatment*, 2020, 40(10):14–20.

Jiwang He, Tao Li, Xiaojian Dang, et al. Zero-discharge technology of flue evaporative crystallization desulfurization wastewater for ultra-supercritical 660MW unit[J]. *Thermal Power Generation* 2019, 48(1):110–114.

Liu Haiyang, Jiang Chengyu, Gu Xiaobing, et al. Progress in zero-discharge treatment technology for wet desulfurization wastewater from coal-fired power plants [J]. *Environmental Engineering*, 2016, 34(4):33–36.

Peng Lian, Kailiang Wang. Analysis and application of flue gas evaporation treatment technology for desulfurization wastewater [J]. *Huadian Technology*, 2018, 40(10):59–62.

Shuangchen Ma, Jiaqi Wen, Zhongcheng Wan, et al. Research progress and standard revision suggestion of desulfurization wastewater treatment technology for coal-fired power plants in China [J]. *Clean Coal Technology*, 2017, 23(4): 19–28.

Zhaowei Yuan. Research progress on zero discharge treatment technology of desulfurization wastewater from coal-fired power plants [J]. *Coal Processing and Comprehensive Utilization*, 2018(10): 49–56.

Zhongmei Zhang. Discussion on zero discharge technology of desulfurization wastewater from coal-fired power plants [J]. *Energy Conservation and Environmental Protection*, 2019(10): 47–48.

Advances in Energy, Environment and
Chemical Engineering – Abdullah & Osman (Eds)
© 2023 The Author(s), ISBN: 978-1-032-36083-6

Stability analysis of landslide under rainfall infiltration

Pengyu Zhang*
School of Highway, Chang'an University, Xi'an, Shaanxi, China

Zhefeng Lin & Hengming Shen
Shaanxi Nuclear Industry Engineering Survey Institute Co., Ltd, Shaanxi, China

Meijuan Bian
Xincheng District Emergency Management Bureau, Xi'an, Shaanxi, China

ABSTRACT: Based on the case of a high soft rock slope in Xixiang County, Hanzhong City, the process of slope instability is analyzed by combining numerical analysis with monitoring data. Through the analysis, it is found that the slope has experienced two stages of accelerated deformation and failure. Under the triggering effect of rainfall, the slope soil softens and finally leads to landslide occurrence. At the same time, combined with the analysis of monitoring data, the strength reduction method is used to analyze the stability of soft rock slopes under the condition of rainfall infiltration. The study of the example slope shows that the instability of the soft rock slope under the condition of rainfall infiltration is the partial collapse of the surface layer of the slope at the initial stage of rainfall. With the increase in rainfall duration, the form of instability is the combination of local layer collapse and overall sliding.

Keywords: cohesive soil; rain infiltration; slope stability; numerical simulation

1 INTRODUCTION

Landslide is one of the most common and harmful natural disasters in the world. Landslide is one of the most common and harmful natural disasters in the world. Rainfall is the most important natural factor leading to landslides, and its relationship with landslides has always been the focus of research by many scholars (Huang 2007; Ying 1996). In fact, the occurrence of landslides usually goes through four stages: creep stage, uniform deformation stage, acceleration stage, and failure stage (Xu 2012, 2009). Through a large number of practical engineering studies, scholars have found that when the slope is in the creep or uniform deformation stage, no matter how much displacement is generated, the slope will not be damaged as a whole; that is, a landslide will not occur (Zheng 2004). Therefore, it is particularly important to judge what stage a landslide is in and its future development trend through practical engineering (Zeng 2009).

Currently, the research methods of rainfall on slope stability mainly include engineering geology analysis, limit equilibrium analysis, numerical analysis, reliability analysis (Wang 2016), and dynamic analysis and prediction combined with monitoring data. In fact, the impact of rainfall on slope stability is also a gradual process. First, the erosion of rainfall on the slope body leads to the softening of slope soil, reduced shear strength, and increased dead weight. Therefore, combined with the monitoring data of rainfall and slope displacement (Cheng 2019), it is safer and more reliable to conduct a dynamic analysis of landslide stability (Hou 2021). In this paper, a high soft rock slope in the Xixiang section of Hanzhong City is taken as an example to analyze the variation of rainfall and landslide displacement, and the stability of the slope is analyzed dynamically with the method of numerical analysis.

---

*Corresponding Author: 2749026513@qq.com

                                     DOI 10.1201/9781003330165-99

2 PROJECT OVERVIEW IN THE STUDY AREA

Tanmu village landslide in Xixiang County, Hanzhong city is located about 1km northeast of Yangpo village, group 1, Tanmu village, Xixiang County, Hanzhong City, Shaanxi Province, and the section from Baotou city to Nanning k1596+725~+850 of G210 national highway. Geographical coordinates E: 107°C 50′37″, N: 32°C 56′12″. Tanmu village landslide is pocket shaped on the whole. It is adjacent to the high terrace of the river in the East and the flood plain of Jingyang River in the West. Pebbles are exposed at the front edge. The landslide is a continuous high, steep and long slope with a slope aspect of 228°, a transverse width of about 220 m, a longitudinal height of about 170 m, a front slope of about 80°, a middle slope of about 50 ~ 60°, and a rear slope of about 40°. The maximum deformation thickness of the slope is about 36 m, and the volume is about 280000 m³. At the same time, bedrock is directly exposed at the landslide location, and due to severe weathering, it is mostly completely weathered, and sporadic diluvium can be seen. Under the influence of such stratum lithology, the strength of the slope is low, and there are many complex weak surfaces. Under the action of rainwater, local softening and strength reduction will occur, which will lead to the overall sliding of the landslide. The overall plan and profile of the landslide are shown in Figures 1 and 2.

Figure 1.   Panorama of landslide.

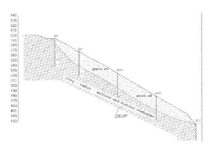

Figure 2.   Profile of landslide.

3 LAYOUT OF MONITORING NETWORK

According to the scale and deformation characteristics of the landslide, the main sliding position is divided into four parts by three profiles. The monitoring equipment shall be installed along the profile. A dragline displacement monitor (LF1, LF2, LF3) was set in section i, section ii and section iii respectively. Two inclinometers were set up in section ii to monitor the deep displacement. The specific arrangement of monitoring equipment is shown in Figures 3 and 4:

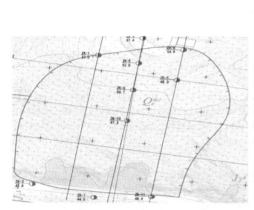

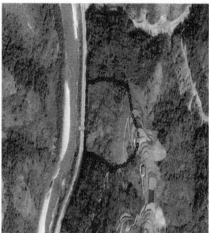

Figure 3. Layout plan of monitoring.   Figure 4. Plane distribution form of landslide.

4   ANALYSIS OF MONITORING DATA

From May to September of each year, it is the rainy season in Shaanxi. During this period, due to the influence of frequent rainfall and heavy rainfall, the saturation and self-weight of the slope soil increase. At the same time, due to the scouring effect of rainfall, the soil softens. Under this series of actions, the stability of the slope becomes worse, leading to the occurrence of landslides. Therefore, the rainy season in Shaanxi is selected for the processing and analysis of monitoring data.

4.1   *Surface displacement analysis*

Through the analysis of the displacement change data and rainfall data of Tanmu village landslide in Xixiang County, ous rainfall, while the displacement increment at LF1 and LF2 is very small (Figure 5 and Figure 6). Therefore, we can infer that LF3 is at the main sliding position, and LF1 and LF2 are at a relatively stable position.

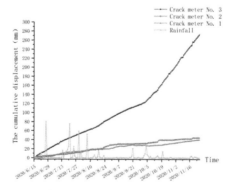

Figure 5.   Cumulative surface displacement time curve.

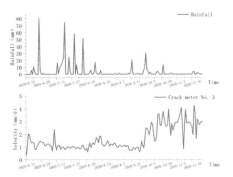

Figure 6. Displacement rate-rainfall diagram.

Therefore, it is necessary to focus on the analysis of displacement changes at LF3. Firstly, the angle between rainfall and displacement change rate is analyzed (Figure 6). At the beginning of June, the displacement rate of the No. 3 displacement meter fluctuated around 1.25mm/d, which was relatively stable. The rate reached the peak value of 2.4 mm/d for the first time on July 11, and there was a continuous rainfall process from June 20 to June 27, of which the rainfall on June 27 reached 80.4mm/d, belonging to the rainstorm level. It can be inferred that the first sudden increase in the displacement rate of the landslide is caused by the lag of the impact of rainfall on the stability of the landslide. This phenomenon is fundamentally caused by the formation lithology of the landslide. The landslide formation contains a large number of mudstone and sandstone, and most of them are interbedded. Most of the bedrock is exposed to severe weathering, which leads to great differences in the permeability of the landslide mass in all directions. In addition, the softening of the rock mass by rainwater is a slow process, so the short-term heavy rainfall does not immediately lead to a sudden increase in the rapid rate of landslide displacement.There was a period of continuous rainfall from July 14 to August 24, when the rainfall reached rainstorm level for three days. However, the landslide displacement rate did not fluctuate significantly within one month after that; that is, the landslide was still in the stage of uniform deformation during this period. Since October 5, the displacement rate of the landslide began to show a large fluctuation and showed no sign of flattening, so it can be inferred that the landslide has entered the stage of accelerated deformation.

4.2   *Deep displacement analysis*

By analyzing the cumulative displacement variation law of the No. 5 inclinometer hole, it can be found that the displacement variation is small. When the landslide is in the uniform deformation stage (before October 5), the maximum cumulative displacement change is only 20.69mm. When the landslide is in acceleration, the maximum deformation rate is only 3mm/d. Therefore, it can be inferred that the displacement change of No. 5 is not within the main landslide location.

By analyzing the cumulative displacement time curve of hole 6 (Figure 8), it can be found that when the landslide is in the constant velocity stage, under the action of long-term continuous rainfall, the displacement growth presents a stepped increase, and with the infiltration of rainwater, the landslide body gradually softens, the self-weight increases and the strength decreases, which makes the landslide enter the accelerated deformation stage. Through the displacement analysis at different depths of the landslide, it can be found that the cumulative displacement change at 18 m and 23 m is small, and the displacement starts to change suddenly from the depth of 13 m. The overall change trend of the displacement above 13m is roughly consistent with that at 13 m. The overall change trend is that the closer to the surface, the greater the displacement change and the greater the displacement.

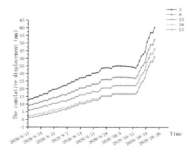

Figure 7.　Cumulative displacement time curve of hole 5.

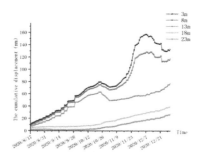

Figure 8.　Cumulative displacement time curve of hole 6.

In order to further judge the deformation of the landslide and the position of the weak surface, the deep displacement of the landslide can be analyzed. Firstly, the horizontal displacement depth curves of No. 5 and No. 6 inclinometer holes can be analyzed. The overall change law of the horizontal displacement depth curve of the No. 5 inclinometer hole (Figure 9) is relatively simple, which shows that the displacement increases with the extension of time. When the landslide is at the junction of the constant deformation stage and the accelerated deformation stage, its displacement basically remains unchanged, and then its displacement change gradually increases.

By comparing the horizontal displacements and depth curves of No. 5 and No. 6 survey holes (Figures 9 and 10), the maximum horizontal displacements of No. 5 survey holes are much smaller than that of No. 6 survey holes. Combined with the analysis of the layout position of monitoring points, it can be found that the sliding surface of the landslide has not developed to the position of No. 5 survey holes, and the landslide is still a local landslide. By analyzing the deep displacement and horizontal displacement of hole No. 6 (Figure 10), it can be seen that its displacement has a mutation at 13m underground, so it can be inferred that its sliding surface is roughly 13m underground.

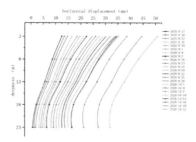

Figure 9.　Horizontal displacement depth curve.

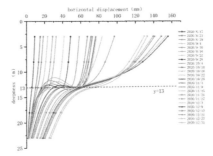

Figure 10.　horizontal displacement depth.

5　NUMERICAL CALCULATION

5.1　*Model establishment*

Through the actual investigation of the slope, a three-dimensional slope calculation model with the actual size of 280m*240m*280m is established, as shown in Figures 11 and 12:

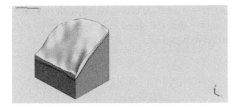

Figure 11. 3D solid model geometry.

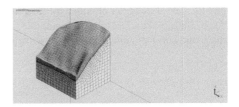

Figure 12. 3D solid mesh generation diagram.

The model calculation parameters are shown in the following Table 1:

Table 1.　List of physical and mechanical parameters of the calculation model.

	Modulus of elasticity (k N//m²)	Poisson's ratio	Natural weight (kN/m³)	Saturated gravity (kN//m³)	Friction angle (°)	Cohesion (kN//m²)
Slip zone	68700	0.3	20.2	20.6	24.5	26.5
Sliding body	210000	0.27	21	21.5	33.8	30
Sliding bed	4609000	0.25	22	22.4	39	50

5.2　*Result analysis*

The calculation results are shown in the figures above (from Figure 13 to Figure 16). Obviously, the displacement of the slope body becomes larger and diffuses to the surrounding under the action of rainfall. The place with large displacement is mainly on the right side of the slope toe, with a thickness of about 14~25m, and the displacement is about 43.3~65.1 cm, accounting for about 0.7% of the whole displacement distribution area. At the same time, it can be seen that the slope displacement of 37.9~27.1cm is mainly distributed at the front edge of the slope, and the distribution thickness is about 20~50m, accounting for about 2.8% of the whole displacement distribution area. In terms of the whole total displacement distribution area, the total displacement of the slope surface and the front edge of the slope surface is the largest, which is also consistent with the slope deformation in the field investigation, proving the reliability of the simulation.

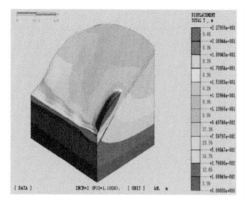

Figure 13. Natural state displacement.

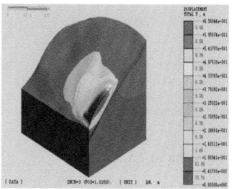

Figure 14. Displacement after rainfall.

Finally, the safety factor obtained by the transfer coefficient method is compared with the safety factor of the simulation results. For example, Table 2 shows that the slope has poor stability after rainfall, and corresponding reinforcement measures need to be taken.

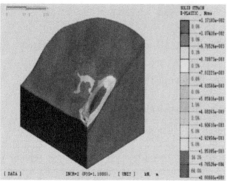

Figure 15. Maximum plastic strain.

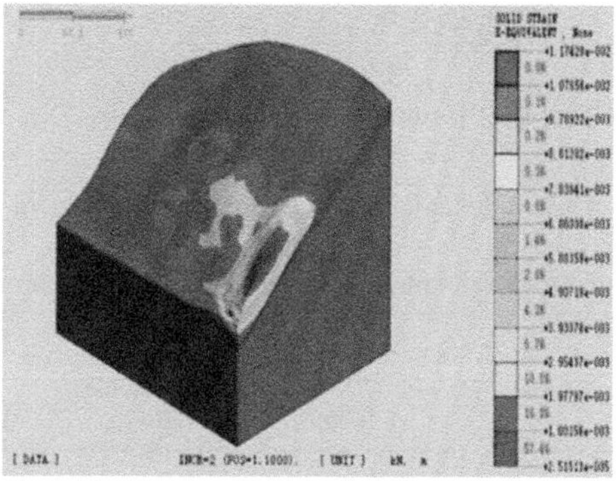

Figure 16. Maximum equivalent strain.

Table 2. Comparison table of stability coefficients of two calculation methods.

| Side slope | Working condition | Stability coefficient | | Evaluation results |
		Transfer coefficient method	Finite element strength reduction method	
Tanmu village slope	Normal situation	1.051~1.074	1.100	Basically stable
	Rainfall	0.999~1.018	1.025	Unstable

6 CONCLUSION

Based on the case of slope instability in Tanmu village, Xixiang, this paper discusses the displacement change and slope stability of the slope under rainfall and natural conditions. Combined with the monitoring data, the main conclusions are as follows:

(1) By studying the displacement distribution law of the slope body, it is found that the displacement of the slope body gradually decreases from the slope surface to the inside of the slope body until the displacement at the bottom of the slope body becomes zero. The displacement at the middle and lower parts of the slope body close to the slope surface deflects towards the slope direction.

(2) By studying the strain distribution law of the slope body, it is found that the distribution state of shear strain is in a strip shape, distributed in the completely weathered sandy mudstone layer of the slope body. The position where the maximum shear strain occurs is near the shear outlet of the slope to where the shear stress mutation occurs, and the maximum shear strain has not yet been penetrated.

(3) The simulation results show that the stability coefficient obtained by the arc length strength reduction method is slightly larger than that obtained by the limit equilibrium method. The slope stability coefficients calculated under natural and rainfall conditions are 1.1 and 1.025, respectively.

REFERENCES

Dongxiujun, Xu Qiang, Tang Chuan, et al. Study on physical simulation test of landslide displacement time curve characteristics [J]*Journal of engineering geology*, 2015(03): 43–49

Fu Hongyuan, Zeng Ling, Wang Guiyao, JIANG Zhongming, Yang Qingxia. Soft rock slope stability analysis under the condition of rainfall infiltration [J]. *Rock and soil mechanics*, 2012(8): 2359–2365. The DOI: 10.16285 / sm j.r. 2012.08.012.

Qi Xing, XU Qiang, ZHENG Guang, Hu Zeming. Dynamic warning mechanics model of bedding rock and soil landslides induced by rainfall. [J]. *Journal of catastrophology*, 2015, 30(03):38–42.

Huang Runqiu. Large-scale landslides and their mechanism in China since the 20th century [J]. *Chinese Journal of Rock Mechanics and Engineering*, 2007, 26(3):22.

Rong Guan, Wang Sijing, Wang Enzhi, Wang Jianxin.Chinese Stability of typical Engineering slope of Yuanmo Highway under Heavy Rainfall [J]. *Chinese Journal of Rock Mechanics and Engineering*,2008(04):704-711.

Saito M. Research on forecasting the time of occurrence of slope failure[J]*Soil Foundation*, 1969, 17:29-38

Wei Ning, QIAN Pingyi, FU Xudong. Effect of rainfall and evaporation on soil slope stability [J]. *Rock and Soil Mechanics*, 2006.

Wu Hongwei, Chen Shouyi, PANG Yuwei. Study on parameters of influence of rainwater infiltration on unsaturated soil slope stability [J]. *Rock and Soil Mechanics*, 1999.

Xu qiang. Deformation and failure behavior and internal mechanism of landslide [J]. *Journal of engineering geology*, 2012, 20(02):145–151.

Xuqiang, Lixiuzhen, Huang Runqiu. Research progress of landslide time prediction [J]*Advances in Geoscience*, 2004, 19(3)

Yin Kunlong, Yan Tongzhen. Landslide prediction and related models [J]*Journal of Rock Mechanics and Engineering*, 1996, 15 (001): 1–8

Zhong Yingan. Study on comprehensive information prediction method of Huanglashi landslide [J]. *Chinese Journal of Geological Disaster and Prevention*, 1995(04):68–74+80.

Advances in Energy, Environment and
Chemical Engineering – Abdullah & Osman (Eds)
© 2023 The Author(s), ISBN: 978-1-032-36083-6

Benthos impact on wastewater purifying capacity of constructed wetlands

Chi-yuan Wei* & Hai-xia Li
Nanjing Tech University Pujiang Institute, Jiangsu, China

Ting Zhang
College of Urban Construction, Nanjing University of Technology, Jiangsu, China

Xu-jian Qi
The IT Electronics Eleventh Design & Research Institute Scientific and Technological Engineering Corporation Ltd Eastern China Branch, Jiangsu, China

An-qi Zhang
Nanjing Tech University Pujiang Institute, Jiangsu, China

ABSTRACT: In this study, the impact of benthos on plants growth and the response of wastewater purifying capacity of constructed wetlands to benthos is investigated through the simulated construction of integrated vertical flow constructed wetlands, with reeds, cannas, and water iris used as test plants, and loach and earthworms used as test benthos. The results show that loach and earthworms obviously promote the growth of plant height, leaf length, and leaf width. The removal rate of TP, COD, ammonia nitrogen, and TN has been improved when loaches and earthworms are added to constructed wetlands. By comparing differently constructed wetlands, loaches increase the removal rate of TN by 5.89%, earthworms increase COD and TN removal rate by 3.18% and 0.61%, respectively, while loaches and earthworms combined help improve TN removal rate by as high as 7.18%. Green sand in the matrix has a great role in promoting the removal of the four indicators mentioned above. Through varied comparative studies, it is found that different benthos has obvious promoting effects on wastewater treatment.

1   INTRODUCTION

The constructed wetland system is a composite system comprising of plants, benthos, matrix (such as soil and gravel), and waters, which purifies pollutants through biological, chemical, and physical collaboration. Compared to conventional wastewater treatment plants, the constructed wetlands system offers selective and pertinent treatment schemes which feature such advantages as little investment, low operation cost, cheap maintenance expense, and no secondary pollution, etc. Based on the pattern of water flow, constructed wetlands may take the form of horizontal flow, surface flow, vertical flow, or composite flow (Liu et al. 2021). With their better water quality and capacity to improve the ecological environment, such wetlands are being increasingly used and developed in wastewater treatment.

Wetland plants may directly absorb pollutants to improve the matrix environment. Selecting easy-to-live, long growth cycle, soiling-resistant, beautiful, and high economic value plants, such as reed, cattail, and Acorus calamus (Parde et al. 2021), etc., helps to secure the constructed wetland system's capability in pollutant purification. Through the study on 19 species of wetland plants, it is found that reed and cattail have shown higher removal efficiency, and they can absorb 70% nitrogen and phosphorus in wastewater, while canna, buttercup, and iris offer a better sense of beauty than other species of wetland plants (Ding 2022). According to studies by Zhu Shijiang et al. (2022),

---

*Corresponding Author: chiyuanweinanjing@ 163.com

DOI 10.1201/9781003330165-100

Acorus calamus, water hyacinth, reed, and water lily suit for low concentration ammonia nitrogen environment, while canna and reed perform better in high ammonia nitrogen concentration, and are more tolerant to ammonia nitrogen. According to experiments by Zhou Riyu et al. (2021), in a vertical flow constructed wetland system planted with canna, Reinecke carnea, iris, and thalia dealbata, and under $0.2 \text{ m}^3 \cdot (\text{m}^2 \cdot \text{d})^{-1}$ influent load, it is shown that such a wetland system removes COD, NH_4^+-N and TP in the water by 84.38%, 65.78% and 74.67%, respectively. According to conclusions by Wang Yunchen et al. (Wang et al. 2021), wetland plants improve the system's oxygen dissolving environment and organics content through their own roots and offer a venue for denitrified microorganisms to grow and breed, thereby improving the microbial community structure, boosting the nitrification-denitrification process and promoting plants' absorption of inorganic nitrogen.

Benthos is an integral part of the wetland ecosystem and mainly inhabits the wetland matrix or sediment. Benthos may damage the original structure of sediment via stir behavior such as predation, crawling, burrowing, etc., to release nitrogen and phosphorus, and may also affect the activity of microorganisms in matrix or sediment, making it possible for the nitrification-denitrification process to be enhanced and nitrogen removal being boosted (Guo et al. 2021). Based on studies by Yang Xiaotong (2021), with the increase of mussel quantity, matrix absorption of phosphorus improved, and as more mussels were added, removal of pollutants such as TN, TP, and COD in the wetland system was also raised. By introducing tubificidae to the constructed wetland system, Kang Yan (2019) observed that in summer, TN and TP removal increased by 22.3% and 27.4%, NO_3^--N removal by 28.4%, and NH_4^+-N removal by 6.3%, respectively. According to studies by Franck Michael Zahui et al. (2018) on large invertebrate communities and the influence of their abundance in treating sediment of synthetic domestic wastewater in the vertical flow constructed wetland, it is disclosed that plants grew quite well during the period, and removal of TN, TP, and TSS in the domestic wastewater of the constructed wetland all improved. In addition, benthos may also be used as an effective bio-indicator for real-time monitoring of constructed wetland operations. Benthos may divert pollutants from a constructed wetland system to other circulatory systems, which to some extent helps reduce the pollution load of a constructed wetland. For instance, earthworms, after absorbing wetland pollutants, may be used as feed and introduced to other recycling systems (Xu 2010).

This study focuses on the changes in plant physical features for a constructed wetland wastewater treatment system in which benthos is provided, as well as on the influence of introducing benthos upon the constructed wetland's wastewater treatment and purification capacity. It offers references for a constructed wetland wastewater treatment system to realize highly efficient and stable operation under the joint functioning of plants, matrix, and benthos, and therefore helps promote the extensive application of benthos in this field.

2 EXPERIMENT MATERIALS AND METHODS

2.1 *Experiment devices*

Comprising a tall pool (intake pool) and a low pool (discharge pool), a composite vertical flow constructed wetland is simulated and built. The tall pool is 60cm long, 60cm wide, and 90cm high, while the low pool is 60cm long, 60cm wide, and 60cm high. The two pools are connected at the bottom by the use of two 5cm-diameter PVC tubes. The intake pool and the discharge pool are each separated vertically into three zones by use of gauze, and four plants of canna, reed, and calamus are planted in each of these three zones.

2.2 *Experiment materials*

2.2.1 *Selection of experiment plants, matrix, and benthos*
1) Experiment plants

Canna: absorbing hazardous substances like hydrogen chloride and sulfur dioxide etc., and featuring rapid growth and huge biomass, therefore capable of removing excessive redundant nutrients.

Reed: top priority plant for wastewater treatment in a constructed wetland, also an aquatic plant with wide application, high biomass, and strong stress resistance, and its developed root system helps for microbial attachment.

Calamus: highly resistant against fouling and cold, an emerged herbaceous plant, and having a developed root system similar to reed, therefore also helpful for microbial attachment.

2) Experiment matrix: fine sand, green sand, and organic matter.

3) Experiment with benthos

Loach: artificially bred and highly adaptive to the environment. 10 pieces are added to the pool where necessary.

Earthworm: Eisenia foetida is selected, and 1.5kg of such earthworms are added to the pool where necessary.

2.2.2 *Wastewater setting*

Add 10.4g ammonium sulfate, 16.0g glucose, 3.2g peptone, and 3.2g potassium dihydrogen phosphate to 100L tap water.

2.2.3 *Water distribution device*

The constructed wetland adopts intermittent inflow. A 20L water bucket is provided in each pool for the purpose of even distribution of wastewater water on a daily basis. This method may maintain a steady inflow and, at the same time, prevent water from evaporating too fast, which will have a negative impact on the experiment.

2.3 *Experiment scheme*

Four types of composite vertical inflow constructed wetlands are simulated and built, and detailed configurations of these wetlands are shown in Table 1 below.

Table 1. Wetland Configuration.

Wetland	Max	Plant	Loach	Earthworm
I	Fine sand, organic matter	Canna, reed & calamus	No	No
II	Fine sand, organic matter	Canna, reed & calamus	Yes (10 pieces)	No
III	Fine sand, organic matter	Canna, reed & calamus	Yes (10 pieces)	Yes (1.5kg)
IV	Fine sand, green sand & organic matter	Canna, reed & calamus	Yes (10 pieces)	Yes (1.5kg)

Hydraulic loading: 0.06 $m^3/(m^2 \cdot d)$
Hydraulic retention time: three days.

2.4 *Method of analysis*

2.4.1 *Water quality measurement*

Four indicators, i.e., total phosphorus (TP), chemical oxygen demand (CODcr), ammonia nitrogen (NH_4^+-N), and total nitrogen (TN), are selected, and analysis is conducted by use of the 4th Edition of *Water and Wastewater Monitoring and Analysis Method* (Editorial Board for *Water and wastewater monitoring and analysis method*, the State Environmental Protection Administration., 2022).

Table 2. Analysis Items and Methods.

Analysis items	Analysis methods
TN	Alkaline potassium persulfate digestion UV spectrophotometric method
NH_4^+-N	Salicylic acid hypochlorite photometric method
TP	Mo-Sb Anti spectrophotometric method
CODcr	COD quick determination instrument - 5B-2H Type

2.4.2 *Measurement of plant growth indicators*

Original plant height data was measured on April 1, 2021, and plant height was measured again on May 15, 2021, for comparison. Plant height: is measured directly by use of tape. Plant leaf length: is measured at the leaf bottom using tape. Plant leaf width: the max width of a middle-height leaf is measured using a vernier caliper.

2.4.3 *Absorption of phosphorus by the matrix*

The steps are as follows: weigh 10g naturally dried matrix samples (three parallel samples for each sample), put them in a 150mL triangular flask, add 100mL KH_2PO_4 solution (using 0.02 mol/L KCl solution) with a phosphorus content of 5 mg·L^{-1}, 10 mg·L^{-1}, 20mg·L^{-1} 40 mg·L^{-1}, 80 mg·L^{-1} and 120 mg·L^{-1}, respectively, and add two droplets of chloroform in each sample to inhibit microbial activity. Put the flask on an oscillator with settings of 25°C and 200 r·min^{-1}, oscillate for 24h, and take the samples out for centrifugation (5000 r·min^{-1}, 10 min). Determine the phosphate concentration of the supernatant equilibrium solution using the Mo-Sb Anti spectrophotometric method. The difference between phosphorus added and phosphorus in equilibrium solution is phosphorus absorbed (mg/kg).

2.5 *Data statistics*

Data sorting and classification and preliminary analysis are conducted using Microsoft Excel, and data mapping is performed using Origin 2018 software.

3 RESULTS AND DISCUSSION

3.1 *Benthos influence on plant growth indicators in a constructed wetland*

3.1.1 *Plant height*

See Table 3 below for variations in plant height of different plants. Given the same external growth environment and conditions, it can be seen that the plant height of reed and cattail increases apparently, save for canna. No matter for the intake pool or the discharge pool, given the same plants, plant height variation in constructed wetland No. III (fine sand, organic matter + loach, earthworm) is higher than that in constructed wetland No. II (fine sand, organic matter + loach), and hence it is apparent that earthworm facilitates plant growth. Plant height variation in constructed wetland No. III (fine sand, organic matter + loach, earthworm) is less than that in constructed wetland No. IV (fine sand, green sand, organic matter + loach, earthworm), but the difference is not so obvious. Plant height variation is compared by use of average growth rate (the value of plant height variation within two adjacent measurements for the same plants, divided by interval measurement time, cm·d^{-1}), and plant growth rates of differently constructed wetlands are given

Table 3. Variation in Plant Height of Different Plants (cm).

Wetland No.	Intake Pool			Discharge Pool		
	Reed	Canna	Calamus	Reed	Canna	Calamus
I(May)	63.20	29.23	56.78	112.45	28.20	63.93
I(April)	22.00	20.50	34.00	40.00	20.00	46.00
II(May)	64.47	36.48	53.35	74.40	23.63	55.38
II(April)	19.00	28.00	21.00	32.00	16.00	27.00
III(May)	93.98	39.48	54.15	87.80	33.80	53.88
III(April)	38.00	30.00	23.00	28.00	25.00	20.00
IV(May)	78.10	34.88	54.60	75.93	29.88	55.78
IV(April)	30.00	26.00	26.00	26.00	20.00	25.00

in Figure 1 below. It can be observed from Figure 1 that, among the four types of constructed wetlands, plant height and average growth rate of reed are the highest, followed by those of water calamus, while those of canna are the lowest. Among constructed wetlands II, IIII, and IV in which benthos are added, the average growth rate of reed, canna, and water calamus is basically higher than constructed wetland I, and the variation in the average growth rate of canna is not notable. The average growth rate of reed in the discharge pool of constructed wetland I have no rules to follow, and the reasons for this remain to be further researched.

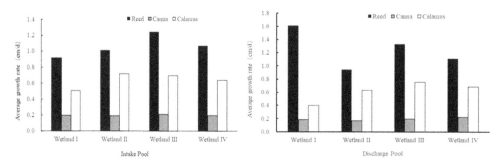

Figure 1.   Average growth rate of plant height.

3.1.2 *Plant leaf length*
For the intake pool, it can be seen from Figure 2 that the leaf length of reed in constructed wetlands II, III, and IV is higher than in constructed wetland I, where benthos is not added. Reed leaf length is 70.98cm and 63.18cm, respectively, for constructed wetlands III (fine sand, organic matter + loach, earthworm) and IV (fine sand, green sand, organic matter + loach, earthworm), much higher than constructed wetland II (fine sand, organic matter + loach), which proves earthworm promotes plant growth greatly, while no apparent variation in leaf length has been observed for canna and water calamus. For the discharge pool, except for canna leaf length, which shows the rules mentioned above, neither reed nor water calamus has specific characteristics.

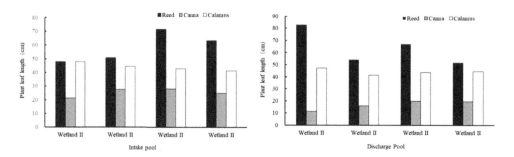

Figure 2.   Leaf length of plants in differently constructed wetlands.

3.1.3 *Plant leaf width*
As variation in plant leaf width is minor, the average values of all measurement data are used for plant leaf widths during the experiment period. Figure 3 shows that the rule for three plants in the intake pool is as follows: among four constructed wetlands, plant leaf width in the constructed wetland where benthos is added is wider. As for the discharge pool, variation in plant leaf width presents no apparent rule. Except for that variation in leaf width of canna is easy to observe, variation in leaf width of reed and water calamus is imperceptible.

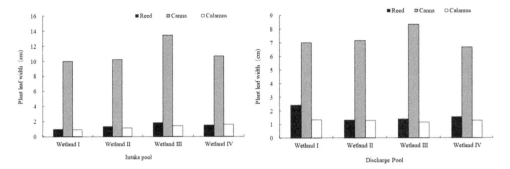

Figure 3.   Leaf width of plants in differently constructed wetlands.

3.2 *Adsorption capacity of matrix to phosphorus*

Isothermal adsorption of phosphorus by a matrix in four types of constructed wetlands is shown in Figure 4 below. It can be seen from Figure 4 that, after green sand is added to the matrix of constructed wetland IV (fine sand, green sand, organic matter + loach, earthworm), its adsorption capacity to phosphorus is higher than the other three constructed wetlands, which shows green sand is quite adsorptive to phosphorous. On the other hand, almost no difference has been observed in the phosphorus adsorption capacity of the matrix in constructed wetlands I (fine sand, organic matter), II (fine sand, organic matter + loach), and III (fine sand, organic matte + loach, earthworm). And the reason may be that green sand has a higher specific surface area and micropore volume, or it has a higher content of such elements as Ca, Fe, Al, and Mg, which performs adsorption reaction and precipitation reaction with phosphorus (Bai et al. 2019).

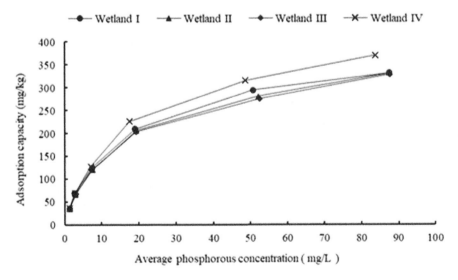

Figure 4.   Adsorption isotherms of phosphorus on different constructed wetland substrates.

3.3 *Purification capacity of constructed wetlands to wastewater*

3.3.1 *Removal of Total Nitrogen (TP) by constructed wetlands*
Removal of phosphorus by constructed wetlands is realized mainly through such mechanisms as plant absorption, filler absorption, filtration, microbial removal, etc. (Wan et al. 2011).

Initial TP concentration in raw water is 1.782 mg/L, and the removal of TP by constructed wetlands is shown in Figure 5 below. The TP removal capacity of all four constructed wetlands is powerful, all above 96%. Constructed wetland IV (fine sand, green sand, organic matter + loach, earthworm) has the highest removal rate of 97.88%, possibly because loach and earthworm added help promote the growth and breeding of microorganisms in constructed wetlands and thereby improve the removal of total phosphorous. Besides, the reasons may also include that benthos, to some extent, promote plant growth and improve phosphorus absorption by plants. According to studies by Zhang Yu et al. (2015), the addition of earthworm and loach to constructed wetlands promotes the conversion of calcium and phosphorus in the matrix into organic phosphorus and enhances TP content in the matrix, therefore reducing phosphorus content in the wastewater. As disclosed by Kang Jiu (2019), by introducing earthworms, the proportion of colloidal phosphorus and particulate phosphorus is boosted, and phosphorus settlement increases, which helps improve phosphorus absorption by plants.

The removal rate of constructed wetland I (fine sand, organic matter) to TP is 97.7%, slightly higher than that of constructed wetland II (fine sand, organic matter + loach), 97.59%, possibly because only loach is added to constructed wetland II, loach disturbance may cause the release of phosphorus in the matrix, which interferes with TP removal by constructed wetlands. The TP removal rate of constructed wetland III (fine sand, organic matter + loach, earthworm), 96.24%, is lower than constructed wetland II, as both loach and earthworm are added to constructed wetland III, which results in further release of phosphorus.

Considering constructed wetlands III and IV, the same benthos (loach and earthworm) are added, but constructed wetland IV has a higher TP removal rate because green sand is added to constructed wetland IV. From Figure 4, it can be noticed that the matrix of constructed wetland IV shows stronger phosphorus adsorption capacity, which contributes to higher phosphorus removal by constructed wetland IV compared to constructed wetland III.

3.3.2 *Removal of COD by constructed wetlands*

COD removal by constructed wetlands relies on microorganisms in the matrix. To be exact, aerobic microorganisms and anaerobic microorganisms synthesize organic pollutants in the wastewater into new cells or just metabolize these pollutants under aerobic conditions and anaerobic conditions, respectively, and in the end, organic pollutants in the wastewater are removed through flocculation, sedimentation, and separation by bacterial organic matter, thereby achieving the purpose of water quality purification (Xiao 2016).

Initial COD concentration in raw water is 284.4mg/L, as shown in Figure 5 below, and the COD removal rate of constructed wetland IV (fine sand, green sand, organic matter + loach, earthworm) is the highest, 85.18%, as compared to constructed wetland III (fine sand, organic matter + loach, earthworm), which proves green sand adsorbs COD. According to Figure 5, after introducing earthworms to constructed wetland III, its COD removal rate is 3.18% higher than constructed wetland II, possibly because plant growth is boosted and plant photosynthesis is enhanced upon introduction of earthworms to wetlands, improving the ability of its root system to secrete oxygen (Su et al. 2015). The COD removal rate of constructed wetlands II, III, and IV where benthos is provided rises gradually (77.77%, 80.95%, and 85.18%).

3.3.3 *Removal of ammonia nitrogen by constructed wetlands*

Initial ammonia nitrogen concentration in raw water is 2.775 mg/L. According to Figure 5, the removal rate of constructed wetlands to ammonia nitrogen increases above 91%. The removal rate of constructed wetlands II, III, and IV where benthos is added is higher than constructed wetland I without benthos, by 0.22%, 0.83%, and 1.12%, respectively. The removal rate of ammonia nitrogen in the water body is positively correlated to the consumption volume of dissolved oxygen (Fu et al. 2021), and the addition of benthos improves the DO content of the wetland system. After the addition of earthworm, the removal rate (92.79%) of constructed wetland III is higher than that (92.18%) of constructed wetland II, which shows earthworm has an obvious influence upon microorganisms and plants, the absorption of nutrient elements by plants gets stronger, vertical transmission of pollutants by plant root system gets deepened, and the abundance of functional bacteria conversion by nitrogen in the matrix gets improved (Kang 2019). Green sand is added to the matrix of constructed wetland IV, causing its removal rate (93.08%) to be ammonia nitrogen higher than constructed wetland III, which proves green sand plays a certain role in adsorbing ammonia nitrogen.

3.3.4 *Removal of Total Nitrogen (TN) by constructed wetlands*

Removal of nitrogen in wastewater by constructed wetlands can be realized through nitrification and denitrification by microorganisms and absorption by plants (Zhang 2017). Initial TN concentration in raw water is 16.721 mg/L. According to Figure 5, the removal rate of constructed wetlands to TN increases in turn. Compared to constructed wetland I, where benthos is not added, the TN removal rate of constructed wetlands II, III, and IV, where benthos is added, is 5.89%, 5.89%, and 10.34% higher, respectively. The removal rate of constructed wetland II is higher than constructed wetland I because loach introduced promotes plant growth, therefore boosting TN removal by constructed wetlands. The removal rate of constructed wetland III is higher than constructed wetland II because earthworm, as well as loach, is introduced to constructed wetland III, while earthworm helps promote microorganism growth and breeding and facilitates nitrification and denitrification. Green sand is added to the matrix of constructed wetland IV as compared with constructed wetland III, resulting in a higher removal rate (86.77%) of constructed wetland IV than that (83.61%) of constructed wetland III, which reveals the addition of green sand is favorable for loach and earthworm to survive, and more favorable to facilitate microorganism and plant growth through their activity in wetlands.

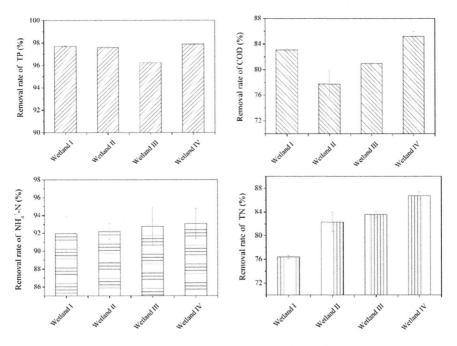

Figure 5. The Purification Ability of Different Constructed wetlands to Pollutants.

4 CONCLUSION

This study focuses on the influence of benthos on plant growth and on the effects of wastewater treatment by adding different varieties of benthos to four types of constructed wetlands. According to the findings of this study, among the four types of constructed wetlands, the reed has the highest plant height and average growth rate, followed by water calamus, while canna has the lowest. Among constructed wetlands II, III, and IV, where benthos is added, the average growth rate of reed, canna, and water calamus is basically higher than constructed wetland I, though the average growth rate difference of canna is not obvious. For the intake pool, the leaf length of reeds in constructed wetlands II, III, and IV where benthos is added is apparently higher than constructed wetland I, while leaf length of the other two plants, i.e., canna and water calamus, shows no apparent difference in the intake pool. For the discharge pool, except for canna leaf length, which conforms to the rules said above, reed and water calamus show no specific features. All three plants follow the same rules for the intake pool: among four constructed wetlands, those where benthos is added have wider plant leaf width. Save for canna leaf width, whose variation is relatively easy to observe, variation in leaf width of reed and water calamus is extremely imperceptible.

In constructed wetlands where benthos (loach and earthworm) is added, the average growth rate of plant height is higher than in constructed wetlands without benthos. Although benthos has a certain promotional function upon the growth of plant leaf length and leaf width, the effects are not so obvious.

It can be observed in wastewater treatment effects that all four constructed wetlands have a strong capacity in TP removal, higher than 96%. However, the removal rate varies as benthos may release phosphorus. The removal rate of 4 constructed wetlands to ammonia nitrogen and total nitrogen keeps increasing, which shows benthos promotes the removal of ammonia nitrogen and total nitrogen. By comparing the removal rate of four indicators in constructed wetlands III and IV, green sand plays a big role in adsorbing phosphorus, COD, and nitrogen. Given the same matrix conditions, adding benthos may improve constructed wetlands' removal rate to TP, COD, ammonia nitrogen, and TN, and a positive correlation is shown. Given the same benthos conditions, green sand in the matrix portrays a strong capacity in adsorbing TP, COD, ammonia nitrogen, and TN.

ACKNOWLEDGMENTS

This research was supported by The Natural Science Foundation of the Jiangsu Higher Education Institutions of China (21KJD560006).

REFERENCES

Bai Junhong, Ye Xiaofei, Hu Xingyun, et al. Adsorption kinetic characteristics of typical reed constructed wetland soil phosphorus in the Yellow River Estuary [J]. *Journal of Natural Resources*, 2019, 34(12): 2580–2587.

Ding Dexin, Tan Guozhi, Zeng Xiaona, Ma Jing, Zhang Yue, Zhang Hui, & Hu Nan. Comparison of remediation effects of three plant-constructed wetlands on uranium in impregnated water from uranium tailings pond [J]. *Industrial Water Treatment*, 2022, 42(01):126–142.

Editorial Board for *Water and wastewater monitoring and analysis method*, the State Environmental Protection Administration. *Water and wastewater monitoring and analysis method: 4th Edition*. China Environmental Science Press, 2002.

Franck M Z, Jean-Marie P O, Lacina C. Vegetation effect upon macroinvertebrate communities in a vertical-flow constructed wetland treating domestic wastewater [J]. *International Journal of Advanced Research* (IJAR), 2018, 6(7):1027-1042.

Fu Yao, Chen Fanli, Jiang Wenqiang, et al. Research on the influence of dissolved oxygen and temperature on the removal of ammonia nitrogen and total phosphorus from micro-polluted water by subsurface flow constructed wetland [J]. *Shandong Chemical Industry*, 2021, 50 (15): 259–262.

Guo Ningning, Li Xi, Zhou Xunjun, et al. Characteristics of zoobenthos community in constructed wetlands in subtropical hilly area [J]. *China Environmental Science*, 2021, 41(02): 930–940.

Kang Yan. *Study on mechanisms of enhanced pollutants removal in constructed wetland added with typical benthic fauna* [D]. Shangdong University, 2019.

Liu Jingyi, Sun Yaosheng & Mo Qiang et al. The application of constructed wetland technologies in industrial wastewater treatment [J]. *Water & Wastewater Engineering*, 2021, 57(S1):509–516.

Parde D, Patwa A, Amol Shukla A, et al. A review of constructed wetland on type, treatment and technology of wastewater J. *Environmental Technology & Innovation*, 2021, 21:101261.

Su Bei, Xu Defu, Wei Chiyuan, et al. Plant growth characteristics and purification capacity in different constructed wetland systems [J]. *Journal of Nanjing University of Information Science and Technology*: Natural Science Edition, 2015, 7 (03): 247–253.

Wan Jinbao, Lan Xinyi, Tang Aiping, & Liu Feng. Combined process of multi-surface flow constructed wetland and oxidation pond for treatment of micro-polluted raw water [J]. *China Water & Wastewater*, 2011, 27(21):11-14.

Wang Yunchen, Cheng Fangkui, Wang Siyu et al. Research advances in influencing mechanism of constructed wetland plants on nitrogen removal [J]. *Water Purification Technology*, 2021, 40 (06):21–27.

Xiao Zhuowen. *Design method and application of IBR integrated biochemical pool in Shiyan Guanyin wastewater treatment plant* [D]. Hubei: China Three Gorges University, 2016.

Xu Defu, Li Yingxue, Fang Hua, Zhao Xiaoli, & Wu Fangfang. Earthworm's optimizing function in constructed wetland system [J]. *Journal of Nanjing University of Information Science and Technology*: Natural Science Edition, 2010, 2(03):242–247.

Yang Xiaotong. *Study on denitrification mechanism of the constructed wetland system under zoobenthos influence* [D]. Shangdong Jianzhu University, 2021.

Zhang Ling. *Experimental study on treatment technologies of concentrated domestic wastewater in rural areas and communities* [D]. Chang'an University, 2017.

Zhang Yu, Xu Defu, Li Yingxue, Li Huili, & Guan Yidong. Effect of two kinds of animals on phosphorus forms in substrate of integrated vertical flow constructed wetland [J]. *Chinese Journal of Environmental Engineering*: Natural Sciences Edition, 2015, 7 (03):247–253.

Zhou Riyu, Zhang Hongtao, Zhou Mingluo, et al. The purification of rural wastewater by vertical flow constructed wetlands and microbiome [J]. *Journal of Safety and Environment*: 1–10. DOI: 10.13637/j.issn. 1009-6094.2021.12.29.

Zhu Shijiang, Li Kaikai, Zhou Mingluo, et al. Experimental study on ammonia nitrogen tolerance of several common aquatic plants in constructed wetlands [J]. *Yangtze River*, 2022, 53(05):94–100.

*Advances in Energy, Environment and
Chemical Engineering – Abdullah & Osman (Eds)
© 2023 The Author(s), ISBN: 978-1-032-36083-6*

Risk assessment of geological hazards in Southern Shaanxi based on machine learning

Xiuwu Zhou*
School of Highway, Chang'an University, Xi'an, Shaanxi, China

Lifei Qiu
Shaanxi Nuclear Industry Engineering Survey Institute Co., Ltd., Xi'an, Shaanxi, China

Song Dong
Baihe County Natural Resources Bureau, Ankang, Shaanxi, China

Haochi Zhang
School of Highway, Chang'an University, Xi'an, Shaanxi, China

Zhengxia Qiu
College of Mathematics and Statistics, Guizhou University, Guizhou, China

ABSTRACT: Due to the complex geological conditions and frequent disasters in southern Shaanxi, the regional geological disaster risk assessment is of great significance for disaster prevention and control in southern Shaanxi. Taking southern Shaanxi as the research object, this paper selects 13 factors such as topography, land use, and vegetation cover as evaluation factors, and uses three machine learning models of BPNP, KLR, and LMT to evaluate the hazard of disasters in southern Shaanxi. Vulnerability is obtained based on AHP. Using the risk calculation model proposed by the United Nations, the vulnerability and hazard are superimposed to calculate the risk zoning map. The results show that the dense road networks and river valley population aggregation areas have a high vulnerability, the area of low-risk area is about 59438.931 km^2, the medium-risk area is about 7915.934 km^2, and the high-risk area is about 2567.653 km^2, and the extremely high-risk area is about 102.456 km^2.

1 INTRODUCTION

Southern Shaanxi is located in the Qinba Mountains, with complex topography, steep peaks, deep valleys, and large elevation differences. The lithology is weak and complex. There are many geological tectonic activities and strong dynamic action. Rainfall intensity is relatively large. Since the 1990 s, engineering activities have been frequent, resulting in a sharp increase in disaster points, which poses a huge threat to the safety of lives and property of residents. Geological disasters have become a major factor restricting the development of southern Shaanxi (Tang 2020). Research on the risk of a single slope has been unable to meet the needs. It is necessary to zoning the landslide in southern Shaanxi to guide the local government's prevention and control work. This paper hopes to evaluate the risk of southern Shaanxi, draw the risk zoning of southern Shaanxi, and provide guidance for the prevention and control of southern Shaanxi.

For the hazard evaluation of disasters, the evaluation models mainly include logistic regression (Lee 2010; Patriche 2016), analytic hierarchy process (Li 2016; Liu 2014), information content

---

*Corresponding Author: 1063235476@qq.com

 DOI 10.1201/9781003330165-101

mode (Zhang 2018), frequency ratio model (Li 2020; Qi 2017;, certainty coefficient (Zhao 2020), evidence weight model (Hu 2020; Yan 2021), artificial neural network (Sadighi 2020), support vector machine (Luo 2019; Yu 2020), random forest (Wu 2021), and multiple methods coupling model (Tan 2018). This paper will comprehensively consider the geological environment and inducing factors, select 13 factors such as topography, vegetation coverage, seismic activity, and human engineering activities, and use BP neural network (BPNN), Kernel logistic regression (KLR), and logical model tree (LMT) to evaluate the risk.

2 SUSCEPTIBILITY EVALUATION MODEL

2.1 *BP Neural Network (BPNN)*

BPNN is a training model widely used at present, and it is a multi-layer feedforward neural network containing an input layer, an output layer, and one or more hidden layers. It can get the corresponding weight of each evaluation factor through the training of landslide samples, which is a simple and practical method in regional landslide evaluation.

2.2 *Kernel Logical Regression (KLR)*

As a classical linear classification algorithm, the limitation of logistic regression is that it cannot be applied to nonlinear data sets. At present, it is common to combine logistic regression algorithm with kernel method to make nonlinear data structured and linearly separable in high dimensional space by mapping nonlinear data to high dimensional space.

2.3 *Logical Model Tree (LMT)*

LMT combines two common machine learning models of logical regression and decision tree. The decision tree is widely used and has good performance in solving classification problems. It can also be applied to machine learning problems of probability estimation, clustering, regression, and ranking. A simple decision tree algorithm usually uses a greedy algorithm to construct recursively from top to bottom. The purpose of the decision tree algorithm is to complete the partition of the subset based on the selected features to improve the purity of the divided subset.

3 EVALUATION PROCESS

3.1 *Selection of evaluation factors*

This section is combined with the geological conditions of the study area and literature. It is concluded that the geological disasters in southern Shaanxi are closely related to topography lithology, and structure. This paper selects 13 factors such as topography, vegetation cover, geological structure, seismic activity, and human engineering activities to analyze the susceptibility (Table 1).

3.2 *Correlation analysis of evaluation factors*

To test the correlation between each evaluation factor and disaster, this paper analyzes the multi-index collinearity of the index system and calculates the variance expansion factor by multivariate statistical analysis method. When the threshold is less than 10, the collinearity is acceptable. The calculation results are shown in Table 2.

It can be seen from the table that the values of each VIF are less than the threshold of 10, indicating that the collinearity is acceptable.

Table 1.   Evaluation factors.

Environmental conditions		Inducing factors	
Terrain conditions	Elevation	Human activities	Land use
Geological conditions	Slope Slope direction Terrain classification Lithology Fault buffer zone Soil texture	Rainfalls	Road buffer zone Rainfalls Soil moisture
Vegetation	Vegetation cover	Earthquakes	Peak seismic acceleration

Table 2.   Value of VIF.

Factor	Tolerance	VIF	Factors	Tolerance	VIF
Elevation	0.866	1.129	Rainfalls	0.336	2.980
Slope	0.652	1.534	Road buffer zone	0.951	1.051
Slope direction	0.334	2.996	Earthquake	0.532	1.879
Soil moisture	0.572	1.748	Terrain classification	0.398	2.513
Vegetation cover	0.409	2.447	Soil texture	0.826	1.210
Land use	0.362	2.764	Fault buffer zone	0.419	2.388
Soil texture	0.870	1.149			

3.3   Classification of evaluation factors

The evaluation factors in the survey area were classified by the natural breakpoint method. Combined with the development characteristics of geological hazard points, each factor is classified Some categories are shown in Figure 1

3.4   Hazard evaluation

In this paper, BPNN, KLR, and LMT are used to study the hazard assessment of southern Shaanxi.

Three machine learning models are built using Python Kersa libraries, ScipPy libraries, and Scikit-learn libraries to train with training sets and validation sets. The Python Gdal library was used to normalize the evaluation factors, input the model completed by training, and carry out hazard zoning according to the density of the predicted disaster. The hazard zoning was divided into four levels: low risk, medium risk, high risk, and extremely high risk (from density 0 to density maximum). The results of hazard zoning were imported into QGIS software and plotted. According to the accuracy, sensitivity, and specificity of the three models, the prediction accuracy of KLR is better.

It can be seen from Figure 2 that the risk areas in Hanzhong and Shangluo are relatively small and scattered, while the risk areas in Ankang are relatively large and concentrated. The extremely high-risk areas are mainly concentrated in Ziyang County, Shiquan County, Shanyang County, and Nanzheng District. High-risk areas are mainly around extremely high-risk areas and are also distributed in Xunyang City, Zhenping County, and Baihe County.

3.5   Vulnerability evaluation

Vulnerability refers to the severity of the geological disaster that the bearing body may be damaged by geological disasters in the region.

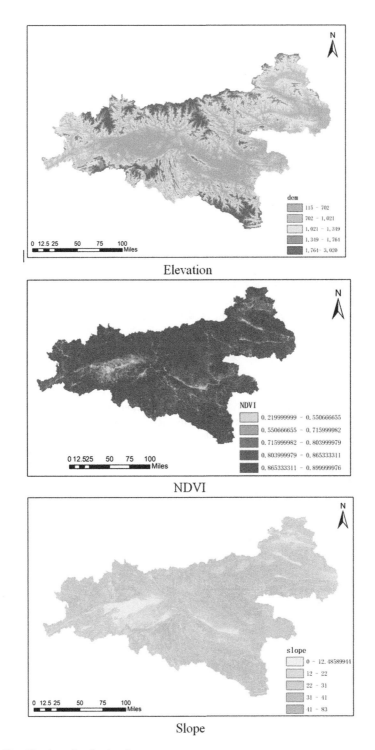

Figure 1.   Classification of evaluation factors.

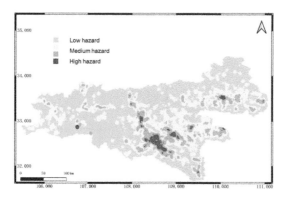

Figure 2.   Zoning results of the KLR model.

The factor calculation is programmed by QGIS and Python. The night light filter is constructed to downscale the economic and demographic factors to improve the spatial resolution of data. The AHP model is used to assign weights according to the vulnerability characteristics such as the value and structure of the index, and the consistency test is carried out. After determining that there is no contradiction in the weight of the model, the vulnerability is calculated, and finally, the vulnerability zoning map is drawn. The calculation results show that the dense road networks and river valley population aggregation areas have high vulnerability.

3.6   *Risk assessment*

The risk assessment of geological disasters is the evaluation of the property losses caused by geological disasters and the expected value of life and casualty of residents in a certain period. The contents include the vulnerability of geological hazard-bearing bodies and risk assessment of disasters. The risk evaluation of this study is calculated using the expression "risk score = risk score × vulnerability score". This calculation was proposed by the United Nations in 1992 to calculate the risk. Firstly, QGIS is used to cut the above zoning data, and the Python GDAL is used to traverse the grid to calculate the risk score. According to the grid score, zoning is carried out, and the region is divided into extremely high, high, medium, and low-risk levels. Finally, QGIS is used to draw the results.

According to the final zoning results, low-risk areas cover an area of approximately 59438.931 km^2, medium-risk areas cover an area of approximately 7915.934 km^2, high-risk areas cover an

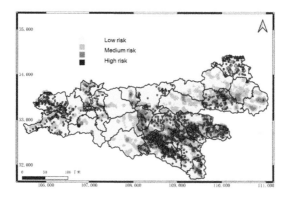

Figure 3.   Risk zoning map.

area of approximately 2567.653 km^2, and extremely high-risk areas cover an area of approximately 102.456 km^2. Lueyang County, Ningqiang County, and Liuba County in Hanzhong City have medium and low-risk areas. The north of Ziyang County in Ankang is a high-risk area, and the middle and low-risk areas in Shiquan County, Zhenping County, and Hanyin County are widely distributed. Shangluo area is mainly the south with low-risk regional distribution, including Shanyang County, Shangnan County, Danfeng County, and Zhen'an County. The results are shown in Figure 3.

4 CONCLUSIONS

The main conclusions are as follows:

(1) Three machine learning models can effectively evaluate the risk of geological disasters in the study area, and the KLR model has the best evaluation effect. The risk zoning map shows that geological disasters occur frequently in Ziyang County and Hanbin District of Ankang City, with a large area of medium and high-risk areas. There are also high-risk areas in Shanyang County and Nanzheng District of Hanzhong City in the Luonan area.
(2) The vulnerability score is calculated based on the AHP model and the zoning map is drawn. The calculation results show that the dense road networks and river valley population aggregation areas have high vulnerability. The area of middle and high vulnerability areas in Hanzhong City is large. The areas along the Hanjiang River are mostly high vulnerability areas.
(3) By adding vulnerability and risk scores, the distribution of risk areas can be obtained. The results show that the area of low-risk area is about 59438.931 km^2, the medium-risk area is about 7915.934 km^2, the high-risk area is about 2567.653 km^2, and the extremely high-risk area is about 102.456 km^2.

In the vulnerability assessment, it is difficult to obtain the data of large facilities such as mines and hydropower stations. In the next vulnerability study, the corresponding factors can be added.

The risk evaluation model used in this paper is simple, and the evaluation results lack credibility. In the follow-up work, it is suggested to use a more complex evaluation model to get a more detailed risk zoning evaluation.

REFERENCES

C Yu, J Chen. (2020) Landslide Susceptibility Mapping Using the Slope Unit for Southeastern Helong City, Jilin Province, China: A Comparison of ANN and SVM [J]. *Symmetry* 12 (6): 1 047.

J. F. Tang, 2020. *Research on kernel logic regression algorithm and its application* [D]. Chongqing University of Posts and Telecommunications.

L. N. Liu, C. Xu, X. W. Xu, etc. (2014), The landslide risk assessment of Lushan earthquake area in 2013 based on AHP method supported by GIS [J]. *Disastery*, 29 (4): 183-191.

Patriche CV, Pirnau R, Grozavu A, et al. (2016), A comparative analysis of binary logistic regression and analytical hierarchy process for landslide susceptibility assessment in the Dobrov River Basin. Romania [J]. *Pedosphere*, 26 (3): 335–350.

Qi X, B. L. Huang, G. N. Liu, etc. (2017), Evaluation of landslide sensitivity of Zigui syncline basin in the Three Gorges area based on GIS technology and frequency ratio model [J]. *Journal of Geomechanics*, 23 (1): 97–104.

R. Z. Wu, X. D. Hu, H. B. Mei, etc. (2021), Spatial susceptibility evaluation of landslides based on random forests: taking Hubei section of the Three Gorges Reservoir area as an example [J]. *Geosciences*, 46 (1): 321–330.

S. H. Tan, J. J. Zhao, L. Yang., etc. (2018), Research on landslide susceptibility based on GIS and information quantity-rapid clustering model – Take Fugong County of Yunnan Province as an example [J]. *Journal of Yunnan University (Natural Science Edition)*, 40 (6): 1148–1158.

S. T. Lee, T. T. Yu, W. F. Peng, et al. (2010) Incorporating the effect of terrain amplification in the analysis of earthquake-induced landslide hazards using logistic regression [J]. *Natural Hazards and Earth System Sciences* 10 (12): 2 475–2 488.

Sadighi M. Motamedvaziri B Ahmadi H, et al. (2020) Assessing landslide susceptibility using machine learning models: a comparisonbetween ANN, ANFIS, and ANFIS-ICA [J]. *Environmental Earth Sciences* 79 (24): 1–14.

W. Y. Li, X. L. Wang. (2020), Application and comparison of frequency ratio and information model in landslide susceptibility evaluation in loess gully region [J]. *Journal of Natural Disasters*, 29 (4): 213–220.

X. Luo, F Lin, S Zhu, et al. (2019), Mine landslide susceptibility assessment using IVM, ANN and SVM models considering the contribution of affecting factors [J]. *PLoS ONE* 14 (4): 1–18.

X. Zhao, W Chen. (2020), GIS-based evaluation of landslide susceptibility models using certainty factors and functional trees-based ensemble techniques [J]. *Applied Sciences* 10 (1): 16.

X. Y. Zhang, C. S. Zhang, H. J. Meng, etc. (2018), Landslide susceptibility evaluation of Beijing-Zhangjiakou high-speed railway based on GIS and information model [J]. *Journal of Geomechanics*, 24 (1): 96–105.

Y. Hu, D. Y. Li, S. Meng, etc. (2020), Susceptibility evaluation of landslide disaster in Badong County based on evidence right method [J]. *Geological science and technology bulletin*, 39 (3): 187–194.

Y. Q. Yan, Z. H. Yang, X. J. Zhang, etc. (2021), Evaluation of landslide susceptibility of Batang fault zone in eastern Tibetan Plateau based on weighted evidence weight model [J]. *Modern geology,* 35 (1): 26–37.

Y. T. Li, H. L. Zhu, S. H. Chen. (2016), Evaluation of landslide susceptibility in the upper reaches of the Yellow River by analytic hierarchy process [J]. *Surveying and mapping science*, 41 (8): 67–70 + 75.

*Advances in Energy, Environment and
Chemical Engineering – Abdullah & Osman (Eds)
© 2023 The Author(s), ISBN: 978-1-032-36083-6*

Research on the technique of reflected in-seam wave detection in coal mine structure exploration

Kun Chen*

Xi'an Research Institute of China Coal Technology Engineering Group, Shaanxi, China

ABSTRACT: The premise of seismic exploration lies in the difference in wave impedance (velocity × density) in the medium. As a special medium, the elastic properties of coal seam are significantly different from its roof and floor. A coal seam is usually a weak intercalated layer with low velocity and density in the coal measure strata. Therefore, the coal-rock interface is a strong reflection interface, which is conducive to seismic wave detection related to coal seam and coal seam structure. When the underground coal seam runs into geological anomalies such as goaf, fault, collapse column and magmatic intrusion, the continuity of the coal body can be destroyed laterally, and its integrity can be weakened due to stress, thus showing different density and elastic wave velocity at the coal rock interface. This provides a basis for seismic exploration. Taking the detection of the reflection channel wave in the goaf of the old kiln as an example, as the goaf is characterized by a cavity or caving zone, reflection occurs when the seismic wave propagates from the coal seam to the boundary of the goaf. The location of the goaf can be analyzed following the time-frequency characteristics of the reflection wave received. This paper adopted the new generation of high-speed and high-resolution mine seismographs to detect the reflection channel waves in a coal face in Shanxi Province. Using the collected data, the location of the goaf was determined, and this location was verified by the subsequent face drive.

1 INTRODUCTION

Mine water disasters are a significant factor restricting the safe and efficient production of coal mines, among which goaf water and abandoned old kiln water have become the main source of water bursting (Hu 2018). In recent years, millions of small coal mines have been integrated or closed. Due to the lack of relevant technical data in original mines and the absence of systematic technical regulations and engineering protection measures during the abandonment and closure of mines, the distribution of mining space and hydrogeological conditions of the abandoned mines is unknown, leading to the frequent exposure of old kiln water in the later mining. Characterized by irregular distribution, large instantaneous water inflow, and strong suddenness, old kiln water could easily lead to flooding and casualties. According to statistics, the water filling of small coal mines due to goaf and abandoned water accounts for 80% of the total water bursting in coal mines every year. So, it is of great significance to find out the distribution location and water abundance of old kilns to prevent water bursting and ensure efficient production in coal mines (Chang 2017; Kang 2018). Geophysical exploration techniques play a crucial role in the detection of old kiln goaf, for instance, 3D seismic, high-density electrical method, mine transient electromagnetic, seismic reflection method, geological radar, and direct current advanced detection have achieved good application results in some areas (Chen et al. 2018; Chen & Xue 2013; Fan 2017; Wang et al. 2018; Zhan et al. 2010). The results of numerous studies and practices demonstrate that roadway seismic advanced detection technology has certain advantages in detecting geological abnormal structures in front of a road, as it is close to the abnormality, will not be affected by ground conditions, and will be less affected by road environment (Chen 2016; Zhang 2009).

*Corresponding Author: ck4106@126.com

2 DETECTION PRINCIPLES

2.1 *Principles of head-on seismic advanced detection*

In combination with roadway characteristics, the advanced detection was designed as an observation system to detect the geological and hydrogeological conditions in front of the roadway by arranging the sources and sensors along the rear of the roadway, as shown in Figure 1. The seismic wave was usually derived from small blasting at a specific location. In general, the shot points were arranged straightly along both sides of the roadway. In this way, a range of slight seismic sources in the regular arrangement was artificially made to form a seismic section. The seismic waves from these sources could generate reflection waves when encountering bad interfaces, such as formation layer, joint surface, especially fractured interface and goaf, karst cave, underground river, karst collapse pillar, and ooze zone. Different from the ground reflective seismic exploration based on horizontal or low dip reflection interface, the reflection interface in front of the roadway had a vertical or high dip spatial relationship with the seismic line, showing a unique time-distance feature of negative apparent velocity. To obtain the reflection wave information in front of the roadway from seismic records, the up and down traveling waves are separated and the traveling waves (negative apparent velocity) are retained in the information processing. In essence, this is to suppress the information from the vertical survey line to retain horizontal reflection information.

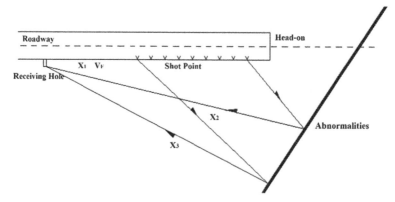

Figure 1.   Principle of head-on seismic advanced detection.

2.2 *Principle of side-roadway reflection seismic detection*

Concerning the geological conditions of the side-roadway, the two-dimensional migration was generally carried out by arranging sources and detectors along the side-roadway-roadways and using the reflection wave method, as shown in Figure 2. The location and extension of the reflection interface could be obtained by using this method to conduct effective phase contrast and tracking.

Seismic migration imaging refers to the occurrence of a reflection phenomenon on the wave impedance interface satisfying Snell's theorem by using the wave impedance difference of different media. The original seismic records include the arrival information about the spatial position of excitation and receiving points and the reflection phase. In the time domain, the reflection phase in original records cannot directly represent the actual shape of the underground geological body. For example, in the common-shot gather, the reflection phase of the horizontal layer is curved in the time domain. So, the reflection wave needs to be reflected in the reflection interface in the subsequent processing of original records. The effects during propagation must be removed, such as attenuation and diffusion. The seismic records showing correct migration and the actual underground reflection interface shall be obtained finally. Based on geometric seismology and the diffraction theory of waves, ray migration can automatically achieve the spatial migration of reflection and diffraction waves on the computer. According to the law of seismic geometry, if the

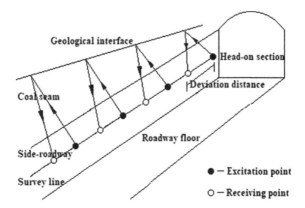

Figure 2. Arrangement diagram of side-roadway survey lines.

input section is measured by a non-zero offset observation system (with an offset section) and the velocity V is constant, the impulse response shall be an ellipse. It takes the reflection point S and the receiving point R as focal points. Its fixed-length, major semi-axis, and minor semi-axis are $V * t, a = V * t/2,$ and $b = \sqrt{a^2 - (l/2)^2}$ respectively, where l is the offset. The elliptic equation is as follows.

$$\frac{(x - x_d)^2}{(\frac{1}{2}V \cdot t)^2} + \frac{Z^2}{(\frac{1}{2}V \cdot t)^2 - (\frac{1}{2}l)^2} = 1$$

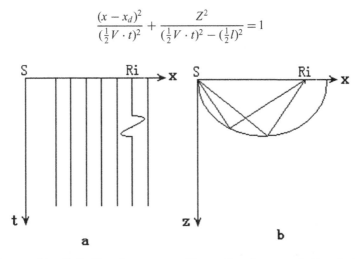

Figure 3. Diagram of the elliptical impulse response. a. Time section of common-shot impulse b. The response section of migration impulse was an ellipse with excitation point S and receiving point R_i as the two focal points.

3 METHOD SELECTION

In accordance with the physical characteristics of the target, the seismic reflection wave method was adopted for this detection. The integrity of the coal seam was destroyed by geological anomalies such as goaf and fault. So, it had different physical characteristics in comparison with raw coal (unmined coal seam), especially in the density of media and the propagation speed of seismic waves. This can provide a physical basis for the seismic wave exploration method. Seismic wave reflection could take place when encountering inhomogeneous geological bodies (with the difference in wave impedance) during seismic wave propagation. By analyzing the kinematics and dynamic characteristics of reflection waves, the position and distance of geological anomalies in the survey area could be obtained. The KDZ3113 mine seismograph, developed by Xi'an Research Institute Ltd of China Coal Technology Engineering Group, was selected as the seismic detecting instrument. This product is an intrinsically safe and portable device for mineral work. Its system is mainly composed of Strata Visor NZ 24 channel explosion-proof acquisition station, seismic central control station, power supply unit, seismic detector, and explosion synchronous trigger device. The main parameters for the on-site collection are as follows. Channels: 24; sampling interval: 0.125 ms; sampling length: 2 s; advanced sampling points.

4 APPLICATION EXAMPLE

4.1 Survey line layout and workload

The data collection was carried out from November 15th to 16th, 2021, near the head-on of the belt roadway (P28+35 m) in the second mining area. The vibration shot was applied as the source for roadway head-on advanced detection and side-roadway detection, using 24-channel seismographs to conduct data collection. The detectors were arranged from front to back with a channel distance of 10 m, and a total of 35 holes were designed, including 21 holes for head-on advanced detection, as shown in P1-P21 (red) in Figure 4 and the space was 2 m; 14 holes for side-roadway detection, as shown in P22-P35 (blue) in Figure 4 and the space was 10 m.

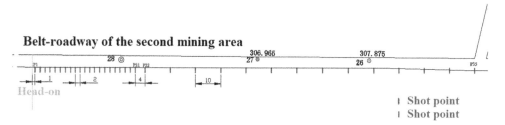

Figure 4. Layout diagram of seismic detecting holes.

4.2 Construction specifications

In this project, the seismic wave advanced detection technology and side-roadway reflection wave method were used to investigate the geological anomalies in the front and sides of the roadway driven. Both the head-on seismic wave advanced detection and the side-roadway advanced detection took explosives in holes as the source. In line with the observation system of seismic wave detection and underground conditions, the following requirements were brought forward for hole construction and blasting based on safe production measures in coal mines. About 100 g of latex explosive is loaded into each hole at a depth of 1.5 m.

4.3 Evaluation of data quality

A total of 35 shots for construction were detected, of which the instrument was not triggered at the excitation of the 34th shot. A total of 34 effective shots, and 816 effective physical points. Two water pumps were used for drainage at the construction site, which had a significant impact on the nearby detectors. In the meanwhile, the blasting caps used for construction were not the same batch (with different delays), leading to differences in the arrival time of seismic data and increasing the difficulty of data processing. The data quality met the exploration requirements overall.

5 DATA PROCESSING AND INTERPRETATION

5.1 *Head-on advanced detection*

The geophysical data collected on-site was transformed into usable physical maps after processing. The head-on seismic wave advanced detection data was processed on the MSP2.0, a self-developed software platform. The specific process was as follows: data preprocessing — spectral analysis — obtaining direct wave — obtaining reflection wave — velocity analysis — depth migration — extracting interface.

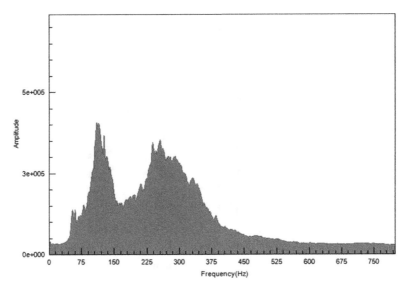

Figure 5.   Spectrum of the seismic record.

In terms of spectral analysis, the FFT tool was used for the frequency-domain analysis of seismic waveforms collected in the roadway. The purpose was to know the main frequency distribution of seismic waves in the roadway in the condition of an explosive source, which was conducive to guiding the subsequent data processing. Figure 5 was the spectral analysis diagram of seismic waveforms, showing that the main frequency range was 60–450 Hz.

Figure 6 shows the pickup of seismic waveforms collected and direct wave velocity. The recorded original offset and the first arrival time were used to simulate the first-arrival straight line of direct waves. As shown in the figure, the direct P-wave velocity was 2.7 m/ms (green) and the direct S-wave velocity was 1.56 m/ms (blue). With the direct wave velocity, the velocity range of the detected area could be determined and used as the velocity background value.

As a core part of MSP processing, depth migration is to migrate the reflected energy from the front medium to a spatial point in the condition of a given velocity model. Based on this resulting map, the reflection interface in front of the roadway was extracted. The detection range was long due to the use of an explosive source, about 200 m overall, including 63.05 m exposed area and 136.95 m unexposed area.

Figure 7 shows the imaging results of depth migration of seismic P-wave and S-wave in front of the roadway. The depth migration section showed the spatial position relation between interfaces with an elastic difference in front of the roadway. In the figure, the amplitudes of reflection waves were expressed in different colors. The darker the warm color was, the stronger the reflection wave energy was. As shown by the section, there were significant reflection phases within 100 m from the head-on to the front of the roadway, among which two groups of reflection phases showed strong

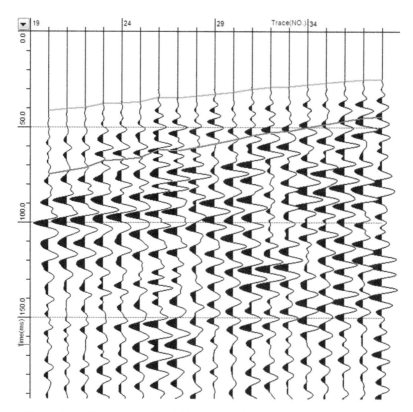

Figure 6. Pickup of waveforms measured and direct wave velocity.

energy. The head-on of the roadway was at X = 63.05 m, mainly explaining the strong reflection interface at the following positions:

The section showed the two groups of reflection interface with strong energy, that is, R1 and R2 marked in the figure. Figure 7 shows the position of the roadway as shown in the black box, and the position of the reflection interface relative to the roadway face (P28+35 m). There was a strong wave impedance reflection interface (R1) within 32–36 m, and a strong wave impedance reflection interface (R2) within 50–66 m in front of the roadway head-on respectively. Combined with the known geological data and the characteristics of borehole water, it was speculated that the anomaly was the roadway abandoned by an old kiln.

5.2 Side-roadway seismic detection

During the data processing of side-roadway detection, the influences of surface waves, roadway sound waves, and other interference factors (Figure 8) were eliminated. In addition, energy compensation was conducted on the seismic signals from the deep layer due to the quick energy decrement of seismic waves. Therefore, the data preprocessing should be conducted before depth migration. The processing was as follows: coordinate setting → delay correction → spectral analysis → bandpass filtering → AGC → inter-roadway balance → velocity analysis → depth migration. Figure 9 shows the result of side-roadway depth migration, of which the origin of coordinates was at the head-on section.

Figure 9 is the section of side-roadway depth migration, showing a northwest geological anomaly between the coordinates (30, 30) and (105, 100) on the left side of the roadway, with an included angle of about 45 degrees with the heading direction of the belt-roadway in the second mining area. It was speculated that the anomaly was caused by the faults and caving zones. As the roadway

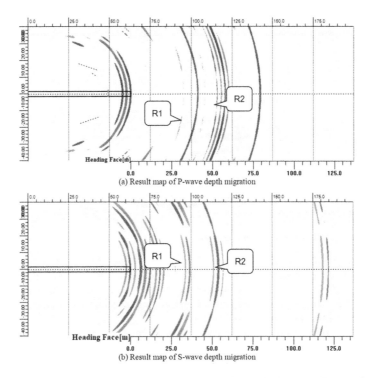

(a) Result map of P-wave depth migration

(b) Result map of S-wave depth migration

Figure 7.   Result map of depth migration of head-on advanced detection. (a) Result map of P-wave depth migration, (b) Result map of S-wave depth migration.

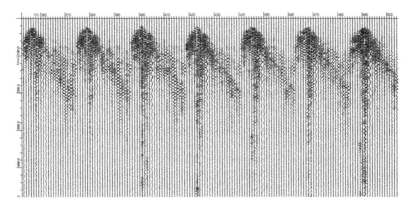

Figure 8.   Original seismic records in the detected side-roadway.

position of the second mining area was designed to pass through this abnormal area, it was suggested to reinforce the hydrogeological observation and conduct further advanced detection in this area before roadway construction.

6   CONCLUSIONS

During the mine seismic detection in the belt-roadway of the second mining area, the actual shots were 34 and a total of 816 physical points were collected. The data quality met the exploration requirements. The following conclusions were obtained through data processing and analysis.

709

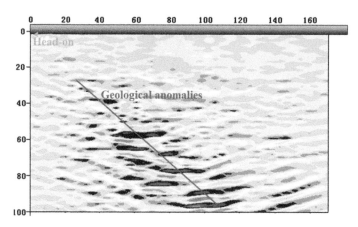

Figure 9.   Section of side-roadway depth migration.

(1) There were two strong wave impedance reflection interfaces at 32–36 m (R1), and 50–66 m (R2) in front of the belt-roadway head-on of the second mining area. Combined with the known geological data and the characteristics of borehole water, it was speculated that the anomaly was the roadway abandoned by an old kiln.
(2) There was a northwest geological anomaly between the coordinates (30, 30) and (105, 100) on the left side of the roadway, with an included angle of about 45 degrees with the heading direction of the belt-roadway in the second mining area. It was speculated that the anomaly was caused by faults and fractured zones.

REFERENCES

Chang J. (2017). *Three-dimensional Numerical Simulation and Application of Mine Transient Electromagnetic Responses of Water-rich Area in Coal Mine*. China University of Mining and Technology.
Chen D. (2016). *Study on Wave Field Characteristics of Drillhole Transient Electromagnetic in Mine Roadway Whole Space Using Numerical Simulation*. China University of Mining & Technology (Beijing).
Chen D., Cheng J. & Wang A. (2018). Numerical Simulation of Drillhole Transient Electromagnetic Response in Mine Roadway Whole Space Using Integral Equation Method [J]. *Chinese Journal of Geophysics*. 61 (10), 4182–4193.
Chen W. & Xue G. (2013). Application on Coal-Mine Voids Detection with Multi-device TEM Technology. *Progress in Geophysics*. 28 (5), 2709–2717.
Fan T. (2017). Experimental Study on the Exploration of Coal Mine Goaf by Dynamic Source and Fixed Reception Roadway-Borehole TEM Detection Method. *Journal of China Coal Society*. 12, 3229–3238.
Hu Y. (2018). Three Dimensional Detection Method of Mine Transient Electromagnetic and Application. *Coal Technology*. 37 (7), 202–204.
Kang J. (2018). Detection of Mine Aquifer and Prevention of Water Disaster based on TEM Method. *China Coal*. 44 (5), 32–35.
Wang Y., Guo G. & Li X. (2018). Application Research of YCS200 Transient Electromagnetic Instrument in Detection of Frontal Rich Water. *Coal Technology*. 37 (10), 170–172.
Zhan W., Wang Q. & Niu X. (2010). Transient Electromagnetic Exploration Technology to Mining Goaf. *Coal Science And Technology*. 38 (8), 115–117.
Zhang J. (2009). *Analysis of Anomalous Character and Study Vector Intersection for Downhole Transient Electromagnetic Method*. China University of Geosciences (Beijing).

Advances in Energy, Environment and Chemical Engineering – Abdullah & Osman (Eds)
© 2023 The Author(s), ISBN: 978-1-032-36083-6

Evaluation index design of Saihanba forest farm

Ruhan Wei* & ShiChen Zhou
Inner Mongolia University, Hohhot, China

ABSTRACT: Located in Weichang Manchu and Mongolian Autonomous County, Chengde City, Hebei Province, Saihanba Ecological Reserve is located at the southeast edge of The Inner Mongolia Plateau and the intersection of the Inner Mongolia Plateau and the mountainous areas of northern Hebei Province. For more than 60 years, the builders of Saihanba Ecological Reserve have worked hard to turn the former desert into a forest. Saihanba Ecological Reserve is the pioneer of ecosystem restoration in China, and its ecological construction has achieved remarkable results. Its great changes show that clear waters and green mountains are invaluable assets. Therefore, a new ecological environment assessment model for Saihanba Ecological Reserve was established in this paper to provide help for related research. This paper establishes an evaluation system by looking for the main evaluation indexes. The main components affecting the forest environment in Saihanba Ecological Reserve were determined by principal component analysis and the forest environment restoration was evaluated. The effects of the Saihanba mechanical forest farm on climate improvement in Beijing and Chengde were also discussed, and the effects of the Saihanba mechanical forest farm on carbon neutrality were analyzed. The results showed that the forest ecological restoration improved steadily from 2004 to 2013, and the forest restoration in Saihanba improved the climate of Beijing and Chengde after 2017 and promoted the carbon neutrality of Beijing and Chengde after 2011.

1 INTRODUCTION

Saihanba Mechanical Forest farm has the largest area of artificial forest in the world. Through years of afforestation, the ecological environment of Saihanba has been greatly improved. As a model of ecological restoration, the Ecological model of Saihanba has also been promoted across China, helping to achieve the goal of "carbon neutrality" and "carbon peak", and contributing to the control of wind-blown sand in The Beijing-Tianjin-Hebei region of China (Wang 2015).

Forest has the function of water and soil conservation, water conservation, and other ecological environment improvements, can provide ecosystem services, and is the basis of sustainable development. Due to the important role of the Saihanba forest, many scholars have evaluated the forest ecological environment of Saihanba at different levels. For example, Zhang Jiandong (2022) and others evaluated the ecological functions of main forest types in Saihanba Forest Farm and found that the level of forest functions in Saihanba was low. Taking North Mandian Forest Farm in Saihanba as the research area, Yin Hailong (2022) et al. selected eight evaluation factors, including forest stock, forest naturalness, forest community structure, and species community, and found that the ecological function of the forest was significantly improved compared with sandy and wasteland based on the technical regulations of national forest resources planning and design survey. Yi Zechuan (2022) et al. selected 16 evaluation factors and constructed an ecological environmental impact assessment model by using TOPSIS. The results showed that the ecological restoration capacity of Saihanba was enhanced. Xinyue Chen (2022) et al. established the evaluation model

---

*Corresponding Author: Eehd07@163.com

DOI 10.1201/9781003330165-103

of Saihanba and realized the quantitative analysis of the impact of Saihanba restoration on Beijing's sand resistance. Zhenbin Zhang (2021) et al. selected 19 indicators reflecting the ecological environment and established an economic trend evaluation system based on a multi-level fuzzy comprehensive evaluation algorithm. The establishment of forest farms has a positive impact on ecology.

There are few studies on the impacts of forest restoration on climate and carbon emissions in the Saihanba region, so a set of scientific and effective evaluation index systems should be established to evaluate the impacts of forest restoration. Taking Saihanba as the research object, this paper extracted the main components affecting forest ecological restoration based on principal component analysis and evaluated the trend of forest restoration. At the same time, climate assessment and carbon neutrality assessment models were established to evaluate the impacts of Saihanba forest on climate and carbon neutrality in Beijing and Chengde, providing a theoretical basis for forest ecological restoration.

2 THE STUDY AREA AND PREPARATION BEFORE MODELING

2.1 *Overview of the study area*

Saihanba Ecological Reserve is located in Weichang Manchu and Mongolian Autonomous County, Chengde City, Hebei Province, hereinafter referred to as Saihanba, with an altitude of more than 1,000 meters. It is a typical semi-arid and semi-humid continental monsoon climate (Qin 2016; Zhou 2021). According to statistics, 369 young people with an average age below 24 have come to the yellow-sand-covered land since 1962. Through the efforts of several generations, Saihanba has regained its vitality. Today, the forest coverage of Saihanba is as high as 80 percent. With the continuous development of China's economy, ecological and environmental problems are becoming increasingly serious. However, many calls and measures for environmental protection have been put forward one after another, providing possibilities for the development of China's ecological civilization construction. As a successful case of ecological environment improvement and restoration in China, Saihanba has valuable experience.

2.2 *Data sources*

We collected the relevant characteristics of Saihanba Forest Farm, as well as the basic weather characteristics of Beijing, and the spatial and temporal distribution map of Beijing affected by the sandstorm from the Data Center of Resources and Environmental Science, Chinese Academy of Sciences. The meteorological data comes from the China Meteorological Data Network (http://data.cma.cn), including the data on rainfall and temperature from 2000 to 2018 of six meteorological stations in Bashang and surrounding areas, and the annual rainfall. The related features of Saihanba Forest Farm, the basic weather characteristics of Beijing, and the spatial and temporal distribution map of Beijing affected by the dust storms were collected. Meteorological data from the Resources and Environmental Sciences Data Center of the Chinese Academy of Sciences, meteorological data from China Meteorological Information Network (http://data.cma.cn) (including the rainfall and temperature data of six weather stations in 2000-2018, and annual rainfall), and the National Bureau of Statistics.

2.3 *Hypothesis of the model*

As there are many factors involved in the forest environment of Saihanba, it is impossible to list all the factors that affect the forest's ecological environment. Therefore, based on the forest ecosystem, we select a part of them to build the model. At the same time, the impact assessment of climate only considers the impact of the Saihanba forest on the incidence of precipitation and severe weather. The construction of forests in Saihanba has an impact on carbon emissions not only in Chengde but also in Beijing. When considering carbon neutrality, we analyzed two cities, Chengde, and Beijing. However, in addition to Saihanba, other factors in the two regions also affect carbon neutrality, so this paper only uses the Saihanba forest to analyze its impact on carbon neutrality.

2.4 *Description of the symbol*

Table 1. Settings and contents of each variable.

$x_{7,1}$	$x_{7,2}$	$x_{8,1}$	$x_{8,2}$	z_1
The occurrence rate of precipitation weather in Beijing	The occurrence rate of precipitation weather in Beijing	The occurrence rate of precipitation weather in Saihanba	The occurrence rate of severe weather in Sanhaiba	Characteristic indexes of carbon dioxide and oxygen
z_2	z_3	z_4	$z_{4,1}$	$z_{4,2}$
Characteristics of water conservation per unit area	Characteristics of tree accumulation capacity per unit area	Climatic indicator indicator	Beijing climatic	Chengde climatic indicator
FEREI	CII	CE	FECN	
Forest ecological restoration evaluation index	Climate improvement index	Impacts of Saihanba forest on carbon emissions	Forest ecological on carbon neutrality	

3 ESTABLISHMENT AND SOLUTION OF THE MODEL

3.1 *Establishment of forest ecological restoration model in Saihanba*

Saihanba is mainly managed through the construction of forest farms, the role of the forest is crucial. After decades of construction, the forest ecological environment in Saihanba has undergone qualitative changes. However, there are few studies on its evaluation at present. At the same time, the ecological environment restoration model of Saihanba forest was established to better study the function of forest water and soil conservation and climate improvement. Six factors and indicators were selected. Based on the relevant data from 1962 to 2021, the main factors reflecting the ecological environment of Saihanba forest were screened out by principal component analysis, and the ecological restoration effect of Saihanba forest was evaluated.

3.1.1 *The research methods*
Principal component analysis (PCA) is a common multivariate analysis method. Usually, to comprehensively analyze problems, many related factors are put forward in some studies, and each factor reflects the information needed for research to varying degrees. The principal component analysis method is to select a few important factors by a linear transformation of multiple factors, which can reduce the dimension of the data set and maintain the maximum characteristics of the different contributions of the data set. While retaining the main information of the original variables, they are not related to each other, which has advantages over the original variables (Kleshchenko 2022; Liu 2015).

3.1.2 *The establishment of model*
(1) Establish a relevant matrix
    The relevant matrix is obtained by collating the obtained data.

Table 2. Relevant matrix.

		Forest coverage rate	Coverage area (10,000 mu)	Forest stock (10,000 cm²)	Water conservation (100 million cm²)	Oxygen release (10,000 t)	Carbon dioxide absorption (10,000 t)
Correlation	Forest coverage rate	1.000	1.000	0.885	0.760	0.885	0.885
	Coverage area (10,000 mu)	1.000	1.000	0.885	0.760	0.885	0.885
	Forest stock (10,000 cm²)	0.885	0.885	1.000	0.869	1.000	1.000
	Water conservation (100 million cm²)	0.760	0.760	0.869	1.000	0.869	0.869
	Oxygen release (10,000 t)	0.885	0.885	1.000	0.869	1.000	1.000
	Carbon dioxide absorption (10,000 t)	0.885	0.885	1.000	0.869	1.000	1.000

There were significant correlations between forest coverage and forest cover area, forest stock, water conservation, oxygen release, and carbon dioxide absorption. This indicates that there is overlap between the selected data and a strong direct correlation between multiple variables. Therefore, it is feasible to use PCA to find the main components of forest restoration.

(2) Analyze the total variance interpretation and calculate the contribution rate.

Table 3. Total variance interpretation.

Composition	Initial eigenvalues			Extract the sum of squares of loads		
	Total	Percentage of variance	Accumulation (%)	Total	Percentage of variance	Accumulation (%)
1	5.485	91.412	91.412	5.485	91.412	91.412
2	0.348	5.798	97.210			
3	0.167	2.790	100.000			
4	6.505E-16	1.084E-14	100.000			
5	−2.196E-16	−3.660E-15	100.000			
6	−6.252E-16	−1.042E-14	100.000			

The table shows the contributions of the first principal component and the second principal component. As can be seen from the table, the proportion of the first characteristic value is 91.412%, indicating that it has a great impact on the ecological environment of the Saihanba forest. After decades of afforestation in Saihanba, the forest has a great impact on its ecological environment. The second eigenvalue accounted for 5.798%, contributing little. Therefore, a principal component was extracted to meet the requirement of the contribution rate of PCA (>80%).

Table 4. Component matrix.

		Forest coverage rate	Coverage area (10,000 mu)	Forest stock (10,000 cm^2)	Water conservation (100 million cm^2)	Oxygen release (10,000 t)	Carbon dioxide absorption (10,000 t)	One component was extracted
Component	1	0.944	0.944	0.984	0.892	0.984	0.984	

As can be seen from the table, the load of six factors such as forest coverage rate, covered area, and forest stock is relatively high, indicating that the first principal component reflects the information of these indicators. Therefore, extracting one principal component can reflect the information of all indicators, so we decided to use one variable to replace the original six variables.

(3) Standardize the data

Since different dimensions will affect the value of each variable, the effect of dimensions can be eliminated through standardization. The standardized formula is:

$$X_i^* = \frac{X_i - \mu_i}{\sqrt{\sigma_{ii}}}, i = 1, 2, \ldots n,$$

(4) Calculate F to evaluate the environment of Saihanba

Table 5. Component score coefficient matrix.

		Forest coverage rate	Coverage area (10,000 mu)	Forest stock (10,000 cm^2)	Water conservation (100 million cm^2)	Oxygen release (10,000 t)	Carbon dioxide absorption (10,000 t)
Component	1	0.172	0.172	0.179	0.163	0.179	0.179

F is calculated according to the corresponding weight and contribution rate of the first principal component.

Table 6. Table of environment principal components.

Year	2004	2005	2006	2007	2008	2009	2010	2011	2012	2013
F	−1.21	−1.1	−0.94	−0.83	−0.59	0.82	0.86	0.97	1.08	0.94

By analyzing the change curve of the extracted principal components over time, we obtained information related to the environmental restoration of the Saihanba forest. As can be seen from the figure, the overall forest ecological environment in Saihanba is in a trend of continuous improvement. From 1962 to 2000, the forest ecological recovery index increased steadily, indicating that the state began to build Saihanba ecological reserve in 1962 and achieved remarkable results. After 2020, the growth rate of the index accelerated.

3.2 Climate impact model of Saihanba forest in Beijing and Chengde

The correlation between precipitation weather and severe weather is explored.

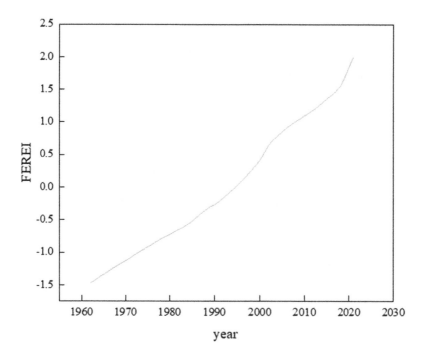

Figure 1.   The trend of forest ecological restoration in Saihanba.

Table 7.   Pearson-related visualization.

	Pearson program	
		Severe weather incidence
Rainy weather incidence	Correlation coefficient	0.286*
	p	0.044

* $p < 0.05$ ** $p < 0.01$

As can be seen from the above table, correlation analysis is used to study the correlation between the proportion of severe weather and the proportion of light rain. The incidence of severe weather and precipitation is at a significant level of 0.044, indicating that there is a significant positive correlation between them.

We selected the severe weather and precipitation weather above to compare the similarity of the climate in Beijing and Saihanba and found that the climate of the two regions was out of balance to some extent. At the same time, combined with correlation analysis, precipitation weather and severe weather in Saihanba could be considered important factors of this model (CII).
z_4:

$$z_{4,1} = x_{7,1} - x_{7,2}$$

$$z_{4,2} = x_{8,1} - x_{8,2}$$

CII:

$$CII = \frac{z_4}{EREI}$$

$x_{8,1}$ and $_{8,2}$ need to be standardized in the calculation process.

By collecting rainy weather incidence data and severe weather incidence data, raw the diagram of CII over time:

716

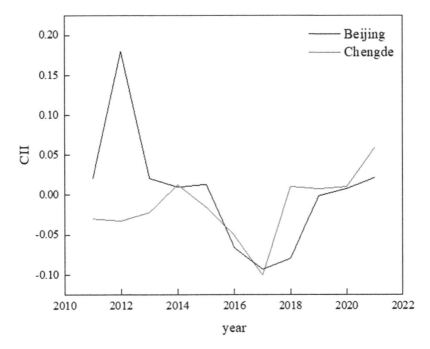

Figure 2. Trends in climate improvement index.

Before 2017, Beijing and Chengde CII showed a trend of decline, proving the woodland ability is crumbling to the improvement of the climate, on the one hand, because both ecological environments after repair to a certain extent, further repair is more difficult; on the other hand, because of the dual resource consumption is increasing, to a certain extent of ecological damage. After 2017, due to the further attention to the construction of this ecological zone, the CIII of this ecological zone for Beijing and Chengde showed an upward trend, and the climate improvement capacity of this ecological zone for both cities was further improved.

3.3 Establish a carbon neutralization model for the Saihanba forest

In the context of the global fight against climate change, China aims to peak its carbon dioxide emissions by 2030 and achieve carbon neutrality by 2060. During the 13th Five-Year Plan period, China's ecological civilization construction will focus on carbon emission reduction, and achieving carbon peak and carbon neutrality will be a profound social and economic change. Therefore, the study of carbon neutralization is of great significance (Su 2021). As a carbon pool, the forest ecosystem in Saihanba is constantly in the process of the carbon cycle. Therefore, it is also important to explore the impact of the Saihanba forest farm on carbon neutralization. As for carbon emission, the data on carbon emission in Beijing and Chengde should be collected and set as CE, and x_5 is the amount of carbon absorbed.

The impact of carbon neutralization on ecosystem FECN is shown below.

$$FECN = CE - x_5$$

According to the graph analysis, the influence curve of Saihanba forest on carbon neutrality in Chengde increased slowly from 2002 to 2005. During the period from 2005 to 2011, the curve was relatively flat, and the impact of the Saihanba forest on carbon neutrality had little change. After 2011, it showed an upward trend, indicating that the Saihanba forest promoted carbon neutrality in Chengde during this period. The carbon neutralization effect of the Saihanba forest on Beijing had

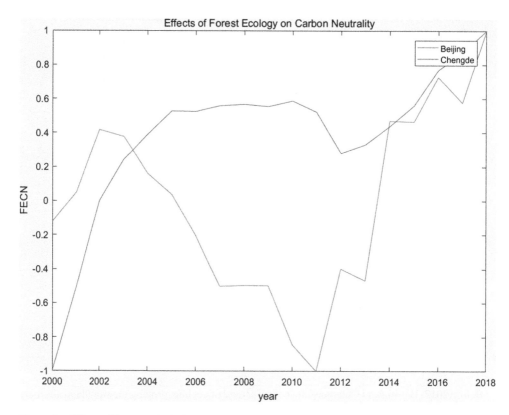

Figure 3.　Effects of forest ecological on carbon neutrality.

a temporary effect from 2000 to 2002, but the curve showed a downward trend from 2002 to 2011. From 2013 to 2018, the curve generally showed an upward trend, and it rose rapidly from 2013 to 2014, indicating that the Saihanba forest played a significant role in promoting carbon neutrality in Beijing. In general, forest ecological restoration in Saihanba has a significant impact on carbon neutrality in Beijing and Chengde.

4　EVALUATION AND IMPROVEMENT OF THE MODEL BASED ON

4.1　*Advantages*

The forest ecological assessment restoration model (FEREI) was established by PCA concerning the ecological environment index and simplified data. Meanwhile, based on FEREI, two characteristics of climate conditions and carbon neutrality were selected to evaluate the impacts of the Saihanba forest on climate and carbon neutrality. For the establishment of a climate improvement model, comparing the climate of Beijing and Saihanba can better analyze the impact of the Saihanba forest on Beijing's climate. The selection of factors should reduce the need for data and simplify the modeling process.

4.2　*Disadvantages*

Due to the limitations of data collection, the period of some data does not match and cannot be collated and compared, leading to certain limitations in model building. The forest ecosystem restoration index only selects several important factors and ignores some other factors, which may affect other indicators. The carbon neutrality assessment model and climate assessment model established in this paper only consider a few factors, there are certain limitations, and there is room for improvement.

5 CONCLUSIONS

(1) To study the changes in forest ecological restoration in Saihanba, the forest ecological restoration evaluation Index (FEREI) was established by PCA based on forest-related factors. The results showed that the forest ecological restoration in Saihanba increased year by year, and the growth rate was relatively stable.

(2) To study the impact of the Saihanba forest on climate, a climate improvement index (CII) model was established to evaluate the severe climate in Beijing. The results show that after 2017, the forest ecological restoration in Saihanba has a certain ability to improve the climate of Beijing and Chengde.

(3) By collecting oxygen and carbon dioxide data, the Carbon neutralization evaluation model (ECI) of Saihanba forest was established to analyze the carbon absorption capacity of the forest and evaluate the impact of the forest on carbon neutralization. The results showed that the Saihanba forest had a significant impact on carbon neutrality in Beijing and Chengde after 2011.

REFERENCES

Kleshchenko A. D. and Savitskaya O. V.. Estimation of Winter Wheat Yield Using the Principal Component Analysis Based on the Integration of Satellite and Ground Information [J]. *Russian Meteorology and Hydrology*, 2022, 46 (12): 881–887.

Liu Xiao, XUE Ying, Ji Yupeng, XU Binduo, Ren Yiping. Water quality assessment of the Yellow River estuary and its adjacent waters based on principal component analysis [J]. *China environmental science* 2015,35 (10): 3187–3192.

Qin Pengyao, Yang Huijuan, Jiang Fengling, Zhang Shubin, Tian Xiaomin, Huang Xuanrui, Zhang Zhidong. Quantitative classification of natural plant communities in Saihanba nature reserve, hebei province [J]. *Chinese journal of applied ecology*, 2016,27 (05): 1383–1392. DOI: 10.13287/j. 1001-9332.201605.007.

Su Jian, Liang Yingbo, Ding Lin, et al. *Journal of Chinese Academy of Sciences*, 2021,36 (9): 1001–1009.

Wang Lijun, Qian Dong, LI Bao-zhu, ZHANG Xiang-bu, Song Jian-biao. Sihanba mechanical forest farm in hebei province reform and development of study [J]. *Journal of forestry economy*, 2015 ((03): 12 + 9-88 DOI: 10.13843 / j.carol carroll nki lyjj. 2015.03.004.

Xinyue Chen et al. Analysis on Restoration Impact Value Model of Saihanba Forest Farm [J]. *International Core Journal of Engineering*, 2022, 8 (3)

Yi Zechuan, TAN Xijun, He Runhui. Based on TOPSIS model of sihanba ecological ecological environment impact assessment [J]. *Energy and energy conservation*, 2022 (6): 26-28 + 45. DOI: 10.16643 / j.carol carroll nki. 14-1360 / td. 2022.06.003.

Yin Hailong, Zhang Zhiwei, Song Wei, Zhang Zewen, Wu Liqin, Deng Ting, Liu Qiang. Sihanba north mann midian forest forest ecological function assessment [J/OL]. *Journal of Beijing forestry university*: 1-10 [2022-07-06]. http://kns.cnki.net/kcms/detail/11.1932. S.20220622.1321.001.html

Yin Xiaohui, Shi Shaoying, Zhang Mingying, Li Jing. Variation characteristics of dust weather in Beijing and analysis of dust source [J]. *Plateau Meteorology*, 2007 (05): 103.

ZHANG Jiandong, SUN He, DENG Ting, LIU Qiang. Evaluation of forest ecological function in Nature Reserve Saihanb Mechanized Forestry Farm [J]. *Forestry and Ecological Sciences*, 2022,37 (2): 185-191.

Zhenbin Zhang1* and Zixuan Ge2 and Bingbing Shi3. Quantitative Study on Environmental Improvement of Saihanba [J]. *International Journal of Computational and Engineering*, 2021, 6 (4)

Zhou Ying, Zhang Zewen, Wen Shuo, Sun He, Liu Qiang. Changes of light response indices of Larix principis-Rupprechtii needles in Saihanba and their influencing factors [J]. *Chinese Journal of Applied Ecology*, 2021,32 (05): 1690-1698. (in Chinese) DOI: 10.13287/ J. 1001-9332.202105.002.

Advances in Energy, Environment and
Chemical Engineering – Abdullah & Osman (Eds)
© 2023 The Author(s), ISBN: 978-1-032-36083-6

Ocean wave period analysis in the North Indian Ocean and prospect of its application on the nautical chart

Xue-hong Li[†], Shi-peng Su[†], Feng-wang Lang, Sun Yao, Zi-ying Li[†] & Cheng-tao Yi[†]
Dalian Naval Academy, Dalian, China

Bo Jiang[†]
National Ocean Technology Center, Tianjin, China

Rui-Ping Yang[†]
No. 91776 of PLA, Beijing, China

Di Wu[†]
Dalian Naval Academy, Dalian, China
Marine Resources and Environment Research Group, Dalian, China

Wei Jiang[†]
Dalian Naval Academy, Dalian, China
No. 92896 of PLA, Dalian, China

Chong-wei Zheng[*]
Dalian Naval Academy, Dalian, China
Marine Resources and Environment Research Group, Dalian, China

ABSTRACT: The wave period has had a significant impact on ocean navigation, ocean engineering, and wave energy resource development. In this study, the temporal and spatial distribution characteristics of the wave period were analyzed, by using the ERA-40 wave reanalysis from the European Centre for Medium-Range Weather Forecasts (ECMWF). The results show that the wave period exhibits a west low and east high spatial distribution. The wave period in most of the Bay of Bengal is above 9.5 s, with a large value center of about 11 s. The wave period in most of the Arabian Sea is below 9.5 s, with a small value center located in the Somali waters (of below 8.0 s). The wave period in spring is highest, while in winter is the smallest. This study also looks forward to embedding the results of the statistical analysis of the wave period into the chart, thereby providing a convenient query and application method for ocean navigation and offshore construction.

1 INTRODUCTION

Among all marine natural disasters, the casualties and property losses caused by waves are at the forefront, posing a serious threat to the safety of marine construction. Studies have shown that there is an average of 70 shipwrecks caused by waves in my country each year, with a loss of about 100 million yuan and about 500 deaths (Xu & Wu 2007; Zheng et al. 2016). In 1944, the U.S. Third Fleet was hit by a typhoon in the waters near the Philippines. The bad sea conditions directly led to the sinking of three destroyers, serious damage to two aircraft carriers, and 146 carrier-based aircraft were thrown into the sea. In 2004, a U.S. Coast Guard helicopter crashed in

---

*Corresponding Author: chinaoceanzcw@sina.cn
†The Authors contributed equally to this work. They are the co-first authors.

DOI 10.1201/9781003330165-104

rough sea conditions during a rescue mission. In 2012, a Seagull helicopter in Taiwan crashed due to bad sea conditions while rescuing people. The impact of waves on ships, offshore platforms, and offshore construction mainly depends on two key elements of the waves: wave height and period, especially waves with a longer period, which are easy for large ships to navigate, moor, load and unload operations, and offshore construction, etc. cause serious harm (Hua et al. 2004). If we can deeply grasp the inherent characteristics of ocean waves and implement clean energy projects such as ocean wave power generation according to local conditions, we can make contributions to alleviating the resource crisis and promoting the sustainable development of human society, thereby supporting the national "dual carbon" goal. The wave period is one of the key elements in wave energy calculation (Zheng 2018).

The predecessors have made a lot of accumulation and contributions to the analysis of the temporal and spatial characteristics of ocean waves. Zhou et al. (2007) conducted a numerical simulation analysis of the wave field in the South China Sea from 1976 to 2005 using the WAVEWATCH-III (WW3) wave model, and found that the significant wave height (SWH) in this sea area was the smallest in the whole year in summer, affected by the northeast monsoon in winter, the SWH is the largest in the whole year. At the same time, it is pointed out that the large center of extreme wave heights in the South China Sea is distributed in the southeast waters of Hainan Island. Wang et al. (2001) used about 6 years of SWH and sea surface wind speed data retrieved by TOPEX/Poseidon (T/P) altimeter and found the sea surface wind field through EOF (Empirical Orthogonal Function analysis) analysis. The first typical field of SWH is the wind field characteristics of the strong monsoon period, and the second type field is the wind field characteristics of the monsoon transition period. The first two modes of SWH are roughly similar to the first two modes of the wind field. Liu et al. (1998) analyzed the characteristics of wind and wave fields in the North Indian Ocean using the ship report data with a spatial resolution of $2° \times 2°$ from 1980 to 1990 and found that the southwest monsoon in this area is much stronger than the northeast monsoon. Zheng et al. (2012) used the simulated wave data to calculate the characteristics of the prevailing wave height and direction of the South China Sea-North Indian Ocean and found that the SWH in most areas of the sea area showed a significant linear increasing trend year by year.

Overall, most of the existing research on ocean waves focuses on the temporal and spatial distribution of effective wave height and wave direction, and the long-term variation law of effective wave height. However, the research on the wave period, which is closely related to the safety of ocean navigation, the safety of offshore construction, and the development of wave energy resources, is extremely scarce. In addition, the North Indian Ocean is one of the key waters of the "Maritime Silk Road" (Zheng 2018; Zheng et al. 2019; 2016), but the current research on the wave period in this area is even rarer, which restricts the national initiative and the safe development of ocean navigation. In this study, the ERA-40 wave reanalysis data from the European Centre for Medium-Range Weather Forecasts (ECMWF) was used to calculate and analyze the temporal and spatial characteristics of the wave period in the North Indian Ocean. The results can be used to query the system for the future and can be published in various forms such as nautical charts, to provide convenient technical support for the security of the marine natural environment for the construction of the "Maritime Silk Road".

2 DATA AND METHODS

The data used in this paper are the ERA-40 wave reanalysis data from ECMWF, which is the first simulation result of the coupled wave model WAM (Wave Modelling, wave simulation) and atmospheric circulation model in the world, and the reanalysis obtained by assimilating the observation data. The biggest feature and biggest advantage of this data is the separation of wind waves and surges. On the whole, the error of the assimilation data from 1991.12 to 1993.05 is slightly larger. For this reason, some researchers have used the buoy observation data to correct the effective wave height of ERA-40 (Caires & Sterl 2005). However, due to the lack of many buoy observation data of wave spectrum, for the time being, it is not possible to make technical improvements to the separation of the model wind wave and swell wave spectrum. To obtain long-term wind, wave, and swell data in the global waters, ERA-40 wave data is still one of the best choices (Semedo & Sušelj 2010). The temporal resolution of this data is 6 hours (that is, one data every 6 hours), and the spatial resolution is $1.5° \times 1.5°$. The spatial range is from 87.5°S to 87.5°N and from 0° to 357.5°E. The time range is from 00:00, September 1, 1957, to 18:00, August 31, 2002. In this paper, the climatic statistical analysis method is used to analyze the temporal and spatial characteristics of the wave period in the North Indian Ocean.

The governing equations of the WAM wave model used in the ERA-40 wave reanalysis data are as follows:

$$\frac{\partial F}{\partial T} + \frac{\partial}{\partial \varphi}(\dot{\varphi}F) + \frac{\partial}{\partial \lambda}(\dot{\lambda}F) + \frac{\partial}{\partial \theta}(\dot{\theta}_g F) = S \tag{1}$$

$$\frac{\partial F}{\partial T} + v \cdot \nabla F = S = S_{in} + S_{nl} + S_{ds} \tag{2}$$

where F is the spectral density, direction, latitude, and longitude; s is the source function term, wind energy input, nonlinear energy transfer due to wave-wave interactions, and dissipation term.

There are various ways to describe the wave period, commonly used include several types of the trans-zero period, characteristic period, maximum period, and spectral peak period, which are expressed as follows:

$$T_z = 2\pi \sqrt{\frac{\mu_0}{\mu_2}} \tag{3}$$

$$T_p = 2\pi \frac{\mu_{-2}\mu_1}{\mu_0^2} \tag{4}$$

$$T_s = 1.2T_z \tag{5}$$

$$T_m = 1.27T_z \tag{6}$$

$$T_p = 1.41T_z \tag{7}$$

In Eqs. (3)-(7), Tz is the trans-zero period, Ts is the characteristic period, Tm is the maximum period, and Tp is the spectral peak period. The spectral peak period is defined as the period corresponding to the maximum spectral value in the wave spectrum by its definition: $\mu_i = \int \omega^i S(\omega)d\omega$; i refers to the zero and second order moments of the waves; $i = 0, 2$. There is a certain conversion relationship between several commonly used periods so that a representative statistical analysis of one of the wave periods can show the general characteristics of the remaining wave periods.

3   SPATIAL AND TEMPORAL CHARACTERISTICS OF THE WAVE PERIOD

Firstly, the wave period from 00:00, March 1, 1958, to 18:00, May 31, 1958, was averaged to obtain the average spring wave period of 1958 on $1.5° \times 1.5°$ per grid point. The same method is used to calculate the average spring wave period for each of the last 45 years, and finally, the average spring wave period for the last 45 years at each grid point of $1.5° \times 1.5°$ is obtained, see Figure 1a. Similarly, the average summer, autumn, and winter wave periods for the last 45 years are obtained (Figure 1b-1d), as well as the average wave period for many years (Figure 2).

In spring, the wave period in the Arabian Sea is significantly smaller than that in the Bay of Bengal, and the North Indian Ocean shows a spatial distribution characteristic of low in the west and high in the east. The wave period contours in the Arabian Sea are roughly in a south-north direction, and the same is true for the east-central part of the Bay of Bengal. It is noteworthy that the wave period contours in the southern tropical waters of Sri Lanka have an east-west trend. The wave period in the Bay of Bengal is basically above 10 s, with large values centered in the eastern part of the Bay of Bengal (up to 11 s). The wave period in the Arabian Sea is basically within 10 s, and the low-value center is in the sea near Somalia, basically within 8.5 s. In summer, the wave period of the whole North Indian Ocean is significantly smaller than that of spring, which is supposed to be caused by the strong southwest monsoon in the North Indian Ocean in summer. Under the

722

influence of the southwest monsoon, the proportion of wind waves in the mixed waves in this sea is higher than that in spring, and the wave period of wind waves is significantly smaller than that of swell waves. The wave period contours in the Arabian Sea are in a northeast-southwest direction, the cycle at the top of the Bay of Bengal is in a northeast-southwest direction, and the cycle in the southern tropical waters of Sri Lanka is in an east-west band. In terms of spatial distribution, the wave period in the northern Indian Ocean still has an overall low west-east distribution. The wave period in most of the Bay of Bengal is above 9 s, and the wave period in most of the Arabian Sea is within 9 s. The low-value center is distributed in the sea near Somalia (within 8 s). In autumn, the wave period in most areas of the Bay of Bengal is above 9.5 s, and the wave period in most areas of the Arabian Sea is within 9.5 s. The contours are roughly oriented north-south. In winter, the wave period in the Bay of Bengal is basically above 9 s, and the wave period in the Arabian Sea is basically within 9 s. The contours in the western part of the Arabian Sea, especially near Somalia, have a northeast-southwest direction, while the northeast part of the Arabian Sea has a northwest-southeast direction. The wave period in most of the Bay of Bengal is in a northwest-southeast direction. It is noteworthy that the wave period within 8 s in the Arabian Sea is the most extensive throughout the year.

Regarding the annual average, the wave period in the North Indian Ocean for the past 45 years is low in the west and high in the east, and the contours are distributed in the shape of north-south bars. The wave period in most of the Bay of Bengal is above 9.5 s, and the large value center is located on the east coast of the Bay of Bengal, which can reach 11 s. The wave period in the Arabian Sea is basically within 9.5 s, and the low-value center is located in the sea near Somalia (within 8 s).

Overall, the wave period in the Bay of Bengal is significantly higher than that in the Arabian Sea, both throughout the year and in all seasons. This should be because the waves in the Bay of Bengal are influenced by the north-borne swell of the westerly zone in the South Indian Ocean all year round (Zheng & Li 2017), and the wave period in the swell-dominated region is usually larger than that in the wind-dominated region (Alves 2006; Zheng et al. 2012; 2016; 2018). The Arabian Sea, on the other hand, is not as significantly influenced by oceanic swells as the Arabian Sea, resulting

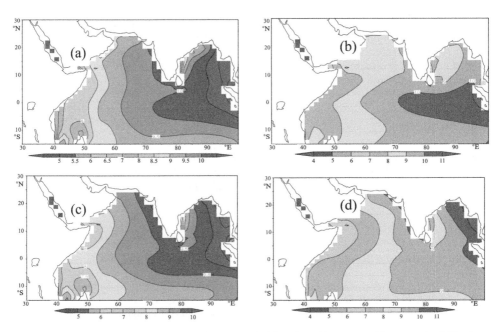

Figure 1.   Wave period in the northern Indian Ocean in (a) spring, (b) summer, (c) autumn, and (d) winter averaged over the last 45 years (unit: s).

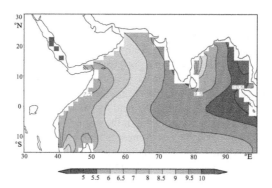

Figure 2.   Annual mean wave period in the northern Indian Ocean over the past 45 years (unit: s).

in its overall low wave period. In the Arabian Sea and most of the Bay of Bengal, the contours are in a south-north trend, and in the tropical East Indian Ocean, the contours are in an east-west band.

4   CONCLUSION AND OUTLOOK

In this paper, the spatial and temporal distribution of the wave period in the North Indian Ocean was statistically analyzed by using the ERA-40 wave reanalysis data from ECMWF for the past 45 years, and the results show that the wave period in this area is generally lower in the west and higher in the east, and the contours are generally distributed in a north-south bar shape. From the spatial distribution of the wave period, the wave period in the Bay of Bengal is significantly higher than that in the Arabian Sea. The wave period in most of the Bay of Bengal is above 9.5 s, and the large value center is located on the east coast of the Bay of Bengal, and the wave period can reach 11 s. The wave period in the Arabian Sea is basically within 9.5 s, and the low-value center is located in the sea near Somalia (within 8 s). From the seasonal difference in wave period, the wave period in the North Indian Ocean is the highest in spring and the lowest in winter. Zheng (Zheng 2020; 2021) created the first Global oceanic wave energy resource application dataset and the first wind energy resource application dataset for the first time. Similarly, we can propose the application dataset of wave period by referring to Zheng.

ACKNOWLEDGMENT

This work was supported by the Open Fund Project of Shandong Provincial Key Laboratory of Ocean Engineering, the Ocean University of China (Grant No. kloe201901). The authors are very grateful to the ECMWF for providing the ERA-40 reanalysis data.

REFERENCES

Alves JH. Numerical modeling of ocean swell contributions to the global wind-wave climate. *Ocean. Model.*, 2006, 11: 98–122.

Caires S, Sterl A. A new nonparameteric method to correct model data: Application to significant wave height from the ERA-40 Re-Analysis. *Atmos. Oceanic Technol.*, 2005, 22, 443–459.

Hua F, Fan B, Lu Y. An Empirical Relation Between Sea Wave Spectrum Peak Period and Zero-crossing Period. *Advances in Marine Science*, 2004, 22(1): 17–20.

Liu JF, Yu MG. Characteristics of wind wave field and optimum shipping line analysis in North Indian Ocean. *Tropic Oceanology*, 1998, 17(1):17–25.

Semedo A, Sušelj K. A Global View on the Wind Sea and Swell Climate and Variability from ERA-40. *Journal of Climate*, 2010. doi:10.1175/2010JCLI3718.1

Wang J, Qi YQ, Shi P. Analysis on the characteristics of the sea surface wind and wave fields over the South China Sea using empirical orthogonal function. *Acta Oceanologica Sinica*, 2001, 23(5): 136–140.

Xu FX, Wu XJ. The harm and distribution of high tiding disaster. *China Maritime Safety*, 2007, 4: 65–66.

Zheng CW, Gao CZ, Gao Y. Climate feature and long term trend analysis of wave energy resource of 21st Century Maritime Silk Road. *Acta Energiae Solaris Sinica*, 2019, 40 (6): 1487–1493.

Zheng CW, Li CY, Li XQ. Temporal and spatial distribution of windsea,swell and mixed wave in Indian Ocean. *Journal of PLA Science and Technology*, 2016, 17(4): 379–385.

Zheng CW, Li CY, Pan J. Propagation route and speed of swell in the Indian Ocean. *Journal of Geophysical Research-Oceans*, 2018, 123: 8–21.

Zheng CW, Li CY, Yang Y, Chen X. Analysis of Wind Energy Resource in the Pakistan's Gwadar Port. *Journal of Xiamen University*, 2016, 55(2): 210–215.

Zheng CW, Li CY. Propagation characteristic and intraseasonal oscillation of the swell energy of the Indian Ocean. *Applied Energy*, 2017, 197: 342–353.

Zheng CW, Li XQ Pan J. Wave climate analysis of the South China Sea and North Indian Ocean from 1957 to 2002. *Journal of Oceanography in Taiwan Strait*, 2012, 31(3): 317–323.

Zheng CW, Pan J, Tian YY, Yan ZZ. *Wave climate atlas of wind sea, swell, and mixed wave in global ocean*. China Ocean Press, 2012.

Zheng CW, Xia LL, Luo X. Characteristics of rough sea occurrence in South China Sea and North Indian Ocean. *Journal of PLA Science and Technology*, 2016, 17(3): 284–288.

Zheng CW. 21st Century Maritime Silk Road: wave energy evaluation and decision and proposal of the Sri Lankan waters. *Journal of Harbin Engineering University*, 2018, 39(4): 614–621.

Zheng CW. Global oceanic wave energy resource dataset—with the Maritime Silk Road as a case study. *Renewable Energy*, 2021, 169: 843–854.

Zheng CW. Temporal-spatial characteristics dataset of offshore wind energy resource for the 21st Century Maritime Silk Road. *China Scientific Data*, 2020, 5(4). DOI: 10.11922/csdata.2020.0097.zh.

Zheng CW. Wind energy evaluation of the 21st Century Maritime Silk Road. *Journal of Harbin Engineering University*, 2018, 39(1): 16–22.

Zheng CW. Wind energy trend in the 21st Century Maritime Silk Road. *Journal of Harbin Engineering University*, 2018, 39(3): 399–405.

Zhou LM, Wu LY, Guo PF. Simulation and study of wave in South China Sea using WAVEWATCH-III. *Journal of Tropical Oceanography*, 2007, 26(5): 1–8.

Advances in Energy, Environment and
Chemical Engineering – Abdullah & Osman (Eds)
© 2023 The Author(s), ISBN: 978-1-032-36083-6

Evaluation of water-driving characteristics and infilling adjustment effect of Chang 6 reservoir in Area M

Ping Liu* & Jiaosheng Zhang*
Exploration and Development Research Institute of Changqing Oilfield Company, Xi'an, Shaanxi, China

Huayu Zhong*
The International Cooperation Administration Department of Changqing Oilfield Company, Xi'an, Shaanxi, China

Ruiheng Wang* & Jing Wang*
Exploration and Development Research Institute of Changqing Oilfield Company, Xi'an, Shaanxi, China

ABSTRACT: The ultra-low permeability reservoir enters the high water cut stage, the distribution of remaining oil is complex, and it is more difficult for the reservoir to stabilize production. Aiming at this problem, infill adjustment is proposed to improve the development effect. In this paper, reservoir engineering, dynamic monitoring, numerical simulation, and horizontal well verification are used to predict the water drive range, thereby laying a foundation for infill adjustment. By evaluating the production decline, formation pressure change, infilling mode, inter-well interference, and recovery change of infill reservoir, the following understandings are obtained. (1) After infill adjustment, due to the reduction of well spacing, there will be some interference to the inter-well seepage field. (2) Two oil wells are infilled between lateral oil wells. When the original angle well is converted to injection, the infilling effect is good. (3) The lateral water drive range can be predicted by using various methods. The lateral water drive width is 80-100 m in 10 years of water injection development.

1 INTRODUCTION

The ultra-low permeability reservoir enters the high water cut stage, well pattern infill adjustment has become an effective means to improve the development effect and tap the potential of remaining oil. From 2010 to 2018, Chang 6 reservoir in Area M implemented scale encryption and the output of infill wells accounted for 54.8% of the whole region. It is necessary to evaluate the adaptability and development effect of the infill well pattern. At present, most studies only focus on basic well pattern water-driving characteristics and well pattern encryption mode, and lack of systematic evaluation of water-driving characteristics, infilling mode, and the effect of infilling adjustment. In this paper, various methods are used to predict the water drive range and remaining oil distribution characteristics at different stages, to further clarify the development technology policy of infilling, which is also a good reference and guidance for similar reservoirs to carry out infilling adjustment. This paper focuses on the research of basic well pattern water-driving characteristics and the evaluation of infilling adjustment effect from production decline, formation pressure change, infilling mode, inter-well interference, and recovery change of infill reservoir.

---

*Corresponding Authors: lp1_cq@petrochina.com.cn; zhangjs_cq@petrochina.com.cn;
zhyu_cq@petrochina.com.cn; wrheng_cq@petrochina.com.cn and wj920_cq@petrochina.com.cn

DOI 10.1201/9781003330165-105

2 WATER DRIVE CHARACTERISTICS OF PRIMARY WELL PATTERN

2.1 *Water drive characteristics*

The basic well pattern in Area M is a rhombic inverted nine-point well pattern, with a well spacing of 480 m and a row spacing of 200 m. The direction of the rhombic long axis is consistent with the direction of the maximum principal stress, which is northeast. The early development characteristics of the oil well show that the water breakthrough of the main oil well is fast, the water breakthrough cycle is 10-12 months, the lateral oil well has low pressure and low production, and the water injection effect is poor. The main lateral pressure difference reaches more than 5 MPa, with an obvious unidirectional banded water line (Wang 2019). In 2002, the main water flooded oil wells were successively switched to injection or shut down, and the drainage water injection was carried out to promote the effectiveness of the lateral wells. The single well production of the lateral wells reached about 3 T, the water cut remained at about 30%, the average annual decline was 0.2%, and the stable production reached 6 years. The drainage water injection achieved good results.

2.2 *Lateral water drive width*

Reservoir engineering, dynamic monitoring, numerical simulation, key well verification, and other methods are used to predict the water drive width on both sides of the waterline in the 10-year water injection development of Reservoir M.

(1) *Reservoir engineering method*

Assuming that the injected water is uniformly pushed from the water injection well to the surrounding areas (Ma Li 2020), then we have

$$W_{inj} - W_p = \pi r^2 h \Phi (1 - S_{or} - S_{wi}) \tag{1}$$

where W_{inj} is cumulative water injection, $(10^4 m^3)$; W_p is cumulative water production $(10^4 m^3)$; r is the radius (m); h is the thickness of the oil layer (m); Φ is the porosity (decimal); S_{or} is the residual oil saturation (decimal); S_{wi} is the irreducible water saturation (decimal).

As Reservoir M has the characteristics that the main water drive speed is faster than the lateral water drive speed, the injection water can be regarded as an ellipse when it is pushed around, then we can obtain

$$S = \pi ab \tag{2}$$

where S is the water drive area (m^2); a is the well spacing (m); b is the lateral water drive width (m).

The lateral water drive width can be predicted by substituting reservoir parameters and actual production data into Formulas (1) and (2). With this method, it can be predicted that the average lateral water drive width of Reservoir M before overall infill adjustment is 72 m.

(2) *Dynamic monitoring method*

From 2010 to 2013, the micro seismic method was used to monitor the water drive in front of 8 well groups in Reservoir M (Guo 2013; Zhao 2010). The results show that the lateral water drive width is 60 to 100 m, and the dominant direction of the water drive is 65 to 80 degrees in the northeast direction.

(3) *Horizontal well verification method*

From 2012 to 2013, the infill test of horizontal wells was carried out in Reservoir M. The horizontal sections of wjp7, wjp2, wjp3, wjp10, and the other four wells were perpendicular to the waterline. The comprehensive analysis based on the combination of horizontal well electrical logging curve and development performance shows that the width of the strong water washing zone on both sides of the fracture is about 60 to 80 m after 10 years of water injection development

(4) *Numerical simulation method*

A high-precision 3D geological model is established and roughened to fit the development history of Area M. From the numerical simulation results, the water drive pattern in this area is oval along the direction of the maximum principal stress, and the lateral water drive range is about 100 m. At the initial stage of development, the streamline along the principal stress direction is relatively dense, indicating that the water drive sweep is mainly along the principal stress direction. Before the overall infill adjustment, the overall water drive sweep is relatively uniform, and the mainstream line is gradually densely distributed laterally.

To sum up, before the overall infilling of Area M, the unidirectional water drive characteristics of the basic well pattern were obvious, and an obvious unidirectional banded water line was formed early. The width of the strong water washing zone was about 60 to 100 m, and the remaining oil was mainly distributed at the side of the water line (Wang 2015).

3 EVALUATION OF ENCRYPTION ADJUSTMENT EFFECT

After fully understanding the water drive characteristics and residual oil distribution characteristics of Reservoir M, the scale infill adjustment has been implemented in Area M since 2010. Three infill methods have been adopted: first, infill two oil wells between lateral oil wells, convert the original angle wells to injection, reduce the well spacing from 480 m to 160 m, and the oil-water well number ratio is 3: 1. Second, two oil wells are infilled between lateral oil wells. The original angle wells are converted to injection, and the well spacing is reduced from 480 m to 160 m. At the same time, the water injection wells are infilled, and the number of oil-water wells is 2: 1. Third, one oil well was infilled between lateral oil wells, the original angle well was converted to injection, and the well spacing was changed from 480 m to 240 m, and the ratio of oil-water wells was 2: 1 (Figure 1) (An 2014). As of 2018, 710 infill wells (547 oil production wells and 163 water injection wells) have been implemented at the waterline side. At the initial stage, the average daily oil production of a single well is 1.70 t, with a water content of 44.7%. At present, the average daily oil production of a single well is 1.16 t, with a water content of 60.4%. The annual oil production reaches 230,600 tons, and the cumulative oil production reaches 1,259,300 tons. The output of infill wells accounts for 54.8% of the region.

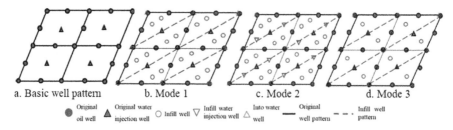

Figure 1.   Schematic diagram of well-infilling patterns.

3.1 Changes in output

At the initial stage of infilling, as the pressure maintaining level reaches about 110% of the original formation pressure, the formation energy is sufficient, the initial decline of the infill well is small, only 12.7% and the water injection takes effect quickly. The water injection takes effect after about 6 months of decline, and the production rises to near the initial production. The decline law of infill well conforms to hyperbolic decline, and the initial decline rate is 5.6 (Figure 2). Therefore, for low permeability reservoirs, energy is the basis to maintain stable production.

3.2 Variation of formation pressure

Before infilling, the formation pressure continued to rise and the pressure remained above 110% due to the transfer of injection or the shutdown of the main oil well and the development of row water injection. After infilling, the injection production ratio decreased from 1.8 to about 1.0 due to the decline of the injection production well number ratio, and the formation pressure-maintained level decreased to 92% of the original formation pressure, which is also the main factor leading to the increase of production decline in the infilling area in the later stage.

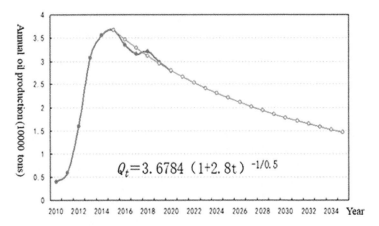

Figure 2. The prediction curve in production decline.

3.3 *Effect comparison of different encryption methods*

After comparing the three infilling methods in Area M, we obtain the following conclusions. In Mode 2, since the water injection wells are supplemented in time, the injection production is balanced, the initial decline of infill wells is small, and the production increases to more than 1.35 times the initial production after the water injection takes effect. At the same time, it has no impact on the old wells, and the water cut remains stable, with the best effect. In Mode 1, the water injection wells were not supplemented in time, resulting in the imbalance between injection and production on the plane, the production decline of the old wells increased, and the interference to the old wells was the most obvious. At the same time, the water cut increased faster than the other two methods (Figure 3).

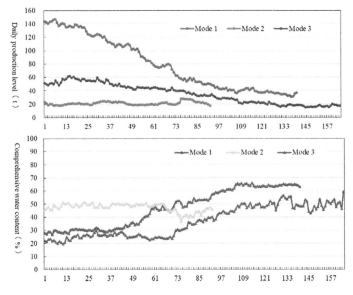

Figure 3. Produced curve of old wells with different infilling patterns.

Therefore, to avoid the decrease in formation pressure and the production of old wells after infilling adjustment, water injection well points should be supplemented while infilling the oil wells to form an injection production balance system of "more points and less injection". At the same time, considering the influence of starting pressure gradient and fracture (Hao 2006), the reasonable water injection intensity is determined to be 1.0 to 1.2 m³/d·m, the daily injection of the single well is 13 to 15 m³/d, and the injection production ratio is maintained at 1.5 to 1.7.

3.4 *Effect of infill adjustment on reservoir development*

There were 95 infilled oil wells in the north of block m from 2010 to 2018. By fitting and predicting the production changes of old wells before and after infilling, the interference degree of infilling on old wells (Chang 2016) and the improvement degree of development effect can be obtained.

On the one hand, the infill adjustment further reduces the well spacing, the injection production balance will be broken in a certain period, and the formation pressure will drop. From 2014 to 2015, the formation pressure will be reduced from 120% before infilling to 92%. At the same time, after the well spacing is reduced, the seepage field between wells will overlap to some extent, resulting in interference. The average daily oil production of an old well will drop by about 1 t. It is predicted that by 2037, a total of 442000 tons will be affected (Figure 4).

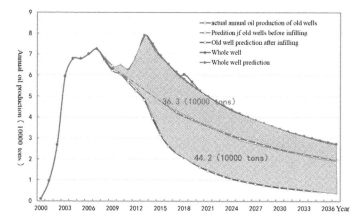

Figure 4.   Predicted curve of production decline before and after well infilling.

On the other hand, it effectively improves the reservoir development effect. Compared with no infill, the oil production rate increased by 0.1 to 0.38%, with an average increase of 0.19 percentage points (Figure 5). It is predicted that by 2037, the recoverable reserves will be increased by 363,000 tons, and the recovery factor will be increased by 5 percentage points (Figure 4 and Figure 6).

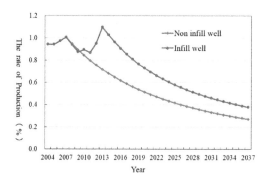

Figure 5.   Contrast curve of oil production rate.

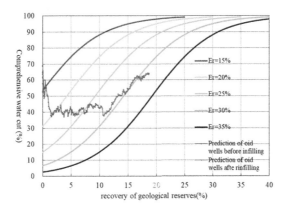

Figure 6.   Curves of the relationship between water-cut and producing degrees.

4   CONCLUSIONS

(1) In the reservoir with obvious unidirectional water drive, the main oil well sees water quickly, the lateral oil well has low pressure and low production, and the water injection effect is poor. The remaining oil is mainly distributed on the water line side.

(2) The lateral water drive width can be predicted by using reservoir engineering, dynamic monitoring, numerical simulation, key well verification, and other methods. The lateral water drive width in this area is 80 to 100 m in 10 years of water injection development.

(3) The infill adjustment has positive and negative effects on reservoir development. On the one hand, due to the reduction of well spacing, the injection production balance will be broken in a certain period, which will interfere with the inter-well seepage field, and the production of old wells will decline to a certain extent; on the other hand, it is a positive side. The oil recovery rate increases by 0.1 to 0.38%, and the recovery factor increases by 5%.

(4) To avoid the decline of formation pressure and the production of old wells after infilling adjustment, water injection well points should be supplemented while infilling the oil wells to form an injection production balance system of "more points and less injection", and determine the reasonable water injection intensity and injection production ratio.

(5) In terms of future work, it is necessary to carry out the prediction of water drive front of multi-direction water-penetrating reservoir, enrich the infilling method and development technology policy under different water drive characteristics, and improve the development effect of medium and high water-cut stage of ultra-low permeability reservoir.

REFERENCES

Fei Hao, Linsong Cheng, Chunlan Li, et al. (2006). Study on threshold pressure gradient in ultra-low permeability reservoir. *J. Journal of Southwest Petroleum Instiute*. 28 (6). 29–32.

Guangtao Chang, Yongchao Xue, Renyi Cao. (2016). Optimizing research of encryption form for rhombus anti nine-spot well pattern in ultra-low permeability reservoir. *J. CHINA SCIENCEPAPER*. 11 (15). 1722–1725.

Li Ma, Jiaosheng Zhang, Ruiheng Wang, et al. (2020). A Method to Determine Water-flood Front Based on Production Performance Data. *J. XINJIANG PETROLEUM GEOLOGY*. 41 (5). 612–615.

Mingsheng An, Qihe Chen, Chao Liu, et al. (2014). The Research of Well Re-infill for ultra-low Permeability Reservoir in Ansai Oilfield. *J. Journal of Oil and Gas Technology* (J. JPI). 36 (7). 133–136.

Wenhuan Wang, Huanhuan Peng, Guangquan Li, et al. (2019). et al. Water-flooding and optimum displacement pattern for ultra-low permeability reservoirs. *J. OIL&GAS GEOLOGY*. 40 (1). 182–189.

Yanfeng Zhao, Bo Tu. (2010). Time lapse monitoring test of water drive front monitoring technology with microseismic method. *J. Inner Mongolia Petrochemical Industry*. 36 (9), 124–125.

Youjing Wang, Xinmin Song, Changbing Tian, et al. (2015). et al. Dynamic fractures are an emerging new development geological attribute in water-flooding development of ultra-low permeability reservoirs. *J. Petroleum Exploration and Development*. 42 (2). 222–228.

Zhihua Guo. (2013). Application and analysis of microseismic water injection front monitoring in SHI 103 well block of low permeability reservoir. *J. Inner Mongolia Petrochemical Industry*. 39 (8). 137–139.

Advances in Energy, Environment and
Chemical Engineering – Abdullah & Osman (Eds)
© 2023 The Author(s), ISBN: 978-1-032-36083-6

Exploring the fate of carbon in saline soils after CO_2 uptake

Yubin Liu, Jiao Yan*, Wenzhu Yang*, Jing Zhang, Yan Wang, Ling Ling & Yingchao Yan
Key Laboratory of Environmental Chemistry of Inner Mongolia Autonomous Region, Hohhot, China
Chemistry and Environmental Science College, Inner Mongolia Normal University, Hohhot, China

ABSTRACT: A large number of studies have now concluded that CO2, CH4 and N2O are the major greenhouse gases, with CO2 having the greatest impact on global warming. However, the problem of lost carbon sinks has not been solved so far, and finding the destination of the missing carbon can help improve our understanding of the lost carbon sinks. Saline soils are mainly divided into arid and semi-arid regions, and most of the research in recent years has focused on (1) the phenomenon of CO2 uptake by saline soils, which is considered a new carbon sink. (2) Studying the influence of salinity, alkalinity, temperature, moisture, and microorganisms on CO2 uptake in saline soils is also the focus. There are three possible destinations of carbon in saline soils after CO2 uptake: the soil is sequestered as carbonate after CO2 uptake, microorganisms in the soil convert DIC to CH4, and it enters groundwater with precipitation or irrigation water in the form of DIC. Studying the fate of carbon is the key to solving the problem of lost carbon sink, but it is not fully verified at present. Future research in this field should focus on the migration and transformation process of saline soils after CO2 uptake and the significance of inorganic carbon in the terrestrial carbon cycle process and its role in carbon sequestration in the Earth's surface system to be further studied and demonstrated, which will provide some new directions to solve the problem of global carbon loss sinks.

1 INTRODUCTION

After the outbreak of the industrial revolution, massive industrial production led to excessive greenhouse gases being emitted into the atmosphere. Many studies have now concluded that atmospheric CO_2, CH_4 and N_2O are the three main greenhouse gases and that these three gases have a major impact on the greenhouse effect with a contribution of nearly 80% (Yan et al. 2016). Although all these greenhouse gases contribute to the greenhouse effect, the IPCC Fifth Assessment Report states that CO_2 remains the largest gas in terms of its contribution to global warming, with a contribution of about 60% (Zhang et al. 2014). The carbon cycle is the most dominant biogeochemical cycle on Earth, and its changes profoundly affect global climate change as well as influence human survival (Schindler D W. 1999; Xie et al. 2010). However, it was found through research that the problem of missing carbon sink (Missing carbon sink; about 1.8 Pg C/a) in the process of accounting for global carbon balance (Bruce J P. 1999; Fang et al., 2007; Joos F. 1994) is still unresolved, which indicates that we still need to improve the understanding of carbon cycle and enhance the research on carbon cycle (Fang et al., 2011; Liu et al., 2007).

Soil is the largest carbon reservoir on the Earth's surface, and the global soil carbon reservoir stores 2.2×10^3 to 3×10^3 Pg, which is 2 to 3 times the storage of the vegetation carbon reservoir and 2 times the storage of the global atmospheric carbon reservoir (Post et al. 1982; Schimel D S. 1995). However, saline soils are widely distributed on the earth, about 9.55×10^8 hm² globally, with Australia, the former Soviet Union, and China being the three countries with the largest saline

---

*Corresponding Authors: Jiaoyan@imnu.edu.cn and yangwzh@imnu.edu.cn

DOI 10.1201/9781003330165-106

soil areas. In the case of China, saline soils are abundant and diverse. The area of various types of saline soils in China is about 99.13 million hm^2, among which the area of saline soils on cultivated land reaches 9.2094 million hm^2, accounting for 6.62% of the national cultivated land area. In recent years, some studies have found that saline soils can absorb CO_2, and this discovery may find a new direction for the problem of "lost carbon sink" (E. L. Yates 2013; R. Stone 2008; Xie et al. 2009). Subsequently, some studies found that the DIC converted from CO_2 uptake by saline soils was eventually converted into carbonate to form a modern carbon sink, rather than simple inorganic carbon migration.

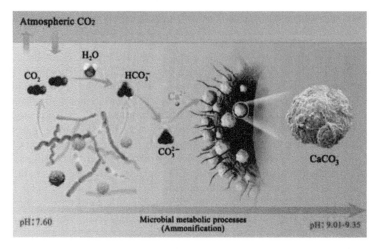

Figure 1.　Schematic diagram of carbon process from different sources, which is involved in the formation of carbonate precipitation (Liu et al. 2014).

2　FACTORS INFLUENCING CO_2 UPTAKE IN SALINE SOILS

The ability of saline soils to absorb CO_2 was first demonstrated by Chinese and foreign scientists in 2014 (Anonymous. 2014). A subsequent study in Nevada, USA found that saline soils play a large role in regulating atmospheric CO_2 at the local level. A study by (Yates E L. 2013) on saline soils in Australia found a nascent carbonate build-up in the soil (Filippi P et al. 2020). These studies show that the ability of saline soils to absorb CO_2 indicates that they have the potential to be called carbon sinks, so many studies have been conducted to investigate whether saline soils can act as carbon sinks. The focus of the research is mainly on the process of carbon-carbon fixation in saline soils after CO_2 uptake, which is the key question to revealing whether carbon sinks exist.

The key to the problem of CO_2 source-sink in saline soils lies in exploring the destination of the absorbed carbon. Based on $\delta^{13}C$ carbon isotope tracing techniques, Fa et al (2016) found in a study in the Mauwusu sands that CO_2 absorbed by saline soils at night was converted in large amounts to Dissolved Inorganic Carbon (DIC) in the liquid phase after excluding the influence of biological factors (Fa et al. 2018). This study has been questioned by some as to whether CO_2 is converted to DIC. To prove that these are DIC formed from CO_2 absorbed by the soil from the atmosphere, some researchers have used $\delta^{14}C$ isotope dating to answer that the newly formed DIC in the soil is modern carbon, while the age of carbonates in the soil is up to 10,000 years, and the two cannot be homologous. What is more, the DIC in the soil is younger than organic carbon, indicating that DIC can only originate directly or indirectly (root respiration) from the atmosphere (Li et al. 2015). These studies confirm that saline soils are not only capable of absorbing CO_2, but also fixing the absorbed carbon in the soil after converting it into carbonate, forming a carbon sink.

Many experiments are now conducted to investigate the factors influencing the conversion of CO_2 uptake into DIC in saline soils. Soil uptake of CO_2 is positively correlated with salinity and alkalinity, and alkalinity has a greater effect on soil uptake of CO_2, which is determined by salinity at the same alkalinity, and alkalinity is the main influencing factor for different salinities (Wang et al. 2019). The solubility of CO_2 in saline water is much higher than that in pure or acidic water, and the solubility of CO_2 in saline water increases linearly with conductivity and increases exponentially as soil alkalinity increases (Wang et al. 2013). The analysis of the above experimental content shows that soil pH has a dominant effect on soil uptake of CO_2, soil pH has a regulatory effect on soil inorganic carbon flux at the same temperature, and conductivity contributes to inorganic carbon flux, but its contribution to inorganic carbon flux is relatively small. Therefore, soil pH is a key factor in our study of CO_2 uptake by soils. Most studies have focused on single-factor studies on pH and conductivity, and there are few studies on the effect of the interaction of pH and conductivity on CO_2 uptake by saline soils. The interaction between pH and conductivity is inherent in saline soils, and single-factor analysis can only explain local problems. To make the experimental conclusions more general and more comprehensive to represent saline soils, we must investigate the interaction between the two.

As the experiment progressed further it was gradually found that temperature also had a significant effect on the CO_2 flux absorbed by the soil. One researcher in a Salt Lake study in the United States found that under controlled conditions, soil temperature was directly related to atmospheric CO_2 concentration. This suggests that soils can also act as a source of CO_2 when daytime temperatures are elevated. By this point, it was shown that temperature has a significant effect on CO_2 concentration where there is a large diurnal temperature difference, and that soil can act as a source of C to release CO_2 at high temperatures so that temperature has a large effect on CO_2 uptake by saline soils. A subsequent study by (Yates E L et al. 2011) in a desert in Ningxia found that the rate of temperature change had a dramatic effect on soil uptake of CO_2, with the soil absorbing a large amount of atmospheric CO_2 at night and being a weak source of CO_2 during the day. Although the effect of temperature on abiotic soil carbon fluxes cannot be explained directly in terms of soil temperature, the rate of change of soil temperature can accurately explain the role of temperature in abiotic soil-atmosphere CO_2 exchange, with soil CO_2 fluxes varying with temperature throughout the day, and with a clear progression from positive to negative soil CO_2 fluxes as temperature decreases (Liu et al. 2015; Xie et al. 2008). Through the above study, the change of temperature affects the change of CO_2 flux, and the saline soils in China are in arid and semi-arid areas, and these areas have the characteristics of diurnal temperature difference, so the temperature has important significance for the study of CO_2 flux. At present, the effect law of temperature change on CO_2 flux is clear, which will provide important help for our subsequent research.

The abiotic process of CO_2 uptake in saline soils is influenced by many factors, among which the influence of microorganisms on soil CO_2 uptake cannot be ignored. Researchers in Xinjiang found that microorganisms significantly affected the balance between soil CO_2, dissolved HCO_3^- and $CaCO_3$ by delaying carbonate recrystallization. $CaCO_3$ recrystallization was higher in sterile solutions than in non-sterile solutions. $CaCO_3$ recrystallization in soils increased with increasing CO_2 concentration, but the presence of microorganisms inhibited recrystallization and uptake of abiotic CO_2 in saline soils. We conclude that increasing CO_2 concentration significantly increases the recrystallization of calcium carbonate and the sorption of CO_2 by the soil. The presence of microorganisms attenuates this effect (Zhao et al. 2020). There are few studies on the effect of microorganisms on soil uptake of CO_2 fluxes, but microorganisms do affect carbon sequestration in saline soils, and some studies have pointed out that microorganisms can promote CO_2 uptake in saline soils, so more studies are needed to refine the effect of microorganisms on soil uptake of CO_2 fluxes and provide a basis for subsequent studies on carbon sinks in saline soils.

3 MIGRATION AND TRANSFORMATION PROCESSES OF INORGANIC CARBON IN SALINE SOILS

There are currently three possible mechanisms for carbon sequestration of inorganic carbon in saline soils: (1) dissolution/precipitation processes in carbonate systems; (2) leaching/absorption processes of atmospheric CO_2; and (3) downhole ventilation processes (Wang et al. 2019). The mechanism of carbon sequestration in saline soils is still under study, and most studies are still unclear about the mechanism of carbon sequestration in saline soils. Currently, with the increasing level of science and technology, isotope tracking technology is becoming increasingly mature, and many studies have started to use isotopes to investigate the source of soil carbon and the carbon content of each soil layer (Campeau Audrey et al. 2017; Cartwright I et al. 2017).

Soil inorganic carbon in a broad sense includes the solid, liquid, and gas phases. The gas phase is CO_2, derived from CO_2 produced by soil respiration and from the atmosphere in the soil; the liquid phase includes CO_2, H_2CO_3, D_2CO_3, and CO_3^{2-}, derived from H_2CO_3 and HCO_3^- rich solutions produced by the reaction of CO_2 with water; and the solid phase is mainly carbonate, derived from lithogenic and occurring carbonates. Lithogenic carbonates, also known as primary or inherited carbonates, refer to carbonates that originate from the parent soil or parent rock and are preserved without weathering soil formation; occurring carbonates, also known as secondary or authigenic carbonates, are carbonates formed during weathering soil formation and occur mostly in arid soils (Pan et al. 1999; Yang et al. 1999; Yu et al. 1990).

Soil inorganic carbon stable isotopes are mainly determined by the $\delta^{13}C$ values of soil CO_2, which ismainly derived from soil respiration, soil organic matter decomposition and ground atmospheric CO_2 exchange (Am Undson R et al. 1998; Pan et al. 2000). Rovira P et al. 2008; Yang et al. 2006). The more lithogenic carbonate predominates in soil inorganic carbon (Soil Inorganic Carbon SIC) sources, the closer the $\delta^{13}C$ value is to 0. The more occurring carbonate predominates in SIC sources, the more the $\delta^{13}C$ value tends to be negative (Krull EG et al. 2005). The results of Wang Na et al. in the desert area of the southern edge of the Junggar Basin showed that the $\delta^{13}C$-SIC values in the surface soil (0-20 cm) of the desert edge were closer to 0, indicating that lithogenic carbonate was dominant in the inorganic carbon composition and occurring carbonate was dominant in the deep soil; the lack of vegetation cover in the surface layer in the desert hinterland, the low content of soil organic carbon, the slow soil formation process, and the lithogenic carbonate was more residual. Therefore, the $\delta^{13}C$-SIC values in the surface layer tend to be negative, indicating the predominance of occurring carbonates in the inorganic carbon composition and the predominance of lithogenic carbonates in the deeper soils (Wang et al. 2017). Therefore, the difference in carbon isotope values used in the study can roughly determine the source of carbon and prove that saline soils can absorb atmospheric CO_2, but this can have certain drawbacks and does not allow continuous observation of the specific destination of carbon, and we need more research methods to refine these elements to clarify the mechanism of CO_2 absorption in saline soils and the destination of carbon.

Soil CO_2 is mainly derived from plant root respiration, organic matter decomposition, microbial respiration, and atmospheric CO_2 (Zhang et al. 2011). The first three effects generate homogenous CO_2 with essentially the same $\delta^{13}C$ values. Modern atmospheric CO_2 has a $\delta^{13}C$ value of about −8%, and the occurring carbonates formed from pure atmospheric carbon sources have higher $\delta^{13}C$ values of about 2% to 4% (Yang et al. 2006). The proximity of the soil surface to the atmosphere, the increased mixing effect of atmospheric CO_2, the increased $\delta^{13}C$ values of soil CO_2, and the ability of soil occurrence carbonates to sequester soil CO_2 during formation and recrystallization may also contain some atmospheric sources of CO_2 (Cui et al. 2013) so that near-surface exchange of atmospheric and soil CO_2 leads to $\delta^{13}C$ enrichment of soil carbonates.

Zhang Lin et al. studied the desert steppe area, in the 1 m depth range, the soil carbonate content increased with the increase of soil depth, until it reached the maximum at about 80 cm depth, and its average content varied from 16.50 to 274.3 g kg^{-1}. The variation pattern of soil carbonate $\delta^{13}C$ value was as follows. With the increase of depth, the $\delta^{13}C$ value gradually decreased to a relative minimum of -8.7% at 20-30 cm; and then, the $\delta^{13}C$ value of soil carbonate increased gradually with further deepening of the soil layer (Zhang et al. 2011). Studies by others for different soil depths found that the same proportion of inorganic carbon storage was 10%, 35%, and 55% for the same level of farmland and wasteland, respectively, for the entire soil profile from 0-100 cm, 100-300 cm, and below 300 cm (Luo et al. 2017; Wang et al. 2013). From these studies, soil is mainly enriched in the surface layer of soil after absorption of atmospheric CO_2, and inorganic carbon in deep soil is composed of primary carbonates, and whether the CO_2 absorbed by soil can be sequestered in soil for a long time or some carbon is sequestered after deep transfer still needs further study.

Although saline soils are capable of absorbing atmospheric CO_2, what is known about the phenomenon of atmospheric CO_2 uptake by soils and carbon isotope tracing studies are only part of the carbon sequestration process, but the process of carbon uptake by soils into dissolved inorganic carbon (DIC), the mechanisms involved in the carbon sequestration and transformation process, and the destination of the absorbed carbon are still uncertain. The fate of the absorbed carbon is still uncertain. Numerous studies have identified three possible destinations for carbon.

The first one is the researcher's study in the Mauwusu sands using isotope tracing technology found that after the soil absorbed $\delta^{13}C$-CO_2, the absorbed $\delta^{13}C$ may be enriched in the soil solid phase in the form of carbonate by observation and form a carbon sink (Liu et al. 2015). Fixed inorganic carbon exists in soil as CO_2, HCO_3^-, CO_3^{2-} and carbonate, and carbonate is the main form of soil inorganic carbon pool in the form of $CaCO_3$ and $CaMg(CO_3)_2$. By further studying the post-CO_2 uptake process in soil, it was found that soil is converted to soluble inorganic carbon after absorbing CO_2 from the atmosphere, after which DIC is converted to soluble inorganic carbon through the interaction with calcium and magnesium ions in saline soils to form carbonate precipitates, which eventually immobilize carbon in the soil as carbonates to form carbon sinks (Li et al. 2015). A recent study was carried out in saline soils in Huanglong County, Shaanxi Province, which used $\delta^{13}C$ isotope tracer properties to reveal the amount of absorbed CO_2 fixed in the soil by the soil, and this study has important implications for quantifying the carbon sequestration potential of saline soils, which found that 33.2% of the absorbed $\delta^{13}C$-CO_2 was preserved in the soil solid phase, indicating that some of the non-energetically absorbed carbon was stably fixed in the soil. In other research (Zhang et al. 2020), it was found that the carbon fixed in the soil accounted for one-third of the total carbon content absorbed, but there is a problem that the soil may have the possibility of carbon exchange, and the exchanged carbon could not be detected in the experimental tests because it was not labeled, which requires more subsequent studies to provide theoretical support for quantifying the carbon sequestration capacity of saline soils. Little is known about the storage capacity of inorganic carbon pools at different depths in various types of saline soils. The study of the formation mechanism of inorganic carbon in different types of saline soils is of great practical importance for reducing atmospheric CO_2 concentration.

The second is that some microorganisms have now been found to convert DIC to CH_4 (Kohl L et al. 2016). Although these microorganisms are mostly found in areas such as hot springs and swamps (Kelley D S et al. 2005; Schrenk M O et al. 2013), and it is not clear whether such microorganisms are also present in saline soils, genetic analysis of soil microorganisms in the Mauwusu sands by researchers found that methanogenic genes are present in these genes, indicating that there may be some microorganisms with methanogenic potential (data not yet published), and therefore, saline soils are endowed with this kind of carbon conversion possibility. However, there is no data to prove that there are methanogenic bacteria in saline soils that convert soil DIC to CH_4, so further research is needed in this area. There are relatively few studies on the conversion of DIC to CH_4, and the focus is on the conversion of DIC to carbonate, which can be tested in future studies for the presence of methanogenic bacteria in the soil. At present, a large amount of DIC in the soil is unaccounted for, so the conversion of DIC to CH_4 is likely to be one of the causes of the problem of the "lost carbon sink".

The third one is that the CO_2 absorbed by the soil is converted to DIC and then enters the groundwater body with precipitation or irrigation water. This deduction was proved by some studies. Researchers found in the karst area of Guizhou that the source of DIC in water is mainly the dissolution of soil CO_2 and carbonate, so we should focus on the dissolution of soil CO_2 and carbonate in our experiments to find the source of DIC (Li et al. 2010; Wen et al. 2016). Li et al. (Zhao et al. 2020) in the Tarim Basin, by using the $\delta^{14}C$ isotope tracing technique, found that a large amount of absorbed carbon appeared in the groundwater body within a short period after irrigation, indicating that the newly formed DIC in the liquid phase of desert soils is likely to undergo rapid longitudinal transport into the groundwater body after a large amount of water injection. Yu Li (Yu et al. 2017) started a study in Shiyang Lake and found that due to strong evaporation, DIC infiltrates into groundwater upstream and precipitates in groundwater in the terminal area. Precipitation or irrigation carries salts out of the soil and moves downward into the saline aquifer throughout the basin, and this saline water, which contains large amounts of dissolved carbon dioxide, is enriched, and precipitated in the terminal area due to strong evaporation. This carbon is sequestered in the groundwater bodies and forms a carbon sink, as the basins studied are closed basins with no outlet for the groundwater bodies. These studies show that carbon entering groundwater in the form of DIC can be fixed and form a carbon sink in groundwater in arid and semi-arid areas that are landlocked because they are not connected to the ocean.

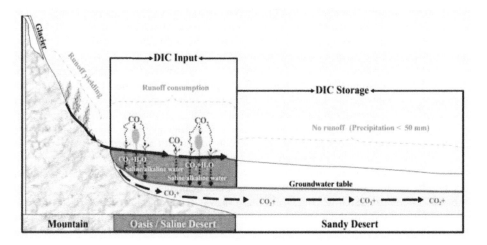

Figure 2.   Leaching and migration of DIC in arid saline-alkali Basin. (Li et al.).

4   CONCLUSIONS

Although most research results in recent years have shown that saline soils can absorb CO_2 from the atmosphere and sequester it in the soil after absorption, the factors affecting CO_2 absorption in saline soils include pH, conductivity, temperature, and soil microorganisms, whether the carbon sequestered in the soil will migrate to groundwater and the significance of inorganic carbon in the terrestrial carbon cycle need to be further studied and demonstrated. The significance of inorganic carbon in the terrestrial carbon cycle needs to be further studied and demonstrated. The following in-depth studies are needed for CO_2 sequestration and carbon transport in saline soils.

If DIC can be converted to carbonate precipitation in groundwater, it can indicate that carbon is sequestered in groundwater, which means that a real carbon sink is formed. Therefore, in the next study of groundwater inorganic carbon sink, we should try to investigate the conversion of DIC to form a complete cycle of carbon after entering groundwater.

In recent years, studies on soil inorganic carbon have gradually increased, and the soil inorganic carbon pool is the main form of soil carbon pool in arid and semi-arid regions, which is generally 2-5 times larger than the soil organic carbon pool. Some scholars have estimated that the global soil inorganic carbon pool is 700 to 1000 Pg (Bhattacharyya & T et al. 2001; Jose L et al. 2003; Yan et al. 1999). Soil carbon pools are of increasing interest and many studies are attempting to elucidate the impact of soils on the global carbon cycle. Although the role of soil organic carbon pools in the global carbon cycle has been studied, relatively little research has been conducted on soil inorganic carbon, especially in the form of soil-occurring secondary carbonates, and the significance of inorganic carbon in terrestrial carbon cycling processes and its role in carbon sequestration in the Earth's surface system is not well understood (Yang et al. 2007).

Especially in saline soils, the effect of the interaction of salinity and alkalinity on the carbon sequestration capacity of soils after CO_2 uptake from the atmosphere and the destination of carbon in soils after CO_2 uptake is still not very clear. Therefore, in future research, we hope that the mechanism of the whole cycle of carbon from the air to soil and then to groundwater can be clarified by using the method of isotope tracing or other research methods, and the destination of carbon can be found. This will provide some new directions to solve the problem of global carbon loss sink, and further exploration to trace the footprint of CO_2 absorbed by the soil in soil and groundwater is the key to clarifying the problem of soil carbon sink in saline areas.

ACKNOWLEDGMENTS

This work was financially supported by the National Natural Science Foundation of China (Grant No. 41865010), the Program for Young Talents of Science and Technology in Universities of Inner Mongolia Autonomous Region, China (Grant No. NJYT-20-A04), Project of Grassland Talent of Inner Mongolia Autonomous Region, China (Grant No. 2020).

REFERENCES

Am Undson R, Stern L, Baisden T, et al. 1998. The isotopic composition of soil and soil-respired CO_2 [J]. *Geoderma*,82(1-3):83–114.

Anonymous. 2014. Chinese and foreign scientists prove for the first time that saline soils can absorb carbon dioxide[J]. *North China Land and Resources*,(3):30–30.(in Chinese)

Bhattacharyya, T., Pal, D. K., Velayutham, et al. 2001. Total carbon stock in Indian soils: issues, priorities and management. *Land resource management for food and environmental security. Soil Conservation Society of India, New Delhi*. 1–46.

Bruce J P. 1999. Carbon sequestration in soils [J]. *Science*, 54(1): 382–389.

Campeau Audrey et al. 2017. Multiple sources and sinks of dissolved inorganic carbon across Swedish streams, refocusing the lens of stable C isotopes.[J]. *Scientific reports*, 7(1): 9158.

Cartwright I, Cendon D, Currell M, Meredith K. 2017. A review of radioactive isotopes and other residence time tracers in understanding groundwater recharge: *Possibilities, challenges, and limitations*. j. Hydrol. 555: 797–811.

Cui Lifeng, Liu Congqiang, Tu Chenglong, et al. 2013. Inorganic carbon distribution and isotopic composition characteristics of soils under different covers in loess areas[J]. *Journal of Ecology*, 32(05):1187–1194.

E.L. Yates, A.M. Detweiler, L.T. Iraci et al.2013."Assessing the role of alkaline soils on the carbon cycle at a playa site," *Environmental Earth Sciences*, vol. 70, no. 3, pp. 1047–1056.

Fa K Y, Liu Z, Zhang Y Q, et al. 2016. Abiotic carbonate dissolution traps carbon in a semiarid desert [J].*Scientific Reports*, 6:23 570. doi:10. 1038/ srep23570.

Fakeyu, Lei Guangchun, Zhang Yuqing, et al.2018. Atmosphere-soil carbon exchange processes in desert areas[J]. *Advances in Earth Sciences*,33(05):464–472.(in Chinese)

Fang Jinyun, Zhu Jiangling, Wang Shaopeng, et al.2011. Global warming, carbon emissions and uncertainty [J]. *China Science*: Series D, 41(10): 1 385–1 395.

Fang, Jing-Yun, Guo, Zhaodi.2007. In search of lost terrestrial carbon sinks [J]. *Journal of Nature*,29(1): 1–6.

Filippi P, Cattle S R, Pringle M J, et al.2020. A two-step modelling approach to map the occurrence and quantity of soil inorganic carbon[J]. *Geoderma*, 371:114382.

J. X. Xie, Y. Li, C. X. Zhai, C. H. Li,et al.2009. "CO_2 absorption by alkaline soils and its implication to the global carbon cycle,"*Environmental Geology*, vol. 56, no. 5,pp. 953–961.

Jiabin Liu et al. 2015. Abiotic CO_2 exchange between soil and atmosphere and its response to temperature[J]. *Environmental Earth Sciences*, 73(5), 2463–2471.

Joos F.1994. Imbalance in the budget [J]. *nature*, 370(6 486):18l–182.

Jose L, Diaz-Hernandez A, Enrique B F. 2003. Organic and inorganic carbon in soils of semiarid regions:A case study from the Guadix-Baza basin (Southeast Spain) [J].*Geoderma*, 114:65–80.

Kelley D S, Karson J A, Früh-Green G L, et al. 2005. A serpentine-hosted ecosystem: the Lost City hydrothermal field [J].*Science*,307(5 714): 1 428-1 434.

Kohl, L., Cumming, E., Cox, A., et al. 2016. Exploring the metabolic potential of microbial communities in ultra-basic, reducing springs at the cedars, ca, usa: experimental evidence of microbial methanogenesis and heterotrophic acetogenesis. *Journal of Geophysical Research Biogeosciences*,121(4), 1203–1220.

Krull EG, Bray SS. 2005. Assessment of vegetation change and landscape variability by using stable carbon isotopes of soil organic matter. *Australian Journal of Botany. Australian Journal of Botany*, 53: 651–661.

Li Dewen,Tang Zhonghua,Liu Ying,et al.2015.Characteristics of inorganic carbon distribution in saline soils of Heilongjiang Province[J]. *Anhui Agricultural Science*,43(15):85–87+89.

Li S L , Liu C Q , Li J , et al. 2010. Geochemistry of dissolved inorganic carbon and carbonate weathering in a small typical karstic catchment of Southwest China: Isotopic and chemical constraints[J]. *Chemical Geology*,277(3-4):301–309.

Li Y , Zhang C , Wang N , et al. 2017.Substantial inorganic carbon sink in closed drainage basins globally[J].*Nature Geoscience*, 10(13).

Li, Yan,Wang, Yu-Gang,Houghton, R A, Tang, et al. 2015. Hidden carbon sink beneath desert[J]. *Geophysical ResearchLetters*, 42(14).

Liu J B, Fa K Y, Zhang Y Q, et al. 2015. Abiotic CO_2 uptake from the atmosphere by semiarid desert soil and its partitioning into soil phases [J]. *Geophysical Research Letters*, 42(14): 5 779-5 785.

Liu Zaihua, Wolfgang Dreybrodt, Wang Haijing.2007.A potentially important CO_2 sink generated by the global water cycle [J].

Liu, Zhen, Zhang, et al. Desert soil bacteria deposit atmospheric carbon dioxide in carbonate precipitates[J]. Catena: An Interdisciplinary Journal of Soil Science.

Luo Qiong, Wang YG, Deng Caiyun, Ma Jian. 2017. Distribution of inorganic carbon in soil profiles in arid zones and its relationship with salinity and alkalinity[J]. *Journal of Soil and Water Conservation*, 31(05):240–246.

Pan G X. 1999. Study on carbon reservoir in soils of China. *Bulletin of Science and Technology*, 15(5): 330–332.

Pan Gen-Xing, Cao Jian-Hua, Zhou Yun-Chao. 2000. Soil carbon and its significance in the carbon cycle of the Earth's surface system[J]. *Quaternary Research*, (04):325–334.

Pan Gen-Xing. 1999. Occurrence of soil carbonate in arid regions of China and its significance on carbon transfer in terrestrial systems[J]. *Journal of Nanjing Agricultural University*, (01):54–60.

Post W M, Emanuel W R, Stangenberger A G. 1982. Soil carbon pools and world life zones. *nature*, 298(5870):156–159.

R. Stone. 2008."Ecosystems: have desert researchers discovered a hidden loop in the carbon cycle?" *Science*, vol. 320, no.5882, pp. 1409–1410.

ROVIRA P, VALLEJO V R.2008. Changesin $\delta^{13}C$ composition of soil carbonates driven by organic matter decomposition in a Mediterranean climate: a field incubation experiment [J].*Geoderma*, 144: 517–534.

Schimel D S. 1995. Terrestrial ecosystems and the carbon cycle. *Global Change Biology*, 1(1): 77–91.

Schindler D W. 1999. Carbon cycling:The mysterious missing sink [J]. *Nature*, 398(6723):105–106.

Schrenk M O, Brazelton W J, Lang S Q.2013.Serpentinization, carbon, and deep life [J]. *Reviews in Mineralogy and Geochemistry*, 75(1): 575–606.

WANG Na, XU Wenqiang, XU Huajun, et al. 2017. Soil carbon distribution and its stable isotope variation in the desert region of the southern edge of the Junggar Basin[J]. *Journal of Applied Ecology*, 28(07):2215–2221.

Wang YG, Wang ZW, Li Y. 2013. Characteristics of inorganic carbon fraction distribution in saline soil profiles in arid areas[J]. *Geography of Arid Regions*, 36(04):631–636.

WANG Zhongyuan,XIE Jiangbo,WANG Yugang,et al.2013. Relationship between inorganic CO_2 fluxes and soil salinity properties in saline soils[J]. *Journal of Ecology*, 32(10):2552–2558.

Wenfeng W , Xi C , Hongwei Z , et al. 2016.Soil CO_2 Uptake in Deserts and Its Implications to the Groundwater Environment[J]. *Water*, 8(9).

Xiaoning Zhao et al.2020.The effect of microorganisms on soil carbonate recrystallization and abiotic CO_2 uptake of soil[J]. *Catena*, 192.

Xiaotong Wang,Zhixiang Jiang,Yue Li,et al.2019. Inorganic carbon sequestration and its mechanism of coastal saline-alkali wetlands in Jiaozhou Bay, China[J]. *Geoderma*, 351.

Xie Gaodi.2010. Global climate change and carbon emission space[J]. *Leadership*, (08):15–31.(in Chinese)

Xie Jingxia et al.2008. CO_2 absorption by alkaline soils and its implication to the global carbon cycle[J]. *Environmental Geology*, 56(5) : 953–961.

Y Gao, Zhang, P. , Liu, J. 2020. One third of the abiotically-absorbed atmospheric co2 by the loess soil is conserved in the solid phase. *Geoderma*, 374(6), 114448.

Yan Cui-Ping,Zhang Y-M,Hu Chun-Sheng et al. 2016.Greenhouse gas exchange and its integrated warming potential in wheat-corn rotation farmland under different tillage practices[J]. *Chinese Journal of Ecological Agriculture*, 24(06):704–715.

Yang L F, Li G T, Zhao X R, Lin Q M. 2007.Profile distribution of soil organic and inorganic carbon in chestnut soils of Inner Mongolia. *Ecology and Environment*.16(1): 158–162.

Yang, L.F., Li, Guitong, Li. 2006. Baoguo. Stable isotope models of soil-occurring carbonate carbon and their applications[J]. *Advances in Earth Sciences*, (09):973–981.

Yang, L.F., Li.1999. Guitong. Progress of soil inorganic carbon research[J].*Soil Bulletinx*,42(04):986–990.

Yang, Qui-Fang, Li, Guitong, Li, Baoguo. 2006. Modeling and Application of Stable Carbon Isotope of Pedogenic Carbonate [J]. *Advances in Earth Sciences*, 21(9): 973–981.

Yates E L, Detweiler A M, Iraci L T, et al. 2013. Assessing the role of alkaline soils on the carbon cycle at a playa site[J]. *Environmental Earth Sciences*, 70(3):1047–1056.

Yates E L , Kathleen S , Max L , et al. 2011. Carbon Dioxide and Methane at a Desert Site-A Case Study at Railroad Valley Playa, Nevada, USA[J]. *Atmosphere*, 2(4):702–702.

Yu T R, Chen C C. 1990. Chemical processes in the soil pedogenesis [M]. *Beijing: Science Press*, 336–365.

Zhang Lin, Sun Xiangyang, Gao Chengda, et al. 2011. CO_2 sequestration during secondary carbonate formation and turnover in desert grassland soils[J].*Journal of Soil Science*, 48(03):578–586.

Zhang Lin, Sun Xiangyang, Gao Chengda, et al. 2011. Sequestration of CO_2 during secondary carbonate formation and turnover in desert grassland soils[J]. *Journal of Soil Science*, 48(03):578–586.

Zhang XH, Gao Y, Qi Y, et al.2014. Analysis of the implications of the main findings of Working Group I of the IPCC Fifth Assessment Report for the United Nations Framework Convention on Climate Change process [J]. *Advances in Climate Change Research*, 10(1):114–19.

*Advances in Energy, Environment and
Chemical Engineering – Abdullah & Osman (Eds)
© 2023 The Author(s), ISBN: 978-1-032-36083-6*

Method design and program development of tailings pond emergency response of water level over limit

Yiyuan Cui*, Guodong Mei, Yali Wang & Sha Wang
BGRIMM Technology Group, National Center for International Joint Research on Green Metal Mining, Beijing, China

ABSTRACT: Water level over the limit is a key problem causing dam breaks. However, there is no method to provide suggestions to tailings pond staff on how and when to lower the water level to safety. In this paper, a method of the emergency response to tailings pond water levels over the limit is introduced to improve safety management and a calculation program is developed by LabVIEW. An application of a tailings pond located in Shaanxi province is conducted to verify the effectiveness of this method. The result indicates that the method can provide suggestions for removing cover plates and applying a water pump.

1 INTRODUCTION

Water is the most important factor causing tailings pond dam-break (Xu 2017; Yaya 2017). There are over 5000 tailings ponds in China (Tang 2020). According to the statistics of dam-break accidents of tailings ponds in China from 2000 to 2020, more than 70% of dam-break accidents of tailings ponds are caused by water levels exceeding design. In recent years, extreme weather has happened frequently in China. On July 20th, 2021, the highest rainfall in Chinese history occurred in Zhengzhou, Henan Province, reaching 201.9 mm/h. Therefore, the safety management of tailings ponds is facing tremendous challenges.

Flood routing is an effective means to predict the maximum water level when the design flood comes (Dimache 2016, Wang 2011). However, there are still many shortcomings in flood routing. For example, it is a professional job and cannot be handled by ordinary workers (Cui 2021). Meanwhile, the lack of emergency response function also leads to this technology being divorced from actual production (Zheng 2011). When an over-designed flood comes, the most important thing for tailings pond workers to consider is how and when to lower the water level. But the traditional flood routing cannot offer this help.

To solve these problems, this paper introduces a method of lowering water levels based on flood routing theory and provides emergency response suggestions for the staff of tailings ponds. At the same time, a program is developed to improve its usability. To prove its applicability, a case of emergency drainage of a tailings pond is taken as an example to demonstrate.

2 METHOD AND PROGRAM DESIGN

2.1 Calculation method

The basic theory of the calculation method is the water balance equation. The increase of water level in the tailings pond equals the volume of flood flowing into the tailings pond minus the volume of

---

*Corresponding Author: cuiyiyuan@bgrimm.com

DOI 10.1201/9781003330165-107

drainage of the discharge system. It can be shown in the formula below.

$$\frac{1}{2}(Q_s + Q_z)\Delta t - \frac{1}{2}(q_s + q_z)\Delta t = V_z - V_s \tag{1}$$

where Q_s and Q_z are the amounts of water that flow into the tailings pond, q_s and q_z are the amounts of water that drained out of the tailings pond; V_s and V_z denote the amount of water stored in the tailings pond at the beginning and end of the period; Δt stands for any period, respectively.

According to the water level storage capacity curve, calculate the storage capacity corresponding to the over-limit water level and the standard water level. The amount of water to be discharged can be obtained by the following formula.

$$\Delta Q = Q_1 - Q_2 \tag{2}$$

where ΔQ is the amount of water to be discharged, m^3, Q_1 stands for the storage capacity corresponding to the over-limit water level, m^3, Q_2 stands for the storage capacity corresponding to the standard water level, m^3.

There are two methods to discharge this water. The common method is to lower the cover plate of the drainage well. However, when lowering the cover plate cannot meet the need of discharging water in a limited time, the water pump should be applied to increase the ability to discharge. Therefore, the ΔQ can be also demonstrated as follows.

$$\Delta Q = (q_1 + q_2)t \tag{3}$$

where q_1 stands for the discharge rate of drainage well, m^3/s, q_2 stands for the discharge rate of water pump, m^3/s, t stands for the limit time of discharging water ΔQ.

q_1 is derived from drainage calculation models related to water level (Cui 2021). The suggested water level can be obtained by the following equation.

$$\frac{Q_1 - Q_2}{t} - q_2 = \frac{\int_h^{h_0} q(H)}{h_0 - h} \tag{4}$$

where $q(H)$ stands for the relation formula between water level and discharge rate, h_0 stands for the highest limit water level, m, h stands for the suggested water level, m. The process of the method is shown in Figure 1.

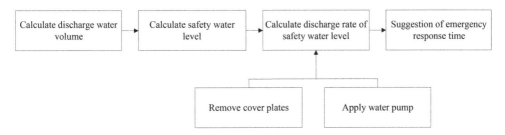

Figure 1. Process of the method.

2.2 *Program design and development*

In this paper, the calculation program is developed by the LabVIEW platform. The calculation program includes storage capacity calculation, discharge rate calculation and suggestion of emergency response time. The function of this program is to receive rainfall forecast data and calculate the possible highest water level. When the possible highest water level is over the limit of the design water level, the program will calculate and output the measures of lower water level to safety, including removing the cover plates and applying water pumps. Then, the shortest discharging time will also be given to guide emergency response.

3 CASE STUDY

3.1 Field conditions

The case demonstrated in this paper is a tailings pond located in Henan province. The basic situation of the tailings pond is shown in Table 1.

Table 1. The basic situation of tailings pond.

Height of beach crest	Current water level	Design flood depth	Total storage capacity
1007.3 m	1004.8 m	1 m	$563 \times 10^4 \text{m}^3$

Water level-storage capacity		Water level-discharge rate	
Water level (H)	**Storage capacity (m³)**	**Water level (H)**	**Discharge rate (m³/s)**
1004.8	0	1004.8	0
1005.2	46522	1005.2	4.68
1005.6	95766	1005.6	12.597
1006	147551	1006	21.951
1006.4	201711	1006.4	31.96
1006.8	258015	1006.8	36.60
1007.2	316239	1007.2	36.73

According to the weather forecast, the maximum rainfall of 12 hours is 59.8 mm/h. The highest water level will reach 1006.95 m exceeding the design flood depth of 0.65 m. The tailings pond staff should lower the water level to safety before the rainstorm. The height of the cover plate is 0.3 m.

3.2 Results and discussion

The design safety water level of the current situation should be below 1006.3 m. According to the relationship between water level and storage capacity, the tailings pond should discharge 92,001 m³. Based on the current situation, the water level should be lower to 1003.9 m. The result calculated by the program is shown in Figure 2.

To ensure the safe water level, 3 cover plates should be removed at least. The discharge curve is shown in Figure 3.

According to Figure 3, the average discharge rate can be calculated at 6.13 m³/s. Therefore, to lower the water level to safety, the program suggests removing 3 cover plates 4.16 hours early before the rainstorm.

In the actual situation, tailings pond staff found that only 2 cover plates could be removed. Therefore, to ensure the water level can be lowered to safety, water pumps should be applied to increase discharge rates.

Two water pumps are applied to discharge water. The total discharge rate is 2,000 m³/h. The discharge curve is shown in Figure 4.

According to Figure 4, the average discharge rate can be calculated as 3.53 m³/s. The result shows when removing 2 cover plates and applying a 2,000 m³/h water pump, the tailings pond should be discharged 7.2 hours early before the rainstorm.

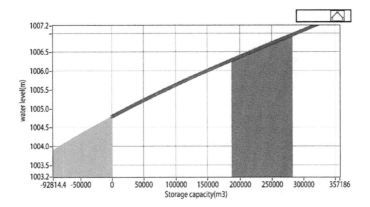

Figure 2.   Discharge to safety water level.

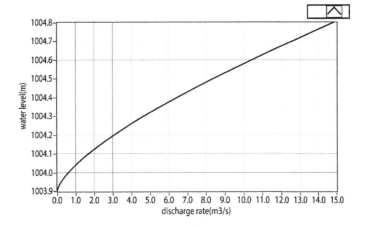

Figure 3.   Discharge curve when removing 3 cover plates.

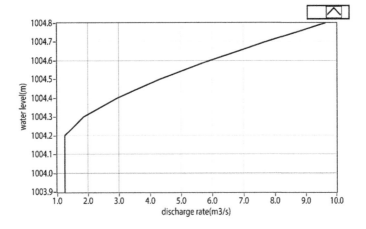

Figure 4.   Discharge curve when removing three cover plates and applying water pumps.

4 CONCLUSION

In this paper, a method to provide emergency response when the water level exceeds design is introduced and a calculation program is developed. It can advise tailings pond staff on when and how to lower the water level before a rainstorm, which can greatly improve the safety management of tailings pond. To verify its availability, a case study of a tailings pond located in Shaanxi province is tested by this program. The result indicates the program performs obvious effects of providing emergency response suggestions fast and accurately.

ACKNOWLEDGEMENT

This work was financially supported by the Key Program of BGRIMM (02-2201) and the Youth Innovation Fund of BGRIMM (04-2134).

REFERENCES

Cui, Y. Y., Mei, G. D., Wang, Y. L. & Wang, S. (2021). Development of tailings pond flood routing simulation system based on LabVIEW. *Journal of Physics: Conferences series*, V2029, n1, 012024.

Dimache, A., Batali, L., Iancu, I., Pantel, G., & Omer, I. (2016). Numerical modelling of exfiltrations from leaching tailing ponds. *Energy Procedia*, 85: 193–200.

Tang, L. & Liu, X., Wang, X., Liu, S., Deng, H (2020). Statistical analysis of tailings ponds in China. *Journal of Geochemical Exploration*, 216: 106579.

Wang, T. & Zhou, Y. (2011). A safety assessment of the new xiangyun phosphogypsum tailings pond. *Minerals Engineering*, 24 (10): 1084–1090.

Xu, F. & Yang, X. (2017). Dam-break flood risk assessment and mitigation measures for the Hongshiyan landslide-dammed lake triggered by the 2014 Ludian earthquake. *Geomatics Natural Hazards and Risk*, 8, 803–821.

Yaya, C., Tikou, B. & Cheng, L. Z. (2017). Numerical analysis and geophysical monitoring for stability assessment of the northwest tailings dam at westwood mine. *International Journal of Mining Science and Technology*, 27 (2017): 701–710.

Zheng, X, Xu, X. & Xu, K. (2011). Study on the risk assessment of the tailings dam break. *Procedia Engineering*, 26: 2261–2269.

*Advances in Energy, Environment and
Chemical Engineering – Abdullah & Osman (Eds)
© 2023 The Author(s), ISBN: 978-1-032-36083-6*

Optimization of treatment scheme for high slope of highway soft rock based on MIDAS

Yikai Wang*
School of Highway, Chang'an University, Xi'an, Shaanxi, China

Zhefeng Lin & Jing Zhang
Shaanxi Nuclear Industry Engineering Survey Institute Co., Ltd., China

Yiming Xu
School of Highway, Chang'an University, Xi'an, Shaanxi, China

ABSTRACT: Highway landslide disaster often occurs, the selection of reasonable support mode, not only to ensure the safety and stability of highway slope but also to save the number of support materials has become an important research direction. Under this background, this paper takes the highway landslide control project of Tanmu Village, Xixiang County, Hanzhong City, Shaanxi Province as an example. Based on MIDAS software displacement cloud map, plastic strain cloud map, truss element axial force distribution map calculation results, combined with construction difficulty and economic factors. The three schemes are optimized, and finally, the third scheme is the best supporting scheme. While stability has been improved, the use of materials has also been greatly reduced, to achieve the premise of ensuring stability and safety to save materials and play a greater effect.

1 INTRODUCTION

China has a vast territory, and there are great differences in rock and soil types, topography, and hydrology conditions, which lead to great differences in highway disaster vulnerability (Li 2015).

In practice, highway slope geological disasters often occur, and adopting an appropriate support structure and landslide control scheme is an effective means to prevent such disasters. Through numerical simulation, combined with the actual situation and experience, optimizing the support scheme, while ensuring the safety of support, and saving workforce and material resources, speeding up the construction period, has very important practical significance.

The common rock slope supporting structures include anchor rods, anchor cables, retaining walls, and anti-slide piles. However, in the preliminary design, there will often be unreasonable construction and waste of materials. In recent years, to solve this problem, geotechnical workers have done a lot of research work.

Xu Yuming (2015), Yang Hongfu (2020), Fang Wenkai (2020), and Mohammad R. A (2021) used finite element simulation software to optimize the supporting schemes using bolt, anchor cable, soil nailing wall and anti-slide pile, respectively, and obtained the final treatment scheme with better safety and economy. Wang Jiangrong (2019), Che Han (2016) and Zhang Tao (2013) respectively put forward a variety of design schemes for the stability of different rock slopes. After comprehensive analysis and in-depth discussion, they put forward a support optimization design scheme that is safer and has better coordination between economy and environment. Wang Yujia (2021), Wang

---

*Corresponding Author: 1414551425@qq.com

Yu (2011), Chen Zesong (2010) studied the optimization of the slope support structure, analyzed the influence of slope support parameters on slope stability through experiments, and proposed an optimization scheme in the construction process based on the experimental data.

This paper uses the finite element numerical simulation method and the theory of slope stability analysis, analysis of real-time monitoring data and the stress-strain and displacement of the slope under the condition of supporting situation, choose highway slope safety can be ensured and the support scheme can save consumption of the supporting materials, for governance program design and site construction to provide theoretical basis and reference.

2 PROJECT SUMMARY

2.1 Engineering geological condition

The landslide is located in Tanmu Village, Xixiang County, Hanzhong City, Shaanxi Province, G210 National highway K1596+725~K1596+850 Xixiang section. The landform where the slope is located belongs to the middle and low mountain landform with a slope of about 42°. At the foot of the slope, the Jingyang River develops here and Tanmu Village Bridge and G210 national road pass through here. The panoramic view of the sloping front is shown in Figure 1.

Figure 1.   Panoramic view of slope front.

In October 2019, under the influence of heavy rainfall, the slope body appeared to large deformation, resulting in the cracking and deformation of tea gardens, roads, and houses in the middle and rear of the slope body. During the investigation, preliminary on-site observation showed that the slope was still in the stage of continuous deformation, which was quite dangerous. The rear edge of the slope presents a creep and tensile failure mode, and there are many tensile cracks. The front edge of the slope has the characteristics of traction landslide failure, and several small collapses have occurred, resulting in the development of multiple ring cracks in the upper part of the front edge of the slope. The tensile cracks at the back edge of the landslide are shown in Figure 2.

According to the drilling results in the study area, the formation lithology is mostly the bedrock of the Middle Jurassic Shaximiao Formation (J2s), which is dominated by massive purple mudstone and yellow-green sandstone, interbedded with feldspar quartz sandstone and coal bands. The mudstone and sandstone are mostly interbedded. The exposed layer of the slope body is mainly the loose layer of residual slope accumulation, and its thickness is larger. The underlying bedrock is sand-mudstone interbedded in the middle Jurassic Shaximiao Formation (J2s), which is seriously weathered and fractured. Most of the bedrock in the exploration area is exposed directly, mostly in the form of weathering, and sporadic slope deposits can be seen in some areas. Figure 3 shows the severely weathered mud sandstone.

Figure 2. Tensile cracks at the rear edge of the landslide.

Figure 3. Severely weathered mud sandstone.

The surface water in the study area is not developed and is mainly atmospheric precipitation in the rainy season. According to the drilling results, the depth of groundwater level in the landslide area is about 10~16 m, the water level height is greatly affected by rainfall, and there is bedrock fissure water but the depth is large.

2.2 Mechanism of landslide formation

According to the survey data of the site and the force state of the landslide, and referring to many engineering experience, the causes of slope deformation and sliding are analyzed as follows: Firstly, the continuous heavy rainfall increases the external load of the slope body, and a large amount of rainwater penetrates the sandstone which is more seriously weathered, thus increasing the gravity, and sliding force of rock and soil body. The second is the formation lithology. The slope lithology is interbedded with mud and sandstone of unequal thickness and has a high degree of weathering. The rock and soil structure are loose, and the thickness of the weathered rock and soil layer is large, which can reach 45 m~54 m. The weathered mudstone has strong water resistance, but weak binding force between weathered layers, so it is easy to deform and slip along the bedding. Therefore, in the slope engineering of Tanmu Village, continuous heavy rainfall and special stratum lithology are the important factors for slope instability.

3 OPTIMIZATION ANALYSIS OF SUPPORTING SCHEME

3.1 MIDAS modeling and parameter selection

At present, the commonly used methods for slope design mainly include the limit equilibrium method, limit analysis method, and numerical analysis method (Wei 2017). In this paper, MIDAS software is used to establish a model for the slope. The effect diagram of grid division and natural finite plastic cloud picture is shown in Figure 4.

In this section, the parameters of supporting structures such as concrete and bolt can be defined according to relevant specifications. The parameters of geotechnical mechanics can be obtained based on in-situ experiments and laboratory experiments, and modified by the engineering analogy method referring to similar strata. Finally, the rock and soil parameters obtained are as follows.

3.2 Comparison and evaluation of supporting schemes

According to the field investigation and measurement results, after comprehensive discussion and analysis, the main measures for the governance of the K1596+710~K1596+880 section of

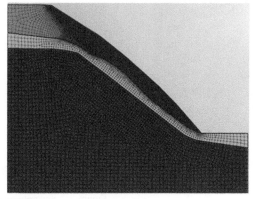

(a) Meshing diagram

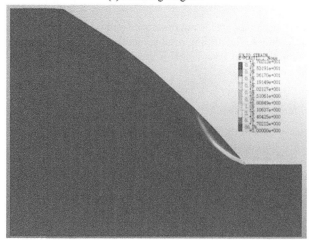

(b) Natural finite plastic cloud diagram

Figure 4. Meshing diagram and natural finite plastic cloud diagram.

Table 1. Physical and mechanical parameters of rock and soil mass.

Rock and soil layer	Modulus of elasticity (MPa)	Cohesive force (kPa)	Friction angle (°)	Poisson ratio	Natural gravity (kN/m³)
Fully weathered sand mudstone	112	26	24	0.273	21
Strongly weathered sand mudstone	700	48	33	0.232	22
Weathered sand mudstone	12000	230	39	0.183	24

G210 national Highway include: temporary back pressure, clearance unloading, slope foot anchorage, comprehensive drainage, combined with the actual situation and environmental needs, three treatment schemes are proposed. The processing ideas of each scheme are as follows.

Scheme 1: At the foot of the slope for sandbag back pressure, on the top of the slope for cutting side unloading, at the top of the landslide after the use of anchor steel pipe pile beam to stabilize, clean up the slope loose gravel, set up anti-slide pile, and then between the sloping beam and anti-slide pile from top to bottom to cut the slope and set up anchor cable frame beam, finally remove the sandbag, make concrete retaining wall. The final design content is sandbagging back pressure + the upper cutting + anchor cable steel pipe pile floor beam + anchor cable anti-slide pile + slope anchor cable frame beam + slope foot concrete retaining wall. The section design is shown in Figure 5.

Scheme 2: There are tea gardens and arable land at the top of the slope. From the perspective of land acquisition and environmental protection, the volume of excavation needs to be controlled. Anchor cable steel pipe pile ground beam is used for reinforcement, without cutting, the ground beam is set in the middle and upper part of the slope, and the rest is treated according to scheme 1. The final design content is sandbag back pressure + anchor cable steel pipe pile floor beam + anchor cable anti-slide pile + slope anchor cable frame beam + slope foot concrete retaining wall. The section design is shown in Figure 6.

Scheme 3: From the economic point of view, to facilitate the construction, the design scheme of the cutting square is adopted to completely remove the landslide, and the unloading of the cutting square is carried out from top to bottom. Sixteen-level slopes are set up, and two wide platforms are set in the middle to make steel pipe piles. An anchor pipe frame beam is applied in the process of slope cutting, and an anchor pipe frame beam is applied in the place of a large shear force in the construction process. The foot of the slope is also provided with a concrete retaining wall. The final design content is cutting unloading + anchor frame beam + wide platform steel pipe pile + concrete retaining wall. The section design is shown in Figure 7.

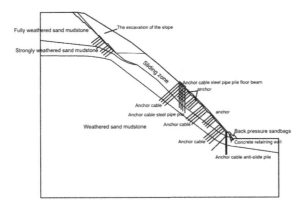

Figure 5.   Schematic diagram of scheme 1.

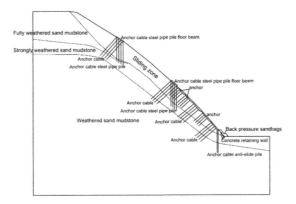

Figure 6.   Schematic diagram of scheme 2.

In the process of engineering management, the design scheme not only relates to whether the quality of engineering management meets the requirements but also affects the size of the economic

751

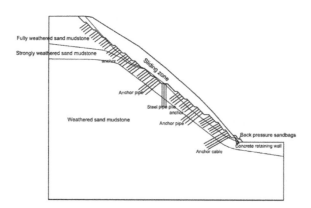

Figure 7.   Schematic diagram of scheme 3.

and environmental problems, construction difficulty and other aspects produced in the process of management. Before determining the final design scheme, it is necessary to compare and optimize several schemes proposed, and then choose a more reasonable scheme for slope treatment.

Scheme 1: unloading the top of the landslide greatly reduces the overall sliding force of the landslide and the safety risk of subsequent construction. Then, on the top of the collapsed part, the anchor cable steel pipe pile ground beam is set to limit the further deformation of the collapsed block and provide a guarantee for the excavation and construction of the anti-slide pile. It is carried out on the sandbag platform with good operation surface and high safety. The advantages of this scheme are high safety and reliability, good stability in construction and later period. The disadvantage is that the construction is difficult, and the upper part of the landslide is about 35.8 acres of land (including 20.4 acres of tea garden), occupying a large land area.

Scheme 2: from the point of view of land acquisition and environmental protection, an anchor cable steel pipe pile is used to reinforce the ground beam without cutting. The overall safety of landslide is guaranteed by a two-stage anchor cable steel pipe pile floor beam a row of anchor cable anti-slide piles and a two-stage anchor cable frame beam. This scheme has the advantage of strong environmental protection, saving land resources, reducing the pressure of land acquisition, and preserving most of the tea gardens. The disadvantage is that the safety is lower than in the first scheme, and the risk in the construction process is larger.

Scheme 3: Adopt the support structure thinking of the first two schemes, but to a certain extent, the support content is divided more carefully, and the design of key parts is more targeted. The construction of an anti-slide pile on a soft rock slope is difficult and dangerous, which is convenient for rapid construction. The solution eliminates a large area of landslides with less risk during construction. The disadvantage is that the amount of square cleaning is large, it is difficult to discard slag, and it will cause certain damage to the surrounding ecological environment, so a series of protection measures such as resource regeneration and ecological protection is needed.

Work safety must always be the top priority. Among the three schemes, scheme three has strong pertinence, high security, and thorough treatment of the hidden dangers of the slope. At the same time, the construction difficulty is less, which is convenient for the smooth progress of the later project. The cultivated land at the top of the slope has also been appropriately sheltered and protected.

4   SLOPE TREATMENT EFFECT

The final support design of the landslide is scheme 3. It can be seen from the simulation results that the slope stability coefficient after optimization increases from 1.0125 to 1.3395. The displacement cloud diagram and slope plastic strain diagram in the X direction are shown in Figure 8. The maximum axial force of the truss element appears at the anchor cable, and the maximum axial force is 66.38 kN, which meets the safety requirements, as shown in Figure 9. Stability has been improved at the same time for the use of the material has been greatly reduced, only the use of bolt saved 53.5%, to ensure stability and safety under the premise of saving materials.

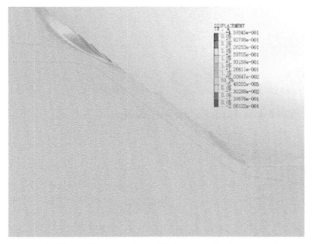

(a) Cloud diagram of the X-direction displacement |

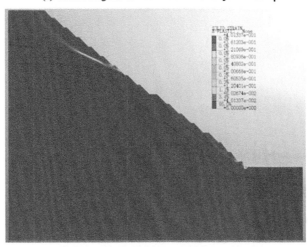

(b) Diagram of the slope plastic strain

Figure 8.   Cloud diagram of the X-direction displacement and slope plastic strain diagram.

After completion, the landslide was monitored for orifices displacement, and the monitoring results for more than one year are shown in Figure 10. In the first 30 days, the displacement rate is large, about 5.0 mm, and then the displacement growth slows down. After half a year of construction, the displacement tends to be stable, and after one year, the pore displacement is stable at about 11.6 mm, indicating that the optimal design of the slope achieves the expected support effect.

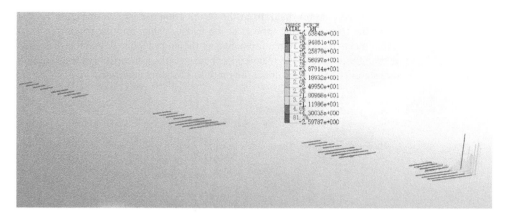

Figure 9. Axial force distribution diagram of the truss element.

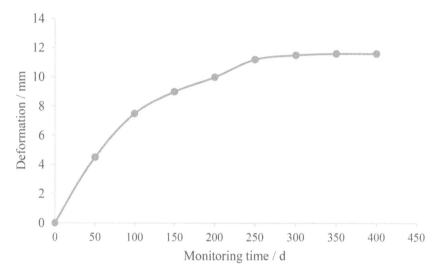

Figure 10. Slope monitoring curve after completion.

5 CONCLUSIONS

In this paper, MIDAS software is used to optimize the highway landslide control scheme in Tanmu Village, Xixiang County, Hanzhong City, Shaanxi Province.

(1) The combined support of cutting square unloading, anchor frame beam, wide platform steel pipe pile, and concrete retaining wall is the best treatment scheme, namely scheme 3. The scheme has strong pertinence, high safety, and thorough treatment of the hidden danger of slope.
(2) The slope stability coefficient is increased to 1.3395, and the maximum axial force is 66.38 kN, which appears at the anchor cable, meeting the need for safety.
(3) The optimization of the landslide control scheme has its complexity. In determining the control scheme, we cannot pursue a single support structure for the sake of simple design or construction. The mixed-use and coordination of various support structures are inevitable. Under the premise of ensuring stability and safety, reducing the number of materials used is also a consideration.

(4) In the study area, the slope is tall, the stratum is divided into many types, and the treatment scheme is complicated. In this paper, due to the limitations of the conditions, the rock and soil body stratum has been simplified. It is hoped that further research can be done to refine the stratum, and the influence of the structural plane and other conditions can be considered in the slope model to reflect the actual situation of the slope body more comprehensively.

REFERENCES

Che Han. (2016). Optimization of Bolt Supporting Design for Soft Rock Slope Based on the Consideration of Changes in Water Table. J. 36 (2): 19–22.

Chen Ze-song. (2010). Stability Evaluation and Application of High Steep Slope in Excavation. *J. Journal of Wuhan University of Technology*. 22 (5): 133–140.

Fang Wenkai. (2020). Study on Granite Residual Soil High Slope Stability and Support Scheme.

Li Jiachun. (2015). Vulnerability Assessment of Natural Disasters of Chinese Highway. *J. Journal of Beijing University of Technology*. 41 (07), 1067–1072.

Mohammad R. A. (2021). Optimization of Nail Inclination Angle in Soil Nail Walls Based on a Prevalent Limit Equilibrium Method. *J. Indian Geotechnical Journal*.

Wang Jiangrong. (2019). Finite Element Analysis of Shitouping Scenic Spot Slope Stability and Study on the Optimization of Support Measures. *J. Industrial Safety and Environmental Protection*. 45 (3): 8–11,20.

Wang Yu. (2011). Response Surface Optimization Design for Cutting Slope Anchoring Parameter. *J. Journal of Yangtze River Scientific Research Institute*. 28 (7): 19–23.

Wang Yujia. (2021). Optimisation of a Slope-Stabilisation System Combining Gabion-Faced Geogrid-Reinforced Retaining Wall with Embedded Piles. *J. KSCE Journal of Civil Engineering*. 25 (12): 4535–4551.

Wei Qibing. (2017). Stability analysis of slope reinforcement of the broadening project of Laoshan road in Qingdao. *J. Coal Geology and Exploration*. 45 (2): 101–104.

Xu Yuming. (2015). Optimization Design of Expressway in Mountain Area Slope Support Scheme Based on FLAC 3D. *J. Highway Engineering*. (6): 145–148,152.

Yang Hongfu. (2020). Optimization Study of Supporting Parameters for Highway Slope Based on Midas/GTS. *J. Highway*. 5 (09): 39–43.

Zhang Tao. (2013). *Stability Analysis and Supporting Design Optimization on Typical Phyllite High Slope of Shiyan to Tianshui Expressway*.

Advances in Energy, Environment and
Chemical Engineering – Abdullah & Osman (Eds)
© 2023 The Author(s), ISBN: 978-1-032-36083-6

Impact of sea surface temperature on the metallic marine materials

Wen-kai Zhang
School of Materials Science and Engineering, Dalian Jiaotong University, Dalian, China

Lin Zhou*
College of Meteorology and Oceanography, National University of Defense Technology, Changsha, China

Xia Zhang
No. 967 Hospital of PLA, Dalian, China

Xue-hong Li, Zi-ying Li & Di Wu
Dalian Naval Academy, Dalian, China

Yu-wei Sui
No. 91937 of PLA, Zhoushan, China

Zhen-yu Zheng
Dalian Naval Academy, Dalian, China

Da Zhang
No. 91526 of PLA, Zhanjiang, China

Guo-jie Chen
No. 91959 of PLA, Sanya, China

De-chuan Yu
School of Textile and Material Engineering, Dalian Polytechnic University, Dalian, China

ABSTRACT: This paper analyzes the influence of sea surface temperature on the corrosion of marine materials, it summarizes the current work progress in this area. The spatial and temporal distribution characteristics and long-term evolution law of SST in the Pacific Ocean are analyzed by using COBE-SST data. It mainly includes an annual variation trend, a seasonal difference of variation trend, and a dominant season of variation trend to provide a scientific basis of SST for anti-corrosion work. The results show that (1) 23°C and 25°C are two inflection points that affect the corrosion rate of marine materials. (2) In February, the 23°C isoline in the North Pacific was mainly along the line of Hainan Island-Taiwan Island-Hawaii-Revia Hihedo Island. The south Pacific 23°C isoline roughly follows the 30°S line. In both the North and South Pacific, the 23°C isoline in August is about 10-20 latitudes higher northward than in February. (3) During the last 50 years (1971–2020), SST in most of the Pacific Ocean increased significantly year by year (larger than 6×10^{-3}°C year on year). SST trends in different sea areas were dominated by different seasons.

1 INTRODUCTION

With the "Maritime Silk Road" and "Maritime community with a shared future" initiatives put forward one after another, a new chapter has opened in win-win cooperation and connectivity for humankind. However, the development and construction of a far-reaching sea are faced with a complex marine environment with high temperature and high salinity. It has a serious impact on marine

---

*Corresponding Author: zhoulin4458@nudl.edu.cn
All the authors contributed equally to this work. They are the co-first authors.

   DOI 10.1201/9781003330165-109

construction, mainly reflected in the corrosion threat to marine materials. marine engineering facilities and equipment, such as ships, offshore wind power platforms, submarine pipeline systems, deep-sea detection devices, as well as its components such as submersible pumps, connecting bolts, articulated columns, ship tail bearings, and anchor chains, are damaged and invalid due to varying degrees of corrosion in the marine environment for a long time, which seriously restricts the application and promotion of marine materials. According to statistics, the direct economic loss caused by corrosion every year accounts for 2-4% of the gross national product of all countries in the world, among which the loss caused by seawater corrosion accounts for about 30% of the total loss (Hou 2004). Therefore, it is of great practical significance to study the anticorrosion of marine materials.

Mastering the corrosion characteristics of marine materials is the premise to reduce the corrosion of marine materials. Feng et al. (Feng 2005) summarized several common marine corrosion forms of metal materials, according to the different corrosion forms, the methods, and indexes for evaluating the corrosion resistance of materials are introduced, and the main corrosion indexes are measured by typical sample sections and metallographic method. Wang et al. (2007) studied the off-site corrosion behavior of 5 kinds of marine engineering steels in the deep-sea environment. They found that the rise in sea temperature would cause the steel corrosion rate to present an extreme phenomenon, with the increase in temperature, the corrosion rate of steel increases first and then decreases, and the maximum value appears at about 23°C. Cheng et al. (2014) introduced the corrosion law of different ocean regions and the influence of dissolved oxygen, temperature, pH value, salinity, and other factors on the corrosion of metal materials in the marine environment. Chen et al. (2018) measured the corrosion potential of aluminum alloy in Sanya seawater with a DNCSDW-24 multi-channel automatic potential acquisition device. The fluctuation of corrosion potential was observed by tidal variation. It was found that at low tide, corrosion potential shifted positively with the increase of time. Shen et al. (2020) introduced the corrosion characteristics of typical metal equipment materials such as high-strength steel, stainless steel, and aluminum alloy in the marine atmospheric environment, and proposed that equipment protection in marine atmospheric environment should start from material selection and environmental adaptability design, effective surface protection, environmental control, regular maintenance, and other aspects.

Many excellent previous studies have been carried out on the corrosion form of metal materials and the factors affecting the corrosion of metal materials in the marine environment, but there is still a lack of effective research on the classification of ocean temperature grades, which is closely related to the anticorrosion work of marine materials. In this paper, according to the SST requirements of marine materials, the SST classification of the South China Sea-Northern Indian Ocean is carried out, which provides a scientific basis for the anticorrosion work of marine materials in the practical application process and assists decision-making.

2   DATA AND METHODS

Firstly, the influence of sea surface temperature on the corrosion of marine materials is analyzed, and then the current work progress in this area is summarized. Using the COBE-SST Data from Japanese Oceanographic Data, this paper analyzed the temporal and spatial distribution characteristics of SST in the Pacific Ocean and the long-term evolution of SST, including the annual trend, a seasonal difference of the trend, and the dominant season of the trend. It provides the scientific basis of SST for anti-corrosion work. COBE-SST Data were provided by Japanese Oceanographic Data with a time series from January 1891 to the present with a time resolution of the month. It covers global seas (89.5° N-89.5 °S, 0.5° E-359.5 °E) with a spatial resolution of $1.0^o \times 1.0°$.

3   RESEARCH PROGRESS OF SEA TEMPERATURE ON CORROSION OF METALLIC MARINE MATERIALS

The sea surface temperature (SST) varies from 0 to 35°C with different seasons and latitudes, and the sea bottom water temperature has relatively little change. As the depth of the water increases, the temperature of the water gradually decreases. Within the depth of 300 m, the water temperature drops rapidly due to the weakening of light, while in the depth of 300-1000 m, the temperature drops relatively slowly. When the water depth exceeds 1,000 m, the seabed temperature is almost constant at about 0°C (Hou et al. 2008).

Seawater temperature is part of the important factors affecting the corrosion of materials, but the effect of seawater temperature on corrosion is very complicated. On the one hand, the increase in seawater temperature will accelerate the reaction rate of cathodic and anodic processes, and accelerate the diffusion rate of oxygen, thus promoting the corrosion process; on the other hand, the solubility of oxygen in seawater decreases with the increase of seawater temperature, which speeds up the formation of protective calcareous scale and slows down the corrosion rate of metal in seawater. The interaction of factors in the actual marine environment is more complex. For example, the rise of seawater temperature will accelerate the reproduction rate of marine microorganisms, and it is easy for microorganisms to form an adhesion layer of a certain thickness on the metal surface, which can block the diffusion channel of oxygen ions and inhibit the uniform corrosion of metals. However, local corrosion of metals may be more serious. Domestic scholars' research on the corrosion of SST on metal materials mainly focuses on copper alloy, steel, and other widely used metal materials.

As important functional materials, copper and its alloys are widely used in communication, electronic components, and outdoor buildings. Similar to other metal materials, their corrosion in the marine environment is a complex physical and chemical process, mainly affected by temperature, humidity, and atmospheric pollutants. As early as last century, Zhu et al. (1998) found that the average corrosion speed of most copper and copper alloy in the full leaching area increased with the increase of seawater temperature by analyzing the experimental data obtained in Yulin, Xiamen, Qingdao, and Zhoushan. They found that most copper and copper alloys are very sensitive to temperature changes in a short period, and the sensitivity gradually decreases with the increase of time, but some alloys, such as BFE30-1-1 and HAL77-2A alloys, are not sensitive to temperature. Fan et al. (2018) studied the relationship between corrosion and scaling performance of B30 copper-nickel alloy at different temperatures. The corrosion and scaling of the copper-nickel alloy were recorded by an online monitoring device of corrosion and scaling. It was found that the corrosion rate of B30 copper-nickel alloy showed an upward trend at the beginning of the experiment. Under the condition of 50°C water, the corrosion rate increase about 70% less than 30°C under seawater, and the scale layer reduced the induction period of 3 hours, which means early experiments in copper and nickel alloy surface do not form a stable scale layer, destroying seawater corrosive chestnuts oxide film, fully exposed to the metal in the water, accelerate the corrosion process. As the temperature increases, the solubility of Ca^{2+} and Co_3^{2-} decreases, leading to more crystallization dirt attached to the metal surface, forming crystal nuclei and crystal embryos of initial dirt, shortening the induction period of the scale layer. During 3-9 h of the experiment, the corrosion rate decreases rapidly mainly because a relatively stable scale layer has been formed on the metal surface, which can play a good protective role on the metal matrix. In addition, the average corrosion rate of the copper and nickel alloys detected by the equipment is 0 during the whole experiment. The change of SST mainly affects the local pitting rate of B30 copper-nickel alloy, and only affects the corrosion rate of the copper-nickel alloy in a certain period.

To get more general and universal conclusions, scholars studied the relationship between the corrosion of marine metal materials and seasonal change and the marine environment on this basis. Gao et al. (2017) applied potentiodynamic polarization curve and electrochemical impedance (EIS) technology in the shipboard laboratory to conduct a preliminary study on the corrosion electrochemical behavior of 45# medium carbon steel, Q235 low carbon steel, X80 pipeline steel, and 316L stainless steel in the seawater of the Yellow Sea and the East China Sea and analyzed the causes of the phenomenon in combination with specific sea conditions. It is found that the corrosion behavior of 45# medium carbon steel, Q235 low carbon steel, and X80 pipeline steel in the Yellow Sea area has obvious seasonal change, the corrosion in winter is more serious than in summer, and the difference of Q235 low carbon steel in winter and summer is more obvious, while the corrosion of 316L stainless steel has no significant seasonal change in the Yellow Sea area. This phenomenon is mainly caused by the cold current in the Yellow Sea with low temperature (6-12°C) and high oxygen content in summer and the warm current in the Yellow Sea with high salt and high temperature (13-20.5°C) in winter. However, for 316L stainless steel in the East China Sea, the influence of seasonal change on its corrosion rate is not obvious. In the whole East China Sea, the corrosion rate of 316L stainless steel is very small.

In many engineering fields such as marine equipment, metal equipment is often corroded and its service life falls short of what should be expected. At present, the effective way to improve the corrosion resistance of the metal structure is surface treatment, and coating is one of the most effective methods of surface treatment. Fang et al. (2018) took amorphous Ni-P coating as the contrast coating and adopted an electrochemical testing method to test the polarization curves and AC impedance spectroscopy (EIS) of Ni-P and Ni-Cu-P coating samples were measured. Then the effects of seawater temperature on the corrosion resistance of amorphous and nanocrystalline Ni-Cu-p coating (treated at 400°C) and their contrast with amorphous Ni-Cu-p coating were analyzed. The corrosion current density (J_{corr}) of amorphous Ni-P, Ni-Cu-p, and nanocrystalline Ni-Cu-p coating increased with the increase of artificial seawater temperature, and the corrosion potential showed a negative trend. It shows that the corrosion resistance of the three coatings decreases with the increase in temperature. At 20, 40, 60, and 80°C, the corrosion current density of amorphous Ni-Cu-p coating is lower than that of amorphous Ni-P coating, and the higher the temperature, the closer the corrosion current density of the two is, indicating that the co-deposition of Cu is beneficial to improve the corrosion resistance of amorphous Ni-P coating, and the lower the temperature, the more obvious improvement effect. Comparing Figure 11 with Figure 10, it is found that the corrosion current density of nanocrystalline Ni-Cu-P coating is significantly lower than that of amorphous Ni-Cu-P coating, and the corrosion potential is significantly positive. The results show that heat treatment at 400°C can improve the corrosion resistance of Ni-Cu-p coating in artificial seawater.

Many foreign corrosion scientists have carried out extensive and in-depth studies on the corrosion behavior of metal materials in different marine environments. Structural steel is often used to manufacture engineering components and machine parts, but most welded steel structures usually have defects that cause fatigue cracks, which seriously affect the service life of structural steel. Vosikovsky et al. (1987) placed the structural steel in artificial seawater at room temperature (25°C) and 0°C, two stress ratios (R = 0.05 and 0.5), and two electrochemical conditions (free corrosion and cathode potential of -1.04 V), respectively, to conduct fatigue crack propagation tests.

Thiosulfate is one of the major sulfide oxidation products in seawater, which poses a serious corrosion threat to marine metal materials. To explore the relationship between temperature, thiosulfate, and metal corrosion, Ezuber (2009) used potentiodynamic polarization measurements (DC) to estimate the corrosion rate of white copper alloys (90 Cu-10Ni) in seawater with and without thiosulfate species (50-650 PPM). The results show that the corrosion current of 90Cu-10Ni alloy increases and the corrosion of white copper alloy accelerates when the seawater temperature increases. When the temperature rises to 25°C, thiosulfate activates the dissolution rate of the alloy, and it is found that the higher the thiosulfate concentration is, the higher the activation degree is, indicating that the corrosion reaction is more active. At 50 or 80°C, thiosulfate promotes the dissolution rate in the early stage, but then interferes with the formation of surface films, producing a black film and effectively reducing the corrosion rate of the alloy. At higher potentials, however, the film becomes nonprotective, leading to accelerated corrosion again. This experiment explains the corrosion behavior of 90-10 white copper alloy in seawater, which provides a foundation for further study of the relationship between the three alloys.

A series of electrochemical tests were carried out on three copper-based alloys using fixed and rotating disk electrodes. It was found that the corrosion rate of brass increased with increasing temperature. The corrosion current density (I_{corr}) increases, while the corrosion current density (I_{corr}) of Cu-5Ni alloy decreases slowly with increasing temperature. It is found that the Cu/Ni ratio in the corrosion products of Cu-5Ni alloy decreases with increasing temperature. Therefore, it can be concluded that selective dissolution of copper occurs on the metal surface. With the increase in temperature, Cu-5Ni alloy forms passivation film in seawater, which reduces the corrosion rate.

4 TEMPORAL AND SPATIAL CHARACTERISTICS AND EVOLUTION OF SST

According to the above literature review, it is found that 23°C and 25°C are two inflection points affecting the corrosion rate of marine materials. Based on the COBE-SST data, the temporal and spatial distribution characteristics and long-term evolution of SST in the Pacific Ocean are analyzed, which provides a scientific basis for corrosion prevention.

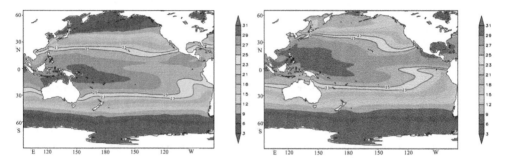

Figure 1. SST in February (left) and August (right) in the Pacific Ocean in recent 50 years (°CC)

Averaging the grid-by-grid SST in February for nearly 50 years (1971−2020), the multi-year average Pacific SST in February is obtained, as shown in Figure 1 on the left. By using the same method, the multi-year average Pacific SST in August is obtained, as shown in Figure 1 on the right. In February, the 23°C isoline in the North Pacific was mainly along the line of Hainan Island-Taiwan Island-Hawaii-Revia-Shihedo Islands, and the 23°C isoline in the South Pacific was roughly along the 30°S line. The spatial distribution of 25°C isoline is roughly the same as that of 23°C isoline, but the range is reduced. In August, the 23°C isoline in the central and western North Pacific Ocean is about 20 latitudes higher than that in February, but only a little higher in the eastern North Pacific Ocean. The 23° C contour in the South Pacific is about 10 latitudes higher northward than in February.

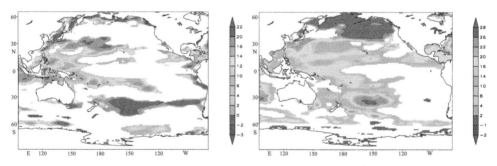

Figure 2. Climatic trend of SST in February (left) and August (right) in the Pacific Ocean in recent 50 years ($\times 10^{-3}$ °C/yr).

Using linear regression, the variation trend of Pacific SST in February and August in the recent 50 years was calculated respectively, as shown in Figure 2. In February, during the past 50 years (1971-2020), SST increased significantly (passing the significance test of 0.05) in the middle and low latitude areas of the Northwest Pacific Ocean and the eastern zone of New Zealand, with the strongest increasing trend in the Sea of Okhotsk, Sea of Japan, Ryukyu Islands-Taiwan Island, and the eastern zone of New Zealand. The trend is above 26×10^{-3}°C/yr. There was no significant change in SST in most of the other sea areas, but the SST decreased significantly in sporadic sea areas. It is worth noting that the SST in most of the offshore areas of China has no significant change. In August, SST increased significantly in most Pacific Ocean, with the trend above 8×10^{-3}°C/yr, especially in the north Pacific Westerly zone, with the strongest trend above 26×10^{-3}°C/yr. The SST in the eastern part of the South Pacific has no significant change, and only the SST in sporadic areas decreases significantly. As for offshore China, the SST in the Yellow Sea and the Bohai Sea has no significant change, the SST in the East China Sea has an increasing trend of $14\text{-}22 \times 10^{-3}$°C/yr, and the SST in most areas of the South China Sea has a significant increasing trend of $4\text{-}10 \times 10^{-3}$°C/yr.

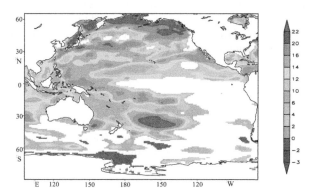

Figure 3.   Annual climatic trend of SST in the Pacific Ocean in recent 50 years ($\times 10^{-3}$°C/yr).

Using linear regression, the annual variation trend of Pacific SST in the past 50 years was calculated respectively, as shown in Figure 3. During the past 50 years, the SST in most parts of the Pacific Ocean increased significantly year by year, with a trend of being larger than 6×10^{-3}°C/yr. The areas with a particularly strong increasing trend were mainly distributed in the offshore areas of the North Pacific Ocean (Ryukyu Islands-Japan-Okhotsk Sea-Ryukyu Islands-Western Alaska sea, arcuate to the north), In the northeast of New Zealand, the trend is over 20×10^{-3}°C/yr. The SST in the eastern part of the South Pacific has no significant change trend, and the area with significant decline is mainly distributed in a small area of the Antarctic. By comparing Figure 2 and Figure 3, it is not difficult to find that the increasing trend in the Ryukyu Islands-Japan area is mainly reflected in February, the increasing trend in the Okhotsk Sea is reflected in both February and August, and the increasing trend of Ryukyu Islands-Alaska west sea area is mainly reflected in August. The trend of SST in the eastern South Pacific was dominated by August.

5   CONCLUSIONS

Through literature review, it is found that 23°C and 25°C are two inflection points affecting the corrosion rate of marine materials. Previous studies have made a great deal of accumulation and contribution to the analysis of SST characteristics in various sea areas around the world. However, overall, few studies are focusing on the analysis of temperature characteristics of 23°C and 25°C, which is not conducive to providing a scientific basis for corrosion prevention. In this paper, COBE-SST Data from Japanese Oceanographic Data is used to analyze the spatio-temporal distribution characteristics of SST in the Pacific Ocean, focusing on the spatio-temporal evolution of 23°C and 25°C. The long-term evolution of SST in the Pacific Ocean was analyzed by linear regression, including annual variation trend, the seasonal difference of variation trend, and the dominant season of variation trend, which provided a scientific basis for corrosion prevention work. The results show that in February, the 23°C isoline in the North Pacific is mainly along the line of Hainan Island-Taiwan Island-Hawaii-Revia-Shihedo Islands, and the 23°C isoline in the South Pacific is roughly along the line of 30°S. In both the North and South Pacific, the 23°C contours in August are about 10-20 latitudes higher northward than in February. During the past 50 years (1971-2020), the regions with significant increasing SST in February were mainly distributed in the middle and low latitude areas of the Northwest Pacific Ocean and the eastern zone of New Zealand. The increasing trend was strongest in the Sea of Okhotsk, Sea of Japan, Ryukyu Islands-Taiwan Island, and the eastern zone of New Zealand, and the trend was above 26×10^{-3}°C/yr. In August, SST increased significantly in most of the Pacific Ocean, with a trend of more than 8×10^{-3}°C/yr, especially in the north Pacific Westerly zone, where the trend was the strongest, with a trend of more than 26×10^{-3}°C/yr. During the past 50 years, SST in most of the Pacific Ocean increased significantly year by year, with a trend of being larger than 6×10^{-3} °C/yr. SST trends in different sea areas were dominated by different seasons. Maritime nursing, rescue, and life-saving also pay great attention to SST. In future work, it is also necessary to find the critical value of SST for maritime nursing and rescue and life-saving to better carry out relevant work

ACKNOWLEDGMENT

This work was supported by the Open Fund Project of Shandong Provincial Key Laboratory of Ocean Engineering, the Ocean University of China (Grant No. kloe201901). The authors are very grateful to the reviewers for providing useful suggestions.

REFERENCES

Chen MD, Zhang F, Liu ZY, Yang ZH, Ding GQ, Li XG. Corrosion potential sequence of metallic materials in Sanya Seawater and effect of Alloy Composition on corrosion resistance [J]. *Acta metallurgica sinica*, 2018,54 (09): 1311-1321.

Cheng P, Huang XQ, Zhang WL, Lang FJ. Research progress on corrosion of metallic materials in Marine environment [J]. *Wuhan iron and steel technology*, 2014,52 (05): 59–62.

Ezuber HM. Effect of temperature and thiosulphate on the corrosion behaviour of 90-10 copper-nickel alloys in seawater [J]. *Anti-Corrosion Methods and Materials*, 2009, 56 (3): 168–172.

Fan XW. *Corrosion and scaling properties of B30 Cu-Ni alloy in Seawater* [D]. Hebei University of Technology, 2018.

Fang XX, Du XJ, Gao SH, Zhang S. Effect of artificial Seawater temperature on the Electrochemical behavior of Ni-P and Ni-Cu-p alloy coatings [J]. *Chinese journal of nonferrous metals*, 2018,28 (06): 1176–1181.

Feng WD. Corrosion of Metal Materials in Marine Environment and its Evaluation Method [J]. *Equipment Environmental Engineering*, 2005, 6: 86–89.

Gao Y. *Study on electrochemical Corrosion behavior of typical Steel in Offshore Seawater of Huang-Donghai Sea* [D]. University of Chinese Academy of Sciences (Institute of Oceanology, Cas), 2017.

Hou BR. Corrosion behavior and control technology in Marine environment [J]. *Science and Management*, 2004, 5: 7–8.

Hou J, Guo WM, Deng CL. Influence of Deep-sea Environmental Factors on Corrosion Behavior of Carbon Steel [J]. *Equipment Environmental Engineering*, 2008 (06): 85–87+104.

Shen J, Ding XX, Song KQ, Zhang Min, Cong DL. Research progress on corrosion and protection of equipment materials in Marine atmosphere [J]. *Equipment environmental engineering*, 2020,17 (10): 103–109.

Vosikovsky O. The Effect of Sea Water Temperature on Corrosion Fatigue-Crack Growth in Structural Steels [J]. *Canadian Metallurgical Quarterly*, 1987.

Wang J, Meng J, Tang X, Zhang W. Corrosion Behavior Evaluation technology of Steel in Deep-sea Environment [J]. *Chinese Journal of Corrosion and Protection*, 2007 (01): 1–7.

Zhu XL, Lin LY, Xu J. Corrosion Behavior and Main Influencing Factors of Copper Alloy in Seawater Environment [J]. *The Chinese Journal of Nonferrous Metals*, 1998 (S1): 215–222.

Advances in Energy, Environment and
Chemical Engineering – Abdullah & Osman (Eds)
© 2023 The Author(s), ISBN: 978-1-032-36083-6

Study on online measurement method and device of rheology of drilling fluids

Jinzhi Zhu
Petrochina Tarim Oilfield Company, Korla, Xinjiang, China

Ren Wang
CNPC Engineering Technology R & D Company Limited, Beijing, China

Hongtao Liu, Tianxing Wei, Shaojun Zhang*, Zhi Zhang, Haiying Lu & Tao Wang
Petrochina Tarim Oilfield Company, Korla, Xinjiang, China

Yushuang Liu
CNPC Engineering Technology R & D Company Limited, Beijing, China

ABSTRACT: Currently, the rheological measurement of drilling fluids is manually performed using the Fann six-speed rotational viscometer, which is time-consuming, labor-intensive, prone to manual operation errors, and leads to drilling safety risks. Given the aforementioned, this research presents an online measuring device for the rheology of drilling fluids. First, the double measuring tube (with different diameters) configuration is used. The two tubes with the same flow rates and yet different flow velocities are connected in a cascade manner, and the differential pressure of the two tubes is measured by the corresponding differential pressure transducers. Then, based on the rheology, fluid mechanics, and the drilling fluid rheological parameters measured by a mass flowmeter, the rheological calculation models for different fluids and flow regimes are developed. Finally, the flow curves measured by the presented online measuring device are compared with the measurements of the Fann six-speed rotational viscometer. The results show that the deviations can be kept within 4% in the presence of the system environmental error, which effectively meets the requirement of the on-site automated measurement.

1   INTRODUCTION

For petroleum drilling and production engineering, the properties and functions of drilling fluids are important conditions for the successful drilling of oil and gas wells (Chen et al. 1998). The functions of drilling fluids include cleaning the bottom hole, transporting cuttings, cooling and lubricating the bit and drill string, transmitting water power, and controlling and balancing the formation pressure. The rheological property as well as performance in water loss and filter-cake forming of drilling fluids, are two basic technical features to ensure normal drilling. These two properties have direct impacts on improving the rate of penetration (ROP), maintaining downhole safety, avoiding damage to oil and gas reservoirs, and reducing drilling costs, so it is vital to monitor the rheological property of drilling fluids in a timely and accurate manner (Zhang et al. 1998).

The Brookfield family in America first invents the rotary viscosity measurement method, which pioneers the measurement of dynamic viscosity based on the correlation of the shear rate and resistance between the uniquely-designed rotor and fluids. The Brookfield family designs the

---

*Corresponding Author: 1272094712@qq.com

DOI 10.1201/9781003330165-110   763

world's first dynamic viscometer under its family name (Brookfield viscometer), namely the rotary viscometer. From the initial dial viscometer, the rotary viscometer has undergone 75 years of modification and upgrading, with continuously improved performance. Later, the digital viscometers DV-I, DV-II, and DV-III emerge, which are considered great progress in viscosity measurement. Recently, in order to realize the online measurement of fluid viscosity that produces more real-time data and facilitates automation, Liu Ximin and Liu Baoshuang develop a straight pipe viscometer (Liu et al. 2016), particularly suitable for the real-time online measurement of drilling fluids. Yet, this invention is not completely automated.

Given the aforementioned, this research presents an online measuring device for the rheology of drilling fluids to enable accurate real-time monitoring of drilling fluids. First, the device adopts double measuring tubes with different diameters. The two measuring tubes with identical flow rates and yet varied flow velocities are connected in a cascade manner, and by doing so, the differential pressure transducers in the two tubes record two differential pressure values. Moreover, based on the rheology and fluid mechanics, combined with the rheological parameters of drilling fluids measured by the mass flowmeter, the calculation models of rheological parameters for different fluids and flow regimes are developed to effectively analyze the rheological properties.

2 SOLUTION OVERVIEW

The solution of double measuring pipes with different diameters is adopted. The two measuring tubes are connected in series and under such circumstances, they have the same flow rates and yet different flow velocities. Therefore, the differential pressure transducers installed in the two tubes deliver two different values of differential pressure. Furthermore, with the parameters measured by the mass flowmeter, such as the density, temperature, and mass flow rate, and the theories of rheology and fluid mechanics, the calculation models of rheological parameters for different fluids and flow regimes are established, which enables the real-time measurement of rheological parameters such as the density, temperature, plastic viscosity (PV), yield point (YP), flow consistency coefficient (k), flow behavior index (n), and gel strength (Li et al. 1998).

2.1 *Drilling fluid performance*

With the continuous development of drilling technologies, various drilling fluids emerge in the industry. At present, there are various classification methods of drilling fluids all around the world. Here present several relatively simple classification methods:

By the density, drilling fluids can be divided into the high-density and low-density drilling fluids and the threshold is typically $1.35 \text{g} \backslash \text{cm}^3$.

By the intensity of induced clay hydration (swelling), drilling fluids can be divided into inhibitive and non-inhibitive drilling fluids. By the nature of the continuous phase, drilling fluids can be divided into water-based, oil-based, and gas-based drilling fluids.

There are nine types of drilling fluids approved by the American Petroleum Institute (API) and the International Association of Drilling Contractors (IADC): (1) non-dispersed drilling fluids; (2) dispersed drilling fluid; (3) calcium-treated drilling fluids; (4) polymer drilling fluid; (5) low-solids drilling fluid; (6) saltwater drilling fluid; (7) oil-based drilling fluid; (8) synthetic drilling fluid; (9) air, mist, foam, and aerated fluids (Tan 2015).

The device presented in this research is mainly applicable to water-based drilling fluids.

The rheological performance of drilling fluids is the fluid mechanics' manifestation of the rheological properties in the flowing state and thixotropy in the static state, which is usually characterized using the flow curve, plastic viscosity, yield point, gel strength, and other rheological parameters of drilling fluids.

(1) Shear rate γ

The shear rate refers to the increment of velocity per unit distance perpendicular to the velocity direction: $\gamma = dv \backslash dx$ (the velocity unit is m/s and the distance unit is m). Hence, the unit of the shear rate is s^{-1}. The greater the flow rate is, the greater the shear rate is.

(2) Shear stress τ

As the velocity of each layer of the liquid flow is different from each other, the interaction between layers is present. Because of the internal cohesion of liquid, the liquid layer with a faster flow velocity drives the adjacent liquid layer with a slower flow velocity, while the liquid layer with a slower flow velocity, in turn, hinders the adjacent liquid layer with a faster flow velocity. In such cases, internal friction occurs between liquid layers with different flow velocities, which results in pairs of internal friction forces (i.e., shear forces) that impede the shear deformation of liquid layers. Usually, the physical property to resist shear deformation during the liquid flow is called the viscosity of liquids (Le & Rasouli 2012).

The shear stress τ is defined as the internal friction per unit area of fluids

The Newtonian friction law defines internal friction as below: When liquid flows, the magnitude of internal friction forces (F) between liquid layers is related to the properties and temperature of liquids. It is in direct proportion to the contact area (S) and shear rate (γ) between liquid layers, and moreover, independent of the pressure on the contact surface:

$$F = \mu S \gamma \tag{1}$$

The internal friction force F divided by the contact area S gives the shear stress in liquid:

$$\tau = F \backslash S = \mu \gamma \tag{2}$$

where F is the internal friction of fluids, N; S is the area, m^2; μ is the viscosity coefficient, Pa·s (mPa·s is more used in practice and 1 cP =1.1mPa·s

In general, the fluid that is associated with the correlation between the shear stress and shear rate following the Newtonian internal friction law is called the Newtonian fluid; the fluid that does not obey the Newtonian internal friction law is called the non-Newtonian fluid. Pure liquids such as water and alcohols, light oil, low molecular compound solutions, and low-velocity gases are mostly Newtonian fluids, while concentrated solutions and suspensions of high molecular polymers are typically non-Newtonian fluids. Most drilling fluids are non-Newtonian fluids.

2.2 *Workflow and principles of measurement*

The electric diaphragm pump (speed-adjustable and constant-rate) is used to extract a certain volume of drilling fluids from the mud tank into the measuring system (Zhang & Li 2001). After filtering out the impurities in drilling fluids through the filter screen, extracted drilling fluids first pass through the damper and then the mass flowmeter, and the measured data is transmitted to the upper computer, which determines whether the flow has reached a stable rate and adjust the pump rate as per the feedback to ensure a constant output flow rate of drilling fluids. At the same time, the temperature, density, and flow rate signals measured by the mass flowmeter are also transmitted to the upper computer. Drilling fluids pass through the two measuring tubes with different diameters (316 stainless steel pipes), and a set of planar membrane film differential pressure transducers is used to measure the differential pressure loss along a certain measuring section. The differential pressure signal is transmitted to the upper computer and the plastic viscosity, yield point, gel strength or flow consistency coefficient, and flow behavior index are calculated in the same way as the results of the conventional six-speed rotational viscometer measurement. Besides, the pH, conductivity, and chloride ion concentration of drilling fluids are measured using the pH, conductivity, and chloride ion online monitoring electrodes, respectively. It takes 65 s to deliver all the results (excluding cleaning), and finally, the extracted drilling fluids return to the mud tank. According to the transmitted signals and measuring pipe geometry, the drilling fluid rheology is computed and recorded continuously (Chen 2011).

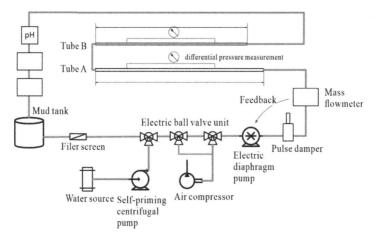

Figure 1.   Schematic diagram of the device workflow.

2.3   *Measurement workflow*

In the measurement process, measuring the flow rate requires maintaining the flow state for a certain period (about 30 s, and the specific time is adjusted according to the actual PID), in order to perform the measurement after the flow rate is stable. The measurement time is 1s (the sampling frequency of the pressure transducer is 1000 Hz, and the average of multiple sampling results (tentatively 100 sampling results) is taken as the measurement result to prevent value fluctuation. The results corresponding to the shear rates of 511 s^{-1} (300 rpm) and 1022 s^{-1} (600 rpm) can be read out at the same time at a certain flow rate, while those of 3 rpm, 6 rpm, 100 rpm, and 200 rpm can be obtained by controlling and adjusting the pump via the computer program (Gao 2002).

It takes about 31 s to complete the measurement of 300 rpm and 600 rpm. Because the rotation speeds, 3 rpm, and 6 rpm, are rather slow, the measurement needs to be completed quickly and the PID adjustment time is changed to 2 s and the measurement time is 1 s. It takes 65 s (excluding cleaning) to obtain the plastic viscosity, yield point, gel strength or flow consistency coefficient, and flow behavior index, calculated in the same way as the results of the conventional six-speed rotational viscometer measurement.

The workflow is shown below:

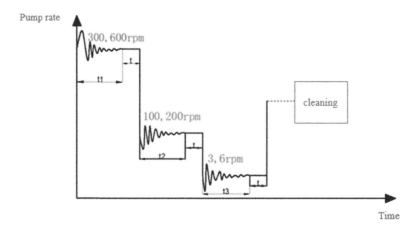

Figure 2.   Time schedule of measurement.

The device frame is built by welding stainless steel rectangular tubes. The measuring tube is a stainless steel round tube. The tube connection uses the flange, which is convenient for disassembling and assembling pipelines and zonal cleaning. The outlet of the electric diaphragm pump is connected with the mass flowmeter using the stainless steel round tube. The inlet of the electric diaphragm pump is connected with the mud tank and the measuring outlet using the hose.

The whole system requires shock absorption, and the pump is not connected to the frame. When skid-mounted, the shock-reducing rubber or other shock absorbers with good performance are placed between the skid base and the pump base. Because the device is close to the vibration source of the well site, the frame vibrates and the shock absorption design is incorporated below the fixed point of each transducer (Zhou et al. 2012).

The viscosity of drilling fluids is greatly affected by temperatures, so the whole measuring system is placed in the skid-mounted cabinet to eliminate this interference. When the temperature fluctuates greatly, extra insulation is necessary to ensure measurement accuracy.

2.4 *Rheological model of non-Newtonian fluids measured by capillary tubes*

(1) Steady-state flow

According to the force balance in a capillary tube, two forces are identified. One is the force on the tube wall F_1 at both ends of the capillary tube applied by the pressure difference ΔP, and the other is the viscous resistance F_2 when fluid flows on the wall. Therefore, in the case of the laminar flow, the formula of the shear stress on the wall is (Li et al 2017):

$$\pi r^2 \cdot \Delta P - 2\pi r L \tau = 0 \tag{3}$$

$$\tau = \frac{\Delta P \cdot r}{2L} \tag{4}$$

where τ is the shear stress on the capillary tube wall. At $r = R$, τ is the largest, corresponding to the maximal flow resistance and zero flow velocity. At the center of the tube ($r = 0$), τ is the least, corresponding to the minimal flow resistance and highest flow velocity. Thus, we have:

$$\frac{\tau}{\tau_B} = \frac{r}{R} \tag{5}$$

The above two formulas are applicable to any fluids flowing in a time-independent steady state.

(2) Considering the flow characteristics of Newtonian fluids in the pipe flow, the Hagen equation can be rearranged as below:

$$Q = \frac{\Delta P \cdot \pi \cdot R^4}{8\mu L} \tag{6}$$

Because $Q = \frac{\pi D^2}{4} \cdot V$, we have:

$$\tau_B = \mu \frac{8v}{D} \tag{7}$$

According to Newton's equation and $\tau_R = \mu\left(-\frac{dv}{dr}\right)$, we have $\left(-\frac{dv}{dr}\right) = \frac{8v}{D}$

(3) Flow characteristics of non-Newtonian fluids in the pipe flow

In order to obtain the flow curve of non-Newtonian fluids and determine their rheological properties, it is necessary to clarify the shear stress τ_B and shear rate $\left(-\frac{dv}{dr}\right)$ in pipe flow. Assuming

$y = v$, then $dy = dv$. For $z = \pi r^2$, then $dz = 2\pi r \, dr$. An integral can be written as below (Yang et al 2016):

$$\int_a^b y \, dz = yz \big|_a^b - \int_a^b z \, dy \qquad (8)$$

When $r = R$, the flow velocity $v = 0$ and Eq. 8 can be written as:

$$Q = \int_0^R \pi r^2 \left(-\frac{dv}{dr} \right) dr \qquad (9)$$

Substituting $Q = v \pi R^2 = \frac{v \pi D^2}{4}$ and $R = \frac{1}{2} D$ into Eq. 9, we have:

$$\frac{8v}{D} = \frac{32}{D^3} \int_0^{\frac{D}{2}} r^2 \left(-\frac{dv}{dr} \right) dr \qquad (10)$$

The shear rate and shear stress can be expressed using $\left(-\frac{dv}{dr} \right) = f(\tau)$ According to Eq. 5, we have $dr = \frac{D}{2\tau_B} d\tau$. By substituting it into Eq. 10, we have:

$$\frac{8v}{D} = \frac{4}{(\tau_B)^3} \int_0^{\tau_B} \tau^2 f(\tau) \, d\tau \qquad (11)$$

Then,

$$Q = \frac{\pi R^3}{(\tau_B)^3} \int_0^{\tau_B} \tau^2 f(\tau) \, d\tau \qquad (12)$$

$\frac{8v}{D}$ is the function of the shear stress of non-Newtonian fluids on the pipe wall. It is assumed that $\frac{8v}{D} = \varphi(\tau_B)$. By substituting it into Eq. 11 and taking the derivatives of both sides of the equation, we have

$$4f(\tau_B) = 3\varphi(\tau_B) + \frac{\tau_B}{d\tau_B} \cdot d\tau_B \cdot \frac{\varphi'(\tau_B)}{\varphi(\tau_B)} \cdot \varphi(\tau_B) \qquad (13)$$

For $\frac{d\tau_B}{\tau_B} = d\ln\tau_B$, we gain $\frac{\varphi'(\tau_B)}{\varphi(\tau_B)} = d\tau_B$. By substituting it into Eq. 13 and introducing the generalized flow behavior index $\frac{d\tau_B}{\tau_B} = d\ln\tau_B$ and introducing $\frac{d\ln\tau_B}{d\ln\frac{8v}{D}} = N$, we have:

$$4\left(-\frac{dv}{dr} \right) = 3\left(\frac{8v}{D} \right) + \frac{d\ln\frac{8v}{D}}{d\ln\tau_B} \cdot \frac{8v}{D} \qquad (14)$$

By substituting $\frac{d\ln\frac{8v}{D}}{d\ln\tau_B} = \frac{1}{N}$ into Eq. 14, the shear rate on the pipe wall can be computed as below:

$$\left(-\frac{dv}{dr} \right) = \gamma = \frac{8v}{D} \left(\frac{3N + 1}{4N} \right) \qquad (15)$$

The above Eq. 15 is the calculation model of the shear rate of time-independent non-Newtonian fluids (including Newtonian fluids) on the pipe wall.

2.5 *Fluid constitutive equation*

Fluids include Newtonian and non-Newtonian fluids. At present, most drilling fluids used in the field are non-Newtonian fluids. The correlation between the shear stress and shear rate of drilling fluids can accurately characterize the rheological properties of drilling fluids, which can be illustrated as the flow curve (Figure 3).

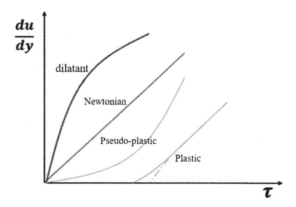

Figure 3.   Schematic diagram of flow curves.

The drilling safety is greatly affected by the rheological property of drilling fluids. Good and stable rheological properties allow for accurate calculation of the bottom-hole pressure and effective cleanup of drilling cuttings and are in favor of high-quality fast drilling. The rheological models commonly used in the field are mainly the two-parameter, three-parameter, and four-parameter rheological models. The two-parameter rheological models include the Bingham model, power-law model, and Casson model. The three-parameter rheological model includes the Herschel-Bulkley (H-B) model (Fan et al. 2010).

1) Bingham fluid model

This rheological model is proposed in 1922 to describe the flow regime of plastic fluids. Such fluids are associated with a gel strength corresponding to the zero shear rate, which is the minimum shear stress required to be exceeded by the external shear force for fluid flow. The gel strength is also referred to as the yield point. Because the two rheological parameters of this model, namely the yield point τ_0 and plastic viscosity η, are calculated within the shear rate range of $511\mathrm{s}^{-1} - -1022\mathrm{s}^{-1}$. Therefore, it describes the rheology of fluids at high shear rates. The constitutive equation of the Bingham model is shown below:

$$\tau = \tau_0 + \eta\gamma \tag{16}$$

Because the apparent viscosity of Bingham fluids decreases with the increasing shear rate, Bingham fluids are shear-thinning fluids.

2) Casson fluid model

With the development of the calculation theory of hydraulics, Casson proposes a new rheological model (Casson model) in the 1950s to accurately calculate the bottom-hole pressure, which can better describe the yield point and shear-thinning behavior of fluids. Compared with the Bingham model applicable to high shear rates, this model is found to have better application performance in cases of low and medium shear rates. The constitutive equation of the Casson model is as below (Dai et al. 2017):

$$\tau^{\frac{1}{2}} = \tau_c^{\frac{1}{2}} + \eta_\infty^{\frac{1}{2}}\gamma^{\frac{nl}{2}} \tag{17}$$

3) Power-law fluid model

This model is applicable to fluids at low shear rates. The two parameters of the model can be calculated at any shear rate. Generally, the power-law model is suitable for both pseudo-plastic and dilatant fluids. The flow behavior index n < 1 indicates pseudo-plastic fluids, while that > 1 suggests dilatant fluids. Since there is no yield stress, such fluids begin to flow, with any non-zero external force (no matter how small it is). There is a nonlinear relationship between the shear rate and shear stress in this model. The constitutive equation can be expressed as:

$$\tau = K\gamma^{n} \tag{18}$$

4) Herschel-Bulkely fluid model (H-B)

This model is applicable to pseudo-plastic fluids demanding a minimum external force for fluid flow (defined as the yield stress). When the yield stress is ignored and the flow behavior index n=1, the model is simplified into the Newton fluid model. With the flow behavior index n < 1 or n > 1, this model is the same as the power-law model. For drilling, the three-parameter H-B model can better describe the rheological properties of drilling fluids, compared with the two-parameter Bingham and power-law models. Moreover, it is applicable to both polymer and water-based drilling fluids. This model is found with higher accuracy particularly at low shear rates, compared with the above three models. Its constitutive equation is shown below (Yuan et al. 2018):

$$\tau = \tau_0 + K\gamma^{n} \tag{19}$$

where τ is the shear stress, Pa; τ_0 is the yield point, Pa; γ is the shear rate, s^{-1}; η is the Bingham plastic viscosity, Pa · s; η_∞ is the Casson plastic viscosity, Pa · s; –Consistency coefficient; K is the fluid consistency coefficient, Pa · s; n is the flow behavior index.

3 EXPERIMENT DATA AND ANALYSIS

5% bentonite slurries are prepared and used as the drilling fluid sample, which is extracted into the drilling fluid rheology real-time measuring device for testing. During the test, the data is recorded after the differential pressure at each pump rate is stable, so as to reduce the transient effect. The measured differential pressure of drilling fluids is shown in Figure 4.

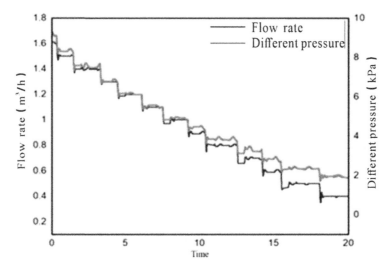

Figure 4. Flow rate vs. differential pressure.

It can be observed that the differential pressure is associated with some fluctuation (yet, within a certain range) and complies with the rheological pattern of drilling fluids. The differential pressure data is processed using the method presented above. Part of the processed differential pressure data is used to plot the viscosity curve, which reflects the characteristics of non-Newtonian fluids in a wide shear rate range from low to high (Figure 5). The flow curve is a must for real-time measurement and calculation of the rheology of drilling fluids.

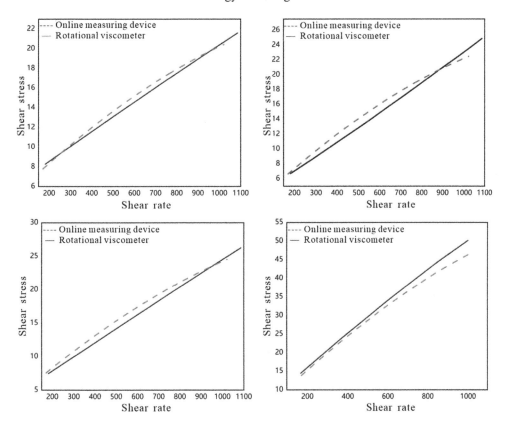

Figure 5. Flow curves, AV 20–46 mPa·s.

It can be seen that for 20 mPa·s, 24 mPa·s, and 38.5 mPa·s, the flow curves measured by the online measuring device are almost identical to those measured by the rotational viscometer, and only a slight deviation is found in cases of higher shear rates of 46 MPa·s, which indicates possible random system environmental error.

In general, the flow curves demonstrate that the measurements of the presented online measuring device are very close to those of the Fann rotational viscometer. They present almost identical variation trends and the deviations are small. Even in the presence of the system environmental error, the deviations are within 4%, which meets the requirement of the on-site automated measurement. $\ln\gamma$ versus$\ln\tau_w$ of each case of viscosity can be used to compute the generalized flow behavior index N in a real-time manner, as required by model calculation. It also shows that the n value is relatively stable, due to the optimization of the flow channel, and the adopted fitting algorithm produces highly matched fitting curves.

The viscosity of drilling fluids can be calculated as long as the generalized flow behavior index N is determined, in accordance with the flow curves corresponding to varied viscosity values. In our research, the rotational viscometer is used for comparison. The measurements of the online measuring device and rotational viscometer are summarized and compared in Table 1.

As shown in Table 1, the discrepancies between the measured data of the presented online measuring device and the rotational viscometer are small and considered within a reasonable error

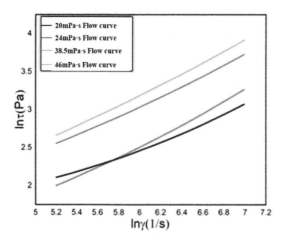

Figure 6. $\ln\gamma$ vs.$\ln\tau_w$, AV 20–46 mPa·s.

Table 1. Summary and comparison of experiment results.

Experimental device			Rotary viscometer			Deviation		
AV	PV	YP	AV	PV	YP	AV	PV	YP
7.76	4.21	0.83	7	4	2.04	0.76	0.21	−1.21
12.43	5.34	5.46	11	5	4.59	1.43	0.34	0.87
15.79	7.85	5.03	14	8	6.13	1.79	−0.15	−1.1
17.33	10.03	4.79	19	9.5	6.13	−1.67	0.53	−1.34
24.19	18.91	4.56	24	17.7	6.43	0.19	1.21	−1.88
38.21	28.64	10.39	40	29.8	10.42	−1.79	−1.16	−0.03
43.78	33.54	12.12	45.25	32	13.54	−1.47	1.54	−1.42
53.3	40.96	12.06	52	39	13.29	1.3	1.96	−1.23
72.02	60.63	11.29	72	59	13.29	0.02	1.63	−2
89.36	71.73	19.15	92	72	20.44	−2.64	−0.27	−1.29

AV is the apparent viscosity, mPa·s; PV is the plastic viscosity, mPa·s; YP is the yield point, Pa; The flow rate unit is m^3/h

range. Thus, the presented online measuring device can be used as an alternative to the six-speed rotational viscometer. It can greatly improve the efficiency of field operations and is expected to make great contributions to the petroleum drilling industry.

4   CONCLUSIONS

The current measurement of the rheology of drilling fluids is mostly manual, which is time- and labor-consuming and suffers from manual operation errors. Given such problems of manual measurement of rheological properties of drilling fluids, this research develops an online automated property measuring device for drilling fluids, so as to accurately monitor the rheology of drilling fluids in a real-time manner.

The measurement adopts two measuring tubes having equal flow rates and different flow velocities, which are connected in a cascade manner. The differential pressures of the two measuring tubes are measured respectively by the corresponding differential pressure transducers. Subsequently, based on the rheology, fluid mechanics, and the drilling fluid rheological parameters measured by a mass flowmeter, the rheological calculation models for different fluids and flow regimes are developed. The flow curves measured by the presented online measuring device are compared with the measurements of the Fann six-speed rotational viscometer. The results of the two devices present nearly identical variation trends and very close values. The deviations can be kept within 4% in the presence of the system environmental error, which meets the requirement of the on-site automated measurement. The measured generalized flow behavior index N is stable and the adopted fitting algorithm produces highly-matched curves. These validate the accuracy of the invented online automated measuring device of drilling fluids. To sum up, the developed online automated measuring device for drilling fluids can greatly boost the progress of automation of well drilling and completion and provide a solid basis for future development of drilling technologies. This invention is of great social and economic value and has high scientific significance.

ACKNOWLEDGMENTS

This work was supported by Forward-looking Basic Strategic Technical Research Topics of CNPC (Grant number 2021DJ4404), "Evaluation of drilling fluid online monitoring technology" project of PetroChina Tarim Oilfield Company (No. 201021102345).

REFERENCES

Haobo Zhou, Honghai Fan, Yinghu Zhai, et al. Accurate calculation method of rheological parameters in the four-parameter model and its application [J]. *Acta Petrolei Sinica*, 2012, 33(01): 128–32.

Honghai Fan, Guangqing Feng, Guo Wang, et al. A new rheological model and its application evaluation [J]. *Journal of China University of Petroleum*(Edition of Natural Science), 2010, 34(05): 89–93+9.

Hui Dai, Xiangchuan Ouyang, Xuan Liao, et al. Discussion and improvement on the calculation method of circulating pressure loss in deep well drilling operation[J]. *Natural Gas and Oil*, 2017, 35(01): 70–3+10.

Jiatian Zhang, Xinhua Li. Software design of an automatic testing system for drilling fluid rheological parameters[J]. *Coal Geology & Exploration*, 2001(05): 63–64.

Junliang Yuan, Jianliang Zhou, Jinyin Deng, et al. Effect of the upper limit of collapse pressure on safety density window in directional well drilling[J]. *Oil Drilling & Production Technology*, 2018, 40(05): 572–6.

Le K, Rasouli V. Determination of safe mud weight windows for drilling deviated wellbores: a case study in the North Perth Basin; proceedings of the Petroleum, F, 2012 [C].

Maobo Tan. *Research on the calculation method of bottom hole pressure based on temperature and pressure influence* [D]; Southwest Petroleum University, 2015.

Ping Chen. *Drilling and completion engineering*[J]. 2011,

Qi Li, Tao Wang, Teng Teng, et al. A new method for calculating rheological parameters of drilling fluid and optimization of rheological model [J]. *Journal of Xi'an Shiyou University*(Natural Science Edition), 2015, 30(02): 84–7+92+11.

Shuangbao Liu, Zhongjie Wang, Yunqian Liu, et al. A new method for online testing of drilling fluid rheology[J]. *Drilling Fluid & Completion Fluid*, 2016, 33(04): 56–59.

Xiumei Zhang, Zhaomo Zhou, Zunli Wang. Pressure Tube viscometer[J]. *Chinese Journal of Scientific Instrument*, 1998(03): 3–5.

Xu Li, Kai Zhang, ZhiFeng Shi, et al. Summary of calculation methods of formation pressure[J]. *West-China Exploration Engineering*, 2017, 29(03): 32–3.

Yingbo Yang, Jun Li, Yanqing Shen, et al. Discussion on the limitation of the traditional method for calculating cycle pressure loss and verification of the improved method [J]. *Petrochemical Industry Application*, 2016, 35(07): 33–6.

Yingjie Chen, Chuanguang Deng, Shoutian Ma. Reliability theory evaluation method of wellbore instability risk [J]. *Natural Gas Industry*, 2019, 39(11): 97–104.

Yukai Gao.Experimental study on online measurement technology of drilling fluid viscosity[J]. *Oil-Gas Field Surface Engineering*, 2002, 27(1): 67–68.

Advances in Energy, Environment and
Chemical Engineering – Abdullah & Osman (Eds)
© 2023 The Author(s), ISBN: 978-1-032-36083-6

Influence of large-scale PV station construction on regional air temperature

Wei Wu, Bo Yuan, Lei Ren & Shengjuan Yue
State Key Laboratory of Eco-hydraulics in Northwest Arid Region, Xi'an University of Technology, Xi'an, P.R. China

Penghui Zou & Ruoting Yang
State Power Investment Group Qinghai PV Industry Innovation Center Co., Ltd., Xining, P.R. China

Xiaode Zhou
State Key Laboratory of Eco-hydraulics in Northwest Arid Region, Xi'an University of Technology, Xi'an, P.R. China

ABSTRACT: This study analyzes the variation characteristics of the air temperature of the Gonghe photovoltaic (PV) station in Qinghai based on the measured data, and discusses the impact of PV development on the temperature. The results show that under the height of 2.5m of the Gonghe PV station in the desert area, the daytime temperature has a warming effect in spring and summer, and the PV power station has a cooling effect in autumn and winter. Under different seasonal conditions, the temperature increase range of PV stations is small during the day, but it has a significant cooling effect at night, which is characterized by the minimum cooling in summer, and the maximum cooling in winter is 2.33°C.

Keywords: Air temperature; PV power station; desert area

1 INTRODUCTION

Global warming is faster than expected and its harm is constantly emerging, forcing the need to adjust the energy structure. Under the pressure of global warming and the energy crisis, vigorously developing clean energy represented by solar energy and wind energy is an important measure to achieve the "double carbon" goal and actively deal with climate change (Wang et al. 2021). How to scientifically and reasonably develop solar energy resources on a large scale while ensuring the sustainable development of regional ecology and climate environment has become a growing concern of experts and policymakers at home and abroad (Rahman et al. 2022). It is also a new challenge for China's long-term development of clean energy and coping with climate change in the future (Al-Dousari et al. 2019; Guo et al. 2019).

As we all know, solar energy will not produce a large number of greenhouse gases and other pollution gases in the process of power generation, but the PV array converts part of solar energy into electric energy, changing the surface radiation balance, which may have an impact on the weather and climate of large-scale PV power stations and surrounding areas (Hosenuzzaman et al. 2015; Pérez et al. 2018). In recent years, the research on the impact of PV stations on climate and the environment has been paid more and more attention (Balakrishnan et al. 2019; Tawalbeh et al. 2021; Zhao et al. 2020). Genchi et al. (2003) studied the impact of large-scale roof PV devices in Tokyo on the urban heat island effect and found that the impact is insignificant. Nemet et al. (2009) studied the impact of the change of albedo caused by the installation of a wide range of PV panels on the global climate and found that the adverse impact on the global climate is very small. Theocharis et al. (Tsoutsos et al. 2005) investigated 32 kinds of effects of the PV electric

DOI 10.1201/9781003330165-111

field in its whole existence stage, determined that these effects have no negative effects except the "local climate effect", and pointed out that the problem of the "local climate effect" needs further observation and research.

The desert area in Northwest China is a typical fragile ecological area and also an area rich in solar energy resources. The construction of solar power stations in desert areas can not only promote the ecological restoration of regional vegetation but also make full use of solar light resources to provide energy guarantees for regional economic development. However, due to the reflection of the solar PV array, some solar radiation is fed back into the air, and the additional energy input will cause a change in the microclimate environment. At present, there are relatively few studies on the impact of large-scale PV power station construction on air temperature in desert areas. Based on this, this study takes Qinghai Gonghe's large-scale centralized PV power station as the research object, uses a long series of monitoring data, analyzes the multi-scale variation characteristics of air temperature in the construction area of the large-scale PV power station, and analyzes the impact of PV power station construction on it, to provide support for the future exploration and in-depth study of the effects of PV development on climate change and ecological environment.

2 STUDY AREAS AND METHODS

2.1 Overview of the study area

Qinghai Gonghe PV power generation industrial park is located in Gonghe County, Hainan Tibetan Autonomous Prefecture, Qinghai Province. The project is 12 km away from Gonghe County. At present, the total planned area of the industrial park is 3028.5 km^2, including 609.6 km^2 of PV power park, 2400 km^2 of wind power park, and 18.9 km^2 of industrial processing and manufacturing zone. Gonghe County has a plateau continental climate, with drought and little rain, a cool climate, sufficient sunshine, and a large temperature difference between day and night. The annual average temperature is 4.1°C, the average annual precipitation is 250-450 mm, the average annual sunshine hours are 2907.8 h, and the average annual solar radiation is 6564.26 MJ/m^2.

2.2 Data sources and analysis methods

To compare and analyze the various characteristics of air temperature inside and outside the PV power station, three monitoring stations are arranged in the PV power station, the transition zone, and outside the PV power station (Figure 1a, b). The first monitoring point is located in the PV power station (Figure 1c, 36.131°N, 100.567°E, altitude 2913 m). The second monitoring point is located in the transition zone from the PV power station to the desert area (Figure 1d, hereinafter referred to as transition zone 36.096°N, 100.507°E, altitude 2924 m). The third monitoring point is located outside the PV power station (Figure 1e 36.187°N, 100.509°E, with an altitude of about 2922 m). The monitoring station outside the PV power station is used as the control point of the experiment. The three monitoring stations use the same monitoring equipment. The data collector adopted by the observation station is cr1000x, which collects and records synchronously every 30 minutes.

The annual average, seasonal average, and daily average values are calculated from the 30-minute raw air temperature data for comparison. The data obtained in this study are the monitoring data from January 2019 to December 2020 for two consecutive years. SPSS 22.0 software (IBM, Montauk, New York, USA) was used for one-way ANOVA, Pearson correlation analysis, and significance test, and origin Pro software was used for mapping (Originpro, Originlab, USA).

3 RESULT ANALYSIS AND DISCUSSION

3.1 Daily variation of the air temperature inside and outside PV station

The annual average daily variation process and daytime and nighttime average values of the temperature at 2.5m inside, in the transition area, and outside the PV power station are shown in Figure 2.

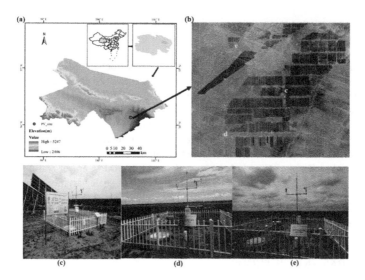

Figure 1.   Location of the sampling sites in the PV station of Gonghe, Qinghai.

It can be seen that the temperature at the three monitoring points reaches the minimum at about 8:00 and the maximum at about 16:30. The temperature of the three monitoring sites tends to be the same during the day, while the difference is significant at night. Considering the significant difference in temperature change between day and night and the main working hours of PV power stations during the day, the temperature is divided into day and night. The daytime temperature is the average value of 08:00-20:00 local time, the nighttime is the average value of 21:00-07:00 local time, and the daily average temperature is the average value of 24 hours a day. The annual average daytime temperature in the PV power station, transition zone, and outside the power station are 6.51°C, 6.41°C, and 6.48°C respectively; The annual average night temperature is - 0.54°C, 0.76°C, and 0.9°C respectively. The annual average daily temperature in the PV power station, the transition area, and outside the power station are 3.14°C, 3.70°C, and 3.81°C respectively. The annual average daily temperature in the PV power station is 0.56°C and 0.67°C lower than that in the transition area and outside the power station respectively.

Figure 3 shows the annual average daily variation process of the air temperature difference in the PV power station, the transition area, and outside the power station, and the average value of the temperature difference between day and night. It can be seen that the air temperature difference between the PV power station and the transition area and inside the PV power station and outside the power station increases first and then decreases. The temperature difference between day and night is above 0°C and below 0°C. The annual average daytime air temperature difference shows that the temperature inside the PV power station increases by 0.04°C and 0.11°C respectively compared with that outside the power station and the transition zone. The annual average night temperature difference is 1.30°C and 1.44°C lower than the transition area and outside the PV power station. Therefore, from the change of annual average daily temperature, the PV power station has a temperature increasing effect during the day, but the temperature increasing effect is not significant, while it has a cooling effect at night, and cooling effect is significant.

3.2   Seasonal and annual variation of air temperature inside and outside PV station

Figure 4 shows the seasonal average daily variation process of air temperature difference in the PV power station, transition area, and outside the power station and the average value of temperature difference between day and night. It can be seen that the fluctuation range of temperature difference is the largest in winter and the smallest in autumn. During the day in spring, the temperature in the

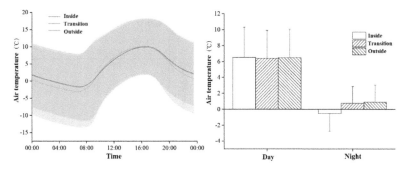

Figure 2.   The annual average daily temperature change process and the daytime and nighttime changes.

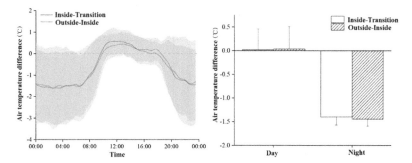

Figure 3.   The annual average daily temperature difference change process and the daytime and nighttime changes.

transition area and outside the power station increases by 0.11°C and 0.22°C respectively, while at night, the temperature in the transition area and outside the power station decreases by 1.34°C and 1.38°C respectively; The temperature in the transition area of the PV power station is 19.0°C in summer, which is the same as that in the transition area of the PV power station and 0.0°C outside the transition area of PV power station in the daytime; In autumn, the temperatures of the three monitoring points are the same. The temperatures in the transition area and outside the PV power station are reduced by 0.02°C and 0.01°C respectively, and 1.01°C and 1.12°C respectively at night. In winter, the temperatures in the transition area and outside the PV power station are reduced by 0.4°C and 0.26°C respectively, and 2.33°C and 2.18°C respectively at night. By comparing the seasonal daily average temperature difference, it is found that the PV power station has the effect of increasing temperature in the daytime of spring and summer; The temperature inside and outside the PV power station is the same in autumn; In winter, the daytime PV power station has a cooling effect. The PV power stations at night in all seasons show a cooling effect, and the cooling at night in winter is the most significant.

Figure 5 shows the annual variation of air temperature difference in the PV power station, transition area, and outside the power station. It can be seen that the temperature variation trend of the three monitoring points is consistent. The maximum temperatures inside, in the transition zone, and outside the PV power station are 15.11°C, 15.12°C, and 15.25°C respectively; The lowest temperature is - 11.20c, - 10.09°C, and - 9.95°C respectively. In the process of annual change, the temperature in the PV power station is lower than that in the transition area and outside the power station. The PV power station has a cooling effect. The maximum temperature difference between the PV power station and the transition area and between the PV power station and the outside of the power station is 1.38°C and 1.32°C respectively, and the minimum value is 0.10°C and 0.14°C respectively. The temperature difference between the inside of the PV power station and

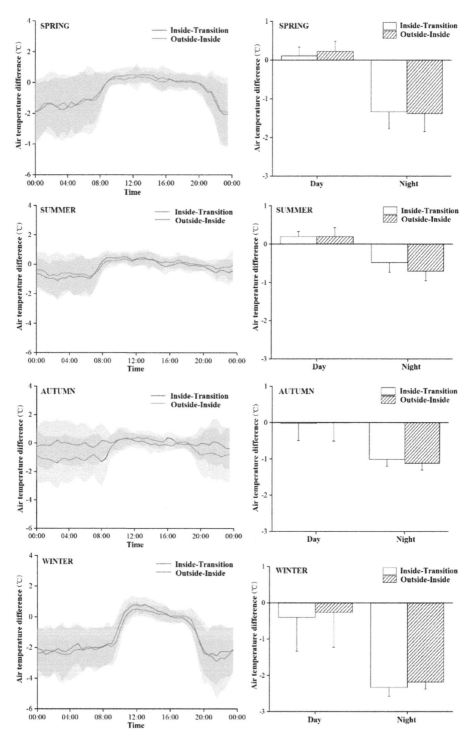

Figure 4.　Seasonal mean daily change process of the temperature difference and daytime and nighttime changes.

the transition zone and outside the power station is small in the high-temperature period, but large in the low-temperature period.

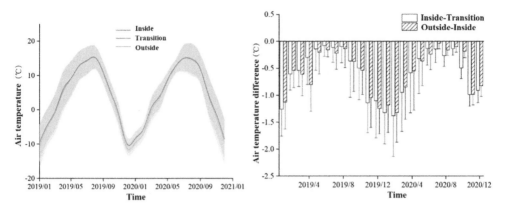

Figure 5. Annual variation of temperature and temperature difference.

3.3 Environmental effect analysis of PV station on air temperature

In this study, the daytime temperature of 2.5m in the Gonghe PV electric field in a desert area has a warming effect in spring and summer (Figure 4). The warming effect is the largest in summer, and the closer it is to the PV center, the greater the warming effect. This may be due to the higher solar radiation intensity in the warm season in the desert area, the longer the operation time of the PV electric field, and the significant impact of PV panels on the warming effect of the surrounding environment. The daytime warming effect has also been reported in the field monitoring research of PV electric fields in other regions at home and abroad. For example, the daytime temperature of the Qinghai Golmud PV electric field increases the most in 2m in summer, with a temperature increase of 0.67°C (Yang et al. 2017). Due to the different layout positions of temperature monitoring instruments, the influence degree of the obtained PV electric field on the temperature is different. For example, the daytime temperature of the PV panel itself increases by 9.7°C compared with the maximum temperature of 2m in the surrounding environment (Chang et al. 2018). With the increase of the height from the ground, the PV electric field produces a cooling effect, and the generated heating heat will completely dissipate into the environment at the height of 5-18m. With the increase in the distance from the PV electric field, the heat energy dissipates rapidly, and the air temperature is close to the ambient temperature (Fthenakis & Yu 2013). The PV electric field has a cooling effect in autumn and winter (Figure 4). The maximum temperature drop at night in winter is 0.4 °, which may be due to the low solar radiation intensity in the cold season. The PV electric field absorbs part of the solar radiation and converts it into electric energy so that the energy obtained by the underlying surface in the electric field is lower than that outside the power station.

In this study, the temperature at night in four seasons has a cooling effect (Figure 4), with the minimum temperature in summer and the maximum temperature in winter, with the maximum temperature of 2.33°C. Different underlying surfaces have different conclusions on the impact of PV electric fields on night temperature. For example, Tucson's PV electric field in the desert area is 2.5m, and the night temperature is 3-4°C higher than that in surrounding areas. In the warm season (spring and summer), the PV heat island effect increases significantly than the urban heat island effect (Barron-Gafford et al. 2016); On the other hand, in the Golmud PV field in the desert area, the temperature at 2m at night also has a warming effect, and the PV panel has a thermal insulation effect on the near-surface layer. In this study, the underlying surface of the Gonghe PV electric field is desert with sparse vegetation. Chang et al. (2018) showed that PV panels at night have a cooling effect, which is consistent with the conclusion of this paper. Based on different climatic conditions, seasons, underlying surface and the scale of the PV electric field, the ambient temperature inside and outside the PV electric field may be cooled differently at night.

In this study, the temperature increase of the PV electric field during the day is small, but the cooling effect at night is significant. The average daily change of PV electric field in four seasons is the cooling effect (Figure 2). The maximum temperature in the PV electric field is 1.38°C and 1.32°C lower than that in the transition area and control area respectively. The PV electric field will not form a "heat island effect" and potentially adverse microclimate, which may be related to the cooling effect of the PV electric field monitoring station close to the roads in the field (Fthenakis & Yu 2013). And the advection of impervious roads may affect the monitored air temperature and humidity (Broadbent et al. 2019).

4 CONCLUSION

This study shows that under the height of 2.5m of the Gonghe PV station in the desert area, the daytime temperature has a warming effect in spring and summer, and the PV power station has a cooling effect in autumn and winter. Under different seasonal conditions, the temperature increase range of PV stations is small during the day, but it has a significant cooling effect at night, which is characterized by the minimum cooling in summer, the maximum cooling in winter, and the maximum cooling of 2.33°C.

REFERENCES

Al-Dousari, A.; Al-Nassar, W.; Al-Hemoud, A.; Alsaleh, A.; Ramadan, A.; Al-Dousari, N.; Ahmed, M. Solar, and wind energy: Challenges and solutions in desert regions. *Energy* 2019, 176, 184–194, DOI:https://doi.org/10.1016/j.energy.2019.03.180.

Balakrishnan, A.; Brutsch, E.; Jamis, A.; Reyes, W.; Strutner, M.; Sinha, P.; Geyer, R.; Ieee. Environmental Impacts of Utility-Scale Battery Storage in California. In Proceedings of the IEEE 46th Photovoltaic Specialists Conference (PVSC), Chicago, IL, 2019 Jun 16-21, 2019; pp. 2472–2474.

Barron-Gafford, G.A.; Minor, R.L.; Allen, N.A.; Cronin, A.D.; Brooks, A.E.; Pavao-Zuckerman, M.A. The Photovoltaic Heat Island Effect: Larger solar power plants increase local temperatures. *Scientific Reports* 2016, 6, 35070, doi:10.1038/srep35070.

Broadbent, A.; Krayenhoff, E.; Georgescu, M.; Sailor, D. The Observed Effects of Utility-Scale Photovoltaics on Near-Surface Air Temperature and Energy Balance. *Journal of Applied Meteorology and Climatology* 2019, 58, doi:10.1175/JAMC-D-18-0271.1.

Chang, R.; Shen, Y.; Luo, Y.; Wang, B.; Yang, Z.; Guo, P. Observed surface radiation and temperature impacts from the large-scale deployment of photovoltaics in the barren area of Gonghe, *China. Renewable Energy* 2018, 118, 131–137, DOI:https://doi.org/10.1016/j.renene.2017.11.007.

Fthenakis, V.; Yu, Y. Analysis of the potential for a heat island effect in large solar farms. In Proceedings of the 2013 IEEE 39th Photovoltaic Specialists Conference (PVSC), 16-21 June 2013, 2013; pp. 3362–3366.

Genchi, Y.; Ishisaki, M.; Ohashi, Y.; Kikegawa, Y.; Takahashi, H.; Inaba, A. Impacts of large-scale photovoltaic panel installation on the heat island effect in Tokyo. In Proceedings of the Fifth Conference on the Urban Climate, Tokyo, Japan, 2003.

Guo, B.; Javed, W.; Khoo, Y.S.; Figgis, B. Solar PV soiling mitigation by electrodynamic dust shield in field conditions. *Solar Energy* 2019, 188, 271–277, doi:10.1016/j.solener.2019.05.071.

Hosenuzzaman, M.; Rahim, N.A.; Selvaraj, J.; Hasanuzzaman, M.; Malek, A.B.M.A.; Nahar, A. Global prospects, progress, policies, and environmental impact of solar photovoltaic power generation. Renewable and Sustainable Energy Reviews 2015, 41, 284–297, DOI:https://doi.org/10.1016/j.rser.2014.08.046.

Nemet, G.F. Net Radiative Forcing from Widespread Deployment of Photovoltaics. *Environmental Science & Technology* 2009, 43, 2173–2178, doi:10.1021/es801747c.

Pérez, J.C.; González, A.; Díaz, J.P.; Expósito, F.J.; Felipe, J. Climate change impact on future photovoltaic resource potential in an orographically complex archipelago, the Canary Islands. *Renewable Energy* 2019, 133, 749–759, DOI:https://doi.org/10.1016/j.renene.2018.10.077.

Rahman, A.; Farrok, O.; Haque, M.M. Environmental impact of renewable energy source based electrical power plants: Solar, wind, hydroelectric, biomass, geothermal, tidal, ocean, and osmotic. *Renewable and Sustainable Energy Reviews* 2022, 161, 112279, DOI:https://doi.org/10.1016/j.rser.2022.112279.

Tawalbeh, M.; Al-Othman, A.; Kafiah, F.; Abdelsalam, E.; Almomani, F.; Alkasrawi, M. Environmental impacts of solar photovoltaic systems: A critical review of recent progress and future outlook. *Sci. Total Environ.* 2021, 759, 143528, DOI:https://doi.org/10.1016/j.scitotenv.2020.143528.

Tsoutsos, T.; Frantzeskaki, N.; Gekas, V. Environmental impacts from the solar energy technologies. *Energy Policy* 2005, 33, 289–296, DOI:https://doi.org/10.1016/S0301-4215(03)00241-6.

Wang, Y.; Guo, C.-h.; Chen, X.-j.; Jia, L.-q.; Guo, X.-n.; Chen, R.-s.; Zhang, M.-s.; Chen, Z.-y.; Wang, H.-d. Carbon peak and carbon neutrality in China: Goals, implementation path, and prospects. *China Geology* 2021, 4, 720–746, DOI:https://doi.org/10.31035/cg2021083.

Yang, L.; Gao, X.; Lv, F.; Hui, X.; Ma, L.; Hou, X. Study on the local climatic effects of large photovoltaic solar farms in desert areas. *Solar Energy* 2017, 144, 244–253, DOI: https://doi.org/10.1016/j.solener.2017.01.015.

Zhao, X.; Huang, G.; Lu, C.; Zhou, X.; Li, Y. Impacts of climate change on photovoltaic energy potential: A case study of China. *Applied Energy* 2020, 280, 115888, DOI: https://doi.org/10.1016/j.apenergy.2020.115888.

*Advances in Energy, Environment and
Chemical Engineering – Abdullah & Osman (Eds)*
© 2023 The Author(s), ISBN: 978-1-032-36083-6

Variations of terrestrial gravity field before Yangbi, Yunnan M_S6.4 earthquake

Qiuyue Zheng, Yunhua Jin, Jiangpei Huang & Dong Liu*
Yunnan Earthquake Agency, Kunming, Yunnan Province, China

ABSTRACT: A M_S6.4 earthquake occurred in Yangbi, Yunnan Province, China on 21 May, 2021, and its epicenter was located at the western boundary of the Sichuan-Yunnan block in the southeastern margin of the Tibetan Plateau. In this study, we adopted the classical adjustment method to calculate the regional terrestrial gravity monitoring data in southwestern China and obtained the dynamic variation characteristics of the regional gravity field in different time scales from 2016 to 2020. The results showed: (1) The variations of terrestrial gravity field near the epicenter of the Yangbi earthquake showed notable changes in general and were related to the regional tectonic activities; (2) An obvious four-quadrant distribution of gravity changes appeared near Dali, and the feature reversed in the next year. (3) Half a year before the Yangbi M_S6.4 earthquake, there was a significant reversal of the high gradient zone of gravity field near the epicenter, and the Yangbi M_S6.4 earthquake occurred near the zero contour line of this high gravity gradient zone.

1 INTRODUCTION

The gravity field is one of the basic geophysical fields of the earth. Earthquakes, groundwater changes, volcanism, tectonic movements, vertical crustal deformation, other material changes, and geodynamic processes in the earth can cause the change of the earth's gravity field to a certain extent (Chen et al. 2015; Kuo & Yue 1993; Zhu et al. 2012). Since an earthquake is a special form of concentrated release of internal energy in the process of crustal tectonic movement, the relationship between earthquake and gravity is closely related to crustal deformation and density change. As early as the 1960s, studies have shown that the change of gravity field was caused by large earthquakes (Barnes 1966; Fujii 1966). And in the past 30 years, there have been more research efforts devoted to investigating possible associations between gravity variations and earthquakes in China (Chen et al. 2016, 1979; Zhu et al. 2010; Zhu & Zhan 2012). Therefore, non-tidal variations of regional gravity field obtained by regular gravity repeated monitoring on the surface are likely to capture gravity precursor information related to changes in source materials (Chen et al. 1979; Li 1983), and scholars have been expecting to obtain information of medium changes in deep crustal seismogenic regions through terrestrial gravity observations (Chen et al. 2015, 2016, 1979; Kuo & Yue 1993; Li 1983; Sun et al. 2014; Van et al. 2017; Zhu et al. 2012, 2010; Zhu & Zhan 2012).

Since the 21st century, satellite gravity monitoring methods represented by GRACE (Gravity Recovery and Climate Experiment) have developed rapidly, and have been providing global time-variable gravity field data for geoscience-related fields (Chen et al. 2007; Han et al. 2006; Tapley et al. 2004; Landerer & Swenson 2012). Different from the satellite gravity observation method, the high-precision surface repeated gravity observation method has the characteristics of being close to the field source in the crust, having strong repeatability of observation position, and high precision

---

*Corresponding Author: doumo@whu.edu.cn

DOI 10.1201/9781003330165-112

of observation instrument, and it is suitable for monitoring the microgravity signal directly related to the field source in the crust. In recent years, the CMONOC (Crustal Movement Observation Network of China) has been established, which includes 101 absolute gravimetric stations and more than 3,000 relative gravimetric stations now. The Yangbi $M_S6.4$ earthquake occurred in the southwestern region of CMONOC, which with a dense distribution of gravity stations, and the average distance of relative stations is about 20km. In this study, we have recalculated the terrestrial gravity data, in addition, the relationship between Yangbi $M_S6.4$ earthquake and the evolution of the gravity field before the earthquake has also been discussed.

2 STUDY AREA AND DATA PROCESSING

2.1 *Structure and gravity network distribution of the study area*

As shown in figure 1a, the seismogenic area in the southeastern margin of the Tibetan Plateau, is an important channel for the eastward escape of materials from the Qinghai-Tibet Plateau, and the background of the regional geological structure is very complicated due to the strong orogeny and large strike-slip faults. Yangbi earthquake occurred as the result of strike-slip faulting at relatively shallow depths. In fact, the seismogenic fault is in the west of the Weixi-Qiaohou-Weishan fault, which is an active fault in late Quaternary with the character of NNW dextral strike-slip, is one of the branches of the NW Red River fault zone extending to the north, connecting with the Red River fault in the south and the Jinshajiang fault in the north (Chang et al. 2016). The regional gravity network (figure 1b) is basically covering the main active faults, and the average station spacing is about 20km around Dali and Kunming region, which can monitor the main tectonic activities in the study area.

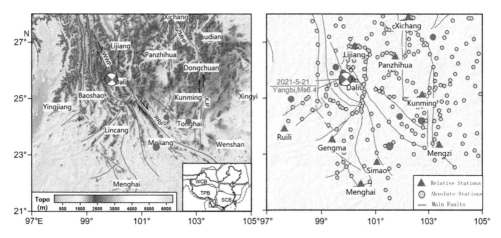

Figure 1. Structure and gravity network distribution of study area ZMHF – Zemuhe Fault, XJF – Xiaojiang Fault, RRF – Red River Fault, WWF – Weixi-Weishan Fault, and the red dots are earthquakes(M≥5) from 2016 to 2021.

2.2 *Data processing*

The relative gravity data is obtained from CG-5 and LaCoste-G relative gravimeter, and the repeated gravity observations are carried out at fixed time intervals (twice a year). The regional observation network is also controlled by 10 absolute gravity stations, obtained by FG-5 and A-10 absolute gravimeters yearly. And the precision of the absolute gravity survey data is higher than 5×10^{-8} m/s^2. In the data processing, the classical adjustment method (Liu et al. 1991) is used and the data is corrected by air pressure, earth tide, oceanic tide, land water load, polar motion, instrument drift, and instrument height. Then the average precision is better than 12×10^{-8}m/s². On the basis of this data processing, the time-variable gravity signals in different spatio-temporal scales are calculated.

3   RESULTS AND DISCUSSION

3.1   *Characteristics of variations of the regional gravity field*

In order to eliminate the interference of seasonal rainfall variation, the data in the second half of each year are used for calculation in this paper. And Kriging interpolation method is used to obtain the characteristics of gravity variations.

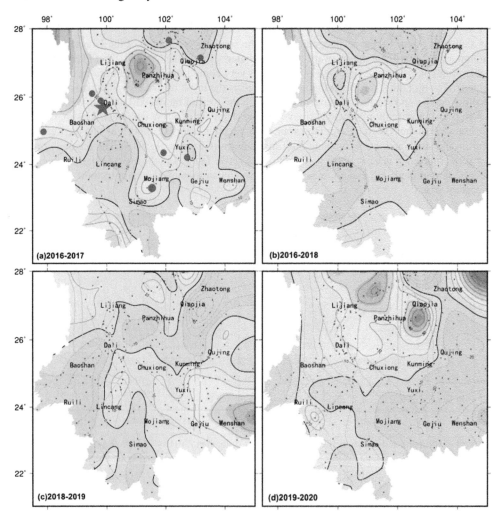

Figure 2.   Characteristics of regional gravity field variations (2016–2020).

2016-2017 (Figure 2a): The changes in the gravity field exhibited a large range of positive variations, accompanied by regional negative variations. There is a gradient zone characteristic of positive and negative gravity variations around the Dali region, and the amplitude is about -10×10^{-8}m/s² to +40×10^{-8}m/s², which shows that the region has a significant background of strong earthquakes. 2016-2018(figure 2b): The cumulative gravity variations from 2016 to 2018 show a four-quadrant gravity distribution of negative (NS) and positive (EW) gravity in the center of Dali. The gravity variation range is about 400km around the Dali region, and the difference between positive and negative gravity is about 50×10^{-8}m/s^2. 2018-2019(figure 2c): Compared with the previous image (figure 2b), there appeared a significant reversal of the four-quadrant feature near Dali in this period, which implied a marked increase in tectonic activity in this area. 2019-2020(figure 2d): The image shows a large region of the positive gravity field in the Baoshan - Dali - Panzhihua area, which may indicate that the materials underground moved into this area before the Yangbi earthquake.

3.2 *Variation of gravity field before the earthquake in the epicenter region*

The accumulated gravity variations show that there is a high gradient zone of positive (East) and negative (West) gravity near the Dali region before the Yangbi earthquake (2016-2020), and Dali is located on the zero contour line of the high gravity gradient zone (figure 3a). However, this high gradient zone reversed rapidly only half a year before the Yangbi earthquake. The variations of gravity field exhibited overall varying characteristics from the southwest to the northeast and alternative distribution of positive and negative variations. And the contour of the gradient zone is basically consistent with that of the Red River fault and Weixi-Weishan fault, which reveals a notable correlation with active structure.

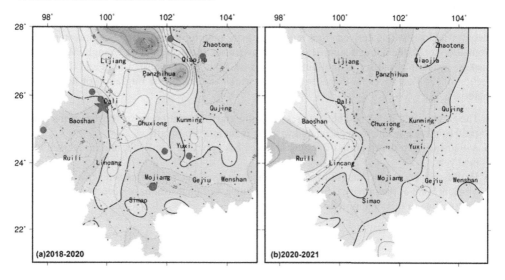

Figure 3.   Variations of gravity field before Yangbi MS6.4 earthquake.

3.3 *Discussion*

Strong earthquakes tend to occur in the high gradient belt of gravity variation related to tectonic activities, the turning part of the contour line of gravity variation, or near the center of the four-quadrant [3, 6]. By analyzing the dynamic evolution images of the regional gravity field before the Yangbi earthquake, it is found that from 2016 to 2017 (Figure 2a), the whole study area presents a large-scale regional of positive gravity changes, which may be large-scale trend gravity changes caused by regional stress enhancement, indicating that this area has a significant background of strong earthquakes. Correspondingly, a M5.0 earthquake occurred in Yunlong, a few dozen kilometers away from Dali, in 2016, and another M5.1 earthquake happened in Yangbi in 2017. The increasing gravitational field may have a relation to the two earthquakes that occurred during this period. However, the cumulative change of gravity field in the two-year period from 2016 to 2018 (Figure 2b) shows a four-quadrant distribution near the epicenter. And this special gravity field variation is related to the main faults in the western boundary of the Sichuan-Yunnan block obviously. The area of the epicenter is a significant tensile deformation area of material extrusion in the southeastern margin of the Qinghai-Tibet Plateau. NNW Weixi-Weishan fault and Red River fault zone on the western boundary of the block both show dextral strike-slip characteristics. The variation pattern of the four quadrants is consistent with the compression and tension region of the focal mechanism, which indicates that the mass migration near the epicenter under the action of tectonic stress may be the cause of the variation of the regional gravity field. In the next period, there was a significant four-quadrant reversal (Figure 2c) from 2018 to 2019, indicating complex tectonic activities around the focal region before the earthquake. The accumulative gravity variation before the Yangbi M_S6.4 earthquake shows a notable high gradient of gravity variation near the epicenter (Figure 3a). However, it changes intensively only half a year before the earthquake, with a significant reverse change of gravity gradient zone (figure 3b), and the epicenter is located on the zero contours of the high gradient zone. In addition, in the southern study area, appearing a four-quadrant distribution in Simao, which is related to the M5.9 Mojiang earthquake that occurred on 8 Sep, 2018. Moreover, the four-quadrant and high gradient character also appeared in the northeast of the study area, related to the Qiaojia M5.0 earthquake on 18 May, 2020.

4 CONCLUSIONS

Gravity field variations derived from regional gravity monitoring data from southwestern China from 2016 to 2020 exhibited significant variations before the occurrence of Yangbi M_S6.4 earthquakes, which may mean that the fault activities near the Weixi-Weishan fault and the northern section of the Red River fault are more active, which leading to the migration of underground materials.

- The gravity field shows notable changes generally, and the variation characteristics of the gravity field near the epicenter are correlated with the main faults basically.
- A four-quadrant distribution of gravity changes has appeared near Dali, and it reversed in the next year.
- Half a year before the Yangbi M_S6.4 earthquake, there was a significant reversal of the high gradient zone of the gravity field near the epicenter, and the earthquake occurred near the zero contour line of this high gravity gradient zone.

ACKNOWLEDGMENTS

Supported by the Special Fund of Yangbi Earthquake of Yunnan Earthquake Agency (grant No: 2021YBZX06) and Seismic Situation Tracing Fund of China Earthquake Administration (grant No: 2021010207).

REFERENCES

Barnes D F. (1966) Gravity Changes during the Alaska Earthquake. *Journal of Geophysical Research-Atmospheres*, 71(2): 451–456.

Chang Z F, Chang H, Zand Y, et al. (2016) Recent active features of Weixi-Qiaohou Fault and its relationship with the Red River Fault. *Journal of Geomechanics*, 22(3): 517–530 (in Chinese).

Chen J L, Wilson C R, Tapley B D, et al. (2007) GRACE detects coseismic and postseismic deformation from the Sumatra-Andaman earthquake. *Geophysical Research Letters*, 34(13): 173–180.

Chen S, Jiang C, Zhuang J, et al. (2016) Statistical Evaluation of Efficiency and Possibility of Earthquake Predictions with Gravity Field Variation and its Analytic Signal in Western China. *Pure and Applied Geophysics*, 173(1): 305–319.

Chen S, Liu M, Xing L, et al. (2015) Gravity increase before the 2015 Mw 7.8 Nepal earthquake. *Geophysical Research Letters*, 43(1): 111–117.

Chen Y T, Gu H D, Lu Z X. (1979) Variations of gravity before and after the Haicheng earthquake, 1975, and the Tangshan earthquake, 1976. *Phys Earth Planet Inter*, 18: 330–338.

Fujii Y. (1966) *Gravity Change in the Shock Area of the Niigata Earthquake*, 16 Jun. 1964. Zisin, 19(3): 200–216.

Han, S C, Shum C K, Bevis M, et al. (2006) Crustal Dilatation Observed by GRACE After the 2004 Sumatra-Andaman Earthquake. *Science*, 313(5787): 658–662.

https://earthquake.usgs.gov/earthquakes/eventpage/us7000e532/executive

Kuo J T, Yue-Feng S. (1993) Modeling gravity variations caused by dilatancies. *Tectonophysics*, 227(1): 127–143.

Landerer F W, Swenson S C. (2012) Accuracy of scaled GRACE terrestrial water storage estimates. *Water Resources Research*, 48(4): 4531.

Li R H. (1983) Local gravity variations before and after the Tangshan earthquake (M=7.8) and dilatation process. *Tectonophysics*, 97(1):159–169.

Liu D, Li H, and Liu S. (1991) *"A management and analysis system of gravity survey data-LGADJ,"* in *Proceedings of the Research Symposium of Application of Seismological Prediction Methods*. Seismological Press, Beijing, China, pp. 339–350.

Sun, S A, Hao H H, et al. (2014) Characteristics of gravity fields in the Jinggu M6.6 earthquake. *Geodesy & Geodynamics*, 5(4): 34–37.

Tapley B D, Bettadpur S, Ries J C, et al. (2004) GRACE Measurements of Mass Variability in the Earth System. *Science*, 305(5683): 503–505.

Van C, de Viron O, Watlet A, et al. (2017) Geophysics From Terrestrial Time-Variable Gravity Measurements. *Reviews of Geophysics*, 55: 938-992.

Zhu Y Q, Liu F, Cao J P, Zhao Y F. (2012) Gravity changes before and after the 2010 Ms7.1 Yushu earthquake. *Geod Geodyn*, 3 (4): 1e7.

Zhu Y Q, Zhan F B, et al. (2010) Gravity Measurements and Their Variations before the 2008 Wenchuan Earthquake. *Bulletin of the Seismological Society of America*, 100(5B): 2815-2824.

Zhu, Y Q, Zhan, F B. (2012) Medium-Term Earthquake Forecast Using Gravity Monitoring Data: Evidence from the Yutian and Wenchuan Earthquakes in China. *International Journal of Geophysics*, 2012( ), 1–6.

*Advances in Energy, Environment and
Chemical Engineering – Abdullah & Osman (Eds)*
© *2023 The Author(s), ISBN: 978-1-032-36083-6*

Analysis of NDVI variations and its climate factors based on geographically and temporally weighted regression: A case study in Inner Mongolia, China

Yuwei Wang*
School of Artificial Intelligence, Jianghan University, Wuhan, China

Xiaoliang Meng
School of Remote Sensing and Information Engineering, Wuhan University, Wuhan, China

Kaicheng Wu
School of Artificial Intelligence, Jianghan University, Wuhan, China

ABSTRACT: Exploring the spatiotemporal trajectory of vegetation growth in response to climate factors is of paramount importance for maintaining global ecosystem stability and sustainable regional development. In this paper, based on remote sensing and geographic information system (GIS) technology, the geographically and temporally weighted regression (GTWR) model was applied to investigate the effect of climate change on normalized difference vegetation index (NDVI) index to capture spatial and temporal heterogeneity. As a case study, the GTWR model was compared with global ordinary least squares (OLS), temporally weighted regression (TWR), and geographically weighted regression (GWR) in terms of modeling NDVI in Inner Mongolia, China from 2014 to 2018. Indicated by the goodness of fit of the model, the results confirmed the effectiveness of the GTWR method and its predominance over OLS, TWR, and GWR models, highlighting the necessity of incorporating both spatial and temporal nonstationarity for modeling spatiotemporal variation in NDVI.

1 INTRODUCTION

Acting as an indicator of climate change, vegetation plays a vital role in global terrestrial ecosystems and is the most sensitive component in response to climate change (Afuye et al. 2021; Cao et al. 2021; Li et al. 2019). Monitoring vegetation variations and their underlying determinants have become a hot topic of current research (Chen et al. 2021; Parker, 2020; Villa et al. 2022). Along with the development of remote sensing and GIS, long-time series remote sensing image data are available to effectively track the vegetation dynamics and thus explore the vegetation status at different spatial and temporal scales (Ivanova et al. 2022; Sizov et al. 2021). Meanwhile, vegetation indices are the most important metrics to study the spatial and temporal changes of vegetation, among which normalized difference vegetation index (NDVI), which is closely related to vegetation coverage, is the most commonly used indicator to reflect the vegetation status. NDVI provides a valuable characterization of vegetation growth and has been employed in many studies to illustrate the effects of climate change on vegetation coverage (Gandhi et al. 2015; Huang et al. 2021; Wang et al. 2022).

Due to the spatial autocorrelation and nonstationarity of geographic data, the direct application of general regression analysis to geographic surveys with spatial structural features often fails to

---

*Corresponding Author: weberwang@jhun.edu.cn

   DOI 10.1201/9781003330165-113

fully portray the true relationship between variables, and the problem is more pronounced in areas with large spatial heterogeneity of environments (Shen et al. 2020; Zhao et al. 2021). Among the researches on spatial heterogeneity and association in GIS, the geographically weighted regression (GWR) model is one of the most widely used regression models. In essence, the method extends the linear regression approach by evaluating local estimated parameters so that the model has the capacity to grasp spatial heterogeneity between multiple factors. Nevertheless, GWR can only model cross-sectional data and excludes temporal dimensions, hence the conclusions obtained during spatial modeling are incomplete (Wang et al. 2020). To cope with this, the GWR model is extended by including temporality into geographically and temporally weighted regression (GTWR) as a way to improve the accuracy and explanatory effect of the model. The model is capable of establishing quantitative relationships between dependent and independent variables over temporal and spatial locations and makes it possible for mining the potential spatiotemporal variations, therefore, this methodology has received considerable attention in recent days. Huang et al. first introduced temporal effects into a GWR model to build GTWR model while used to deal with the spatiotemporal nonstationarity of real estate data in Calgary, and their results showed significant benefits of modeling both spatial and temporal nonstationarity, the GTWR model reduced the absolute error in the test samples (Huang et al. 2010). Fotheringham et al. proposed an effective model calibration method for the local effects in time and space considered by the GTWR model, and applied GTWR using 19 years of house price data in London, the empirical results confirmed the validity and superiority of the method (Fotheringham et al. 2015). Based on NDVI and climate data, Yang et al. used GTWR to address the evolution of NDVI trends and responses to climate change, the study highlighted that GTWR had better adaptation in the driving regimes of vegetation (Yang et al. 2021). These scholars provide a tremendous contribution to the knowledge of modeling spatiotemporal variations.

Our aim in this study was to analyze the impacts of climate factors on vegetation trajectory in Inner Mongolia from spatial and temporal perspectives. First, based on a series of remote sensing data from 2014 to 2018, the vegetation cover status and climatic conditions were explored in the study area. Second, the GTWR model was implemented in terms of climate factors, namely precipitation and temperature, to reveal the changing characteristics of the climate pattern and its impact on vegetation productivity. Third, the OLS, TWR, GWR, and GTWR models were examined and compared to validate the effectiveness of including temporal variation information. The results provide a scientific basis for ecological protection and sustainable development of the local livestock industry in Inner Mongolia.

2   STUDY AREA, DATA AND METHODOLOGY

2.1   *Study area*

The Inner Mongolia Autonomous Region is located in northern China (Figure 1), with a total area of about 1,183,000 square kilometers and an average elevation of approximately 1,000 meters (Bai et al. 2021). The average annual temperature ranges from 0°C to 8°C, whereas the annual precipitation fluctuates between 100 and 450 mm. Dominated by a temperate continental monsoon climate, the vegetation cover types in this region ranked in order of forest-steppe, grassland, desert steppe, and desert from east to west, which serves as an important ecological security barrier in northern China (Chen et al. 2022).

Since Inner Mongolia is situated in the transition belt from the humid zone to the arid and semi-arid zone in the north of China, therefore the region is one of the ideal areas to investigate the influence of regional climate changes on vegetation productivity. At the same time, as its ecological environment is more fragile due to the arid climate and the destruction of human reclamation, it is urgent to undertake research on the vegetation change in Inner Mongolia for the formulation of policies related to the conservation and management of Inner Mongolia grasslands (Guo et al. 2021).

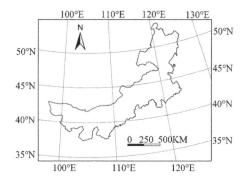

Figure 1. The location of Inner Mongolia.

2.2 *Data*

The data selected for this study consist of NDVI and climatological data. NDVI data were obtained using the 16-day synthetic MOD13A1 product from NASA's EOS/MODIS for the period January 2014 to December 2018, with a spatial resolution of 500m. The product was pre-processed with geometric, radiometric, and atmospheric correction. It was projected and mosaicked through MODIS reprojection tools (MRT) and was clipped by the Inner Mongolia boundary. Then the maximum value composite (MVC) method (Holben, 1986) was utilized to further reduce the effects of clouds, atmosphere, and solar altitude angle to yield monthly NDVI data. The climatic datasets were originated from the National Meteorological Information Center (http://data.cma.cn/), including precipitation (mm) and temperature (°C) data. The daily precipitation data were recalculated as monthly cumulative values and temperature data as monthly average values. Then the recomputed data were interpolated into a continuous raster surface with a spatial resolution of 500 m using ArcGIS 10.2 software for consistency with the NDVI data.

The workflow of the study is shown in Figure 2.

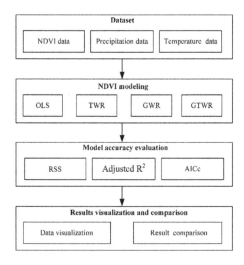

Figure 2. Description of the study workflow.

2.3 Methodology

2.3.1 Geographically and temporally weighted regression

The traditional GWR model accommodates spatial nonstationarity by constructing a weight matrix based on spatial distances (Yellow Horse et al. 2022). Therefore, GWR can only interpret local impacts in space and is inefficient in mining the regression relationship between the dependent and independent variables in the temporal dimension, leading to the omission of some potential local features. An extension of the GWR model, the GTWR model simultaneously considers temporal and spatial nonstationarity by constructing spatiotemporal distances. In detail, the GTWR model introduces temporal dimension so that the coefficients of the local regression equation are a function of geographic location and time scale, which provides powerful assistance in analyzing the temporal and spatial features of the regression relationship. The coordinates of sample i in the spatiotemporal coordinate system are defined as (u_i, v_i, t_i), the general expression of the GTWR model is as follows (Huang et al. 2010):

$$Y_i = \beta_0(u_i, v_i, t_i) + \sum_k \beta_k(u_i, v_i, t_i)X_{ik} + \varepsilon_i \qquad (1)$$

Where Y_i denotes the value of the explanatory variable at sample point i; t_i is the temporal coordinate of the ith sample point; $\beta_0(u_i, v_i, t_i)$ denotes the temporal intercept term of sample point i; X_{ik} denotes the value of the kth explanatory variable at sample point i; $\beta_k(u_i, v_i, t_i)$ denotes the regression coefficient of the kth variable as a function of the spatiotemporal coordinate; and ε_i denotes the residual.

Similar to the GWR model, the estimation of $\beta_k(u_i, v_i, t_i)$ relies on the spatiotemporal weight matrix. The spatiotemporal weight $W(u_i, v_i, t_i)$ is a reflection of the spatiotemporal distance (d_{ST}) from sample point i to other sample points. When calculating the distance d_{ST}, the parameter λ and μ can be multiplied as the balance factor of the spatiotemporal weights.

$$d^{ST} = \lambda d^s \otimes \mu d^T \qquad (2)$$

Where d^S denotes the spatial distance and d^T denotes the temporal distance, operator $\otimes$ represents the operation on spatiotemporal distances. In particular, when the scale factor μ equals 0, only spatial distance and spatial heterogeneity are taken into account, and GTWR becomes the GWR model; when the scale factor λ equals 0, only temporal distance and temporal nonstationarity are focused, which becomes the TWR model.

2.3.2 Model accuracy evaluation

In order to evaluate the influence of climatic factors on NDVI variations, the NDVI data were taken as the dependent variable, while the climate factors were regarded as the independent variables. Subsequently, four categories of regression models, namely, OLS, TWR, GWR, and GTWR were tested and compared with the goodness of fit of the model. Diagnostic indicators of the model, residual sum of squares (RSS), Adjusted R^2, and corrected Akaike information criterion (AICc) are adopted to measure the model's overall accuracy (Permai et al. 2021). RSS estimates the variance in the residuals and represents the effect of random error, the smaller the value, the better the model fits the data. Adjusted R^2 provides a relative measurement of the percentage of explained variance for the dependent variable by the model, reflecting the goodness of fit of the model, ranging from 0 to 1, with the closer to 1.0 the better the fit. AICc builds on the concept of entropy and can examine the complexity of the model and the goodness of the fitted data. It is generally accepted that model with lower AICc is the preferred alternative model to better illuminate the relationship between response variables and explanatory predictions.

3 RESULT

3.1 *Spatiotemporal NDVI and climate variations*

During the period from 2014–2018, the temporally averaged NDVI across Inner Mongolia was calculated and shown in Figure 3. A significant zonal difference was observed in the distribution of NDVI. The higher values were mainly found in the eastern parts of the study area and the lower values were concentrated in the central and western regions. This is attributed to the fact that Inner Mongolia spans a large longitude from east to west, and the distance from the ocean varies in different regions, resulting in a considerable variation in vegetation cover from east to west, which is dominated by forest, steppe and desert respectively. In addition, Figure 4 depicts the temporally averaged precipitation and temperature in Inner Mongolia from 2014 to 2018. The mean precipitation ranged from 95mm to 543.24mm while the mean temperature fluctuated from 3.26°C to 11.14°C. It has been found that the distribution trend of precipitation is consistent with the distribution pattern of NDVI values. It is clear from the figure that the eastern part of the study area is covered with dense vegetation and that these areas experience high levels of precipitation. Meanwhile, the sparse vegetation and bare areas were mainly found in the western sections with lower precipitation (Figure 4a). As opposed to precipitation, the average temperature elevated from northeast to southwest (Figure 4b).

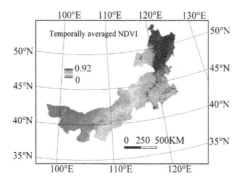

Figure 3.   Spatial patterns of temporally averaged NDVI in Inner Mongolia during 2014–2018.

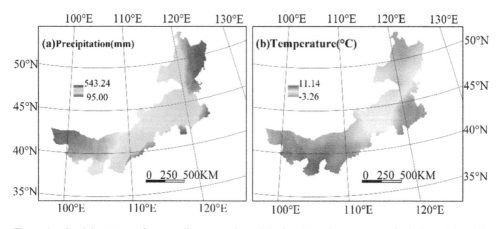

Figure 4.   Spatial patterns of temporally averaged precipitation (a) and temperature (b) in Inner Mongolia during 2014–2018.

792

3.2 Comparison of model accuracy for OLS, TWR, GWR and GTWR

The model accuracy indices for OLS, TWR, GWR, and GTWR models, in terms of RSS, Adjusted R^2, and AICc, were summarized in Table 1.

As seen in Table 1, the value of Adjusted R^2 had a significant enhancement, which was 0.74 in OLS, 0.76 in TWR, and 0.81 in GWR, compared to 0.85 for the GTWR model. Without loss of generality, the GTWR model possesses the maximum Adjusted R^2 value among all the models, indicating the optimum goodness of fit of the model. In addition, the reduction of RSS 40.24 in GTWR, in contrast to 43.45 in GWR, 60.48 in TWR, and 64.57 in OLS, proved that the GTWR model outperformed the other models in grasping spatiotemporal variations in NDVI. Equivalent to the RSS value, the GTWR model with the minimum value is considered to be the most optimal when the difference in AICc value exceeds 3. The above results highlighted that with added temporal information, the GTWR tended to better model NDVI variation than OLS, TWR, and GWR.

Table 1. Model accuracy results between OLS, TWR, GWR, and GTWR models.

Model	RSS	Adjusted R^2	AICc
OLS	64.57	0.74	−2996.45
TWR	60.48	0.76	−3177.46
GWR	43.45	0.81	−4157.6
GTWR	40.24	0.85	−4351.00

3.3 Model result from GTWR

Given that the GTWR model has a parametric set of local estimates for each sample, only the minimum (Min), lower quartile (LQ), median (Med), upper quartile (UQ), and maximum (Max) values of the coefficients for two climatic variables are presented in Table 2 to imply the variability of the estimated parameters. The regression coefficients of precipitation, i.e., 0.09 to 0.98, were consistently positively correlated with NDVI across the whole study area during all study years. Correspondingly, the regression coefficients of temperature i.e., -0.46 to 0.11, transited from negative to positive. The LQ value of -0.46 and UQ value of -0.05 indicated that the coefficient of temperature was negative for most Inner Mongolia areas in different periods.

Table 2. GTWR parameter estimate summaries.

Parameter	Min	LQ	Med	UQ	Max
Intercept	−1.24	−0.37	−0.14	0.24	0.95
Precipitation	0.09	0.44	0.57	0.71	0.98
Temperature	−0.46	−0.21	−0.15	−0.05	0.11

Figure 5 further portrays the spatiotemporal variability of the regression coefficients of the climate factors. Overall, the climate factors manifested different effects on NDVI in different Inner Mongolia regions through different periods. Precipitation displayed a persistent positive effect on NDVI in the whole study area, implying that a higher amount of precipitation would yield greater NDVI values in different spatiotemporal locations in Figure 5(a). Moreover, the negative coefficients for the temperature were observed in most of the study areas and time intervals in Figure 5(b), confirming that increasing temperature would decrease NDVI, which was a sign of a negative impact on vegetation productivity. It is noteworthy that in the northwestern part of Alxa League and Chifeng City in the region, the opposite effect of temperature was detected with an enhancing effect on NDVI. In these localized areas, a higher temperature tended to contribute to more vegetation productivity.

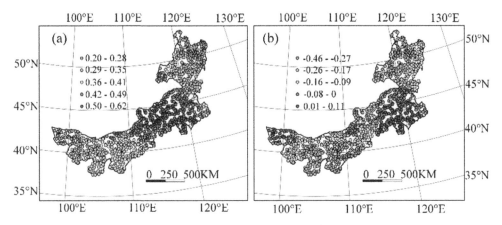

Figure 5. Regression coefficient of the climatological variables against NDVI from GTWR. Coefficient for (a) Precipitation, (b) Temperature.

4 DISCUSSION

In order to explore and analyze the trends of vegetation trajectory in time and space, the two-dimensional GWR model was extended to a three-dimensional GTWR model in this paper, and the related regression analysis was conducted with Inner Mongolia as the sample area. The GTWR model investigated the climatic factors associated with the spatial and temporal variation of NDVI during 2014-2018. Our emphasis on the GTWR model was based on the consideration that incorporating temporal dimension would convey more information in the modeling process, which can better reflect the spatial and temporal correlations.

Inner Mongolia features a wide span and complex landscape types, hence the climatic conditions vary substantially for different regions. Most of the grassland areas in Inner Mongolia feature an arid and semi-arid climate, and the growth of vegetation depends primarily on climatic factors. Significant explanatory factors, namely precipitation and temperature, have been identified as key forces changing vegetation in the study area (Wang et al. 2022). The results of the comparison between different models further verify the existence of spatiotemporal heterogeneity in the effects of temperature and precipitation on NDVI. Climate change has the capacity to modify the composition and structure of vegetation communities, thus affecting the biomass of vegetation, and being a major cause of triggering changes in vegetation cover (Pérez et al. 2021). GWR can provide quantitative results that enable visual and detailed analysis of local spatial relationships. Furthermore, GTWR captures the variability of the effects of climate factors on vegetation NDVI at different time intervals based on GWR, which in turn can provide NDVI analysis at different locations at different periods. This conclusion was also confirmed in the model results from the GTWR model, which presented a positive effect of precipitation on vegetation NDVI and the complex relationship between temperature and NDVI. The analysis revealed that the climate parameters tended to vary over the study area from a spatiotemporal perspective. This deserves to be expected because GTWR can handle both spatial and temporal heterogeneities. The result reinforces the importance of including temporal dimension in the spatial modeling process.

5 CONCLUSION

To conclude, based on GIS spatial analysis technology and the GTWR model, the current study conducted modeling analysis for NDVI variations and climate data of Inner Mongolia from 2014-2018. Our study confirmed that the spatiotemporal heterogeneity should be taken into consideration when modeling vegetation productivity because, it was found that the GTWR model had a better overall model accuracy compared to the OLS, TWR, and GWR models.

For Inner Mongolia, where NDVI distribution exhibits typical spatial and temporal effects, GTWR can be applied to model the spatial and temporal fluctuation trends of vegetation more effectively, thus providing policymakers with more accurate and detailed analysis results to assist decision-making and achieve the purpose of promoting sustainable development of grassland ecosystems. Future work needs to pay attention to compiling more potential variables to explain more NDVI variations.

ACKNOWLEDGEMENTS

This research was funded by the National Natural Science Foundation of China (NSFC): 41971352.

REFERENCES

Afuye, G.A., Kalumba, A.M., Orimoloye, I.R. 2021. Characterization of vegetation response to climate change: a review. *Sustainability* 13, 7265.

Bai, Y., Deng, X., Cheng, Y., Hu, Y., Zhang, L. 2021. Exploring regional land use dynamics under shared socioeconomic pathways: A case study in Inner Mongolia, *China. Technol. Forecast. Soc. Change* 166, 120606.

Cao, D., Zhang, J., Xun, L., Yang, S., Wang, J., Yao, F. 2021. Spatiotemporal variations of global terrestrial vegetation climate potential productivity under climate change. *Sci. Total Environ.* 770, 145320.

Chen, A., Mao, J., Ricciuto, D., Lu, D., Xiao, J., Li, X., Thornton, P.E., Knapp, A.K. 2021. Seasonal changes in GPP/SIF ratios and their climatic determinants across the Northern Hemisphere. *Glob. Chang. Biol.* 27, 5186–5197.

Chen, K., Ge, G., Bao, G., Bai, L., Tong, S., Bao, Y., Chao, L. 2022. Impact of Extreme Climate on the NDVI of Different Steppe Areas in Inner Mongolia, *China. Remote Sens.* 14, 1530.

Fotheringham, A.S., Crespo, R., Yao, J. 2015. Geographical and temporal weighted regression (GTWR). *Geogr. Anal.* 47, 431–452.

Gandhi, G.M., Parthiban, S., Thummalu, N., Christy, A. 2015. Ndvi: Vegetation change detection using remote sensing and gis–A case study of Vellore District. *Procedia Comput.* Sci. 57, 1199–1210.

Guo, X., Chen, R., Thomas, D.S.G., Li, Q., Xia, Z., Pan, Z. 2021. Divergent processes and trends of desertification in Inner Mongolia and Mongolia. *L. Degrad. Dev.* 32, 3684–3697.

Holben, B.N. 1986. Characteristics of maximum-value composite images from temporal AVHRR data. *Int. J. Remote Sens.* 7, 1417–1434.

Huang, B., Wu, B., Barry, M. 2010. Geographically and temporally weighted regression for modeling spatio-temporal variation in house prices. *Int. J. Geogr. Inf. Sci.* 24, 383–401.

Huang, S., Tang, L., Hupy, J.P., Wang, Y., Shao, G. 2021. A commentary review on the use of normalized difference vegetation index (NDVI) in the era of popular remote sensing. *J. For. Res.* 32, 1–6.

Ivanova, N., Fomin, V., Kusbach, A. 2022. Experience of Forest Ecological Classification in Assessment of Vegetation Dynamics. *Sustainability* 14, 3384.

Li, Y., Xie, Z., Qin, Y., Zheng, Z. 2019. Estimating relations of vegetation, climate change, and human activity: A case study in the 400 mm annual precipitation fluctuation zone, *China. Remote Sens.* 11, 1159.

Parker, G.G. 2020. Tamm review: Leaf Area Index (LAI) is both a determinant and a consequence of important processes in vegetation canopies. *For. Ecol. Manage.* 477, 118496.

Pérez, L., Correa-Metrio, A., Cohuo, S., González, L.M., Echeverría-Galindo, P., Brenner, M., Curtis, J., Kutterolf, S., Stockhecke, M., Schenk, F. 2021. Ecological turnover in neotropical freshwater and terrestrial communities during episodes of abrupt climate change. *Quat. Res.* 101, 26–36.

Permai, S.D., Christina, A., Gunawan, A.A.S. 2021. Fiscal decentralization analysis that affects economic performance using geographically weighted regression (GWR). *Procedia Comput. Sci.* 179, 399–406.

Shen, Y., Zhang, C., Wang, R., Wang, X., Cen, S., Li, Q. 2020. Spatial heterogeneity of surface sediment grain size and aeolian activity in the gobi desert region of northwest China. *Catena* 188, 104469.

Sizov, O., Ezhova, E., Tsymbarovich, P., Soromotin, A., Prihod'ko, N., Petäjä, T., Zilitinkevich, S., Kulmala, M., Bäck, J., Köster, K. 2021. Fire and vegetation dynamics in northwest Siberia during the last 60 years based on high-resolution remote sensing. *Biogeosciences* 18, 207–228.

Villa, M.V.E.D., Bruzzone, O.A., Goldstein, G., Cristiano, P.M. 2022. Climatic determinants of photosynthetic activity in humid subtropical forests under different forestry activities. *Remote Sens. Appl. Soc. Environ.* 100735.

Wang, H., et al. 2020. Urban expansion patterns and their driving forces based on the center of gravity-GTWR model: A case study of the Beijing-Tianjin-Hebei urban agglomeration. *J. Geogr. Sci.* 30, 297–318.

Wang, S., Li, R., Wu, Y., Zhao, S. 2022. Vegetation dynamics and their response to hydrothermal conditions in Inner Mongolia, China. *Glob. Ecol. Conserv.* 34, e02034.

Wang, W., et al. 2022. Prediction of NDVI dynamics under different ecological water supplementation scenarios based on a long short-term memory network in the Zhalong Wetland, China. *J. Hydrol.* 608, 127626.

Yang, L., Guan, Q., Lin, J., Tian, J., Tan, Z., Li, H. 2021. Evolution of NDVI secular trends and responses to climate change: A perspective from nonlinearity and nonstationarity characteristics. *Remote Sens. Environ.* 254, 112247.

Yellow Horse, A.J., Yang, T. C., Huyser, K.R. 2022. Structural inequalities established the architecture for the COVID-19 pandemic among native Americans in Arizona: a geographically weighted regression perspective. *J. Racial Ethn. Heal. Disparities* 9, 165–175.

Zhao, X., Liu, J., Bu, Y. 2021. Quantitative analysis of spatial heterogeneity and driving forces of the thermal environment in urban built-up areas: A case study in Xi'an, China. *Sustainability* 13, 1870.

Advances in Energy, Environment and
Chemical Engineering – Abdullah & Osman (Eds)
© 2023 The Author(s), ISBN: 978-1-032-36083-6

Effect evaluation on release and proliferation of *Gymnocypris eckloni eckloni* in the Qinghai section of the Upper Yellow River

Shihai Zhu*, Maoji Dou, Xiangyu Huang, Jiansheng Wang, Yanming Zhao, Shiyan Wang & Ting Ma
College of Eco-Environmental Engineering, Qinghai University, Xining, Qinghai Province, China

ABSTRACT: The effects of release and proliferation were evaluated by constructing the age structure of Longyangxia, Laxiwa, Guide, and Lijiaxia of the Upper Yellow River, and the comparative study of biomass and fish diversity of spotted naked carp before and after proliferation and release was conducted. The results showed that under high fishing pressure, the highest diversity index was 2.05 in the Guide section, 2.02 in the Longyangxia section, 1.80 in the Lijiaxia section, and 1.22 in the Laxiwa section. The species diversity index in the section with more artificial proliferation and release was significantly higher than that in the section with less release. Artificial proliferation and release in the Upper Yellow River can effectively increase the number of released fish, and the species diversity of the community, which play an important role in the recovery of local fishery resources, improvement of the water ecological environment, and protection of endangered species biodiversity.

1 INTRODUCTION

There are 35 species of fish in the Qinghai section of the Upper Yellow River, which belong to 23 genera, 6 families, and 3 orders, including 22 indigenous species and 13 exotic species (Wang et al. 1988; Wu et al. 1992; Zhan 1995). The composition of fish is relatively simple, including the Schizothoracine fishes, nemacheiline fishes of the Cobitidae (loach family), etc. With the rapid development of society and the economy, human activities such as wading project construction create a lot of pressure on the ecosystem of the Yellow River. Dam construction for hydropower development is an important reason for the loss of fish population diversity and the decline or even extinction of fish resources (Chen et al. 2011). Meanwhile, environmental pollution and man-made fishing have greatly changed the fish composition and resource status in the Qinghai section of the Upper Yellow River, and the indigenous fish population in the Upper Yellow River has sharply decreased.

Artificial proliferation and release are currently the means widely used in the world to protect rare and endangered species and restore fishery resources (Pan et al. 2020). In the past ten years, Qinghai Fishery Environmental Monitoring Station has artificially released *Gymnocypris eckloni eckloni* in Longyangxia, Laxiwa, and Guide Reach of the Upper Yellow River. There is no research report on the proliferation and release of *Gymnocypris eckloni eckloni* in the Qinghai section of the Upper Yellow River. We conducted age identification on 15 sections of the *Gymnocypris eckloni eckloni* captured at four sampling points, Longyangxia, Laxiwa, Guide, and Lijiaxia, in the Upper Yellow River. By constructing the age structure of the community, we calculated the proportion of fish biomass and fish species diversity in the breeding section of the corresponding river. Based on the statistical data of aquatic biological economic species proliferation and release in Qinghai

*Corresponding Author: zsh696000@163.com

DOI 10.1201/9781003330165-114

province from 2010 to 2019, the effects of proliferation and release of spotted naked carp were evaluated. The effects of proliferation and release on fish resources were analyzed to provide basic data for the future proliferation and release of indigenous fish in the upper reaches of the Yellow River and to carry out scientific and accurate proliferation and release.

2 MATERIALS AND METHODS

2.1 Sampling site

Longyangxia reservoir area, Laxiwa district, Guide county, Lijiaxia reservoir of the Upper Yellow River.

2.2 Sampling time and method

From October 2019 to September 2020, samples were collected at Longyang Gorge, Laxiwa, Guide, and Lijiaxia in the Upper Yellow River, 4 times at each site.

2.3 Sample processing method

Fish morphological identification and measurement: In the field, the basic data such as length, weight, the number of individuals, and most of the back after the original acquisition waters were determined for species identification.

2.3.1 Age determination of *Gymnocypris eckloni eckloni*: Age determination of *Gymnocypris eckloni eckloni* was carried out using otolith and spine (Zhang et al. 2009)

2.4 Data processing

Data processing software: Excel 2010

3 RESULT ANALYSIS

3.1 Analysis of age structure of Gymnocypris eckloni eckloni in major river sections

Table 1. Age structure of carpio tabularise from Longyangxia to Lijiaxia in the Upper Yellow River.

Age Ratio(%)	2^+	3^+	4^+	5^+	6^+	7^+	8^+	9^+	10^+	11^+	13^+	Number
Longyangxia	3.11	21.40	28.02	26.07	19.46	1.17	0.78					257
Laxiwa	5.69	64.51	29.41	0.39								1530
Guide	0.43	4.69	10.02	18.55	35.18	20.89	6.82	1.70	1.07	0.43	0.21	469
Lijiaxia				2.70	62.16	29.73	2.70		2.70			37

The age proportion of *Gymnocypris eckloni eckloni* in Longyangxia, Laxiwa, Guide, and Lijiaxia sections was analyzed: the total number of individuals of *Gymnocypris eckloni eckloni* in the Longyangxia sample area was 257, aged from 2 to 8 years. The four years old individuals accounted for 28.1%, followed by the five years old individuals (26.07%), the sixth years old (19.46), the two years old individuals (3.03%), the three years old individuals (21.4%), the seven years old individuals (1.17%) and the eight years old individuals (0.78%). In the Laxiwa sample area, there were 1530 individuals, and the age group was 2–5 years. In each age composition, the two years old individuals accounted for 5.69%, and the three years old fish had the largest proportion, accounting for 64.51% of the total. The four years old individuals accounted for 29.41% of the total, while the five years old individuals accounted for only 0.39% of the total. There were 469 individuals in the sample area of Guide, and the age group was 2–13 years. The two years old individuals accounted for 35.18%, and the five years old individuals accounted for 35.18% of the total. The second and seventh-aged individuals accounted for 20.89% of the total. There were fewer individuals over 10 years of age, accounting for only 1.71% of the total. In the Lijiaxia sample area, the total number of fish was 37, and the age group was 5–10 years. Among all age components, the six years old individuals accounted for the largest proportion, accounting for 62.16% of the total. The second was seven years old individuals, accounting for 29.73% of the total. 5, 8, and 10 years old individuals accounted for only 8.1% of the total.

3.2 Analysis of community diversity in different river reaches

Table 2. Species diversity analysis from Longyangxia section to Lijiaxia section of the Upper Yellow River.

Site	Pielou's Index (E)	Margalef Index (D_m)	Shnong-weinaer Index (H)	Simpson Index
Longyangxia	0.6466	2.5861	2.0275	0.9212
Laxiwa	0.63	0.8784	1.2259	0.6967
Guide	0.7833	2.1797	2.0593	0.8721
Lijiaxia	0.3536	1.5922	1.8035	0.8538

As can be seen from Table 2, Pielou's Evenness Index is 0.7833 in the Guide section, 0.3536 in Lijiaxia, and 0.3536 in Longyang and Laxiwa. The Margalef Richness Index was 2.5861 in Longyangxia, 2.1797 in Guide, 1.5922 in Lijiaxia, and 0.8784 in Laxiwa. Shannon-Weiner Index was 2.0593 in Guide, 2.0275 in Longyang Gorge, 1.8035 in Lijiaxia, and 1.2259 in Laxiwa. Simpson Dominance Index shows that the longyangxia section has the highest value of 0.9212, Guide Section 0.8721, Lijiaxia section 0.8538, and Laxiwa section 0.6967.

3.3 Weight analysis of catch

Table 3. Fish catch of *Gymnocypris eckloni eckloni* from Longyangxia section to Lijiaxia section in the Upper Yellow River.

Site	Production (g)	Ratio (%)
Longyangxia	49356.3	33.8
Laxiwa	111582.1	71.85
Guide	143406	77.47
Lijiaxia	14937.9	39.34

The maximum catch was 143.406 kg 111.5821 kg, 49.3563 kg, and 14.9379 kg in the Guide section, Laxiwa section, Longyangxia section, and Lijiaxia section respectively. The weight proportion of naked carp caught was 71.85%, 77.47%, 39.34%, and 33.8% in the , Laxiwa section, Guide section, Lijiaxia section, and Longyangxia section respectively.

4 RESULTS AND DISCUSSION

4.1 *Age structure analysis of Gymnocypris eckloni eckloni from Longyangxia to Lijiaxia in the Upper Reaches of the Yellow River*

The four years old fish accounted for 28.02% of the total, followed by the five-year-old individuals (26.07%), the three-year-old individuals (21.4%), the six-year old individuals (19.46%), the two-year-old individuals (3.11%), the seven-year-old individuals (1.17%), and the eight-year-old individuals (less), which accounted for 0.78% of the total. In 2019, 100,000 *Gymnocypris eckloni eckloni* were released artificially in the Longyangxia section. In 2020, 8 two-year-old individuals of *Gymnocypris eckloni eckloni* were captured, accounting for 3.11% of the total. The captured *Gymnocypris eckloni eckloni* were mainly in the group of 3-, 4-, 5-, and 6-year old, with an average weight of 190g. There are very few large fish with economic value in Longyang Gorge, and the *Gymnocypris eckloni eckloni* is one of the main targets of local fishermen. The fishing intensity is large, which makes it difficult for the *Gymnocypris eckloni eckloni* in Longyang Gorge to be more than six years old, with a small body size and younger age (Mao et al, 2011). The total number of fish caught in the Laxiwa section was 1530, and the age group was 2–5 years. In each age composition, the two-year-old individuals accounted for 5.69%, and the three-year-old individuals accounted for 64.51% of the total. The four-year-old individuals accounted for 29.41% of the total number, while the five-year-old individuals accounted for only 0.39% of the total number. In the Laxiwa section, the terrain on both sides of the Yellow River is very dangerous, and the sampling area is extremely limited. The sampling section is located in the gentle water flow area of Laxiwa, and the water is relatively deep. The total number of carps in the guide sample area was 469, and the age group was 2–13 years. In all age components, the six-year-old individuals accounted for the largest proportion, accounting for 35.18% of the total number of individuals, followed by the seven-year-old individuals, accounting for 20.89% of the total number of individuals, and the number of individuals above age ten was less, accounting for only 1.71% of the total number of individuals. The Guide section accounts for the largest number among the four sampling sections, and the number of captured carps is more, and the age of *Gymnocypris eckloni eckloni* in the Guide section is significantly higher than that in the Longyangxia section. The effect of artificial proliferation and release on the increase of fishery resources is very obvious, it has a good promotion effect on the recovery of natural fishery resources, and plays an important role in the protection of endangered species and biodiversity (Yan 2016). In the Lijiaxia sample area, the total number of carps was 37, and the age group was 5–10. Among all age components, age-six accounted for the largest proportion and the largest number of individuals, accounting for 62.16% of the total. The second was age-seven individuals, accounting for 29.73% of the total. Individuals in the age group of 5, 8, and 10 accounted for only 8.1% of the total. In 2015, 200,000 fish were released in the Lijiaxia section. Judging from the age, the fish released artificially in 2015 were exactly the six-year-old, and the artificial proliferation and release effectively replenished the population of Spotted naked carp.

4.2 *Analysis of community diversity in different river segments*

According to Table 3, the Shannon-Weiner Index is 2.0593 in Guide, 2.0275 in Longyang Gorge, 1.8035 in Lijiaxia, and 1.2259 in Laxiwa. According to the Shannon-Weiner Index, the guide and Longyangxia sections belong to grade iii level and have good fish diversity. The diversity of lijiaxia and Laxiwa members generally belongs to the class ii level. Through visits, expensive and aggressive fish were captured in Lijiaxia. This dominance of nudibranchs, which has led to the discharge of 2.4 million nudibranchs in the DE period over a decade, has effectively compensated for the fishery resources, reduced the pressure on other fish catches (Liu et al. 2011), and maintained the biodiversity of the section. The evenness index of Pielou was 0. 7833, the Margalef Richness Index of the Guizhou section was 2.1797, and the Simpson Dominance Index of the Guizhou section was 0.8721. The evenness index, richness index, and dominance index are significantly higher than those in the Lijiaxia and Laxiwa sections, and the effect of proliferation and discharge is obvious (Han 2016). Although the fishing intensity in the Longyang Gorge section is also high, the fishing pressure of fish is scattered due to the large water area (Huang 20020, and the fish diversity is relatively rich. The Margalef Richness Index in Longyang Gorge Section is 2.5861, and the Simpson Dominance Index in Longyang Gorge Section is 0.9212. Shannon-Wiener Index, evenness index, richness index, and dominance index also confirm this point; the evenness index, richness index, dominance index, and Shannon-Weiner Index in the Lijiaxia section are much lower than those in the Guide and Longyang Section, and the evenness index is only 0.3536, indicating the lack of fish resources.

4.3 *Catch quality analysis*

A total of 469 (143.406 kg) were caught in the Guide section, 1530 (111.5821 kg) in the Laxiwa section, 257 (49.3563 kg) in the Longyangxia section, and 37 (14.9379 kg) in the Lijiaxia section. According to the analysis of the catch number and catch quantity, the maximum catch of Guide was 143.406 kg, and the average weight was 300g. The population number of Pied naked carp was effectively increased by artificial breeding and releasing (Yan 2016). The catch of the Longyangxia section was 49.3563 kg, with an average tail weight of 192g, and that of the Lijiaxia section was 14.9379 kg, with an average tail weight of 403g. The fishing intensity in these two sections was large, and the number of breeding and releasing was small. Only 100,000 fish were released artificially in the Longyangxia section in 2019, and 10,000 fish were released artificially in the Lijiaxia section in 2012. In 2015, 200,000 fish were released, but it was difficult to replenish the number of spotted naked carp captured. The population of spotted naked carp in the two reservoir areas decreased significantly, resulting in a low catch amount.

5 CONCLUSION

Through the analysis of the population age structure of longyangxia, Laxiwa, Guide, and Lijiaxia in the Qinghai section of the Yellow River, artificial proliferation and release can effectively improve the population number of Naked carp (Hong et al. 2009) and maintain a reasonable population age structure of naked carp. Diversity of four sample points and comparison to fishing pressure, high-intensity discharge section of the species diversity of artificial proliferation index (Shannon Nasdaq number, evenness index, richness index, and dominance index) is significantly higher than the dischargeamount.. Proliferation discharge can improve river fish community biodiversity (Zhu et al. 2009) and maintain the ecological balance of the river. Through the comparison of the catch amount, it can be seen that the population quantity and biomass of naked carp in the artificially breeding and releasing section are significantly higher than those in the less-releasing section.

In conclusion, the artificial proliferation and release of fish seedlings in the Upper Reach of the Yellow River can increase the number of fish resources in the water area, effectively increase the number of the released fish population, increase the diversity of community species, and play an important role in the recovery of local fishery resources, improving the ecological environment of the water area, and protecting endangered species and biodiversity.

ACKNOWLEDGMENTS

This work was financially supported by the Open Project of Qinghai Agriculture and Rural Department, KLPA-2019-01.

REFERENCES

Bin Zhu, Haitao Zhen. Artificial reproduction and release of freshwater fishes in the Yangtze River Basin and its ecological effects [J]. *Chinese fishery economy*, 2009, 27 (2):74–87.

Bingyi Zhan.*Assessment of fishery resources* [J]. Beijing: China Agriculture Press: 1995, 18–38.

Bo Hong, Zhenzhong Sun. Evaluation of fishery resources proliferation and release effect in the upper reaches of Huangpu River [J]. *Fisheries science and technology information*, 2009, 36 (4):178–181.

Jilin Wang, Zhuoqun Jiang. *Fishery Resources and Fishery Regionalization in Qinghai Province* [M]. Xining: Qinghai People's Publishing House, 1988.

Jinguo Huang.Conservation and sustainable utilization of wetland biodiversity in Dongting Lake region [J]. *Chongqing environmental science*, 2002, 24(6), 18–20.

Long Chen, Gaodi, Xie, Chunxia Lu, et al. Effects of water conservancy projects on the fish ecological environment: A case study of fish changes in Baiyangdian Lake in recent 50 years [J]. *Resources Science*, 2011, 33 (8) : 1475–1480.

Qing Liu, Ran Xu. *Fisheries Environmental assessment and ecological restoration* [M]. Beijing: China Ocean Press:2011, 173–184.

Ruiyi Chen, Bao Lou.Discussion on artificial proliferation and discharge technology [J]. *Hebei Fisheries*. 2014, (5):50–53.

Sili Yan.*Studies on biological characteristics, reproductive characteristics, embryo development, and artificial culture of Carpio polygoni* [D]. China West Normal University, 2016

Xiumei Zhang, Xijie Wang. Current situation and the prospect of fishery resources proliferation and release in Shandong Province [J]. *China fishery economy*, 2009, 27 (2):51–58.

Xuwei Pan, Linlin Yang, Weiwei Ji, et al.Research progress of proliferation and discharge technology[J].*Jiangsu Agricultural Sciences*, 2020, 4: 236–240.

Yong Han. *Age, growth, and genetic characteristics of Gymnocypris eckloni eckloni in the Upper Reaches of the Yellow River* [D]. Shanghai Ocean University, China West Normal University, 2016

Yunfei Wu, Cuizhen, Wu. *Fish of Qinghai-Tibet Plateau* [M]. Chengdu: Sichuan Science and Technology Press, 1992.

Zhigang Mao, Xiaohong Gu, Qingfei Zeng, et al.. Fish community structure and diversity in Taihu Lake [J]. *Chinese Journal of Ecology*, 2011, 30 (12): 2836–2842.

Advances in Energy, Environment and
Chemical Engineering – Abdullah & Osman (Eds)
© 2023 The Author(s), ISBN: 978-1-032-36083-6

Characteristics of shale gas in thrust belt of Chengkou basin margin in Chongqing

Weijun Zhao*
SDIC Chongqing Shale Gas Development and Utilization Co., Ltd, Chongqing, China

Xinying Yang
Exploration and Development Research Institute of Tarim Oilfield Company, Korla, China

Ping Li
SDIC Chongqing Shale Gas Development and Utilization Co., Ltd, Chongqing, China

Yanbo Nie
Kela Department of Oil and Gas Development, Tarim Oilfield Company, Korla, China

Yinghuan Luo
SDIC Chongqing Shale Das Development and Utilization Co., Ltd, Chongqing, China

Qing Xia
Exploration and Development Research Institute, PetroChina Southwest Oil & Gasfield Company, Chengdu, China

ABSTRACT: To study the geological and engineering characteristics of shale gas in the mountainous area of the basin margin, the static geological index parameters of organic shale in the work area, such as organic carbon, maturity, porosity, and permeability, are counted, showing that the indicators are good; however, the organic-rich shale gas reservoirs have limited dispersion and concentrated sections of true thickness and poor preservation conditions, mainly residual adsorbed gas. The mountains are high and the valleys are deep, the topography is undulating, and the dip of the strata is large. It is difficult to accurately determine the structural strata due to the large geological exploration risk; since the occurrence of the strata is changeable and the lithology changes are complex, it is difficult to carry out horizontal wells and other conditions, so it is difficult to effectively develop the block at present.

1 INTRODUCTION

At present, shale gas research is mainly aimed at shale formations with flat landform, good preservation conditions, relatively slow formation, few faults, and good organic matter conditions (Bowker 2007; David 2007; Dong et al. 2009; Li et al. 2007; Martineau 2007; Richard 2007; Ross & Bustin 2007; Zhang et al. 2007, 2009), while shale gas in Chengkou block, which is located at the basin edge, with large formation angle, landform dominated by medium and high mountains and relatively developed faults, is less involved. For this problem, based on surface outcrops, geological and geochemical data, the geological characteristics of shale gas in the Chengkou area are analyzed. The surface and tectonic conditions of the study area are complex, and the stratigraphic dip angle is large. Two deep faults are developed to destroy the integrity of the formation. Fold faults are

---

*Corresponding Author: zhaoweijun@sdic.com.cn

803

developed, and strata inversion is common, which is not conducive to shale gas preservation. The topography is undulated with the high mountains and deep valleys regional, the stratigraphic dip angle is large, and the geophysical exploration is difficult. The organic-rich shale sections in the profile are too scattered, have large spacing, thin single-layer thickness, and are unable to drill horizontal wells, so the actual available resources are very limited.

2 GEOLOGICAL SETTING

The Chengkou District of Chongqing is located at the junction of the Dabashan trough fold area and the Dabashan platform edge depression, and there are two large faults mainly developed, namely, the Chengba Fault and the Wuping Fault. The structure is composed of a series of NW-trending tight linear compound folds and oblique thrust faults, forming an imbricate thrust nappe structure. The syncline is relatively complete, and the anticlines are generally damaged by faults. The intensity and density of folds in the northeastward direction of the Chengba fault zone change from strong to weak, from dense to sparse in-plane distribution. On the profile, the closer to the surface, the stronger the intensity of folds, and the strata with a high dip angle of 60°C to 70°C are common in the field. The fault trend is consistent with the stratigraphic trend, with the undulating section and the great occurrence changes. And the overall dip is northerly, and the dip angle varies between 40°C and 70°C. The local fault in the Chengkou area has a dip of 220°C and a dip of 65°C. The strata on the north side of the fault are all thrust to the south, and the maximum fault displacement is more than 2000m (Figure 1).

Secondary structures and derived small faults have developed well in the area. In the southern part of the Chengba fault (the main target layer), the derivative small faults and small folds are very developed. The fault density is $14/10 \times 10$ km^2, and there are more than 10 faults with a length of more than 10 km. Small folds are also relatively developed, making the stratum complex and changeable. The dip of the same structural belt varies greatly, and it is a typical thrust-type stratum with developed faults, developed secondary structures, and large strata dip.

3 GEOLOGIC FEATURES

Drilling confirmed that there is no exploration & development potential in the Longmaxi Formation, and the main target layer is the Cambrian Shuijingtuo Formation.

3.1 *Basic geological parameters*

The analysis of organic carbon shows that the average TOC of the target shale is 3.11%, the organic matter type is type I, the average vitrinite reflectance is 2.81%, the average effective porosity is 3.4%, and the average permeability is 0.0133 mD. Natural fractures in shale are very developed, mainly including natural open fractures, weathered shale fractures, X shear joints, sedimentary discontinuity unconformity, and early extrusion joints filled with calcite veins.

3.2 *Limited true thickness of shale gas reservoirs with rich organic matter and dispersed*

The overall gas content of the Chengdi 4 well is not high, 86% of which is less than 1m^3/t and 14% is more than 1m^3/t (Figure 2). Considering the poor preservation conditions near the surface, to obtain the real situation, the gas content of the on-site desorption samples was counted from below 1000 m. A total of 100 samples were counted, of which 92 samples were less than 1 m^3/t and 8 samples were more than 1 m^3/t. The total gas content is $0.02322 \sim 1.6313$ m^3/t, with an average of 0.5011 m^3/t. The total thickness of the strata with gas content greater than 1 m^3/t is only 1.6 m, and is separated by the strata with gas content less than 1 m^3/t; it shows that shale heterogeneity is very strong (Figure 3).

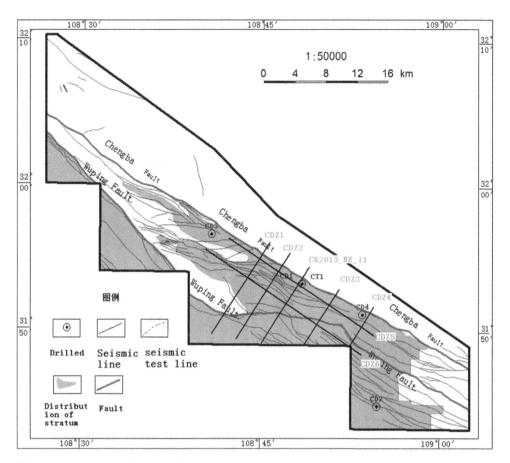

Figure 1.   Structural outline map of the study area.

The resource abundance of CT1 well, CD1 well, and CD4 well is $4.6 \times 10^8 m^3$ / km², and it is low. The organic layer in the profile is too scattered, large interval, thin single-layer thickness (Table 1), coupled with the lack of an alpine landform well site, unable to drill horizontal wells, and other reasons, the effective resources that can be exploited are very limited and difficult to effectively develop.

3.3   *Poor storage conditions and residual adsorbed gas dominated.*

The study area is located in the basin margin thrust zone. After several periods of strong transformation, thrust faults are developed, and the strata are broken and mainly high and steep (inclination angle is 40°C–80°C). The target layer in most areas is exposed to the surface, and the preservation conditions are poor.

The liquid depth in the CT1 wellbore is 1468 m, and the salinity of formation water is about 100,000 mg / L, corresponding to the formation water density of 1.07 g / cm³. Therefore, the liquid pressure in the wellbore is 15.39 MPa when the dynamic liquid level is balanced. The corresponding reservoir depth is 1762 m, and the hydrostatic column pressure is 17.26 MPa, so the shale reservoir pressure coefficient is 0.89, which is a low-pressure shale gas layer. The above data show that the free gas content of the well is low, and the adsorbed gas is dominated there. During the drainage process, the pressure continues to decline while the flame height does not rise, and the times of gas lifts have no effect, which proves that the free gas content of the reservoir is insufficient.

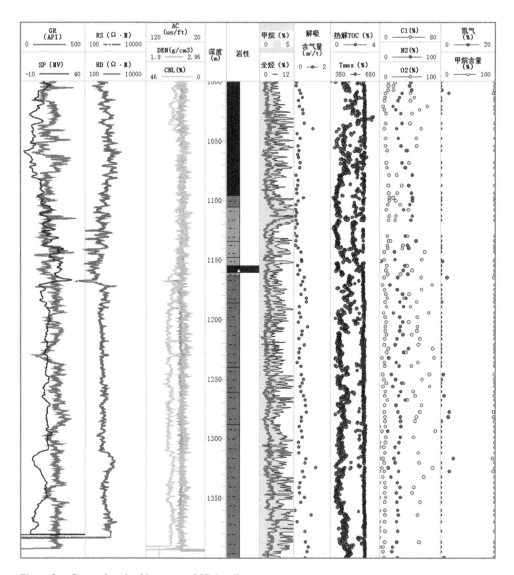

Figure 2. Comprehensive histogram of CD4 well.

3.4 *The difficulty of geophysical exploration where high mountains and deep valleys*

The study area is a double complex area, with high and steep terrain, large stratum dip, dense faults, chaotic strata, complex geological structure, and great changes in the quality of single-shot recorded data in the whole area. The reflected wave group has poor continuity, low signal-to-noise ratio, chaotic geological information, prominent static correction problems, difficult reflection homing, and difficult interpretation (Figure 4). Through processing and interpretation, we can identify the changing trend of the profile structure, and the continuity of the event axis is poor, but the micro-amplitude structure can be identified. The fault breakpoint is unclear, the horizon traceability is poor, the wave group characteristics of each horizon are disordered, and the stratigraphy and

806

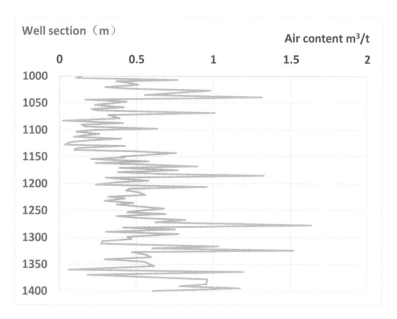

Figure 3.   Profile distribution of desorption gas content at depth of 1000 m in Well CD4.

basement top boundary cannot be identified, which is of little guiding significance for exploration and deployment.

The study area has experienced multi-stage complex tectonic movements, faults, and folds developed, and the relative elevation difference of topography and geomorphology in the reserved area is 1300 m (Figure 4). 64 coring statistics of CD4 well show that 39 samples with formation angle of 60°C∼ 80°C, accounting for 50. 9%; six samples with stratum angle of 80°C – 90°C layers, accounting for 9.4%; 12 samples with stratum angle of 50°C – 60°C, accounting for 18.8%; seven samples with stratigraphic angles of 40°C ∼ 50°C, accounting for 10.9 %. The proportion of high and steep strata (60°C – 80°C) is more than half (Figure 5). Seismic exploration construction is difficult due to high and steep strata, high surface lithology, and high mountain landform. Geophysical exploration data quality is poor, underground structure and target layer distribution characteristics cannot be accurately identified, plane prediction is difficult, and engineering operation and reserves evaluation are difficult.

3.5   *The changeable lithology makes horizontal wells difficult to carry out*

Drilling shows that the geological conditions are complex. The straight-line distance between Well CD1 and Well CT1 is only 150m, but the lithology revealed by the drilling of the two wells is very different: Well CD1 is dominated by siliceous shale, while Well CT1 is dominated by carbonaceous shale and black mudstone. The lithology of Well Chengdi 3, which is about 13 km apart, is also different; Well Chengdi 2 is deployed in the target layer with a complete surface profile, large thickness, and no surface fractures (over 100m), but is affected by underground fractures, only the target layer is revealed 38m.

The drilling also showed that the stratum is very fragmented, the lateral tracking is difficult, the difference between wells is large, the stratigraphic correlation is difficult, the underground structure is complex, and it is difficult to carry out horizontal good drilling.

Table 1. Vertical distribution of shale gas resource abundance in wells CT1, CD1, and CD4.

Well	Original depth (m)	Maximum cutting depth (m)	Thickness (m)	Total gas density (g/cm³)	Resource content (m³/t)	abundance (10⁸m³/km²)	Interpretation results
CD4	610	618.2	8.2	2.6	2.89	0.6161	Type II shale gas reservoir
	629.2	630.7	1.5	2.6	3.25	0.1268	Type II shale gas reservoir
	641.3	648.2	6.9	2.6	2.96	0.5310	Type II shale gas reservoir
	814.5	818.2	3.7	2.6	2.08	0.2001	Type II shale gas reservoir
	1254.3	1260	5.7	2.6	2.93	0.4342	Type II shale gas reservoir
	1271.2	1275.8	4.6	2.6	2.51	0.3002	Type II shale gas reservoir
	1280.6	1283.1	2.5	2.6	2.75	0.1788	Type II shale gas reservoir
	1295.2	1296.7	1.5	2.6	3.11	0.1213	Type I shale gas reservoir
CD1	610	618.2	8.2	2.6	2.89	0.6161	Type II shale gas reservoir
	629.2	630.7	1.5	2.6	3.25	0.1268	Type II shale gas reservoir
	641.3	648.2	6.9	2.6	2.96	0.5310	Type II shale gas reservoir
	814.5	818.2	3.7	2.6	2.08	0.2001	Type II shale gas reservoir
	1254.3	1260	5.7	2.6	2.93	0.4342	Type II shale gas reservoir
	1271.2	1275.8	4.6	2.6	2.51	0.3002	Type II shale gas reservoir
	1280.6	1283.1	2.5	2.6	2.75	0.1788	Type II shale gas reservoir
	1295.2	1296.7	1.5	2.6	3.11	0.1213	Type I shale gas reservoir
CT1	1624.9	1627.2	2.3	2.6	7.8	0.4664	Type I shale gas reservoir
	1627.2	1634	6.8	2.6	4.9	0.8663	Type I shale gas reservoir
	1678.4	1679.5	1.1	2.6	3.6	0.1030	Type II shale gas reservoir
	1682.5	1683.6	1.1	2.6	2.9	0.0829	Type II shale gas reservoir
	1685.4	1691.5	6.1	2.6	4.6	0.7296	Type II shale gas reservoir
	1725.8	1732.7	6.9	2.6	4.4	0.7894	Type II shale gas reservoir
	1734.8	1744.4	9.6	2.6	6	1.4976	Type II shale gas reservoir
	1748.4	1760	11.6	2.6	8.6	2.5938	Type I shale gas reservoir
	1790.6	1794.3	3.7	2.6	7.6	0.7311	Type I shale gas reservoir
	1801.5	1803.4	1.9	2.6	7.8	0.3853	Type II shale gas reservoir
	2196.9	2200	3.1	2.6	6.3	0.5078	Type I shale gas reservoir

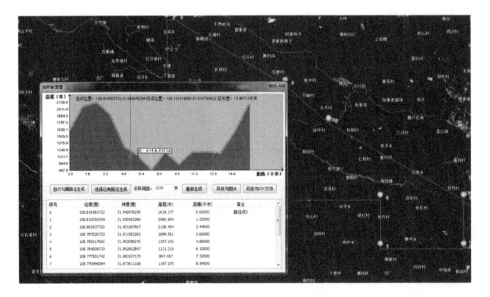

Figure 4. Profile of Topography elevation.

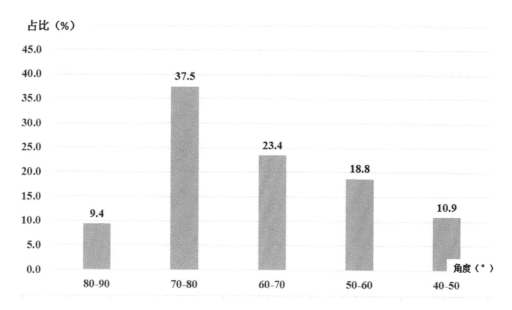

Figure 5. Statistical chart of formation dip angle measured by logging in Well CD4.

4 CONCLUSION

To determine the geological and engineering characteristics of shale gas in the mountainous area of the basin margin, the static geological index parameters of organic shale in the work area were investigated and discussed. Conclusions and suggestions are as follows:

(1) The surface and tectonic conditions of the study area are complex, and the stratigraphic dip angle is large. Two deep faults, the Chengba fault, and the Wuping fault are developed to destroy the integrity of the formation. Fold faults are developed, and strata inversion is common, which is not conducive to shale gas preservation.
(2) The topography is undulated with the high mountains and deep valleys regional, the stratigraphic dip angle is large, and the geophysical exploration is difficult.
(3) The organic-rich shale sections in the profile are too scattered, have large spacing, thin single-layer thickness, and are unable to drill horizontal wells, so the actual available resources are very limited.

ACKNOWLEDGMENTS

This research was financially supported by Chongqing Science and Technology Commission Project "Key Technologies and Research Application of Middle-Deep Shale Gas Exploration and Development in Chongqing Area" (Grant No. cstc2017zdcy-zdyfX0040).

REFERENCES

Bowker K A. Barnett shale gas production Fort Worth basin: Issues and discussion [J]. *AAPG Bulletin*, 2007, 55(1):523–524.
David F Martineau. History of the Newark East field and the Barnett shale as a gas reservoir [J]. *AAPG Bulletin*, 2007, 55(1):399–403.

Dong Dazhong, Cheng Keming, Wang Shiqian, etc. Evaluation method of shale gas resources and its application in Sichuan Basin [J]. *Natural gas industry*, 2009, 29(5): 33–39.

Li Xinjing, Hu Suyun, Cheng Keming.Inspiration from exploration and development of fractured shale gas in North America [J]. *Oil exploration and development*, 2007, 34(4): 392–400.

Ross D J K, Bustin R M. Impact of mass balance calculations on adsorption capacities in microporous shale gas reservoirs [J]. *Fuel*, 2007, 86:2696–2706.

Richard M P. Total petroleum system assessment of undiscovered resources in the giant Barnett Shale continuous (unconventional) gas accumulation, Fort Worth Basin, Texas [J]. *AAPG Bulletin*, 2007, 91(4): 551–578.

Zhang Jinchuan, Xu Bo, Nie Haikuan, etc. Two important areas of natural gas exploration in China [J]. *Natural gas industry*, 2007, 27 (11): 1–6.

Zhang Linye, Li Zheng, Zhu Rifang. Formation and development of shale gas [J]. *Natural gas industry*, 2009, 29 (1): 124–128.

*Advances in Energy, Environment and
Chemical Engineering – Abdullah & Osman (Eds)
© 2023 The Author(s), ISBN: 978-1-032-36083-6*

Grey correlation analysis on photovoltaic construction in the embedded port area and "dual carbon" target

Xumin Wu*
Anhui Transport Survey & Design Institute Co., Ltd., Anhui, China

ABSTRACT: This paper proposes for the first time the concept of embedding photovoltaic energy storage in the construction of the port area. This paper opens up a new field of new applications for clean energy. This paper analyzes the research on the gray correlation between clean energy (solar energy) and the "dual carbon" goal and proposes relevant countermeasures. It provides important references and new ideas for relevant government and group investment and pollution control work.

1 INTRODUCTION

Climate change is one of the most serious challenges faced by humanity. Since the Industrial Revolution, a large amount of carbon dioxide emitted from the burning of fossil energy by human activities, industrial processes, and changes in agriculture, forestry, and land use has remained in the atmosphere, which is the main cause of climate change. To cope with climate change and promote the sustainable development of human society, efforts must be made to reduce greenhouse gas emissions.

2 DEVELOPMENT STATUS OF CLEAN ENERGY APPLICATIONS UNDER THE "DUAL CARBON" TARGET

2.1 *"Dual carbon" target*

On 22 September 2020, General Secretary Xi Jinping pointed out at the general debate of the 75th United Nations General Assembly: China will increase its nationally determined contribution, adopt more powerful policies and measures, and strive to peak carbon dioxide emissions before 2030 and aim for carbon neutrality by 2060.

The carbon peak refers to the carbon emission of the global, country, city, enterprise, and other subjects in the process of rising and falling, and the highest point of carbon emission is the carbon peak. Most developed countries have achieved carbon peaks, and carbon emissions have entered a downward channel. Although China's current carbon emissions have slowed down compared with the rapid growth period of 2000–2010, they are still increasing and have not yet peaked. Carbon neutrality refers to balancing anthropogenic sources of emissions with anthropogenic sinks through afforestation, carbon capture, and storage (CCS) technologies, etc. Carbon neutrality goals can be set at different levels such as global, national, city, corporate activities, etc. It refers to carbon dioxide emissions in a narrow sense, and can also refer to all greenhouse gas emissions in a broad sense.

2.2 *Status of clean energy applications*

In the era of grid parity, the proportion of non-fossil energy sources will become an important guiding indicator of photovoltaic prosperity. According to the previous policy plan, the development

*Corresponding Author: shade2003127@163.com

target of non-fossil energy is 15% by 2020, not less than 20% by 2030, and not less than 50% by 2050.

The most important form of consumption of non-fossil energy is power generation. In 2019, non-fossil energy consumption accounted for 14.6%, of which 14.2% was provided by non-fossil energy power generation.

Photovoltaic is the non-fossil energy with the highest power generation potential. The anchor of the industry's prosperity will be determined by the proportion of non-fossil energy. In short, if the proportion of non-fossil energy is increased to 17.5% (14.6% in 2019), it will bring photovoltaic energy to 60%, that is, -70 GW/year of new installations, 1.3–1.6 times this year's 45 GW (expected). In the long run, the new infrastructure will also bring a certain degree of benefits and development space to photovoltaics.

3 GRAY CORRELATION ANALYSIS AND RESEARCH ON PHOTOVOLTAIC CONSTRUCTION IN THE EMBEDDED PORT AREA AND "DUAL CARBON" TARGET

The impact of the application of new energy on the "dual carbon" goal is multifaceted. Because the influencing factors are more complex, specific, and quantitative analysis is required. This paper will use the gray system theory to carry out a gray relational analysis on the influence of the "two-carbon" target. In this way, the impact of the embedded photovoltaic construction in the port area on the "dual carbon" target of the port area can be quantitatively determined. Since the "dual carbon" target is greatly affected by the total amount of electric energy storage, the relevant data on the total amount of electric energy consumed and stored in the port area will be used as the impact parameter of the system.

3.1 Gray relational model

Gray relational analysis is a sequence of gray relational grades (referred to as the gray relational order) to describe the strength of the relationship between factors, the size of the order of the method, through the gray relational grade system to analyze and determine the extent of the impact factors of a method. The basic idea is: to factor in the data sequence based on the method of mathematical geometric factors relationship.

Analysis of correlation, in general, includes the following: raw data transformation, calculation of correlation coefficient, and correlation, scheduling related procedures are listed in incidence matrix. In practical applications, these calculations are not necessarily required.

Transformation of original data. Including two methods: 1) mean transformation, 2) initial value transformation. While analyzing the correlation degree of the socio-economic system, most of the series show an increasing trend, so this paper adopts the initial value transformation.

Maintaining the Integrity of the Specifications.

Transform the original data. Includes two ways: means of transformation, initial transformation. Socioeconomic system is the analysis of correlation, the number out of the increasing trend in the majority; it transforms the initial value in this paper.

[1] Transformation of original data. It includes two methods: 1) mean transformation and 2) initial transformation. While analyzing the correlation degree of social and economic system, most of the series show an increasing trend, so this paper adopts the initial value transformation.

[2] Seeking gray correlation difference information $\Delta_{0i}(k)$, $\Delta_{0i}(k)$ Is the absolute difference between the two comparison sequences at time k, is $\Delta_{0i}(k) = |x_0(k) - x_i(k)|$; $1 \leq i \leq m$; $\Delta_{0i}(\max)$, $\Delta_{0i}(\min)$ are the maximum and minimum of the absolute difference of all comparison sequences at each time. Because the comparison sequence is transformed into the same starting point, $\Delta_{0i}(\min) = 0$ is generally taken.

[3] Calculate the correlation coefficient. The parent sequence after data transformation is recorded as $x_0(t)$, and the sub sequence is $x_i(t)$, I = 1,2,..., M. Then at time t = k, the correlation

Month	202X Total energy used	Year-on-year growth rate	202X 1–11 month Total stored power	202X year 1–11 month Total stored power
6	219.34	14.90	136.21	83.13
7	193.55	5.51	105.2	88.35
8	230.52	9.47	138.8	91.72
9	239.85	8.19	141.19	98.66
10	251.19	12.73	144.75	106.44
11	256.39	3.64	150.79	105.6
12	267.87	7.77	159.31	108.56
1	270.49	5.33	158.68	111.81
2	290.01	9.85	173.74	116.27
3	271.75	6.69	166.78	104.97
4	252.31	5.31	135.83	116.48

Figure 1. Energy use table of an inland river port area in Anhui unit: (10^4kwh,%).

coefficient between the parent sequence and the sub sequence is:

$$\gamma(x_0(k), x_i(k)) = \frac{\min\limits_i \min\limits_k \Delta_{0i}(k) + \xi \max\limits_i \max\limits_k \Delta_{0i}(k)}{\Delta_{0i}(k) + \xi \max\limits_i \max\limits_k \Delta_{0i}(k)}$$

Where ξ It is called the resolution coefficient, $\Delta_{0i}(\max)$ which is used to weaken the influence of too large value and distortion, and improve the significance of the difference between correlation coefficients, $\xi \in (0,1)$, generally 0.1–0.5 is appropriate. It can be seen from the above formula that the correlation coefficient reflects the closeness (proximity) of the two groups of comparison sequences at a certain time. For example, at time $\Delta_{0i}(\min)$, the correlation coefficient $\gamma(x_0(k))_i(k)) = 1$ is the largest, while at time $\Delta_{0i}(\max)$, the correlation coefficient is the smallest. Therefore, the variation range of the correlation coefficient is $0 < \gamma(x_0(k))_i(k)) \leq 1$.

[4] Find the correlation degree. The correlation degree of the two comparison sequences can be quantitatively expressed by the average of the correlation coefficients of the two sequences at each time. The calculation formula is:

$$\gamma(x_0, x_i) = \frac{1}{N} \sum_{k=1}^{N} \gamma(x_0(k), x_i(k))$$

[5] Arrange association order. In the analysis of correlation degree, the more practical significance is to compare the correlation degree of each sub sequence for different parent sequences. Therefore, the correlation degree of M sub sequences to the same parent sequence is arranged in order of magnitude to form the correlation order. It directly reflects the "advantages and disadvantages" or "primary and secondary" relationship of each sub sequence.

[6] Column incidence matrix.

3.2 Correlation calculation between power consumption and storage in port area and "double carbon" target

Used in the following calculations ω_1, ω_2, ω_3, ω_4 Respectively represent the total energy used, the total stored electricity, the total electricity consumption and the year-on-year growth rate of each index in the economic system, then the influence space of the system is:

$\omega_1 = \{159.04, 118.30, 145.59, \ldots\ldots, 156.26, 147.35\}$
$\omega_2 = \{83.13, 88.35, 91.72, \ldots\ldots, 104.97, 116.48\}$

813

Month	202X year Total energy used	Total power consumption (total energy used) × 0.86)
6	184.93	159.04
7	137.56	118.30
8	169.29	145.59
9	174.02	149.66
10	176.08	151.43
11	175.94	151.31
12	196.8	169.25
1	207.33	178.30
2	200.4	172.34
3	181.7	156.26
4	171.34	147.35

Figure 2. Calculation table of total energy consumption unit: (10^4 kwh,%).

$\omega_3 = \{136.21, 105.2, 138.8, ..., 166.78, 135.83\}$
$\omega_4 = \{14.90, 5.51, 9.47, ..., 6.69, -10.10\}$
The initial value is transformed into: x_1

$$x_i = (x_i(1), x_i(2), x_i(3), ..., x_i(10), x_i(11)) = \left(\frac{\omega_i(1)}{\omega_i(1)}, \frac{\omega_i(2)}{\omega_i(1)}, \frac{\omega_i(3)}{\omega_i(1)}, ..., \frac{\omega_i(10)}{\omega_i(1)}, \frac{\omega_i(11)}{\omega_i(1)} \right)$$

After calculation, the following data results are obtained:

	X1	X2	X3	X4	Δ12	Δ13	Δ14
1	1.0000	1.0000	1.0000	1.0000	0.0000	0.0000	0.0000
2	0.7438	1.0628	0.7723	0.3700	0.3189	0.0285	0.3739
3	0.9154	1.1033	1.0190	0.6354	0.1879	0.1036	0.2801
4	0.9410	1.1868	1.0366	0.5494	0.2458	0.0956	0.3917
5	0.9521	1.2804	1.0627	0.8543	0.3283	0.1106	0.0978
6	0.9514	1.2703	1.1070	0.2442	0.3189	0.1557	0.7072
7	1.0642	1.3059	1.1696	0.5216	0.2417	0.1054	0.5425
8	1.1211	1.3450	1.1650	0.3577	0.2239	0.0438	0.7634
9	1.0837	1.3987	1.2755	0.6611	0.3150	0.1919	0.4226
10	0.9825	1.2627	1.2244	0.4492	0.2802	0.2419	0.5333
11	0.9265	1.4012	0.9972	-0.6777	0.4747	0.0707	1.6042

Figure 3. Data calculation.

The correlation between total energy consumption and total stored electricity is: $\gamma(x_1, x_2) = \frac{1}{11}\sum_{k=1}^{11} \gamma(x_1(k), x_2(k)) = 0.1480$.

The correlation between total energy consumption and total electricity consumption is: $\gamma(x_1, x_3) = \frac{1}{11}\sum_{k=1}^{11} \gamma(x_1(k), x_3(k)) = 0.1745$.

The correlation between the total energy used and the year-on-year growth rate of the previous year is: $\gamma(x_1, x_4) = \frac{1}{11}\sum_{k=1}^{11} \gamma(x_1(k), x_4(k)) = 0.1212$.

	$\gamma(x_1(k), x_2(k))$	$\gamma(x_1(k), x_3(k))$	$\gamma(x_1(k), x_4(k))$
1	1.0000	1.0000	1.0000
2	0.7155	0.9657	0.6821
3	0.8102	0.8856	0.7412
4	0.7654	0.8935	0.6719
5	0.7096	0.8789	0.8913
6	0.7155	0.8375	0.5314
7	0.7684	0.8839	0.5965
8	0.7818	0.9482	0.5124
9	0.7180	0.8070	0.6550
10	0.7411	0.7683	0.6006
11	0.6282	0.9190	0.3333

Figure 4. Calculation results.

3.3   Influence relationship between electricity and "double carbon" target

3.3.1   Standard coal conversion

Every 1 kWh of electricity saved = 0.404 kg of standard coal (the equivalent value of electricity converted to standard coal).

1 ton of standard coal = 1000 kg of standard coal = 2475.2 kWh.

For every 1 degree (1 kWh) point saved, 0.404 kg of standard coal is saved accordingly; at the same time, pollution emissions are reduced by 0.272 kg of carbon dust, 0.997 kg of carbon dioxide (co2), 0.03 kg of sulfur dioxide (SO2), and 0.015 kg of nitrogen oxides (NOx).)

The above two values can be calculated

Saving 1 kWh of electricity = emission reduction of 0.997 kg of carbon dioxide = emission reduction of 0.272 kg of carbon

Saving 1 kg of standard coal = emission reduction of 2.493 kg of carbon dioxide = emission reduction of 0.68 kg of carbon

That is, every kilogram of carbon emission reduction = 3.676 kWh of electricity saved

Every kilogram of carbon emission reduction = 1.471 kilograms of standard coal saved

Every 100 million tons of coal emission reduction = 100 billion kg of carbon emission reduction = 147.1 billion kg of standard coal saved = 367.6 billion kWh of electricity saved.

3.3.2   Data derivation

According to the above calculation, the gray correlation sequence is arranged as: $x_3 > x_2 > x_4$, the corresponding data are:0.1745, 0.1480, 0.1212, which is: $\gamma(x_1, x_3) > \gamma(x_1, x_2) > \gamma(x_1, x_4)$, the sequence relationship is R(3,2,4) = R(Total foreign trade exports, total foreign trade imports, growth rate of total foreign trade imports and exports).From this data analysis, the absolute value coefficient of the correlation between the various indicators of Shenzhen's foreign trade import and export economic system and Shenzhen's road container freight volume is not large, and the absolute number is not much different, but the relative comparison result is that the total foreign trade export The gray correlation with road container freight volume is the largest, followed by total foreign trade imports, and finally the growth rate of total foreign trade imports and exports.

The above calculation results show that the total amount of stored electric energy is the most important factor affecting the "double carbon" goal. In addition, the total amount of energy used and the growth rate of the total electricity consumption are also important factors affecting the completion of the "double carbon" goal.

4 CONCLUSION

The author uses the gray relational model to analyze and calculate the gray relational degree of the relevant indicators between the photovoltaic construction of the port area and the "dual carbon" target, and compares the degree of the relational degree to do some basic work for the related research work.

REFERENCES

Zhang L, Qin QD, Wei YM. China's distributed energy policies: Evolution, instruments and recommendation [J]. *Energy Policy*, 2019(125):55–64.

Wang S Q, Tiong R L K, Ting S K, et al. Evaluation and management of foreign exchange and revenue risks in China's BOT projects [J]. *Construction Management & Economics*, 2000, 18(2):197–207.

Liu ZJ, Wu D, He BJ, et al. Using solar house to alleviate energy poverty of rural Qinghai-Tibet region, China: A case study of a novel hybrid heating system [J]. *Energy & Buildings*, 2018 (178):294–303.

Sam G, Shen W, Gongbuzerenc. Solar energy for poverty alleviation in China: Stateambitions, bureaucratic interests, and local realities [J]. *Energy Research & Social Science*, 2018(41): 238–248.

Advances in Energy, Environment and
Chemical Engineering – Abdullah & Osman (Eds)
© 2023 The Author(s), ISBN: 978-1-032-36083-6

Vegetation restoration in rare earth tailings in Northeast Guangdong

Qihe Yang*, Jiaoqing Li*, Hesheng Yang*, Xiongjun Liu* & Lihui Mou*
Jiaying University, Meizhou, China

ABSTRACT: According to the microterrain, idle time, man-made interference degree, and the rare earth tailings in Huangnijiao were divided into 4 areas: near the summit, around the acid pool, slope, and entrance area of the tailings. The area near the mountain summit has been idle for about 40 years with mainly natural recovery and without artificial restoration measures. In the area around the acid pool, the main shrubs were removed in order to build an acid pool, but the surface soil was not seriously affected. The slope area has been idle for over 20 years, *Pinus elliottii, Elaeocarpus sylvestris*, and *Miscanthus floridulus* were planted in 2015 for artificial restoration, but there was hardly any natural vegetation before. In the entrance, the original vegetation was not seriously damaged, but due to the filling of mine slag and in the lowest position, this area was a confluence of wastewater of tailings, changing the physical and chemical nature of the soil to a certain extent. The vegetation investigation revealed that there were more plants, higher Shannon diversity, Pielou uniformity, and Margalef richness near the summit and in the entrance, while there were fewer individuals around the acid pool and in the slope area, but higher Simpson dominance. The Jaccard and Sorensen similarity indexes were higher between the area around the acid pool and the slope, while the Cody dissimilarity index was low. However, the Jaccard and Sorensen similarity indexes between the area near the summit and the entrance were low, and also lower between both and the area around the acid pool or the slope, and the Cody dissimilarity was large. Areas with similar environmental conditions and similar idle time had the highest plant species similarity. The site conditions play an extremely important role in plant restoration in rare earth mines, and sufficient water and nutrients are the key factors to ensure rapid vegetation recovery in rare earth tailings.

1 INTRODUCTION

Pingyuan County in Northeast Guangdong Province of China is extremely rich in rare earth resources distributed in all 12 towns in this county, especially Renju Town and Huangshe Town. There is a reserve of nearly 86822 t, nearly 30 mining sites, an opening capacity of 800 t in 1987 (by REO), annual output of 640 t, to nearly 1000 t in 1990. The rare earth mine in Pingyuan County is an ion adsorption type with the characteristics of many types, good distribution, high quality, easy mining and extraction. It has been the key rare earth mining area in Southern China for many years. In the 1990s, all towns in Pingyuan County had large and small rare earth mines. In the past 15 years, with the adjustment of the national rare earth strategy, many small and medium-sized mining factories have been shut down. Rare earth mining has brought wealth, but also caused many ecological environment problems (Li et al. 2015, 2018; Yang et al. 2013; Zhang et al. 2016a, b).

Ecological restoration in tailings is a complete ecosystem engineering including restoration and reconstruction of all species of animals, plants, and microorganisms and their constituent communities and ecosystems, while the exchange of material, energy, and information occurs within and between adjacent systems with strong dynamics. When ecological processes change, the species distribution pattern shifts from one model to another. Monitoring the process of ecological

---

*Corresponding Authors: yangqh@jyu.edu.cn; lijiaoqing12@126.com; yhs@jyu.edu.cn; 609449126@qq.com and moulihui@jyu.edu.cn

restoration requires frequent attention to its plant community structure and composition changes. The ecological restoration and reconstruction of rare earth tailings in South China is actually also a process of ecological succession (Li et al. 2015; Yang et al. 2013; Zhang et al. 2016a, b). Due to the superior climate, rich water, and heat in Northeast Guangdong, if some suitable plants can be selected in the tailings, supplemented by some beneficial artificial measures, vegetation restoration and reconstruction can be accelerated.

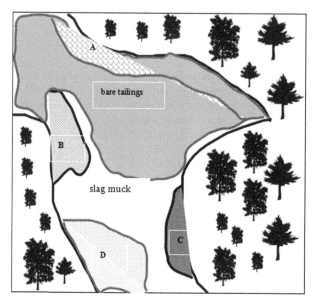

Figure 1.   Topographic graphical diagram of the surveyed tailings area. A: near the summit; B around the acid pool; C: slope; D: entrance.

Places with tree patterns indicate the surrounding woodland.

In the tailings, due to different topographic conditions, abiotic factors are also different, the distribution and growth of plants are also different, and plant communities in different areas are also different. Habitat heterogeneity has been thought to be an important factor in the maintenance of ecosystem biodiversity, and natural selection pressures in complex and diverse niches, leading to species composition, community structure, and diversity. The selection of resistant or suitable plants in tailings is the key and difficulty of taking artificial measures for ecological restoration. Yang et al. (2013) combined the study of habitat heterogeneity with the improvement of soil fertilizer, studied the composition of plant community and diversity distribution pattern of mining areas under different micro-environments, and further understood the ecological succession law of plant communities. Zhang et al. (2016a) investigated the vegetation of rare earth tailings in Northeast Guangdong, analyzed the changes in plant diversity in natural restoration areas in 5 years after plant restoration on different slopes, and found that the slope direction and position were important determinants of community composition, species distribution and ecosystem process. Different vegetation covering in the tailings affect the physical and chemical properties of soil, and the change, in turn, restricts vegetation recovery and succession. Different areas of the same tailings show different plant growth, distribution, and vegetation cover due to differences in microhabitat conditions such as soil substrate, light, and even temperature. Therefore, constructing the plant diversity of stable plant communities can effectively improve the ecological restoration effect, and play an important role in improving the ecological environment quality. Based on the survey of rare earth tailings in Pingyuan County, this paper will mainly discuss the effect of the scheme to restore the ecology of rare earth tailings, and provide theoretical and technical guidance for the ecological restoration of these tailings.

2 MATERIALS AND METHODS

2.1 Overview of the study area

The rare earth tailings in Huangnijiao in this study are located in Huangshe Village, Renju Town, Pingyuan County. The village is rich in forest resources, and the main cash crops are sweet corn, Chinese mesona herb (*Mesona chinensis*) and flue-cured tobacco. In recent years, the planting of Chinese herbal medicine, such as *Pogostemon cablin*, has been vigorously promoted. However, in the past 20-30 years, a large area of deforestation for rare earth mining has caused serious soil erosion. The tailings surveyed are in hilly areas, on the middle and upper position on the southeast slope of a mountain, covering an area of about 100 hm². The study area is a subtropical monsoon humid climate with geographic coordinates of 115°51'54"E, 24°48'25"N, 255-415m altitude, the summer maximum temperature is about 30°C, winter minimum temperature 0-7°C, average annual temperature 18.7°C, annual precipitation above 1480 mm, 75% concentrated in Apr.-Sept., frost-free period of 298 d.

In the non-mining area of Pingyuan County, in general, the vegetation types are mostly coniferous broad-leaved mixed forest, evergreen broad-leaved forest, shrub tussock, etc. Around the tailings, the forest types near its entrance are *Pinus massoniana* forest and *Cunninghamia lanceolata* forest, belonging to the artificial forest, but on the other side of the mountain summit is the evergreen broad-leaved forest in northeast Guangdong, and Fagaceae, Theaceae, and Hamamelidaceae are the dominant families. The soil types were mainly montane partial acidic yellow soil (pH5.1-5.9), with the highest elevation of 420m and the lowest of 130 m. The mine was started in 1983 (near the mountain summit) and abandoned in July 2010. Small parts of the mine area were re-mined when approved by the state in 2021. Surface desertification was serious, it was very difficult for plants to grow, and above 80% of the ground was still a bare wasteland.

Table 1. Altitude, plant cover and soil characteristics of different areas in the tailings.

Area	Slope	Altitude	Vegetation cover (%)	Soil gravel content (%)	Soil fertility and water retention capacity	Soil acidity
Near summit	10–15°	330-420	25-30	30	1	1
Around acid pool	5–10°	250	50	5–8	3	4
Slope	25–30°	200	35-40	15–20	2	2
Entrance	5–10°	150–200	50–60	10-15	4	3

Soil acidity indicates sorting: the smaller the value, the smaller the acidity, the larger the value, the higher the acidity. The same is soil fertility and water retention capacity.

According to the different micro-topography, idle time, and degree of artificial disturbance, the surveyed tailings were divided into 4 areas: near the summit, around the acid pool, slope and entrance, that is, A, B, C, and D areas in Figure 1. An area was near the mountain summit, idle for over 40 years, whose gravel content was about 30%, with poor water protection performance, but the gravel was relatively large and its vegetation cover was 25-30%. As for the B area around the acid pool, its position was slightly 2-3m higher than the bottom of the acid pool, about 10 years ago, in order to build the acid pool, the adult trees were removed. Most shrubs and herbs were retained, and the vegetation cover (mainly *Miscanthus floridulus*) on the soil surface was over 50%, but due to the acid pool leakage, the deep soil was acidified, so the trees and shrubs grew poorly due to their deeper root system. C area was on a slope, idle for over 20 years, with about 15% gravel content in the soil, and the gravel was mostly small. In 2015, the area was transformed into bench terraces and *Miscanthus floridulus* was replanted, in 2020, *Elaeocarpus sylvestris* and *Pinus elliottii* were planted with a vegetation cover of 35-40%. D area was the entrance area of tailings to the main highway and at the lowest position, with a vegetation cover of 50~60%, close to the forest edge of *Cunninghamia lanceolata* and *Pinus massoniana* at the far downstream of tailings, filled with slag (usually below the height of 1m) or household waste, the soil composition was complex and very different. In the rainy season, the water flow in the A, B, and C areas flew into the stream in the D area.

2.2 *Community surveyrame*

In Jan-May 2021, 10 large samples in 4 different areas, each with 10m×10m, and all trees (DBH≥5 cm) were surveyed. After random selection of 10-14 shrub quadrats of 2 m×2 m and herb quadrats of 1m×1m in the large sample quadrats, all individuals in the shrub and herb layer in these quadrats were investigated and recorded respectively. The species name, number, DBH and height of the individuals in the tree layer were recorded; the species name, number, coverage of individuals in the shrub (DBH<5 cm, height≥50 cm) and herb layer were recorded (herbs, ferns, and seedlings of tree and shrub species at a height below 50 cm were all classified into individuals in herb layer).

2.3 *Data processing*

Referring to relevant investigating methods (Fang et al. 2004; Hou & Lu 2019; Yang et al. 2013), the surveyed data were processed as follows.

Growth index: average height, DBH, base diameter, etc.

Community characteristic index: the relative advantage of the tree layer was calculated by cross-sectional area at breast height, and that of the shrub and herb layer was calculated by the coverage.

Important value = Relative density+relative advantage+relative frequency (1)

Relative Density (RD) = 100× (Density of a certain species/density of all species) (2)

Relative advantage = 100× Advantage of a certain species/advantages of all species (3)

Relative frequency = 100× Frequency of a certain species/frequencies of all species (4)

Plant diversity index:

Margalef index = $(S-1)/\log_2 N$ (5)

Shannon–Wiener Diversity Index (H)= $-\sum_{i=1}^{S} (P_i)\, ln\, (P_i)$ (6)

Community evenness =H/lgS (7)

Simpson index= $1-\frac{\sum_{i=1}^{s} n_i(n_i-1)}{N(N-1)}$ (8)

Pielou evenness inde $\frac{H}{\ln S}$ (9)

In the above formulas, N is the total number of individuals of a certain species; n_i is the number of individuals of species i; P_i is the ratio of n_i to the total number of individuals of all species; N is the total number of species within all the survey sample quadrats.

Jaccard similarity index: Cj=c/(a+b-c) (10)

Sorensen similarity index: Cs=2c / (a+b) (11)

Cody dissimilarity index: βc=(g+l)/2, g=a-c, l=b-c. (12)

In Formulas (10), (11) and (12), a and b are the respective number of plant species of the two quadrants, and c is the number of species common to the two quadrats.

3 RESULTS

3.1 *Plants in different areas of the tailings*

There were 76 plant species in the tailings (all 4 areas) in Huangnijiao, belonging to 63 genera in 43 families. Among them, Poaceae was the family having the most species, had 7 species which were *Miscanthus floridulus, Neyraudia reynaudiana, Cynodon dactylon, Digitaria sanguinalis, Paspalum thunbergii, Microstegium vimineum* and *Indocalamus tessellatus. Rubiaceae* had 5 species, with the 2nd most species which were *Paederia scandens, P. cavaleriei, Morinda umbellata, Mussaenda Pubescens* and *Gardenia jasminoides*. Compositae and Theaceae both had 4 species, with the 3rd most species, the former were *Ageratum conyzoides, Crepidiastrum lanceolatum, Conyza canadensis* and *Ageratum conyzoides*, the latter were *Schima superba, Eurya groffii, E. japonica* and *Adinandra millettii*. Other families, Camphaceae, Rosaceae and Verbenaceae, all had 3 species.

Of the 4 areas, the species richness in the entrance area was the most, with 61 species, followed by the area near the summit, with 36 species, that around the acid pool and in the slope area were the least, with only 11 or 10 species, corresponding, the areas with more species, also with more genera and families. From the perspective of life type, there were more shrubs and herbs in the whole tailings, as in the entrance and near the summit, however, in the area around the acid pool and slope area, with short idle time, there were more ferns occupying a higher proportion. Although there were young individuals and seedlings of trees, there were no adult trees and no vines. The more species near the summit were likely due to a long idle time for natural recovery, while that in the entrance was mainly because the original vegetation was not seriously damaged, although the surface soil was contaminated and filled with some slag, the physical structure was relatively good, retaining some species from original vegetation.

Table 2.   Plant statistics in Huangnijiao rare earth tail mining area.

Area	NO. families	NO. genus	NO. species	NO. trees	NO. shrubs	NO. lianas	NO. herbs (excluding herbal ferns)	NO. ferns
Near the summit	24	31	36	4	16	6	6	4
Around the acid pool	11	11	11	4	2	0	1	4
Slope	9	10	10	3	2	0	1	4
Entrance	37	52	61	10	19	8	17	7
Total	43	63	76	15	24	11	18	8

In the whole tailings, the herb species (including herbal ferns) were the most, 26 species, accounting for about 34%, followed by shrubs, 24 species, about 32%, 15 trees, about 20%, and lianas were the least, 11, about 14%. In the slope area, the area around the acid pool, and near the summit, the original vegetation was directly damaged, in this order, their idle time was gradually extended, and plant species increased. The original vegetation in the entrance area was not directly damaged, but the slag scattered accumulation here, and because of the lowest location, the sewage also flew downstream, probably because the deep soil structure was not completely destroyed, so the plant species were relatively abundant.

In the area near the summit and entrance area, the vertical structure of the community could be divided into 3 layers: tree, shrub, and herb layer. The dominant species of tree layer in the area near the summit were *Pinus massoniana* and *Schima superba*, which were similar to the surrounding coniferous and broad-leaved mixed forest, while those in the entrance area were *Pinus massoniana, Melia azedarach*, and *Cunninghamia lanceolata*, which were likely affected by the surrounding plantations of *Pinus massoniana* and *Cunninghamia lanceolata*. In the area around the acid pool and the slope area, there were only shrub and herb layers, without trees with a height above 5m, with only their saplings or seedlings. However, in the 4 areas, Miscanthus *floridulus* was the dominant species in the herb layer, with only relatively good soil structure and water preservation performance, which was often the 1st dominant species. In the area near the summit, around the acid pool, and the slope area, the original vegetation was destroyed, *Miscanthus floridulus* and *Dicranopteris dichotoma* were all the dominant species in the herb layer, which also shows that the 2 plants had strong resistance to the tailings. However, although there were few plants in the area around the acid pool, the vegetation cover was second only to the entrance area, reaching 50% (Table 1), mainly because of relatively many individuals and the high cover of resistant plants, such as *Miscanthus floridulus* and *Dicranopteris dichotoma. Ageratum conyzoides, Solanum photeinocarpum*, and *Digitaria sanguinalis* were more distributed in the entrance area and generally grew in places without slag accumulation in good water and fertilizer condition, especially some domestic garbage dump places. However, the dominant species of shrub layer in the area near the summit and in the slope area were the saplings of *Elaeocarpus sylvestris*, *Pinus elliottii*, and *Schima superba*, but in the former area, the 1st dominant species was *Pinus elliottii*, while in the latter, that was *Elaeocarpus sylvestris*.

Table 3. Species and degree of plant communities in rare earth mines.

Area	Level	Species name	Relative frequency	Relative density	Relative dominance	Importance value
Near the summit	Tree	*Pinus massoniana*	55.56	62.92	77.90	0.65
		Schima superba	22.22	24.72	16.95	0.21
		Diospyros morrisiana	11.11	8.99	3.19	0.08
	Shrub	*Litsea cubeba*	14.29	25.86	22.34	0.21
		Morinda umbellata	14.29	10.34	29.67	0.18
		Pinus massoniana (sapling)	11.90	12.07	12.09	0.12
	Herb	*Dicranopteris dichotoma*	18.00	32.95	54.92	0.35
		Gahnia tristis	16.00	13.29	15.20	0.15
		Miscanthus floridulus	10.00	13.29	7.05	0.10
Around acid pool	Shrub	*Pinus elliottii* (sapling)	27.27	40.00	39.49	0.36
		Elaeocarpus sylvestris (sapling)	31.82	30.00	24.84	0.29
		Schima superba (sapling)	18.18	13.33	14.65	0.15
	Herb	*Miscanthus floridulus*	33.33	41.82	45.85	0.40
		Dicranopteris dichotoma	30.00	31.82	29.07	0.30
		Pteris vittata	26.67	17.27	19.03	0.21
Slope	Shrub	*Elaeocarpus sylvestris* (sapling)	53.85	61.11	34.97	0.50
		Pinus elliottii (sapling)	23.08	22.22	25.14	0.23
		Schima superba (sapling)	7.69	5.56	15.30	0.10
	Herb	*Miscanthus floridulus*	40.00	54.43	53.95	0.49
		Dicranopteris dichotoma	24.00	21.52	10.61	0.19
		Pteris vittata	12.00	3.80	26.64	0.14
Entrance	Tree	*Pinus massoniana*	47.37	66.67	79.62	0.65
		Melia azedarach	10.53	10.71	7.39	0.10
		Cunninghamia lanceolata	10.53	8.33	1.77	0.07
	Shrub	*Litsea cubeba*	9.84	11.49	10.95	0.11
		Trema tomentosa	6.56	8.05	17.24	0.11
		Rubus corchorifolius	6.56	6.90	6.49	0.07
	Herb	*Miscanthus floridulus*	16.33	20.30	33.17	0.23
		Neyraudia reynaudiana	8.16	5.58	10.40	0.08
		Ageratum conyzoides	4.08	13.17	6.19	0.08

3.2 Comparison of plant community diversity in different areas of tailings

By measuring species diversity and uniformity, the community structure of different areas in the tailings can be comprehensively analyzed. The indexes of Shannon diversity, Pielou uniformity, Margalef richness and the number of individuals in the entrance area were the largest, followed by the area near the summit, the area around the acid pool ranked the 3rd, and the slope area was the smallest. The Simpson dominance index was exactly the opposite of the aforementioned indexes (Table 4). The field investigation found that vegetation cover in the area around the acid pool and the entrance was larger, both reaching 50% and even above, and plant distributions were more uniform, the entrance area was in a lower position, the trees such as *Pinus massoniana*, *Melia azedarach* and *Cunninghamia lanceolata* grew well, few individuals of *Pilea japonica*, *Alternanthera philoxeroides* and *Dioscorea japonica* were distributed near puddles or ditches. The vegetation cover on the slope was 35-40%, since it was an artificial restoration area, the plants were cultivated by a fixed strain distance, so the plant distribution was relatively sparse, but overall, it was more uniform, the saplings of *Pinus elliottii* and *Elaeocarpus sylvestris* were not thriving. Plant communities were the least homogeneous and in a typical plaque-like distribution near the summit, *Trema tomentosa*, *Rubus corchorifolius*, *Adinandra millettii*, *Dicranispinis dichotoma* and *Miscanthus floridulus* were mainly under the canopy of *Pinus massoniana*, *Gahnia tristis* was concentrated in the gullies, and at the flat terrain, the vegetation cover was larger, while at the larger slope with more gravel, that was less.

Table 4.   Diversity index of plant communities in different regions in rare earth tailings.

Area	Shannon index	Pielou uniformity index	Simpson dominance index	Margalef enrichment index	Individual number (Convert into within 1,000 m^2)
Near the summit	2.5391	0.7085	0.1334	3.6439	14840
Around acid pool	1.5741	0.6564	0.2726	1.0671	11750
Slope	1.4463	0.6281	0.3286	0.9967	8350
Entrance	3.1804	0.7737	0.0711	5.9713	23110

3.3   *Similarity or dissimilarity of communities in different areas in the tailings*

It can be seen from Table 5 that in addition to the high similarity between the area around the acid pool and the slope area, the Jaccard, Sorensen similarity index exceeded 0.9, and the similarities between the other pairs were below 0.5, indicating that plant diversity characteristics of the area around the acid pool was close to the slope area and had similar community composition, which may be related to natural elements such as topography, pollution status and lighting conditions in the 2 areas. It also can be seen from Tables 3 and 4 that the hierarchy, dominant species, diversity, uniformity and richness of plant communities in the 2 areas were all very close, combined with Table 5, the vegetation cover and individual numbers in the two areas were greatly different, they were still very similar.

Table 5.   Index of Jaccard similarity, Sorensen similarity and Cody dissimilarity of vegetation in 4 areas in the tailings.

Index	Area	Near summit	Around acid pool	Slope	Entrance
Jaccard similarity	Near the summit	1			
	Around the acid pool	0.2222	1		
	Slope	0.2309	0.9091	1	
	Entrance area	0.2576	0.0588	0.05971	1
Sorensen similarity	Near the summit	1			
	Around the acid pool	0.3636	1		
	Slope	0.3750	0.9524	1	
	Entrance	0.4096	0.1111	0.1127	1
Cody dissimilarity	Near the summit	0			
	Around the acid pool	10.5	0		
	Slope	10.0	0.5	0	
	Entrance	24.5	32.0	31.5	0

The original vegetation damage in the area around the acid pool and the slope area was more serious, the recovery time was short, and the species composition and hierarchical structure were simpler. In the area near the summit, at the highest terrain, but after natural recovery for nearly 40 years, the vegetation had a certain degree of recovery, while the entrance area was at the lowest terrain, because most of the original vegetation and soil were not destroyed, just filled with some slag, so the water and fertilizer condition was relatively suitable, so there were more plants growing and distributing here, some plants could hardly grow in the other 3 areas, such as *Melia azedarach*, *Pilea japonica*, and *Dioscuri japonica*, etc., but they appeared in this area.

4 DISCUSSION

In the rare earth tailings in Ping yuan County because the early mining process seriously damaged soil texture, long-term wind and rain erosion, soil desertification, high gravel content, poor soil and water conservation capacity, low organic content, soil acidification, it is difficult for plants to grow normally. The crown density in the surrounding forests, even in secondary forests, could reach above 35%, but because of the low natural ecological restoration process, in the area near the summit in the tailings, after nearly 40 years of natural recovery, it was still below 20%, but in the relatively complete geomantic forest in the surrounding area it could usually reach above 40% (Li et al. 2018; Yang et al 2013).

In the surveyed tailings, there were only 76 plant species, within the sample plots of the same area size in geomantic forests around the tailings there were usually over 100 species. Therefore, the tailings as a whole had low species richness, and the composition and hierarchical structure were also simple. In terms of community succession, the plant communities of rare earth tailings in Hovannisian were still in the first-middle stage of community succession, namely, from the shrub and herb community to the coniferous and broad-leaved mixed forest community, and then it took a long time to become the mature forest–evergreen broad-leaved forest in Northeast Guangdong. The plant diversity in the area around the acid pool and the slope area was significantly lower than in the area near the summit, which was caused by different idle times. The area around the acid pool and the slope area were less idle, while the area near the summit was idle for nearly 40 years, and there were inevitably more settled plants, the community composition was naturally more complex. In the area near the summit, *Pinus masoning* distribution was relatively sparse, the scrub tussock was usually patchily distributed below the canopy of *Pinus masoning*. In the soil in this area, the gravel reached a high and was unevenly distributed, *Pinus masoning* often grew in soil with low gravel content and good quality. Of course, the plants and the soil also interact, plant litter and root secretions can improve the soil quality, and the soil is more suitable for plant growth later. Meanwhile, these plants tend to be acidic soil plants, such as *Pinus elliptic*, *Micropterus dichotomy*, *Blechnum Orientale*, etc. Because the gullies often collect rainwater, the water flow also spreads plant seeds, in the locations around the gullies, the soil water, and fertilizer conditions are relatively good, so *Microseism iminium*, *Trema tomentosa*, *Callicarpa kachina* are often distributed here, but in other locations with high gravel content, it is difficult for plants to take root and grow, some bald spots are formed.

The area around the acid pool and the slope area had low species richness and simple composition structure, their original vegetation had been damaged, the former area around the acid pool, namely close to the acid pool, the surface soil was not obviously affected, but the deeper soil was seriously affected by the acid pool leakage, high acidity and heavy metal pollution, therefore, in the surface, *Micropterus dichotomy*, *Blechnum Orientale*, *Pteris vittate*, *Miscanthus florid ulus* grew well, but the young individuals of shrubs and trees did not grow vigorously. In the slope area, the original vegetation was seriously damaged, and the existing plants, such as *Elaeocarpus sylvestris*, *Pinus elliptic*, *Miscanthus florid ulus* were artificially planted as resistant plants. Due to the large slope, the ground was converted into stepped soil beds, during the rainy season, the rainwater was discharged from the ditch in the soil beds, so the soil had relative stability and small loss volume, and the saplings of *Elaeocarpus sylvestris* and *Pinus elliptic* took root and grew, *Miscanthus florid ulus* also grew relatively prosperously, but because the low soil fertility and short planting time, the vegetation cover was not large. In the entrance area, the vegetation composition structure was the most complex among the 4 areas, because the original vegetation and soil were not directly destroyed. This area was in the lowest position in the tailings, water and fertilizer conditions in the soil were better than in other areas, therefore, there were *Verbena officinalis*, *Paspalum thunbergia*, *Pilea japonica*, *Solanum thunbergia*, and other plants, however, due to sewage confluence and slag stack to change the partial topsoil structure, so *Miscanthus florid ulus*, *Blechnum Orientale* and other acidic soil plants occupied a certain advantage, but the plants that like neutral or alkaline soil were rarely distributed.

In the process of vegetation recovery, the growth of shrubs and herbs presents different degrees of enrichment for soil organic carbon, whole carbon, nitrogen, whole phosphorus, effective nitrogen, and effective phosphorus, also has obvious "seed" and "fertile island" effect (Liu et al. 2019; Xia et al. 2015). The shrubs and herbs improve soil quality in tailings, while the soil improvement accelerates the growth of shrubs and herbs, which can produce more seeds and is conducive to the spread of plant populations. In this study, the plaque distribution of vegetation in the summit top also confirms the conclusion that if vegetation recovery in tailings is to be accelerated, the existing shrubs and herbs must be protected, because the improvement effect of these plants on the soil is not only conducive to its own growth and expansion but also the settlement of other plants, thus making the community structure increasingly complete. The plants that can settle and grow near tailings with short idle time are more adapted to acidic, arid, and poor habitats, and promote the growth of these tolerant plants through some measures such as enhanced water and fertilizer management, or soil improvement. Although in tailings, especially in tailings with short idle duration, there are few adaptive plants, vegetation diversity should also be considered during artificial restoration or reconstruction. In addition to these existing resistant plants, other resistant plants, such as *Vetiver pivanilides*, *Miscanthus sinensis*, *Paspalum notatum*, *Pennisetum Americanum*, *purpureum*, *Medicago sativa*, can be selected and introduced. The shrub and herb community are the vegetation type in the early stage of ecological restoration and has important value in tailings (Zhao et al. 2015).

This study has initially confirmed that in the ecological restoration of tailings, the selection of plant species should be combined with the soil's physical and chemical characteristics and microclimatic characteristics in different areas. In Hovannisian tailings, the settled plants were mainly herbs from Pinaceae and Asteraceae, with only 1 species, *Disodium heterokaryon* from Leguminosae. In contrast to other rare-earth tailings in South China, there were few legumes. In other rare earth tailings in Ping yuan, Southern Jiangxi, there are also other legumes, such as *Macrophile thyroids*, *Albizia kilorad*, and *Kummerian striata*, because their root rhizobia have a nitrogen fixation function, for these plants, the soil nitrogen deficiency will not be an obvious limiting factor. Other relevant studies have also confirmed that in the early restoration of rare earth tailings, it is most likely that because of too high acidity in the soil, the species and number of legumes are less than those of Pinaceae, and generally distribute in less polluted areas in tailings, legumes are important species of soil improvement in the tailings. Therefore, these legumes can be replanted when the soil acidity is reduced to some extent. This is consistent with the conclusions of others in other rare earth tailings in South China. In addition, it should also be noted that in Hovannisian earth tailings, landslides often occur, so it is necessary to prevent landslides. The soil contains more gravel and sand, and has a loose structure with poor adhesion, especially in the rainy season, soil and water loss is particularly serious. Soil physicochemical properties and climate factors are the potential key factors for vegetation recovery (Wang 2011; Yang et al. 2013). Ping yuan County is located in Lingnan, with superior climate conditions, but in the rare earth tailings, unfavorable soil conditions, such as too high acidity, less organic matter content, poor fertility, loose structure, poor viscosity, and poor water retention, are the key factors limiting plant growth and distribution. Therefore, in the early stage of recovery in rare earth tailings, some acid-resistant, drought-resistant, barren-resistant herbs, such as *Pteris vittate*, *Blechnum Orientale*, *Miscanthus spp.*, should be selected to be planted (Yang et al. 2013; Wu 2018), after a period of time, the acid of soil decrease, the physical structure of the soil is improved, then some legumes with strong adaptability to per acidic, poor, arid soils, such as *Sesbania cannabina*, *Tephrosia candida*, *Pueraria lobata*, should be then planted (Yang et al. 2013; Zhang et al. 2016b). The tailings will turn green faster with barnyard manure or other organic fertilizer. In previous surveys, it has been found that the vegetation cover reached more than 50% after 3-5 years. At present, there are other chemical and microbial fertilizer methods for soil improvement, which have also achieved good results.

5 CONCLUSION

In the 4 areas in the Hovannisian rare earth tailings, near the summit, around the acid pool, slope and entrance, the vegetation structure and the diversity index were different among them. The larger number of individual plants was found in the area near the summit and in the entrance area, moreover, the Shannon diversity index, Pileous uniformity index and Marg alef richness index were higher than those in the slope area and around the acid pool, but the Simpson dominance index was exactly the opposite of the aforementioned index in different areas. The Jaccard and Sorensen similarity index of plant community in the area around the acid pool and the slope area were as high as above 0.9, but the Cody dissimilarity was as low as 0.5, which indicated a great similarity between the plant communities in the 2 areas. But, both the 2 areas were very different from the other 2 areas. Areas with similar environmental conditions and similar idle time usually have the highest plant species similarity, the site condition plays an extremely important role in plant restoration in rare earth tailings, and sufficient water and nutrients are the key factors to ensure the rapid recovery of vegetation. Natural restoration of vegetation in rare earth tailings is a longer process. Rare earth mining not only seriously destroys the physical structure of the soil, but also changes the physical and chemical nature of the soil. Early settled plants must usually be acid resistant, drought resistant and barren resistant soil. In rare earth tailings in northeast Guangdong, these resistant plants are usually *Pteris vittate*, *Blechnum Orientale*, *Miscanthus florid ulus*, etc. Plants that can settle and grow near tailings with short idle time are more adapted to acidic, arid and poor habitats, and promote the growth of these tolerant plants, through measures such as enhanced water and fertilizer management, or soil improvement.

ACKNOWLEDGEMENTS

This study is financially supported by the Special Fund for Guangdong Provincial Science & Technology Innovation and Rural Revitalization Strategy in 2021 (2021A0305), Guangdong Base Hakka Research Institute Bidding Project in 2020 (20KYKT09), Characteristic Innovation Project of Guangdong General Universities in 2019 (2019KTSCX171), and Guangdong Provincial Special Projects in Key Areas in 2020 (2020ZDZX1036).

REFERENCES

Fang JY, Shen ZH, Tang ZY, et al. (2004). The protocol for the survey plan for plant species diversity of China's mountains. *Chin Biondi*, 12(1), 5–9.

Hou J, Lu JG. (2019). Investigation and analysis of plant community characteristics of Hiking trails in Zijin Mountain of Forest Park. *North Horti*, 23, 86–92.

Li HD, Gao YY, Yan SG, et al. (2018). Supervisory countermeasures of ecological restoration of abandoned mine areas in the ecological conservation redline area. *J Eco Rural Environ*, 34(8), 673–677.

Li HD, Shen WS, Jia M, et al. (2015). Economic losses assessment for ecological destruction and environmental pollution in large-scale opencast mine. *J Nanjing for Univ* (Nat Sci Ed), 39(6), 112–118.

Liu MQ, Zhang XX, Ma JH, et al. (2019). The effect of different restorative treatments on the diversity of soybean rhizobia in the abandoned land of Ionic rare earth mine. *Soybean Sci*, 2019, 38(1), 77–83.

Wang JH. (2011). *Study on characteristics of soil microbe in the exogenous rare earth accumulation area of Baogang tailings dam*. Hohhot: Inner Mongolia Normal Univ.

Wu JW. (2018). *Geological mineral characteristics and prospecting indicator of Pinghu REE deposit of ion-adsorption type in Meizhou*, Guangdon. World non-ferrous metals, 1(2018)

Xie LN, Zhou JW, Xu W. (2015). Current situation and advanced technology of tailings management in Australia. *Environ Eng*, 33(10), 72–76.

Yang QH, Lin QY, Lai WN, et al. (2013). Investigation on vegetation restoration in Pingyuan rare earth tailing. *Guangdong Agri Sci*, 40, 150–154+166.

Yang QH, Pan SF, Lai WN, et al. (2015). Investigation of the community characteristics and species diversity of Qiaoxi Geomantic Forest in Meizhou. *Guihaia*, 35(6), 833–841.

Zhang L, Liu SH, Zhou LY, et al.(2016a). Characteristics of plant community in the restored rare earth mine area. *Guangdong Agr Sci*, 43(7), 73–80.

Zhang L, Liu W, Liang H. (2016b). Plant diversity under different site conditions of rare earth mines in northeastern Guangdong. *Guangdong Agri Sci*, 43(10), 82–88.

Zhao ZQ, Wang LH, Bai ZK, et al. (2015). Development of population structure and spatial distribution patterns of a restored forest during 17-year succession (1993-2010) in Pingshou opencast mine spoil, China. *Environ Monit Assess*, 187(7), 431–443.

Author index

831

For Product Safety Concerns and Information please contact our EU
representative GPSR@taylorandfrancis.com Taylor & Francis Verlag GmbH,
Kaufingerstraße 24, 80331 München, Germany

Printed and bound by CPI Group (UK) Ltd, Croydon, CR0 4YY
01/05/2025
01858470-0005